Great Lakes Fisheries
Policy and Management

Great Lakes Fisheries Policy and Management

A Binational Perspective

William W. Taylor

C. Paola Ferreri

Michigan State University Press

East Lansing

Michigan State University Press
East Lansing, Michigan 48823-5202

04 03 02 01 00 99   1 2 3 4 5 6 7 8 9

LIBRARY OF CONGRESS CATALOGING-IN-PUBLICATION DATA
Great Lakes fisheries policy and management: a binational perspective /
edited by C. Paola Ferreri, William W. Taylor.
p. cm. — (Canadien series)
Includes bibliographical references and index.
ISBN 0-87013-483-3 (clothbound: alk. paper)
1. Fishery policy—Great Lakes. 2. Fishery management—Great Lakes.
I. Taylor, William W. II. Ferreri, C. Paola, 1966– III. Series.
SH219.6 .G725 1999
333.95'6'0977—dc21
99-050480

Cover design by Michael Smith, View Two Plus.
Book design/typography by Sharp Des!gns, Inc.

Visit Michigan State University Press on the World-Wide Web at:
www.msu.edu/unit/msupress

Contents

Preface

The Laurentian Great Lakes are a unique natural resource that has directly affected the lives of the citizens of the United States and Canada. Their presence and importance to society are evident in historical and modern-day records of nearly every aspect of civilization. This book concentrates on the Great Lakes fisheries, a highly visible and important product of the Great Lakes ecosystem. The history of these fisheries is directly related to the cultural values of the people within the basin, and therefore, reflects the ever changing socioeconomic landscape of the Great Lakes basin and the political forces within both countries that have greatly influenced the physical, chemical and biological makeup of the Great Lakes themselves.

I [Bill] grew up on the shore of Lake Ontario at a time when the lake was considered a wasteland for industrial and municipal uses. The idea that Lake Ontario was a valuable fishery resource would have seemed ludicrous to me at this point in my life. It was a time when beaches were closed due to high bacterial counts, stench was high due to dead alewives floating in the water and resting on the beaches, and people viewed the inland oceans as places where societal waste could be diluted. Fishing and boating on these bodies of waters would never have been considered a serious option. Marinas were poorly developed and the main ships on the lake were iron ore transports. Sadly, perhaps because of my adolescent age, I accepted the state of the lakes and did not have the vision that they would ever be much of a fisheries resource. I had implicitly accepted the fact that the degraded aquatic system that I lived adjacent to was okay and that I could do nothing to change its state or value to myself or society. Thus, living only two houses away from Lake Ontario, my family chose to build a pool in our backyard rather than put our efforts towards improving one of the largest and most unique aquatic resources that exists today. However, by the time I arrived in Michigan in 1980, the Great Lakes fisheries were magnificent and valuable to citizens of the Great Lakes basin and the two great countries which border its shores. What a change from the time I left Pultneyville, New York in 1968! The waters were clean, the recreational fisheries thriving, and the beaches free of the stench of dead and decaying alewives. Society had come together and mandated that a clean and healthy environment was important, and a new generation of individuals had been mobilized to actively maintain our vigilance in protecting and enhancing our Great Lakes and their fisheries

Our intent in producing this book was to provide, in a central place, the history and current status of the Great Lakes fisheries and the their relationship to the physical, chemical, biological, and social components of the basin. This volume is divided into four major sections. The first section, Historical Perspective, provides a general background relating to the physical, chemical, social, institutional, and biological aspects of the Great Lakes fisheries. The second section, Current Issues Facing Fishery Management, describes many of the complex issues that fishery managers in the Great Lakes are currently dealing with. These include biodiversity considerations, sustainability of Great Lakes fisheries, effects of non-indigenous species, defining and conserving fish habitat, and understanding the role that contaminants play in shaping the fisheries of the Great Lakes. The third section, Allocation of Fishery Resources, provides an overview of historical and current allocation to the commercial and recreational fishery. This section also provides an overview of the forage base available in the Great Lakes and how it might impact allocation decisions in the future. The fourth section, Case Studies, provides detailed chapters regarding historical and current management of both single species and multiple species fisheries. Finally, we close the volume with a chapter calling for the collaborative management of Great Lakes fisheries at an ecosystem level.

We are greatly indebted to the many people that helped this book become a reality. First and foremost, are the authors of the various chapters. Without their vision, insight, and patience, none of this would have been possible. In addition, we thank the many professionals who have interacted with us over our careers to help us better understand the Great Lakes ecosystem and their fisheries. These include not only the authors and reviewers of this volume, but also the many people with the state, provincial, tribal and federal resource management agencies, universities, and binational commissions who have dedicated their time and energy to making the Great Lakes a sustainable resource for current and future generations to enjoy. Their outstanding professionalism and contributions to bettering the Great Lakes and their fisheries are noted throughout the chapters in this book. Given the skill and dedication level of these professionals, we believe that the Great Lakes ecosystem is in good hands for the future.

Each chapter in this volume was externally peer reviewed by at least two individuals, and the comments of the reviewers were carefully considered by both the authors and the editors. For the time and care they invested as peer reviewers, we thank: James Bence, Peggy Blair, Ed Brown, Tom Brydges, Tom Busiahn, Ed Crossman, Tracy Dobson, Doug Dodge, Randy Eshenroder, John Gannon, Bob Haas, Michael Hansen, Dan Hayes, Roger Knight, John Lehman, Wayne MacCallum, Chuck Madenjian, Doran Mason, Ed Mills, Stephen Nepszy, Robert O'Gorman, Jim Peck, Phil Pister, Charles Rabeni, Richard Santer, Paul Seelbach, Gerald Smith, Thomas Todd, Steven Yeo, and James Zorn.

Friends are particularly important to those individuals who attempt to produce a book of such magnitude. We especially appreciate the mentoring of Dr. Victor Howard, Professor Emeritus and former Director of the Canadian Studies Center at Michigan State University, who in so many ways stimulated us to take on this project and gave us the confidence to proceed when stumbling blocks seemed insurmountable. His sense of humor, never failing support and friendship has been greatly appreciated and will never be forgotten. Really understanding the Great Lakes and their fisheries requires one to interact with the resource in a very personal and meaningful way. Two individuals that have assisted us in this endeavor are John Robertson, former Chief of the Fisheries Division and current Chief of the Forest Management Division of the Michigan Department of Natural Resources, and Denny Grinold, past President of the Michigan Charter Boat Association. Through their efforts and insight, we have seen the Great Lakes fisheries within the context of a complex ecosystem and have learned firsthand the importance of a well functioning productive environment to society.

Finally, we would like to express our gratitude to Julie L. Loehr of the Michigan State University Press who has graciously assisted our every request and has expended much effort to make this book the very best that it can be. Her dedication to this project was essential to its completion. The Michigan Sea Grant College Program and the Canadian Embassy, through its Canadian Studies Research Grant Program, provided financial assistance that facilitated the production of this book.

Our goal was to create a volume that would be useful to fishery managers, researchers, and students who were interested in the fisheries of the Great Lakes. We felt that by bringing together an understanding where we have been, how we got there, and where we currently are, we could provide a resource useful to those involved with Great Lakes fisheries issues. Let us never forget the lessons we have learned from abusing our environment, and stay active in providing the stewardship for these resources so that current and future generations are able to enjoy the benefits that our Great Lakes offer humankind on a sustainable basis.

William W. Taylor and C. Paola Ferreri
East Lansing, Michigan

PART 1

Historical Perspective

CHAPTER 1

An Introduction to the Laurentian Great Lakes Ecosystem

Alfred M. Beeton, Cynthia E. Sellinger, and David F. Reid

Great Lakes Physical Processes

Physical processes that occur in the Great Lakes influence the distribution of nutrients, geographic location, and relative abundance of the biological communities. Processes such as water movement, which are important to the distribution of benthos and plankton, are ultimately responsible for the location of Great Lakes fish, which feed heavily on these biota. Additional physical processes, such as light penetration and water temperature are also responsible for the location and abundance of Great Lakes biota. Since light is the important factor in photosynthesis, the quantity and quality of light penetrating the water column controls the nourishment and growth of phytoplankton. Light is also important for distribution of zooplankton (e.g., vertical migrations). Temperature, which varies due to latitude and depth of a particular lake, influences the distribution and abundance of Great Lakes invertebrate fauna. Simply put, the upper deep Great Lakes have different zooplanktonic and zoobenthic fauna than the shallow waters of Green Bay, Saginaw Bay, and Lake Erie due partially to cooler temperatures.

Water level changes, although not as dynamic as water movement, indirectly affect the biota's habitats. Variations in water levels on a seasonal, long period, or geologic-time-scale basis affect coastal wetlands, shore erosion, resuspension of sediments, and the depth in the water column.

Underlying these physical processes is the basin morphology. Not only do latitude and depth determine water temperature, but climatic patterns and water abundance are geographically specific. Therefore, this section will briefly describe basin morphology, or watershed dimensions; provide an explanation of how water is maintained in the Great Lakes; (i.e., the water budget) describe the optical properties of the water column; and conclude with examples of processes that are responsible for water movement.

Watershed Dimensions and Lakes

The Great Lakes basin covers approximately 754,100 km^2 of North America and Canada, and it encompasses an ecosystem which includes a combination of massive water bodies and

TABLE 1

Morphometric data from the Laurentian Great Lakes

Coordinating Committee on Great Lakes Basic Hydraulic and Hydrologic Data, 1977

LAKE	AREA (KM²)		DEPTH (M)		VOLUME[b] (KM³)	DIMENSIONS (KM)			RANKING[c]
	LAND	WATER	MEAN[a]	MAX		LENGTH	BREADTH	SHORELINE	
Superior	127,700	82,100	149	405	12,233	563	257	4,795	2
Michigan	118,000	57,800	85	281	4,913	494	190	2,633	5
Huron	131,300	59,600	59	229	3,516	332	295	6,157	4
St. Clair	12,430	1,114	4	6	5	42	39	413	124
Erie	58,800	25,700	19	64	488	388	92	1,402	13
Ontario	60,600	18,960	86	244	1,631	311	85	1,146	17

[a] From Herdendorf (1982)   [b] Calculated   [c] Ranking among the large lakes in the world with respect to surface area (Herdendorf 1982)

watersheds that are populated with human beings, farms, wildlife, and forests, to name a few. The water bodies are composed of lakes, embayments, rivers, and littoral zones (an interface zone between the land of the watershed and the open water of the lake). The five Laurentian Great Lakes rank among the seventeen largest lakes of the world, ranging in size from Lake Ontario, with a surface area of 18,960 km², to Lake Superior, with a surface area of 82,100 km² (table 1). Among the freshwater lakes, Lake Superior is the largest in surface area.

Of the five Laurentian Great Lakes and their connecting channels, Lakes Superior, Michigan, Huron, and Ontario all have maximum depths >200m. Furthermore, of the large lakes of the world, Lake Superior ranks among the fifty lakes with depths of 283m or greater (Herdendorf 1982).

Lake Superior has the highest elevation and is the largest and deepest of the Great Lakes. Its basin area covers 28% of the total Great Lakes basin and its water accounts for 54% of all of the basin water. Water flows from Lake Superior to Lakes Michigan and Huron through the St. Mary's River (fig. 1).

Lake Michigan is the only Great Lake that is located solely within the United States, affording the title as the largest lake in the continental United States. Its basin area covers 23% of the total Great Lakes basin, and its water volume accounts for 22% of the total water supply. Flows out of Lake Michigan are through the Chicago Diversion and the Straits of Mackinac. Major water exchange occurs through the Straits of Mackinac's broad, deep channel with net easterly flows into Lake Huron (Saylor and Sloss 1976).

Lakes Michigan and Huron are hydraulically connected through the Straits of Mackinac, and have the same water level (fig. 2). Lake Huron is the second largest Great Lake in surface area and is located in the central portion of the Great Lakes Basin. Lake Huron accounts for 25% of the total Great Lakes basin area, as well as 15% of the total water volume. Outflows from Lake Huron are discharged into the St. Clair River and, subsequently, to Lake St. Clair.

Although not considered a Great Lake, Lake St. Clair receives outflows from Lakes Michigan and Huron, and therefore, is part of the Great Lakes system. It accounts for 0.1% of the total basin area and less than 1% of the total volume of water. Lake St. Clair outflows into another connecting channel, the Detroit River, which flows into Lake Erie.

FIG 1. *The Laurentain Great Lakes*

Lake Erie is the shallowest of the Great Lakes; it is the only one in which the greatest depth is above sea level (Great Lakes Basin Commission 1975). It accounts for 11% of the total basin area and 2% of the entire Great Lakes water volume. Lake Erie discharges to Lake Ontario through the Niagara River and the Welland Canal.

Lake Ontario is second in depth to Lake Superior. The smallest of the Great Lakes, Lake Ontario accounts for 11% of the total basin area and 7% of the total water volume. Outflows are into the St. Lawrence River to the Atlantic Ocean.

Great Lakes Water Supply

The Great Lakes water volume is a function of the hydrologic cycle. The earth's hydrologic cycle is a representation of the circulation of water from land and water masses into the atmosphere and back again. Basically, water enters land and water surfaces in the form of precipitation. Lakes receive precipitation either directly from the atmosphere or indirectly through either tributary runoff or groundwater seepage. Water then leaves the land, and water surfaces through the process of evapotranspiration. This water exchange is accounted for by a water budget, which is an application of the conservation of mass law expressed by the equation of continuity (equation 1)

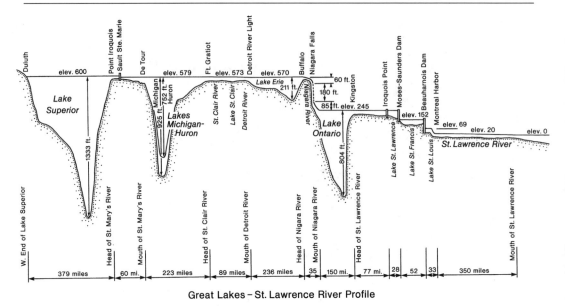

Great Lakes – St. Lawrence River Profile

FIG 2. *Lake surface elevations*

(Brooks et al. 1991), where the difference between inflows (*I*) and outflows (*O*) is equal to the change of storage (ΔS) in the system.

$$I - O = \Delta S$$

In the Great Lakes region, this can be expressed more specifically with terms of the hydrologic cycle (equation 2).

$$Q_i + P \pm R \pm G_w - E_t - D_i - Q_o = \Delta S$$

Where inflows are from flows into the system through connecting channels (Q_i), precipitation (*P*), stream runoff (*R*), and groundwater seepage—which, in some cases, may be considered an outflow (G_w); outflows from the system are from evapotranspiration (E_t), diversions (D_i), and water flowing out through connecting channels (Q_o). These inflows and outflows account for the volumetric change in the lakes' storage as well as the linear change in the lakes' levels.

Optical Properties

Most of the energy necessary to produce metabolic processes in fresh water is derived from solar radiation. Energy derived from solar radiation provides a source of heat to the aquatic environment, provides light to plants for photosynthetic processes, and influences the behavior of zooplankton and fish.

Light incident on the lake surface is reflected, absorbed, and scattered. The amount of radiation that penetrates the lake's surface is dependent upon many factors, some of which include the angle of incident, spectral distribution of light with depth, and the amount and characteristics of dissolved and suspended materials. Wetzel (1983) reported that on a clear, calm lake surface 5% to

6% of incident light is reflected away from the water's surface, allowing most of the light to be absorbed; when the surface is disturbed by wave action, reflection is increased by 20%; and an average of 75% of light is reflected from frozen lakes covered by snow.

Penetration of light with respect to depth is also dependent upon the spectral distribution of light. For example, studies by Beeton (1962) showed that orange light, with wavelengths of 590–610μm, penetrated the deepest in Lakes Erie and Ontario, while Lakes Superior and Michigan had their greatest transparency in the green spectrum with wavelengths of 490–540μm, and Lake Huron in the blue, with wavelengths of 440–90μm.

Optical properties, such as water clarity, control the amount of light that enters the lakes and, subsequently, the photosynthetic nature of its biota. Research has been scarce with respect to quantifying the depth of light penetration in the Great Lakes. Researchers (Beeton 1958; Charlton et al. 1993) have employed either Secchi disks, or photometers, or a combination of the two, to quantify the areal variations in Great Lakes transparency.

Water Movement

Understanding water movement to the Great Lakes is important not only for navigational, hydroelectrical, and shoreline protection purposes, but the turbulence resulting from water movements is of major importance to the lakes' biota. Water movement influences both the gathering and distribution of nutrients and food, and the distribution of microorganisms and plankton. Due to the scope of this section, only a few major water movement processes are presented. Information on more specific processes can be found in Wetzel (1983).

Water movement does not only consist of wind-induced waves. Various processes, such as the thermal variations, free and forced lake oscillations, internal waves, wind-generated surface waves, and short and long-term lake-level variations are categorized as water movement in the Great Lakes. Additionally, since the Great Lakes are a large body of water, circulation patterns are affected by the earth's rotation, commonly known as the Coriolis force (Rousmaniere 1979).

Thermal Variations

Located within a temperate zone between the equator and the Arctic Circle, the Great Lakes exhibit large seasonal changes in insolation, resulting in cold winters and mild summers (Irbe 1992). Solar radiation absorbed by the Great Lakes is the driving force controlling the lakes' biology, chemistry, heat storage, and circulation patterns (Great Lakes Basin Commission 1975). Solar radiation affects each lake differently according to geographic location and water depth.

Geographically, lakes that are closer to the equator receive more solar radiation; Lake Erie receives more radiation than Lake Superior. Likewise, the deeper a lake is and the larger its volume, the longer it takes to warm up and cool down. Again, Lake Superior takes longer than Lake Erie to warm up, and due to Superior's large volume, it has a larger heat-storing capacity; consequently, it cools more slowly. Therefore, due to location and depth, there is a temperature-phase difference between lakes.

Not only do these lakes exhibit temperature variations due to location and depth, but they also exhibit vertical variation within themselves due to volume, depth and geographic location. The Great Lakes are monomictic, since they mix freely from fall through spring and then become

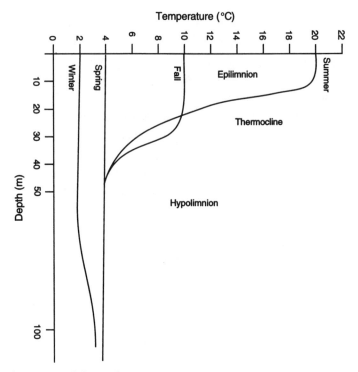

FIG. 3. *Great Lakes seasonal thermal structure*

stratified in the summer. These lakes, on rare occasions, may become dimictic if they develop 100% ice cover during winter. The ice cover prevents solar and wind energies from mixing the lakes' waters, thus interrupting the mixing pattern. Each Great Lake attains its maximum normal ice cover in February, with a range of 45% ice cover on Lake Michigan to 90% on Lake Erie. Of the five Great Lakes, Lakes Superior and Erie have historically achieved a maximum 100% ice cover (Assel et al. 1983).

The relationship between temperature and depth for the four seasons of the year is illustrated in figure 3. As water reaches its maximum density of 4°C it tends to sink to the lake's bottom while colder (less than 4°C), less-dense water rises to the lake's surface, resulting in inverse thermal stratification. The waters continue to cool and temperatures at or below 1°C occur. Thus, the coldest water is at the surface, where ice forms, usually in the near shore and the bays of the Great Lakes. By spring, the shallower, near shore area warms up more quickly than the deep area offshore. Near shore, waters that warm up to 4°C sink beneath the colder, less dense surface waters. This phenomena is most pronounced in Lakes Michigan, Huron and Ontario during the spring, when the near shore warms faster than the central basin (Brooks and Lick 1972). The area where the water temperature is exactly 4°C (temperature of maximum density) is called the thermal bar. It separates warmer (> 4°C) near shore water from colder (< 4°C) offshore water. As the lake warms around the edges, the thermal bars migrate from the shore toward the center of the lake. At this point, warmer water can cover the entire surface of the lake, with colder water below. As summer progresses, the water column becomes stratified with the upper (warmer), more turbulent water (the epilimnion) overlying the colder layer (the hypolimnion). These two layers are separated by the metalimnion (or, the thermocline) (fig. 3), a water layer of large thermal gradient. Vertical mixing continues once again during the late summer and fall as surface waters are cooled. Again, when this colder water reaches its maximum density, it sinks, and is mixed by a combination of convection currents and wind-induced circulation (Wetzel 1983). With time, the entire lake is involved in this mixing, known as the fall turnover. In monomictic lakes, the mixing continues until the next thermal stratification.

Currents, Seiches, and Waves

Water motion in the form of currents is caused by many factors, some of which are still under investigation. Currents can result from water-level oscillations or long-standing waves (Great Lakes Basin Commission 1975). Long-standing waves result from displacement of the water mass of the whole lake. These waves have wavelengths the same order as the basin's dimension. These long waves are reflected at the basin boundaries, and combine into standing wave patterns; these patterns, both internal and surficial, are known as seiches.

Surges

Surges, or short-term variations that occur on the Great Lakes, are independent of the volume of water available. These events occur as a result of atmospheric pressure differences, or winds blowing over the surface. A steady wind blowing over the lake's surface can produce a sloshing, or tilting, of water toward one end of the lake. Certain lakes are more susceptible to this sloshing, or wind setup, than others. Lake Erie is notorious for wind setups. For example, water piles up on the western end, as a result of easterly gales. This pileup of water on the western end often results in failing of hydroelectric power generated at the Niagara River (Gresswell and Huxley 1965).

A 1972 flood that caused enormous damage on the lower Great Lakes resulted from both differences in barometric pressure and high winds. A quote from Brazel and Phillips (1974, p. 57) illustrates the sequence of events that led to this flood:

> The major flood damage was experienced on 14 November 1972, when a short wave trough in the middle troposphere over Oklahoma on the thirteenth moved north to position over southern Ohio on the next day. Associated with this trough was a large surface low, deepening rapidly as it moved to just south of the lower Great Lakes. Surface pressure falls up to 7.5mb in six hours, along with widespread showers, preceded the advancing low.

Brazel and Phillips (1974) continued to describe a high pressure with a line through Minnesota, northern Michigan, and central Ontario that kept the low pressure stationary over the southern lakes. Additionally, they reported that the surface pressure gradient was so high that winds were clocked at $33ms^{-1}$ (74mph). The combination of pressure gradient differences and extremely high winds produced extreme lake-level changes in specific areas and damaging waves; thus, the flood of November 1972.

TABLE 2

Great Lakes seasonal highs and lows

Adapted from Great Lakes Basin Commission, 1975

LAKE	HIGH	LOW
Superior	August or September	March
Michigan - Huron	July	February
Erie	June	February
Ontario	June	January

Seasonal Lake-Level Variations

Seasonal water-level variations occur primarily due to the hydrologic cycle. A hydrograph of Lakes Michigan and Huron illustrates a typical seasonal cycle (fig. 4). In spring, when the snow melts and evapotranspiration is decreased, the lakes approach their maximum seasonal elevation due to increased runoff. In summer, since evapotranspiration is increased and runoff is decreased, the lakes begin their seasonal water-level decline. In fall, evapotranspiration reaches a maximum, which hastens the seasonal decline. Finally, in winter, the lakes remain low because freezing temperatures keep runoff low. Due to factors that include differences in latitude and depth of the lake bottom, each lake reaches its peak and low levels in different months of the year (table 2).

Long Period Lake-Level Variations

Variations that occur over long periods are the result of a cumulative change in normal hydrologic parameters, such as temperature or precipitation. Figure 5 illustrates an area weighted, five-year averaged index of Great Lakes precipitation correlated with Lakes Michigan, Huron, and Erie's water levels (International Joint Commission 1993). This figure, which was used to identify years

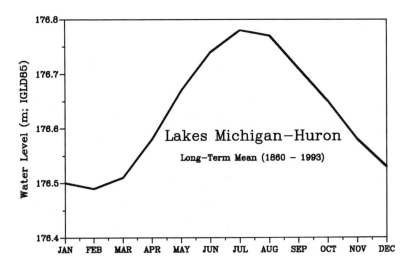

FIG. 4. *Lakes Michigan and Huron seasonal water levels*

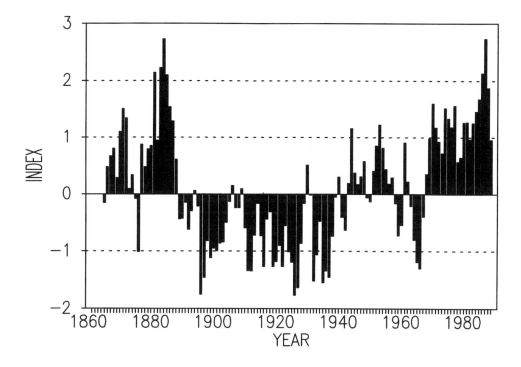

FIG. 5. *Great Lakes precipitation index*

of high and low lake-level variability, shows distinctly high precipitation for the years 1878–83, 1947–52, 1964–69, and 1981–86. Distinct lows are also apparent for 1890–95, 1920–25, 1929–34, and 1960–65. Long period variations in precipitation have been known to correspond to variations in lake-levels. These long-term variations were responsible for record high levels in 1986 and record low levels in 1964.

Patterns of longer duration have also been suggested. It can be interpreted from this precipitation index that a high precipitation regime dominated in 1878–83, then a low regime dominated from 1884–1940.

Geologic Variation

Since the glacial retreat of the Pleistocene time period, land masses in the Great Lakes region have been slowly uplifting (Horton and Grunksy 1927). This uplifting is based upon the principle of isostasy, which states that land masses must maintain an equilibrium within the earth's crust. A tremendous weight, such as a glacier, can theoretically force localized land masses downward into the earth's crust once a compensatory equilibrium is achieved. Once that weight is removed, associated with glacial retreat, land masses will rebound upward to regain their preglacial elevations (Quinn and Sellinger 1990). Due to the varying thickness of the ice cover over the Great Lakes region, differential isostatic rebound rates are location specific (table 3). Uplifting over thousands of years has been partially responsible for changes in the drainage patterns of the Great Lakes and, subsequently, in the present biota habitats.

TABLE 3

Isostatic rebound rates

Adapted from Clark and Persoage 1970

LAKE	AVERAGE RATE (CM PER 100 YEARS)
Superior	34.3
Michigan	11.4
Huron	17.3
Erie	0.0
Ontario	22.86

Since land masses are steadily uplifting in this region, the datum, or elevation reference system, must be changed every twenty-five to thirty-five years. In 1992, the datum was updated to reflect the crustal movement during the last thirty years; water levels were adjusted to reflect these increased elevations (Coordinating Committee on Great Lakes Basic Hydraulic and Hydrologic Data 1992).

Water Chemistry of the Great Lakes

The chemistry of surface waters is determined by a combination of natural and cultural, or anthropogenic, contributions. When the Great Lakes were formed during and after the last glacial advances, their chemistries were determined by the natural chemistry of the glacial meltwaters. As the postglacial lakes evolved, their chemistries were modified by natural contributions from the weathering of drainage-basin soils and bedrock. The bedrock under the Great Lakes basin ranges from the predominantly igneous and metamorphic rocks of the Canadian Shield in the north and west, to massive deposits of carbonates and shales in the south and east (Sloss et al. 1960). These bedrocks are overlain by surficial deposits of sand, gravel, and clays. Table 4 summarizes the common rock types and the major ions that are their primary weathering products.

Because bicarbonate is a major product of weathering, regardless of rock type, it is the predominant anion in natural waters. Calcium is the only other ion that is a weathering product of all four rock types. In addition to the ions listed in table 4, other inorganic components, such as phosphorus, nitrogen compounds, and trace metals, are also present in unmodified natural waters, but usually only in trace amounts. These occur in trace amounts because either they have a low solubility, they are used very quickly by biota, or they occur in only trace amounts in the rocks that are being weathered in the watershed. For example, lead is relatively insoluble and therefore occurs in low concentration in most natural waters. Radium is soluble in natural waters, but its abundance in common crustal rocks is quite low, and this is reflected by its low concentration in most natural waters. Phosphorus and nitrogen, although generally quite soluble, are essential nutrient elements for plant growth and therefore are rapidly taken up by aquatic plants. Based on table 4, natural waters without anthropogenic modification will contain K^+, Na^+, Ca^{+2}, Mg^{+2}, Cl^-, HCO_3^-, and SO_4^- as the predominant ions.

TABLE 4

Common rock and soil types and their primary chemical weathering products

Upchurch 1975

TYPE OF ROCK OR SOIL	K+	Na+	Ca+2	Mg+2	MAJOR WEATHERING PRODUCTS Fe+2,+3	Cl-	HCO3-	SO4-2	H4SiO4	SiO2 quartz	DEGRADED CLAY MINERALS
Basic igneous & metamorphic rocks and soils	X	X	X	X	X	X	X	X	X	X	X*
Acid igneous & metamorphic rocks and soils	X	X	X				X		X	X	X
Shales & clay-rich soils	X	X	X	X			X	X	X		X
Limestones, dolomites, & associated soils			X	X			X	X			X*

* Minor

General Chemical Characteristics of the Great Lakes

The reported average concentration of major ions, and other dissolved chemical constituents in each of the Great Lakes, based on data obtained during the 1960s, are shown in table 5, along with representative (average) concentration in rainwater, river water, average North American fresh waters, and seawater, for comparison. These data must be interpreted loosely since they are not all from the same year, nor is the quantity of data that went into each average the same. However, these data are suitable for qualitative analysis and provide an indication of what was happening chemically in the Great Lakes by the 1960s. The 1960s were a pivotal period in the history of water chemistry of the Great Lakes, since this is the period when the extent of degradation of water quality in the Great Lakes, primarily by overenrichment of nutrients, became a topic of intense focus (Beeton 1961, 1966; IJC 1970).

The dominant anion in the waters of the Great Lakes is bicarbonate, while calcium is the major cation, as expected, based on the previous discussion. In general, the major ion concentration in Great Lakes waters compare well to those of average North American fresh waters, and to average river water, with some exceptions:

1. The concentration of major ions and other dissolved constituents in Lake Superior are below the average for North American waters, although still the same order of magnitude. The Lake Superior basin is underlain primarily by Precambrian igneous and metamorphic rocks (Hough 1958), which are highly resistant to weathering, and therefore, the loading of weathering products (dissolved ions) to Lake Superior is low. In addition, the human population surrounding Lake Superior has been, and remains, the least of all the lakes: 607,000 people in the Lake Superior basin compared to 2.7 million to 11.7 million people in each of the other Great Lakes basins (Environment Canada 1995); and because many dissolved constituents are also added by human activities (see below), the anthropogenic loadings of these ions to Lake Superior are also lower than they are elsewhere in the Great Lakes basin.

TABLE 5

Reported average concentrations (in mg/L) of major ions and other dissolved chemical constituents in various natural waters

Data from the 1960s

DISSOLVED CONSTITUENTS	LAKE SUPERIOR[1]	LAKE HURON[1]	LAKE MICHIGAN[1]	LAKE ERIE[1]	LAKE ONTARIO[1]	RAIN WATER[3,5]	RIVER WATER[3,4]	NORTH AMER. FRESH WATER[6]	SEA WATER[3]
K^+	<1	1	1	1.4	1.3	—	2	1	380
Na^+	1	3	4	11	12	0.4	6	9	10,500
Ca^{+2}	12	27	33	38	38	1.4	15	21	400
Mg^{+2}	3	6	11	10	8	—	4	5	1,350
Cl^-	2	6	7	23	27	0.2	8	8	19,000
HCO_3^-	51[2]	96[2]	130[2]	113[2]	113[2]	—	58	—	142
SO_4^{-2}	2	13	20	21	27	2.1	11	20	2700
pH	7.8[2]	8.0[2]	8.0[2]	8.1[2]	7.9[2]	variable	variable	variable	—
Dis. Solids	52[2]	118[2]	150[2]	201[2]	194[2]	—	90	—	—
Total P (µg/L)	5	10	13	61	24[7]	—	—	—	—
Nitrate–N	0.5	—	0.1	0.1	—	—	—	—	—
SiO_2	2	2	2.5	0.8	0.3	—	13	9	—

[1]Beeton 1971a.   [2]Upchurch 1975.   [3]Hem 1970.   [4]Mean composition of world river water.   [5]One year average for inland sampling stations in the United States.
[6]Livingston 1963.   [7]Robertson & Scavia 1984.

2. The concentration of chloride in Lakes Michigan and Huron are close to the North American average, while Lakes Erie and Ontario have a significantly higher-than-average concentration. Human population centers are major sources of chloride in the Great Lakes, primarily from industrial wastes and from land runoff, the latter reflecting the extensive use of salt (NaCl) on roadways (Ownbey and Willeke 1965). The chloride content of lakes downstream from Lakes Michigan and Superior reflect the integration of additional inputs as the water moves toward the St. Lawrence Seaway. The chloride content of Lake Huron waters is determined by the mixing of inflowing waters from Lakes Superior and Michigan. In addition, Saginaw Bay provides a significant internal source of water with a higher-than-average chloride concentration, caused by leakage from the extensive brine fields in the Saginaw Bay watershed associated with oil fields there (Beeton et al. 1967). The chloride concentration in Lake Erie is more than three times that of Lake Huron. This is a result of Lake Erie's shallowness and small water volume, combined with significant urban and industrial areas that are sources of chloride. Ownbey and Kee (1967) identified the Detroit River and the Grand River as providing 90% of the chloride load to Lake Erie, and Upchurch (1975) added the urban areas of Toledo, Cleveland, and the industrial corridor from Lorain, Ohio, to Erie, Pennsylvania, as source areas. Lake Ontario's chloride level can be attributed to the inflow of high-chloride Lake Erie water, plus additional chloride, added by the urbanized areas of Toronto, Rochester, and Oswego (Upchurch 1975).

3. Magnesium and bicarbonate peak in Lake Michigan, and Lakes Michigan, Huron, Erie, and Ontario are consistently higher in alkalinity, calcium, and magnesium when compared to

average North American waters. Lakes Michigan and Huron follow the perimeter of the Michigan structural basin, which dominates the geologic structure of the lower peninsula of Michigan and is composed of Paleozoic limestones, dolomites, shales, and sandstones. The western shore of Lake Michigan, and the northern shore and part of the eastern drainage basin of Lake Huron, are underlain by massive dolomite deposits (Hough 1958). Weathering of these massive calcium-magnesium carbonate deposits provide a source for the high concentration of calcium, magnesium, and bicarbonate found in Lakes Michigan and Huron. Since alkalinity is driven by the carbonate system, the alkalinity of these lakes also reflects their proximity to carbonate deposits. According to Upchurch (1975), agricultural areas provide large amounts of calcium and magnesium to runoff. He noted that Lakes Erie and Ontario also lie in limestone and dolomite basins, as do Lakes Huron and Michigan. They are also highly agricultural, thus adding to the concentration of these ions coming from the upper lakes.

4. The silica concentration is low in all of the Great Lakes. Silica is a common weathering product of igneous and metamorphic rocks, and is one of the primary nutrients used by diatoms. There are no apparent anthropogenic sources of silica to the Great Lakes, so the primary source to the lakes is the slow weathering of the Precambrian igneous and metamorphic rocks that form the basement under large portions of the Lake Superior, northwestern Lake Michigan, and northern Lake Huron drainage basins (Hough 1958). The depletion of silica in Lakes Erie and Ontario, relative to the upper lakes, is probably a reflection of the much higher biological productivity of these lakes.

5. The concentration of sulfate increases from west to east in the Great Lakes. Upchurch (1975) identified four major sources of sulfate to the environment: atmospheric precipitation; runoff from areas containing deposits of gypsum and anhydrite; industrial wastes; and domestic sewage. Compared to average North American fresh waters, Lakes Superior and Huron have low sulfate concentrations, while Lakes Michigan, Erie, and Ontario are at, or above, the average. Lake Superior, which has the lowest average sulfate concentration, does not have significant sulfate deposits and, because it is the least populated of all the lake basins, industrial and domestic sources are also at a minimum compared to the rest of the Great Lakes basin. Beeton (1971a) concluded that the sulfate levels in Lake Michigan represented a long-term buildup. Lake Michigan supports urban centers at Chicago-Gary, Milwaukee, and Green Bay, which all appear to be significant sources of sulfate to the lake (Upchurch 1975). The concentration in Lake Huron reflects the mixing of low-sulfate Lake Superior water with Lake Michigan water, and the addition of sulfate internally, primarily from the Saginaw Bay watershed. The increases found in Lakes Erie and Ontario were attributed by Beeton (1971a) to the discharge of municipal and industrial wastes from the urban and industrial centers located on these lakeshores. In addition, there are anhydrite and gypsum deposits underlying Lake Huron and the Province of Ontario, which separate Lakes Huron, Erie, and Ontario (Hough 1958). Weathering of these deposits would provide a source of increased sulfate in the lower lakes. Hough (1958) postulated that leaching of sulfate bearing shales underlying the lower lakes might also be contributing to the increases in sulfate found in Lakes Erie and Ontario.

In summary, the concentration of the major weathering-related ions in the Great Lakes tends to increase from west to east, from Lake Superior to Lake Ontario. This trend reflects the nature of the rock and soil types in the drainage basins of the lakes, as well as the integration of natural and anthropogenic inputs as the water flows from Lake Superior and Lake Michigan into Lake Huron, and then from Lake Huron to Lake Erie and Lake Ontario.

Trophic Status and the Great Lakes in 1970

There have been many attempts to develop an acceptable limnologic classification scheme for lakes, but the most widely used is based on their trophic status: oligotrophic, mesotrophic, or eutrophic (Beeton 1965; Vollenweider 1968; Rodhe 1969; IJC 1970). These classifications are based on the nutrient supply and the resulting impact on the lake: oligotrophic lakes have low-nutrient concentrations and low productivity; eutrophic lakes have high-nutrient concentrations and high productivity; mesotrophic lakes have characteristics between oligotrophic and eutrophic. Eutrophic lakes are generally characterized by high concentrations of algae and may suffer from poor water quality—depletion of oxygen in the water column, caused by excessive organic matter production and subsequent decomposition in the water column and bottom sediments, which can result in loss of fish populations, and foul smelling water with a residual bad taste, even after treatment (Beeton 1971b).

In theory, lakes undergo a natural aging process, wherein progressive changes occur in the supply of nutrients and organic matter to the lake, caused in large part by the change in the surface-to-volume ratio of the lake as the lake ages and fills with sediments. As the basin fills, its volume decreases, lowering the surface to volume ratio and increasing the supply of nutrients and organic matter through internal recycling. An increase in nutrients and organic matter caused by natural processes, such as those just described, can result in increased productivity, and a gradual change in the trophic status of the lake from oligotrophic to mesotrophic to eutrophic. The aging of a lake by enrichment of its nutrient or organic matter content is called eutrophication (Beeton 1971b; Likens 1972), and when this occurs through the natural process of aging, it is called natural eutrophication. However, human activities can significantly accelerate the rate of aging of a lake, most often by significantly increasing the supply of nutrients over a short period of time, particularly through the discharge of domestic sewage. Hasler (1947) called this process cultural eutrophication, while Edmondson et al. (1956) and Edmondson (1969a, 1969b) referred to the same process interchangeably as artificial eutrophication, artificial enrichment, and cultural eutrophication.

Vollenweider (1968) provided one of the most comprehensive attempts to develop quantitative criteria for the trophic classification of lakes. He found that nitrogen and phosphorus levels appear to be the most important, though not the only, factors for determining trophic status, and that trophic status can be affected by morphometric, hydrologic, optical, and climate factors, as well as other nutrients. He developed log-log plots of (Annual P Loading v. Mean Depth) and (Annual N Loading v. Mean Depth), in which he was able to differentiate between oligotrophic, naturally eutrophic, and culturally eutrophic lakes. In addition, McIntosh et al. (1977) also proposed a detailed criteria for trophic classification of lakes (table 6).

Scientific studies of the chemistry of the Great Lakes were scattered and irregular until the early 1960s, when attention was drawn to the declining water quality of Lake Erie (Beeton 1961;

TABLE 6

Trophic classification criteria

Developed by McIntosh et al., 1977

	EUTROPHIC	MESOTROPHIC	OLIGOTROPHIC
Total Dissolved Solids (mg/l)	>150	100–150	<100
Mean Depth (m)	<10	<30	>15
Secchi-disk depth (m)	<2.5	2.5–6	>6
Chlorophyll (µg/L)	>8	4–8	<4
Total P (µg/L)	>24	16–24	<16
Dis-O_2^- (descriptive)	Usually depleted in bottom waters	Clinograde to anaerobic at depth	Orthograde, or if clinograde, >60% saturation at depth
Annual Primary Productivity (gC/m^2/yr)	>300	150–250	<100

Ayers 1962; Carr 1962). By the mid to late 1960s, the scientific community had documented several occurrences of cultural eutrophication in the lower lakes, and parts of Lakes Huron and Michigan (Beeton 1969). Lake Superior was the only lake that appeared to be undisturbed. Beeton (1965, 1971a) described the trophic status of the lakes in the 1960s:

- Lake Superior: oligotrophic throughout
- Lake Michigan: oligotrophic in general, but high total dissolved solids normally associated with eutrophic conditions indicate a trend toward mesotrophy; dissolved oxygen depletion in the waters of southern Green Bay indicates local eutrophy.
- Lake Huron: oligotrophic, except inner Saginaw Bay, which is eutrophic.
- Lake Erie: oligotrophic to mesotrophic in the eastern basin, eutrophic throughout the central basin and the western end.
- Lake Ontario: mesotrophic (morphometrically oligotrophic and the biota of an oligotrophic lake, but the physicochemical characteristics are eutrophic).

Cultural/Anthropogenic Contributions to the Changing Water Chemistry of the Great Lakes

Cultural or anthropogenic contributions consist of chemicals that are added as the result of human activities: municipal sewage (phosphorus, organic compounds, heavy metals, particulates), storm sewers (fertilizers and pesticides used by homeowners and businesses for grounds keeping, domestic pet wastes), agricultural wastes (fertilizers containing nitrate, phosphate, ammonia, lime, animal wastes containing urea and other organic compounds), industrial wastes (inorganic and organic compounds, heavy metals), inputs originating with commercial and recreational boating (oil, gasoline, kerosene, human wastes), and chemicals transported through the atmosphere and deposited on land and in surface waters (several organic contaminants including PCBs, Dieldrin, DDT, Toxaphene, and Cd, Pb, Zn, Hg, SO_4^{-2}). These types of inputs are determined by land use, population density, and prevailing wind direction, not by the underlying bedrock or soil type, and their routes of input include direct discharge to the lakes and rivers flowing into the lakes, discharge and seepage into groundwater, and volatilization into the atmosphere, followed by precipitation in rainwater.

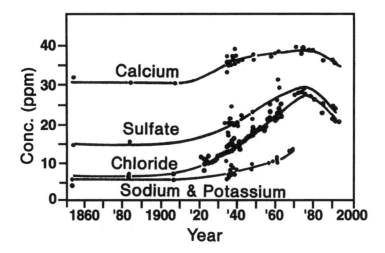

FIG. 6. *Changes in calcium, chloride, sulfate, and (sodium + potassium) ion concentrations in Lake Ontario from 1854 to 1991 (data from 1854 to 1963 after Beeton 1965, 1969; data from 1970 to 1991 from Kwiatkowski 1982, G. Warren, USEPA, personal communication; S. L'Italien, Environment Canada, personal communication).*

Direct, identifiable discharge into lakes and rivers are called point sources, which, once identified, can usually be controlled. Examples of point sources are municipal sewer outfalls and factory waste discharges via pipes flowing directly into a body of water. Inputs from groundwater seepage, atmospheric transport, and precipitation are much more difficult to identify or control, and are usually dispersed over wide areas, rather than at individual discharge points. These are called nonpoint sources, and include chemicals used on agricultural lands and suburban lawns (phosphorus and nitrogen fertilizers, pesticides, and herbicides), and chemicals transported through the atmosphere and deposited with rainwater (nitrogen compounds, organic chemicals).

Beeton (1965) compiled records from as far back as 1854, to the early 1960s. He showed that there has been a progressive increase in the concentration of total dissolved solids, SO_4, and Cl^- in all of the lakes except Lake Superior. Magnesium, on the other hand, did not change significantly in any of the lakes over this time period, which Beeton (1965) interpreted as evidence that the increases observed in the other ions were not caused by a change in the weathering and erosion of the underlying base rock. Beeton's most complete record was from Lake Ontario (fig. 6), and showed that the concentrations of calcium, sulfate, chloride, and (sodium + potassium) started increasing after approximately 1920, and by the early 1960s, when his records ended, the concentrations of all four ions were in an upward trend.

Beeton (1965) attributed these changes to the population growth around southern Lake Michigan, where the population of Chicago exceeded one million people by 1890. Until the Chicago Sanitary and Shipping Canal was completed in 1900, sewage from Chicago was discharged directly into Lake Michigan. Beeton (1965) also cited general census data for the region, which showed that the population in the region increased from 4.5 million to 16 million between 1850 and 1900, and was over 36 million by 1960. During the 1880s, the promise of agricultural land attracted

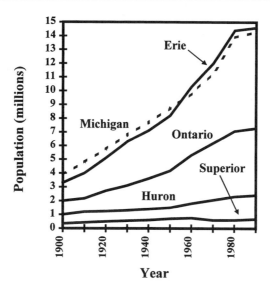

FIG. 7. *Population growth in the Great Lakes basin since 1900 (after Botts and Krushelnicki 1987).*

thousands of immigrants to the Great Lakes region, and by the mid 1800s, most of the land suitable for farming had been settled (U.S. EPA and Environment Canada 1988). The rate of population growth increased sharply around Lakes Michigan, Erie, and Ontario in the early 1900s (fig. 7), as agricultural expansion was replaced by urbanization and industrialization. Urban-industrial corridors grew along southern Lake Michigan (Chicago-Gary, Milwaukee, Green Bay), Lake Erie (Detroit, Toledo, Cleveland-Lorain, Erie, Buffalo), and Lake Ontario (Toronto, Buffalo, Rochester, Syracuse). In addition, the Saginaw Valley became a center for oil and salt/brine mining in southeast Michigan, with concurrent population growth in the Saginaw Bay watershed. The population and industrial growth of these areas, beginning in the early 1900s, corresponds to the period of greatest increase in the chemical content in the waters of the lower lakes (compare figs. 6 and 7).

We were able to update the Lake Ontario ion concentration record with additional data from 1970 through 1991 (G. Warren, USEPA, personal communication; S. L'Italien, Environment Canada, personnel communication). Figure 6 reveals that the upward concentration trends of chloride, sulfate, and calcium continued into the mid 1970s. However, during the late 1970s, calcium, sulfate, and chloride concentrations all peaked and then began to decrease. The downward trend in concentrations has been maintained at least through 1991. The changes coincided quite well with the implementation of the Great Lakes Water Quality Agreements of 1972 and 1978, and the resulting actions to control and reduce anthropogenic inputs to the lakes, which are discussed in the next section.

There are two categories of anthropogenic inputs of specific importance and concern to the water quality and chemistry of the Great Lakes: nutrients, such as phosphorus and nitrogen, which stimulate phytoplankton growth and can result in cultural eutrophication; and toxic substances (substances known to have an adverse impact on the living components of the ecosystem, including humans), such as heavy metals and complex, usually synthetic, organic compounds, such as polychlorobiphenyls (PCBs), pesticides, herbicides, and polycyclic aromatic hydrocarbons (PAHs), which are linked to severe and persistent environmental degradation.

Nutrients

Chemicals required for the growth and reproduction of plants are called nutrients. Those nutrients that are required at relatively high concentrations (parts per million and greater) for plant growth are called macronutrients; and those that, while essential, are required and occur in only trace, or barely detectable, concentrations, are called micronutrients. The primary macronutrients are phosphorus, nitrogen, calcium, potassium, magnesium, sodium, sulfur, carbon, and carbonates, while the primary micronutrients are silica, manganese, zinc, copper, molybdenum, boron, titanium, chromium, and cobalt (NAS 1972). Most of the other nutrient elements are found in natural concentrations that are sufficient to support the productivity of the lake.

In lakes unaffected by human activity phosphorus is usually a limiting nutrient. That is, the available phosphorus in such lakes is used up by phytoplankton growth before the maximum sustainable productivity for the lake is achieved, making phosphorus the chemical nutrient that limits additional productivity. In lakes receiving large amounts of phosphorus, nitrogen may be depleted before the phosphorus, thus become the limiting nutrient for most phytoplankton. However, silica may be a limiting nutrient for diatom production, since diatoms use silica to build their shells (Schelske et al. 1986). In the Great Lakes, phosphorus and nitrogen are the most important chemical nutrients and the ones subject to most scrutiny with respect to water quality.

As noted above, the 1960s were pivotal in the history of water quality in the Great Lakes. The scientific community noted, with increasing concern, the expansion of areas and locations in the Great Lakes, where the waters were turning eutrophic (Beeton 1961, 1965, 1966; Carr 1962). The most severe problems were found in water bodies of restricted circulation and/or small volume, where significant human population growth, intensified agricultural activity, and/or industrial expansion had occurred (Beeton 1970; IJC 1970). For example, although Lakes Michigan, Erie, and Ontario all received large influxes of human populations into their basins during the first half of the twentieth century (fig. 7), it was western and central Lake Erie, Saginaw Bay, and inner Green Bay that suffered the greatest water quality degradation by the 1960s. The effects of the increase in human activities adjacent to Lake Erie were magnified relative to those experienced in other lake basins, such as Lake Michigan, because of Lake Erie's shallow depths and small water volume. The average depth of Lake Erie is only 19m v. 85m for Lake Michigan, and the volume of Lake Erie is only 483km^3 v. 4,920km^3 for Lake Michigan (Michigan Sea Grant 1990a, b), yet the population increase in the Lake Erie basin approximated that of the Lake Michigan basin in both rate of growth and numbers, throughout the first half of the twentieth century (fig. 7). Saginaw Bay and Green Bay are both restricted embayments, and were sites of major urban and industrial growth during the early twentieth century. The loadings of nutrients and organic material brought about by the population growth and industrial activity in their watersheds exceeded their capacity to assimilate these materials, and eutrophic conditions rapidly developed (Beeton 1970).

The International Joint Commission (IJC) was created as a Binational Commission under the Boundary Waters Treaty of 1909 between the United States and Canada to address issues concerning water levels, water flows, and water quality of the shared boundary waters between the two

TABLE 7

1978 GLWQA target phosphorus loads and spring total phosphorus open-lake concentrations

After Neilson et al., 1994

LAKE/LAKE BASIN	TARGET PHOSPHORUS LOAD (10^3 KG/YR)	TARGET SPRING PHOSPHORUS CONCENTRATION (μg/L)
Superior	3400	5
Michigan	5600	7
Huron	4300	5
Erie	11000	
Western Basin		15
Central Basin		10
Eastern Basin		10
Ontario	7000	10

countries. In 1964, in response to the growing evidence of accelerated eutrophication in the lower Great Lakes, the IJC was given a Reference by the two governments to study the water quality conditions in the lower Great Lakes and St. Lawrence River. The results of that study (IJC 1970) led to the development and signing of the 1972 Great Lakes Water Quality Agreement (GLWQA) . The intent of the 1972 GLWQA was to retain (upper lakes) or restore and enhance (lower lakes) the water quality of the Great Lakes. It focused on the growing problem of eutrophication, and, based on the scientific evidence (IJC 1970), identified reduction and control of phosphorus loadings from point sources (municipal and industrial waste outlets) to the lakes as the primary way to control and reverse eutrophication. It also identified control of phosphate in detergents as another effective means of reducing phosphorus loadings.

Five years later, in 1978, after extensive reviews and technical studies by the IJC Water Quality Board, the GLWQA was revised, and included updated phosphorus loading targets for each lake (table 7). These revised loadings were based on achieving average in-lake phosphorus concentrations (table 7) proposed as guidelines by a panel of scientists (Vallentyne and Thomas 1978). This was subsequently changed to "spring open-lake phosphorus concentrations," since the spring phosphorus supply determines the summer phytoplankton biomass (IJC 1970). The 1978 GLWQA also recognized the significance of nonpoint-source phosphorus loadings (primarily from agricultural activities and storm water runoff), and expressed a growing concern over the many toxic organic chemicals being found in the Great Lakes ecosystem. Since 1972, the United States and Canada have spent billions of dollars for municipal wastewater treatment plant improvements and have implemented various nonpoint-source control strategies basinwide. These actions, taken under the GLWQA, have resulted in significant reductions in total phosphorus loadings throughout the basin. Phosphorus loads to all five lakes decreased during the period 1978–81 and have generally declined slightly or been approximately level since then (figs. 8a and 8b; Neilson et al. 1994). By 1985, the phosphorus load to Lake Superior had been reduced to below the target level (fig. 8a) and has remained below that level since, with open-lake total phosphorus concentration always below the target spring concentration (fig. 9a). During the 1990s, open-lake Lake Superior water quality has remained good and Lake Superior continues to be classified as oligotrophic, but

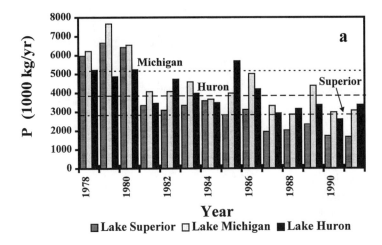

FIG. 8a. *Phosphorus loading to the Upper Great Lakes, 1978–91.*

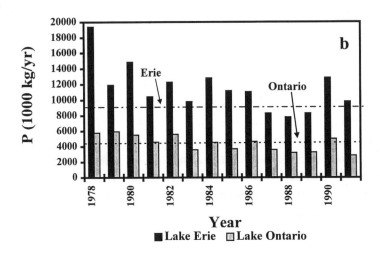

FIG. 8b. *Phosphorus loading to the lower Great Lakes, 1978–91 (both after Neilson et al. 1994). Dashed lines show target phosphorus loading levels established by the GLWQA of 1978.*

there is some local water quality degradation associated with major harbors (Rogers et al. 1993; Neilson et al. 1994).

Except for the years 1982, 1985, and 1987 in Lake Huron, both Lakes Michigan and Huron achieved the target phosphorus loads by 1981 (fig. 8a) and have had open-lake phosphorus concentration below the target levels during most of the period from 1978 through 1991 (fig. 9a). Within Lake Michigan, Green Bay, while remaining classified as eutrophic, showed some improvement in the bottom-water dissolved oxygen concentration, according to the Wisconsin Department of Natural Resources (Rogers et al. 1993); and Saginaw Bay, although still eutrophic, also showed some improvements after the implementation of phosphorus load reductions (Bierman et al. 1984;

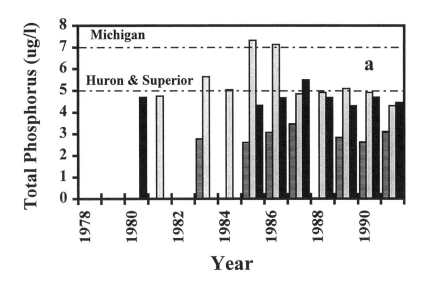

FIG. 9a. *Open-lake spring average total phosphorus concentrations in the upper Great Lakes, 1978–91.*

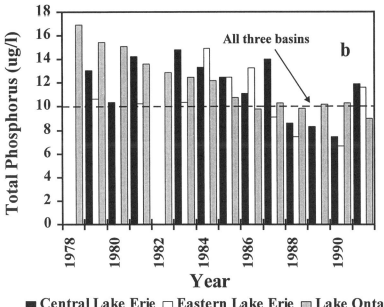

FIG. 9b. *Open-lake spring average total phosphorus concentrations in the lower Great Lakes, 1978–91 (both after Neilson et al. 1994). Dashed lines show the target spring total phosphorus concentrations established by the GLWQA of 1978.*

Bierman and Dolan 1986). Both Lakes Michigan and Huron remain classified as oligotrophic, with local areas of eutrophy (inner Green Bay and inner Saginaw Bay). Lake Michigan has experienced a continuing rise in chloride concentration, attributed, in part, to the weathering of chloride-containing minerals in some areas of its watershed, and in large, to the discharge of municipal and industrial wastes, and the use of salt for deicing roads, all of which contain substantial amounts of chloride (Sonzogni et al. 1983b; Moll et al. 1992). Although Lake Michigan continues to have a total dissolved-solids level high enough to suggest mesotrophy, especially in the southern basin, where the urban-industrial corridor of Milwaukee-Chicago-Gary is located, the bottom waters of Lake Michigan remain well oxygenated.

Lake Erie was the most severely degraded of the Great Lakes at the time the GLWQA phosphorus reductions were implemented. Phosphorus loads to Lake Erie (fig. 8b) were as much as three times what they were to Lakes Michigan and Superior, and up to four times that to Lake Huron. Oxygen depletion of near-bottom waters, a result of oxygen consumption caused by the decomposition of the large amounts of organic matter produced in the eutrophic water column and deposited in the bottom sediments, was first reported in the shallow western basin of Lake Erie in the 1950s (Carr 1962). Studies conducted in Lake Erie in the 1920s generally showed no evidence of severe oxygen depletion in Lake Erie's central basin until 1929, when some depletion was observed in a large portion of the central basin's near-bottom water. Studies in the early 1950s in the same general area showed oxygen concentration consistently lower than in 1929 (Carr 1962). By the late 1950s, large areas of both the western and central basins showed severe oxygen depletion and water quality had become extremely poor in both (Carr 1962; Beeton 1969; IJC 1970). Subsequent to the 1978 load reductions, phosphorus loads to Lake Erie dropped dramatically (fig. 8b) and, since 1981, have hovered near or below the target phosphorus load for the whole basin of 11,000mt per year ($11,000 \times 10^3$ kg/yr). Exceptions were 1982, 1984, and 1990, but in each of these years the exceedance was less than 10%. Based on chlorophyll, dissolved oxygen, and total phosphorus (fig. 9b), the eastern and central basins of Lake Erie have been restored to oligotrophic to mesotrophic conditions, and western Lake Erie is generally considered mesotrophic. However, depletion of dissolved oxygen in the central basin continues to be a problem (IJC 1989a).

Because Lake Erie provides a majority of the phosphorus load to Lake Ontario, the substantial load reduction to Lake Erie has benefited Lake Ontario. The target load of 7,000mt was achieved by 1981, and since then the load to Lake Ontario has hovered near that level, exceeding it as often as it has been below it. Phosphorus concentration was near the target in Lake Ontario starting in 1985, and, similar to the loading, have hovered slightly above to slightly below the target level since then. Lake Ontario is now considered to be oligo-mesotrophic to mesotrophic (IJC 1989a; Leach and Herron 1992).

The decreases in chloride and sulfate concentrations in Lake Ontario after about 1980 (fig. 6) provide additional evidence that the actions taken under the GLWQA have been effective in reducing the loadings, and therefore concentrations of anthropogenic pollutants to the lakes.

Toxic Substances

In addition to nutrients, numerous other chemical contaminants have found their way into the Great Lakes ecosystem. By the time the United States and Canada met in 1978 to renegotiate the

Great Lakes Water Quality Agreement, there was growing concern about the presence of toxic substances, and the incidence of areas in the Great Lakes basin that were experiencing severe environmental degradation due to toxic chemicals. The resulting 1978 GLWQA expanded the focus of the Agreement from eutrophication and nutrient overenrichment to include and emphasize toxic substances and Areas of Concern.

Article I of the 1978 GLWQA defines a toxic substance as one which "can cause death, disease, behavioral abnormalities, cancer, genetic mutation, physiological or reproductive malfunctions or physical deformities in any organism or its offspring, or which can become poisonous after concentration in the food chain, or in combination with other substances."

Annex 2 of the 1978 GLWQA defines an Area of Concern as any geographic area that fails to meet the General or Specific Objectives of the Agreement, where such failure has caused, or is likely to cause, impairment of beneficial use or of the area's ability to support aquatic life. "Impairment of beneficial use(s)" is defined as a change in the chemical, physical, or biological integrity of the Great Lakes system sufficient to cause any of the following:

- restrictions on fish and wildlife consumption;
- tainting of fish and wildlife flavor;
- degradation of fish wildlife population;
- fish tumors or other deformities;
- bird or animal deformities or reproduction problems;
- degradation of benthos;
- restrictions on dredging activities;
- eutrophication or undesirable algae;
- restrictions on drinking water consumption, or taste and odor problems;
- beach closings;
- degradation of aesthetics;
- added costs to agriculture or industry;
- degradation of phytoplankton and zooplankton populations; and
- loss of fish and wildlife habitat.

The IJC has identified 43 Areas of Concern in the Great Lakes basin; 26 in U.S. waters, and 17 in Canadian waters (fig. 10). Almost all 43 areas include toxic substances as a source of major impairment.

Although chemical contamination of the environment has occurred since the industrial revolution, the widespread presence of toxic substances found in the Great Lakes in the 1960s and 1970s was primarily a result of the increased commercial production and widespread use of organic chemicals and heavy metals that started after World War II. Reconstruction of environmental records from radiometrically-dated coherent sediment cores from Lake Ontario revealed the presence of several organic chemical contaminants starting as early as 1915, but increasing sharply after World War II, and peaking in the early 1960s (Durham and Oliver 1983). The records for individual chemicals also showed a clear relationship between the date of the first production of a given substance by the chemical industry and its appearance in the sedimentary record. There are over sixty-five thousand chemicals registered for use in the United States (U.S. EPA 1992), and by

1987, the list of chemicals verified as being present in the Great Lakes ecosystem had grown to 362 (IJC 1987), 329 of which are organic. One third of these chemicals may have toxic effects on life (IJC 1991a).

In 1985, the Water Quality Board of the IJC identified eleven substances which it considered to be Critical Pollutants in the Great Lakes (IJC 1985), including nine synthetic organics and two metals:

- tetrachlorodibenzo-p-dioxin (2,3,7,8—TCDD)
- tetrachlorodibenzo-furan (2,3,7,8—TCDF)
- polychlorinated biphenyls (PCBs—all forms)
- hexachlorobenzene (HCB)
- benzo(a)pyrene (BaP)
- DDT and its metabolites
- Mirex
- Toxaphene
- Dieldrin
- mercury
- alkylated lead

Critical Pollutants are chemicals that are present in the Great Lakes ecosystem, are highly toxic, are persistent in the environment (i.e., they do not break down easily in the environment), can bioaccumulate (i.e., they are concentrated in an organism faster than they are excreted from the organism) and can biomagnify (i.e., their concentration in organisms increases as the contaminant moves up the food chain) to levels which can threaten human health and the aquatic ecosystem (IJC 1985).

The concentration of several of the Critical Pollutants is high enough in some Great Lakes fish, especially lake trout and salmon, that state and provincial governments have issued advisories for people consuming these fish, suggesting that pregnant women and nursing mothers avoid them altogether, and that others in the population not consume Great Lakes fish more than once a week (IJC 1985; Environment Canada 1991a). Some species of fish can no longer be sold commercially because they contain high levels of PCBs, mercury, or other pollutants.

In addition to the potential for human health effects, there is increasing evidence linking persistent toxic substances to a variety of diseases and impairments found in organisms in the ecosystem, such as fish (especially bottom-dwelling species), mink, otter, snapping turtle, and several fish-eating birds (NRC and the Royal Society of Canada 1985; Environment Canada 1991c; IJC 1991b; U.S. EPA 1992, U.S. EPA and Environment Canada 1995). The most consistently documented effects on biota include reproductive failure, declining populations, developmental abnormalities, adult and embryonic mortality, malignancies, carcinogenic effects, neurobehavioral deficiencies, and genetic effects such as teratogenic or embryo deformities (NRC and the Royal Society of Canada 1985; Muir and Sudar 1987; Environment Canada, 1991c, IJC 1991b; U.S. EPA 1992).

Chemical contaminants can enter the Great Lakes ecosystem via several pathways: point sources (direct discharges), tributary loadings, atmospheric deposition, seepage of groundwater contaminated by agricultural runoff and leakage from old land fills and chemical disposal sites,

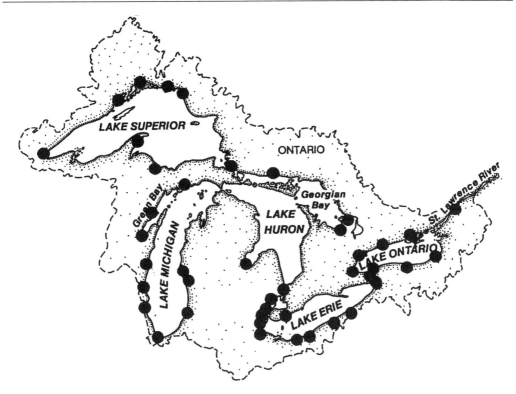

FIG. 10. *The 43 Areas of Concern as identified by the International Joint Commission as of 1995.*

and by remobilization of contaminants attached to sediments. In an attempt to control or eliminate the discharges of the contaminants of greatest concern at the time, the manufacture and/or
use of key chemicals was restricted or banned in the Great Lakes region during the 1970s and
1980s (Environment Canada 1991a; DeVault et al. 1994):

- Aldrin and Dieldrin (insecticides)—banned from use in Ontario in 1969 and in the United
  States in 1974;
- PCBs—manufacture and importation of PCBs prohibited by law in both the United States
  and Canada from 1979–80;
- DDT—banned in the United States in 1971; banned in Canada in 1985, but used until 1989
  under permit;
- Toxaphene—removed from general use in 1974 in Canada; manufacture banned and use
  restricted in the United States starting in 1982.

After implementation of these restrictions and bans, the concentrations of target compounds,
like DDT and PCBs, began to decline dramatically in predator fish and bird populations in the
Great Lakes. However, the decline, while rapid at first, was not sustained, and slowed significantly.
After five to ten years, the concentration trends became essentially level or only slightly downward (Environment Canada 1991b; DeVault et al.1994). Although the restrictions and bans either
reduced or eliminated new loadings of these contaminants to the lakes, there continue to be

significant basinwide contaminant inputs from tributary loadings, atmospheric transport, ground-water seepage, and contaminated sediments in the basins. These sources serve to buffer the lakes by maintaining the concentrations of contaminants at low, but detectable levels.

Tributary loadings can include contaminants provided by atmospheric deposition, point source inputs of newly discharged chemicals, and nonpoint source inputs from groundwater leakage, or from existing sediment-bound pollutants remobilized from contaminated sediments within the tributary watershed. There is, however, scant reliable data concerning tributary contaminant load-ings to the Great Lakes. The average concentration of contaminants in tributaries, at any particu-lar time, may be very low and, therefore, difficult to measure accurately, but the volume flow of the tributary may make the net annual input considerable. Recent studies suggest that the load of contaminants carried by tributaries may be significantly different than that measured by many routine monitoring programs, which often do not sample during and shortly after short-term, high-energy, weather-related events, in particular, storms. Johengen and Beeton (1992) found that a few storms can produce over 50% of the total annual pollutant load carried by a tributary to the Saline River in Michigan, even though such loading spikes lasted only a few days at a time, and would easily have been missed by a monthly sampling program; Ludwig et al. (1993) documented the rapid collapse of Caspian tern reproduction in Saginaw Bay following the apparent injection, in only a few days, of a large spike of contaminants into Saginaw Bay, the result of a 100+ year flood of the Saginaw River in 1986. Thus, tributary loading may be severely underestimated by routine monitoring programs, unless sampling frequency is sufficient to include storm effects, which could account for a significant percentage of the annual load.

Atmospheric transport and deposition are believed to be significant sources of lead, arsenic, cadmium, mercury, polycyclic aromatic hydrocarbons (PAHs), Lindane, Chlordane, DDT, Dield-rin, Toxaphene, and PCBs to the Great Lakes basin today (Keeler et al.1993; Hoff 1999). Atmo-spheric transport can carry contaminants originating internally in the Great Lakes basin, primarily from direct point-source discharges and recycling through volatilization, as well as from sources outside the basin. Internal sources include just about any human activity that involves heat pro-cessing or use of volatile compounds, such as:

- fossil fuel combustion (heavy metals, PAH, dioxins, furans)
- dry and wet process kilns (heavy metals, PAH)
- petroleum refineries (PAH)
- municipal incinerators (heavy metals)
- production of dry cell batteries (heavy metals)
- refineries and chemical plants (various chemicals)
- paper and pulp production (chlorine, PCBs)
- agricultural applications (pesticides)

Potentially large sources of atmospherically-transported contaminants exist in almost all areas of the Northern Hemisphere (Keeler et al. 1993). For example, in the United States, Missouri, Louisi-ana, and Texas generate the largest PAH emissions to the atmosphere, but are not the only sources of such emissions; Missouri is also a leading emitter of arsenic, cadmium, lead, mercury, and other heavy metals that may be transported to the Great Lakes basin by air masses and wind; PCBs appear to originate from spills and leaks from inventories existing all over the United States and Canada, especially from old electrical products still in use, in storage, or previously disposed of in landfills; Mexico is believed to be a large source of atmospheric emissions of heavy metals and persistent organic compounds. In addition, some compounds found in the Great Lakes, such as Lindane, are banned from use in Canada and the United States, leading to the conclusion that they may originate from as far away as India, China, and the former Soviet Union.

Groundwater seepage occurs throughout the basin, wherever the water table intersects the shoreline or lake basin edge. By itself, groundwater is not a problem. However, throughout the basin there are old landfills and chemical dump sites, especially associated with heavily industrialized urban sites that have and are leaking unknown amounts of contaminants into the local groundwater, which can then leak into tributaries leading to the Great Lakes, or leak directly into the Great Lakes. For example, a major source of contamination to the Niagara River, and Lake Ontario by extension, is groundwater contaminated by leakage from chemical and municipal dumps and land-fills adjacent to the Niagara River, and within its drainage basin (NRTC 1984; Environment Canada 1991a).

Many organic and heavy metal contaminants strongly sorb to sediment particles in the water column, which then settle out and accumulate in depositional areas. Particle residence times in the Great Lakes water column are generally much less than one year (Eadie and Robbins 1987). The combination of these processes—sorption and settling—provides a major mechanism by which contaminants are rapidly removed from the water column.

Contaminants removed to, and persisting in, the sediment zone will be removed from further interactions with the ecosystem if they are buried deeply enough, or if they decompose. However, research has shown that upon reaching the bottom, contaminants may be transferred back into the water column by diffusion and biotransfer, and that sediment particles are actively redistributed by bioturbation and repeated cycles of resuspension and redeposition (Robbins 1982, Eadie and Robbins 1987):

Biotransfer: the recent sediments of the Great Lakes generally support a diverse and abundant population of benthic organisms, including freshwater shrimp, amphipods, oligochaete worms, and freshwater clams. Many of these organisms occur at high population densities in Great Lakes sediments, from which they scavenge food, and through which they may come in contact with sediment-associated contaminants. Benthic organisms can be exposed by contact with sediment pore water, which may contain dissolved contaminant concentrations several orders of magnitude higher than those in overlying waters (Eadie and Robbins 1987), or by ingestion of contaminated particles. These organisms are also food (prey) for small fish, which are themselves exposed to the contaminants contained in the prey. This process tends to repeat itself up the food chain, resulting in a transfer and biomagnification of contaminants into the higher trophic level biota of the ecosystem (fig. 11). For example, large predator fish, such as lake trout, often have the highest level of organic contaminants in their tissue of all the organisms below them in their food chain, and fish-eating aquatic birds, such a herring gulls, are found to have higher concentrations in their bodies than the fish they eat (Environment Canada 1991a).

Bioturbation: while scavenging for food and ingesting particles, benthic organisms also mix the upper layers of sediments. The amphipod *Diporeia,* for example, lives at the sediment-water

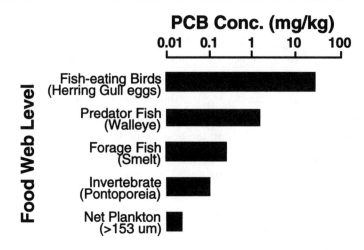

FIG. 11. *Total PCB concentrations in the components of the Lake Erie food web (IJC 1989b).*

interface and is a shallow burrower, typically homogenizing the upper 1–3cm of sediment. Oligochaetes burrow several centimeters into the sediment, and feed in a head-down position. The particles they ingest are excreted through their tails, which usually protrude through the sediment surface. Their feeding behavior brings material from several centimeters below the sediment-water interface back to the interface, where it is resuspendible. The feeding and burrowing of oligochaetes also homogenizes the upper layers of sediment. The thickness of the sediment-mixed layer can range from a few centimeters in the profundal sediments of the open lakes, to as much as 20cm in shallow sediments in bays (Robbins 1982; Eadie and Robbins 1987).

Diffusion: when sediment-associated contaminants are deposited in the sediment zone, they partition between the solid phase and dissolved phase (pore water), ideally moving towards chemical equilibrium between the phases. Measured concentrations of dissolved hydrophobic organic compounds in pore waters are often several orders of magnitude higher than in overlying waters (Eadie and Robbins 1987). Pore water diffusion into the overlying lake water will transport these dissolved contaminants back into the water column.

Resuspension: near-surface sediments (< 0.5 cm, Robbins and Eadie 1991) are periodically subject to resuspension by storms, especially during the winter, when the water column is unstratified. This provides a mechanism by which contaminants from the top of the sedimentary zone can recycle into the water column. Eadie et al. (1984, 1990), Eadie and Robbins (1987), and Robbins and Eadie (1991) discuss this topic thoroughly, and provide insight into the rates and time constants associated with the recycling and depositional processes in the Great Lakes.

The combination of these processes determine if, and how long, it takes for sediment-associated contaminants to become buried deeply enough to be removed from contact with the ecosystem. Based on measurements of the rate of removal of ^{137}Cs from the water column, Robbins (1982) and Robbins and Eadie (1991) estimated that the long-term contaminant removal rates for Lakes Huron and Michigan are on the order of 0.05/year. Thus, under the hypothetical situation of no additional contaminant loading, it would take about 15 years in these lakes for 50% of the existing inventory of contaminated sediments to be buried deeply enough to be out of contact with the ecosystem, and about 60 years for 95% of the present-day contaminants to be buried.

The Plankton and Benthos of the Great Lakes

The flora and fauna of the Laurentian Great Lakes have many species which are cosmopolitan, and likely became part of the biota at various times since glaciation. Some deep, cold-water organisms probably entered the Great Lakes region through proglacial lakes during the Pleistocene period (Henson 1966). A number of non-indigenous species have been introduced to the lakes in recent times (Mills et al. 1993). Some of these have become important in the lake communities (e.g., *Dreissena polymorpha, Eurytemora affinis*). The Great Lakes differ from the smaller inland lakes because of greater species diversity and the proglacial lake species. The biota of the Great Lakes and other large glacial scour lakes in North America, Europe, and Asia are closely similar. For example, Lakes Ladoga and Onega of Russia have the species associations similar to the Laurentian Great Lakes and the large lakes of Canada (Beeton, 1984). These large glacial scour lakes of the northern temperate region are all geologically young. Consequently, they do not have as great a species diversity and endemism found in truly ancient large lakes, such as Baikal in Siberia, Orchid in the Balkans, and the African rift lakes. The large glacial scour lakes are complex systems with a diversity of habitats (e.g., deep profundal regions, bays, marshes, extensive shallows, many kinds of tributaries, and independent water masses). Because of their large size, the Great Lakes provide spatial heterogeneity, vertically and horizontally, which contributes to the stability of zooplankton populations (McNaught 1978).

Phytoplankton

Phytoplankton are the small plants floating at the mercy of currents. These microscopic plants are the basis of the food web. The phytoplankton communities of the upper Great Lakes (Huron, Michigan, and Superior) and of Lake Ontario have many similarities in species composition and abundance (table 8). Some kinds of blue-green algae (cyanophyta) occur in each of the lakes, including the nutrient-poor waters of Lakes Huron and Superior, since blue-green algae are not only associated with nutrient-rich waters. Some species of green algae (chlorophyta) are abundant in each of the four lakes, but a greater diversity of green algae species have been recorded for Lake Ontario. Diatoms are always a major component of the plankton in each lake. Cyclotellas and some other diatoms are usually associated with nutrient-poor waters (Stoermer and Yang 1969), such as Lakes Huron and Superior (table 8). *Fragilaria crotonensis* is a cosmopolitan species abundant in all the Great Lakes (table 8 and 9). Phytoflagellates (Chrysophceae, Cryptophyceae, Euglenophyta) are especially abundant in Lakes Superior and Huron. At times, they dominate the phytoplankton communities of all four lakes. Euglenophyta has been abundant only in Lake Ontario. The phytoplankton communities of Lakes Superior and Huron have the closest similarities among the four lakes. Both lakes are nutrient poor. Furthermore, the major inflow to Lake Huron is from Lake Superior. Very small algae, picoplankton, being <34µm in size, have accounted for 40% to 50% of the carbon assimilation in Lake Superior (Munawar et al. 1987).

Green Bay, Saginaw Bay, and Lake Erie, especially the western basin, and many nearshore areas of the upper lakes, have phytoplankton communities associated with nutrient rich waters

TABLE 8

Common species of phytoplankton (≥5%) in the Great Lakes

Lakes Huron, Ontario, and Superior from Munawar and Munawar 1982, Lake Michigan data from Claflin 1975, Bowers et al. 1986, Holland 1980.
SP = spring, S = summer, F = fall.

	LAKE SUPERIOR			LAKE HURON			LAKE MICHIGAN			LAKE ONTARIO		
	SP	S	F	SP	S	F	SP	S	F	SP	S	F
Cyanophyta												
Anabaena spiroides						X						
A. subcylindrica				X	X	X						
Anabaena sp.	X		X	X				X	X			
Anacystis sp.								X	X			
Aphanocapsa sp.	X		X	X								X
Aphanothece clathrata			X	X								
Aphanothece sp.				X				X				
Chroococcus dispersus				X	X					X	X	X
C. limneticus				X								
C. minutus										X		
Gomphosphaeria aponina				X	X	X						X
G. naegeliana				X	X							
Gomphospheria sp.						X		X	X			
Lyngbya sp.	X		X	X								
Microcystis firma					X	X						
Microcystis sp.					X	X						
Oscillatoria agardhii				X	X	X						
Oscillatoria sp.	X	X				X						
O. tenuis		X										
O. limnetica	X									X	X	X
Chrysophyceae and other flagellates												
Chromulina sp.				X	X	X		X		X	X	
Chrysamoeba sp.	X											
Chrysocapsa sp.	X											
Chrysochromulina parva	X	X	X	X	X	X				X	X	X
Dinobryan bavaricum	X		X									
Dinobryon sp.							X	X	X			
D. divergens	X	X	X	X		X						
D. sertularia	X	X										
D. sociale	X							X		X	X	
Mallomonas sp.	X	X	X		X	X						
Ochromonas sp.	X	X	X	X	X		X					X
Dinophyceae												
Ceratium hirundinella					X				X		X	
Glenodinium sp.					X		X			X	X	X
Gymnodinium helveticum	X		X		X	X				X	X	X
G. uberrimum	X		X	X								
G. varians	X	X	X	X	X							
Gymnodinium sp.	X	X	X	X	X	X	X			X	X	
Peridinium sp.	X		X									
Chlorophyta												
Ankistrodesmus falcatus	X			X			X	X		X	X	
Asterococcus sp.												X
Carteria cordiformis											X	X
Chlamydomonas globosa										X	X	X
Chlamydomonas sp.	X			X	X							

	LAKE SUPERIOR			LAKE HURON			LAKE MICHIGAN			LAKE ONTARIO		
	SP	S	F	SP	S	F	SP	S	F	SP	S	F
Chlorella vulgaris				X	X	X						
Chlorella sp.										X	X	X
Closterium limneticum												X
Coelastrum microporum				X	X						X	X
C. proboscideum											X	
Cosmarium sp.								X			X	X
Crucigenia spp.									X			
Gloecystis ampla										X	X	X
Gloeocystis sp.	X		X		X			X				
Golenkinia radiata				X								
Gyromitus cardiformis										X	X	X
Lagerheimia ciliata											X	X
Mougeotia sp.				X	X						X	
Oedogonium sp.											X	X
Oocystis borgei	X										X	X
Oocystis sp.		X							X		X	
O. lacustris					X	X						
Pediastrum spp.								X			X	X
Pedinomonas minutissima										X	X	X
Phacotus lenticularis											X	X
Scenedesmus bijuga	X	X	X	X	X	X				X	X	X
S. spp.								X				X
Staurastrum paradoxum											X	X
Staurastrum sp.											X	X
Tetraedron sp.												X
T. minimum					X	X						
Ulothrix subtilissima											X	
Cryptophyceae												
Chroomonas acuta	X											
Cryptomonas erosa	X	X	X	X	X	X				X	X	X
C. phaseolus			X									
C. marssonii	X	X	X	X	X	X						
Cryptomonas sp.		X	X		X	X	X	X	X			
C. ovata										X		
Katablepharis ovalis	X	X	X		X					X	X	
Rhodomonas lens			X									
R. minuta	X	X	X	X	X	X	X			X	X	X
Rhodomonas sp.			X				X					
Diatomeae												
Asterionella formosa	X	X	X	X	X	X		X	X	X		X
A. gracillima	X		X	X	X	X				X	X	X
Cyclotella bodanica	X	X	X		X							
C. comta	X	X	X	X	X	X						
C. comensis					X							
C. glomerata	X		X									
C. kutzinginana					X				X			
C. michiganiana					X							
C. ocellata	X	X	X	X	X	X						
C. stelligera	X	X	X	X	X	X		X				
C. striata					X	X						
Cyclotella sp.	X	X	X		X	X	X	X				
Cymbella ventricosa					X						X	
Cymbella sp.					X						X	
Diatoma spp.						X		X				

	LAKE SUPERIOR			LAKE HURON			LAKE MICHIGAN			LAKE ONTARIO		
	SP	S	F	SP	S	F	SP	S	F	SP	S	F
Fragilaria capucina	X				X				X			
F. crotonensis	X	X	X	X	X	X	X	X	X	X	X	
Fragilaria sp.			X					X	X			
M. granulata	X			X	X	X						
M. islandica				X	X		X	X				
Melosira sp.				X		X	X	X				
M. varians										X	X	
M. italica							X		X			
Navicula											X	
Nitzschia palea					X					X	X	X
N. vermicularis				X								
Nitschia sp.						X						
Rhizosolenia eriensis			X	X			X	X				
Stephanodiscus astraea	X	X	X		X					X		
S. binderanus				X						X		X
Stephanodiscus niagarae				X						X		X
S. tenuis					X					X	X	X
S. hantzchii							X	X	X	X		X
Surirella angustata			X									
S. ovalis										X		
Synedra acus			X	X	X							
S. acus var radians	X	X	X	X		X	X					
S. ulna				X			X			X	X	X
Tabellaria fenestrata	X	X	X	X	X	X			X	X		
T. flocculosa	X	X	X	X	X		X	X	X	X	X	X
Tabellaria sp.										X		
Euglenophyta												
Lepoclinclis sp.											X	
Phacus sp.											X	

(table 9). Diatoms usually predominate, although blue-green algae, green algae, and chrysomonads are abundant, in Saginaw Bay. The blue-green algae population sometimes have large blooms resulting in taste and odor problems at water intakes.

The phytoplankton of southern Green Bay is similar to that of Saginaw Bay with diatoms, blue-green and green algae being abundant (table 9). The diatom community of northern Green Bay is closely similar to that of adjacent Lake Michigan (Holland and Claflin 1975). The species associations of the inshore areas of Lake Michigan are closely similar to those of lower Green Bay (Holland 1969).

Lake Erie differs from all the lakes, most likely a consequence of high-nutrient loading from many urban areas, extensive agricultural lands, and its shallow depth. Consequently, it has close similarities to Green and Saginaw Bays. Lake Erie is usually divided into the western, central, and eastern basins because of differences in depth, (increases from west to east, and especially in biota and productivity). The greatest phytoplankton densities are in the western basin and extending into the south central basin. The phytoplankton of the western basin includes many eutrophic species. Usually diatoms, phytoflagellates, and green algae are abundant in spring. The diatoms decrease as green and blue-green algae increase in summer. Diatoms dominate in the fall (Munawar and Munawar 1982).

TABLE 9

Common phytoplankton species (≥5%) in Lake Erie, Green Bay, and Saginaw Bay

Lake Erie data, Munawar and Munawar 1982; Green Bay, Holland 1969, Holland and Claflin 1975, Richman et al. 1984, Stoermer and Stevenson 1979; Saginaw Bay data from GLERL[1]
SP = spring, S = summer, F = fall

	LAKE ERIE WEST			LAKE ERIE CENTRAL			LAKE ERIE EAST			GREEN BAY			SAGINAW BAY		
	SP	S	F	SP	S	F	SP	S	F	SP	S	F	SP	S	F
Cyanophyta															
Agmenellum quadruplicatum														X	
Anabaena sp.		X			X						X			X	
Anacystis incerta											X	X		X	X
Aphanizomenon sp.		X			X			X			X				
Chroococcus dispersus				X			X	X			X				
Glocotricha											X				
Gomphosphaeria lacustris													X	X	X
Microcystis		X	X								X				
Oscillatoria spp.													X	X	X
Merismopedia tenuissima														X	
Chlorophyta															
Ankistrodesmus sp.														X	
Chlamydomonas globosa				X					X						
Chlorella sp.				X											
Cosmarium sp.								X							
Crucigenia quadrata														X	
Gloeocystis planctonica											X	X		X	X
Oedogonium sp.					X			X							
Oocystis											X	X		X	
Pandorina morum															X
Pediastrum boryanum														X	
P. duplex													X	X	
P. simplex		X				X		X	X						
P. tetra													X		
Scenedesmus sp.														X	
Scenedesmus quadricauda										X				X	X
S. denticulatus											X				
Sphaerocystic sp.								X							
Staurastrum paradoxum	X				X			X	X					X	
Chrysophyceae and other flagellates													X	X	X
Chrysochromulina parva				X				X							
Ochromonas sp.							X								
Diatomeae															
Amphora ovalis														X	
Asterionella formosa											X			X	X
A. gracillima															X
Coscinadiscus spp.		X	X												
Cyclotella atomus											X				
C. comensis											X	X	X		X
C. meneghiniana											X		X		
C. kutzighiana														X	
C. ocellata														X	
C. stelligera														X	
Diatoma tenne elongatum													X		
Fragilaria capucina					X			X			X	X	X	X	X
F. construens													X		
F. crotonensis	X		X	X			X	X	X	X	X	X	X	X	X

	LAKE ERIE WEST			LAKE ERIE CENTRAL			LAKE ERIE EAST			GREEN BAY			SAGINAW BAY		
	SP	S	F	SP	S	F	SP	S	F	SP	S	F	SP	S	F
F. pinnata													X	X	X
Melosira ambigua										X	X				
M. granulata										X	X		X	X	X
M. islandica			X										X	X	X
M. italica													X		
Nitzschia fonticola															X
Stephanodiscus astraea											X				
S. binderanus	X			X							X		X		
S. hantzchii				X							X				
S. minutus											X		X		
S. niagarae		X	X	X	X	X	X		X	X	X	X			
S. tenuis	X	X	X	X			X				X		X		
Synedra filiformis										X				X	
Tabellaria fenestrata											X			X	
T. flocculosa											X				
Thalassiosira fluviatilis											X				
Cryptophyceae															
Chroomonas spp.										X					
Cryptomonas erosa	X	X	X	X	X	X	X	X	X						X
Cryptomonas spp.											X	X	X		
Rhodomonas spp.	X	X		X	X	X	X	X	X	X			X	X	X
Dinophyceae															
Ceratium hirundinella		X			X		X								
Dinobryon sp.												X		X	X
Gymnodinium helveticum	X			X			X								
G. uberrimum				X			X						X		
G. spp.										X					
Peridinium aciculiferum				X			X								

[1]Great Lakes Environmental Research Laboratory (GLERL) phytoplankton data from 13 locations sampled in 1991.

In the central basin of Lake Erie, phytoflagellates are especially important for the productivity and biomass in spring, whereas diatoms dominate in the fall, (Munawar and Munawar 1982).

Three biomass peaks have been observed for the eastern basin. Phytoflagellates dominate in the spring; phytoflagellates, green and blue-green algae peak in the summer, while diatoms dominate the fall phytoplankton (Munawar and Munawar 1982).

Significant spatial differences have been shown in the distribution of nutrients and phytoplankton in Lakes Michigan and Ontario. Lake Ontario has pronounced inshore-offshore differences in productivity and species composition of the phytoplankton (Munawar and Munawar 1982). The inshore biomass of Lake Ontario may be 25% greater than offshore. Inshore, two peaks occur in the productivity/biomass, whereas only one peak occurs offshore. Eutrophic species dominate the inshore community. The greater productivity/biomass of inshore water versus offshore is a consequence of nutrient-rich waters from river, runoff, and sewage plants. The open lake phytoplankton has oligotrophic species. Inshore waters of Lake Michigan, less than 16km from shore, may have greater populations of diatoms, species favored by nutrient rich waters, and greater concentrations of nutrients (Holland and Beeton 1972). Holland (1968) found the distribution of

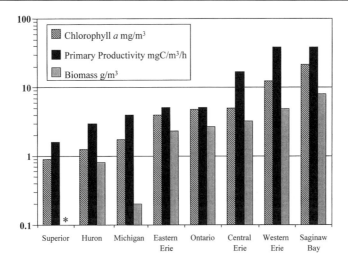

FIG. 12. *Concentrations of chlorophyll a, primary productivity, and biomass in the five Great Lakes including the three basins of Lake Erie and the Saginaw Bay. Chlorophyll from Dobson et. al. (1974), biomass and productivity from Munawar and Munawar (1982) except for lake Michigan. Lakes Huron and Michigan chlorophyll and Lake Michigan productivity calculated from Farnenstiel et. al. (1989). Lake Michigan biomass calculated from Fahnenstiel and Scavia (1987).*

species of *Melosira* reflected differing trophic conditions in Green Bay, inshore and offshore waters of Lake Michigan.

The nutrient-poor, more oligotrophic lakes, have only one pulse in phytoplankton abundance, usually in spring. In the ultra-oligotrophic waters of open Lake Superior, no clear seasonal trend has been observed in phytoplankton abundance (Munawar and Munawar 1982). Lake Superior has very low chlorophyll a, primary production, and biomass (fig. 12). Lake Huron is somewhat more productive with peak plankton abundance in spring, but seasonal peaks are not well developed. The offshore waters of Lake Michigan have one annual phytoplankton pulse (Holland 1969); these waters appear to be more productive than those of Lake Huron (fig. 12). The mesotrophic waters of offshore Lake Ontario and Central Lake Erie also have only one major phytoplankton pulse (Munawar and Munawar 1982).

Great Lakes waters which are more productive (i.e., western Lake Erie, inshore Lake Ontario) (Munawar and Munawar 1982), inshore Lake Michigan and Green Bay (Holland 1969), have two or more peaks of phytoplankton abundance, probably reflecting seasonal temperature changes and availability of nutrients. Western Lake Erie and Saginaw Bay (fig. 12) are the most productive waters of the Great Lakes. Primary productivity is estimated to be about 35mgC/m^3/hr in these ecosystems about 30–33mg/m^3/hr greater productivity than in the other lake waters. Chlorophyll a values up to 25.4mg/m^3 were reported for western Lake Erie and biomass concentrations up to 17.2g/m^3 were reported for Saginaw Bay (Munawar and Munawar 1982). Lower Green Bay is also productive with chlorophyll a values of 10.4mg/m (Holland 1969).

Scavia and Fahnenstiel (1987) proposed a hypothesis to explain seasonal phytoplankton dynamics in Lake Michigan, a hypothesis which probably can be applied to the deep offshore waters of all the Great Lakes, except Lake Erie. The early spring diatom production in a thoroughly mixed system is a function of gradual warming and increasing light. Once the lake stratifies, nutrient concentrations are reduced and limit growth in the epilimnion. As the lake stratifies, diatoms sink from the epilimnion and in-situ growth is insufficient to replace the population loss. The sinking phytoplankton becomes an important food for hypolimnetic and benthic animals, and grazing becomes an important factor for controlling algal abundance. During stratification, the euphotic zone usually extends below the thermocline and a two-layer phytoplankton community develops, giving rise to a deep chlorophyll layer. The lower, sub-thermocline layer is nutrient-light controlled, whereas the upper layer is phosphorus limited. Much of the epilimnetic phosphorus is recycled, and thus, controlled by zooplankton, which effectively graze the algae. This two-layer system begins to break down in the fall, as thermal stratification is disrupted by increased mixing due to falling temperatures and strong autumn winds. The increased mixing may occur early enough to permit a fall plankton pulse because of increased nutrients mixed into euphotic waters. In deep waters of the open lakes this mixing of nutrients occurs too late to be effectively used by the algae because the system is becoming light and temperature limited.

Zooplankton

The zooplankton, those small animals with limited mobility, of the Laurentian Great Lakes consist of fourteen major species of copepods and twelve major species of cladocerans (i.e., Crustacea, that are usually abundant sometime during the year). Of these, some are clearly cosmopolitan species which are abundant in the cold, oligotrophic waters of the upper lakes, as well as in the warmer, more eutrophic waters of Lake Erie, Saginaw Bay, and Green Bay (e.g., *Cyclops bicuspidatus thomasi* and *Bosmina longirostris*) (table 10).

The upper Great Lakes zooplankton is dominated by diaptomids, (i.e., *Diaptomus ashlandi*, D. *minutus*, and D. *sicilis*, and the cyclopoid *C. bicuspidatus*) (table 10). Cladocerans are present in the warmer months, especially *Daphnia retrocurva* and *B. longirostris*. The deep water copepod, (*Senecella calanoides*) is recorded for Lake Superior, Lake Michigan, and the North Channel and Georgian Bay of Lake Huron. It is likely in open Lake Huron. *Limnocalanus macrurus* is reported for all the lakes. These two proglacial lake species are hypolimnetic and may be more abundant than indicated.

The lower lakes, Lakes Erie and Ontario, share a number of species with the upper lakes (e.g., *C. bicuspidatus, Daphnia retrocurva, D. longiremis*, and *B. longirostris* are abundant species). Lake Ontario has two species which are important in one of the upper lakes, but not in Lake Erie, (i.e., *Tropocyclops prasinus* and *Ceriodaphnia lacustris*). Lake Erie has a mixture of species common to the upper lakes as well as species which are unique. The western basin has all of the major upper lakes species, probably because the major inflow is from Lake Huron via the St. Clair and Detroit Rivers. In addition, *Daphnia galeata mendotae, Eubosmina coregoni*, and *Eurytemora affinis* are abundant species. The central and eastern basins do not have all the upper lakes species since these basins are not as greatly influenced by the Detroit River. The species that are shared by Lake Erie, Saginaw Bay, and Green Bay include *Alona sp., Eucyclops agilis*, and *Canthocamptus sp.* The latter two are bottom forms which are found in plankton samples. *Canthocamptus sp.* occurred in 31% of the Great Lakes Environmental Research Laboratory samples collected by Ohio State

TABLE 10

Occurrence of crustacean plankton in Lake Huron (LH), Georgian Bay (GB), North Channel (NC), Saginaw Bay (SB), Lake Superior (LS), Lake Michigan (LM), Green Bay (GrB), eastern Lake Erie (EE), Central Lake Erie (CE), Western Lake Erie (WE), and Lake Ontario (LO)

Data for LH, LS, EE, CE, and LO from Patalas (1972), Gannon (1981), and Watson (1974). Data for GB and NC from Carter and Watson (1977). Data for LM and GrB from Gannon (1972). Data for SB and WE from unpublished data for 1991 and 1992 from A. M. Beeton. X indicates occurrence > 1%, p = present, but < 1%.

SPECIES	LS	LH	GB	NC	SB	LM	GRB	EE	CE	WE	LO
Limnocalanus macrurus	X	p	p	p		X	p	p	p	p	p
Diaptomus ashlandi	X	X	X	X	p	X	p	p	p	X	p
D. minutus	X	X	X	p	p	X	p	p	p	X	p
D. oregonensis	p	X	p	p		X	X	X	X		p
D. sicilis	X	X	p	p	p	p	p	p	p	X	p
D. siciloides	p	X			p	p	X	X	X		p
Senecella calanoides	p		X	X		p					p
Epischura lacustris	p	X	X	X		X	p	p	p	p	p
Eurytemora affinis	p	X		X	X	X	p	p	p	X	p
Cyclops bicuspidatus thomasi	X	X	X	X	X	X	X	X	X	X	X
C. vernalis	p	X	X	X	X	p	X	p	X	p	p
Mesocyclops edax	p	X	X	X	p	X	X	X	X	p	p
Tropocyclops prasinus		X		X	p	p	p	p	p	p	X
Canthocamptus sp.						p	p	p		X	
Alona spp.				p	p	X				p	p
Bosmina longirostris	X	X	X	X	X	X	X	X	p	X	X
Eubosmina coregoni		X	p	p		X	X	p	X	X	X
Ceriodaphnia lacustris	p	X	X	X	p	p	p		p	p	X
Chydorus sphaericus	p	X	X	X	X	X	X	p	p	p	p
Daphnia galeata mendotae	X	p	X	p	p	X	X	X	X	X	p
D. longiremis		X	p	X		X	p	X	X		X
D. retrocurva	X	X	p	p	p	X	X	X	X	X	X
Diaphanosoma leuchtenbergianum	p	X	X	X		p	p	p	p	p	p
Holopedium gibberum	p	X	p	p		p	p	p	p		p
Leptodora kindtii	p	X	X	X		p	X	p	p	p	p
Polyphemus pediculus	p	X	X	X		X	p			p	p

University in 1992 in western Lake Erie. Presumably, these bottom-living forms are mixed into the plankton by storms affecting the shallow waters of western Lake Erie and the bays.

Saginaw Bay is closely similar in its productivity to western Lake Erie, but its zooplankton assemblage is different, since there is major inflow of Lake Huron water. *Cyclops vernalis* is an important species. Diaptomids are not important, and occur infrequently. *E. affinis* is established in the Bay, but it is not especially abundant. The major cladoceran is *B. longirostris*, although *Chydorus sphaericus* is almost as important and frequently dominates the zooplankton.

The spiny water flea, (*Bythotrephes cederstroemi*) and *E. affinis* are exotic species which have become important in the zooplankton. *B. cederstroemi* is abundant at times in Lakes Huron and

Michigan. *E. affinis* is especially abundant in western Lake Erie, and it is an important species in Lake Huron (table 10).

Patalas (1972) explored the possible relationship between species and abundance of crustaceans with temperature and nutrient enrichment. He concluded, "a declining proportion of diaptomids and an increasing abundance of cyclopoids and cladocerans are the more characteristic trends apparently related to increasing nutrients." Thus, we observe that the zooplankton is dominated by diaptomids in the upper more nutrient-poor lakes, whereas cyclopoid copepods and cladocerans are the major groups in the nutrient-rich lower lakes, Saginaw Bay and Green Bay (table 10).

Pronounced differences occur between the species composition and abundance of in-shore and open-lake zooplankton communities in Lake Michigan. Some species are very abundant in-shore, and at the same time, they will not be present in the open-lake plankton. For example, Beeton and Barker (1974) found that *C. bicuspidatus, D. oregonensis* and cladocerans were abundant inshore, but few were found offshore in late fall of 1971 in Lake Michigan. *Limnocalanus macrurus* was abundant offshore but never inshore. Differences of this nature can affect species lists for the lakes. It is more useful to differentiate between the upper lakes open-water community, and the community in Green Bay, Saginaw Bay, and the near shore of all the lakes. Consequently, we find a calanoid dominated zooplankton community in the open waters of the upper lakes (table 10). The species association, which has more cyclopoid copepods and cladocerans, is more characteristic of Lake Erie, Saginaw and Green Bay, and the inshore waters of the other four lakes.

Seasonal trends in the open-lake planktonic Crustacea reported by Gannon (1972) for Lake Michigan may be representative of the offshore waters of the upper lakes. He found that *D. ashlandi, D. minutus, D. oregonensis,* and *C. bicuspidatus* were important all year with maximum numbers in July through September and lowest numbers in October and November. A secondary peak in abundance occurred in winter. Immature diaptomids were abundant from June to October, while immature cyclopoids were present all year and especially abundant from July to September. Cladocera, primarily *D. retrocurva, B. longirostris,* and *E. coregoni,* were abundant July–September with low numbers in October and November. They were essentially absent the remainder of the year.

It appears that both the Cladocera and copepods have two periods of peak abundance, late spring and midsummer, in western Lake Erie (Gannon 1981). The dominant Cladocera have been *B. longirostris* and *E. coregoni.* The dominant copepods were *C. bicuspidatus, C. vernalis,* and *D. siciloides.* The latter two have decreased in importance in the 1990s and *C. bicuspidatus* and *E. affinis* now dominate the zooplankton. *Cyclops bicuspidatus, C. vernalis, B. longirostris, E. coregoni* and *D. siciloides* were also abundant species in May and July in Green Bay (Gannon 1972). *Mesocyclops edax, D. retrocurva,* and *C. sphaericus* were also abundant. The seasonal trends in abundance of cyclopoid copepods and cladocerans observed in western Lake Erie may represent trends in the other more eutrophic waters of the Great Lakes.

The communities of planktonic Crustacea are affected by fish and invertebrate predation, as well as in abundance of their phytoplankton food, which in turn is determined by the availability of nutrients. The importance of fish predation was well demonstrated by major size-related changes in zooplankton populations in Lake Michigan between 1954 and 1966 (Wells 1970). The largest cladocerans and copepods, which were easy prey, underwent major decline, while some medium-sized or small species increased in numbers after large populations of alewife, which feed on plankton, became established. *Daphnia retrocurva* decreased in abundance, and also in size from 1954 to 1966, but the average size increased after an alewife die-off in 1967. Mature *D. retrocurva* were 1.26–1.70mm in 1954, 0.90–1.35mm in 1966, and 1.14–1.56mm in 1968 (Wells 1970).

It appears that all the planktonic Crustacea undergo daily vertical migrations (Wells 1960). The amplitude of migration of copepods, (e.g., *L. macrurus*), can be at least 40m in migrating from the hypolimnion into the epilimnion, whereas cladocerans are usually favored by the warmer temperatures of the epilimnion and their migrations are less than 24m amplitude (McNaught 1966). Diurnal changes in light intensity (Wells 1960) coupled with narrow-band visual pigments (McNaught 1966) appear to be the key environmental and physiological components for vertical migration behavior. Water temperatures, especially sharp thermal gradients affect the extent of migrations. Upward migrations at night are advantageous because of abundant food in the upper strata, but downward movement and dispersal during daylight reduces predation by fish and larger invertebrates.

Rotifers have always been abundant in the zooplankton of the Great Lakes (Davis 1966). Numerically they are important, although their biomass is small in comparison to Crustacea. Nevertheless, they are important in the food web. Rotifers have not been extensively studied in recent years, but Stemberger et al. (1979) found seventy-five species in samples taken in Saginaw Bay and southern Lake Huron in 1974. The greatest numbers of rotifer occurred in late spring and early summer samples. Up to 3,146 rotifers l^{-1} were found in Saginaw Bay. The predominant planktonic genera in the Bay and southern Lake Huron were *Keratella, Polyarthra, Synchaeta, Notholca, Filinia,* and *Conochilus.* These are the same genera which are important in western Lake Erie. Several species found only near the Saginaw River may be eutrophic indicators, (i.e., *Anuraeopsis fissa, Brachionus* spp., *Conochiloides dossnarius,* and *Keratella, cochlearis* f. *tecta*).

Polyarthra, Keratella, Synchaeta and *Notholca* were the main genera found by Stemberger (1973) in Lake Michigan. These genera are probably widespread and common in the Great Lakes. He found significant inshore-offshore differences in abundance and species composition. Rotifers were much less abundant off-shore where less phytoplankton occur. *Conochilus unicornis* was a major offshore species. Inshore species were *Filinia longiseta, Trichocerca multicrinus, Kellicottia longispina,* and others.

Benthos

The benthic communities of the Great Lakes are closely similar in species composition throughout the system. The amphipod *Diporeia* sp. (*Pontoporeia hoyi*), the oligochaete *Stylodrilus heringianus,* small fingernail clams (e.g., *Pisidium conventus*), and midges (*Heterotrissocladius subpilosus*) dominate the profundal zone. Cook and Johnson (1974, 765) concluded that, "the lakes differ from each other quantitatively, not qualitatively; that is in abundance, not kind, of animals inhabiting the benthic environment." This statement certainly applies to the profundal zones of the five lakes, but less clearly to the inshore and bays which are subjected to various natural and human-caused perturbations. Thus, we find that *Diporeia* comprise 70% to 80% of the benthic communities in Lakes Huron and Superior, oligochaetes about 20%, fingernail clams 14%, and midges around 2% to 4% (Schuytema and Powers 1966; Cook and Johnson 1974). *Diporeia* are less abundant in Lake Michigan (ca 65%), oligochaetes are about 20% to 24%, fingernail clams 5% to 15%, and midges around 1% (Cook and Johnson 1974). *Diporeia* comprise about 50% of the benthos in Lake Ontario, whereas oligochaetes, primarily *S. heringianus,* are 30% (Cook and Johnson 1974), and midges and fingernail clams are present, but not in large numbers (Brinkhurst et al. 1968). In Lake Erie, the profundal zone is primarily the eastern basin where amphipods *(Diporeia* and *Gammarus)* make up 27% of the benthos, oligochaetes 34%, midge larvae 24%, and fingernail clams 9% (Veal and Osmond 1968).

The relative importance of *Diporeia* and oligochaetes progressively changes from oligotrophic Lake Superior into the more productive lower lakes. *Diporeia* comprises less of the total benthos, and oligochaetes a greater percentage as one progresses from the oligotrophic, to the mesotrophic, into the eutrophic waters. Furthermore, the profundal benthos is primarily *Diporeia, S. heringianus, H. subpilosus,* and *P. conventus* in the upper lakes, whereas other species of oligochaetes, midges, and fingernail clams become important in the nutrient rich lower lakes. Also, *Gammarus* shares prominence with *Diporeia* in eastern Lake Erie. The change in importance of oligochaetes relative to the trophic state is shown by Lake Erie benthos where oligochaetes increase from 34% of the benthos in the eastern basin to 55% in the central basin to 86% in the western basin (Veal and Osmond 1968). The eastern basin is considered oligo-mesotrophic, the central basin mesotrophic, and the western basin eutrophic.

The benthic productivity of the lakes is reflected in the biomass data (table 11). Benthic biomass is < 1.0 g/m^2 in Lake Superior, whereas the western Lake Erie benthic biomass is > 4 g/m^2.

A shrimp-like organism, *Mysis relicta,* is sometimes included as part of the benthos or zooplankton. It is a free-swimming organism, and therefore may more appropriately be called nekton. Nevertheless, these animals, especially the larger ones, are on the lake bottom some of the time. They also make extensive diel vertical migrations to the water surface, or only to the thermocline if there is a sharp thermal gradient (Beeton and Bowers 1982). Because of this migration, they are important to the exchange between pelagic and benthic animals (i.e., pelagic-benthic coupling, since they feed on phytoplankton and zooplankton in the water column, and on detritus in the sediments). They can also feed on larger animals such as *Diporeia* (Parker 1980). The mysids are important in the food chain, especially in the diets of many Great Lakes fish.

Both *M. relicta* and *Diporeia* are especially important to secondary production in the profundal zone of the Great Lakes. Secondary production of *Diporeia* ranges from 2.9 to 6.0g/m^2/yr, and the *M. relicta* secondary production in Lake Michigan is estimated to be between 0.25 and 3.2g/m^2/yr (Sell 1982). Estimates for *M. relicta* productivity in Lake Huron are only 1.5g/m^2/yr.

As mentioned above, mysids are important in pelagic-benthic coupling in the food web because they feed in the water column and on the bottom. Pelagic-benthic coupling is more complex for the profundal benthos, since these organisms do not migrate into the water column to feed, and must rely on the organic material settling out (i.e., the phytoplankton from the photic zone). It appears that the spring bloom of lipid-rich diatoms is the major source of organic material for the profundal benthos, especially *Diporeia,* since summer inputs of organic material are not sufficient to account for the annual production of *Diporeia* (Gardner et al. 1990). The weight-

TABLE 11

Mean abundance of profundal benthos and biomass estimates

LAKE	MEAN ABUNDANCE #M^{-2}	BIOMASS g/m^2
Superior	172[1]	0.09[3]
	314[2]	
Huron	720[4]	1.48[3]
Michigan	4265[5]	3.4[3]
Ontario	2600[6]	1.84[7]
Erie	4600[8]	4.63[3]

[1]Hiltunen l969   [2]Schelske and Roth 1973   [3]Alley and Powers 1970   [4]Shrivastava l974
[5]Powers and Alley l967   [6]Hiltunen l969b   [7]Johnson and Brinkhurst l97l   [8]Veal and Osmond l968

specific lipid content of *Diporeia* can double in a few weeks after the spring diatom bloom, especially *Melosira*.

The benthic communities of the shallow inshore areas, bays, and harbors have a large number of species, many of them cosmopolitan, and found widely in small lakes and rivers. The major groups are clams and snails, insects, crustaceans, and oligochaetes. The clams include the fingernail clams, but the unionidae have been especially abundant in some areas (e.g., Lake St. Clair). The major insect groups in many shallow areas are the chironomids, ephemeropterans, and Tricoptera, although other groups (e.g., Odonata and Plecoptera) may be especially important in some shallow water areas. Amphipods and crayfish are the major Crustacea. A wide variety of oligochaetes occur in bays and harbors. The species distribution and abundance of oligochaetes closely reflect the nature of the habitat (Brinkhurst et al. 1968). In grossly polluted areas oligochaetes are especially abundant, and the most common species is *Limnodrilus hoffmeisterii*, (e.g., Toronto Bay, western Lake Erie, southern Green Bay, and southern Saginaw Bay). Species which do well in nutrient-rich areas are *Peloscolex multisetosus*, *Tubifex Tubifex*, and the *Aulodrilus* and *Potamothrix* species. It is likely that nematodes are abundant (Loveridge and Cook 1976), but commonly used methods do not adequately sample these animals. A variety of other taxa, (e.g., aschelminthes, tardigrades, and protozoans) are found in the sediments (Cook and Johnson 1974).

The zebra mussel (*Dreissena polymorpha*), which apparently entered Lake St. Clair in ballast water from northern Europe in 1986 (Griffiths et al. 1991), has become the dominant benthic species in shallow water environments of Lake St. Clair, Saginaw Bay, and western Lake Erie. This organism has had a major impact on these ecosystems. The unionid clam community of Lake St. Clair has been decimated by the heavy growth of zebra mussels on their shells (Nalepa 1994). Water transparency greatly increased in western Lake Erie as the zebra mussels filtered out particulate matter, which resulted in an 86% decrease in planktonic diatoms (Holland 1993). The decreased abundance of phytoplankton has affected the water chemistry so that nutrients (i.e., soluble reactive phosphorus, ammonia, nitrate, silica) have increased in western Lake Erie (Holland et al. 1995). Analyses of zooplankton samples from 1990–92 at the same site show large decreases in abundance of copepods and rotifers since the 1970s and 1980s (Beeton and Hageman

1994). Closely similar changes have been observed in Saginaw Bay, where phytoplankton abundance has decreased, nutrients increased, and water transparency increased (Fahnenstiel et al.1995). Furthermore, the near-bottom algal community, metaphyton, increased as did the abundance of amphipods, while oligochaetes decreased (T. Nalepa, Great Lakes Environmental Research Lab., personal communication). Blooms of blue-green algae (*Microcystis*), occurred in Saginaw Bay in 1994 and 1995 (H. Vanderploeg, Great Lakes Environmental Research Lab., personal communication). A *Microcystis* bloom also occurred in the island region of western Lake Erie in late summer, early fall of 1995 (J. Hageman, Ohio State University, personal communication). It appears that the filtering activity of the zebra mussels reduces the phytoplankton, and thereby makes more nutrients available for a bloom. Furthermore, experiments conducted by H. Vanderploeg (Great Lakes Environmental Research Laboratory personal communication) show the mussels do not readily ingest and actually reject *Microcystis*, while they continue to filter out small nonblue-green alga.

The shallow water environments are the areas most severely impacted by pollution, habitat changes, and other stresses. Changes in the benthos of western Lake Erie between 1930 and 1961 show the severity and impact of organic pollution (Carr and Hiltunen 1965). In these thirty-one years, oligochaetes increased nine-fold, chironomids four-fold, fingernail clams two-fold, and gastropods six-fold. *Hexagenia* populations were less than 1% of their previous abundance. The greatest changes were at the mouths of the Detroit and Raisin Rivers and in Maumee Bay. *Hexagenia* had been an abundant food for fish. It is unlikely the absence of *Hexagenia* has been compensated by increases in other animals. Closely similar changes in the benthic community were documented for southern Green Bay for the period between 1952 and 1969 (Howmiller and Beeton 1971). The *Hexagenia* population also collapsed in Saginaw Bay, but large increases in other organisms were not observed (Schneider et al. 1969). The benthic community had small increases in amphipods and oligochaetes between 1955 and 1965. The populations of *Hexagenia* and sphaeriid clams collapsed between 1955 and 1956, suggesting some catastrophic event. The clam population had recovered by 1965, but not the *Hexagenia*.

The benthos, zooplankton, and phytoplankton of Green Bay reflect the environmental quality of the system as determined by the inflow of the high-quality Lake Michigan water into the northern part and the inflow of Fox River water, polluted with industrial and domestic waste, in the southern end (Holland and Claflin 1975). A counterclockwise circulation persists, with Lake Michigan water flowing along and down the western shore and Fox River water moving upward along the eastern shore (Modlin and Beeton 1970). The extreme southern end is dominated by the river and has seriously degraded sediments. The planktonic diatoms and zooplankton of northern and western Green Bay are a Lake Michigan species complex. The plankton of the extreme southern bay are those associated with eutrophication and/or pollution. Plankton along the eastern shore is a mixture of the eutrophic and oligotrophic species. The northern bay benthos is similar to Lake Michigan with *Stylodrilus heringianus* an important species. The extreme southern bay is populated with the pollution tolerant *L. hoffmeisteri*. Two species of *Peloscolex* are abundant, *P. ferox* occurs primarily along the western shore in low organic sediments, whereas *P. multisetosus* is along the eastern shore where sediments have a higher organic content due to the Fox River (Howmiller and Beeton 1970).

Dynamics of the Great Lakes

The Laurentian Great Lakes and their basins comprise a dynamic system which has undergone many major and minor changes since their formation. During most of their history, natural changes in water levels, size of lakes, connections between lakes, and outflow have occurred. Changes in the last two hundred years are largely anthropogenic, (e.g., draining and filling of wetlands, damming tributaries, construction of canals, alteration of shoreline for ports, discharge of domestic and industrial waste, toxic substances).

Major changes occurred in the formation of the lakes, changes in levels, connections, and drainage systems associated with advance and retreat of the glaciers (Hough 1958). The last glacial retreat was around six thousand years ago, but present day lake levels probably were not established until 2,500 years ago. The glacial lakes (e.g., Algonquin, Iroquois, and Agassiz) covered large areas of the present basin and adjacent lands, and had a great influence on the zoogeography of the region. These lakes were harsh environments, especially if they were located near the glaciers. Paloeontological data suggests that conditions were high arctic. For example, the early Champlain Sea, which connected the Atlantic Ocean with Lake Ontario probably initially had summer temperatures less than $5^{\circ}C$ and salinity greater than 25% (Elson 1969). Subsequently, crustal uplift due to glacial retreat must have reduced exchange with the ocean, and permitted warming and reduced salinity.

The glaciers were the major force in determining the present day lakes, but the glacial and proglacial lakes provided for invasion and dispersal of major components of the biota of the Great Lakes. The organisms that are frequently referred to as glaciomarine relicts, (i.e., species that may have evolved in arctic seas and invaded North American waters by marine inundations or by dispersal by proglacial waters in front of ice sheets, *Mysis relicta, Diporeia, Limnocalanus,* and *Senecella*), all entered the Great Lakes via the glacial-lake systems (Dadswell 1974).

A period of stability probably persisted for over two thousand years once the climate moderated and the forests had become established. The present-day aquatic flora and fauna, except for recent non-indigenous species, had been established. Immigration of humans into the region probably had little, if any impact on the lakes initially.

Anthropogenic changes in the basin did not occur until within the past two hundred years. These changes, initially were the result of physical structures built in the lakes, (i.e., locks and dams, or canals in the drainage basin) to facilitate transportation. Exploitation of the natural resources in the basin, (i.e., the forests for lumber and mining of mineral deposits) resulted in erosion and early pollution which affected water quality and fish habitat.

Structural changes to the Great Lakes system not only affected the natural water levels and flows, but also affected the Great Lakes ecosystem. River damming, for example, is thought to be responsible for decline in Atlantic salmon spawning prior to 1900 in Lake Ontario tributaries. Although considered an economic benefit to the shipping industry, construction of the Welland Canal, between Lakes Erie and Ontario, which began in 1829 (Jackson 1991), also impacted the fishing industry. Before the canal was constructed, Niagara Falls presented a natural barrier to exotic species which entered the St. Lawrence River from the Atlantic Ocean. The construction of the Welland Canal, provided access to the upper lakes. The sea lamprey was thriving in Lake Ontario by the 1880s and was first observed in Lake Erie in 1921, Lake Huron in 1932, Lake Michigan in 1936, and Lake Superior in 1946 (Smith 1972). The sea lamprey probably migrated through the canal. Similarly, the alewife was also successfully thriving in Lake Ontario by the 1860s and was first noted in Lake Erie in 1931, Lake Huron in 1933, Lake Michigan in 1949, and Lake Superior in 1953 (Smith 1972). The alewife may have migrated through the Welland or Erie Canals. By the 1950s, the sea lamprey had killed nearly all the lake trout in Lakes Huron, Michigan, and Superior. Changes such as regulating water levels on Lakes Superior and Ontario through the introduction of extensive lock systems at the outflows probably caused change in indigenous species' migration patterns, (e.g., the extent of migration by lake trout in the St. Marys River). Changes such as the dredging the St. Clair River for gravel, as well as deepening its channel for navigation purposes lowered Lakes Michigan and Huron's water levels by 0.88ft (Quinn et al. 1993). This lowering may have resulted in the dewatering of fragile wetlands.

Native Americans made use of the extensive copper deposits in the Lake Superior basin. These small mining activities became large-scale with the industrial revolution and the immigration of Europeans into the region. Large quantities of mining tailings, refuse material separated from ground ore, persist today in the basin and along the Lake Superior shore in some areas. The impact on water quality, and on the biota of materials leached from the tailings is unclear, although it could be of some significance locally.

The Great Lakes region had a resource valuable to a developing nation, (i.e., the extensive forests, especially red and white pine). After the Civil War the region became especially important for lumber. Large tracts of old growth forest were lumbered during 1870–1900. This activity must have resulted in soil erosion, large amounts of sawdust, changes in runoff and stream flow, and in tributary water quality. The overall impact of these terrestrial changes on the water quality and biota of the Great Lakes is unknown, except for reports on readily visible gross pollution from sawdust (Smiley 1882). Spawning runs of whitefish ceased by 1880 in the Oconto River, a tributary to Green Bay, presumably because sawdust covered the river bottom and extended two miles into the Bay. Whitefish had been migrating twenty miles upstream to spawn in the 1840s. Many of the forest lands were converted to agriculture, and thus, changed permanently, others have been reforested.

Drainage of Great Lakes wetlands likely had a large impact on water quality, especially on the biota in some shallow water ecosystems of the Great Lakes. Decreases in water levels adversely impact fish spawning, food producing areas for fish, ducks and wildlife (Beeton and Rosenberg 1968). Shallow wetland areas of Lake Erie have been very important nursery and feeding sites for fish (Krecker 1931). For example, the drainage of the Black Swamp, a vast swamp region in northwest Ohio, which extended 120 mi in length and average 40 mi in breadth, 4800 mi^2, presumably had a large impact on the turbidity of western Lake Erie (Langlois 1965). Removal of the swamp forest and establishment of a network of ditches changed the watershed so that streams changed from clear to muddy and turbid. The former swamp has become a major agricultural area.

Undoubtedly the most significant impact on the Great Lakes is associated with the growth of major population centers, and change in land use from forest and prairies to agriculture and industry (Beeton 1969). Pollution from major cities was evident by 1900, (e.g., concern over water borne diseases in Chicago and other southern Lake Michigan cities) led to the construction of the Chicago Sanitary and Barge Canal to divert sewage away from the lake. Elsewhere, pollution was evident in the near shore of Lake Erie, and major limnological studies were undertaken because of collapse of fisheries such as the cisco (Wright 1955). Problems with pollution were increased in the early 1900s by construction of sewerage systems, which used water as a vehicle, to carry municipal and industrial wastes to the lakes and/or tributaries. Little was done to regulate and treat sewage from the major population centers or discharge from the many Great Lakes ships. Many coastal areas, river mouths, and associated environments were radically altered by dredging, disposal of dredged sediments in wetlands, and construction of breakwall, wharfs, and dikes. Consequently, inshore areas were greatly changed physically and by deteriorated water quality. These inshore areas were usually the parts of the lakes important for water supply, recreational use, fishing, wildlife, and fish spawning.

The development and widespread use of various organic chemicals such as pesticides, herbicides, plastics, and others, beginning in the 1940s, brought about subtle changes in the environment and biota (Sonzogni et al. 1983a). Most of these chemicals have relatively low solubility, and become adsorbed to particles which settle into the sediments. Consequently, sediments in some inshore areas, harbors, tributaries, and even some offshore areas, are severely contaminated with exotic compounds which persist for many years. Furthermore, it was discovered that these compounds, which may be at very low concentrations in the environment, can and do bioaccumulate to toxic levels which may be lethal, limit reproduction, or otherwise adversely affect life. It is well known that reproduction of fish eating birds, (e.g., eagles) was seriously impacted. Some fish have tumors, other aquatic organisms may have very low tolerances to toxic organic compounds.

The growth of large cities, industries, and agriculture resulted in ever-increasing loads of nutrients, from sewage, industrial wastes, and fertilizers, entering the lakes. Changes had been observed in benthic communities, algal and fish populations, and in water quality, but it was not clearly established until the 1960s that these changes and others were symptoms of excessive nutrient enrichment, (i.e., cultural eutrophication) (Beeton 1961). The progression of eutrophication is probably best illustrated by information for Lake Erie, since we have a reasonably good historical record of past environmental conditions and the consequences of nutrient enrichment were first documented for this lake. Major changes in the benthos of western Lake Erie first directed our attention to accelerated eutrophication (Beeton 1961). The once abundant mayfly, *Hexagenia*, over five hundred nymphs m^{-2} in 1940s, disappeared from their sedimentary environment in all but a few locations by 1961. Pollution tolerant oligochaetes (e.g., *Limnodrilus hoffmeisterii*) became the dominant organisms. Other pollution tolerant species of fingernail clams, midges, and snails also became more abundant (Carr and Hiltunen 1965). Blooms of blue-green algae occurred in the 1960s and l970s. The abundance of planktonic algae increased threefold between 1919 and 1963 (Beeton 1969). Zooplankton also increased and eutrophic species became abundant.

Severe dissolved oxygen depletion appears to be a major direct factor in demise of *Hexagenia*. Low-dissolved oxygen concentrations occurred during protracted calm periods in the western basin (Britt 1955), and concentrations of 3ppm or less occurred in about 75% of the central basin in 1959 and 1960 (Beeton 1963). Other changes were observed in water quality. Calcium, chloride, sodium, sulfate, and total dissolved solids all increased (Beeton l969). Large increases were observed in nitrogen and phosphorus compounds.

Environmental changes associated with pollution and eutrophication of the Great Lakes generally fall into three main categories: pollution of inshore areas including harbors and tributaries; long-term changes in the open waters; and long-term changes in the sediments. The first category has been, and is a problem in all the Great Lakes, and represents conditions maintained by continued pollution. These areas should show fairly rapid substantial improvement with pollution abatement. Long-term changes in the open waters, which are subtle and may be recognized primarily in changes in nutrients and in phytoplankton, were observed especially in Lakes Erie, Michigan, and Ontario (Beeton 1969), although some of these changes have been reversed as phosphorus loadings were decreased. Major changes observed in Lake Erie, southern Green Bay, and southern Saginaw Bay appear to be closely related to long-term changes in the sediments. This is certainly true for the benthos, and the dissolved oxygen depletion likely has been due to the accumulation of organic materials with a high oxygen demand. Some biological changes may be due to direct activities of man, such as overfishing and introduction of non-indigenous species. It is likely that many of the changes have been entirely or partially the consequence of increased nutrient loading, but it is difficult to sort out others which may be the result of toxic substances, overfishing, exotic species, habitat change (Sonzogni et al. 1983a). For example, some of the changes in the sedimentary environment could be due to accumulation of toxic substances.

We have seen large changes in fish communities. Some of the changes probably are due to overfishing, while others are the consequence of exotics entering the lakes, and some may be due to loss of critical habitat. Major changes in fish populations have occurred in Green Bay, Saginaw Bay, and Lake Erie, which are ecosystems most severely impacted by eutrophication and contaminants. Certainly the contaminated sediments and areas of low dissolved oxygen present a hostile environment for fish eggs. Furthermore, important fish food (e.g., *Hexagenia*) have been impacted. Changes in zooplankton must also affect food and feeding of larval and juvenile fish. Sorting out factors, which are important for the success of open lake fishes (e.g., salmonids) is more difficult, although contaminants and altered spawning reefs may be important.

Non-indigenous species have brought about large changes in the communities of fish, plankton, and benthos, and even in water quality. The impacts of alewife, carp, and sea lamprey are well known to fishery biologists and resource managers. Less well known are the consequences of invasion and establishment of viable populations of other exotic species. A number of non-indigenous invertebrates have successfully colonized the Great Lakes, e.g., *Eurytemora affinis*, Japanese snail, Asiatic clam, *Bythotrephes cederstroemi*, and the zebra mussel. We have little information on the importance of the first three, although the Asiatic clam has been of concern at some water intakes. The latter two organisms have impacted Great Lakes ecosystems, especially the zebra mussel, *Dreissena polymorpha*, as described in a previous section.

The zebra mussel and the quagga mussel, *Dreissena bugensis*, may be enhancing energy transfer to the benthos via filtering biotic and abiotic materials and depositing the material on the bottom as feces and pseudofeces (Stewart and Haynes 1994). Several invertebrates were more abundant in Lake Ontario after the mussels became established. Elsewhere, the zebra mussel has had a negative impact on native clams, especially in Lake St. Clair, where the clam population has been decimated (Nalepa 1994). The importance to the fish communities remains uncertain, but it is possible that increased water clarity and increased nutrient supply will benefit the macrophytes

(Stuckey and Moore 1995), and in turn provide refuges for fish. The decrease in planktonic invertebrates may affect feeding by very small fish, although this may be compensated for by the increased benthos.

References

Alley, W. P., and C. F. Powers. 1970. Dry weight of the macrobenthos as an indicator of eutrophication of the Great Lakes. Proc. 13th Conf. Great Lakes Res., 595–600.

Ayers, J. 1962. Great Lakes waters, their circulation and physical and chemical characteristics. American Association for the Advancement of Science Publication No. 71:71–89.

Assel, R. A., F. H. Quinn, G. A. Leshkevich, and S. J. Bolsenga. 1983. Great Lakes Ice Atlas. U.S. Department of Commerce, Great Lakes Environmental Research Laboratory, Ann Arbor, Michigan 48105.

Beeton, A. M. 1958. Relationship between Secchi Disc readings and light penetration in Lake Huron. Transactions of the American Fisheries Society 87:73–79.

Beeton, A. M. 1961. Environmental changes in Lake Erie. Transactions of the American Fisheries Society 90:153–159.

Beeton, A. M. 1962. Light Penetration in the Great Lakes. University of Michigan, Great Lakes Research Division Publication 9:68–76

Beeton, A. M. 1963. Limnological survey of Lake Erie 1959 and 1960. Great Lakes Fishery Commission Technical Report No. 6.

Beeton, A. M. 1965. Eutrophication of the St. Lawrence Great Lakes. Limnology and Oceanography 10:240–254.

Beeton, A. 1966. Indices of Great Lakes eutrophication. University of Michigan, Great Lakes Research Division Publication 15:1–8.

Beeton, A. M. 1969. Changes in the environment and biota of the Great Lakes. Pages 150–187 *in*: Eutrophication: Causes, consequences, correctives. National Academy of Science/National Research Council, Publication 1700.

Beeton, A. 1970. Statement on Pollution and Eutrophication of the Great Lakes. Special Report No. 11, Center for Great Lakes Studies, University of Wisconsin, Milwaukee.

Beeton, A. 1971a. Chemical Characteristics of the Laurentian Great Lakes. Bulletin of the Buffalo Society of Natural Sciences 25(2):1–29.

Beeton, A. 1971b. The phenomenon of lake eutrophication. Water Resources Engineering Educational Series, University of California, Program VIII: Freshwater Lakes and their Management 7–1 7–12.

Beeton, A. M. 1984. The world's great lakes. J. Great Lakes Res. 10:106–113.

Beeton, A. M., and H. B. Rosenberg. 1968. Studies and research needed in regulation of the Great Lakes. Proceedings of Great Lakes Water Regulation Conference, American Society of Civil Engineers, Toronto. 1968:311–342.

Beeton, A. M., and J. M. Barker. 1974. Investigation of the influence of thermal discharge from a large electric power station on the biology and near shore circulation of Lake Michigan-Part A: Biology. Center for Great Lakes Studies, University of Wisconsin-Milwaukee. Special Report No. 18.

Beeton, A. M., and J. A. Bowers. 1982. Vertical migration of *Mysis relicta* Loven. Hydrobiologia 93:53–61.

Beeton, A. M., and J. Hageman, Jr. 1994. Impact of Dreissena polymorpha on the zooplantkon community of western Lake Erie. Verhandlugen International Vereinigung Fur Theoretische und Angewandte Limnologie 25:2349.

Beeton, A., S. H. Smith, and F. F. Hooper. 1967. The physical limnology of Saginaw Bay, Lake Huron. Great Lakes Fishery Commission Technical Report No. 12, Ann Arbor, Michigan.

Bierman, V. J. Jr., and D. M. Dolan. 1986. Modeling of phytoplankton in Saginaw Bay: II. Post-audit phase. Journal of Environmental Engineering 112(2):415–429.

Bierman, V. J. Jr., D. M. Dolan, R. Kasprzyk, and J. L. Clark. 1984. Retrospective analysis of the response of Saginaw Bay, Lake Huron, to reductions in phosphorus loadings. Environmental Science and Technology 18(1):23–31.

Botts, L., and B. Krushelnicki. 1987. The Great Lakes: An Environmental Atlas and Resource Book. Environment Canada, U.S. Environmental Protection Agency, Brock University, and Northwestern University, Chicago, IL., and Toronto, Canada.

Bowers, J. A., R. Rossmann, J. Barres, and W. Y. B. Chang. 1986. Phytoplankton populations of southeast Lake Michigan 1974–1982. Pages 141–69 *in* R. Rossman, editor. Southeastern Near shore Lake Michigan: Impact of the Donald C. Cook Nuclear Plant. University Michigan ,Great Lakes Research Division Publication 22.

Brazel, A. J., and D. W. Phillips. 1974. November 1972 Floods on the Lower Great Lakes. Weatherwise 27(2):56–62.

Brinkhurst, R. O., A. L. Hamilton, and H. B. Herrington. 1968. Components of the bottom fauna of the St. Lawrence, Great Lakes. University Toronto, Great Lakes Institute. No. PR 33:1–49.

Britt, N. W. 1955. Stratification in western Lake Erie in summer of 1953; effects on the Hexagenia (*Ephemeroptera*) population. Ecology, 36:239–44.

Brooks, I., and W. Lick. 1972. Lake currents associated with the Thermal Bar. J. of Geophysics Research 77(30):6000–13.

Brooks, K. N., P. F. Ffolliott, H. M. Gregersen, and J. L. Thames. 1991. Hydrology and management of watersheds. Iowa State University Press, Ames.

Carr, J. 1962. Dissolved oxygen in Lake Erie, past and present. University of Michigan, Great Lakes Research Division Publication 9:1–14.

Carr, J. F., and J. K. Hiltunen. 1965. Changes in the bottom fauna of western Lake Erie from 1930 to 1961. Limnology and Oceanography 10:551–69.

Carter, J., and N. Watson. 1977. Seasonal and horizontal distribution patterns of planktonic Crustacea in Georgian Bay and North Channel, Lake Huron in 1974. Journal of Great Lakes Research 3:113–22.

Charlton, M. N., J. E. Milne, W. G. Booth, and F. Chiocchio. 1993. Lake Erie offshore in 1990: Restoration and resilience in the central basin. Journal of Great Lakes Research 19(2):291–309.

Claflin, L. W. 1975. A multivariate data analysis of Lake Michigan phytoplankton. Doctoral dissertation. University of Wisconsin-Madison.

Clark, R. H., and N. P. Persoage. 1970. Some implications of crustal movement in engineering planning. Canadian Journal of Earth Sciences 7:628–33.

Cook, D. G., and M. E. Johnson. 1974. Benthic macroinvertebrates of the St. Lawrence Great Lakes. Journal of Fisheries Research Board Canada 31:763–782.

Coordinating Committee on Great Lakes Basic Hydraulic and Hydrologic Data. 1977. Report from Great Lakes Physical Data Subcommittee. 12 pp.

Coordinating Committee on Great Lakes Basic Hydraulic and Hydrologic Data, 1992. IGLD 1985 Brochure on the International Great Lakes Datum 1985.

Dadswell, M. J. 1974. Distribution, ecology, and postglacial dispersal of certain crustaceans and fish in eastern North America. National Museum of National Science Publication Zoology No. 11.

Davis, C. C. 1966. Plankton studies in the largest Great Lakes of the world. University of Michigan Great Lakes Research Division Publication 14:1–36.

DeVault, D., P. Bertram, D. Whittle, and S. Rang. 1994. Toxic Contaminants in the Great Lakes. State of the Lakes Conference Working Paper, Environment Canada and United States Environmental Protection Agency, State of the Lakes Conference (SOLEC), October, 1994, Dearborn, Michigan.

Dobson, H., M. Gilbertson, and P. Sly. 1974. A summary and comparison of nutrients and related water quality in Lakes Erie, Ontario, Huron, and Superior. J. Fish. Res. Bd. Can. 31:731–738.

Durham, R. W., and B. G. Oliver. 1983. History of Lake Ontario contamination from the Niagara River by sediment radiodating and chlorinated hydrocarbon analysis. Journal of Great Lakes Research 9:160–68.

Eadie, B., R. Chambers, W. Gardner, and G. Bell. 1984. Sediment trap studies in Lake Michigan: resuspension and chemical fluxes in the southern basin. Journal of Great Lakes Research 10:307–321.

Eadie, B., and J. Robbins. 1987. The Role of Particulate Matter in the Movement of Contaminants in the Great Lakes. Pages 319–64 in R. A. Hites editor.Sources and Fates of Aquatic Pollutants, American Chemical Society, Washington D.C.

Eadie, B., H. Vanderploeg, J. Robbins, and G. Bell. 1990. The significance of particle settling and resuspension in the Laurentian Great Lakes. Pages 169–209 in M. Tilzer and C. Serruya, editors. Large lakes: ecological structure and function. Springer Verlag, New York.

Edmondson, W. 1969a. Cultural eutrophication with special reference to Lake Washington. Mitteilungen. Internationale Vereinigung fuer Theoretische und Angewandte Limnologie 17:19–32.

Edmondson, W. 1969b. Eutrophication in North America. Pages 124–149 in Eutrophication: causes, consequences, correctives, National Academy Press, Washington, D.C.

Edmondson, W. T., G. C. Anderson, and D. R. Peterson. 1956. Artificial eutrophication of Lake Washington. Limnology and Oceanography 1:47–53.

Elson, J. A. 1969. Radiocarbon dates, Mya arenaria phase of the Champlain Sea. Canadian Journal of Earth Science 6:367–372.

Environment Canada. 1991a. Toxic chemicals in the Great Lakes and Associated Effects-Synopsis. Department of Fisheries and Oceans, Environment Canada.

Environment Canada. 1991b. Toxic Chemicals in the Great Lakes and Associated Effects-Volume I: Contaminant Levels and Trends. Department of Fisheries and Oceans, Environment Canada.

Environment Canada. 1991c. Toxic Chemicals in the Great Lakes and Associated Effects-Volume II: Effects. Department of Fisheries and Oceans, Environment Canada.

Environment Canada. 1995. Great Lakes Factsheet No. 1, Physical Features and Population.

Fahnenstiel, G. L., and D. Scavia. 1987. Dynamics of Lake Michigan phytoplankton. Can. J. Fish and Aquatic Sci. 44:509–514.

Fahnenstiel, G. L., J. Chandler, H. Carrick, and D. Scavia. 1989. Photosynthetic characteristics of phytoplankton communities in Lakes Huron and Michigan: P-I parameters and end-products. J. Great Lakes Res. 15:394–407.

Fahnenstiel, G. L., G. A. Lang, T. F. Nalepa, and T. H. Johengen. 1995. Effects of zebra mussel (Dreissena polymorpha) colonization on water quality parameters in Saginaw Bay, Lake Huron. Journal of Great Lakes Research 21:435–448.

Gannon, J. E. 1972. A contribution to the ecology of zooplankton crustacea of Lake Michigan and Green Bay. Doctoral dissertation, University of Wisconsin.

Gannon, J. E. 1981. Changes in zooplankton populations in Lakes Erie and Ontario. Bulletin of Buffalo Society of Natural Sciences 25:21–40.

Gardner, W. S., M. A. Quigley, G. L. Fahnenstiel, D. Scavia, and W. A. Frez. 1990. *Pontoporeia hoyi*—a direct trophic link between spring diatoms and fish in Lake Michigan. Pages 632–44 in M. A. Tilzer and C. Serruya editors. Large Lakes: Ecological Structure and Function. Springer-Verlag. Berlin.

Great Lakes Basin Commission. 1975. Great Lakes basin framework study. Great Lakes Basin Commission, Ann Arbor. 441 pp.

Gresswell, R. K., and A. Huxley. 1965. Standard encyclopedia of the world's rivers and lakes. G. P. Putnam and Sons, New York. 383 pp.

Griffiths, R. W., D.W. Schloesser, J. H. Leach, and W. P. Kovalak. 1991. Distribution and dispersal of the zebra mussel (*Dreissena polymorpha*) in the Great Lakes region. Canadian Journal of Fisheries and Aquatic Sciences 48:1381–1388.

Hasler, A. 1947. Eutrophication of lakes by domestic drainage. Ecology 28:383–395.

Hem, J. D. 1970. Study and interpretation of the chemical characteristics of natural water. U.S. Geological Survey Water-Supply Paper 1473:363 pp.

Henson, E. B. 1966. A review of Great Lakes benthos research. University of Michigan, Great Lakes Res. Div. 14:37–54.

Herdendorf, C. E. 1982. Large Lakes of the World. J. Great Lakes Res. 8(3):379–412.

Hiltunen, J. K. 1969a. Invertebrate macrobenthos of western Lake Superior. Michigan Academician 1:123–133.

———, 1969b. The benthic macrofauna of Lake Ontario. Great Lakes Fishery Commission Technical Report 14:39–50.

Holland, R. 1968. Correlation of *Melosira* species with trophic conditions in Lake Michigan. Limnology and Oceanography 13:555–557.

———, 1969. Seasonal fluctuations of Lake Michigan diatoms. Limnology and Oceanography 14:423–436.

Holland, R. E. 1980. Seasonal fluctuations of major diatom species of five stations across Lake Michigan, May 1970–October 1972. EPA–600/3–80–066, 85 PP.

———, 1993. Changes in planktonic diatoms and water transparency in Hatchery Bay, Bass Island area, western Lake Erie since the establishment of the zebra mussel. Journal of Great Lakes Research 19:617–624.

———, A. M. Beeton. 1972. Significance to eutrophication of spatial differences in nutrients and diatoms in Lake Michigan. Limnology and Oceanography 17:88–96.

———, L. W. Claflin. 1975. Horizontal distribution of planktonic diatoms in Green Bay, mid July 1970. Limnology and Oceanography 20:365–78.

———, T. H. Johengen, and A. M. Beeton. 1995. Trends in nutrient concentrations in Hatchery Bay, western Lake Erie before and after Dreissena polymorpha. Canadian Journal of Fisheries and Aquatic Sciences 52:1202–9.

Horton, R. E., and C. E. Grunsky. 1927. Hydrology of the Great Lakes. Report of the Engineering Board of Review of the Sanitary District of Chicago on the lake lowering controversy and a program remedial measures. Part III; Appendix II.

Hough, J. L. 1958. Geology of the Great Lakes. University Illinois Press, Urbana 313 pp.

Howmiller, R. P., and A. M. Beeton. 1970. The oligochaete fauna of Green Bay, Lake Michigan. Proceedings of the 13th Conference on Great Lakes Research, International Association for Great Lakes Research, Ann Arbor, Michigan.

———, 1971. Biological evaluation of environmental quality, Green Bay, Lake Michigan. Journal of Water Pollution Control Fed. 43:123–133.

IJC (International Joint Commission). 1970. Pollution of Lake Erie, Lake Ontario, and the International Section of the St. Lawrence River. International Joint Commission, Windsor, Ontario, 174 pp.

———, 1985. Report on Great Lakes Water Quality. Great Lakes Water Quality Board, Report to the International Joint Commission, Windsor, Ontario, 212 pp.

———, 1987. Report on Great Lakes Water Quality. Great Lakes Water Quality Board, Report to the International Joint Commission, Windsor, Ontario, 236 pp.

———, 1989a. Fourth Biennial Report. International Joint Commission, Windsor, Ontario.

———, 1989b. 1987 Report on Great Lakes Water Quality, Appendix B: Great Lakes Surveillance, Volume I. Great Lakes Water Quality Board, Report to the International Joint Commission (D. E. Rathke and G. McRae, eds.), International Joint Commission, Ottawa, Canada.

———, 1991a. Cleaning up the Great Lakes. International Joint Commission, Windsor, Ontario.

———, 1991b. Persistent Toxic Substances: Virtually Eliminating Inputs to the Great Lakes. Interim Report of the Virtual Elimination Task Force. International Joint Commission, Windsor, Ontario.

———, 1993. Level reference study, Phase II. Climate, climate change, water level forecasting and frequency analysis. Volume 1; Water supply scenarios.

Irbe, G. J. 1992. Great Lakes surface water temperature climatology. Climatological studies number 43. Atmospheric Environment Service Canada, Environment Canada.

Jackson, J. N. 1991. Construction and operation of the first, second and third Welland Canals. Canadian Journal of Civil Engineering 18(3):472–483.

Johengen, T., and A. Beeton. 1992. The effects of temporal and spatial variability on monitoring agricultural nonpoint source pollution. Pages 89–95 in Proceeding of the National Rural Clean Water Program Symposium. EPA Seminar Publication, EPA/625/R–92/006, Washington, D.C.

Johnson, M. G., and R. O. Brinkhurst. 1971. Production of benthic macroinvertebrates of Bay of Quinte and Lake Ontario. Journal of Fisheries Research Board of Canada 28:1699–1714.

Keeler, G., J. Pacyna, T. Bidleman, and J. Nriagu. 1993. Identification of sources contributing to the contamination of the Great Waters by toxic compounds. Report prepared for the Great Waters Program, USEPA, Durham, North Carolina, by the University of Michigan, Air Quality Laboratory, Ann Arbor, Michigan.

Krecker, F. H. 1931. Vertical oscillations or seiches in lakes as a factor in the aquatic environment. Ecology 12:156–163.

Kwiatkowski, R. E. 1982. Trends in Lake Ontario surveillance parameters, 1974–1980. Journal of Great Lakes Research 8 (4):648–659.

Langlois, T. H. 1965. Portage River watershed and fishery. Ohio Department of National Resources, Wildlife Division Publication W–130. 22 pp.

Leach, J. H., and R. C. Herron. 1992. A review of lake habitat classification. Pages 27–58 in W.–D.N. Busch and P.G. Sly, editors. The development of an aquatic habitat classification system for lakes. CRC Press, Boca Raton, Florida.

Likens, G. 1972. Eutrophication and aquatic ecosystems. Pages 3–13 in Proceedings of the Symposium on Nutrients and Eutrophication: The Limiting Nutrient Controversy, Kellogg Biological Station, Michigan State University. American Society of Limnology and Oceanography Special Symposia Volume 1:3–13.

Livingston,e D. A. 1963. Chemical composition of ruvers and lakes. Data of Geochemistry, U.S. Geol. Surv. Prof. Paper 44-G, 64 pp.

Loveridge, C. C. and D. G. Cook. 1976. A preliminary report on the benthic macroinvertebrates of Georgian Bay and North Channel Environment Canada, Fisheries and Marine Service, Technical Report 610.

Ludwig, J., and ten coauthors. 1993. Caspian tern reproduction in the Saginaw Bay ecosystem following a 100-year flood event. Journal of Great Lakes Research 19:96–108.

McIntosh, R., and ten coauthors. 1977. Report of the Classification Task Group, Experimental Ecological Reserves, TIE-NSF Project, Kellogg Biological Station, Michigan State University.

McNaught, D. C. 1966. Depth control of planktonic cladocerans in Lake Michigan. University of Michigan, Great Lakes Research Division Publication 15:98–108.

———, 1978. Spatial heterogeneity and niche differentian in zooplankton in Lake Huron. Verhandlungen-Internationale Vereinung fuer Theoretishce und Angewandte Limnologie 20:341–346.

Michigan Sea Grant. 1990a. Lake Michigan. Michigan Sea Grant College Program, Michigan State University, Extension Bulletin E–1867.

———, 1990b. Lake Erie. Michigan Sea Grant College Program, Michigan State University, Extension Bulletin E–1869.

Mills, E. L., J. H. Leach, J. T. Carlton, C. L. Secor. 1993. Exotic species in the Great Lakes: A history of biotic crises and anthropogenic introduction. Journal of Great Lakes Research 19:1–54.

Modlin, R. F. and A. M. Beeton. 1970. Dispersal of Fox River water in Green Bay, Lake Michigan. Proceedings of the 13th Conference on Great Lakes Research. International Association for Great Lakes Research, Ann Arbor, Michigan.

Moll, R., R. Rossmann, J. Barres, and F. Horvath. 1992. Historical trends of chlorides in the Great Lakes. Pages 303–22 in F. D'Itri, editor. Chemicals Deicers and the Environment. Lewis Publishers, Chelsea, Michigan.

Muir, T., and A. Sudar. 1987. Toxic chemicals in the Great Lakes basin ecosystem: some observations. Environment Canada, Burlington, Ontario, 126 pp.

Munawar, M., and I. Munawar. 1982. Phycological studies in Lakes Ontario, Erie, Huron, and Superior. Canadian Journal of Botany 60:1837–1858.

———, W. Norwood, and C. Mayfield. 1987. Significance of autotrophic picoplankton in the Great Lakes and their use as early indicators of contaminant stress. Archiv fuer Hydrobiologie, Beiheff: Ergebnisse de Limnologie 25:141–155.

Nalepa, T. F. 1994. Decline of native unionid bivalves in Lake St. Clair after infestation by zebra mussel, (Dreissena polymorpha). Canadian Journal of Fisheries and Aquatic Sciences, 51:2227–33.

NAS (National Academy of Sciences). 1972. Water Quality Criteria 1972. A Report of the Committee on Water Quality Criteria, Environmental Studies Board. Washington, D.C.

NRC (National Research Council) and the Royal Society of Canada. 1985. The Great Lakes Water Quality Agreement: An Evolving Instrument for Ecosystem Management. National Academy Press, Washington, D.C.

Neilson, M., S. L'Italien, V. Gluman, D. Williams, and P. Bertram. 1994. Nutrients: Trends and System Response. State of the Lakes Conference Working Paper, Environment Canada and United States Environmental Protection Agency, State of the Lakes Conference (SOLEC), October 1994, Dearborn, Michigan.

NRTC (Niagara River Toxics Committee). 1984. Report of the Niagara River Toxics Committee. Burlington, Ontario, Inland Waters Directorate.

Ownbey, C., and Kee. 1967. Chlorides in Lake Erie. Proceedings of the Tenth Conference on Great Lakes Research. International Association for Great Lakes Research, Ann Arbor, Michigan.

———, G. Willeke. 1965. Long-term solids buildup in Lake Michigan waters. University of Michigan, Great Lakes Research Division Publication 13:141–152.

Parker, J. I. 1980. Predation by Mysis relicta on Pontoporeia hoyi: A food chain link of potential importance in the Great Lakes. Journal of Great Lakes Research 6:164–166.

Patalas, K. 1972. Crustacean plankton and the eutrophication of St. Lawrence Great Lakes. Journal of the Fisheries Research Board of Canada 29:1451–1462.

Powers, C. F., and W. P. Alley. 1967. Some preliminary observations on the depth distribution of macrobenthos in Lake Michigan. University of Michigan, Great Lakes Research Division Publication 30:112–125.

Quinn, F.H., and C. E. Sellinger. 1990. Lake Michigan record levels of 1838, a present perspective. Journal of Great Lakes Research 16(1):133–138.

Quinn, F.E., J. A. Derecki, and C. E. Sellinger. 1993. Pre-1900 St. Clair River flow regime. Journal of Great Lakes Research 19(4):660–664.

Richman, S., P. E. Sager, E. Banta, T. R. Harvey, and B. T. Destasio. 1984. Phytoplankton standing stock, size distribution, species composition and productivity along a trophic gradient in Green Bay, Lake Michigan. Verhandlungen-Internationale Vereingung fuer Teoretische und Angewandte Limnologie 22:460–469.

Robbins, J. 1982. Stratigraphic and dynamic effects of sediment reworking by Great Lakes zoobenthos. Hydrobiologia 92:611–622.

——, B. Eadie. 1991. Seasonal cycling of trace elements ^{137}Cs, ^{7}Be, and $^{239+240}$Pu in Lake Michigan. Journal of Geophysical Research 96(C9):17,081–17,104.

Robertson, A., and D. Scavia. 1984. North American Great Lakes. Pages 135–177 *in* F. B. Taub, editor, Lakes and Reservoirs. Elsevier Science Publishers, Amsterdam.

Rodhe, W. 1969. Crystallization of eutrophication concepts in Northern Europe. Pages 50–64 *in* Eutrophication: causes, consequences, correctives. National Academy of Sciences, Washington, D.C.

Rogers, P., T. Heidtke, V. Bierman, Jr., D. Dilks, and K. Feist. 1993. Great Lakes Environmental Assessment. A report prepared for the National Council of the Paper Industry for Air and Stream Improvement (NCASI), LTI LimnoTech, Inc., Ann Arbor, Michigan.

Rousmaniere, J. 1979. The enduring Great Lakes: A natural history book. American Museum of Natural History 112 pp.

Saylor, S. H., and P. W. Sloss. 1976. Water volume transport and Oscillatory Current Flow through the Straits of Mackinaw. Journal of Physical Oceanography 6(2):229–237.

Scavia, D., and G. Fahnenstiel. 1987. Dynamics of Lake Michigan phytoplankton: mechanisms controlling epilimnetic communities. Journal of Great Lakes Research 13:103–120.

Schelske, C. L., and J. C. Roth. 1973. Limnological survey of Lakes Michigan, Superior, Huron, and Erie. University of Michigan, Division of Great Lakes Research Publication 17:1–108.

——, E. F. Stoermer, G. L. Fahnenstiel, and M. Haibach. 1986. Phosphorus enrichment, silica utilization, and biogeochemical silica depletion in the Great Lakes. Canadian Journal of Fisheries and Aquatic Sciences 43:407–415.

Schneider, J. C., F. F. Hooper, and A. M. Beeton. 1969. The distribution and abundance of benthic fauna in Saginaw Bay, Lake Huron. Proceedings of the 12th Conference of Great Lakes Research. International Association for Great Lakes Research, Ann Arbor. Michigan.

Schuytema, G. S., and R. E. Powers. 1966. The distribution of benthic fauna in Lake Huron. University of Michigan, Division of Great Lakes Research Publication 15:155–163.

Sell, D. W. 1982. Size-frequency estimates of secondary production by *Mysis relicta* in Lakes Michigan and Huron. Hydrobiologia 93:69–78.

Shrivastava, H. N. 1974. Macrobenthos of Lake Huron. Fisheries Research Board. of Canada Technical Report 449:145 p.

Sloss, L.L., E. C. Dapples, and W. C. Krumbein. 1960. Lithofacies Maps: An Atlas of the United States and Southern Canada, John Wiley and Sons, Inc., New York.

Smiley, C. W. 1882. Changes in the fisheries of the Great Lakes during the decade 1870–1880. Transactions of the American Fisheries Society 11:28–37.

Smith, S. H. 1972. Factors of Ecologic Succession in Oligotrophic Fish Communities of the Laurentian Great Lakes. Journal of the Fisheries Research Board of Canada 29:717–730.

Sonzogni, W. C., A. Robertson, and A. M. Beeton. 1983a. Great Lakes Management: ecological factors. Environmental Management 7:531–542.

Sonzogni, W. L., P. W. Rodgers, W. L. Richardson, and T. J. Monteith. 1983b. Chloride pollution of the Great Lakes. Journal of Water Pollution Control 55(5):513–521.

Stemberger, R. 1973. Temporal and spatial distributions of rotifers in Milwaukee Harbor and adjacent Lake Michigan. Masters Thesis. University of Wisconsin, Milwaukee.

——, J. Gannon, and F. Bricker. 1979. Spatial and seasonal structure of rotifer communities in Lake Huron. U.S. Environmental protection Agency. EPA–600/3–79–085 pp. 160

Stewart, T. W., and J. M. Haynes. 1994. Benthic macroinvertebrate communities of southwestern Lake Ontario following invasion of *Dreissena*. Journal of Great Lakes Research 20:479–93.

Stoermer, E. F., and R. J. Stevenson. 1979. Green Bay phytoplankton composition, abundance, and distribution. EPA–905/3–79–002, 104 pp.

——, J. J. Yang. 1969. Plankton diatom assemblages in Lake Michigan. University of Michigan, Division of Great Lakes Research Special Report 47, 268 pp.

Stuckey, R. L., and D. L. Moore. 1995. Return and increase in abundance of aquatic flowering plants in Put-in-Bay Harbor, Lake Erie, Ohio. Ohio Journal of Science 95:261–266.

USEPA (U.S. Environmental Protection Agency). 1992. Exposure and Effects of Airborne Contamination. Great Waters Program Report, Chap 2. 2, 201 pp.

USEPA and Environment Canada. 1988. The Great Lakes: An Environmental Atlas and Resource Book. USEPA, Great Lakes National Program Office, Chicago, Illinois, and Environment Canada, Burlington, Ontario, 44 pp.

USEPA and Environment Canada. 1995. State of the Great Lakes—1995. USEPA, Great Lakes National Program Office, Chicago, Illinois, and Environment Canada, Burlington, Ontario, 56 pp.

Upchurch, S. 1975. Chemical Characteristics of the Great Lakes. Great Lakes Basin Framework Study, Appendix 4, Section 7:151–238. Great Lakes Basin Commission, Ann Arbor, Michigan, 441 pp.

Vallentyne, J., and N. Thomas. 1978. Fifth-year review of Canada-United States Great Lakes Water Quality Agreement. Report of Task Group III, A Technical Group to review Phosphorus Loadings to the Parties of the Great Lakes Water Quality Agreement of 1972, International Joint Commission, Windsor, Ontario.

Veal, D. M., and D. S. Osmond. 1968. Bottom fauna of the western basin and near-shore Canadian waters of Lake Erie. Proceedings of the 11th Conference on Great Lakes Research. International Association for Great Lakes Research, Ann Arbor, Michigan.

Vollenweider, R. A. 1968. Scientific fundamentals of the eutrophication of lakes and flowing waters, with particular reference to nitrogen and phosphorus as factors in eutrophication. Technical Report DAS/CSI/68. 27, Organization for Economic Co-Operation and Development, Directorate for Scientific Affairs, Paris, pp. 193.

Watson, N. H. F. 1974. Zooplankton of the St. Lawrence Great Lakes-species composition, Distribution, and Abundance. Journal of the Fishery Research Board of Canada 31:783–794.

Wells, L. 1960. Seasonal abundance and vertical movements of planktonic *Crustacea* in Lake Michigan. U.S. Fish and Wildlife Service Fisheries Bulletin 172:343–369.

———, 1970. Effects of alewife predation on zooplankton populations in Lake Michigan. Limnology and Oceanography 15:556–565.

Wetzel, R. G. 1983. Limnology; Second Edition. Saunders College Publishing, Philadelphia.

Wright, S. 1955. Limnological survey of western Lake Erie. U. S. Fish and Wildlife Service Special Report 1939. 341 p.

CHAPTER 2

Ichthyofauna of the Great Lakes Basin

Thomas G. Coon

The fish fauna of the Great Lakes basin is a relatively recent assemblage of fish species, most of which entered the Great Lakes basin from neighboring watersheds as the Great Lakes evolved. Its uniqueness is not so much in the presence of endemic species, such as in the African Great Lakes, but rather in the combination of species that are present. Because the lakes are relatively recent features, evolutionary processes have not had adequate time to produce more than a few endemic species and subspecies. Rather, the fauna consists of borrowed species (i.e., fish species that originated much earlier than the Laurentian basin and moved into the basin during and following lake formation). In addition, the Laurentian fish fauna has been altered by human activities over the past 150 years, including extinction and extirpation of some fish and introduction of species new to the basin.

This chapter provides an overview of the Great Lakes fish fauna and its origins. Several more detailed reviews are available on the origin of the Great Lakes fauna and its recent changes (Hubbs and Lagler 1964; Christie 1974; Bailey and Smith 1981; Emery 1985; Underhill 1986; Mills et al. 1993). The intent of this chapter is to describe the extant fauna, and to explain how it is different from the fauna that was present in the presettlement state.

Because the Great Lakes are only part of the aquatic ecosystem within the Great Lakes basin, and because fish can move freely among many of these ecosystems, this description encompasses the five Great Lakes, their tributaries, any inland lakes within the Great Lakes watershed, and the St. Lawrence River and its tributaries. In addition, the status of each species in the extant and the historic fauna is classified according to whether it was native or introduced (intentional and unintentional introductions), and whether it is extant or extirpated. Status is indicated separately for the watershed of each of the five Great Lakes, Lake Nipigon, and the St. Lawrence River. Species which have been recorded in the basin, but which have never been documented to be self-sustaining, are assigned the status of being a failed introduction, whether the introduction was intentional (e.g., *Oncorhynchus clarki*) or unintentional (e.g., *Platichthys flesus*).

Composition of the Fauna

The extant fish fauna of the Great Lakes basin includes 179 species (table 1), which represent 29 families in 18 orders and 2 classes of fish. Seven of the extant species are found only in the St. Lawrence River and its tributaries, and 172 are found in one or more of the Great Lakes and their tributaries.

The current fauna includes 157 species that were native to the basin, seven species that have been introduced from the Mississippi River watershed, one species introduced from other Atlantic watersheds, and fourteen species introduced from a source beyond eastern North America (Pacific Ocean watersheds of North America or Eurasia). In addition, three species have been introduced to the basin in the past, either by human intent or by accident, but have failed to become established, *Oncorhynchus clarki* (Hubbs and Lagler 1964), *Platichthys flesus* (Crossman 1991), and *Ameiurus catus* (Van Meter and Trautman 1970).

The native fauna also included six species which have been extirpated from the basin (*Polyodon spathula, Notropis boops, Lagochila lacera, Thymallus arcticus, Etheostoma chlorosomum, Percina evides*), and three species that were endemic to the lakes and now are extinct (*Coregonus johannae, C. nigripinnis nigripinnis, C. reighardi*; Bailey and Smith 1981, Underhill 1986, Webb and Todd 1995). Two more species (*C. kiyi, C. zenithicus*) have been extirpated from all lakes except Lake Superior (Smith and Todd 1984, Todd and Smith 1992).

Fourteen species have expanded their range within the watershed, either by intentional or accidental introductions. The most noted of these are the introductions of *Petromyzon marinus, Alosa pseudoharengus, Osmerus mordax*, and *Alosa sapidissima* from the St. Lawrence River, and possibly Lake Ontario into the upper lakes above Niagara Falls by means of human activities, such as canal construction, shipping and intentional stocking (Hubbs and Lagler 1964; Bailey and Smith 1981; Underhill 1986). Whether *A. sapidissima* was ever native to Lake Ontario remains uncertain. Greeley (1940) suggested they had been native, but no records confirm this (Smith 1985). That they were stocked and then failed to become established in Lakes Erie and Ontario is certain (Hubbs and Lagler 1964; Scott and Crossman 1973).

The species counts reported here differ from those of Bailey and Smith (1981) and Underhill (1986) for several reasons. Taxonomic changes, changes in status, and new introductions account for most of the differences. Underhill (1986) listed 177 species for the five lake basins combined, and eight more species found only in the St. Lawrence River. That list (Underhill 1986) included *Coregonus alpenae*, which has been synonymized with *C. zenithicus* (Todd et al. 1981), and *Ameirus catus*. *A. catus* has been introduced to Lake Erie on repeated occasions in the past, but has not become established and is no longer being introduced (Van Meter and Trautman 1970; Emery 1985). It is included in table 1, but as a failed introduction, rather than as an established species in the fauna. Five species have become established in the basin since Underhill's list was compiled: *Alosa chrysochloris* (Fago 1993), *Scardinius erythrophthalmus* (Crossman et al. 1992), *Gymnocephalus cernuus* (Simon and Vondruska 1991), *Neogobius melanostomus*, and *Proterorhinus marmoratus* (Jude et al. 1992). Nine species included in Underhill's count are either extirpated from the basin or extinct, (*Polyodon spathula, Coregonus johannae, C. nigripinnis, C. reighardi, Thymallus arcticus, Notropis boops, Lagochila lacera, Etheostoma chlorosomum, Percina evides*).

TABLE 1

Fish fauna of the Laurentian Great Lakes and the St. Lawrence River

Distributional status is given for each species in each lake and its tributaries, and the St. Lawrence River, including its tributaries. Status designations are: A = Absent, N = Native, P = Possibly native or introduced from nearby drainages, Ep = Extirpated, Ex = Extinct, In = Introduced, native elsewhere in the Laurentian Basin, Im = Introduced, native to Mississippi Basin, Ia = Introduced, native to Atlantic drainages, Ie = Introduced, not native to eastern North America, If = Introduced, but no evidence of self-sustaining populations (Failed introductions, either intentional or unintentional).

SCIENTIFIC NAME	COMMON NAME	LAKE NIPIGON	LAKE SUPERIOR	LAKE MICHIGAN	LAKE HURON	LAKE ERIE	LAKE ONTARIO	ST. LAWRENCE RIVER
Ichthyomyzon castaneus	chestnut lamprey	A	A	N	A	A	A	A
Ichthyomyzon fossor	northern brook lamprey	A	N	N	N	N	A	N
Ichthyomyzon unicuspis	silver lamprey	A	N	N	N	N	N	N
Lampetra appendix	American brook lamprey	A	N	N	N	N	N	N
Petromyzon marinus	sea lamprey	A	In	In	In	In	P	N
Acipenser fulvescens	lake sturgeon	N	N	N	N	N	N	N
Acipenser oxyrhynchus	Atlantic sturgeon	A	A	A	A	A	A	N
Polyodon spathula	paddlefish	A	A	Ep	Ep	Ep	A	A
Lepisosteus oculatus	spotted gar	A	A	N	A	N	A	A
Lepisosteus osseus	longnose gar	A	N	N	N	N	N	N
Lepisosteus platostomus	shortnose gar	A	A	N	A	A	A	A
Amia calva	bowfin	A	A	N	N	N	N	N
Hiodon tergisus	mooneye	A	A	N	N	N	N	N
Anguilla rostrata	American eel	A	In	In	In	In	N	N
Alosa chrysochloris	skipjack herring	A	A	Im	A	A	A	A
Alosa pseudoharengus	alewife	A	In	In	In	In	P	N
Alosa sapidissima	American shad	A	A	A	A	P / If	Ep[4]	N
Dorosoma cepedianum	gizzard shad	A	N	N	N	N	N	N
Campostoma anomalum	central stoneroller	A	A	N	N	N	N	A
Campostoma oligolepis	largescale stoneroller	A	A	N	A	A	A	A
Carassius auratus	goldfish	A	A	Ie	Ie	Ie	Ie	Ie
Clinostomus elongatus	redside dace	A	A	N	A	N	N	A
Couesius plumbeus	lake chub	N	N	N	N	A	N	A
Ctenopharyngodon idella	grass carp	A	A	Ie	A	Ie	A	A
Cyprinella analostana	satinfin shiner	A	A	A	A	A	N	A
Cyprinella spiloptera	spotfin shiner	A	A	N	N	N	N	N
Cyprinus carpio	common carp	A	Ie	Ie	Ie	Ie	Ie	Ie
Erimystax x-punctatus	gravel chub	A	A	A	A	N	A	A
Exoglossum laurae	tonguetied minnow	A	A	A	A	A	N	A
Exoglossum maxilingua	cutlips minnow	A	A	A	A	A	N	N
Hybognathus hankinsoni	brassy minnow	A	N	N	N	N	N	N
Hybognathus regius	eastern silvery minnow	A	A	A	A	A	N	N
Luxilis chrysocephalus	striped shiner	A	A	N	N	N	N	N
Luxilis cornutus	common shiner	A	N	N	N	N	N	N
Lythrurus umbratilis	redfin shiner	A	A	N	N	N	N	N
Macrhybopsis storeriana	silver chub	A	A	A	A	N	N	A
Margariscus margarita	pearl dace	N	N	N	N	N	N	N
Nocomis biguttatus	hornyhead chub	A	N	N	N	N	N	A

SCIENTIFIC NAME	COMMON NAME	LAKE NIPIGON	LAKE SUPERIOR	LAKE MICHIGAN	LAKE HURON	LAKE ERIE	LAKE ONTARIO	ST. LAWRENCE RIVER
Nocomis micropogon	river chub	A	A	N	N	N	N	A
Notemigonus crysoleucas	golden shiner	A	N	N	N	N	N	N
Notropis amblops	bigeye chub	A	A	A	A	N	N	A
Notropis anogenus	pugnose shiner	A	A	N	N	Ep	N	A
Notropis ariommus	popeye shiner	A	A	A	A	Ep	A	A
Notropis atherinoides	emerald shiner	N	N	N	N	N	N	N
Notropis bifrenatus	bridle shiner	A	A	A	A	A	N	N
Notropis blennius	river shiner	A	A	N	A	A	A	A
Notropis boops	bigeye shiner	A	A	A	A	Ep	A	A
Notropis buccatus	silverjaw minnow	A	A	N	A	N	A	A
Notropis buchanani	ghost shiner	A	A	A	Im	Im	A	A
Notropis chalybaeus	ironcolor shiner	A	A	N	A	A	A	A
Notropis dorsalis	bigmouth shiner	A	N	N	A	N	N	A
Notropis heterodon	blackchin shiner	A	N	N	N	Ep	N	N
Notropis heterolepis	blacknose shiner	N	N	N	N	Ep	N	N
Notropis hudsonius	spottail shiner	N	N	N	N	N	N	N
Notropis photogenis	silver shiner	A	A	A	A	N	A	A
Notropis procne	swallowtail shiner	A	A	A	A	A	N	A
Notropis rubellus	rosyface shiner	A	A	N	N	N	N	N
Notropis stramineus	sand shiner	A	N	N	N	N	N	N
Notropis texanus	weed shiner	A	A	N	N	A	A	A
Notropis volucellus	mimic shiner	N	N	N	N	N	N	N
Opsopoeodus emiliae	pugnose minnow	A	A	N	A	N	A	A
Phenacobius mirabilis	suckermouth minnow	A	A	A	A	Im	A	A
Phoxinus eos	northern redbelly dace	N	N	N	N	N	N	N
Phoxinus erythrogaster	southern redbelly dace	A	A	N	A	N	A	N
Phoxinus neogaeus	finescale dace	N	N	N	Ep	A	N	N
Pimephales notatus	bluntnose minnow	A	N	N	N	N	N	N
Pimephales promelas	fathead minnow	N	N	N	N	N	N	N
Pimephales vigilax	bullhead minnow	A	A	N	A	In	A	A
Rhinichthys atratulus	blacknose dace	A	N	N	N	N	N	N
Rhinichthys cataractae	longnose dace	N	N	N	N	N	N	N
Scardinius erythrophthalmus	rudd	A	A	A	A	A	Ie	Ie
Semotilus atromaculatus	creek chub	A	N	N	N	N	N	N
Semotilus corporalis	fallfish	A	A	A	A	A	N	N
Misgurnus anguillicaudatus	oriental weatherfish	A	A	A	Ie	A	A	A
Carpiodes cyprinus	quillback	A	A	N	N	N	N	N
Catostomus catostomus	longnose sucker	N	N	N	N	N	N	N
Catostomus commersoni	white sucker	N	N	N	N	N	N	N
Erimyzon oblongus	creek chubsucker	A	A	N	A	N	N	A
Erimyzon sucetta	lake chubsucker	A	A	N	N	N	N	A
Hypentelium nigricans	northern hog sucker	A	A	N	N	N	N	A
Ictiobus cyprinellus	bigmouth buffalo	A	A	N	A	N	A	A
Ictiobus niger	black buffalo	A	A	N	In	A	A	A
Lagochila lacera	harelip sucker	A	A	A	A	Ep	A	A
Minytrema melanops	spotted sucker	A	A	N	N	N	A	A
Moxostoma anisurum	silver redhorse	N	N	N	N	N	N	N

SCIENTIFIC NAME	COMMON NAME	LAKE NIPIGON	LAKE SUPERIOR	LAKE MICHIGAN	LAKE HURON	LAKE ERIE	LAKE ONTARIO	ST. LAWRENCE RIVER
Moxostoma carinatum	river redhorse	A	A	N	A	N	A	N
Moxostoma duquesnei	black redhorse	A	A	A	N	N	A	A
Moxostoma erythrurum	golden redhorse	A	A	N	N	N	N	A
Moxostoma hubbsi	copper redhorse	A	A	A	A	A	A	N
Moxostoma macrolepidotum	shorthead redhorse	N	N	N	N	N	N	N
Moxostoma valenciennesi	greater redhorse	A	N	N	N	N	N	N
Ameiurus catus	white catfish	A	A	A	A	If[7]	A	A
Ameiurus melas	black bullhead	A	N	N	N	N	N	A
Ameiurus natalis	yellow bullhead	A	A	N	N	N	N	N
Ameiurus nebulosus	brown bullhead	A	N	N	N	N	N	N
Ictalurus punctatus	channel catfish	A	A	N	N	N	N	N
Noturus flavus	stonecat	A	N	N	N	N	N	N
Noturus gyrinus	tadpole madtom	A	A	N	N	N	N	N
Noturus insignis	margined madtom	A	A	A	A	A	N	In
Noturus miurus	brindled madtom	A	A	A	A	N	N	A
Noturus stigmosus	northern madtom	A	A	A	A	N	A	A
Pylodictis olivaris	flathead catfish	A	A	N	A	N	A	A
Esox americanus	grass & redfin pickerels	A	A	N	N	N	N	N
Esox lucius	northern pike	N	N	N	N	N	N	N
Esox masquinongy	muskellunge	A	N	N	N	N	N	N
Esox niger	chain pickerel	A	A	A	A	In	N	N
Umbra limi	central mudminnow	A	N	N	N	N	N	N
Osmerus mordax	rainbow smelt	In	In	In	In	In	P	N
Coregonus artedi	lake herring	N[1]	N	N	N	N	N	N
Coregonus clupeaformis	lake whitefish	N	N	N	N	N	N	N
Coregonus hoyi	bloater	N	N	N	N	A	Ep	A
Coregonus johannae	deepwater cisco	A	A	Ex	Ex	A	A	A
Coregonus kiyi	kiyi	A	N	Ep	Ep	A	Ep	A
Coregonus nigripinnis	blackfin cisco	A[1]	A	Ex	Ex	A	A	A
Coregonus reighardi	shortnose cisco	A	A	Ex	Ex	A	Ex	A
Coregonus zenithicus	shortjaw cisco	N[2]	N	Ep[3]	Ep[3]	Ep[3]	A	A
Oncorhynchus clarki	cutthroat trout	A	A	If	A	A	A	A
Oncorhynchus gorbuscha	pink salmon	A	Ie	Ie	Ie	Ie	Ie	A
Oncorhynchus kisutch	coho salmon	A	Ie	Ie	Ie	Ie	Ie	A
Oncorhynchus mykiss	rainbow trout	A	Ie	Ie	Ie	Ie	Ie	Ie
Oncorhynchus nerka	sockeye salmon	A	A	If	Ie	A	If	A
Oncorhynchus tshawytscha	chinook salmon	A	Ie	Ie	Ie	Ie	A	Ie
Prosopium coulteri	pygmy whitefish	A	N	A	A	A	A	A
Prosopium cylindraceum	round whitefish	N	N	N	N	A	A	A
Salmo salar	Atlantic salmon	A	In	In	In	A	N	N
Salmo trutta	brown trout	A	Ie	Ie	Ie	Ie	Ie	Ie
Salvelinus alpinus	Arctic char	A	A	If[5]	A	A	If[5]	N
Salvelinus fontinalis	brook trout	N	N	N	N	In	N	N
Salvelinus namaycush	lake trout	N	N	N	N	N	N	N
Thymallus arcticus	Arctic grayling	A	Ep[6]	Ep[6]	Ep[6]	A	If[6]	A
Percopsis omiscomaycus	trout-perch	N	N	N	N	N	N	N
Aphredoderus sayanus	pirate perch	A	A	N	N	N	N	A

SCIENTIFIC NAME	COMMON NAME	LAKE NIPIGON	LAKE SUPERIOR	LAKE MICHIGAN	LAKE HURON	LAKE ERIE	LAKE ONTARIO	ST. LAWRENCE RIVER
Lota lota	burbot	N	N	N	N	N	N	N
Microgadus tomcod	Atlantic tomcod	A	A	A	A	A	A	N
Fundulus diaphanus	banded killifish	A	A	N	A	N	N	A
Fundulus dispar	starhead topminnow	A	A	N	A	A	A	A
Fundulus heteroclitus	mummichog	A	A	A	A	A	A	N
Fundulus notatus	blackstripe topminnow	A	A	N	A	N	A	A
Gambusia affinis	western mosquitofish	A	A	Ie	A	Ie	A	A
Labidesthes sicculus	brook silverside	A	A	N	N	N	N	N
Apeltes quadracus	fourspine stickleback	A	In	A	A	A	A	N
Culaea inconstans	brook stickleback	N	N	N	N	N	N	N
Gasterosteus aculeatus	threespine stickleback	A	A	In	In	A	N	N
Gasterosteus wheatlandi	blackspotted stickleback	A	A	A	A	A	A	N
Pungitius pungitius	ninespine stickleback	N	N	N	N	A	N	N
Cottus bairdi	mottled sculpin	N	N	N	N	N	N	N
Cottus cognatus	slimy sculpin	N	N	N	N	N	N	N
Cottus ricei	spoonhead sculpin	N	N	N	N	Ep	Ep	N
Myoxocephalus thompsoni	deepwater sculpin	N	N	N	N	N	Ep	A
Morone americana	white perch	A	A	A	In	In	N	N
Morone chrysops	white bass	A	N	N	N	N	N	N
Morone mississippiensis	yellow bass	A	A	A	Im	A	A	A
Morone saxatilis	striped bass	A	A	A	A	A	A	N
Ambloplites rupestris	rock bass	A	N	N	N	N	N	N
Enneacanthus gloriosus	bluespotted sunfish	A	A	A	A	A	Ia	A
Lepomis cyanellus	green sunfish	A	N	N	N	N	N	A
Lepomis gibbosus	pumpkinseed	A	N	N	N	N	N	N
Lepomis gulosus	warmouth	A	A	N	N	N	A	A
Lepomis humilis	orangespotted sunfish	A	A	A	A	Im	A	A
Lepomis macrochirus	bluegill	A	N	N	N	N	N	N
Lepomis megalotis	longear sunfish	A	A	N	N	N	N	N
Lepomis microlophus	redear sunfish	A	A	Im	A	Im	A	A
Micropterus dolomieu	smallmouth bass	In	N	N	N	N	N	N
Micropterus salmoides	largemouth bass	A	N	N	N	N	N	N
Pomoxis annularis	white crappie	A	A	N	N	N	N	A
Pomoxis nigromaculatus	black crappie	A	A	N	N	N	N	N
Ammocrypta clara	western sand darter	A	A	N	A	A	A	A
Ammocrypta pellucida	eastern sand darter	A	A	A	A	N	A	N
Etheostoma blennioides	greenside darter	A	A	N	N	N	N	A
Etheostoma caeruleum	rainbow darter	A	A	N	N	N	N	A
Etheostoma chlorosomum	bluntnose darter	A	A	Ep	A	A	A	A
Etheostoma exile	Iowa darter	N	N	N	N	N	N	N
Etheostoma flabellare	fantail darter	A	N	N	N	N	N	N
Etheostoma microperca	least darter	A	N	N	N	N	N	A
Etheostoma nigrum	johnny darter	N	N	N	N	N	N	N
Etheostoma olmstedi	tessellated darter	A	A	A	A	A	N	N
Etheostoma spectabile	orangethroat darter	A	A	A	A	N	A	A
Etheostoma zonale	banded darter	A	A	N	A	A	A	A
Gymnocephalus cernuus	ruffe	A	Ie	A	Ie	A	A	A

SCIENTIFIC NAME	COMMON NAME	LAKE NIPIGON	LAKE SUPERIOR	LAKE MICHIGAN	LAKE HURON	LAKE ERIE	LAKE ONTARIO	ST. LAWRENCE RIVER
Perca flavescens	yellow perch	N	N	N	N	N	N	N
Percina caprodes	logperch	N	N	N	N	N	N	N
Percina copelandi	channel darter	A	A	A	N	N	N	N
Percina evides	gilt darter	A	A	A	A	Ep	A	A
Percina maculata	blackside darter	A	A	N	N	N	N	A
Percina phoxocephala	slenderhead darter	A	A	N	A	A	A	A
Percina shumardi	river darter	A	A	N	N	N	A	A
Stizostedion canadense	sauger	N	N	N	N	N	N	N
Stizostedion vitreum	walleye & blue pike	N	N	N	N	N	N	N
Aplodinotus grunniens	freshwater drum	A	A	N	N	N	N	N
Neogobius melanostomus	round goby	A	A	Ie	Ie	Ie	A	A
Proterorhinus marmoratus	tubenose goby	A	A	A	A	Ie	A	A
Platichthys flesus	European flounder	A	If	A	If	If	A	A

1   *Coregonus nigripinnis regalis* and *C. nipigon* are included with *C. artedi* for Lake Nipigon (Bailey and Smith 1981), and *C. hubbsi* from Ives Lake, Michigan is included with
    C. artedi for the Lake Superior basin (Bailey and Smith 1981).

2   *C. bartletti* from Lake Siskiwit on Isle Royale is included with *C. zenithicus* for the Lake Superior basin (Bailey and Smith 1981).

3   *C. alpenae* formerly reported from Lakes Michigan, Huron, and Erie (Scott and Smith 1962), are included here as *C. zenithicus* (Todd et al. 1981).

4   *A. sapidissima* may have been native to Lake Ontario (Greeley 1940), but the historic record is not sufficient to determine. Efforts to introduce it have been unsuccessful
    (Hubbs and Lagler 1964; Scott and Crossman 1973).

5   *S. alpinus* introduction attempts have been to Lake Ontario and to inland waters within the Lake Michigan and Lake Ontario basins (Emery 1985).

6   *T. arcticus* introductions have been attempted in Lakes Michigan, Huron, Superior, and Ontario drainages, without success (Emery 1985).

7   *A. catus* was introduced repeatedly into Lake Erie and its tributaries, without ever establishing reproducing populations (Van Meter and Trautman 1970).

Although these were important species in the native fauna, they are not included in the count of extant species reported here.

Bailey and Smith (1981) only listed species for the five lakes and their tributaries, and did not include the St. Lawrence River and its tributaries. Their number of extant species for the five lakes and tributaries (174) included seven species that were listed as being extirpated or extinct and two species that are listed as being extinct here. In addition to the five species added since Underhill (1986) compiled his list, *Ctenopharyngodon idella* has become established (by illegal stocking) in the basin, and *Apeltes quadracus* has become established in the upper lakes (Holm and Hamilton 1988).

Prior to documented extinction and extirpation, the native fauna of the five lakes and their tributaries included at least 158, and as many as 162 species, depending on the presence of *P. marinus, A. pseudoharengus, A. sapidissima,* and *O. mordax* in the Lake Ontario fauna. Underhill (1986) included three species as natives, *Notropis buchanani, Phenacobius mirabilis,* and *Lepomis humilis,* which are considered here as being introduced. All three species were first documented in the Great Lakes basin in the twentieth century (Hubbs and Lagler 1964; Trautman 1981), and may have entered the basin from the Mississippi basin via introduction or as natural colonists (Lodge 1993). Following their introduction to the basin, they have dispersed further within the basin, again either by means of introduction or natural dispersal mechanisms (Holm and Coker 1981; Holm and Houston 1993; Trautman 1981; Noltie and Beletz 1984; Noltie 1990).

The most diverse fauna is in the Lake Michigan watershed (136 species; table 2), followed by Lake Erie (129 species), Lakes Ontario and Huron (119 and 117, respectively), the St. Lawrence River (105 species), Lake Superior (83 species) and Lake Nipigon (39 species). This pattern reflects

TABLE 2

Species diversity and status of fishes by sub-basin in the Laurentian Great Lakes and their tributaries

STATUS	LAKE NIPIGON	LAKE SUPERIOR	LAKE MICHIGAN	LAKE HURON	LAKE ERIE	LAKE ONTARIO	ST. LAWRENCE RIVER	TOTAL 5 LAKES & TRIBUTARIES	TOTAL BASIN
Native	37	69	118	96	106	108	98	145	157
Possibly Native	0	0	0	0	0	4	0	4	0
Extirpated	0	1	5	5	9	4	0	6	6
Extinct	0	0	3	3	0	1	0	3	3
Introduced from elsewhere in basin	2	7	6	8	8	0	1	1	0
Introduced from Mississippi Basin	0	0	3	2	5	0	0	7	7
Introduced from Atlantic drainages	0	0	0	0	0	1	0	1	1
Introduced: not from eastern N. America	0	7	9	11	10	7	6	14	14
Failed introductions: intentional and accidental	0	1	3	1	2	4	0	4	3
Total Extant Species	39	83	136	117	129	119	105	172	179
Total Original Native Species	37	70	126	104	115	116	98	158	166

the presettlement distribution of species among lakes: 126 species were native to Lake Michigan and its tributaries, followed by Lake Ontario (116), Lake Erie (115), Lake Huron (104), St. Lawrence River (98), Lake Superior (70) and Lake Nipigon (37).

Native Fauna

Because the Laurentian basin is relatively young, most of the fish fauna originated outside of the basin and colonized the basin after the retreat of the Wisconsin glaciers. Bailey and Smith (1981) and Underhill (1986) have carefully documented the likely avenues of colonization, including passage from the Mississippi River basin (into Lakes Superior, Michigan and Erie) and from Atlantic coastal river systems (Susquehana and Mohawk-Hudson) into Lake Ontario. Underhill (1986) estimated that 10.3% (19 of 185) of the species were of Atlantic origin, 14.1% (26) were of Atlantic and Mississippi origin and 64.9% (120) were of Mississippi origin. The remainder are either endemic (*Coregonus reighardi, C. nigripinnis, C. kiyi, C. johannae, C. hoyi*) or introduced. It is likely that the endemic coregonines evolved within the Great Lakes basin from as few as two ancestral species (*C. artedi* and *C. zenithicus*) that entered the basin from the Mississippi refugium (Smith and Todd 1984, Todd and Smith 1992). Thus, the Great Lakes fish fauna is largely a subset of the Mississippi basin fauna, suggesting that the Chicago-Illinois River outlet, the Maumee-Wabash River outlet, and to a lesser extent, the Lake Duluth-St. Louis River and the Fox-Wisconsin River outlets were the major avenues for colonization following the retreat of the glaciers. The importance of the two southern outlets to the Mississippi basin, combined with the greater diversity of habitats in the Lakes Michigan and Erie basins account for the high ranks of these basins in species richness.

The most distinctive elements of the Great Lakes fish fauna are those species that are endemic to the basin. The greatest example of this is the species flock of *Coregonus* species, including *C. hoyi, C. johannae, C. kiyi, C. nigripinnis,* and *C. reighardi* (Smith and Todd 1984). The only other endemic species in the basin is *Moxostoma hubbsi*, which is restricted to the St. Lawrence River. Several other groups reflect some degree of endemism; however, their status as species or subspecies remains open to debate. The Lake Superior population of *Prosopium coulteri*, has no subspecific designation to separate it from the other populations of this species found in the Columbia River basin (Eschmeyer and Bailey 1955). The Lakes Superior, Michigan and Huron populations of *Thymallus arcticus* were disjunct from populations in the headwaters of the Missouri and Columbia Rivers and in northern Canada and Alaska, and were extirpated early in the twentieth century (Taylor 1954). Hubbs and Lagler (1964) argued against considering this a separate subspecies from the other populations of arctic grayling. However, Hubbs and Lagler (1964) did consider the endemic blue pike to be a distinctive subspecies, *Stizostedion vitreum glaucum*. This form of walleye was concentrated primarily in Lake Erie, but also occurred in western Lake Ontario (Campbell 1987). Aside from the difference in color, it was distinct from the yellow form in that it had larger and more closely-set eyes, and did not grow to be as large as the yellow form (Hubbs 1926). Hubbs (1926) speculated that blue pike spawned at deeper sites, and at later dates than the yellow walleye.

A variety of extant species show evidence of genetic differentiation within the Great Lakes basin. Some are recognized as morphologically distinct subspecies (e.g., *Rhinichthys atratulus, Fundulus diaphanus, Erimyzon oblongus, Margariscus margarita, Etheostoma flabellare, E. blennioides*; Bailey and Smith 1981). Others have been distinguished on the basis of genetic data, supplemented with morphological data in some cases; some examples are, *Salvelinus namaycush* (Burnham-Curtis and Smith 1994); *Coregonus clupeaformis*, (Ihssen et al. 1981; Bernatchez and Dodson 1990) *Perca flavescens*, (Billington 1993; Todd and Hatcher 1993) *S. v. vitreum*, (Billington and Hebert 1988; Todd and Haas 1993).

These distinctive populations or stocks, along with the recognized subspecies and the coregonine species flock suggest that the speciation process had begun in the short time (approximately ten thousand to twelve thousand years) that the Great Lakes have existed, yet isolating mechanisms had not evolved to the extent of forming true species, except in some of the coregonines (Smith and Todd 1984). In contrast, the cichlid species flocks of the African Great Lakes have evolved in much more ancient systems, facilitated not only by the age of the lakes, but also by the distinctive courtship and reproductive behavior of haplochromine cichlids (Greenwood 1984). Spawning philopatry is strong in many of the species that have evolved distinctive stocks and this plays a more important role in the distinctiveness of these stocks than true geographic isolation (Bailey and Smith 1981).

Perhaps the best documented examples of philopatry and its role in forming distinctive stocks is in the reef spawning populations of *Salvelinus namaycush* in Lake Superior (Eschmeyer 1955; Eschmeyer and Phillips 1965; Rahrer 1965; Horrall 1981; Ihssen et al. 1988; Burnham-Curtis and Smith 1994). Three distinct phenotypes have been documented, including the "lean" trout, which occupy shallow, near shore areas and spawn in late October and early November. The siscowet trout is more deep-bodied, and occurs in deeper, offshore habitats, spawning earlier than the lean trout. The third form, humpers, spawn on shallow shoals which are surrounded by deep water from August to October (Burnham-Curtis and Smith 1994). Within these phenotypes, distinctive forms are found associated with different reefs within Lake Superior. It is likely that similar degrees of genetic and phenotypic diversity of *S. namaycush* were extant in the other lakes, however, in most of these, the mid-twentieth century decline (Lakes Huron and Ontario) or extirpation (Lake Michigan) of lake trout eliminated the original within-lake diversity (Brown et al. 1981). Subsequently, a diversity of hatchery strains of lake trout have been used in an effort to rehabilitate lake trout populations, with the result that these rehabilitated populations exhibit modest within-lake genetic diversity (Grewe et al. 1994). These hatchery strains have caused the decline or elimination of distinctive wild strains as well (Evans and Willcox 1991).

Koelz (1929) also documented a great degree of phenotypic variation among Great Lakes coregonines, and concluded that many of the different forms represented true species. Many of these also exhibited strong philopatry to spawning reefs. However, much of the phenotypic variation among the coregonines reflected their developmental plasticity. Bailey and Smith (1981) and Smith and Todd (1984) concluded that the original list of more than forty species and more than twenty subspecies more likely represented four to eight species with some incipient isolation among stocks within these species. Bailey and Smith (1981) considered *C. nigripinnis regalis* and *C. nipigon* from Lake Nipigon, and *C. hubbsi* from Ives Lake, Michigan, to be variants of *C. artedi*, and *C. bartletti* from Siskiwit Lake, Isle Royale to be a variant of *C. zenithicus*. Although these conclusions have not been confirmed for synonymy, they are likely to be validated, and are reflected in table 1.

Another important aspect of the native faunal diversity was the presence of phenotypically distinctive river-spawning (potamadromous) populations of *Salvelinus namaycush* (Loftus 1958; Scott and Crossman 1973) and *Stizostedion vitreum* (Spangler et al. 1977a; Schneider and Leach 1977). Many other species had potamadromous strains as well, including *C. artedi* and *C. clupeaformis, Acipenser fulvescens, Micropterus dolomieu, Esox lucius,* and many of the sucker species (*Catostomus catostomus, C. commersoni, Moxostoma* spp.), but there is no evidence that any of these are or were genetically distinct. Native Lake Ontario populations of *S. salar* in the nineteenth century apparently included some fish that matured in the lake, and others that migrated down the St. Lawrence River to mature in the Atlantic Ocean (Parsons 1973). Because these were extirpated by the late nineteenth century, it is not possible to determine if these represented distinct genetic stocks, or if environmental factors determined which fish remained in the lakes and which migrated to the ocean. Like the Lake Ontario *S. salar,* many of the potamadromous strains of other species have been extirpated, either as a result of overharvest or habitat degradation (Spangler et al. 1977b; Smith 1995). Current stocks of *S. salar* in Lake Ontario are maintained by stocking of hatchery-reared fish, and do not originate from the ancestral stocks that were in Lake Ontario (Lange et al. 1995).

Introduced Fauna

The fish fauna of the Great Lakes has been altered dramatically over the past 150 years, and one of the major agents of change has been the introduction of fish species. Known introduction of at least twenty-four species that were not native to the Great Lakes basin have been documented (table 2). In addition, fourteen species have been introduced from one part of the basin to other locations within the basin where they were not native. The twenty-four species that have been introduced from outside the Laurentian basin represent four families that were not in the native fauna (Cobitidae, Poeciliidae, Gobiidae, Pleuronectidae). Salmonids and cyprinids each account for six of these introduced species. Three of the introduced species are centrarchids, and two are gobiids.

Fish species have entered the Great Lakes basin and have been redistributed within the basin over the past 150 years by several means, including intentional introduction by humans, accidental releases of fish held in captivity, introductions resulting from release of live bait fish, passage around dispersal barriers (e.g., waterfalls and watershed boundaries) through shipping canals, and introductions from ballast water of cargo ships (Mills et al. 1993). *Cyprinus carpio* is the most ubiquitous of the intentionally introduced species. Four salmonid species have been intentionally stocked into one or more of the lakes, including *Salmo trutta, Oncorhynchus mykiss, O. tshawytscha,* and *O. kisutch.* In all cases, hatchery-produced fish continued to be stocked for recreational fisheries, even though naturalized populations of each persist in all of the lakes except Lake Erie (Biette et al. 1981). *Gambusia affinis* also was introduced intentionally, ostensibly for control of mosquito larvae in small lakes and wetlands in the basin (Emery 1985). However, severe winter conditions limit the extent to which this species has become naturalized in the basin.

Species which have been introduced by accidental release from captive populations include *Ctenopharyngodon idella, Misgurnus anguillicaudatus, Oncorhynchus gorbuscha,* and *Osmerus mordax. C. idolla* has not established reproducing populations within the basin, partially due to restriction on their introduction in states and provinces in the Great Lakes basin. However, they have become naturalized in the Mississippi River basin (Brown and Coon 1990), and may be able to make passage through the Chicago River-Illinois River Canal. In addition, illegal transport and stocking of these into inland lakes of the Lake Michigan and Lake Erie basins are likely to persist (Bailey and Smith 1981; Emery 1985). *M. anguillicaudatus* escaped from a captive population used for producing fish for the aquarium trade in Lake Huron drainage as early as 1939, and persists within the Saginaw Bay watershed (Schultz 1960). *O. gorbuscha* escaped from a captive population in the Lake Superior basin in 1956, and has become naturalized in Lakes Superior, Huron and Michigan. *O. mordax* was initially stocked into Crystal Lake within the Lake Michigan basin in 1912, and quickly escaped to Lake Michigan, and became naturalized throughout the Great Lakes basin (Van Oosten 1937).

Several baitfish species have become established in the Great Lakes or their tributaries, apparently as a result of angler releases of live bait. *S. erythrophthalmus* is a bait fish that has been introduced into some waters within the Lake Ontario basin, and it has been documented from Lake Ontario (Mills et al. 1993). Holm and Coker (1981) suggested that *Notropis buchanani* was introduced to Lake Huron tributaries as a result of bait fish transfers, and Rubec and Coad (1974) suggested a similar means of introduction for *Noturus insignis.*

The most significant introductions for Great Lakes fisheries have been those that occurred as a result of passage through shipping canals. Whether *Petromyzon marinus* were native to Lake Ontario remains debatable, but their introduction into the upper lakes is clear, and most likely occurred by passage of adults through the Erie and Welland Canals. Clearly, their impact has been devastating for *Salvelinus namaycush* fisheries (Lawrie 1970). *Alosa pseudoharengus* also found passage around Niagara Falls through the Erie and Welland Canals and now is one of the most prevalent prey species in the upper lakes (Emery 1985). *Anguilla rostrata* has invaded the upper lakes, probably through two avenues: the Welland and Erie Canals into Lake Erie, and the Chicago River-Illinois River Canal. Numerous early attempts at introducing them to the upper lakes also contributed to their range extension (Emery 1985). *Morone americana* has become most abundant in Lakes Erie and Ontario, which it entered by passage through the Erie Canal (Mills et al. 1993). Other incidental introductions may have been facilitated by inter-basin canals such as the Chicago River–Illinois River Canal, (e.g., *Alosa chrysochloris*), (Fago 1993), and Lake St. Mary's, Ohio, which has constructed outlets to the Mississippi and Great Lakes basins (*Phenacobius mirabilis, Lepomis humilis*, (Hubbs and Lagler 1964; Trautman 1981)).

Ballast water from ships has become an important vector of species introductions into and among the Great Lakes over the past twenty years (Mills et al. 1993). At least four European species have been introduced by this mechanism, including the gobies, *Proterorhinus marmoratus, Neogobius melanostomus* (Jude et al. 1992), the flatfish, *Platichthys flesus* (Crossman 1991) and the percid, *Gymnocephalus cernuus* (Simon and Vondruska 1991). However, there is no evidence of reproduction in the Great Lakes. In addition, several species appear to have been transferred from the St. Lawrence River estuary to locations as far upstream as Lake Superior (*Apeltes quadracus*, Holm and Hamilton 1988) and Lake Michigan (*Gasterosteus aculeatus*, Stedman and Bowen 1985).

Although efforts have been made to reduce transfer of species by ballast water through stricter controls on ship operations (Mills et al. 1993), it is likely that more introductions will occur. For example, Great Lakes shippers voluntarily agreed to exchange ballast water in mid-Lake Superior on all ships disembarking from the Duluth-Superior Harbor (original site of *G. cernuus* introduction). Yet this species was reported from Thunder Bay, Lake Huron and Thunder Bay, Lake Superior, in 1995, two ports which receive a sizable volume of shipping traffic from the Duluth-Superior Harbor.

The combination of introductions has had dramatic effects on the Great Lakes fish fauna, far beyond the addition of twenty-four species to the native fauna of 162 species. The original food web in pelagic habitats consisted of prey species, such as the deepwater ciscoes (*Coregonus* spp.), sculpins (*Myoxocephalus thompsoni, Cottus* spp.), and suckers (*Catostomus catostomus*), and the predominant piscivore, (*Salvelinus namaycush*). Currently, this web has been altered to include predominant prey, such as *A. pseudoharengus, O. mordax*, and *M. americana*, as well as the invasive *P. marmoratus* and *G. cernuus*, and predominant piscivores, including *P. marinus*, the *Oncorhynchus* species, and *S. trutta*.

Endangered Species

In addition to the introduction of species, the native Great Lakes fauna has suffered loss or reduction of some native species (table 3). Eight species have been extirpated from the basin completely, two of which were endemic species and thus, are extinct. In addition, several other species are at risk of extinction, and have been listed as potential candidates for listing as threatened or endan-

TABLE 3

Status of species listed or proposed for listing as threatened, endangered, or extinct/extirpated in U.S. or Canada waters of the Great lakes.

Status codes are given at bottom of table.

SCIENTIFIC NAME	GREAT LAKES STATUS	U.S.	CANADA	SCIENTIFIC NAME	GREAT LAKES STATUS	U.S.	CANADA
Ichthyomyzon castaneus			V	*Pylodictis olivaris*			ID
Ichthyomyzon fossor			V	*Esox americanus*			PV
Acipenser fulvescens		C2		*Esox niger*			PV
Polyodon spathula	Ep	C2	Ep	*Coregonus artedi*			PE
Lepisosteus oculatus			V	*Coregonus clupeaformis*			PT
Anguilla rostrata			PV	*Coregonus johannae*	Ex	Ex	Ex
Campostoma anomalum			V	*Coregonus kiyi*		C2	V
Clinostomus elongatus			V	*Coregonus nigripinnis*	Ex	Ex	T
Erimystax x-punctatus			Ep	*Coregonus reighardi*		C2	T
Exoglossum maxilingua			PV	*Coregonus zenithicus*		C2	T
Hybognathus regius			PV	*Prosopium cylindraceum*			PV
Macrhybopsis storeriana			V	*Salmo salar*			C2
Notropis anogenus			V	*Thymallus arcticus*	Ep		
Notropis boops	Ep			*Microgadus tomcod*			PV
Notropis dorsalis			V	*Fundulus diaphanus*			V
Notropis heterodon			PV	*Fundulus notatus*			V
Notropis photogenis			V	*Myoxocephalus thompsoni*			T
Notropis rubellus			PV	*Lepomis gulosus*			PV
Notropis texanus			PV	*Lepomis humilis*			V
Opsopoeodus emiliae			V	*Ammocrypta pellucida*		C2	PV
Pimephales notatus			PV	*Etheostoma blennioides*			V
Erimyzon oblongus			PX	*Etheostoma chlorosomum*	Ep		
Erimyzon sucetta			PV	*Percina copelandi*			T
Ictiobus cyprinellus			V	*Percina evides*			Ep
Ictiobus niger			V				
Lagochila lacera	Ep						
Minytrema melanops			V				
Moxostoma carinatum			V				
Moxostoma duquesnei			T				
Moxostoma hubbsi			T				
Moxostoma valenciennesi		C2					
Noturus insignis			T				
Noturus miurus			V				
Noturus stigmosus			ID				

STATUS CODES
IS   Insufficient Data (Canada)
V   Vulnerable
Ep   Extirpated
Ex   Extinct
E   Endangered
T   Threatened
PE   Proposed Endangered (Canada)
PT   Proposed Threatened (Canada)
PV   Proposed Vulnerable (Canada)
PX   Proposed Extirpated (Canada)
C2   Proposed Endangered or Threatened (U.S.)

gered species. The U.S. list of endangered and threatened species does not include any species in the Great Lakes Basin (U.S. Fish and Wildlife Service 1994). However, several Great Lakes species are listed as potential candidate species. These include *Acipenser fulvescens, Coregonus kiyi, C. zenithicus, Salmo salar, Moxostoma valenciennesi,* and *Etheostoma pellucida. Polyodon spathula* also is on this list, although it is already extirpated from the Great Lakes fauna. For each of these species, the U.S. Fish and Wildlife Service has determined that it may be appropriate to list the species as endangered or threatened, but that data on biological vulnerability and threat are not currently adequate to support listing.

The classification of endangered and threatened species in Canada is slightly different from that used in the United States, however, a number of species from the Great Lakes fauna are on the Canadian lists of vulnerable, threatened, or endangered species (Campbell 1992, 1993). In addition to the species listed in the U.S., one more is listed as extirpated from Canadian waters (*Erimystax x-punctata*), one is proposed for listing as endangered (*Coregonus artedi*), and one is proposed for listing as threatened (*Coregonus clupeaformis*, in Lakes Erie and Ontario, only). Twenty species whose distributions include the Great Lakes are listed as Vulnerable in Canada, one of which is on the U.S. list for consideration of listing (*C. kiyi*). Eight of the Great Lakes species are listed as threatened in Canada, including one species that is on the U.S. list for consideration (*C. zenithicus*) and two that are considered extinct on the U.S. list (*C. nigripinnis nigripinnis* and *C. reighardi*). One species is proposed as being listed extirpated (*Erimyzon oblongus*) and fourteen species are proposed as being listed vulnerable, including one that is on the U.S. list for consideration (*Ammocrypta pellucida*). For two other species (*Noturus stigmosus* and *Pylodictis olivaris*), insufficient data are available for determining if listing is appropriate in Canada.

Future Prospects

It is likely that the Great Lakes fish fauna will continue to change dramatically over the next few decades. The factors that have led to the critically important introductions have not been eliminated. Efforts to manage ballast water introduction to prevent introduction of non-native species need to be accelerated, and regulations need to be enacted and strictly enforced, or more species are likely to enter from other temperate regions, primarily from Europe and east Asia. New introductions may be as devastating to the native fauna as *P. marinus* and *A. pseudoharengus* have been.

In addition to ship-borne introductions, canals continue to pose a threat to the Great Lakes fauna. For example, improved water quality in the Chicago River–Illinois River Canal has made it more likely that other species will enter the Great Lakes from the Mississippi Basin (Underhill 1986). Of particular concern are the Asian carp species, (e.g., *Aristichthys nobilis* and *Ctenopharyngodon idella*) that have become established in the Mississippi River System and have spread relatively quickly throughout the watershed. It is extremely difficult to restrict movement of fish through this and other inter-basin canals. New technologies are needed to restrict inter-basin transfer of fish. One alternative to this would be to establish and maintain oxygen-depleted zones in these canals as a barrier to fish dispersal.

Intentional introductions, either of species or of genetic stocks that are new to the basin need to be restricted to prevent further destabilization of the Great Lakes fish fauna and to prevent further dilution of the genetic resources extant in the fish of the basin. Potentially attractive game species, (e.g. *Morone saxatilis*) are likely to cause more damage to the Great Lakes fish fauna and its fisheries than they counteract by increasing the diversity of fishing opportunities. Interstate and international agreements are needed to ensure against introductions that can have severe impacts beyond the single state or province of initial introduction.

Finally, some species are likely to become extinct or extirpated from the Great Lakes fauna. The most vulnerable species and genetic stocks are likely to be those that have already declined severely, particularly among the coregonines. Other species that are likely to face risk of extirpation are those that require migration into Great Lakes tributaries for spawning. Dams and poor water quality continue to threaten the sustainability of these species, (e.g. the larger *Moxostoma* species). In addition, loss of headwater stream habitats to channelization or degraded water quality threaten some species in the basin that may never occupy any of the Great Lakes, but contribute to and reflect the overall diversity of aquatic ecosystems connected with the Great Lakes, such as *Notropis anogenus, Clinostomus elongatus, Etheostoma microperca,* and *Ichthyomyzon fossor.*

The unique assemblage of fish in the Great Lakes basin is a legacy generated by the melting waters of glaciers. Human activities have dramatically altered this assemblage, yet much of the fauna that was present in the presettlement basin remains extant. The challenge to resource managers and to human populations within the Great Lakes basin is to find ways of living adjacent to the Great Lakes and using their resources without further taxing what remains of this valuable legacy.

Literature Cited

Bailey, R. M., and G. R. Smith. 1981. Origin and geography of the fish fauna of the Laurentian Great Lakes Basin. Canadian Journal of Fisheries and Aquatic Sciences 38 (12): 1539–1561.

Bernatchez, L.and J. J. Dodson. 1990. Allopatric origin of sympatric populations of lake whitefish *(Coregonus clupeaformis)* as revealed by mitochondrial-DNA restriction analysis. Evolution 44 (5): 1263–1271.

Biette, R. M., D. P. Dodge, R. L. Hassinger, and T. M. Stauffer. 1981. Life history and timing of migrations and spawning behavior of rainbow trout *(Salmo gairdneri)* populations of the Great Lakes. Canadian Journal of Fisheries and Aquatic Sciences 38: 1759–1771.

Billington, N. 1993. Genetic variation in Lake Erie yellow perch (*Perca flavescens*) demonstrated by mitochondrial DNA analysis. Journal of Fish Biology 43: 941–943.

——, and P. D. N. Hebert. 1988. Mitochondrial DNA variation in Great Lakes walleye (*Stizostedion vitreum*) populations. Canadian Journal of Fisheries and Aquatic Sciences 45: 643–654.

Brown, D. J., and T. G. Coon. 1990. Grass carp larvae in the lower Missouri River and its tributaries. North American Journal of Fisheries Management 11: 62–66.

Brown, E. H., Jr., G. W. Eck, N. R. Foster, R. M. Horrall, and C. E. Coberly. 1981. Historical evidence for discrete stocks of lake trout (*Salvelinus namaycush*) in Lake Michigan. Canadian Journal of Fisheries and Aquatic Sciences 38: 1747–1758.

Burnham-Curtis, M. K., and G. R. Smith. 1994. Osteological evidence of genetic divergence of lake trout (*Salvelinus namaycush*) in Lake Superior. Copeia 1994: 843–850.

Campbell, R. R. 1987. Status of the blue walleye, *Stizostedion vitreum glaucum*, in Canada. Canadian Field-Naturalist 101 (2): 245–252.

——. 1992. Rare and endangered fishes and marine mammals of Canada: COSEWIC Fish and Marine Mammal Subcommittee Status Report VIII. Canadian Field-Naturalist 106 (1): 1–6.

——. 1993. Rare and endangered fishes and marine mammals of Canada: COSEWIC Fish and Marine Mammal Subcommittee Status Report IX. Canadian Field-Naturalist 107 (4): 395–401.

Christie, W. J. 1974. Changes in the fish species composition of the Great Lakes. Journal of the Fisheries Research Board Canada 31: 827–854.

Crossman, E. J. 1991. Introduced freshwater fishes: a review of the North American perspective with emphasis on Canada. Canadian Journal of Fisheries and Aquatic Sciences 48 (Suppl. 1): 46–57.

——, E. Holm, R. Cholmondeley, and K. Tuininga. 1992. First record for Canada of the rudd, *Scardinius erythrophthalmus,* and notes on the introduced round goby, *Neogobius melanostomus.* Canadian Field-Naturalist 106 (2): 206–209.

Emery, L. 1985. Review of fish species introduced into the Great Lakes, 1819–1975. Great Lakes Fishery Commission Technical Report 45. Ann Arbor, MI.

Eschmeyer, P. H. 1955. The reproduction of lake trout in southern Lake Superior. Transactions of the American Fisheries Society 84: 47–74.

——, and R. M. Bailey. 1955. The pygmy whitefish, *Coregonus coulteri*, in Lake Superior. Transactions of the American Fisheries Society 84: 161–199.

——, and A. M. Phillips, Jr. 1965. Fat content of the flesh of siscowets and lake trout from Lake Superior. Transactions of the American Fisheries Society 94: 62–74.

Evans, D. O., and C. C. Willcox. 1991. Loss of exploited, indigenous populations of lake trout, *Salvelinus namaycush*, by stocking of non-native stocks. Canadian Journal of Fisheries and Aquatic Sciences 48 (Suppl. 1): 134–147.

Fago, D. 1993. Skipjack herring, *Alosa chrysochloris*, expanding its range into the Great Lakes. Canadian Field-Naturalist 107 (3): 352–353.

Greeley, J. R. 1940. Fishes of the watershed with annotated list. Pages 42–81 *in* Anonymous, editor. A biological survey of the Lake Ontario watershed. Supplement 29th Annual Report New York State Conservation Department.

Greenwood, P. H. 1984. African cichlids and evolutionary theories. Pages 141–154 *in* A. A. Echelle and I. Kornfield, editors. Evolution of Fish Species Flocks. University of Maine Press, Orono.

Grewe, P. M., C. C. Krueger, J. E. Marsden, C. F. Aquadro, and B. May. 1994. Hatchery origins of naturally produced lake trout fry captured in Lake Ontario: temporal and spatial variability based on allozyme and mitochondrial DNA data. Transaction of the American Fisheries Society 123: 309–320.

Holm, E, and G. A. Coker. 1981. First Canadian records of the ghost shiner (*Notropis buchanani*) and the orangespotted sunfish (*Lepomis humilis*). Canadian Field-Naturalist 95 (2): 210–211.

——, and J. G. Hamilton. 1988. Range extension for the fourspine stickleback, *Apeltes quadracus*, to Thunder Bay, Lake Superior. Canadian Field-Naturalist 102 (4): 653–656.

——, and J. Houston. 1993. Status of the ghost shiner, *Notropis buchanani*, in Canada. Canadian Field-Naturalist 107 (4): 440–445.

Horrall, R. M. 1981. Behavioral stock-isolating mechanisms in Great Lakes fishes with special reference to homing and site imprinting. Canadian Journal of Fisheries and Aquatic Sciences 38: 1481–1496.

Hubbs, C. L. 1926. A check-list of the fishes of the Great lakes and tributary waters, with nomenclatural notes and analytical keys. University of Michigan Museum Zoological Miscellaneous Publication No. 15.

——, and K. F. Lagler. 1964. Fishes of the Great Lakes region. University of Michigan Press, Ann Arbor. 213 pp.

Ihssen, P. E., J. M. Casselman, G. W. Martin, and R. B. Phillips. 1988. Biochemical genetic differentiation of lake trout (*Salvelinus namaycush*) stocks of the Great Lakes region. Canadian Journal of Fisheries and Aquatic Sciences 45: 1018–1029.

——, D. O. Evans, W. J. Christie, J. A. Reckahn, and R. L. DesJardine. 1981. Life history, morphology, and electrophoretic characteristics of fish allopatric stocks of lake whitefish (*Coregonus clupeaformis*) in the Great Lakes region. Canadian Journal of Fisheries and Aquatic Sciences 38: 1790–1807.

Jude, D. J., R. H. Reider and G. R. Smith. 1992. Establishment of Gobiidae in the Great Lakes basin. Canadian Journal of Fisheries and Aquatic Sciences 49 (2): 416–421.

Koelz, W. 1929. Coregonid fishes of the Great Lakes. Bulletin U.S. Bureau of Fisheries 43: 297–643.

Lange, R. E., G. C. LeTendre, T. H. Eckert, and C. P. Schneider. 1995. Enhancement of sportfishing in New York waters of Lake Ontario with hatchery-reared salmonines. Pages 7–11 *in* H. L. Schramm, Jr., and R. G. Piper, editors. Uses and effects of cultured fishes in aquatic ecosystems: Proceedings of the international symposium and workshop held in Albuquerque, New Mexico. American Fisheries Society Symposium No. 15, American Fisheries Society, Bethesda.

Lawrie, A. H. 1970. The sea lamprey in the Great Lakes. Transactions of the American Fisheries Society 99: 766–775.

Lodge, D. M. 1993. Biological invasions: lessons for ecology. Trends in Ecology and Evolution 8 (4): 133–137.

Loftus, K. H. 1958. Studies on river-spawning populations of lake trout in eastern Lake Superior. Transactions of the American Fisheries Society 87: 259–277.

Mills, E. L., J. H. Leach, J. T. Carlton, and C. L. Secor. 1993. Exotic species in the Great Lakes: a history of biotic crises and anthropogenic introductions. Journal of Great Lakes Research 19: 1–54.

Noltie, D. B. 1990. Status of the orange-spotted sunfish, *Lepomis humilis*, in Canada. Canadian Field-Naturalist 104 (1): 69–86.

——, and F. Beletz. 1984. Range extension of the orangespotted sunfish, *Lepomis humilis*, to the Canard River, Essex County, Ontario. Canadian Field-Naturalist 98 (4): 494–496.

Parsons, J. W. 1973. History of salmon in the Great Lakes, 1850–1970. U.S. Fish and Wildlife Service Technical Paper No. 68. 80 pp.

Rahrer, J. F. 1965. Age, growth, maturity, and fecundity of "humper" lake trout, Isle Royale, Lake Superior. Transactions of the American Fisheries Society 94: 75–83.

Rubec, P. J., and B. W. Coad. 1974. First record of the margined madtom (*Noturus insignis*) from Canada. Journal of the Fisheries Research Board Canada 31 (8): 1430–1431.

Schneider, J. C., and J. H. Leach. 1977. Walleye fluctuations in the Great Lakes and possible causes, 1800–1975. Journal of the Fisheries Research Board Canada 34: 1878–1889.

Schultz, E. E. 1960. Establishment and early dispersal of a loach, *Misgurnus anguillicaudatus*, in Michigan. Transactions of the American Fisheries Society 89: 376–377.

Scott, W. B., and E. J. Crossman. 1973. Freshwater fishes of Canada. Fisheries Research Board Canada Bulletin 184: 1–966.

———, and S. H. Smith. 1962. The occurrence of the longjaw cisco, *Leucichthys alpenae*, in Lake Erie. Journal Fisheries Research Board Canada 19 (6): 1013–1023.

Simon, T. P., and J. T. Vondruska. 1991. Larval identification of the ruffe, *Gymnocephalus cernuus*, in the St. Louis River estuary, Lake Superior drainage basin, Minnesota. Canadian Journal of Zoology 69: 436–442.

Smith, C. L. 1985. The inland fishes of New York state. New York State Department of Environmental Conservation. Albany, New York.

Smith, G. R., and T. N. Todd. 1984. Evolution of species flocks of fishes in north temperate lakes. Pages 45–68 *in* A. A. Echelle and I. Kornfield, editors. Evolution of Fish Species Flocks. University of Maine Press, Orono.

Smith, S. H. 1995. Early changes in the fish community of Lake Ontario. Great Lakes Fishery Commission Technical Report Number 60. 38 pp.

Spangler, G. R., N. R. Payne and G. K. Winterton. 1977a. Percids in the Canadian waters of Lake Huron. Journal of the Fisheries Research Board of Canada 24: 1839–1848.

———, N. R. Payne, J. E. Thorpe, J. M. Byrne, H. A. Regier, and W. J. Christie. 1977b. Responses of percids to exploitation. Journal of the Fisheries Research Board of Canada 34: 1983–1988.

Stedman, R. M., and C. A. Bowen II. 1985. Introduction and spread of the threespine stickleback (*Gasterosteus aculeatus*) in Lakes Huron and Michigan. Journal of Great Lakes Research 11: 508–511.

Taylor W. R. 1954. Records of fishes in the John N. Lowe collection from the Upper Peninsula of Michigan. Miscellaneous Publication Museum Zoological University Michigan 87: 1–50.

Todd, T. N., G. R. Smith, and L. E. Cable. 1981. Environmental and genetic contributions to morphological differentiation in ciscoes (*Coregoninae*) of the Great Lakes. Canadian Journal of Fisheries and Aquatic Sciences 38: 59–67.

———, and R. C. Haas. 1993. Genetic and tagging evidence for movement of walleyes between Lake Erie and Lake St. Clair. Journal of Great Lakes Research 19 (2): 445–452.

———, and C. O. Hatcher. 1993. Genetic variability and glacial origins of yellow perch (*Perca flavescens*) in North America. Canadian Journal of Fisheries and Aquatic Sciences 50: 1828–1834.

———, and G. R. Smith. 1992. A review of differentiation in Great Lakes ciscoes. Polskie Archiwum Hydrobiologii 39: 261–267.

Trautman, M. B. 1981. The fishes of Ohio. 2nd edition. Ohio State University Press, Columbus. 782 pp.

Underhill, J. C. 1986. The fish fauna of the Laurentian Great Lakes, the St. Lawrence lowlands, Newfoundland and Laborador. Pages 105–136 *in* C. H. Hocutt and E. O. Wiley, editors. The zoogeography of North American freshwater fishes. Wiley & Sons, New York.

United States Fish and Wildlife Service. 1994. Endangered and threatened wildlife and plants; animal candidate review for listing as endangered or threatened species. The Federal Register, November 15, 1994.

Van Meter, H. D., and M. B. Trautman. 1970. An annotated list of the fishes of Lake Erie and its tributary waters exclusive of the Detroit River. Ohio Journal of Sciences 70 (2): 65–78.

Van Oosten, J. 1937. The dispersal of smelt, *Osmerus mordax*, in the Great Lakes region. Transactions of the American Fisheries Society 66: 160–171.

Webb, S. A., and T. N. Todd. 1995. Biology and status of the shortnose cisco *Coregonus reighardi* Koelz in the Laurentian Great Lakes. Archiv fur Hydrobiologie. Special Issues. Advances in Limnology 46: 71–77.

CHAPTER 3

Demographic and Economic Patterns in the Great Lakes Region
Richard Groop

The human landscape of the Great Lakes region is characterized by three dominant themes. First, there is considerable diversity across the region in terms of the demographic characteristics of the people and the economic activities these people pursue. That diversity is evident in population density, age structure, income, urban and rural distribution, and employment in farming, manufacturing and other occupations. Second, the region has changed considerably since the 1950s in terms of these characteristics: people have redistributed themselves and occupational structure and employment opportunities have been altered. While some of these changes reflect national trends in the United States and Canada, some are unique to the Great Lakes region. Third, these human changes have greatly influenced the environmental characteristics of the region. For example, growing populations in the early 1900s spurred the expansion of commercial fishing in the Great Lakes; industrialization and the growth of cities increased pollution risks in both water and air quality; and spreading human settlements and recreational demands have encroached on shoreline habitats. This chapter explores demographic and economic patterns in the region, and highlights some of the more important changes that have taken place in relation to their impact on the Great Lakes environment.

Demographic Characteristics

Population Distribution

Human migration into the region began with Native American populations, probably around 10,000 B.C. These peoples used the Great Lakes for transportation routes and as a food source. Their settlements would have been small and widely scattered, and the population density of the region would have been minimal by today's standards. Their only significant impact on the environment might

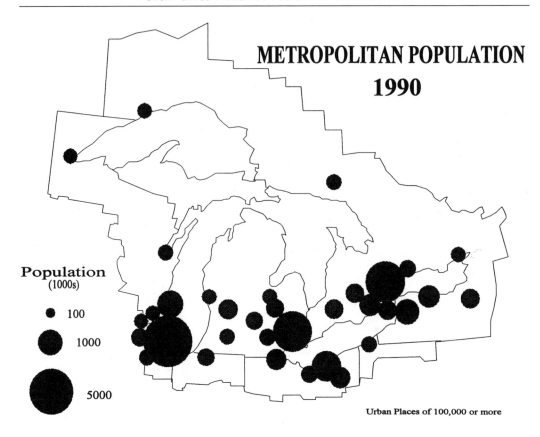

FIG. 1. *Metropolitan population in the Great Lakes Region, 1990.*

have been an increase in forest burning with the use of fire, but that would have had small consequence and, for the most part, Native Americans did not greatly alter the region.

Europeans began arriving in the Great Lakes in the 1600s in the form of French and English fur traders, missionaries, and military expeditions. They constructed small settlements and military outposts, some of which later would grow into permanent villages and cities, such as Detroit, which was founded in 1701. Like the Native Americans, these early explorers used the Great Lakes as water transportation and subsisted on the fish and game that was plentiful in the region. Since their numbers were small, impacts on water, air, and land environments were minimal. But human impacts on those environments were coming.

Like most of Canada and the United States, the Great Lakes began to be more densely settled by European-derived peoples migrating from east to west beginning in colonial times (Getis and Getis 1995). In the late 1700s, settlers moved into the Lake Ontario region from the St. Lawrence and Mohawk River valleys to the east. By 1820, farming and lumbering populations began to spread across the countryside, and permanent settlements began to appear in present-day Cleveland, Detroit, Chicago, and elsewhere to the west. This trickle of population migration increased dramatically with the opening of the Erie Canal in 1825 in upstate New York; by 1840, the farming settlement frontier had been pushed to Indiana, central Michigan and western Ontario. Railroad building from the 1850s onward facilitated the influx of migrants, many directly from Europe in the great migration streams of the mid to late 1800s. This stream of migrants slowed to a trickle by 1900 but by then, the Great Lakes region was full: rural population had reached its maximum, and a natural population increase (the excess of births over deaths) became the dominant demographic characteristic of the area (U.S. Census 1960).

Today, most people in the Great Lakes region are concentrated in the south, stretching from the southern half of Lake Michigan to the eastern end of Lake Ontario, south of a line running from Milwaukee to Ottawa (fig. 1). The greatest concentration is near the southwestern shore of Lake Michigan (Chicago to Milwaukee), the western end of Lake Erie (Detroit to Cleveland), and the western end of Lake Ontario (Toronto to Buffalo). These concentrations obviously result from urban clusters and their associated metropolitan sprawl, but here, the rural areas, reflecting a long-established productive agricultural base, are also densely settled. To the north, in northern Minnesota, Wisconsin, Michigan and Ontario, population is far more sparse. There are some medium-sized cities, such as Duluth, but these are few, and the more scattered rural population reflects few agricultural or other economic opportunities.

Density of population in the Great Lakes region varies from less than three people per square mile in some northern counties to some of the most concentrated urban densities in North America (fig. 1). The urban corridor across the southern Great Lakes has long been predicted as a future megalopolis, where large cities and their suburbs begin to merge and crowd out intervening rural territory. While a megalopolis has yet to arrive in the southern Great Lakes region, trends in population growth and redistribution support the concept of urban sprawl, where central-city people migrate to suburban and nearby rural areas. To the north, no such urban concentration threatens the rural landscape, although selected areas have experienced increased rural population settlement. The following sections detail these different population growth scenarios.

This general pattern of population distribution has not changed greatly in the past fifty years: a 1950 map of Great Lakes population would look very similar to figure 1 (U.S. Census, 1950; Census of Canada, 1951). Growth and decline have taken place, local migration has occurred, and population characteristics have changed; however, most of these changes have affected local areas only, and not surprisingly, the overall distribution of people in the Great Lakes region has remained remarkably consistent over time.

Population Change

Figure 2 illustrates the growth of population in the region, in Canada, and the United States from 1900–90. While some of the Great Lakes population change can be attributed to immigration, particularly in Ontario, most of the growth resulted from natural increase. Two things should be noted on figure 2: (1) As a whole, the Great Lakes region did not grow as rapidly as Canada and the rest of the United States. This is not caused by lower birth rates than in other parts of the two countries, but rather by lower rates of immigration into the Great Lakes. (2) The U.S. portion of the Great Lakes has grown considerably slower than Ontario, Canada, and the rest of the United States since 1970. This is a reflection of changing economic conditions in the Great Lakes states, resulting in higher outmigration from the region. The Canadian part of the region, less dependent on the vagaries of industrial change, has experienced more consistent growth.

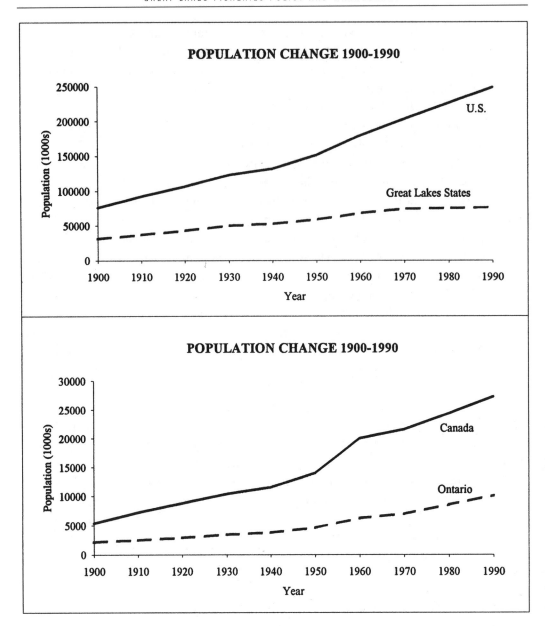

Sources: Historical Statistics of the U.S., 1960; Statistical Abstract of the U.S., 1950-1990; Census of Canada, 1951-1991; Historical Statistics of Canada, 1957.

FIG. 2. *Population change in the Great Lakes Region, 1900–90.*

Considerable change in population numbers occurred across the Great Lakes region from 1950–90, both in terms of the entire region, and in terms of local areas. The Great Lakes region grew from twenty-eight million in 1950 to nearly forty-four million by 1990, an increase of 55% (table 1). Almost all of this population increase is attributable to natural increase, an excess of births over deaths. Current national birth rates for Canada and the United States are fourteen and fifteen

TABLE 1

Ppoulation change 1950–1990 for counties within the Great Lakes region

Sources: Statistics Canada, 1951–1991; U.S. Census Bureau, 1950–1990.

	1950	1990	% CHANGE
Ontario	4,166,618	10,851,057	160.4
Indiana	1,195,535	1,771,670	48.2
Michigan	6,371,766	9.295,297	45.9
Wisconsin	1,436,687	3,473,260	42.5
Illinois	5,282,724	7,429,181	40.6
Ohio	4,178,906	5,370,536	28.5
New York	3,656,368	4,542,353	24.2
Pennsylvania	526,405	591,642	12.4
Minnesota	259,550	263,019	1.3
Total U.S. Portion	23,907,941	32,736,958	36.9
Total Great Lakes Region	28,074,559	43,588,015	55.3
All Canada	14,009,429	27,296,859	94.8
All U.S.	151,325,798	248,709,873	64.4

births per thousand population respectively, while death rates are seven and nine deaths per thousand population respectively. These birth and death rates result in the current annual growth rate of 0.7% for the two countries (Population Reference Bureau 1995). The Great Lakes region as a whole does not vary significantly from these national rates and, when adjusted to age, local areas have little variation as well.

Some people have been added through international migration from foreign countries, but that proportion of the total growth is small. More important is migration to and from other parts of Canada and the United States. In the last twenty years, average outmigration from Great Lakes states has exceeded immigration by nearly 10%. This net migration loss has been attributed to loss of employment opportunities, particularly manufacturing jobs, and to retirement migration to warmer climates. This combination of demographic change factors has resulted in a lower % change in population than that of the United States and Canada, as reflected in table 1. Most of the region's population increase occurred from 1950, with considerable slowing of the rate after 1970, again reflecting national trends and increased outmigration.

For individual states/provinces within the Great Lakes region, growth rates varied significantly. Ontario's growth far surpassed other areas in the United States and other parts of Canada. Much of this growth results from the rapid urban sprawl of the Toronto metropolitan area, but most of the counties (or equivalent areas) in Ontario also grew at higher-than-average rates. Most of this growth can be attributed to a large influx of foreign immigration into Ontario, particularly the Toronto area. In the United States, selected local areas experienced significant increases but overall, growth lagged far behind the rest of the United States and Ontario.

Growth and decline in local areas from 1970–90 exhibited considerable diversity within the region (fig. 3). There are four kinds of population change worthy of note on this map. First, counties containing large cities either lost population or were among the slowest growing in the region, particularly those counties in which Chicago, Detroit, Cleveland and Buffalo were located. Loss of

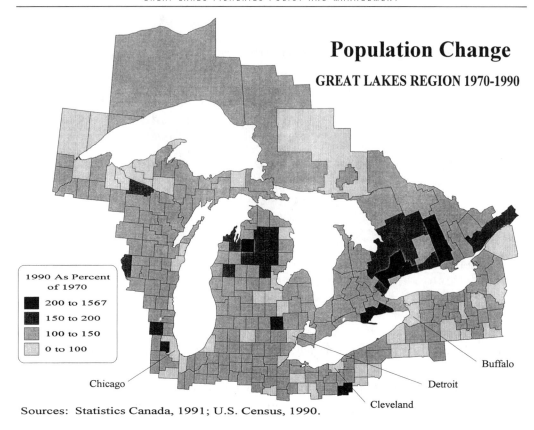

Population Change

GREAT LAKES REGION 1970-1990

1990 As Percent
of 1970

■ 200 to 1567

▨ 150 to 200

▦ 100 to 150

☐ 0 to 100

Buffalo

Chicago

Detroit

Cleveland

Sources: Statistics Canada, 1991; U.S. Census, 1990.

FIG. 3. *Population change by county in the Great Lakes Region, 1970–90.*

population due to outmigration from these central cites to suburban and rural areas is a phenom-enon that has been widely recognized nationwide in the United States since the 1960s. In Canada, because of fewer negative push factors such as high crime rates and poor schools, central cities, particularly Toronto, have not experienced this decline in population (Getis and Getis 1995). Sec-ond, suburban ring counties around large cities have experienced the highest growth in popula-tion throughout the region. As suburban areas have grown, highway systems have become clogged by commuters between central city and suburb. Business offices, factories, retail stores, and other large employers have followed their labor force outward to the suburbs and new edge cities have begun to sprout on the outskirts of Chicago, Detroit, Toronto, and on a smaller scale, around small and medium-sized cities as well. Suburbanization is not unique to the Great Lakes region, but its impact on central city decline, urban sprawl, population and wealth redistribution, and social change, are perhaps most apparent within the region: Detroit, Buffalo, Chicago, and Cleveland are perhaps the best examples of the suburbanization phenomenon (Birdsall and Florin 1992). Third, some counties in the northern part of the Great Lakes region have experienced high per-centage rates of growth, particularly since 1970. These include the north-central parts of the lower peninsula of Michigan, Wisconsin and Ontario. Unlike suburban growth areas to the south, these rural areas have experienced what is called a "population turnaround," gaining people after earlier long-term outmigration and population decline (Groop and Manson 1987). This type of growth can be found in a few other areas of Canada and the United States, but is best illustrated in the northern Great Lakes region. Here, the high rate of growth is caused by modest numbers of migrants moving into sparsely-populated areas. Most of these people are older retirees seeking recreational fishing, hunting, skiing, and scenic benefits found in the northern part of the region. These migrants have created employment demands in service industries, and younger job-seekers have added to this migration stream. They work as waitpersons in restaurants, golf course workers, retail sales persons, and other similar occupations. While population experts are uncertain as to the long-term continuation of this trend, and numbers of migrants are not large, their economic, political and social impact on the local receiving areas is significant. Fourth, counties in the northern part of the Great Lakes region have consistently lost population through outmigration. These are usually areas that have undergone economic decline associated with mining, forestry, farming or other primary industries. The outmigrants are often younger job seekers who move to metropolitan areas farther south, leaving behind older people in declining small towns and rural areas. This long-term trend is representative of rural depopulation in other parts of Canada and the United States and, like rural growth, has significant social and economic impact on the local areas left behind (Manson and Groop, 1992).

Population Characteristics

In addition to population numbers, people in the Great Lakes region exhibit considerable diversity in terms of noneconomic, cultural characteristics, such as ethnic and racial background, age structure, religious adherence, political affiliation, and language association. Unfortunately, differences in census collection between Canada and the United States prohibit the examination of comparable detailed data for the region, and few cultural patterns can be directly compared.

In general, the people of the Great Lakes region are midwestern in cultural characteristics: they are predominantly a mixture of Protestant and Catholic religion, with Catholics more often found in cities; a majority of conservative political opinions with pockets of liberal voters, particularly in cities; a majority of European-derived populations in rural areas, with other ethnic peoples more commonly represented in cities.

Two characteristics that are comparably collected for both countries are age and gender. County male-female ratios are fairly uniform across the region, ranging from 92 males per 100 females to 116 males to 100 females, with far more counties having more women (Santer 1993). However, age structure varies considerably across the region. Figure 4 shows the distribution of older people in the Great Lakes region. In the United States, the proportion of people 65 years or older is 12.6% (1990), while in Canada the comparable figure is 11.6% (1991). Using these benchmarks, many counties in northern Minnesota, Wisconsin and Michigan, and fewer in Ontario, New York and Pennsylvania have concentrations of older population. In some places these concentrations have resulted from inmigration of older people; in some counties, it is the result of outmigration of younger population; and, in a few extreme cases, both migration phenomena have occurred simultaneously. Younger populations can be found most often in counties surrounding larger cities, particularly the suburban areas of Chicago, Detroit, and Toronto. While some counties can

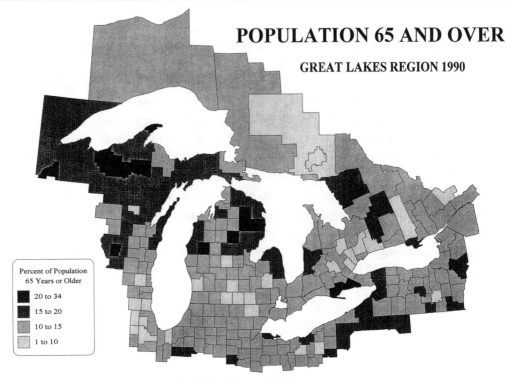

Sources: Statistics Canada, 1991; U.S. Census, 1990.

FIG. 4. *Distribution of older population in the Great Lakes Region, 1990.*

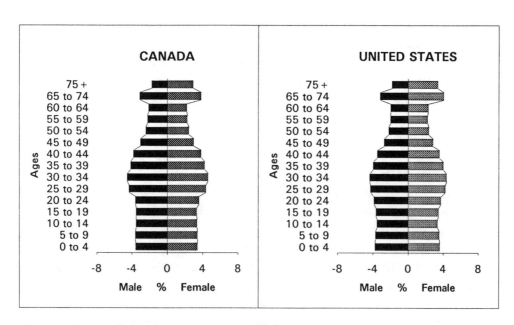

Sources: Statistics Canada, 1991; U.S. Census, 1990.

FIG. 5. *Age-gender composition for the United States and Canada, 1990.*

AGE-GENDER COMPOSITION

Sample Counties, 1990

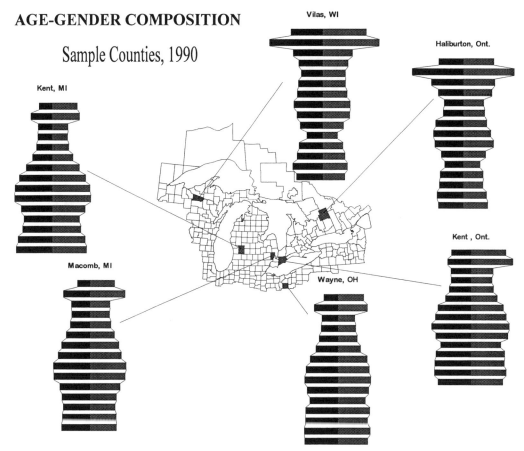

Sources: Statistics Canada, 1991; U.S. Census, 1990.

FIG. 6. *Sample county age-gender composition in the Great Lakes Region, 1990.*

have aberrant age structures caused by small population numbers, in general, the northern part of the Great Lakes region has a significantly higher average age, and a greater concentration of older people than the southern part of the region. Overall age distribution of the Great Lakes region is quite comparable to the national norm.

Age-gender composition, or pyramids, showing proportions of gender and age categories also demonstrate the diversity of the region. Figure 5 illustrates that the United States and Canada have nearly identical age-gender structures typical of most industrialized, modern, "first-world" countries. The bulges in middle-age categories are baby-boomers. As years pass, these bulges will move upward in the pyramid causing a top-heavy aging population along with the attendant problems of health care needs, social security costs, retirement relocation, and other social and economic changes. The pyramids also illustrate that females live relatively longer than males, and thus, make up a higher proportion of the population. The straight-sided younger age categories caused by low birth rates suggest a stabilized and low population growth rate.

In comparison to these national profiles, figure 6 shows some selected county age-gender compositions for different parts of the Great Lakes region. These profiles illustrate regional differences,

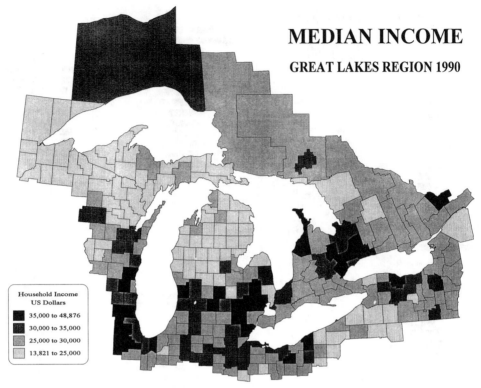

MEDIAN INCOME
GREAT LAKES REGION 1990

Household Income
US Dollars

- 35,000 to 48,876
- 30,000 to 35,000
- 25,000 to 30,000
- 13,821 to 25,000

Sources: Statistics Canada, 1991; U.S. Census, 1990.

FIG. 7. *Distribution of income in the Great Lakes Region, 1990.*

generally from north to south. Vilas, Wisconsin, and Haliburton, Ontario, exhibit typical age-gender profiles for northern parts of the region. The proportion of older-age population is larger, and young adult categories are relatively small. Kent and Macomb, Michigan show typical urban-suburban profiles with larger proportions of middle-age people, modest numbers of children being added, and far fewer older people. Kent, Ontario, and Wayne, Ohio, have relatively straight-sided pyramids, typical of farming areas in the southern part of the region. These are relatively stable populations, and serve as examples of what the Great Lakes, Canada, and the United States are likely to move toward in the next fifty years.

Economic Characteristics

Income and Employment

Just as demographic variables exhibit diversity from north to south across the Great Lakes, so do economic characteristics. A map of income for the region (fig. 7) is typical of these general trends. (In this map and the following discussion, Canadian income has been converted to U.S. dollars). High-income areas are found throughout the urban corridor in the south, with most suburban counties in the highest category. As noted above, the movement of large employers (as well as high-wage employers) to these areas has relocated employment income along with population.

UNEMPLOYMENT

GREAT LAKES REGION 1990

Percent of
Labor Force

- 10 to 21
- 8 to 10
- 5 to 8
- 3 to 5

Sources: Statistics Canada, 1991; U.S. Census, 1990.

FIG. 8. *Distribution of unemployment in the Great Lakes Region, 1990.*

DuPage, Illinois, Oakland, Michigan, and York, Ontario, are examples of counties with above-average income and many of the socioeconomic characteristics those incomes imply: larger professional populations, higher property values, better schools, higher property taxes, longer commuting distances, infrastructure growing pains, and urban sprawl. To the north, because of few urban and suburban employment opportunities, income declines sharply. In these counties, particularly those with older populations, employment income is supplemented with two types of nonemployment income: investment sources such as stock dividends, rental property, retirement benefits; and transfer payments such as social security benefits and welfare income (Manson and Groop 1988, 1990). These nonemployment income sources comprise nearly one-third of all local income in the United States, and in northern counties in the Great Lakes region, that proportion can exceed 50% of all income.

A nearly mirror image of income distribution is found in the pattern of unemployment (fig. 8). Although there are differences in method of calculation between the two countries, in general, Canadian unemployment rates are higher than those of the United States. Canada's overall rate of unemployment was approximately 11% in 1990, while the United States had between 5% and 6% of its work force unemployed. As figure 8 illustrates, the northern Great Lakes region exceeds these national averages while the southern part of the region is at or below those averages.

These patterns of income and unemployment have remained fairly persistent for at least the last twenty years. Overall rates have fluctuated with national economic conditions of recession, inflation, or economic growth, and some places may have gone up or down with changing local situations. However, income and employment differences between northern and southern parts of the region have long been in place and will likely remain so in the foreseeable future because they are tied to the economic activities taking place there.

Economic Activities

The Great Lakes region is highly segmented in terms of physical environment, land use, and the economic functions carried out in those different settings. Extractive industries such as mining and forestry are found scattered in the northern part of the region. While many primary industries survive as important employers and sources of economic stability for local areas, many of these industries reached peak production in earlier years. Much of the mining and forestry has declined in the last fifty years coinciding with unemployment, lower income, and outmigration. These declines have been caused by resource depletion and by more competitive sources of minerals or timber. Of particular importance has been the growth of overseas mining activity with relatively low labor costs driving market prices downward, and making mines in the Great Lakes region uncompetitive (Birdsall and Florin 1992). Agriculture has never been an important part of the northern Great Lakes region because of short growing seasons, cold winters, and poor agricultural soils. As a result, there are vast tracts of land virtually unused for human economic activity.

Hunting, trapping, and fishing are practiced widely in the north. Hunting activity is primarily for sport and recreation, and occurs on a seasonal basis, depending on state or province conservation law. Fishing can be for sport, commercial, or subsistence purposes. Sport and commercial fishing are controlled seasonally by law; subsistence fishing is practiced primarily by Native American tribes who depend on their catches for food. Most trapping is subsistence in nature, as well. These activities have only small impact on the environmental landscape or land use, and in economic terms, have relatively small importance compared to industry and commerce.

In the southern part of the Great Lakes region, a great variety of economic activity is present with agriculture, commercial sales, and manufacturing dominant on the landscape (Patterson 1989). Figure 9 shows the relative importance of agriculture between north and south in the region. In northern Minnesota, Wisconsin, Michigan, and Ontario, most counties have less then 20% of the landscape in agricultural land. The highest concentration is in the south, with the exception of urban and suburban counties where agriculture cannot compete economically with urban land uses. Crop combinations and productivity levels also vary across the region. In Illinois, Indiana, western Ohio, the southern tier of counties in Michigan, and southern Ontario, cash grain and livestock predominate with corn, wheat, sorghum, cattle, and hogs as major products. These have long been high productivity areas with large farms, high yields per acre, and commercially-successful farmers. And, while urban sprawl continues to displace valuable farm land at a rapid pace, these areas, along with others in the agricultural Midwest should continue to be some of the most productive, surplus-producing breadbaskets of the world.

To the north, product combinations change to those traditionally associated with the dairy belt: hay and other fodder crops used to feed milk cows, along with wheat and other small grains dominate those areas used for agriculture (Patterson 1989). Concentration of specialized agriculture, primarily fruit and vegetable crops such as apples and cherries, are found in the Door Penin-

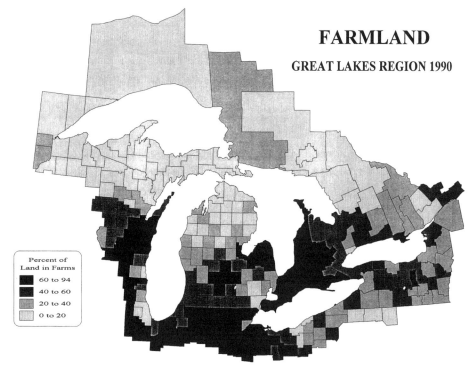

FARMLAND

GREAT LAKES REGION 1990

Percent of
Land in Farms

- 60 to 94
- 40 to 60
- 20 to 40
- 0 to 20

Sources: Statistics Canada, 1986; U.S. Dept. of Agriculture, 1992.

FIG. 9. *Distribution of farmland in the Great Lakes Region, 1990.*

sula of Wisconsin, the eastern shore of Lake Michigan, and other areas, where the climate is moderated by nearness to the lakes.

The southern part of the Great Lakes region began to rapidly expand its industrial base at the beginning of the twentieth century with the movement of steel plants and other associated industries to port locations such as Lorain, Ohio, Buffalo, New York, Gary, Indiana, and Hamilton, Ontario (Patterson 1989). These locations were profitable transshipment points between iron ore and other minerals coming down the Great Lakes in ore carriers and coal moving northward by rail. After the turn of the century, the automotive industry, begun in Detroit, spread throughout the southern part of the region, bringing with it an increase in related manufacturing activities. World War II spurred this growth even further, and by 1950, the region was emerging as one of the leading industrial areas in the world. As industrial growth continued, more and more plants were established in the Great Lakes area to take advantage of nearness to raw materials, such as steel, produced in the region. Called "industrial inertia," this self-sustaining growth, generated primarily by the steel and automotive industries, dominated the economic landscape from 1920 to 1960. From 1950–70, the region enjoyed leadership in the manufacture of automotive, electrical, steel, and other products, particularly those in heavy industry categories (Birdsall and Florin 1992).

In environmental terms, industrial expansion placed great stress on the natural environment. Water quality in lakes and rivers within the region declined because of the following: (1) industrial waste discharge; (2) air pollution from industrial smoke stacks increased; (3) fish populations declined because of deteriorating environments; (4) prime agricultural land was converted to urban-industrial uses; (5) long-term solid wastes and chemical pollutants from factories began to accumulate. In sum, the industrial boom in the region coincided with, and was the primary cause of, a significant decline in environmental quality. Perhaps the environmental nadir was reached in the early 1970s, when Lake Erie was declared dead and industrial pollutants were cited as the primary cause.

Since that time, a number of major industrial location factors have converged to change that pattern. In 1950, most cities had a large proportion of employment in "heavy" manufacturing industries such as steel and automobiles (fig. 10). Growth continued through 1970 (fig. 11) as these industries continued to thrive, and were bolstered by growth in light manufacturing sectors, such as electronics. However, by 1990 (fig. 12), the proportion of manufacturing employment had declined considerably. These declines can be attributed to any number of causes, but some combination of the following was usually involved: (1) aging of physical plant facilities and the high cost of replacing those facilities in central cities; (2) relocation of factories to lower labor cost areas of the country; (3) decline of mineral quantity and quality, particularly iron ore, being shipped on the lakes; (4) relocation of factories to overseas destinations where labor costs, raw material supply and markets are all improved; (5) displacement in the North American economy of manufacturing industries with quaternary kinds of activities, such as information technology or service sector functions. While some of these changes have occurred in Canada and the United States as national trends, the Great Lakes area, because of its dependence on heavy industry, has been particularly affected by these relocations, as illustrated in figures 11 and 12. This decline in manufacturing and other industries, particularly in the last twenty years, has corresponded with decaying central cities, lower-than-average population growth, and a relative decline in the general economic health of the region.

Transportation

The Great Lakes have long been one of the world's great inland water transportation routes. Entirely navigable throughout Lakes Erie, Huron, and Michigan, this portion of the Great Lakes has long served as a highway for transporting goods and people within the region. The transportation function was greatly enhanced with the opening of the Welland Canal in 1829, bypassing Niagara Falls and connecting Lake Ontario and the St. Lawrence River to the system. Access to Lake Superior was added in 1870, with the construction of locks around the rapids in the St. Marys River at Sault Ste. Marie. Various improvements to the system occurred periodically, culminating in the St. Lawrence Seaway Project completed in 1957. This opened the Great Lakes to the Eastern Seaboard, ocean-going traffic, international trade, and increased connectedness to the outside world.

Figures 13 and 14 illustrate the history of shipping on the Great Lakes. Figure 13 shows the net tonnage of all Great Lakes commercial carriers prior to the opening of the St. Lawrence Seaway, and indicates that carrying capacity greatly increased around the turn of the century. This coincides with the advent of larger, steam-powered ships, the increase of industrialization nationwide, and the concomitant demand for bulk natural resources, such as coal and iron ore. Figure 14 shows cargo tonnage since 1900, and illustrates that while individual years may vary considerably, depending on national economic fluctuations, the overall pattern has been fairly constant since World War II. In general, bulk cargoes shipped internally on the Great Lakes have declined

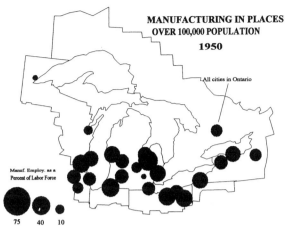

FIG. 10. *Distribution of urban manufacturing population in the Great Lakes Region, 1950.*

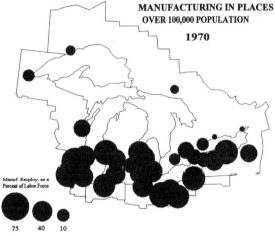

FIG. 11. *Distribution of urban manufacturing population in the Great Lakes Region, 1970.*

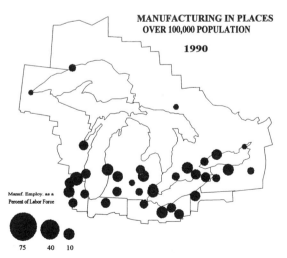

FIG. 12. *Distribution of urban manufacturing population in the Great Lakes Region, 1990.*

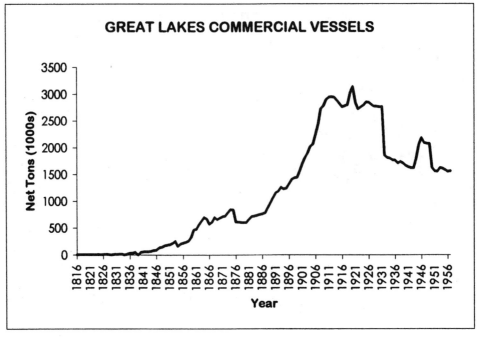

Source: Lesstrang, 1977.

FIG. 13. *Ships on the Great Lakes, 1816–1957.*

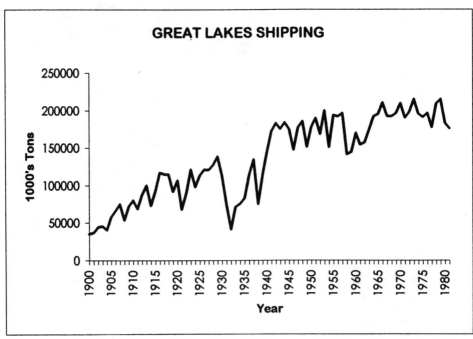

Sources: Lake Carriers Association, 1970,1980; Seaway Maritime Directory, 1992.

FIG. 14. *Shipping on the Great Lakes, 1900–81.*

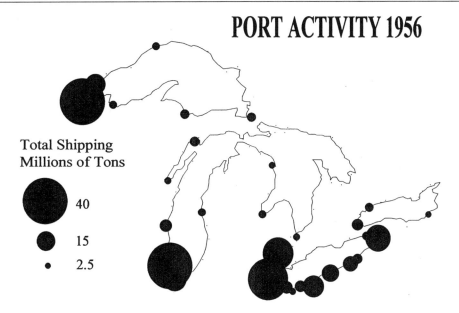

PORT ACTIVITY 1956

Total Shipping
Millions of Tons

40

15

2.5

Only ports larger than 2.5 million shown
Source: Great Lakes Commission, 1957.

FIG. 15. *Shipping at Great Lakes ports, 1956.*

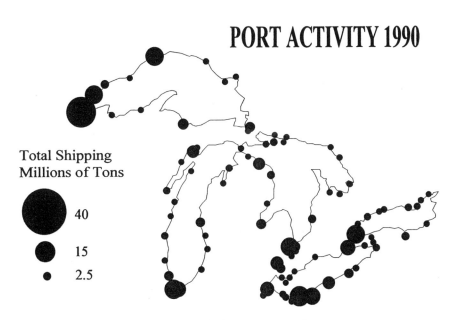

PORT ACTIVITY 1990

Total Shipping
Millions of Tons

40

15

2.5

Source: Statistics Canada, 1992; St. Lawrence Seaway
Development Corp., 1990.

FIG. 16. *Shipping at Great Lakes ports, 1990.*

since the 1960s, and have been replaced by international trade through the St. Lawrence Seaway since 1957.

This growth in the diversity of shipping is perhaps best exemplified by port activity on the Great Lakes. Larger ports are depicted in figure 15 just prior to the opening of the St. Lawrence Seaway. At that time, most cargo transport was concentrated at larger port facilities and consisted of bulk goods, such as iron ore, coal, stone, and grain. By 1990, this pattern had changed considerably (fig. 16). Volume of goods had declined somewhat, particularly at iron-related ports; but, more importantly, cargo handling spread to more ports and consisted of more diversified cargoes, generally of higher value goods such as manufactured imports and exports.

Figure 16 also shows the relative importance of ports around the Great Lakes and, as with other economic variables, there are differences across the region. Most of the tonnage from northern ports is outbound to urban areas down the Lakes or to foreign export markets. These are mostly bulk commodities, primarily grain (passing through the region from further west), iron ore, and stone products with low per-ton value. Southern ports, coinciding primarily with large urban areas are more complex. Here, raw materials, particularly coal, arrive by rail to be loaded on bulk carriers for export; bulk minerals, particularly iron ore, arrive by boat to feed factories; manufactured products depart, many destined for foreign ports; and manufactured goods from other parts of the world arrive to be transshipped to markets in or near the Great Lakes (Birdsall and Florin 1992). Dollar value of goods in and out of southern ports is considerably higher than those to the north. As manufacturing relocations and mineral depletion continue to take their economic toll, it is likely that commercial port activity, at least in terms of bulk commodities and total shipping tonnage, may well decline in coming years.

These port cities interact with their hinterlands throughout the Great Lakes primarily by rail and truck. The last forty years has seen the decline of rail traffic throughout both countries nationally, and the Great Lakes is no exception. Railroads have been consolidated, low-traffic routes have been abandoned, and only those high-volume lines connecting to major ports continue to function. Many abandoned railroad right-of-ways have been converted to recreational uses. Truck carriers have replaced rail transport in many places. Growth and improvements in interstate-type highways in the region has encouraged that change along with larger trucks and carrying capacities.

People and the Environment

In the last forty years, the Great Lakes region has undergone significant change in its socioeconomic landscape. For the region as a whole, population has grown significantly, and the overall economy has remained relatively strong, despite recessionary and inflationary ups and downs. The relocation of manufacturing plants outside the region has been the largest economic change, and has undoubtedly slowed overall population growth. Within the region, local areas have experienced more pronounced change. Some northern rural areas have continued long-term declines in population and economic opportunities, while others have experienced a "turnaround" in growth. Central cities, at least in the United States, have declined in population, and economic activities have followed. Suburban areas have sprawled into rural farmland in the southern part of the region, and urbanization continues at a rapid rate. In general terms, these changing socioeconomic conditions in the Great Lakes region have had a major impact on many aspects of the physical environment, particularly the rivers and lakes of the region as well as the diverse life forms dependent on that environment. And, that impact has probably been both positive and negative.

On the negative side, a rapidly growing population in the Great Lakes region over the last fifty years has put pressure on various aquatic systems in a variety of ways. Currently, the region is home to forty million people, about 13% of the U.S. population and 30% of the Canadian population. In addition, the Great Lakes region houses half of Canada's industry, and about 40% of U.S. industry (Miller 1994). To support the growth of these concentrations, freshwater supplies for human consumption are in greater demand, rivers and lakes are exposed to more pollution from human sewage systems, industrial waste, and other toxic sources, and recreational uses increase. Extractive activities, industrialization, agricultural intensification, and urban growth and land use are all changes that require water resources; and various forms of pollution, and/or biological disruption seem to be the inevitable result. A well-known example is Lake Erie in the 1960s, which experienced eutrophication, large fish kills, and contamination from bacteria and other wastes. Another is the Cuyahoga River in downtown Cleveland, which briefly "burned," its surface on fire, fueled by industrial chemicals. And, while these may be extreme examples of the pressure of human populations and activities on water resources, they are also indicative of the kinds of environmental damage that generally accrue from human and economic growth. A different example comes from the opening of the canal systems in the early 1800s, and the subsequent invasion of sea lampreys and other alien life forms. The devastation of certain fish species, such as the lake trout was spectacular, and had tremendous impact on the biology and commercial use of the Lakes. A similar scenario occurred in the 1980s, when zebra mussels migrated to the Great Lakes via ocean going vessels; their financial and biological impact has yet to be fully recognized. In short, population increase and growth in almost any segment of the economy can (and most often do) have negative impacts on the aquatic environment. While the Great Lakes region has not undergone the phenomenal growth and attendant environmental problems of places such as southern California or coastal Florida, the negative impacts are easily observed.

On the positive side, the Great Lakes, including Lake Erie, have been at least partially resuscitated. This has come about because people and their governments on both sides of the border have recognized that environmental quality was on the decline, that change was needed, and that environmental improvement was dependent on human management of that environment. Various programs have been implemented in recent years, including a $20 billion joint United States-Canada venture to control toxic agents in the Great Lakes (Miller 1994). As a result, toxic chemicals have declined, dissolved oxygen levels have increased, fishing has improved, and, in general, the Lakes are a healthier environment than they were twenty-five years ago. Ironically, population and economic growth of the region have allowed this improvement to occur: Great Lakes environmental initiatives and improvements have come about because the local population and their governments in the region can afford to do them. Vibrant populations with healthy, growing economies provide the monetary and technological resources necessary for the maintenance of good environmental conditions. The relatively high value of the rivers and lakes of the region, in terms of water quality, habitat, and human uses, can be attributed, at least in part, to healthy economic activities, and the people that pursue them.

Human activity in a healthy environment is a balancing act. Growth in population and economic activity puts pressure on environmental quality and too-rapid growth can outpace the ability to manage land and water resources effectively. Conversely, regions in economic decline are seldom able to support environmental quality because unhealthy economies do not have the monetary resources necessary for environmental maintenance and improvement.

Both Government and individuals should pursue the goal of economic diversification, but not at the expense of environmental quality. Losses in manufacturing employment in the Great Lakes region must continue to be offset with jobs in other sectors of the economy but the resultant economic growth must be paced. Basic agricultural productivity on prime farm land must be maintained, but competing economic activities should be encouraged. The importation of tourist dollars into the region must continue to grow but recreational pressures must be controlled. Fishing and hunting demands must be met, but depletion of fish and wildlife populations must be curtailed. The list of human activity versus environmental concern dilemmas goes on. The future of the Great Lakes region is dependent on our ability to maintain this delicate balance.

Literature Cited

Anonymous. 1992. Seaway Maritime Directory. Fourth Seacoast Publishing Co., Inc., St. Clair Shores, Mich.

Birdsall, Stephen S. and John W. Florin. 1992. Regional Landscapes of the United States and Canada, 4th edition. John Wiley and Sons, Inc., New York.

Bureau of the Census. 1990, 1980, 1970, 1960, 1950. Census of Population and Housing. U.S. Department of Commerce, Washington, D.C

Bureau of the Census. 1990, 1980, 1970, 1960, 1950. Statistical Abstract of the United States. U.S. Department of Commerce, Washington, D.C.

Bureau of the Census. 1960. Historical Statistics of the United State to 1957. U.S. Department of Commerce, Washington, D.C.

Department of Agriculture. 1992. Census of Agriculture. U.S. Department of Agriculture, Washington, D.C.

Dominion Bureau of Statistics. 1960. Canada Yearbook. Government of Canada, Ottawa.

Getis, A., and J. Getis. 1995. The United States and Canada: The land and the people. Wm. C. Brown Publishers, Dubuque, Iowa.

Great Lakes Commission. 1957. Projects and Developments. Great Lakes Commission, Ann Arbor.

Groop, R. E. and G. Manson. 1987. Nonemployment income and migration in Michigan. East Lakes Geographer 22:103–109.

Lake Carriers Association. 1980, 1970, 1960. Annual Report. Lake Carriers Association, Cleveland.

Lesstrang, J. 1977. Lake carriers: The saga of the Great Lakes fleet—North America's fresh water merchant marine. Salisbury Press, Seattle.

Manson, G., and R. E. Groop. 1988. Concentrations of nonemployment income in the United States. Professional Geographer 40:444–450.

———, 1990. The geography of nonemployment income. The Social Science Journal 27:317–325.

———, 1992. Michigan state-to-state migration patterns, 1976–1988. Michigan Academician 24:443–451.

Miller, G. T. 1994. Living in the environment, 8th edition. Wadsworth Publishing Company, Belmont, CA.

Patterson, J. H. 1989. North America: A geography of the United States and Canada, 8th edition. Oxford University Press, New York.

Population Reference Bureau. 1995. 1995 world population data sheet. Population Reference Bureau, Inc., Washington, D.C.

Santer, R. A. 1993. Geography of Michigan and the Great Lakes Basin. Kendall/Hunt Publishing Co., Dubuque, Iowa.

Statistics Canada. 1992. Shipping in Canada. Government of Canada, Ottawa.

Statistics Canada. 1991, 1981, 1971, 1961. Census of Canada. Government of Canada, Ottawa.

Statistics Canada. 1991. Revised intercensal population and family estimates. Government of Canada, Ottawa.

Statistics Canada. 1986. Census of Agriculture. Government of Canada, Ottawa.

St. Lawrence Seaway Development Corporation. 1990. U.S. Great Lakes ports statistics for overseas and Canadian waterborne commerce: Report for 1990. U.S. Department of Commerce, Washington, D.C.

Urquhart, M. C. 1983. Historical Statistics of Canada. Statistics Canada, Ottawa.

CHAPTER 4

Authorities, Responsibilities, and Arrangements for Managing Fish and Fisheries in the Great Lakes Ecosystem

Margaret Ross Dochoda

Introduction

The breadth and complexity of the Great Lakes ecosystem is matched only by that of its institutional infrastructure and arrangements. Canadian and American agencies undertake many actions that affect fish in the Great Lakes, often operating in the context of Canada–Ontario agreements, of U.S. federal interagency and federal–state agreements, of U.S.–Canada agreements, such as the 1954 Convention on Great Lakes Fisheries, and of the multi-agency Joint Strategic Plan for Management of Great Lakes Fisheries.

Canadian federal authority over Great Lakes fish comes directly from the Constitution Act, 1982 (formerly the British North America Act, 1867), whereas that of the United States is limited to case-by-case legislation in support of interstate commerce, treaty-making, and other authorities granted the U.S. federal government by the U.S. Constitution (1788). In addition to managing fish, Great Lakes states and the province of Ontario manage recreational, commercial, and certain tribal fisheries, in particular, setting the conditions for harvest and stocking. Some U.S. tribes have management and self-regulatory authority related to treaty-reserved fishing rights, both on and off-reservation. In Canada, aboriginal rights, including off-reservation fishing in portions of the Great Lakes, are protected in the 1982 Constitution Act.

In pursuit of effective, efficient, and equitable management of that remarkable resource, the Great Lakes ecosystem, agencies increasingly wield their respective authorities and responsibilities in a coordinated stewardship. The first part of this chapter will summarize the source and nature of authorities and responsibilities which inform and flavor all such attempts to coordinate fish management in the Great Lakes, first for Canada, then the United States, and finally, for treaty and related authorities from both countries. The second part of this chapter will consider attempts to coordinate management[1] in a convention negotiated by Canada and the United States, through interstate arrangements, and in a more informal agreement among federal, state, provincial, and intertribal agencies.[2]

Authorities for Management of Great Lakes Fish and Fisheries

Canadian Authorities

Dual Authorities

Great Lakes fish and fisheries are managed by Ontario and Canada under intricately interwoven authorities arising from the division of powers in the Constitution Act (1982). As proprietor or owner of the lakebed, the Province of Ontario has authority in licensing or otherwise permitting access to the fishes in Ontario waters of the Great Lakes.[3] The federal government has authority for conservation of Great Lakes fish, which can override the province's proprietary authority in certain instances. The federal government, however, tends to delegate its authority to Ontario, with the resulting execution of federal and provincial authority often proving a very complicated matter:

> The management of the Canadian (inland) fisheries is under the authority of both federal and provincial governments. While it is common to the Canadian confederation that both levels of government have dual roles in many areas of Canadian life, the situation with respect to the fisheries tends to be more complicated in the intricacies of federal-provincial jurisdiction. The result has led at times to a genuine confusion as to the management prerogatives over the Canadian fisheries. (Thompson 1974)

In fact, the Constitution Act (1982) assigned to the federal government exclusive legislative authority for the inland (and sea coast) fishery (including the Great Lakes), but assigned the Provinces authority over "Property and Civil Rights." At the same time, common law vests the Province of Ontario with ownership of the bed of the Great Lakes, with the possible exception of unsurrendered Indian lands and waters. Under common law, the right to fish, or the ownership of the fishery is directly connected to the ownership of the bed of the watercourse, except in tidal waters. Because of provincial authority over "Property and Civil Rights," the Crown in the right of Ontario as owner of the bed of the Great Lakes, is thus responsible for managing fisheries of the Canadian waters of the Great Lakes (Thompson 1974). In Dixon v. Snetsinger, 23 U.C.C.P. 235 (1873); and R. v. Moss, 26 S.C.R. 322 (1896), the courts specifically declared that the title to soil of navigable waters was vested in the Crown in the right of the province, and not in the right of the Dominion (Piper 1967).

Ontario's Proprietary Rights v. Canada's Protection Authority

Canadian federal laws take precedence over provincial laws for the Great Lakes fishery where these are in conflict, but only to the extent necessary to protect the fishery. In response to an 1898 provincial challenge to the federal government's assumed responsibility over inland fisheries, the Judicial Committee of the Privy Council ruled that provinces may exercise their proprietary rights by legislating with respect to the conveyance, disposal, and rights of succession of several fisheries within the province, including the conditional granting, leasing, and licensing of provincial fisheries. However, the Privy Council determined that provincial acts must be consistent with federal legislation (Thompson 1974). Unsurrendered Indian lands and waters, for which exclusive federal jurisdiction was not at issue, were exempted from the Privy Council's consideration.

According to the Privy Council, the federal government may legislate for the regulation, protection, and preservation of the sea coast and inland fisheries, but they may not grant any right to fish in waters where the exclusive right of fishing exists "in the Crown in right of the province" (as with the Great Lakes). The Privy Council recognized that federal power might amount to a partial confiscation of property rights (e.g., closing down a fishery entirely), but left it to the courts to ensure that such an action was necessary for the conservation of the inland fishery, and did not constitute a takeover of provincial property rights (Thompson 1974).

Provincial Management Activities

Thus, Ontario has authority in licensing or otherwise permitting access to the Great Lakes fishery. Where aboriginal and treaty rights are involved, Ontario's authority is subject to treaty rights due to provisions of the Indian Act, s. 88, and is administered under federal, not provincial regulations, under delegated authority from the federal government. Up until the last two decades, licensing and allocation issues in the Great Lakes had a commercial-fishery focus. Sport and aboriginal fishermen, however, have shown increased interest as near shore pollution has been curbed, and as offshore fisheries recovered from sea lamprey predation and overfishing.

While the federal government has legislative authority for conservation of the Great Lakes fishery, it has largely delegated its authority to Ontario. Thus, Ontario prepares the Ontario Fishery Regulations that the federal government subsequently imposes under the Fisheries Act of Canada. Both orders of government may enforce the regulations, but in practice, enforcement has been undertaken by Ontario. Periodically, a Canada–Ontario Fisheries Agreement is negotiated to facilitate implementation of the dual roles in managing the Great Lakes fishery. The way is not always smooth, however. At this writing (1997), the Province discontinued its (unreimbursed) management of aquatic habitat, ultimately a federal responsibility, and Fisheries and Oceans Canada must now handle tens of thousands of permit requests that were once processed by the Ontario Ministry of Natural Resources annually.

Federal Management Activities

The federal government's authority for conservation, but not for use of Great Lakes fish is reflected in the focus of its research (ecosystem rehabilitation), in its environmental and habitat interests, and in its establishment with the United States of the Great Lakes Fishery Commission, in part, to control the sea lamprey (Convention on Great Lakes Fisheries 1954). The invasion of the highly predacious sea lamprey was seen as a conservation (Canadian) responsibility and not as a proprietary (Ontario) issue. Federal contribution to Great Lakes fish research and management is most visible as the Department of Fisheries and Oceans' Great Lakes Laboratory for Fisheries and Aquatic Sciences as well as for its Sea Lamprey Control Center. (The latter delivers sea lamprey control under contract to the U.S.–Canadian Great Lakes Fishery Commission.)

United States Authorities

State Authority over Great Lakes Fish

Amendment 10 (1791) of the Constitution of the United States of America states that: "The powers not delegated to the United States by the Constitution, nor prohibited by it to the States, are reserved to the States respectively, or to the people." Early court decisions described states as having supreme control over fish in their waters:

> Over fish found in its waters, and over wild game, the State has supreme control (Bayside Fish Co. v. Gentry, 297 U.S. 422, 426 (1936)). It may regulate or prohibit fishing and hunting within its limits (Manchester v. Massachusetts, 139 U.S. 240 (1891); Geer v. Connecticut, 161 U.S. 519 (1896)).

In 1949, however, in a South Carolina case, the Supreme Court qualified that:

> The whole ownership theory, in fact, is now generally regarded as but a fiction expressive in legal shorthand of the importance to its people that a State have power to preserve and regulate the exploitation of an important resource (Toomer v. Witsell, 334 U.S. 385, 402 (1948)).

State control (or state power to preserve and regulate exploitation of Great Lakes fish) is, however, subject to the terms of various Indian treaties negotiated by the U.S. federal government, and to legislation that Congress implements under its constitutional powers, including, in particular, "To regulate commerce with foreign nations and among the several states, and with the Indian tribes."

There are hundreds of court decisions on a myriad of issues throughout the United States which redress the balance between state and (Indian) treaty-derived authority, and, in particular, disallow state legislation which conflicts with federal authority. This judicial turmoil has its parallel in the combative history among various U.S. agencies with responsibility for Great Lakes fish, particularly in the first half of the century. Coupled with the binational character of the Great Lakes, it is the states' efforts to protect their fish management prerogatives from federal encroachment that largely shaped the interjurisdictional institutions discussed in the second half of this chapter.

State Ownership of Lakebeds

A notable difference between the Great Lakes and marine coastal waters is that the Great Lakes lakebeds are the property of either a state or Ontario (with some tribal ownership), with no national or international ownership. This accident of geography, plus the binational character of the Great Lakes and the Great Lakes states' willingness to contest federal encroachment, no doubt was responsible for the U.S. federal government's greater reliance on its treaty making, rather than its interstate commerce authority in facilitating desired improvements in the management of an interjurisdictional fishery resource. State pressure was such, however, that the resulting U.S.–Canadian Convention on Great Lakes Fisheries (1954) confined itself to managing only one species, the exotic sea lamprey.

As in Ontario, title to the waters and the beds of the Great Lakes rests on the common-law principle that the state owns the seabed under the tidal waters within its jurisdiction. Although the lakes are non-tidal, they are public, navigable waters and the same principle of public ownership applies (Piper 1967 19):

> In Pollard v. Hagan (3 How. (44 U.S.) 212,223 (1845)) the Court held that the original States had reserved to themselves "the ownership of the shores of navigable waters and the soils under them, and

that under the principle of equality the title to the soils of navigable water passes to a new State upon admission" to the Union.

In the United States, the eight riparian states (New York, Pennsylvania, Ohio, Illinois, Indiana, Michigan, Minnesota, and Wisconsin) possess, in their respective public capacities, the waters and lakebeds within their territorial limits. (The principle of state ownership of the lakebeds is affirmed in the Submerged Lands Act of 1953. 67 Stat.29. See also 65 C.J.S., Navigable Waters, sec. 92; Illinois Central Railroad v. Illinois, 146 U.S. 387 (1892); and Hilt v. Weber, 233 N.W. 159 (1929); Moore, Digest, I, 672; and 56 Am. Jur., Waters, sec. 52.). (The territorial limits of each of the mentioned states extends into the waters of the adjacent lake or lakes. (See for example the Ohio Revised Code annotated (Page, 1959 Supp.), sec.123.03, wherein it is stated that the State of Ohio possesses, for public use, the waters and the bed of Lake Erie within the boundaries of the state, extending to the international boundary. See also Illinois Central Railroad v. Illinois, 146 U.S. 387, 452 (1892); Bowes v. City of Chicago, 120 N.E. 2d 709 (1948)) (Killian and Beck 1987).

Federal Supremacy over State Law

When acting within an area of its constitutional authority, legislation passed by the U.S. Congress overrides state authorities. "Even where Congress has not entirely displaced state regulation in a specific area, state law is preempted to the extent that it actually conflicts with federal law" (Ray v. Atlantic Richfield Co., 435 U.S. 151, 158 (1978)). Such a conflict arises when "compliance with both federal and state regulations is a physical impossibility" (Florida Lime and Avocado Growers v. Paul, 373 U.S. 132, 142–143 (1963), or where state law "stands as an obstacle to the accomplishment and execution of the full purposes and objectives of Congress" (Hines v. Davidowitz, 312 U.S. 52, 67 (1941)).

Federal Authority over Interstate Commerce

Federal authority over interstate commerce is the basis for much (but not all) federal legislation which potentially can preempt state control of Great Lakes fish, (e.g., Endangered Species Act of 1973).[4] Drawing on its constitutional authority to regulate interstate commerce, the U.S. Congress may assign specific authorities to the federal government at the expense of state authority. Article 1, Section 8 of the U.S. Constitution states that, "The Congress shall have Power . . . to regulate Commerce with foreign Nations, and among the several States, and with the Indian Tribes." According to Killian and Beck (1987), the authority to regulate interstate commerce is:

> . . . the direct source of the most important powers which the Federal Government exercises in peacetime, and, except for the due process and equal protection clauses of the Fourteenth Amendment, it is the most important limitation imposed by the Constitution on the exercise of state power . . . Of the approximately 1,400 cases which reached the Supreme Court under the clause prior to 1900, the overwhelming proportion stemmed from state legislation.

Regulation of interstate commerce is the intent for exercise of this Congressional power within the United States. Chief Justice Marshall interpreted this clause to include action applied "to those internal concerns which affect the states generally; but not to those which are completely within a particular state, which do not affect other states, and with which it is not necessary to interfere, for the purpose of executing some of the general powers of the government" (9 Wheat. (22U.S.) 1, 194, 195 (1824)).

The power to regulate commerce was interpreted broadly in Brooks v. the United States (267 U.S. 432, 436–437 (1925), in which the Court ruled that "Congress can certainly regulate interstate commerce to the extent of forbidding and punishing the use of such commerce as an agency to promote immorality, dishonesty, or the spread of any evil or harm to the people of other states from the state of origin." Killian and Beck (1987) note that the power to regulate interstate commerce has been exercised to enforce majority conceptions of morality, to ban racial discrimination, and to protect the public against evils both natural and manmade, and is, therefore, "rightly regarded as the most potent constitutional grant of authority of the powers of Congress."

U.S. Federal Role

In contrast with Canadian federal responsibility for conservation and preservation of Great Lakes fish, the U.S. federal role under the interstate commerce clause in Great Lakes fish management is continually being defined in legislation (appendix). Its legislated role in Great Lakes fish management is most authoritative in matters of pollution, habitat protection, and endangered species, and increasingly of non-indigenous aquatic nuisance species. U.S. federal resource agencies have played a welcome support role to the states and tribal governments, for example, in research, rehabilitative stocking, and law enforcement. The U.S.–Canadian Great Lakes Fishery Commission has contracted with the U.S. Fish and Wildlife Service and the U.S. Geological Survey (as well as Fisheries and Oceans Canada) in discharging its sea lamprey control commitments.

Except for species such as lake trout, where U.S. federal agencies have an interest under the Endangered Species Act, federal research tends not to focus on the top-predator fish species targeted by fishermen, but rather on forage fishes, such as alewife and sculpin. Increasingly, however, state, tribal, and provincial fish managers have come to rely on research and other information assembled by federal researchers on forage fish and lake trout. Research and monitoring is accomplished through labs such as the Great Lakes Science Center (U.S. Geological Survey) and the Fisheries Resource Stations of the U.S. Fish and Wildlife Service.

State Activities

As with the U.S. federal agencies, states redirected the focus of their natural resource management programs from business development to resource protection as the great abundance of Great Lakes fishes declined under the market fishing ethos of the 1800s and early 1900s. Hatcheries were built and restrictions imposed to maintain the commercial fishery. Overfishing, followed by invasion by the sea lamprey in the upper lakes, and chemical and phosphorus pollution prompted interjurisdictional discussion, and eventually, cooperative action in addressing common problems. Technological fixes were developed in response to various crises in the mid-twentieth century that made it possible for managers to develop important, new sport fisheries (e.g. lampricide control, control of alewife via stocked salmon, toxic, and chemical control programs). Today, state fishery managers work closely with federal, provincial, and tribal fishery managers (and increasingly with their environmental counterparts) to sustain their sport and commercial fisheries, and to influence other activities with potential to disrupt those communities.[5]

Treaty Fishing Rights

Historical differences in American and Canadian relations with their native peoples shape native claims to the Great Lakes fishery, depending on whether one is north or south of the border. On the United States side of the Great Lakes, fishing rights are rooted in negotiated treaties with Indian peoples, who, incidentally, were only fully extended American citizenship in the Citizen Act of 1924. Long considered subjects of the Crown, Canadian aboriginal people of the Great Lakes may claim off-reservation fishing rights as aboriginal rights protected by the Canadian Constitution Act of 1982.

United States

In the United States, certain treaties preserved tribal fishing rights in the Great Lakes. The Treaty of Washington, 28 March 1836, U.S.–Ottawa and Chippewa nations, 7 Stat. 491, preserves tribal fishing rights in northern Michigan waters of Lakes Superior, Michigan, and Huron. The Treaty of 1842 with the Chippewa, 7 Stat. 591, preserves tribal fishing rights in portions of Lake Superior in Wisconsin, Michigan, and Minnesota. For some tribes having a reservation on Lake Superior (such as Red Cliff and Bad River Tribes), fishing rights are guaranteed by the Treaty of 1854 with the Chippewa, 10 Stat. 1109 (as litigated in State v. Gurnoe, 53 Wis. 2d 390 (1972)).

Thus, U.S. treaty-making authority has preserved tribal fishing rights in certain ceded territories relative to fish management authority of the states. Where guaranteed by treaty, the courts have ruled (Table 1) that Indian fishing rights in ceded territories may not be qualified by the state, but such items as the manner of fishing, the size of the take, and the restriction of commercial fishing may be regulated by the state[6] for conservation reasons (such as irreparable harm), if, for example, the regulation does not discriminate against Indians, and represents the least restrictive alternative available. The preference is for tribal regulation of Indian fishermen.

Canada

In 1982, section 35 of the new Canadian Constitution Act recognized and affirmed "the existing aboriginal and treaty rights of the aboriginal people." As David Milne (1982) explained soon thereafter, section 35 "raises the most ambiguity about the future rights of native peoples. It can be read in two distinct ways: the first is that section 35 offers constitutional recognition of the admittedly weak position of aboriginal rights in the statute and common law as of the date of proclamation; the second is that aboriginal and treaty rights themselves are accorded constitutional protection in the sense that no statute may infringe them. It was certainly the view of the federal government that only the first meaning applied." In the fifteen years since, however, the courts appear to have adopted the second reading—that section 35 rights may not be infringed.

The existing aboriginal rights affirmed by the Constitution Act (1982) are now being characterized by the highest Canadian courts on cases throughout Canada, including the Great Lakes (for example, R. v. Jones [1993]14 D.R. (2d) 421). In the Sparrow decision, the Canadian Supreme Court

TABLE 1

Examples of court decisions pertaining to regulation of U.S. tribal fishing in ceded territories.

Cal.—People v. McCovery, 205 Cal.Rptr. 643, 685 P.2d 687, 36 C.3d 517, certiorari denied California v. McCovey, 105 S.Ct. 544, 469 U.S. 1062, 83 L.Ed2d 432.

Or.—State v. Jim, 725 P2d 372, 81 Or.App. 189.

Or.—State v. Smith, 625 P.2d 1321, 51 Or.App 223, review denied 631 P.2d 341, 291 Or. 118.

U.S.—Puyallup Tribe, Inc. v. Department of Game of State of Washington, Wash., 97 S.Ct. 2616, 433 U.S. 165, 53 L.Ed.2d 667.

U.S.—Lac Courte Oreilles Band of Lake Superior Chippewa Indians v. State of Wisconsin, D.D.Wis., 668 F.Supp. 1233.

U.S.—U.S. v. Eberhardt, C.A.9 (Cal.), 789 F.2d 1354.

U.S.—U.S. v. State of Michigan, C.A.Mich., 653 F.2d 277, certiorari denied 102 S.Ct. 971, 454 U.S. 1124, 71 L.Ed.2d 110.

U.S.—U.S. v. State of Michigan, D.C.Mich., 505 F.Supp. 467.

U.S.—U.S. v. State of Michigan, D.C.Mich., 520 F.Supp. 207.

U.S.—U.S. v. State of Oregon, C.A.9 (Or.), 769 F.2d 1410.

Wash.—State v. Courville, 676 P.2d 1011, 36 Wash.App. 615.

Wash.—State v. Miller, 689 P.2d 81, 102 Wash.2d 678.

Wis.—State v. Lemieux, 327 N.W.2d 669, 110 Wis.2d 158.

found that aboriginal rights existed unless extinguished by the Crown with clear intention. The Court developed tests to determine whether aboriginal (fishing) rights had been unreasonably interfered with, and whether interference was justified (i.e., consistent with duty toward aboriginal peoples and in pursuit of a valid legislative objective) (R. v. Sparrow [1990] 1 S.C.R.).

While aboriginal fishing rights are assigned a very high priority and traditional commercial fishing some priority, the recreational fishery has much less priority. The Sparrow decision stated that "Any allocation of priorities after valid conservation measures have been implemented must give top priority to Indian food fishing" (R. v. Sparrow [1990] 1 S.C.R.). In the Gladstone decision, the Supreme Court found that, "With regards to the distribution of the fisheries resource after conservation goals have been met, objectives such as the pursuit of economic and regional fairness, and the recognition of the historical reliance upon, and participation in, the fishery by non-aboriginal groups are the type of objectives which can (at least in the right circumstances) satisfy this standard" (R. v. Gladstone [1996] 2 S.C.R.).

Aboriginal rights is a fairly inclusive concept. Aboriginal rights were defined as integral to the precontact community and culture (i.e., not tied to a particular piece of land) (R. v. Gladstone [1996] 2 S.C.R.). In Van der Peet the Court stated that "The concept of continuity is the means by which a "frozen rights" approach to s. 35(1) will be avoided" and that, "A practice existing prior to contact can be resumed after an interruption." "Section 35(1) must be given a generous, large and liberal interpretation, and ambiguities or doubts should be resolved in favour of the natives" (R. v. Van der Peet [1996] 2.S.C.R.). A federal requirement in British Columbia for a free, easily available fishing license, as distinct from its conditions, was found to not constitute an infringement of s.35(1), the aboriginal rights section of the Constitution Act, 1982 (R. v. Nikal [1996]1 S.C.R.).

Native Management Activities

Most U.S. tribes with off-reservation fishing rights in the Great Lakes have established intertribal agencies to which they have delegated certain responsibilities to help regulate their members' activities. The Chippewa-Ottawa Treaty Fishery Management Authority and Great Lakes Indian Fish and Wildlife Commission also represent the tribes in interagency technical and management committees of the Great Lakes Fishery Commission (GLFC). In the 1980s, tribal biologists joined those of other agencies in interagency technical committees developing a common understanding of stocks to be managed in Lakes Superior, Michigan, and Huron. In 1989, the two intertribal agencies were invited by the GLFC onto the management committees for the three upper lakes, where they joined with fish managers from the states and province. The inclusion of the intertribal agencies has proven to be a constructive step. For example, a lake trout rehabilitation plan subsequently developed by the management committee for Lake Huron was taken as the basis for a consent agreement implementing the 1836 Treaty of Washington (U.S. v. Michigan, no. M-26–73CA (W.D. Mich. Apr. 10, 1985)). The consent agreement established the framework for all fisheries in the treaty area, thereby strengthening efforts toward common objectives for the resource, and easing a potentially volatile situation.

Canadian native people exercising aboriginal rights, rights which were only recognized in 1982, have made less progress than some U.S. tribes in seeking a recognized role in managing Great Lakes fisheries. It is not yet clear the degree to which Canadian native people will manage their own fishing activities. As of this writing (1997), Canadian native people have not participated in interagency technical committees. Apparently, some native lawyers advise that such participation might be construed as consultation, which is necessary for a province to impose a management restriction on off-reservation aboriginal fishing.

Coordination and Joint Programs

In the twentieth century, federal, state, and provincial authorities pursued both informal and formal mechanisms for coordinating their management of Great Lakes fish and fisheries. In the 1960s, individuals such as Howard Tanner, Michigan proponent of salmon stocking, greatly influenced Great Lakes fishery management in all eight Great Lakes states and the Province of Ontario. In 1954, the Great Lakes Fishery Commission (GLFC) was established by the convention between the United States and Canada, and, in 1981, fishery agencies[7] themselves negotiated a Joint Strategic Plan for Management of Great Lakes Fisheries. The strategic plan provided rules of engagement to further the cooperation that had evolved to that date in the lake committees of the Great Lakes Fishery Commission.

Arrangements for coordinating Great Lakes fishery management are largely shaped by such forces as the binational profile of the resource, states rights issues (relative to U.S. federal power), and the interplay of a complex array of federal, tribal, state and provincial management authorities for Great Lakes fish, as discussed in the first half of this chapter. Briefly, the province owns the lakebed and the fish, and determines access; states own defined portions of the lakebed, and have power to preserve and regulate access to fish. Both the provinces' and the states' powers are subject to important federal and treaty limitations in their respective countries. Canada has a constitutional responsibility to conserve and preserve Great Lakes fish, which it tends to delegate to Ontario. The U.S. Congress legislates federal authorities primarily under the interstate commerce clause of the U.S. Constitution. U.S. and Canadian federal governments have created new fish management players through treaties with each other and with various Indian tribes. Some U.S. tribes have fishing rights that they did not relinquish in treaties with the United States. And many Canadian aboriginals have constitutional fishing rights in the Great Lakes, which are being upheld by the courts.

As a consequence of this complex array of authorities, management positions do not tend to line up in rigid alliances (e.g. along national lines), science is valued as a common language, and discussion tends to focus on common objectives, needed action, and monitoring progress. In practice, equity is assumed for the more or less equal state, provincial, and tribal players on management committees of the Great Lakes Fishery Commission. The potentially powerful commission and federal fishery agencies provide vitally important support such as sea lamprey control, research, and in the United States, hatchery trout to rebuild stocks decimated by the sea lamprey. After many false starts, the two most important mechanisms which emerged in the twentieth century for rationalizing management of shared fisheries of the Great Lakes are the Great Lakes Fishery Commission (by U.S.–Canadian convention in 1954) and a Joint Strategic Plan for Management of Great Lakes Fisheries (an interagency agreement signed in 1981).

Evolution and Attributes of Mechanisms for Coordinated Management
(A Joint Strategic Plan for Management of Great Lakes Fisheries)

State Compacts

Early in the twentieth century, at the 1937 New York Conference, the Great Lakes states recommended consideration of an interstate compact for bringing about agreements for conservation of their fisheries. Recognizing that the Province of Ontario must be a party to any initiative determined to improve regulation of the Great Lakes fisheries, New York Congressman, William T. Byrne, introduced H.R. 4096 seeking advance approval for a compact entered into between two or more states and "with a contiguous sovereignty or dominion or any state or province thereof for the purpose of conserving the wildlife resources of the several states."[8] Asked to comment, then-Secretary of State Hull:

> was unable to concur in the proposal to authorize states to enter into compacts concerning wildlife with the contiguous countries and political divisions thereof . . . If the officials of the interested states will bring to my attention . . . any subject matter that is not covered by existing treaties, I assure you that I shall be glad to give most careful consideration to the possibility of adopting additional measures on the subject or subjects (Gallagher et al. 1943).

In 1955, an interstate commission came into being that afforded the states an opportunity to work together on natural resource and shipping issues.

> In 1955, the riparian American states entered into the Great Lakes Compact to achieve greater cooperation and uniformity in the use and development of the water and natural resources of the Great

Lakes . . . In the initial compact, there was provision for the membership of the provinces of Ontario and Quebec. The State Department objected to this provision when the compact was presented to Congress for approval on the ground that with provincial participation the commission entered the field of international relations,[9] moreover, the compact contained some provisions that would enable the member states to contradict the policies of the U.S. federal government . . . As a voice of riparian states (the Great Lakes Commission) is in a position to influence the international law concluded by the two federal governments (Piper 1967).

Congressional consent legislation was enacted in 1968 approving a revised Great Lakes Compact (which contained no allowance for provincial membership), and the (U.S.) Great Lakes Commission.

Management Committees of the Great Lakes Fishery Commission

The year prior to the drafting of the first Great Lakes Compact, the Great Lakes Fishery Commission was created by the Governments of Canada and the United States in a Convention on Great Lakes Fisheries (1954). The new commission had, in addition to sea lamprey control responsibilities, a duty to study and advise the two countries on issues affecting sustainable benefits from fish stocks of common concern, and on remedial or preventative measures. On the U.S. side, enabling legislation (Great Lakes Fishery Act of 1956, Appendix) provided for governor-nominated advisors for the U.S. section of the new commission, in recognition of state interests and authorities in management of Great Lakes fish and fisheries. Advisors represented agencies, anglers, commercial fishermen, and the public-at-large.

Prior to the formation of the Great Lakes Fishery Commission, state fish management agencies invested in informal, management-coordination lake forums that included their Ontario counterparts. In 1965, the states and the province abandoned the last of these informal forums for lake committees which the GLFC established to address a broader array of fishery problems (Eshenroder 1987). Lake committees initially focused on rehabilitating such species as lake trout in Lake Superior (Pycha and King 1975), a deepwater cisco known as bloater in Lake Michigan (Brown et al. 1985), and walleye in Lake Erie (Hatch et al. 1987).

In 1972, the Great Lakes Fishery Commission established a Great Lakes Fish Disease Control Committee (now the Great Lakes Fish Health Committee) at the recommendation of the Lake Michigan Committee. Peopled by agency fishery pathologists and hatchery administrators, the Great Lakes Fish Health Committee is a forum to prevent introduction, and to reduce the range of serious fish disease.

A Joint Strategic Plan for Management of Great Lakes Fisheries

State and provincial satisfaction with progress under lake committees was such that the states rejected an opportunity to form a U.S. Regional Fishery Management Council under the Magnuson Fishery Conservation and Management Act of 1976 (Appendix). Instead, with their provincial counterparts, U.S. natural resource agencies requested the assistance of the Great Lakes Fishery Commission in drafting a strategic plan in which fishery agencies would formalize their commitment to lake committees as their major action arm (Great Lakes Fishery Commission 1979, 1980). Directors of fishery agencies reiterated their commitment in a 1985 review of the resulting strategic plan (Dochoda 1988), and again in 1995, when responding to a proposal by Rollie Harmes, then Director of the Michigan Department of Natural Resources, to review implementation of the 1981 Joint Strategic Plan for Management of Great Lakes Fisheries.

The Joint Strategic Plan for Management of Great Lakes Fisheries (1981, revised in 1997) is an agreement signed by agency directors, in which fishery agencies, while ceding none of their authority, agree to wield it in a coordinated, binational, science-based fashion for the common good. Fish community objectives are set by lake committees composed of fish management agencies (i.e., natural resource agencies of the Great Lakes States, Ontario, and certain tribes), and advised by technical subcommittees where federal agency researchers often play a pivotal role. For the most part, agencies have been successful back home in implementing common objectives and strategies, crafted with their fellow managers of Great Lakes fish. Progress on objectives is assessed by the technical subcommittees which report to lake committees in annual public meetings.

The heart of the Strategic Plan is a common goal statement and four strategies or rules of engagement. In the Strategic Plan, fishery agencies agreed,

> To secure fish communities, based on foundations of stable self-sustaining stocks, supplemented by judicious plantings of hatchery-reared fish, and to provide from these communities an optimum contribution of fish, fishing opportunities and associated benefits to meet needs identified by society for: wholesome food, recreation, cultural heritage, employment and income, and a healthy human environment.

The agencies' four strategies or rules of engagement for achieving their common goal are as follows:

- Consensus must be achieved when management will significantly influence the interests of more than one jurisdiction.
- Fishery management agencies must be openly accountable for their performance.
- The Parties (to the Plan) must exercise their full authority and influence in every available arena to meet the biological, chemical, and physical needs of desired fish communities.
- Fishery agencies must cooperatively develop means of measuring and predicting the effects of fishery and environmental decisions (Great Lakes Fishery Commission 1997).

The strengths of the Strategic Plan include mechanisms for collective use of information in decision making, and procedures for ongoing and coordinated management (Dochoda and Koonce 1994). The Strategic Plan gives explicit procedural guidance regarding the application of the Strategic Plan's strategies. For example, "What happens if a consensus decision cannot be achieved? The problem will be taken to the Great Lakes Fishery Commission for mediation and (nonbinding) arbitration at the request of one or more of the parties in the dispute at the lake committee level."

A challenge for fishery agencies is that the Strategic Plan, typical of informal learning-led[10] ecosystem-management institutions, tends to be limited in fiscal resources. Fishery agencies cover the expenses of their representatives on commission committees guided by the Strategic Plan. In addition, the Great Lakes Fishery Commission provides staff and modest financial support as part of its responsibility to advise governments on issues and on measures which will secure sustained benefits from the fishery. Individual agencies pool data and analytical capability, but continuation of these in-kind contributions are threatened as government cutbacks[11] force agencies into reactive postures.

For the Strategic Plan to realize its promise, mechanisms must be found that will fund science-based, coordinated research to address new and continuing challenges in rebuilding ecosystems devastated by overfishing, exotics, and pollution. In addition, without adequate resources, learning-led lake committees will find it difficult to constructively engage the relatively well-funded lakewide, environmental management plans created by the U.S.–Canadian Great Lakes Water Quality Agreement (as amended by protocol in 1987).

Evolution and Attributes of Mechanisms for Coordinated Management (Great Lakes Fishery Commission)

Agency successes under the Strategic Plan through the lake committees are all the more remarkable when one considers that the Canada–U.S. Convention on Great Lakes Fisheries was signed in 1954, only after nearly sixty years of unsuccessful attempts to establish a joint fishery commission or another mechanism for producing effective, standardized regulation of the Great Lakes fishery. An early difficulty,

> . . . precluding the conclusion of a treaty regulating the fishery was the claim by the state governments of exclusive control over the fishery and Canadian uncertainty as to the authority of the United States federal government to enforce the proposed regulations. To clarify the federal government's authority, Attorney General Griggs, in 1898, stated that the regulation of the fishery in the boundary waters between the United States and Canada was a proper subject for a treaty, and that although a treaty might supersede state laws, its legal validity would not be impaired (Piper 1967).

Also, early negotiators withheld Georgian Bay a tributary water, and Lake Michigan, from draft fish treaties, the Canadians presumably to gain leverage, and the Americans in order to avoid raising the question of U.S. federal treaty making authority over waters surrounded by U.S. territories. Both waters were included in the 1909 Boundary Waters Treaty, thereby confirming U.S. federal treaty making authority in Lake Michigan (Piper 1967).

In fact, prior to 1954, two agreements were signed by the two governments, but were terminated. A Treaty between the United States and Great Britain, Fisheries in the United States and Canadian waters, was signed, ratified, and proclaimed in 1908, only to be terminated in 1915 because of the failure of the U.S. Congress to approve the regulations. A Convention between the United States of America and Canada for the Development, Protection, and Conservation of the Fisheries of the Great Lakes was signed in 1946, but not ratified by the U.S. Congress due to opposition to the transfer of regulatory authority from states to an international body (Fetterolf 1980). Finally, negotiators dropped regulatory authority in their treaty in order to be able to respond to the invading sea lamprey, which had decimated stocks of lake trout and whitefish in the 1940s and 1950s[12].

In his book, "The Joint Organizations of Canada and the United States," Willoughby (1979) points out that the Great Lakes fish management situation is unique in that the Great Lakes

Fishery Commission created for that water body is the only U.S.-Canadian fishery commission without regulatory authority. The principal reason, according to Willoughby, was the "vehement objections raised by the vested American interests—particularly by spokesmen of the Ohio commercial fishermen."

In retrospect, it was perhaps the absence of regulatory authority in the Great Lakes Fishery Commission that allowed states to be comfortable in using its binational forum. A possible indication of continuing sensitivity to federal encroachments on state authority is that signatories were careful to specify in the Strategic Plan that they looked to the commission for arbitration that was nonbinding.

From Ontario's perspective, the commission doubtless provides a comforting 1:1 binational venue under which the province may coordinate its management in forums where it is otherwise always outnumbered by state and (U.S.) tribal counterparts. Each agency's veto power under the Strategic Plan's consensus strategy helps allay the Canadian perception of being outnumbered, as does organization around lakes (where the U.S./Canadian ratio is somewhat lower), and the binational sharing of committee leadership[13].

The commission's study-and-advise mandate is broad and open relative to the sustainable Great Lakes fishery, and it is a flexible instrument which is administratively acceptable in both countries as a partner or vehicle for various initiatives:

> . . . the commission shall, in so far as feasible, make use of private or other public organizations, including international organizations, or any person . . . The commission may seek to establish and maintain working arrangements with public or private organizations for the purpose of furthering the objectives of the convention (Convention on Great Lakes Fisheries, 1954).

The various missions and components of the commission balance and enhance each other. While nonregulatory, the commission's sea lamprey control mission is an essential prerequisite to most, if not all, fish management activities in the Great Lakes, and thus, the commission is a significant member of the fish management community.

Through lake committees, fish management agencies have found significant opportunities to influence and benefit from the commission's more basic study-and-advise mandate, and sea lamprey control mission. Finally, the commission explicitly relies on its lake committees, and related management committees for sea lamprey control targets and, in part, for delivery of its study-and-advise mandate (Dochoda 1991, Great Lakes Fishery Commission 1992).

Within the commission, the practical work of sea lamprey control is given vision by the study-and-advise mandate, while science is focused on the very real management and policy needs of the basin. Because the states and the province are both proprietors of Great Lakes fish, positions of commissioners tend not to become stratified along national lines. Often, a commissioner representing a state has as much in common with a commissioner from the province as with one from the U.S. federal government. The process of issue-resolution returns repeatedly to the needs of the ecosystem, and to the language of science, both touchstones with which all agree.

Future Challenges for Great Lakes Management

Resource and fiscal challenges for fishery management (as well as for environmental management) are now such that no jurisdiction or discipline can progress on its own. Great Lakes fishery managers must coordinate and leverage each others' actions through investment in coordinated planning, management, and assessment. The most promising initiatives and institutions for coordinated, effective management are those negotiated with recognition of our numerous, varied sources of authority for managing fish and fisheries in the Great Lakes ecosystem.

Acknowledgments

This overview relies heavily on others' compilations and analyses for Canada, the United States, and regional and binational mechanisms, which the author gratefully acknowledges. Mary Lynn Becker, Mike Conlin, Tom Gorenflo, Bill James, Bob Kavetsky, Bob Lange, Ron Poff, John Robertson, and Jim Zorn kindly provided many additional references. The following people provided valuable guidance in their review of drafts: Jim Zorn, Peggy Blair, an anonymous reviewer, Doug Dodge, Chris Goddard, Chuck Krueger, Bill Horns, and Mike Conlin. Any errors or omissions in describing the complex institutional arrangements for management of Great Lakes fish are the responsibility of the author.

Notes

1. Fishery managers have only a few tools available to them for manipulating the fish community and fisheries in the Great Lakes Ecosystem, e.g., (1) managers can add new species to the ecosystem, or supplement the relative numbers of certain existing species, or prevent harmful invasions, e.g., through stocking decisions and regulation of private sector activities, ballast management, and decisions on water or transportation projects; (2) managers can reduce the relative numbers of certain existing species, e.g., through harvest regulations and chemical control; (3) managers can improve or create physical and chemical habitat.

2. Government agencies with responsibility for the Great Lakes fishery: Canada Department of Fisheries and Oceans; Indiana Department of Natural Resources; Michigan Department of Natural Resources; Minnesota Department of Natural Resources; National Marine Fisheries Service (U.S.); New York State Department of Environmental Conservation; Ohio Department of Natural Resources; Ontario Ministry of Natural Resources; Pennsylvania Fish and Boat Commission; U.S. Fish and Wildlife Service; U.S. Geological Survey; Wisconsin Department of Natural Resources; Chippewa-Ottawa Treaty Fishery Management Authority (U.S.); Great Lakes Indian Fish and Wildlife Commission (U.S.). (Note: each member tribe of COTFMA and GLIFWC has delegated certain authorities to its intertribal agency in its ceded territory context.)

3. There is question whether this proprietary interest extends to the lakebed of unsurrendered Indian lands, e.g. portions of Lake Huron. (Peggy Blair, personal communication)

4. Water management, too, has its roots in the interstate commerce clause. The federal government has regulatory power related to the navigable waters of the Great Lakes in that waterborne interstate commerce depends upon navigation. Partly to further navigation interests, the U.S.–Canadian Boundary Waters Treaty of 1909 created the International Joint Commission (IJC). One of the IJC's responsibilities is to decide matters relative to Great Lakes water levels, which are of primary concern to navigation interests, but also have importance for fish management. Health concerns were also behind the Boundary Waters Treaty. The Great Lakes Water Quality Agreements, executive agreements between U.S. and Canadian agencies operating in the forum of the IJC, that followed in the 1970s and 1980s, greatly modified the nutrient and toxic chemical profile of the Great Lakes, again affecting the makeup, production, and use of fish communities.

5. In 1988, fishery managers on Lake Superior, responding to reports of Eurasian ruffe in the St. Louis River, successfully called for management of ballast in intercontinental shipping to prevent further introductions.

6. In addition, U.S. federal regulations 25 CFR Part 249 provide that either the affected tribe or state may request U.S. federal regulation to protect a stock from overfishing.
7. Department of Fisheries and Oceans-Canada, Illinois Dept. of Natural Resources, Indiana Dept. of Natural Resources, Michigan Department of Natural Resources, Minnesota Department of Natural Resources, National Marine Fisheries Service, New York Department of Environmental Conservation, Ohio Department of Natural Resources, Ontario Ministry of Natural Resources, Pennsylvania Fish and Boat Commission, Wisconsin Department of Natural Resources, and U.S. Fish and Wildlife Service signed in 1981; Chippewa-Ottawa Treaty Fishery Management Authority and Great Lakes Indian Fish and Wildlife Commission signed in 1989; U.S. Geological Survey signed in 1997.
8. Section 10 of the U.S. Constitution states that, "No State shall enter into any Treaty, Alliance or Confederation."
9. See U.S. Congress, Senate, Subcommittee of Foreign Relations, Hearings, The Great Lakes Basin, 84th Cong., 2d Sess., 1956, pp. –, 14, 17, 3–2; idem, Subcommittee on the Judiciary, Hearings, Great Lakes Basin Compact, 85th Cong., 2d Sess., 1958, p.5.
10. Human collaborations have been categorized as planning-led, vision -led, and learning-led. All have their advantages and limitations. (Westley 1995) The lakewide management plans of the Great Lakes Water Quality Agreement are examples of a planning-led enterprise. Howard Tanner's introduction of the salmon might be thought of as a vision-led enterprise. The lake committees of the Great Lakes Fishery Commission have many characteristics of learning-led assemblages.
11. The International Joint Commission reported to the Governments of Canada and the United States (1996) in its Eighth Biennial Report on Great Lakes Water Quality that 31 Great Lakes (environmental and fishery) research organizations, responsible for 80% of 199–2 research, were projecting reductions of 23% to 53% percent of total operating funds, and 31% to 45 % of salaries by 1997.
12. U.S. senate, Subcommittee of the Committee on Foreign Relations, hearing, Great Lakes Fisheries Convention, 84th Cong.,1st Sess., 1955, p. 10
13. There are Commission committees, however, such as the basin-focused Great Lakes Fish Health Committee and the Council of Great Lakes Fishery Agencies, where Canadians are greatly outnumbered and must rely on committee chairmen to prevent agendas from becoming overly burdened with exclusively internal U.S. concerns. (With 8 states and 2 intertribal agencies and 3 federal agencies, some exclusively internal U.S. discussions are, of course, unavoidable.)

Literature Cited

Brown, E. H., R. W. Rybicki, and R. J. Poff. 1985. Population dynamics and interagency management of the bloater (*Coregonus hoyi*) in Lake Michigan, 1967–1982. Technical Report. 44. Great Lakes Fishery Commission, Ann Arbor, Michigan.

Burkett, D. P. 1995. Great Lakes Fishery Resources Restoration Study. U.S. Fish and Wildlife Service Report to Congress.

Dochoda, M. R. (ed.). 1988. Committee of the Whole Workshop on Implementation of the Joint Strategic Plan for Management of Great Lakes Fisheries (reports and recommendations from the 18–20 February 1986 and 5–6 May 1986 meetings). Great Lakes Fishery Commission Special Publication 88–1. 170 p.

———, 1991. Meeting the challenge of exotics in the Great Lakes: the role of an international commission. Canadian Journal of Fisheries and Aquatic Sciences 48 (Suppl. 1): 171–176.

———, and J. F. Koonce. 1994. A perspective on progress and challenges under A Joint Strategic Plan for Management of Great Lakes Fisheries. University of Toledo Law Review 25 425–442.

Dodge, D. P., and R. Kavetsky. July 1994. Aquatic Habitat and Wetlands of the Great Lakes. State of the Great Lakes Ecosystem Conference. Working Paper.

Eschenroder, R. L. 1987. Socioeconomic Aspects of Lake Trout Rehabilitation in the Great Lakes. Trans. Am. Fish. Soc. 116:309–313.

Fetterolf, C. M. 1980. Why a Great Lakes Fishery Commission and why a Sea Lamprey International Symposium. Canadian Journal of Fisheries and Aquatic Sciences 37:1588–1593.

Gallagher, H. R., A. G. Huntsman, D. J. Taylor, and J. Van Oosten. 1943. International Board of Inquiry for the Great Lakes Fisheries, Report and Supplement. U.S. Government Printing Office. Washington, DC.

Great Lakes Fishery Commission. 1979. Minutes of the annual meeting of the Council of Lake Committees. Great Lakes Fishery Commission, Ann Arbor, MI.

Great Lakes Fishery Commission. 1992. Strategic Vision of the Great Lakes Fishery Commission for the Decade of the 1990s. Great Lakes Fishery Commission, Ann Arbor, MI.

Great Lakes Fishery Commission. 1997. A Joint Strategic Plan for Management of Great Lakes Fisheries. Great Lakes Fishery Commission, Ann Arbor, MI. (Supersedes 1981 version)

Hatch, R. W., S. J. Nepszy, K. M. Muth, and C. T. Baker, 1987. Dynamics of the recovery of the western Lake Erie walleye (*Stizostedion vitreum vitreum*) stock. Canadian Journal of Fisheries and Aquatic Sciences 44 (Suppl. 2): 15–22.

Killian, J. H., and L. E. Beck. 1987. The Constitution of the United States of America Analysis and Interpretation: Annotations of Cases Decided by the Supreme Court of the United States to July 2, 1982.

Koonce, J. F. 1994. Ecosystem partnership coordination final report. Great Lakes Fishery Commission, Ann Arbor, MI.

Milne, D. A. 1982. The New Canadian Constitution. James Lorimer and Co., Publishers. 1982.

Piper, D. C. 1967. The International Law of the Great Lakes: A Study of Canadian-United States Cooperation. Duke University Press, Durham, NC.

Pycha, R. L., and G. R. King. 1975. Changes in the lake trout population of southern Lake Superior in relation to the fishery, the sea lamprey, and stocking, 1950–70. Technical report 28. Great Lakes Fishery Commission, Ann Arbor, MI.

Stanley, J. G., R. A. Peoples Jr., and J. A. McCann. 1991 U.S. federal policies, legislation, and responsibilities related to importation of exotic fishes and other aquatic organisms. Canadian Journal of Fisheries and Aquatic Sciences 48 (Suppl.1): 162–166.

Thompson, P. C. 1974. Institutional constraints in fisheries management. Journal of the Fisheries Research Board of Canada 31:1965–1981.

USFWS. 1992. Digest of Federal Resource Laws of Interest to the U.S. Fish and Wildlife Service. U.S. Department of the Interior, U.S. Fish and Wildlife Service, Office of Legislative Services. April 1992.

Westley, F. 1995. Governing design: the management of social systems and ecosystems management *in* Gunderson, L.H., C.S. Holling, and S.S. Light (eds.). 1995. Barriers and Bridges to the Renewal of Ecosystems and Institutions. Columbia University Press, New York, NY.

Willoughby, W. R. 1979. The Joint Organizations of Canada and the United States. The University of Toronto Press, Toronto, ON.

Appendix. Some U.S. Federal Resource Laws of Importance to Fish Management in the Great Lakes. (Burkett 1995, Dodge and Kavetsky 1994, Stanley et al 1991, USFWS 1992).

Anadromous Fish Conservation Act of 1965 (as amended, 16 USC 757a-757g) Authorizes the Secretaries of the Interior and Commerce to enter into cooperative agreements with the states and other non-federal interests for conservation, development, and enhancement of anadromous fish, including those in the Great Lakes, and to contribute up to 50% as the federal share of carrying out such agreements.

Coastal Zone Management Act of 1972 (as amended, 16 USC 1451–1464) Established a voluntary national program within the Department of Commerce to encourage coastal states to develop and implement coastal zone management plans. Funds were authorized for cost-sharing grants to states

Comprehensive Environmental Response Compensation and Liability Act of 1980 (Superfund) (As amended, 26 USC 4611–4682) Established liability to the U.S. Government for damage to natural resources over which the U.S. has sovereign rights. The Department of Interior is a trustee for natural resources.

Endangered Species Act of 1973 (as amended, 16 USC 1531–1543) Provides for the conservation of threatened and endangered species of fish, wildlife, and plants by U.S. federal action and by encouraging the establishment of state programs.

Estuary Protection Act of 1968 (16 USC 1221–1226) authorizes the Secretary of the Interior, in cooperation with other federal agencies and the states, to study and inventory estuaries of the United States, including land and water of the Great Lakes, and to determine whether such areas should be acquired by the Federal Government for protection.

Federal Aid in Sport Fish Restoration Act of 1950 (Dingell-Johnson and Wallop-Breaux) (As amended, 16 USC 777–777k) Provides funding from a 10% excise tax on sport fishing tackle to states for management of sport fish, (i.e., land acquisition, research, development and management projects).

Federal Power Act of 1920 (as amended, 16 USC 791–828c) Incorporates fish and wildlife concerns in licensing, relicensing, and exemption procedures. (e.g., the Department of Interior is represented on board overseeing the Great Lakes Fishery Trust established in 1995 as mitigation for fish loss due to Ludington Pumped Storage Plant).

Federal Water Pollution Control Act Amendments (33 USC 1251–1365, 1281–1292, 1311–1328, 1342–1345, 1361–1376) In 1987 a "National Estuary Program" was established which provides a mechanism by which the Governor of any State could nominate to the administrator of EPA, an estuary within its boundaries as an estuary of national significance and request a management conference to develop a comprehensive management plan for the estuary.

Fish and Wildlife Act of 1956 (as amended, 16 USC 742a-742j) Directs the Secretary of the Interior to (1) develop measures for "maximum sustainable production of fish"; (2) make economic studies of the industry and recommend measures to insure stability of the domestic fisheries; (3) undertake promotional and information activities to stimulate consumption of fishery products; (4) take any necessary steps to develop, manage, protect, and conserve fishery and wildlife resources, through research, acquisition of refuge lands, and development of existing facilities, and other means. Confirmed U.S Fish and Wildlife Service in the Department of Interior.

Fish and Wildlife Coordination Act of 1934 (as amended, 16 USC 661–667e) Authorizes Secretaries of Agriculture and Commerce to provide assistance to and cooperate with federal and state agencies to protect, rear, stock, and increase the supply of game and fur-bearing animals, as well to study the effects of sewage, trade wastes, and other polluting substances on wildlife. e.g. the USFWS raises 70% of all lake trout stocked into the Great Lakes.

Fishermen's Protective Act of 1954 (as amended, 22 USC 1971–1979) Section 8 (the Pelly Amendment) upon expansion authorizes the President to embargo fish and wildlife products whenever the Secretary of Commerce or Interior certifies that nationals of a foreign country are engaging in trade or taking that diminishes the effectiveness of an international conservation program for the conservation of endangered or threatened species.

Fish and Wildlife Improvement Act of 1978 (16 USC 7421) Authorizes the Secretaries of the Interior and Commerce to establish, conduct, and assist with national training programs for state fish and wildlife law enforcement personnel. Also authorizes funding for research and development of new or improved methods to support fish and wildlife law enforcement.

Great Lakes Fish and Wildlife Restoration Act of 1990 (16 USC 941a-941g) Authorizes 1) a comprehensive study of the status, and the assessment, management, and restoration needs, of the fishery resources of the Great Lakes basin; 2) proposals to implement recommendations resulting from the study; and 3) assistance to the Great Lakes Fishery Commission, States, Native Americans, and other interested entities in encouraging cooperative conservation, restoration, and management of Great Lakes fish and habitat.

Great Lakes Fishery Act of 1956 (16 USC 93–39c) Authorizes any agency of the U.S. Government to cooperate with the U.S. Section of the Great Lakes Fishery Commission in conduct of research programs, and to enter into agreements with the U.S. Section for assisting in carrying out the sea lamprey control program.

Interjurisdictional Fisheries Act of 1986 (16 USC 4101–4107) Authorizes grants by the Secretary of Commerce to states and to interstate compacts in support of management of interjurisdictional fishery resources throughout their range.

Lacey Act Amendments of 1981 (as amended, 16 USC 3371–3378) Provides enforcement authorities and penalties for violation of various wildlife related acts and regulations. Prohibits the transportation of wildlife or their parts or products taken or possessed in violation of federal, state, or foreign laws or regulations.

National Aquaculture Development Act of 1980 (16 USC 2801–2810) Directed the Secretaries of Commerce, Agriculture, and Interior to develop a National Aquaculture Development Plan to identify those aquatic species that could be cultured on a commercial or other basis and to set forth for each species a program of necessary research and development, technical assistance, demonstration, education and training activities. Established a Joint Committee on Aquaculture chaired by Secretary of Agriculture.

National Environmental Policy Act of 1969 (as amended, 42 USC 4323–347) Requires that all federal agencies prepare detailed environmental impact statements on proposed federal legislation or major action, or federally funded state projects. e.g. winter navigation

National Wildlife Refuge System Administration Act of 1966 (as amended, 16 USC 668dd-668ee) Provides for administration and management of areas in the National Wildlife System including wildlife refuges, areas for the protection and conservation of fish and wildlife that are threatened with extinction, wildlife ranges, game ranges, wildlife management areas, and waterfowl production areas.

Non-indigenous Aquatic Nuisance Prevention and Control Act of 1990 (16 USC 4701–4741) Authorizes establishment of a national program to reduce the risk of unintentional introductions of non-indigenous aquatic species and to control, when warranted, aquatic nuisance species that have become established. Also authorized are coordination of non-indigenous species activities in the Great Lakes and a review of policies for addressing intentional introductions.

Sikes Act of 1960 (16 USC 670a-6700) Provides for cooperation by the Departments of the Interior and Defense with state agencies in planning, development and maintenance of fish and wildlife resources on military reservations.

Wild and Scenic Rivers Act of 1968 (as amended, 16 USC 1271–1287) A 1984 amendment, Public Law 98–444, permits installation and operation of facilities to control the lamprey eel (*sic*), as prescribed by the Secretary of Agriculture, for the protection of water quality and the value of the river.

PART 2

Current Issues Facing Fishery Management

CHAPTER 5

Embracing Biodiversity in the Great Lakes Ecosystem

Richard A. Ryder and Judith A. Orendorff

Introduction

Biological diversity within the Great Lakes ecosystem has long been recognized as an ecological attribute worthy of preservation (Agassiz 1850), but the full implication of its value has often been overlooked until quite recently. Biodiversity is now often regarded as an ecosystem attribute to be conserved, sustained, and rationally utilized, according to ecological principles and human mores. From these perspectives, we will take a fleeting look at the massive literature developed around the biodiversity theme, and attempt to equate it to ecosystem matters directly relating to the Great Lakes. In essence then, we may only skim the surface of this vast topic and hope that what is left unsaid, will provide a stimulus in the future for new initiatives of this genre.

In recent years, the increasing emphasis placed on biological diversity, both within the biosphere as well as within the multiplicity of ecosystems, communities and organisms comprising it, has proven to be a confusing, if not intractable concept. At least part of the confusion stems from the fact that biodiversity means different things to different people (Noss 1990). Some authors (Watt 1973) have considered diversity to be a primary ecological resource in the same sense that matter, energy, space, and time, are ecological resources. Biological diversity, therefore, may be conceived as a fundamental ecological resource that should be protected and conserved into perpetuity through the agency of wise use. E. O. Wilson (1992) interprets biodiversity as the variety of life forms on earth. Accordingly, for the Great Lakes basin ecosystem, we might apply the definitions of both Wilson and Watt, and regard biodiversity as a valuable resource, composed of the multiplicity of organisms that occupy the basin proper, including its contiguous catchment. This definition is somewhat arbitrary, as it is dependent upon ecosystem boundaries, but its potential for application is eminently pragmatic.

In essence then, but one prerequisite to our understanding of patterns and processes within the Great Lakes is the detailed knowledge of the diversity of indigenous organisms, communities, and ecosystems naturally occurring there. Accordingly, biodiversity, or the diversity of life, ultimately may provide a metric by which we might measure the sustainability of healthy ecosystems

(Rapport 1995), at least into the foreseeable future. As such, we should apply the appropriate management measures required to ensure the perpetuation of the biodiversity resource.

Biodiversity: A Definition

The term biodiversity, through incessant usage, covering many topics and concepts, comes close to becoming an imprecise, and in certain instances, ambiguous designation. Yet, its fundamental meaning of "the multiplicity of life forms" (Wilson 1992) is so rich in ecological implication, that no synonyms, however brilliantly contrived, might supplant biodiversity in terms of its directness, candidness, and even useful ambiguity. Inherent within "the diversity of life forms" is the implied meaning of the complexity of the milieu required to support this diversity, not to mention their sustainability, their productivity, and other key ecological attributes. Suffice it to say, that in consideration of the richness of information embodied in the term biodiversity, its use should be limited to carefully considered, and proven ecological concepts.

The Global Status and Value of Biodiversity

E. O. Wilson (1988) has emphasized that biological diversity is a global resource, and as such it must be indexed, used, and preserved. There is some urgency in this matter because of exploding human populations, lax environmental standards, and the accelerated expansion of culturally-altered environments. Furthermore, science has discovered many innovative uses for biodiversity that might alleviate human suffering, or relieve the current levels of environmental destruction. Among such substances from the past are the antimalarial properties of the bark of the chincona tree (quinine), and the cardiotonic preparations from the foxglove plant (digitalis), which have been in use since medieval times (Eisner 1991). More importantly, perhaps, is the continual discoveries of new miracle drugs, which have the propensity for curing many seemingly incurable diseases in the future. In addition to the possible loss of these potentially valuable drugs even before their discovery, much biodiversity has already been lost. This has been due mainly to the continuing and sequential degradation and encroachment of habitat, or overexploitation of the resources therein. Other values of biodiversity may be more subtle, or at least, not obviously evident. For example, McNeely et al. (1990) have classified the values of biological diversity using the principle of Watt (1973), namely, that biodiversity is a resource. They have broken down the categories into direct values to humankind, as opposed to indirect values. The direct values are defined as those concerned with the enjoyment or satisfaction received directly by consumers of the resource. These included consumptive uses, such as the nonmarket value of firewood or game, for example; and productive use, such as the commercial value of timber and fish.

Indirect values included nonconsumptive uses, such as scientific research and bird watching; option value, namely, the value of maintaining options available for the future; and existence value, or the value of ethical feelings of existence of wildlife. The global status of biodiversity also occurs in microcosm within the Laurentian Great Lakes, albeit at varying scales and rates. As a generalization, with many notable exceptions at local scales, a cline in biodiversity in terms of species and phenotypic stocks (especially salmonids) now exists from Lake Superior and its relatively pristine headwater catchment, to Lake Ontario at the downstream terminus, generally conceded by many environmentalists to be the most degraded lake of the five major basins. The distinction between

TABLE 1

Five major categories of biodiversity and some specific types of biodiversity within each category

Note that the first four categories are spatially differentiated while the last category is temporally determined.

TAXONOMIC	Species, stocks, communities
DEMOGRAPHIC	Cohorts, ages, year-classes, size-classes
ECOLOGICAL–PHYSIOLOGICAL	Guilds, niches, fecundities, behaviors, strategies (r or K)
GENETIC	Genotypic, phenotypic
TEMPORAL	Evolutionary (long), ontogenetic (short)

phenotypes and genotypes in salmonids may be moot because of their inherent polyploidy, but we will make this distinction in our many references to phenotypic stocks or stock flocks, in order to emphasize the following critical concept to our understanding of biodiversity. That is, we regard phenotypic stocks as those that retain distinctive morphological features, behaviors, or growth rates, and spawn at discrete spawning sites and at different times of the year, thereby retaining reproductive isolation from one another. In each of these attributes, and wherever possible, we equate phenotypic stocks to species for the assessment of biodiversity.

Biodiversity as a Descriptor of Fish Assemblages

Biodiversity within fish assemblages may be classified in a multiplicity of ways. The most fundamental (in a conventional Darwinian sense) is species and stock diversity or richness, a measure of heterogeneity (table 1), or more simply, the number of different fish species (or stocks) present within an assemblage or community, regardless of their individual abundance. A further step that might be regarded as ecological egalitarianism is the determination of relative abundance of the different species comprising an assemblage or community, that is, an evenness index (Pielou 1966, 1975; Perkins 1983). Evenness addresses the abundance of an organism within a community relative to other species, and accordingly, the potential impact or ecological role it might play in a community. For example, abundant terminal predators play important roles in retaining prey fish populations in check, or maintaining ecological dynamic equilibrium (Ryder and Kerr 1990). Overabundant prey species may exceed the limitation of the trophic opportunity that exists and change an efficiently functioning harmonic community into an unbalanced astatic one. The management role in this instance would be to regulate the fishery in such a manner as to harvest both the predator and the prey in ecologically apposite numbers. Because of the dynamic nature of predator-prey systems, this objective may be difficult to achieve. Alternatively, close attention could be paid to the harvest levels of predator species, to ensure that they are not fished down to a size that would preclude their ecological role as terminal predator. Consequently, both diversity and evenness are important factors in assessing the biodiversity of any community.

Other types of biodiversity categories exist, and may be conveniently classified as taxonomic, demographic, ecological-physiological, or genetic (table 1). For example, behavioral biodiversity (such as homing of spawners to a natal site) from an ecological point of view, may be equally important in the ecosystemic scheme of things as species diversity per se (Keenleyside 1979). Both size-class diversity and body size spectra are species independent, and mirror trophic processes in aquatic ecosystems. These have a potential for deriving practical criteria for community stability and integrity (Kerr 1974a).

Implicitly, each category contains within, a specific measurement of biodiversity that is spatially distributed, in the present instance, among the various fish assemblages and communities of the Laurentian Great Lakes. The heterogeneity and evenness attributes of the organisms from any two of the Great Lakes within the same system may differ substantially, and this difference, in some instances, would be masked if these two attributes were averaged for the whole system. Each measure of biodiversity represents an observation at a particular time, and in a particular location, and is subject to change over various ecological time scales, such as the seasons and the years. Hence, each element of biodiversity that is spatially distributed, may be used in a comparative sense to indicate whether or not one lake is ecologically healthier (Rapport 1995) than another, using numerical indices of heterogeneity and evenness as a principal means of comparison.

However, the real strength of biodiversity measurements come into play when their relative values are projected over time to indicate whether applied management measures have been effective or not. Similarly, hindcasts of biodiversity measurements may be usefully made in order to calibrate predictive diversity models.

Another category of biodiversity measurements include those that are inherently temporal, such as evolutionary (long-term), and ontogenetic (short-term) measurements of time. The notion of biodiversity occurring within a single organism over time (ontogenetic diversity), supports the thesis that *Natura vacuum abhorret*, and complements the occasionally sparse spatial elements of biodiversity, where these fall short of ecological efficiency. For example, the fry and fingerlings of most piscivorous fish are planktivorous, and thereby fill a void in systems lacking the usual prey species particularly adapted to a lifetime of planktivory. In monocultures, these juvenile stages fall prey to the adults of the same species, and this, in turn, adds a measure of stability to depauperate fish assemblages.

We hasten to mention that the categories of biodiversity measurements for fishes described herein are both arbitrary, and nonexhaustive, and are presented in order to discourage the use of biodiversity exclusively, only in its narrowest interpretation, that is, as the diversity of species. We have, in a sense, for the sake of orientation and operational practicality, forced our various categories of biodiversity into a procrustean bed. Despite this, and for the sake of uniformity throughout the remainder of this document, we will focus mainly on species and stock diversity. Many of the concepts and principles invoked herein will, implicitly, apply equally well to the many other types of biotic diversity listed in table 1.

Biodiversity and the Hutchinsonian Niche

The Hutchinsonian niche may be loosely defined as a multidimensional envelope surrounding an organism, in which each dimension of the envelope represents one environmental zone of tolerance for that particular organism (Hutchinson 1957). Fry (1947), in a remarkable anticipation of the Hutchinsonian niche concept, provided both a methodology and a metric whereby the niche dimensions of any fish species might practically be measured (Kerr and Ryder 1977). Intuitively, these methods may be applied to the concept of biodiversity as being dependent upon the integration of specific dimensions for the various scopes for activity of a particular species within the Hutchinsonian niche envelope.

This Hutchinsonian niche application has been further enhanced for the Great Lakes in a demonstration of the need for the preservation of genetic diversity (Berst and Simon 1981). It was shown therein, that genotype could be regarded as a surrogate for fundamental or potential niche, while phenotype was representative of realized, or operational niche (Ryder et al. 1981). Accordingly, preservation of a variety of genetic materials is essential, *inter alia*, in order to permit efficient use of the various resources of the environment, and thereby confer survival advantage onto an organism, stock, or community.

Cultural Stresses on Biodiversity

Reduction in levels of biodiversity may occur due to natural events, such as the massive eruption in Krakatau in 1883, followed by an enormous tsunami that traveled most of the way around the world (Wilson 1992). More subtly, losses of biodiversity may occur through the agency of emigration and extinction processes (MacArthur and Wilson 1967; MacArthur 1972; Magnuson 1976). These processes may reach a state of dynamic equilibrium over time. The anthropogenic effects on aquatic ecosystems as exemplified in the Great Lakes, however, loom large when compared to any natural effects. For centuries past, these have been limited to events both periodic and episodic, and of relatively moderate amplitudes by global standards, such as twenty-five or one hundred year storms. However, the stresses of millennial time scales, which only recently have seemed important, include ice ages, which human-generated disturbances do not yet rival in terms of level of effect. The principal cultural intrusions which either directly or indirectly affect biodiversity levels include: selective exploitation of aquatic resources; exposure of aquatic resources to a variety of xenobiotic and other toxic contaminants; excessive nutrient loadings; and introductions or invasions of non-indigenous fishes, invertebrates, and plants (Loftus and Regier 1972; Mills et al. 1993). One other major intrusion of humankind involves either the erection of, or destruction of, physical structures that tend to significantly alter habitat. These include the construction of dams, berms, groins, locks, or various modifications of the shoreline or bottom structures, and deforestation. Tributary streams to the Great Lakes (Hynes 1970) are also modified through channelization and other physical alterations that affect the normal patterns of water movement. Each of these physical environmental incursions have a tendency to simplify diverse habitats in such a way that they no longer serve as centers of organization (Regier et al. 1980), but rather become uniformly austere, and essentially inhospitable to the existence of fish life. Regardless of the nature of the physical incursions, under an increasing stress regime, environments may eventually be expected to reach similar states of equifinality, through parallel reductions in complexity (von Bertalanffy 1968).

The Great Lakes Basin Ecosystem

The study of biodiversity, or the spectrum of diversity of the various life forms within the Great Lakes basin ecosystem, has formed the basis for some of the earliest observations of fishes within the vast expanse of the five lake basins and their contiguous waters (Agassiz 1850). Early organized studies of biodiversity on the Great Lakes first emanated from universities (Bocking 1990) or national institutions such as the Smithsonian (Jordan and Evermann 1896), and later, from federal governments. Virtually all of these studies implicitly addressed biodiversity as they usually bore a strong systematic component (Koelz 1921; Dymond 1922). Later studies were generally initiated by state or provincial agencies (Nash 1908), and were largely restricted to the limits of their jurisdictional boundaries. These studies, for the most part, focused on the demographics of fish assemblages as a practical requirement for regulation setting, with the emphasis placed on the establishment of meaningful regulations for the perpetuation of viable fisheries (Hile et al. 1951). During this period, methodologies developed on small inland lakes (Fry 1949), where experimental design was easier to implement, were exported and applied to the more complicated Great Lakes situation. Despite the general emphasis on demographic attributes of fish assemblages, implicitly, the measurement of biodiversity provided both a fundamental and critical dimension to the studies. Over time, the measurement of biodiversity became explicit, that is, not only was it an essential component of the demographic studies, but also, natural biodiversity had intrinsic values of its own, necessitating policies and practices not previously recognized as essential to the perpetuation of a healthy fishery. Specifically, we refer to the policies of the 1960s, which acknowledged the need for the protection of indigenous species. Protection from the unintentional introduction of exotic species was also recognized as desirable at that time, but planned introductions were generally acknowledged as being an acceptable practice if they formed part of a management scheme approved through a process of biological consensus. While consensus was eagerly sought, unanimity was but rarely attained amongst the many state and provincial agencies, and exotic species introductions continued, albeit often under strong protest.

Before elaborating on the topic of biodiversity in the Great Lakes per se, it becomes necessary to place certain historical events in perspective, inasmuch as they may have influenced the biodiversity patterns of today. Accordingly, we will first present a brief descriptive overview of the previously pristine environments and resources of the Laurentian Great Lakes, dealing especially with fish assemblages and cultural stresses that have occurred over the last two hundred years or so. For further details relating to historical events, refer to the documented evidence which has been cited.

The Pristine Environment

The pristine environments of the five Laurentian Great Lakes and their forested catchments are often treated as a single system, although practically, any effective management in the past, has usually been achieved at the individual lake level. While it is appropriate from an ecosystem point of view, to regard the Great Lakes as but a single system, a more pragmatic approach would treat them as five separate subsystems, albeit contiguous. Obviously, the outflow from each lake forms the influent to each lake or river immediately downstream, and therefore, these effects of downstream loadings must be taken into account during any ecosystem studies.

The pre-European status of the Great Lakes and their forested catchments probably approached our concept of near-pristine conditions, and typically, had clearer and colder waters, with substantially less nutrient content than currently exists (Beeton 1965; Smith 1972; Sly 1990). The catchments for each of the five basins were generally well forested, but the areas of the individual catchments were small, relative to the large surface areas of the lakes. The three upper Great Lakes (Superior, Michigan, and Huron), with substantial drainage from the Precambrian Shield, were typically oligotrophic, with low phosphorous levels (Beeton 1965; Dobson 1981), while Lake Ontario had substantially higher nutrient levels, because it derived most of its water from the fertile Lake Erie. Accordingly, Lake Ontario was considered to be morphometrically oligotrophic because of its relatively great depths, which were believed to overpower the edaphic effects of a rich nutrient supply (Rawson 1952). Lake Erie, because of its geological setting off the Precambrian Shield, also had relatively high nutrient levels when compared to the three Upper Lakes, and was substantially shallower than any of the other Great Lakes, another factor contributing to its high productivity. In its near primal condition, the western basin of Lake Erie was, undoubtedly, mesotrophic, with a clinal gradation through a marginally mesotrophic, and sometimes anoxic central basin (Carr 1962), terminating at the eastern basin, which was probably best described as oligotrophic. However, when the lakes are categorized in terms of nutrient concentration, there exists a generally strong trophic cline, increasing in a downstream direction from the headwaters of Lake Superior to the outflow of Lake Ontario (Beeton 1965).

Primal Fish Assemblages

The primal fish assemblages and communities occurring in the Great Lakes circa 1800 (Loftus and Regier 1972) varied from lake to lake because of their dependency on the habitat zonation in the individual lakes.

In the three upper Great Lakes, salmonine and coregonine "stock flocks" predominated in the pelagic and benthipelagic zones, while fish species belonging to most other taxa, with few exceptions, were more often found in the near shore littoral zones (Berst and Spangler 1973; Lawrie and Rahrer 1973; Wells and McLain 1973). This ordination of species was also true for Lake Ontario (Christie 1973) which, in addition to the aforementioned taxa, was also home to an indigenous salmonid, the Atlantic salmon (*Salmo salar*). The eastern basin of Lake Erie, despite its relatively high nutrient content, was otherwise more closely similar to the upper Great Lakes and Lake Ontario in many respects (Hartman 1973). However, it contained substantial numbers of coolwater species (preferring mesotrophic conditions), probably because of its contiguity with the central basin of Lake Erie. The latter basin was formerly dominated by the endemic blue pike (*Stizostedion vitreum glaucum*) considered to have been a subspecies of walleye (*S. vitreum vitreum*) according to Hubbs and Lagler 1970. The blue pike, which is now extinct, at one time constituted one of the largest fish stocks of commercial value in all of the Great Lakes (Trautman 1957; Parsons 1967; Scott and Crossman 1973; Baldwin et al. 1979).

As a generalization, the outer limnetic waters of the upper Great Lakes, Lake Ontario, and the eastern basin of Lake Erie may be classified as suitable habitat offering the appropriate trophic opportunities for cold-water (oligotrophic) species, such as the chars, whitefishes and especially the cisco complex. On the other hand, the many embayments, such as Black Bay (Lake Superior), Green Bay (Lake Michigan), Saginaw Bay (Lake Huron), Bay of Quinte (Lake Ontario), and the western basin, and to a lesser extent, the central basin of Lake Erie, may more appropriately, be designated as habitats for cool-water species of fish such as the yellow perch (*Perca flavescens*), the walleye, the northern pike (*Esox lucius*), and the white sucker (*Catostomus commersoni*).

Near shore zones, small embayments, and river deltas were usually inhabited by warm-water fish species characterized by the centrarchids (sunfishes), the cyprinids (minnows), and the ictalurids (catfishes).

The following descriptions and generalized classification herein, of species and their preferred habitats in pristine times, have been inferred from data obtained from the commercial fisheries in the late 1800s and early 1900s (Baldwin et al. 1979), other anecdotal evidence (Sly 1990), and the exhaustive works of Hubbs and Lagler (1970) and Scott and Crossman (1973).

Biodiversity in Great Lakes Fisheries—Stresses and Changes

Given that the current basins of the Great Lakes were covered with continental glaciers about twelve-thousand to eight-thousand years ago (Flint 1957; Crossman and McAllister 1986), the initial biodiversity of the reinvading fishes was limited, to a large extent, by the existing biodiversity at their source, that is, the glacial refugia from which they originated. Accordingly, the habitat in each refugium had to be ecologically amenable to its occupants, and they, of necessity, had to be coadaptive if mutual survival were to occur. Additionally, contiguous water routes had to be available in order to permit refugia biota access to the newly formed Great Lakes basins.

Reinvasion of the Great Lakes, which initially provided very different habitats than are present today, required species that could adapt to rather austere environmental conditions (Ryder et al. 1981), such as extreme cold, winter ice cover, low nutrient levels, moderate to high levels of glacial flour suspended in the water column, and a dearth of appropriate foods. Accordingly, early invaders had to have adaptive survival mechanisms such as phenotypic plasticity in order to exist in this harsh and austere environment. Such plasticity may also confer a degree of capability for ecological character displacement (Grant 1994). In the sense that the ontogeny of a species recapitulates the phylogeny of its forebears, the history of the reinvasion of the Great Lakes by various fish stocks emanating from glacial refugia might, to a certain degree, mimic the present-day spatial occurrence of fish stocks present within the five Great Lakes. One might ponder as well, the nature of the constituents of the native biota present in each of these lakes, and attempt to determine what particular characteristic permitted certain phenotypic stocks or species to be among the avant-garde of the reinvaders. The bulk of the first reinvading fishes into Lake Superior were probably composed of only two "stock flocks," that is, two single species that had differentiated into a multiplicity of phenotypes, each adapted to a particular set of environmental conditions (Koelz 1929; Eschmeyer 1957). The two stock flocks comprise two of the existing subfamilies of salmonids, namely, the native salmonines, and coregonines (formerly a single thymalline existed, but is now extirpated). These stocks were admirably adapted to the sparse and somewhat austere environment presented by the open, limnetic waters of the Great lakes, as well as the cold, bathypelagic depths. Obviously, to survive in such an austere environment, where physiological stresses were extreme, and food resources sparse, rapid adaptation was desirable in order to withstand competitive stresses. That is, it is unlikely that sufficient time was available for genetic adaptation to these new and fluctuating milieux, the existence of which is measured in increments of thousands of years, rather than the millions usually required for adaptive evolution. Accordingly, the salmonids were doubly fortified against the capriciousness of the environment, having phenotypic response to short-term variability, with genotypic adaptation to geologically-timed events. Rapid adaptation to differing, and ever-changing habitats within the Great Lakes basin was effectively achieved through the phenotypic adaptation of all species having this inherent capability (fig. 1). This resulted in the formation of the two principal "stock flocks," which, in a relatively short time segregated into separate stocks of many morphs and behaviors, that functioned

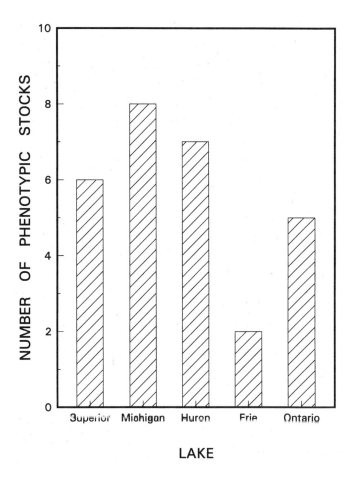

FIG. 1. *A "stock flock" likely derived from a prototypical lake herring* (Coregonus artedi)*, was composed of eight species in the Great Lakes proper (Koelz 1929). The phenotypic plasticity inherent in these variable stocks probably provided an adaptive mechanism for survival during their invasion of the cold, austere environments that existed following glacial recession. Lake Erie, which is relatively warm during the open water season, and is also the shallowest of the Great Lakes, provided the least favorable conditions for this suite of cold-water fish.*

ecologically, as if they were several different species. From a genetic point of view, many of these "species" within a stock flock may prove to be genetically homogeneous (Ryder et al. 1981); however, the possession of a broad genetic base permitted rapid phenotypic differentiation of stocks in different habitats, particularly at early ages during imprinting.

There may be some exceptions, however, and this aspect of Great Lakes fisheries requires further study. Foremost, among any possible exceptions may be the blue pike, which appeared to possess a particular adaptation to the rather unique environment provided by the central basin of Lake Erie, where it achieved its greatest abundance. It also occurred in lesser numbers in the east and west basins of lake Erie, as well as in the east end of Lake Ontario, where it was not known to spawn (Trautman 1957; Scott and Crossman 1973). The blue pike appeared to be particularly adapted to feed in the limnetic waters of the central basin of Lake Erie. There, its large eyes (proportionately larger than the walleye), gray-blue dorsum and flanks, and its immaculate white underbelly and pelvic fins provided it with apposite camouflage for such conditions.

Our conclusion of maximum likelihood derived from these conjectures is that taxon diversity applies not only at the species level, but rather extends to phenotypic stocks as well. Stock flocks within the Great Lakes contain suites of phenotypically differentiated forms that from an ecological perspective, should probably be assigned equal weight with species in any biodiversity assessment.

Native Peoples' Subsistence Fisheries

It may be inferred from the primitive nature of many subsistence fisheries that exist today in northern Canada, including parts of the Great Lakes, that in pre-European times, native peoples had probably reached a degree of dynamic equilibrium with available fish stocks, and that much of the harvest was opportunistic, or unintrusive at best. Traditional passive gears such as wicker-type basket traps, and other primitive entrapment gears, concentration devices such as weirs, or active projectiles such as spears, were sufficiently ineffective that substantial levels of escapement were virtually guaranteed. Additionally, fishing effort was distributed over a wide-range of species, making the likelihood of overfishing much less likely, and in some instances, spawning areas were even set aside as exempt from exploitation, because they had been designated as sensitive religious areas (Russel Barsch, University of Alberta, personal communication).

Opportunistic fishing took place most frequently during spawning runs, when fish were concentrated and preoccupied with prespawning behavior, thereby permitting relatively easy capture. These fisheries, for the most part, were concentrated towards adfluvial spawning runs, or otherwise in protected embayments. Intensive fishing in the open lake or in the bathypelagic depths probably occurred but rarely, as even the largest birch-bark canoes were ill equipped for fishing the open, wind swept waters of the Great Lakes. Large catches were often dried or smoked for preservation and storage in order to provide sustenance for leaner times. In the case of the Algonquian peoples (Ojibwa, Potawatomi) whose various bands inhabited a substantial portion of the shoreline of the Great Lakes, fishing was a seasonal occupation, and most of their fishing effort was also directed at spring or fall spawning runs. Winters were usually devoted to trapping and hunting (Honigmann 1962; Bishop 1974), and the spoils from these constituted virtually all of their winter food (Rogers and Black 1976). During the moderately short summers, most Algonquian peoples congregated at summer gathering places, usually on a waterfront, where fish provided a substantial part of the food. During these periods, it was traditional for the women to fish, and in general, provide the summer sustenance on a daily basis.

Overall, the prevailing evidence would support the notion that the previously sparse distribution of aboriginal peoples (particularly those belonging to the Algonquian linguistic group), combined with their nomadic fishing, hunting and gathering lifestyles (Jenness 1958), were not conducive to the overharvest of native biota in much of the Great Lakes basin and its tributaries. This observation may, or may not apply to the Iroquoian peoples who inhabited the more southerly portion of the Great Lakes. These aboriginal peoples derived most of their food staples from agriculture, and because of this, had a relatively sedentary nature, which permitted only short fishing and hunting excursions (Jenness 1958). Therefore, fishing effort, which was concentrated in only a few locations adjacent to their villages, may have created opportunities for some un-documented local incidences of overexploitation of specific stocks, particularly on adfluvial spawning runs in small streams. On the other hand, this supposition may be countered with the fact that agricultural peoples normally store excess harvests to utilize in lean times, and the need to rely on fishing as a source of their staple diet may not have existed. Accordingly, any assumed overfishing that might have occurred by native peoples is purely conjectural, and whatever their supposed negative effect on fish stocks more than two-hundred years ago might have been, they likely pale into insignificance when compared with the commercial levels of exploitation currently experienced on the Great Lakes (Baldwin et al. 1979). It would seem that the development of more efficient gears and other new technologies, plus changing demographics, have been at the root of current levels of overexploitation. For example, the walleye stocks of the Thames River, a tributary of Lake St. Clair, in particular, may be at risk from excessive harvests by native peoples during the spawning season. Spearing is often highly selective of the largest, and most experienced, and therefore, the most valuable brood stock, and is often disruptive of previously deposited eggs because of disturbances to the substrate. Spotlight-assisted spearing, coupled with any accompanying disturbance of the bottom materials, may also be detrimental to the spawning activities of many species, but particularly the light sensitive walleyes.

Currently, native peoples' fisheries using modern gears, and assuming that reasonable regulations are applied and enforced, should not have greater proportional effect on the fish stocks than does the rest of the fishery. However, the use of spears on spawning grounds (a native prerogative in some jurisdictions), may have had a particularly onerous effect on shallow spawning fish. Some streams in southern Ontario may be particularly vulnerable to overfishing because they are adjacent to high population densities in Indian reservations or local communities. Overall, native peoples' fisheries were directed at the most abundant species in the Great Lakes, which would tend to mask any deleterious practices that may have occurred. The heterogeneous component of biodiversity, under these circumstances, would be little affected, and the evenness component may have been slightly affected. Where noticeable effects did occur, they would almost always be intermittent and on a local scale.

The Effect of Modern Fishing Gears on Biodiversity

The relative effectiveness of fishing gears during the harvest of fish have been discussed by numerous authors (Hamley 1975; Gulland 1983), and need not be repeated here. Modern fishing gears, used primarily for commercial exploitation, have provided a seemingly insolvable enigma to the fisheries managers of the day (see Ryder and Scott 1994 for elaboration). While the use of most gears is subjected to rigorous regulation, the creation of regulations per se, often assumes that a complete demographic knowledge of all of the species comprising a fishery is known, as well as the fish communities' responses to other stresses, and interaction with other species and environments. Such an assumption, unfortunately, is rarely justified in Great Lakes fisheries, where obtaining this kind of information for all major fisheries would be both costly and logistically impractical.

Almost always, in new fisheries, the quota restrictions are based on the behavior of the fishery during the fishing-up phase (Regier and Loftus 1972), rather than on any real understanding of the projected demographics likely to occur once a substantial fishing effort has been established. Accordingly, most new fisheries tend to be overexploited by fishing up the capital (broodstock), but the overfishing is not readily apparent at first. Reduced quotas come into effect when this fact becomes known to the manager, usually only after the evenness attribute of biodiversity has been severely disrupted. Often, the evenness index is radically distorted by a fishery through directed effort towards a single valuable species, rather than the attempt to retain ecologically appropriate production/biomass ratios within the fish community as advocated by Ryder and Henderson (1975).

We emphasize that inordinately high levels of exploitation using efficient fishing gears is usually due either to outright mismanagement, or through a lack of understanding of the fishing-up process. The consequent density-dependent, growth compensation following fishing up tends to mask its effects. This conclusion derives from the fact that the gears, of themselves, are not inherently destructive when wisely utilized. Yet, certain gears do have an inherent potential to be more exploitive or damaging than others. Most impounding gear, however, under favorable conditions, provide an opportunity to dispose of living, rather than dead bycatches (Gulland 1983). Entangling gears such as gill nets (Hamley 1975), may be extremely destructive to bycatches, as virtually all of the captured fish are eventually killed, if not in the net, then through delayed mortalities. These may result from the inconspicuous injuries of rough handling during the removal of a fish from a net prior to its release, or through the inadvertent removal of some of the protective mucous covering of the epidermis. Accordingly, fish released after gillnet capture may be extremely vulnerable to viral, bacterial, and fungal infections.

Another negative effect from fishing gillnets is the potential provided to vary the mesh size efficiently in order to take advantage of a particular sized year-class of a valuable species (Hamley 1975). One mesh size may be effectively used in this fashion until most of a year-class of a valuable species is harvested, at which time a reduced mesh size may be used, and effort directed toward another year-class of smaller-sized fish. Using these tactics, individual size classes of some stocks may be fished out over time. This type of fishery was at least partially responsible for the demise of some of the lake trout stocks of Lake Superior in the 1950s and 1960s, which were sequentially fished-up, one stock after another (Lawrie 1978). In general, the more efficient the gear, the more rigorous should be the regulations limiting its use.

Impounding gears such as pound nets and trap nets, are generally more benign, and therefore, less selective of a single species or size-class, although this is not always the case. Impounding gears usually provide an opportunity to release alive, designated sports species, or undersized commercial species, although survival rates may be low on fragile species with deciduous scales, such as rainbow smelt (*Osmerus mordax*) or alewives (*Alosa pseudoharengus*). On the other hand, the speculated cause of the demise of an extremely valuable whitefish (*Coregonus clupeaformis*) fishery in Lakes Huron and Michigan in the 1930s and 1940s, was attributed (at least in part), to the fishing of deep trap nets (Van Oosten et al. 1946).

Angling Effects on Biodiversity

As recently as fifty years ago, it could be said with impunity, that the extent of angling on the Great Lakes, except perhaps for some local exceptions, had little or no marked effect on the biodiversity of fish communities. However, a burgeoning population of anglers now fishing in all of the Great Lakes has resulted in a fishery conservatively valued in excess of one billion dollars, and perhaps, the source of several billions of dollars of economic activity (Talhelm 1988). Some of the technologies now used for sport fishing in the Great Lakes, plus some that have been adapted from marine sport fisheries, make anglers more efficient, at first locating, and then selecting for individual species or stocks. Among these gears are included downriggers and outriggers, planer boards, sidescan sonar, and a variety of modern devices and space-age lures.

The increase in the number of sport fishing vessels adapted to Great Lakes angling in the 1970s and 1980s has been only somewhat short of phenomenal, although more recently it has been waning. This, coupled with a plethora of fishing derbies occurring in each of the five Great Lakes, makes sport fishing an item of sufficient magnitude to require the continued concern of managers. The recognition of the possibility for overexploitation, coupled with the economic benefits perceived to be either circulated, or derived from sport fisheries, have caused some of the management agencies on the Great Lakes (e.g., the State of Michigan) to permanently close large segments of their commercial fisheries.

Sport fisheries are usually directed towards the capture of terminal predators, which are often keystone species in the sense of Paine (1966). The effects emanating from an improperly managed sport fishery could contribute to the near extirpation of one of these native keystone organisms, such as the lake trout or walleye, and may ultimately cause a decomposition of naturally-integrated fish assemblages within the Great Lakes (Ryder and Kerr 1990). We expect that in the future, there will be tighter controls on sport fishing yields, sizes and seasons, coupled with a redirection of effort towards other species not currently sought by anglers. This redirection of effort, if successful, would tend to maintain the evenness aspect of biodiversity. However, changing an angler's preference from heavily exploited walleyes to the relatively underexploited burbot (*Lota lota*), for example, is easier said than done, and marketing burbot for sport or food may be a very difficult proposition indeed. Apparently, before such a transformation may be made, there is, of necessity, an appropriate interval of ecological marketing necessary. Many similar initiatives to utilize a variety of edible biomass from the Great Lakes have failed in the past, due to traditional prejudice against certain unglamorous species.

Other Stress Effects on Biodiversity

Besides exploitation, the fish assemblages of the Laurentian Great Lakes are subjected to the debilitating effects of at least four other major stress categories. These include cultural eutrophication, toxic contaminant loadings, physical alteration of the land-water interface or catchment, and invasion or introduction of non-indigenous species. Any two or more of these stresses may act in concert, and result in devastating synergistic effects on native fish assemblages, thereby making effective management intractable. The ultimate consequences of each of these exogenous stresses will affect negatively, the heterogeneity and evenness components of biodiversity.

Contaminants

Most fishes occupy the lower reaches, or termini of environments, such as lakes and oceans, and often, are also terminal predators of the food web. Hence, they are doubly vulnerable to all of the human insults, not only to the aquatic compartments of ecosystems, but also to their contiguous terrestrial and atmospheric compartments as well (Ryder and Scott 1994). Because of this, fishes are usually the last direct recipients in the cycling of toxic waste material and accordingly, are often highly vulnerable to the effects of both biomagnification and bioaccumulation. Toxic waste enters fish in various ways, including directly through the gills, or indirectly through food intake. The effects of assimilated toxic materials vary, and often include a reduction in reproduction rates (Burdick et al. 1964, 1972; Willford et al. 1981), and many other sublethal behavioral effects bridging a cline leading to outright mortality. Whatever effect there may be, there is invariably an accompanying loss of biodiversity at one or more levels (table 1), but particularly of those items associated with reproduction or other physiological processes.

Toxic contaminants, combined with other environmental impacts, have the potential to cause the loss of fish species, or in extreme instances, whole communities. Some contaminants are being reevaluated on the basis of new knowledge, and have been found to have damaging effects, far beyond what was previously thought (Birnbaum 1993). Despite the dedication of hundreds of highly-skilled personnel who have devoted thousands of hours, and spent millions of dollars towards the solution of the toxic waste problem in the Great Lakes, new dimensions of the problem continue to emerge (Anonymous 1994).

Cultural Eutrophication

As with virtually any disruption of the natural system, the fish community suffers in terms of both the quantitative and qualitative aspects of biodiversity. Cold-water species, such as lake trout, produce their largest yields in oligotrophic systems. Any departure from the optimum environment may contribute to ecological instability and ultimately a loss of revenues in the case of a commercial fishery, or loss of quality in a sport fishery.

Massive loadings of phosphorus from municipal wastes, agriculture, and other specific and nonpoint sources on the Great Lakes have had dire effects on the heavily populated Lakes Erie and Ontario, as well as southern Lake Michigan, but have been generally less harmful to the other regions of the Great Lakes, except perhaps, for local effects near cities, river mouths or embayments. To a large degree, rehabilitative effects have been at least partially successful due to upgrading of municipal waste treatment facilities and control of detergent use (Vallentyne 1974). Cultural eutrophication may seem outwardly benign compared to toxic waste inputs, yet too much phosphorus pumped into a water body can drastically change its productive nature, not to mention its fundamental limnology. As an oligotrophic body of water receives excessive phosphorus loadings, it may change over time into a mesotrophic condition. Opportunistic cool-water species, which are better adapted to mesotrophy may invade from the shallow bays and proliferate in the open waters of the formerly oligotrophic lake. Meanwhile, the original cold-water, oligotrophic species, may become progressively scarce over time, and constitute but a vestigial remnant of the new assemblage, or else are completely extirpated. Accordingly, biodiversity may be severely compromised.

In this instance, the heterogeneity aspect of biodiversity may increase initially, as species preferring oligotrophic and mesotrophic environments are aggregated, and may even seem to coexist for an indeterminate period. But eventually, the ecological fulcrum upon which they were balanced, will permit a swing irrevocably to one side or the other, favoring the species most benefited by the direction of the environmental change (Svärdson 1976). Accordingly, while cultural eutrophication may have the capability of extirpating species and stocks, its most likely long-term effect will be to distort the evenness indices of biodiversity (Pielou 1966, Perkins 1983).

Structural Modifications of the Environment

In the Great Lakes, structural modifications to the near shore environments have been few to moderate in number, except in the vicinities of the major cities and metropolitan areas, such as Toronto, Chicago, Milwaukee, Detroit, Cleveland, and Buffalo. The greatest single effect of these structural modifications is probably the amount of feeding or spawning near shore habitat that is altered or displaced. In terms of the total shoreline of the Great Lakes, the effects may be significant, but perhaps exaggerated because of its concentration in urban areas, where people are more prone to view, or in other ways to experience the environmental consequences.

Historically, the single largest structural modification on the Great Lakes has almost certainly been the removal of the forests from the catchments, which is substantial for the lower lakes, but has had relatively less impact, to date, in the three upper Great Lakes. Forest removal affects the temperature, turbidity, and sedimentation rates in tributary streams directly, and one or more of these environmental changes may have been the principal cause in the loss of the Atlantic salmon from Lake Ontario (McCrimmon 1950).

Conversely, the increase in temperature, turbidity, and sedimentation rates in these streams, could have improved their suitability for the spawning, and subsequent rearing of sea lamprey ammocoetes (*Petromyzon marinus*), which may have been contemporary marine invaders according to S. H. Smith, (retired) U.S. Fisheries and Wildlife Service. While sea lampreys require clean gravel shoals on which to spawn, adjacent patches of flocculent sediments are necessary for the feeding, growth, and general well being of the ammocoetes. Beyond the cutting of the forests and modification of the shoreline due to human settlement and other cultural activities, many modifications of the physical environment cannot yet be proven to be a current major threat to the maintenance of historical levels of biodiversity in fishes of the Great Lakes. However, continual shoreline modification and deforestation, as well as progressive loss of wetlands, coupled with expanding cities and agricultural and industrial inputs auger ill for the future. Other potential effects on biodiversity from the alteration of the physical structure of the environment include, but are not restricted to: channelization of lower stream reaches, destruction of dunes, construction of mill dams and locks, water diversions from outside the Great Lakes catchment, and various other physical intrusions on the land-water interface of the Great Lakes coastal zones. Intuitively, the effects of each of these is potentially harmful to the maintenance of the natural levels of heterogeneity and evenness. However, the environmental jury is out on the relative ecosystemic incursion effected by each of these land-water abuses when compared with massive inputs of toxic wastes or extreme levels of overexploitation.

Introduction and Invasion of Exotic Species

The introduction and fate of exotic species in the Great Lakes have been discussed exhaustively (Emery 1985; Mills et al. 1993). Their effects range from the seemingly benign influence of the rainbow trout (*Oncorhynchus mykiss*), which is often said to be naturalized (Ryder and Kerr 1984), to the rapacious depredations of the sea lamprey, which undoubtedly has created a reduction in biodiversity in the upper Great Lakes (Pycha 1980) proportionate to that of the Nile perch (*Lates niloticus*) in Lake Victoria (Lowe-McConnell 1992). Even some of the exotic species usually perceived to be ecologically benign, may make strong inroads into natural communities. They may displace indigenous species during the course of invasion, especially those species or stocks which possess closely similar niche envelopes to those of the invading species. Hence, the naturalized rainbow trout is now well established in the lower reaches of most tributary streams in the Upper Great lakes, while at the same time it has effected a displacement of the native brook trout (*Salvelinus fontinalis*) upstream, towards the less productive headwaters of these same stream systems (Ryder and Kerr 1984). In fact, the proclivity of large rainbow trout to inhabit the near shore zone of a lake, rather than the tributary streams for much of the year, has probably accounted for the low levels of "coaster" brook trout in recent years. The adfluvial coasters are known to occupy the near-shore littoral zone of the Upper Great Lakes for the most part, and enter streams only during late summer in preparation for fall spawning.

In general, the interpretation of the effects of the introductions of exotic species have been ambiguous. Two invertebrates, the zebra mussel (*Dreissena polymorpha*) and the Cladoceran (*Bythotrephes cederstroemi*), have made massive intrusions into the native, but much altered communities of some of the Great Lakes. Yet, no dire direct effects to fishes from an ecological perspective can be attributed with confidence to either invading species at this time. However, the potential for ecological disaster is rife, and future studies will undoubtedly reveal underlying disaster that will affect the integrity of the native communities. It would seem that the intrinsic strategy of ecosystems is to buffer the intrusions of exotic species as best they can. This may be part and parcel of the self-organizing and self-sustaining capabilities of not only ecosystems (Kay and Schneider 1994), but also communities. Arrival at dynamic equilibrium with the native community may occur only after considerable interaction has taken place over a substantial period of time, especially when the introduced species has a closely similar niche envelope to a dominant native species. The model for the effect of an invading species into a natural community probably assumes the pattern of a monotonic decay, with the greatest disturbance coming during the post-invasion stage, with steady state being the end condition, arrived at only over substantial time and considerable iteration.

Biodiversity within the Context of the Ecosystem Approach

The establishment of appropriate fishery policy by which to address the biodiversity issue on the Laurentian Great Lakes, such as the Strategic Great Lakes Fishery Management Plan (Anonymous 1980), and its subsequent application through the various provincial and state management agencies has, over time, become both an anachronistic and insufficient approach, insofar as the conservation of biodiversity is concerned, given the rapt attention that is now being focused on ecosystem management (Allen et al. 1993). The thesis engendered by the ecosystem approach from a fisheries perspective, presupposes that fishes constitute but a single component of a highly complex and integrated system, which is composed of a multiplicity of organisms, each with its respective proprietary habitat, and all of this circumscribed by an ecosystem boundary that is somewhat indeterminate at best (fig. 2). For management purposes, an arbitrary boundary is usually assigned, on the basis of practical or operational considerations.

While for the Great Lakes, the original strategic plans (Anonymous 1980) acknowledged the ecosystem approach, traditionally, they have been directed more towards the management of one or more commercially valuable fish species at the individual lake level (e.g., Lake Erie or Lake Superior), although other binational protocols, such as the Great Lakes Water Quality Agreement (GLWQA; Anonymous 1978) of 1978 (modified in 1987; Anonymous 1987), have included all five lakes and their catchments under the collective cognomen of "Great Lakes Basin Ecosystem". An understanding of this ecosystem emphasis requires knowledge of effects even beyond the arbitrarily designated, and somewhat fuzzy hydrospheric boundaries and hence, includes perforce, atmospheric and lithospheric considerations, among others.

Taking all of these ecosystem components, and their respective behaviors into simultaneous account, and empirically observing the individual changes in each over time presents a seemingly intractable problem of enormous magnitude for the journeyman ecosystem manager in the field. These beleaguered souls are conscientiously attempting to ensure the sustainability of "healthy" ecosystems into perpetuity, as proposed by the World Commission on Environment and Development (Brundtland 1987), which appears to be the order of the day for most resource agencies within the Great Lakes basin. Inasmuch as ecosystem health is an integrative science in the process of defining itself (Rapport 1995), we will leave the preferred definition to the individual disposition of the readers.

A potentially effective approach to the solution of this dilemma, in terms of timeliness of application, might be a directed study of surrogate species (Ryder and Edwards 1985; Edwards and Ryder 1990) found at sensitive nodes of ecosystems. The species selected would be representative of the relative health of the total system inasmuch as they would be community surrogates (Ryder and Kerr 1990). However, before these surrogate organisms can be designated, appropriate expert insight into the working components of the ecosystem is required, particularly as these affect the surrogate's current, relative abundance, and the degree of their fluctuations over time. Other indices and models, such as Karr's (1981) index of biotic integrity (IBI) have either proven to be robust in certain applications, or to have a strong potential for future use, such as the Strategic Choice Model (SCM), as it might be adapted to the Great Lakes basin ecosystem (Steedman and Haider 1993). Further pursuit of these types of approaches is highly encouraged. Most management agencies on the Great Lakes, as well as the two binational commissions, have been recently committed, in varying degrees, to managing the uses of the lakes within the context of an ecosystem approach. The implication of this commitment requires that ecosystems and communities be recognized as "wholes," having distinctive emergent properties (Kerr 1974) that may be observed and measured. It also implies that in the future, the hierarchic level of address by the various management agencies will transcend any previous jurisdictional boundaries and responsibilities, and in turn, will require collaborative efforts among the agencies, both within and outside of the

Great Lakes catchment. Within this context, the biodiversity of fishes and other aquatic organisms will play a meaningful role as a metric against which management successes and failures might be measured.

A Measure of Sustainability

Biodiversity, as an attribute of natural systems, may be measured at predetermined stations and times in order to ascertain whether or not a healthy ecosystem has been retained. In the face of a variety of anthropogenic incursions, such as exploitation and cultural eutrophication, a healthy condition is but rarely attained in the Great Lakes Basin.

It is presumed that healthy ecosystems and communities have developed through eons of co-evolution, and that any marked departures from that state may be attributed to human perturbations of various types. Healthy ecosystems, through their inherent resilience, will at first compensate for the effects of many of the abuses of the system, but will eventually succumb to accumulative stresses. Within this context, we interpret succumb as a transformation to one of a number of possible states that are not highly valued by humans. Often, many would naively hope that the Great Lakes will continue to mirror their former pristine conditions as outlined in the SCOL Symposium (Loftus and Regier 1972), while ironically, they ensure that they will not through inappropriate, or excessive uses. Monitoring biodiversity, then, is one of the simple and forthright measures that might be implemented, in order to track a successful return to a healthy condition following release from a suite of anthropogenic stresses. Unfortunately, as a return to pristine conditions is unlikely to ever occur, the next best alternative is to ensure that the stressed system is at least on a recovery trajectory that might terminate ideally, at a near pristine benchmark, that is, the Great Lakes as they were about two hundred years ago (Ryder and Edwards 1985). Measurements of biodiversity provide a means for doing this at little expense, particularly if indices of heterogeneity and evenness are used (Pielou 1966, 1975; Perkins 1983). If, over a period of a decade or more, biodiversity indices show relatively minor fluctuations in heterogeneity or evenness, without the near extirpation or extinction of any native organisms, the ecosystem may be perceived as retaining an acceptable state of health, despite a substantial level of use (fig. 2).

Unfortunately, past experience would tend to preclude the persistence of a constant healthy state in most ecosystems, at least within the grasp of human control through the application of appropriate management policies. For example, in the realm of fisheries management, we have failed miserably to provide sustainable yields of desirable fish species despite the mobilization of legions of binational, federal, provincial, and state agencies on the Great Lakes, all armed to the teeth with the latest technologies to apply to this problem (Larkin 1977). The potential for managing sustainable development as proposed by the World Commission on Environment and Development (Brundtland 1987) seems even more dubious, given the burgeoning nature of the human population, as well as the various external influences, including global increases in greenhouse gases and the reduction of the stratospheric ozone. Especially disconcerting is the proliferation of new virulent diseases such as AIDs and Legionnaires' disease (Levins et al. 1991; Holling 1994), or new strains of old diseases, such as the mutation of the ubiquitous *E. coli* bacterium (Service 1994). All of these factors, and many others are compounded by human avarice, a characteristic that unfortunately appears to be universal. In fact, these somewhat dubious attributes have

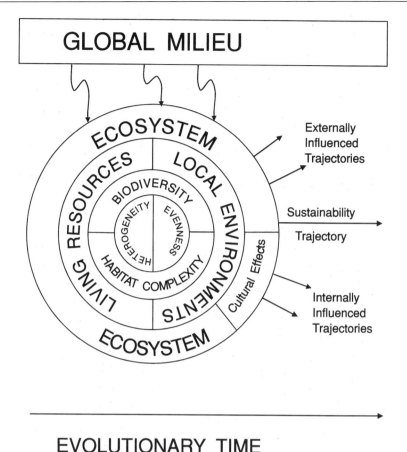

EVOLUTIONARY TIME

FIG. 2. *Schematic diagram of an ecosystem and its predicted optimum trajectory over evolutionary time, that is, the "sustainable" ecosystem as viewed by Utopians. More realistically, perhaps, are the externally and internally influenced trajectories, determined by the vagaries of the global milieu and compounded by cultural effects. In both instances, measures of the two components of biodiversity, heterogeneity and evenness, should provide a metric by which the relative level of sustainability of ecosystems may be reasonably assessed.*

contributed not only to ecosystem distress, but have been observed as contributing factors to the collapse of complex human societies as well (Tainter 1988).

In order to navigate the waters of uncertainty in terms of the sustainability of optimal biodiversity, Ludwig et al. (1993) have proposed some principles for effective ecosystem management. These include: consider human motivation and responses as part of the system to be studied and managed; act in advance of scientific consensus; expect scientists to recognize problems but not necessarily to remedy them; distrust any claims of the sustainability of biodiversity; and finally, confront uncertainty. To this list might be added the profitable practice of assessing biodiversity in terms of change of heterogeneity and evenness over time. Marked variability beyond normal homeostatic boundaries may be evidence enough that the goal of a sustainable ecosystem has not likely been attained.

Caveats to the Use of Biodiversity Measurements

The term biodiversity is sometimes taken to mean any level of biotic goodness within an ecosystem that could be improved by adding more species (see Koshland 1994, for a humorous commentary on such misinformation). Such careless interpretations compromise reasonable applications of biodiversity indices. Pielou (1966, 1975), and others have repeatedly pointed out the double-faceted nature of biodiversity, namely, not only the number of different organisms or ecological attributes present in a system, but also, the natural levels of evenness of their distribution. An array of fishes, for example, may be appropriately ordered ecologically within a community, if the relative number between any two species reflects the level of their trophic interaction, that is, their relative energy exchange. This condition may be deemed to be a healthy community. Another array with the same number of species, but occurring in ecologically inappropriate proportions, may be on the brink of losing species, or worse yet, flipping to another, less desirable ecosystemic state by human standards. Yet a third community may have the highest diversity index of all, but because an inordinate number of nonnative species have invaded or been introduced, may be completely unpredictable as to its future state, and thereby, lose much of its potential value to man. In the latter instance, any perceived benefits, as in the introduction of the chinook salmon (*Onchorynchus tshawytscha*) into Lake Superior, for example, are short term, and are usually accompanied by a serious level of ecological incursion into the native community.

Despite all of this, it is implicitly understood that trophic interaction among fishes are not mutually exclusive, but must be extended to all of the faunal, floral, and microbial inhabitants of an ecosystem in order to provide a fundamental ecological understanding of all of its processes.

Also, caution must be taken in invoking the stability-diversity hypothesis (Woodwell and Smith 1969), which has come under some strong ecological criticism (Goodman 1975). While it is known that monocultures as a rule are inherently unstable, too many species in a mix may evoke niche contention (Ryder and Kerr 1990), an equally undesirable state from an anthropogenic point of view. It would seem that stability-diversity relationships assume an optimality curve in natural systems as with human societies (Tainter 1988). That is, the optimum number of species for maximum stability and persistence is dependent upon the individual community interrelationships, and its particular ecological orientation within the ecosystem as a whole.

Biodiversity, Perturbations, and Heterogeneity

Recent concepts of ecosystem structure and function, in the Great Lakes and elsewhere, have indicated that ecosystems reside at some distance from equilibrium (Kay 1991). Continued maintenance, and therefore the ensured persistence of these systems, relies upon self-organizing properties that create increasingly complex molecules, which in turn, evolve into increasingly complex systems with a greater biodiversity. Retention of these systems within homeostatic boundaries depends upon periodic or episodic events (i.e., resets) which rejuvenate the systems through the mechanism of perturbations that set back the successional state of the ecosystem (Holling 1985). Major perturbations increase the biodiversity of an ecosystem at first, by increasing its heterogeneity (Reice 1994). Accordingly, in almost all instances, observed ecosystems are constantly recovering from their last disturbance, and evolving inevitably towards greater complexity. Put in context, the Great Lakes are in the process of recovering from Pleistocene glaciation. Accordingly, its natural level of biodiversity may be reasonably deduced to be that of the invading species from Pleistocene refugia.

Evaluation of Biodiversity

The evaluation of biodiversity is but one metric that may practically be implemented under the aegis of the ecosystem approach. It calls for the assessment of numbers of species or stocks (heterogeneity), and their relative abundance within the total fish assemblage (evenness). Data suitable for these evaluations have been collected on the Great Lakes for many decades, but more often than not, with some other purpose in mind. However, the vast data banks that have been accumulated at universities, or governmental and quasi-governmental institutions, may permit both qualitative and quantitative assessments of biodiversity over at least the last one hundred years. Of course, much of the earliest information will be anecdotal, offering only slight expectation for quantification. Such information is valuable, however, for the assessment of biodiversity as it might have been under a regime of only minimal stress. Interpretation of such data requires the application of skilled analysis, and verification of the results may be impossible. Nonetheless, future paleolimnological analysis of the various strata of bottom sediments (Lagler and Vallentyne 1956) may, indeed, provide verification for estimated levels of biodiversity previously based only on anecdotal evidence. This will be especially true for the heterogeneity component of biodiversity, and less so for the evenness component. Even rare species have a role to play in the biodiversity drama. Sometimes they proliferate, with a favorable change in milieu, and may exploit a trophic opportunity of another declining species, which has been decimated by the same environmental change (Welch 1967). This dynamic may be set in motion from anthropogenic sources, or more benignly, from natural episodic events.

Other quantitative methods are possible, of course, and to this end we refer to some of the indexing methods elaborated in Karr (1981), Ryder and Edwards (1985), and Steedman and Haider (1993). In using an index method for the assessment of different levels of biodiversity, precision may be sacrificed for timeliness of application (Henderson et al. 1973), and ultimately, a substantial economy of effort and funding may be experienced. Indices are neither a panacea nor an artifact in the assessment of biodiversity, but rather, provide a quick, moderately precise, and economically feasible evaluation.

As an example of a simple, relative index (while temporarily eschewing more complex ones) that may be used for a rapid biodiversity assessment for a body of water, we propose further experimentation with the Natex Index (number of native fish/number of exotic fish; fig. 3). This simple index calculates the ratio of native to exotic species, multiplied by a dimensionally correct scaling coefficient, which retains the index values (in the Great Lakes) to integers less than ten for ease of comparison. As there is a correlation between the area of a lake and the number of fish species that it harbors (Barbour and Brown 1974), other scaling coefficients may be used for smaller lake areas, but for the Great Lakes, a factor of four satisfies our requirements for a Natex Index with comparative utility for lakes of that order of magnitude.

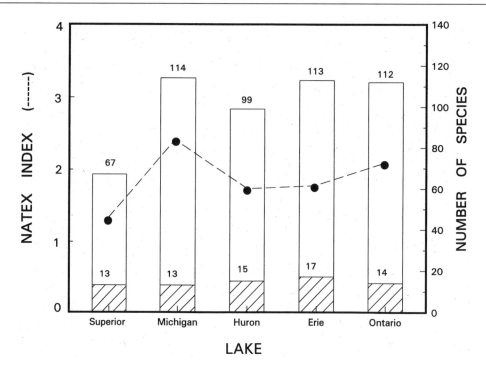

FIG. 3. *Approximate number of indigenous fish species in each of the Laurentian Great Lakes circa 1800 (upper numbers in histogram) after Ryder (1972). The lower numerals represent the number of exotic fish species that have since invaded or been stocked over approximately the last two hundred years (estimated from Mills et al. 1993). The Natex Index, the number of native fishes divided by the number of exotic fishes times a dimensionally correct scaling coefficient (in this case 0.25), is a comparative measure of the relative impacts on native fish communities by exotic species, with an arbitrary upper level of four, representing no introductions and therefore, no intrusive effects. From this figure, it may be assumed that the native fish communities of each of the Great Lakes have been substantially encroached upon, and present a different community profile from that of the early 1800s.*

The Natex Index may be utilized either for spatial comparison among a suite of lakes, as is illustrated in figure 3, or alternatively, as an indicator of changing values of biodiversity over time within a single body of water. In the latter case, for Lake Erie (fig. 4), the upper limit for the sake of convenience, has been arbitrarily set at ten.

In the interpretation of both figures 3 and 4, it should be emphasized that the higher the index value, the better is the heterogeneity factor of biodiversity, that is, there are relatively more native species, and fewer exotic or naturalized species within the lake's total fish assemblage. To some, it may prove surprising to see that Lake Superior, generally conceded to have more pristine attributes than the other four Great Lakes basins, has the lowest Natex Index, an indication of relatively low numbers of native species, and proportionately high numbers of exotic species. This observation leads to an interesting ecological phenomenon that suggests that the more native species a lake has, the less likely it is that successful invasion, and ultimately, naturalization, will occur (Christie

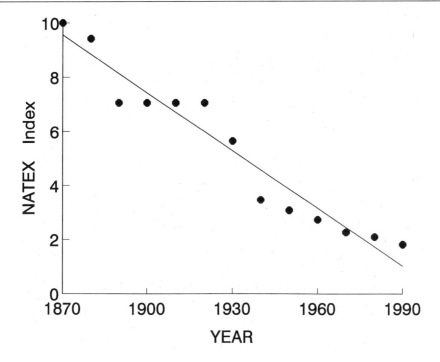

FIG. 4. *Changing values for another version of an index of biodiversity (Natex Index) in fishes of Lake Erie between 1870 and 1990. This index was calculated by dividing the number of native species in the lake by the sum of the introduced and invading species, then multiplying by a dimensionally correct scaling factor, in this case, four. An upper level of ten was arbitrarily established as the highest index possible for Lake Erie. The negative slope of the Natex Index line indicates substantial increases of exotic species and a corresponding decline in native species, hence, a general decline in biodiversity and general ecosystem degradation.*

et al. 1972). This is because more trophic opportunities exist in lakes with a depauperate fish fauna through a relative lack of niche contention.

The Natex Index will respond not only to all of the lake stresses that affect native species, but also to ecosystem stresses, including toxic contaminants, cultural eutrophication, structural modifications, and overfishing. The presence of exotic species, however, is essential for the Natex Index to function. The Natex Index also provides a means to disclaim the oft held proposition that increases in diversity due to invading or stocked exotic species contributes to maximum biodiversity, and therefore, may be (wrongly) perceived as being ecologically sound. Alternatively, it supports the notion that optimum diversity levels are most indicative of a healthy system comprised of mostly indigenous and endemic species, with but few exotic or naturalized species in the mix. Also, the Natex Index as currently applied, deals only with the heterogeneity aspect of biodiversity. It should be noted, that in the case of Lake Superior, if lake trout phenotypic stocks were counted as individual species, the index value provided in figure 3 would be slightly higher, dependent upon how many of these stocks have already been extirpated. Despite the fact that an absolute standard does not exist against which the relative values in figure 3 might be assessed, it may be assumed that the 10 to 20 percent occupancy range for exotic species in the five lakes may be considered as much too high to avoid undue levels of niche contention (Ryder and Kerr 1990).

The Natex Index may be used with either ecologically deteriorating, or improving conditions. It provides only a relative measure of ecosystem health based on biodiversity. However, certain ecological inferences might be drawn from this simple index. For example, low index values may suggest that a lake is more vulnerable to further invasion than other lakes with higher indices. Accordingly, the planned management strategies intended to protect the fisheries of the two lakes would be different. A strategy to address the management needs of the lake with the high index would attempt to maintain the status quo, while the lake with the low index should not only be subjected to rehabilitation measures, but special efforts should be devised to ensure that no further invasions occur.

Management of Biodiversity in the Great Lakes

The management of biodiversity in the Great Lakes is a compelling problem because of the multiplicity of biomes and political jurisdictions that are embraced. Accordingly, the application of appropriate management methods requires not only an ecosystem understanding of a degree not easily obtained, but also needs a sufficiently wide jurisdictional mandate to effectively address ecosystem problems. Aquatic communities have no regard for political boundaries (Regier and Grima 1984), and move freely from jurisdiction to jurisdiction, wherever contiguous waters make it possible. Effective management implies application of holistic science at an ecosystem level which may transcend several jurisdictions. The mandate required for an ecosystem approach using biodiversity indicators on the Laurentian Great Lakes and their contiguous waters would involve, at minimum, eight U.S. states, two Canadian provinces, two federal governments, two binational commissions, a multitude of municipal and city governments, quasi-governmental agencies, tribal agencies, non-governmental organizations, and others. Caldwell (1994) has estimated that there are at least 650 jurisdictional units that come into play on the Great Lakes, excluding municipalities. Efficient meshing of all these governmental, quasi-governmental, and private bodies into a rubric that will optimize information flow leading to an ecosystem understanding, presents an almost humanly insurmountable task. The actual implementation of a coordinated management program to conserve biodiversity under these circumstances becomes even more intractable. And given that ideally, efficient integration of all of these agencies is affected, many of the larger problems including international atmospheric transportation of waste products, subsurface photolytic effects in lakes due to increasing ultra-violet radiation, groundwater infiltration, and climate change, as but four examples, will not likely have been adequately addressed.

International Joint Commission

The International Joint Commission (IJC) was originally created by Canada and the United States to oversee the Boundary Waters Treaty of 1909 (Caldwell 1990). Its effective entrance into Great Lakes water quality issues (implicitly included biodiversity concerns) occurred with the signing of the 1972 Great Lakes Water Quality Agreement, which dealt, for the most part, with the control of persistent toxic substances entering the Great Lakes and contiguous waters. A subsequent agreement in 1978 (Anonymous 1978) embraced ecosystem concerns as well as the concern for toxic substances per se, and included expanded responsibilities. This new strategy, termed the ecosystem approach, was welcomed by the IJC in its newly expanded role (Vallentyne et al. 1978). It implicitly involved the measurement of biodiversity at different trophic levels in order to determine the relative health of the Great Lakes basin ecosystem.

Complementing this approach was the subsequent identification of forty-three areas of concern within the Great Lakes basin. These areas are some of the specific locations within the Great Lakes and contiguous waters that do not meet the objectives of the Great Lakes Water Quality Agreement of 1978. Subsequently, there was an implementation of Remedial Action Plans to address the principal environmental needs within each Area of Concern. Indicators of relative success in the various Remedial Action Plans included before and after biodiversity measurements of a plethora of organisms at several trophic levels (Hartig 1993).

In 1987, the 1978 Water Quality Agreement was amended by protocol to transfer some of the responsibilities previously conferred on the IJC, back to the contracting parties, that is, the federal governments of the United States and Canada. The jury is still out on the effectiveness of this measure, however, Caldwell (1994) observed that:

> In brief, the parties cannot co-ordinate their commitments for the Great Lakes because their parts are not committed to the policies of the whole. The whole is divided against itself, simultaneously pursuing policies that are diverse and sometimes discordant, and limited by constitutions which offer no real protection for the environment where private rights and public interests are in conflict.

One of the principal problems in the Great Lakes, as outlined succinctly above by Caldwell (1994), may not all be subsumed under a biodiversity perspective; still, biodiversity remains an effective measure of ecosystem health, sustainability, resilience, and several other emergent properties of ecosystems.

Great Lakes Fishery Commission

A 1954 Convention on Great Lakes Fisheries resulted in a coordinated effort between Canada and the United States to sustain fisheries productivity in the Great Lakes. This Convention established the Great Lakes Fishery Commission as a binational commission (Anonymous 1992) charged with a range of duties that included: research on the productivity of fisheries in the Great Lakes; coordination of research among the eight states and Ontario; recommendation to the Contracting Parties on the application of new science generated; design and implementation of a program to eliminate sea lampreys in the Great Lakes; and publication of scientific, or other information obtained by the Commission in the course of exercising its mandate.

The implication of these charges to the conservation of biodiversity is obvious. Early attempts to control the invading sea lamprey were somewhat less than successful, but as more information and effort was directed toward this problem, proportionately greater success was obtained.

In the interim, however, major stocks of lake trout and coregonines were lost to the sea lamprey, resulting in a corresponding loss of biodiversity (Pycha 1980; Pycha and King 1975). More recently, the Great Lakes Fishery Commission has opted for an ecosystem approach for both research and management of the Great Lakes fisheries, and has enlarged its mandate within the intent of the original convention. The new direction calls for partnerships to accomplish fish community objectives for each of the Great Lakes, and include state of the lake reports, quantifiable environmental objectives, and priorities for fisheries research.

In retrospect, the Great Lakes Fishery Commission has proven to be a major force in the preservation of biodiversity in the Great Lakes because of the cooperative integration of its stakeholders and clients. Perhaps even greater success could be achieved if the Commission moved rapidly towards greater autonomy.

Policy Development for Biodiversity

While a plethora of information exists on biodiversity from which policy might be derived (McNeely et al. 1990; Noss 1990; Ryder and Scott 1994; Anonymous 1994, 1995, 1995a), in total, these represent only the tip of the iceberg in terms of what will eventually be required to sustain biological diversity over the next few generations. Accordingly, much of what is recommended herein, will depend on the appropriate level of supportive research to back up our recommendations. In this regard, we will focus on the recent recommendations of Ryder and Scott (1994), which were subjected to peer review at a three-day workshop, and have been appropriately modified according to majority consensus. While these recommendations apply generally to Canadian fisheries, we have modified a subset as applicable specifically to the Laurentian Great Lakes and their contiguous waters as follows:

- Initiate studies to determine how aquatic communities respond to exploitation;
- Determine the life history for species of little economic importance, but with potentially critical ecological roles;
- Study the ecological roles of scarce, endangered, or fragile species, and how they might be sheltered from the effects of fishing and habitat loss;
- Implement control of multispecies fisheries, whereby they may be retained indefinitely in the fishing-up stage in order to preserve maximum biodiversity;
- Reduce bycatch to ecologically tractable levels (i.e., sufficiently small portion of standing stocks to allow for retention of heterogeneity and evenness values of biodiversity);
- Identify and use the genre of fishing gears best suited for the retention of apposite levels of biodiversity;
- Evaluate the effects of various extant fisheries regulations designed to preserve the biodiversity of aquatic communities;
- Determine the effects of nonfishery related stresses on the biodiversity of aquatic communities;
- Determine ecologically appropriate levels of heterogeneity and evenness as indicative of healthy fish communities;
- Develop and apply indices for a rapid, first approximation assessment of biodiversity.

Epilogue

Ultimately, we might ponder the future status of biological diversity in the Laurentian Great Lakes basin ecosystem. The prevailing human condition dictates that we remain ever optimistic about the relative health of the biosphere, as we hurtle from one Panglossian illusion to the next. The bare facts would suggest that rather than sustained optimism about continuing deterioration of our extensive aquatic ecosystems, a large portion of reasoned introspection and hind-casting may be more appropriate.

Despite Herculean efforts from various agencies, parties, commissions, and boards, the reality that we face on the Great Lakes would seem to be an ever declining biodiversity over the long term, due ultimately to constantly burgeoning human populations, the latter synergized by the ubiquitous human conditions we have come to know as indifference and avarice. If these primal attributes of mankind are seen as insufficient cause to progressively plunder our resources and raze our landscapes, then we must also consider the confounding effects of some modern day Judeo-Christian religions, the doctrines of which advocate "dominion of mankind over all of nature." The sustainability crutch currently proffered by world authorities is flawed, and provides but little hope in aspiring to a state of sustainable development, as proposed by the Brundtland World Commission on Environment and Development (Brundtland 1987). The gradual accumulation of many interfering variables of global magnitude make the prospect of the sustainability of biodiversity an improbable pipe dream, even for optimistic ecologists, as global externalities preclude any real progress in this regard.

In a desperate last surge, humanity has grasped at ecosystem science and interagency partnerships as the solution to the biodiversity problems of the Great Lakes. Unfortunately, many of these seemingly intractable problems occur at hierarchic levels that are neither easily comprehended, nor readily accessed by mankind. Recently, we have entered a period of rapidly changing conditions in just about every aspect of life. We gleefully ride the roller coaster of improved living conditions tentatively, only to be rapidly plunged into the troughs of ecological and economic despair. The cycle times between halcyon apogees are becoming ever shorter. We exist in a truly boom-or-bust syndrome. Perhaps human existence on the Great Lakes is destined to mirror the historical record of the hapless blue pike, which was swept into extinction almost immediately after experiencing a period of historic maximum abundance and well being.

It is not surprising to learn that the loss of biodiversity has sometimes been called the "ecology of grief" (Windle 1992). Perhaps the "ecology of desperate measures" would be more appropriate! Collectively, we will eventually have to leave the metaphoric roller coaster for a firm ecological foundation, and define the human values that are most precious to us. Only then may we assume a rational approach to the ever-encroaching problems of the Laurentian Great Lakes, which threaten to inundate us all.

Acknowledgments

We gratefully acknowledge the various contributions and constructive criticisms of S. R. Kerr, J. H. Leach, J. Pesendorfer, E. P. Pister, H. A. Regier, R. Steedman, and W. W. Taylor, as well as anonymous reviewers. We appreciate their forbearance, as well as the tolerance shown to our infrequent editorial comments, but exempt them from having any responsibility for the same. J. H. Leach was also a principal collaborator in the development and application of the Natex Index. We thank B. Pond, K. Ing, and Y. Allen for their assistance with the graphic figures.

We are grateful to W. W. Taylor and C. Paola Ferreri for the opportunity to make a contribution to this worthwhile endeavor on Great Lakes biodiversity.

This chapter is contribution No. 94-10 of the Ontario Ministry of Natural Resources Aquatic Ecosystem Science Section, P.O. Box 7000, 300 Water St., St. Petersborough, Ontario, Canada K9J 8M5.

Literature Cited

Agassiz, L. 1850. Lake Superior. Robert E. Krieger Publishing Company, Huntington, New York.

Allen, T. F. H., B. L. Bandurski, and A. W. King. 1993. The ecosystem approach: theory and ecosystem integrity. Report to the Great Lakes Science Advisory Board, Windsor, Ontario. 64 pages.

Anonymous. 1978. Great Lakes Water Quality Agreement of 1978. International Joint Commission, Ottawa, Canada. 52 pages.

———, 1980. A joint strategic plan for management of Great Lakes Fisheries. Great Lakes Fishery Commission, Ann Arbor, Michigan. 23 pages.

———, 1987. Revised Great Lakes Water Quality Agreement of 1978. International Joint Commission, USA and Canada.

———, 1992. Strategic vision of the Great Lakes Fishery Commission for the decade of the 1990s. Great Lakes Fishery Commission, Ann Arbor, Michigan. 39 pages.

———, 1994. Seventh Biennial Report. International Joint Commission, Ottawa, Canada. 58 pages.

———, 1995. The conservation of biological diversity in the Great Lakes ecosystem: issues and opportunities. The Nature Conservancy Great Lakes Program, Chicago, Illinois.

———, 1995a. Canadian biodiversity strategy. Biodiversity Convention Office, Environment Canada, Hull, Quebec.

Baldwin, N. S., R. W. Saalfeld, M. A. Ross, and H. J. Buettner. 1979. Commercial fish production in the Great Lakes 1867–1977. Great Lakes Fishery Commission, Technical Report Number 3:1–187.

Barbour, C. D., and J. H. Brown. 1974. Fish species diversity in lakes. American Naturalist 108:473–489.

Beeton, A. M. 1965. Eutrophication of the St. Lawrence Great Lakes. Limnology and Oceanography 10:240–254.

Berst, A. H., and R. C. Simon. 1981. Introduction to the Proceedings of the 1980 Stock Concept International Symposium (STOCS). Canadian Journal of Fisheries and Aquatic Sciences 38:1497–1506.

———, G. R. Spangler. 1973. Lake Huron-the ecology of the fish community and man's effects on it. Great Lakes Fishery Commission, Technical Report 21. 41 pages.

Birnbaum, L. 1993. Reevaluation of dioxin. Presented to the Great Lakes Water Quality Board's 102nd meeting, Chicago, Illinois. International Joint Commission, Windsor, Ontario. 19 pages.

Bishop, C. A. 1974. The Northern Ojibway and the fur trade: an historical and ecological study. Holt, Rinehart, and Winston of Canada, Toronto.

Bocking, S. 1990. Stephen Forbes, Jacob Reighard, and the emergence of aquatic ecology in the Great Lakes Region. Journal of the History of Biology 23:461–498.

Brundtland, G. H. [Chairman] 1987. Our common future. World Commission on Environment and Development. Oxford University Press, New York.

Burdick, G. E., E. J. Harris, H. J. Dean, T. M. Walker, J. Skea, and D. Colby. 1964. The accumulation of DDT in lake trout and the effect on reproduction. Transactions of the American Fisheries Society 93:127–136.

———, H. J. Dean, E. J. Harris, J. Skea, R. Karcher, and C. Frisa. 1972. Effect of rate and duration of feeding DDT on the reproduction of salmonid fishes reared and held under controlled conditions. New York Fish and Game Journal 19:97–115.

Caldwell, L. K. 1990. Between two worlds: Science, the environmental movement, and policy choice. Cambridge University Press, New York.

———, 1994. Disharmony in the Great Lakes Basin: Institutional jurisdictions frustrate the ecosystem approach. Alternatives 20:26–31.

Carr, J. F. 1962. Dissolved oxygen in Lake Erie, past and present. Great Lakes Research Division, Institute of Science and Technology, Publication No. 9:1–14.

Christie, W. J. 1973. A review of the changes in the fish species composition of Lake Ontario. Great Lakes Fishery Commission, Technical Report 23:1–65.

————, J. M. Fraser, and S. J. Nepszy. 1972. Effects of species introductions on salmonid communities in oligotrophic lakes. Journal of the Fisheries Research Board of Canada 29:969–973.

Crossman, E. J., and D. E. McAllister. 1986. Zoogeography of freshwater fishes of the Hudson Bay drainage, Ungava Bay and the Arctic Archipelago. Pages 5–04 in C. H. Hocutt and E. O. Wiley, editors. The zoogeography of North American freshwater fishes. John Wiley and Sons, Toronto.

Dobson, H. F. H. 1981. Trophic conditions and trends in the Laurentian Great Lakes. World Health Organization, Water Quality Bulletin 6:14–51, 158 and 160.

Dymond, J. R. 1922. A provisional list of the fishes of Lake Erie. Publication of the Ontario Fisheries Research Laboratory. Number 4:57–73.

Edwards, C. J. and R. A. Ryder. 1990. Biological surrogates of mesotrophic ecosystem health in the Laurentian Great Lakes. Report to the Great Lakes Science Advisory Board, International Joint Commission, Windsor, Ontario. 69 pages.

Eisner, T. 1991. Chemical prospecting: a proposal for action. Pages 196–202 in F. H. Bormann and S. R. Kellert [ed]. Ecology, Economics, Ethics: The Broken Circle Yale University Press, London.

Emery, L. 1985. Review of fish species introduced into the Great Lakes, 181–974. Great Lakes Fishery Commission, Technical Report No. 45:1–31.

Eschmeyer, P. H. 1957. Note on the subpopulations of lake trout in the Great Lakes. United States Fish and Wildlife Service, Special Scientific Report, Fisheries Number 208: pp. 129.

Flint, R. F. 1957. Glacial and Pleistocene geology. John Wiley and Sons, Incorporated, New York, N.Y. 553 pages.

Fry, F. E. J. 1947. Effects of the environment on animal activity. University of Toronto Studies in Biology Series 55, Publication of the Ontario Fisheries Research Laboratory 68:62 pages.

————, 1949. Statistics of a lake trout fishery. Biometrics 5:27–67.

Goodman, D. 1975. The theory of diversity-stability relationships in ecology. Quarterly Review of Biology 50:237–366.

Grant, P. R. 1994. Ecological character displacement. Science 266:746–747.

Gulland, J. A. 1983. Fish stock assessment: A manual of basic methods. John Wiley and Sons, Toronto.

Hamley, J. M. 1975. Review of gillnet selectivity. Journal of the Fisheries Research Board of Canada 32:1943–1969.

Hartig, J. H. 1993. Toward integrating remedial-action and fishery management planning in Great Lakes Areas of Concern. Great Lakes Fishery Commission, Ann Arbor, Michigan. 34 pages.

Hartman, W. L. 1973. Effects of exploitation, environmental changes, and new species on the fish habitats and resources of Lake Erie. Great Lakes Fishery Commission, Technical Report Number 22:1–43.

Henderson, H. F., R. A. Ryder, and A. W. Kudhongania. 1973. Assessing fishery potentials of lakes and reservoirs. Journal of the Fisheries Research Board of Canada 30:2000–2009.

Hile, R., P. H. Eschmeyer, and G. F. Lunger. 1951. Status of the lake trout fishery in Lake Superior. Transactions of the American Fisheries Society (1950):278–312.

Holling, C. S. 1985. Resilience of ecosystems: local surprise and global change. Pages 228–269 in T. F. Malone and J. G. Roederer, editors, Global Change. Proceedings of a Symposium sponsored by the International Council of Scientific Unions, Ottawa, Canada. Cambridge University Press, Cambridge.

————, 1994. New science and new investments for a sustainable biosphere. Pages 57–73 in A. Jansson, M. Hammer, Carl Folke, and Robert Constanza, editors. Investing in natural capital. International Society for Ecological Economics, Island Press, Washington, D.C.

Honigmann, J. J. 1962. Foodways in a muskeg community. Northern Coordination and Research Centre, Department of Northern Affairs and National Resources, Ottawa.

Hubbs, C. L., and K. F. Lagler. 1970. Fishes of the Great Lakes Region. University of Michigan Press, Ann Arbor, Michigan.

Hutchinson, G. E. 1957. Concluding remarks. Cold Spring Harbor Symposium on Quantitative Biology 22:415–427.

Hynes, H. B. N. 1970. The ecology of running waters. University of Toronto Press, Toronto.

Jenness, D. 1958. The Indians of Canada. Anthropological Series Number 15, Bulletin 65, Ottawa, Canada.

Jordan, D. S., and B. W. Evermann. 1896. The fishes of North and Middle America: A descriptive catalogue, Parts I–IV. Government Printing Office, Washington, D.C.

Karr, J. R. 1981. Assessment of biotic integrity using fish communities. Fisheries 6:21–27.

Kay, J. J. 1991. A nonequilibrium thermodynamic framework for discussing ecosystem integrity. Environmental Management 15:483–495.

Kay, J. J., and E. Schneider. 1994. Embracing complexity-the challenge of the ecosystem approach. Alternatives 20:3–9.

Keenleyside, M. H. A. 1979. Diversity and adaptation in fish behaviour. Springer-Verlag, Berlin.

Kerr, S. R. 1974. Structural analysis of aquatic communities. Pages 69–74 in W. H. Van Dobben and G. R. Gradwell, editors. Structure, functioning and management of ecosystems. Proceedings of the 1st International Congress of Ecology.

————, 1974a. Theory of size distribution in ecological communities. Journal of the Fisheries Research Board of Canada 31:1859–1862.

————, R. A. Ryder. 1977. Niche theory and percid community structure. Journal of the Fisheries Research Board of Canada 34:1952–1958.

Koelz, W. 1921. Description of a new cisco from the Great Lakes. Occasional Papers of the Museum of Zoology, Number 104:1–4.

———, 1929. Coregonid fishes of the Great Lakes. Bulletin of the United States Bureau of Fisheries, 43(2):1–643.

Koshland, D. E. Jr. 1994. The case for diversity. Science 264:639.

Lagler, K. F. and J. R. Vallentyne. 1956. Fish scales in a sediment core from Linsley Pond, Connecticut. Science 124:368.

Larkin, P. A. 1977. An epitaph for the concept of maximum sustained yield. Transactions of the American Fisheries Society 106:1–111.

Lawrie, A. H. 1978. The fish community of Lake Superior. Journal of Great Lakes Research 4(3–4):513–549.

———, J. F. Rahrer. 1973. Lake Superior: A case history of the lake and its fisheries. Great Lakes Fishery Commission, Technical Report 19:1–69.

Levins, R. and nine coauthors. 1991. The emergence of new diseases. American Scientist 82(1):52–60.

Loftus, K. H. and H. A. Regier. 1972. Introduction to the proceedings of the 1971 symposium on salmonid communities in oligotrophic lakes. Journal of the Fisheries Research Board of Canada 29:613–616.

Lowe-McConnell, R. 1992. The changing ecosystem of Lake Victoria, East Africa. Freshwater Forum 4:76–88.

Ludwig, D., R. Hilborn, and C. Walters. 1993. Uncertainty, resource exploitation, and conservation: Lessons from history. Science 260:17, 36.

MacArthur, R. H. 1972. Geographical ecology. Harper and Row, New York.

———, E. O. Wilson. 1967. The theory of island biogeography. Princeton University Press, Princeton, New Jersey.

Magnuson, J. J. 1976. Managing with exotics: A game of chance. Transactions of the American Fisheries Society 105:1–9.

McCrimmon, H. R. 1950. The reintroduction of Atlantic salmon into tributary streams of Lake Ontario. Transactions of the American Fisheries Society 78:128–132.

McNeely, J. A., K. R. Miller, W. V. Reid, R.A. Mittermier, and T. B. Werner. 1990. Pages 25–35 in Conserving the World's Biodiversity. The World Bank, World Resources Institute, Conservation International, World Wildlife Fund United States, Washington D.C., IUCN, Gland, Switzerland.

Mills, E. L., J. H. Leach, J. T. Carlton, and C. L. Secor. 1993. Exotic species in the Great Lakes: a history of biotic crises and anthropogenic introductions. Journal of Great Lakes Research 19:1–54.

Nash, C. W. 1908. Vertebrates of Ontario. L. K. Cameron, Publisher, Toronto, Ontario.

Noss, R. F. 1990. Indicators for monitoring biodiversity: A hierarchical approach. Conservation Biology 4:355–364.

Paine, R. T. 1966. Food web complexity and species diversity. American Naturalist 100:65–75.

Parsons, J. W. 1967. Contributions of year-classes of blue pike to the commercial fishery of Lake Erie, 194–9. Journal of the Fisheries Research Board of Canada 24:1035–1066.

Perkins, J. L. 1983. Bioassay evaluation of diversity and community comparison indexes. Journal of the Water Pollution Control Federation 55:522–530.

Pielou, E. C. 1966. The measurement of diversity in different types of biological collections. Journal of Theoretical Biology 13:131–144.

———, 1975. Ecological diversity. Wiley-Interscience Publication, John Wiley and Sons, New York.

Pycha, R. L. 1980. Changes in mortality of lake trout (*Salvelinus namaycush*) in Michigan waters of Lake Superior in relation to sea lamprey (*Petromyzon marinus*) predation, 1968–78. Canadian Journal of Fisheries and Aquatic Sciences 37:2063–2073.

———, G. R. King. 1975. Changes in the lake trout population of southern Lake Superior in relation to the fishery, the sea lamprey, and stocking, 195–970. Great Lakes Fishery Commission, Technical Report Number 28:1–34.

Rapport, D. J. 1995. Ecosystem Health: Principles and practice Blackwell Science Inc., Cambridge, Massachusetts.

Rawson, D. S. 1952. Mean depth and the fish production of large lakes. Ecology 33:513–521.

Regier, H. A. and K. H. Loftus. 1972. Effects of fisheries exploitation on salmonid communities in oligotrophic lakes. Journal of the Fisheries Research Board of Canada 29:959–968.

———, T. H. Whillans, and A. P. Grima. 1980. Rehabilitation of the Long Point ecosystem: initiating a process. Contact 12:125–149.

Regier, H. A. and A. P. Grima. 1984. Nature of Great Lakes ecosystems as related to the management of transboundary pollution. Environmental Law, International Business Lawyer, London, United Kingdom.

Reice, S. R. 1994. Nonequilibrium determinants of biological community structure. American Scientist 82:424–435.

Rogers, E. S., and M. B. Black. 1976. Subsistence strategy in the fish and hare period, Northern Ontario: the Weagamow Ojibwa, 188–920. Journal of Anthropological Research 32(1):1–43.

Ryder, R. A. 1972. The limnology and fishes of oligotrophic glacial lakes in North America (about 1800 A.D.). Journal of the Fisheries Research Board of Canada 29:617–628.

———, H. F. Henderson. 1975. Estimates for potential fish yield for the Nasser Reservoir, Arab Republic of Egypt. Journal of the Fisheries Research Board of Canada 32: 2137–2151.

———, S. R. Kerr, W.W. Taylor, and P. A. Larkin. 1981. Community consequences of fish stock diversity. Canadian Journal of Fisheries and Aquatic Sciences 38:1856–1866.

———, S. R. Kerr. 1984. Reducing the risk of fish introductions: a rational approach to the management of integrated cold-water communities. European Inland Fisheries Advisory Commission (EIFAC) Technical Paper 42 Supplement to Volume 2:510–533.

———, C. J. Edwards. 1985. A conceptual approach for the application of biological indicators of ecosystem quality in the Great Lakes Basin. International Joint Commission and Great Lakes Fishery Commission, Windsor, Ontario. 169 pages.

———, S. R. Kerr. 1990. Harmonic communities in aquatic ecosystems: a management perspective. Pages 59–23 in W. L.T. van Densen, B. Steinmetz and R. H. Hughes editors. Management of freshwater fisheries. Proceedings of a Symposium Organized by the European Inland Fisheries Advisory Commission (EIFAC), Göteberg, Sweden. Centre for Agricultural Publishing and Documentation (Pudoc), Wageningen, The Netherlands.

———, W. B. Scott. 1994. Effects of fishing on biodiversity in Canadian waters. Pages121–144 *in* Biodiversity in Canada: A science assessment for Environment Canada. Environment Canada, Ottawa, Canada.

Scott, W.B. and E.J. Crossman. 1973. Freshwater fishes of Canada. Fisheries Research Board of Canada, Bulletin 184.

Service, R. F. 1994. E. coli scare spawns therapy search. Science 265:475.

Sly, P. G. 1990. The effects of land use and cultural development on the Lake Ontario ecosystem since 1750. Rawson Academy of Aquatic Science, Ottawa. 135 pages.

Smith, S. H. 1972. Factors in ecologic succession in oligotrophic fish communities of the Laurentian Great Lakes. Journal of the Fisheries Research Board of Canada 29:717–730.

Steedman, R. and W. Haider. 1993. Applying notions of ecological integrity. Pages 47–60 *in* S. Woodley, J. Kay, and G. Francis, editors. Ecological Integrity and the management of ecosystems. St. Lucie Press, U.S.

Svärdson, G. 1976. Interspecific population dominance in fish communities of Scandinavian Lakes. Institute for Freshwater Research, Drottningholm, Rep. No. 55:144–171.

Tainter, J. A. 1988. The collapse of complex societies. Cambridge University Press, Cambridge, United Kingdom.

Talhelm, D. R. 1988. The international Great Lakes sport fishery of 1980. Great Lakes Fishery Commission, Special Publication 88(4):1–70.

Trautman, M. B. 1957. The fishes of Ohio. Ohio State University Press, Columbus, Ohio.

Vallentyne, J. R. 1974. The algal bowl lakes and man. Department of the Environment Canada, Miscellaneous Special Publication 22.

———, et al. 1978. The ecosystem approach. Great Lakes Research Advisory Board, International Joint Commission, Windsor, Ontario. 47 p.

Van Oosten, J., R. Hile, and F. W. Jobes. 1946. The whitefish fishery of Lakes Huron and Michigan with special reference to the deep-trap-net fishery. United States Fish and Wildlife Service, Fishery, Bulletin 50:297–394.

von Bertalanffy, L. 1968. General system theory. George Braziller, New York. 295 pp.

Watt, K. E. F. 1973. Principles of environmental science. McGraw-Hill, Toronto.

Welch, H. 1967. Energy flow through the major macroscopic components of an aquatic ecosystem. Ph.D. Thesis, University of Georgia, Athens, Georgia.

Wells, L. and A. L. McLain. 1973. Lake Michigan–man's effects on native fish stocks and other biota. Great Lakes Fishery Commission, Technical Report 20:1–55.

Willford, W. A., and eight co-authors. 1981. Introduction and summary. Pages 1–7 in Chlorinated hydrocarbons as a factor in the reproduction and survival of the lake trout (*Salvelinus namaycush*) in Lake Michigan. United States Fish and Wildlife Service Technical Papers 105.

Wilson, E. O. [ed] 1988. Biodiversity. National Academy Press, Washington, D.C.

Wilson, E. O. 1992. The diversity of life. Belknap Press of Harvard Univ. Press, Cambridge, Massachusetts.

Windle, P. 1992. The ecology of grief. Bioscience 42:363–366.

Woodwell, G. M. and H. H. Smith [ed.] 1969. Diversity and stability in ecological systems. Report of Symposium held May 2–8, 1969. Brookhaven National Laboratory, Upton, New York.

CHAPTER 6

Species Succession and Sustainability of the Great Lakes Fish Community

Randy L. Eshenroder and Mary K. Burnham-Curtis

Introduction

Sustainability, an elusive concept (Levin 1993), can be hard to define in a way that leads to clear-cut fishery preferences for the Great Lakes. Larkin (1977) showed that the concepts of maximum sustained yield and optimum yield, precursors to sustainability, were ambiguous. Much of the emerging literature on sustainability deals with issues about the quantity of extraction (Levin 1993). Issues about the composition of what is expected to be extracted from the Great Lakes is equally, if not more important because the salmonine communities in four of the Great Lakes are mostly hatchery dependent.

One well-accepted feature of sustainability is intergenerational equity (i. e., the transfer of resources unimpaired to the next generation) (Meyer and Helfman 1993). However, the native fish assemblages in parts of the Great Lakes have been so altered by human activity that the remnants are not much of a legacy. Any reasonable concept of sustainability should incorporate an obligation to restore, as well as to maintain. We view sustainability, therefore, as a type of husbandry that seeks to establish or maintain fish assemblages that are best adapted to the ecology of the lakes, as evidenced by their diversity as well as by their structural and functional characteristics. Diversity includes differentiation within and among species. How well such assemblages are achieved is a measure of sustainability.

This article concentrates on the sustainability of the offshore pelagic and deepwater fish communities that were historically dominated by lake trout (scientific names of fish in table 1). The causes of alterations in these fish communities (i. e., overfishing, introductions, and cultural eutrophication) were identified by Loftus and Regier (1972). Here we look at the ecology of these altered communities in relation to sustainability and discuss the need for restoration.

Conceptual Approach

Harvest regulation, stocking (including introduction), and habitat protection/modification are major fishery management practices that greatly influence the ecology of the Great Lakes. The key

TABLE 1

Scientific names of fish mentioned in this article.

COMMON NAME	SCIENTIFIC NAME	COMMON NAME	SCIENTIFIC NAME
Sea lamprey	*Petromyzon marinus*	Coho salmon	*Oncorhynchus kisutch*
Alewife	*Alosa pseudoharengus*	Rainbow trout	*Oncorhynchus mykiss*
Emerald shiner	*Notropis atherinoides*	Kokanee	*Oncorhynchus nerka*
Rainbow smelt	*Osmerus mordax*	Chinook salmon	*Oncorhynchus tshawytscha*
Lake herring	*Coregonus artedi*	Atlantic salmon	*Salmo salar*
Lake whitefish	*Coregonus clupeaformis*	Lake trout	*Salvelinus namaycush*
Bloater	*Coregonus hoyi*	Burbot	*Lota lota*
Deepwater cisco	*Coregonus johannae*	Slimy sculpin	*Cottus cognatus*
Kiyi	*Coregonus kiyi*	Spoonhead sculpin	*Cottus ricei*
Blackfin cisco	*Coregonus nigripinnis*	Deepwater sculpin	*Myoxocephalus thompsoni*
Shortnose cisco	*Coregonus reighardi*	Yellow perch	*Perca flavescens*
Shortjaw cisco	*Coregonus zenithicus*	Walleye	*Stizostedion vitreum*
Pink salmon	*Oncorhynchus gorbuscha*		

question is: which management practices are sustainable and which are not? Before particular practices can be assessed for sustainability, an understanding of how fish communities are likely to respond to management initiatives is needed. We use the concept of species succession to show that various associations of species are characteristic of different stages of ecosystem development or maturity. Acceptance of the idea that fish communities are more than random assemblages of species, and that they organize (in a given habitat) on a successional trajectory in response to species interactions (Evans et al. 1987), is necessary to understand this concept. Smith (1968, 1972) recognized successional patterns of fish species change among each Great Lake subjected to the same stresses. The most pronounced change offshore was the extinction or depletion of large-bodied native planktivores, and replacement with small-bodied, invasive species from the Atlantic Ocean. Replacement of marine-origin planktivores with native planktivores recently in Lake Superior (Hansen 1994) supports the idea that succession can be reversed, and is a basic property of the Great Lakes.

Our conceptual approach is based on models developed by Christie et al. (1987), Kay (1990), and Johnson (1994), who discuss ecological succession in relation to energy flow or to ecosystem configuration. Although there is no agreement among these authors on what system property is manifested by certain configurations, the predicted patterns of organization are similar. Their models associate mature communities (those that develop in the absence of perturbation) with increased dissipation of energy and/or increased resistance to change. Succession is not viewed as a strictly linear process, but more mature and resilient communities are associated with higher levels of organization or complexity, increased biomass, and/or low production to biomass (P/B) ratios.

Christie et al. (1987) inferred a positive relationship between resilience and system maturity, and characterized more mature systems as having a dominance of long-lived piscivores. Johnson (1994) defined dominant species as those whose abundance is least controlled by other species, but instead, are largely constrained by the physical boundaries of the ecosystem, their food supply, and their genetic constitution. Kay (1990) recognized that ecosystems are dynamic, and exist in various states that reflect abiotic and biotic properties of the system. His model is also successional, but with the caveat that major changes in ecosystems may not be linear (reversible). Integrity, the ability of communities to resist change, was seen by him to increase with increased connectedness among species.

Johnson (1994) viewed community organization as a response to entropy reduction (i. e., minimization of tendency for disorder). Under any set of conditions, a population accomplishes entropy reduction by tending towards configurations that resist the rate at which energy is turned over or dissipated. In lakes, entropy for the whole community is minimized when a top piscivore becomes the dominant species. Top piscivores achieve larger body sizes and consume larger prey, causing more of the available energy to be metabolized to support storage at the expense of growth and turnover (Christie et al. 1987; Johnson 1994). Thus, a population of dominant piscivores in a lake can be viewed as a successional end point where community turnover as measured by P/B ratios is at a minimum. If this end point has either maximum biotic inertia (Christie et al. 1987), maximum integrity (Kay 1990), or minimum entropy (Johnson 1994), then it also represents a state of maximum sustainability, except that piscivore yields would have to be low for the end point to persist.

Sustainability can be viewed as a characteristic of the successional stage of the fish community, and can be qualified by the amount the community is set back from an end-point that would, in theory, be reached if the piscivore with the broadest niche and longest life span was restrained only by its food supply. Other piscivores could exist under such conditions, but, by definition, their abundance would be constrained by the dominant piscivore. The challenge is to identify and maintain a successional stage that appears to have acceptable prospects for sustainability. Fish harvests and other management actions that set back successional state can be viewed as tradeoffs against sustainability (i. e., they diminish the ability of the community to persist). The question is how much of a tradeoff is ecologically prudent for the Great Lakes.

To accomplish this task of identifying sustainability tradeoffs, we explore how species succession (i. e., change in species composition) affected structural and functional characteristics of the offshore Great Lakes fish community). We examine three specific issues that become concerns when successional state is set back too far. These issues are: (1) a loss of connectedness among trophic components; (2) the potential for unintended species to dominate the community; and (3) loss of fitness in stocked fish. We conclude by contrasting the sustainability of alternative community configurations and by identifying management practices that are important for its achievement.

Compositional Changes in Offshore Fish

European settlement of the Great Lakes basin in the last two centuries has resulted in a process of species succession within the offshore fish community of the Great Lakes. Changes in the diversity and abundance of fish species in each of the Great Lakes have occurred at different times, but all have common patterns and consequences. Eutrophication, overfishing, and the proliferation of introduced species, beginning already in the late 1800s, reduced or eliminated several indigenous species (Smith 1995). In this section, we review compositional changes that resulted in major structural and functional changes to be discussed in subsequent sections.

Early Introductions

The alewife, an anadromous species indigenous to lakes and streams of the Atlantic coastal drainage, was probably the earliest non-native fish species to proliferate in the Great Lakes. It was present in abundance in Lake Ontario as early as 1873 (Bean 1884; Koelz 1926). The successful establishment of the alewife in Lake Ontario probably resulted from migration through the Erie Canal (fig. 1) (Smith 1892; Smith 1970, 1995) as genetic studies have linked Great Lakes alewife populations to those of the Mohawk River (Hudson River) drainage rather than the St. Lawrence River drainage (Ihssen et al. 1992). Alewives could have entered Lake Ontario as early as 1819, but large populations of piscivorous fish may have kept them rare until 1873. They were abundant elsewhere in the Lake Ontario watershed as early as 1868 (Smith 1970). In the mid 1800s, piscivores in Lake Ontario were undergoing a rapid decline in abundance, possibly giving the alewife the opportunity to become well established (Smith 1970). Passage of alewives into the upper Great Lakes was facilitated by the Welland Canal (fig. 1), which opened in 1829 (Smith 1970). Alewives were first reported in Lake Erie in 1931 (Dymond 1932), but never became dominant there. The first reports of alewives in the upper lakes were: Lake Huron in 1933 (MacKay 1934), Lake Michigan in 1949 (Wells and McClain 1973), and Lake Superior in 1954 (Miller 1957). No abundant populations were established in Lake Superior possibly because of thermal limitations (Bronte et al. 1991).

Alewives became well established in Lakes Huron and Michigan in the mid to late 1940s, when lake trout were in decline (Hile 1949; Miller 1957). The most conspicuous change in these populations occurred when the Lake Michigan alewife population suffered a massive die-off in 1967 (Brown 1972). This population recovered by the early 1970s, but populations in both lakes were reduced by intensive predation and unfavorable climatic conditions in the late 1970s to early 1980s (Stewart et al. 1981; Eck and Wells 1987).

Rainbow smelt are also native to the Atlantic coastal drainage, but became established in the Great Lakes after an intentional introduction in Crystal Lake, Michigan, in 1912 (Van Oosten 1937). Smelt spread from Lake Michigan, and became well established in the other Great Lakes during the 1920s and 1930s (Van Oosten 1937), but the original source in Lake Ontario was probably from introduction into the Finger Lakes of New York, which began in 1917 (Bergstedt 1983). An unexplained, severe mortality of smelt caused temporary population declines in Lakes Huron and Michigan in 1942–3 (Van Oosten 1947). The establishment of smelt coincided with the decline in abundance of top piscivores, such as lake trout in Lakes Ontario and Superior, and walleyes in Saginaw Bay and Green Bay (Christie 1974). Unlike the alewife, the smelt proliferated in Lakes Michigan and Huron before lake trout populations collapsed (Baldwin et al. 1979). After a period of maximum smelt abundance in the lakes, density leveled off in the 1970s (Christie 1974).

Undoubtedly, the most destructive invading species to become established in the Great Lakes has been the sea lamprey. The sea lamprey is the largest and most predaceous of the lampreys, and is native to both the eastern and western coasts of the Atlantic Ocean (Leim and Scott 1966; Scott and Crossman 1973). Landlocked populations of smaller-sized sea lampreys were known to inhabit the upper St. Lawrence River, the Finger Lakes, Lake Champlain, and Lake Ontario prior to the 1920s (Pearce et al. 1980). They, like the alewife, probably invaded Lake Ontario through the Erie Canal (fig. 1) (Wright et al. 1985; Mandrak and Crossman 1992; Smith 1995). Genetic evidence links Great Lakes sea lampreys most closely to sea lampreys in the Finger Lakes (Brussard et al.

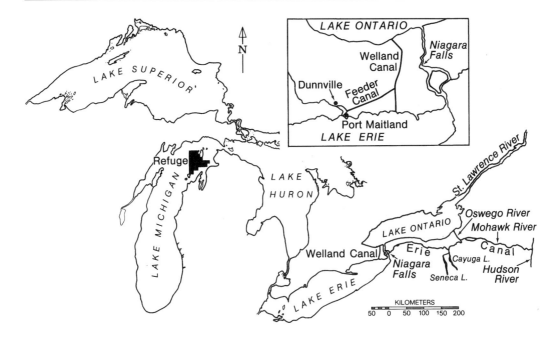

FIG. 1. *The Laurentian Great Lakes, the natural and artificial waterways between lakes and between the lakes and the Atlantic Ocean, and the location of a lake trout refuge in northern Lake Michigan (modified from Smith 1972).*

1981), which they may have invaded just before they entered Lake Ontario. Mitchell (1815), in his description of New York fish made four years before the Erie Canal opened, did not list the sea lamprey among Lake Ontario fish, but he did recognize it as occurring in salt water. The first authenticated record of a sea lamprey in Lake Ontario was in 1835 (Lark 1973; Smith 1995). Sea lamprey attack marks were observed on lake trout and other species in Lake Ontario in the late 1800s (Dymond et al. 1929), but sea lampreys were not thought at that time to pose a threat to fish (Parsons 1973).

The first sea lamprey was observed in Lake Erie in 1921, eighty-six years after the first sighting in Lake Ontario, and ninety-two years after the construction of the first Welland Canal (Smith 1970, 1995). Until the Canal was improved the last time (there were four canals in all), passage of sea lampreys into Lake Erie was apparently inhibited. Originally, the Welland Canal flowed into both Lakes Ontario and Erie because one section of the canal (the Deep Cut) had an elevation above the surface of Lake Erie, necessitating a diversion of water from a feeder canal (fig. 1) constructed off a reservoir on the Grand River at Dunville, Ontario (J. Burtniak, Brock University, personal communication). Adult sea lampreys entering the Welland Canal from Lake Ontario, and seeking an upstream area for spawning, would soon be heading downstream (as they approached Lake Erie) unless they entered the 34km Feeder Canal (Ashworth 1987). Two locks on the Feeder Canal, which ceased to operate as a regional waterway in 1881, would have impeded free passage of sea lampreys from the Welland Canal into the Grand River, a Lake Erie tributary. Once in the

Feeder Canal, sea lampreys may have entered various temporary flowages that drained low-lying agricultural lands. Construction of the fourth Welland Canal, begun in 1913, resulted in a deeper channel which flowed directly from Lake Erie into Lake Ontario, and in 1919 the feeder from the Grand River was cut off, leaving only Lake Erie upstream from Lake Ontario (Ashworth 1987). Two years later, the sea lamprey was reported from Lake Erie (Mandrak and Crossman 1992).

Once sea lampreys became established in Lake Erie, proliferation in the upper Great Lakes was rapid (Smith and Tibbles 1980). They were discovered in Lake St. Clair in 1934, Lake Michigan in 1936, Lake Huron in 1937, and Lake Superior in 1938, although 1946 has been a well-accepted date for this Lake (Hubbs and Pope 1937; Smith and Tibbles 1980). Chemical control was first implemented in Lake Superior in 1958 with the use of TFM (3-triflouromethyl-4-nitrophenol), a larval lampricide (Applegate et al. 1961). By 1978, catches of spawning sea lampreys in Lake Superior were 8% of the precontrol average (Smith and Tibbles 1980). TFM treatments were extended to Lakes Michigan and Huron in the 1960s, to Lake Ontario in 1972, and Lake Erie in 1986 (Smith and Tibbles 1980; Pearce et al. 1980; Cornelius et al. 1995). Chemical treatment has eliminated spawning runs from many streams, but major producers require continued treatments (Smith and Tibbles 1980).

Piscivore Succession

Before the 1900s, the offshore fish community in each of the Great Lakes was dominated by two piscivores, lake trout and burbot, and in Lake Ontario by a third top predator, the Atlantic salmon (Berst and Spangler 1973; Christie 1973; Parsons 1973; Wells and McLain 1973; Leach and Nepszy 1976; Lawrie 1978). The Atlantic salmon probably became established from indigenous anadromous east coast populations (Parsons 1973), and was prevented from establishment in the other Great Lakes by Niagara Falls. In Lake Ontario, Atlantic salmon supported one of the greatest freshwater fisheries in the world before they declined in the mid 1800s and became extinct at the turn of the century (Smith 1892; Fox 1930; Hunstman 1931; Parsons 1973).

During the decline of the Lake Ontario Atlantic salmon, several facilities were established in the United States and Canada for the specific purpose of developing artificial-propagation programs for salmonines (Parsons 1973). These early efforts were not successful at saving the Atlantic salmon or establishing Pacific salmon in the Great Lakes. The rainbow trout was, however, successfully introduced into Lake Huron in 1876, and the brown trout into Lake Michigan in 1883 (Emery 1985). Both species were subsequently planted in watersheds throughout the lakes, but neither species was an important predator in the lakes until massive stocking programs began in the 1970s.

Although three top predators existed in the offshore waters of the Great Lakes, the lake trout was dominant because of its size, distribution, and abundance (Smith 1972). Lake trout are widely distributed throughout northern North America, and the southern edge of their native range lies within the limits of the Pleistocene glaciation (Lindsey 1964). Within the Great Lakes basin, their distribution has been described "from top to bottom and from shore to shore" (Smith 1972). The earliest changes in lake trout abundance occurred in Lake Ontario, concurrent with declines in abundance of Atlantic salmon, and were mirrored in similar patterns of depletion in the other lakes later in the twentieth century (Christie 1973). In Lake Erie, the abundance of lake trout was relatively low compared to that of other indigenous Lake Erie fish, probably due to limited amounts of suitable habitat (Hartman 1973). The pattern of lake trout abundance, exploitation, and decline are similar for Lakes Huron and Michigan, even though Lake Michigan is reported to have produced more lake trout than all the other Great Lakes combined (Koelz 1926). Commercial exploitation of lake trout began in these two lakes around the 1830s, and continued steadily through the early 1900s, after which populations endured a steady decline until their collapse in the 1950s (Brown et al. 1981; Eshenroder et al. 1995a; Holey et al. 1995). In Lake Superior, the lake trout supported a stable commercial fishery from the 1870s through the 1950s (Lawrie and Rahrer 1973).

Though lake trout are morphologically similar outside of the Great Lakes, within the basin they evolved several distinct morphological types yet present in Lake Superior (Agassiz 1850; Goode 1884; Brown et al. 1981; Goodier 1981). The most common form of lake trout was the inshore lean, which occupied all of the Great Lakes. Lake Superior historically contained up to twelve distinct phenotypes recognized on the basis of skin color, body fat, facial characteristics, and habitat (Goodier 1981). Siscowet and humper lake trout, both deepwater forms, are unique to the Great Lakes, and are hypothesized to have evolved within the Lake Superior basin (Burnham-Curtis 1993). Deepwater siscowet-like lake trout were referred to in historical records of fish communities in Lake Michigan (Koelz 1926; Brown et al. 1981) and Lake Huron (Eshenroder et al. 1995a), though none have been seen since the 1920s. A genetic basis for the observed phenotypic variation among Great Lakes lake trout populations has been supported in breeding studies and investigations of morphological and physiological differences (Eschmeyer and Phillips 1965; Khan and Qadri 1971; Burnham-Curtis and Smith 1994). Genetic diversity among lake trout populations is modest (Dehring et al. 1981; Ihssen et al. 1981a; Grewe and Hebert 1988; Burnham-Curtis 1993); however, specific genetic contributions to phenotypic differentiation are insufficient to support the designation of these phenotypes as distinct subspecies.

By the mid 1900s, lake trout were commercially extinct in all of the lakes except Superior, although populations there had seriously declined (Christie 1974). Overfishing and sea lampreys were involved in these losses to various degrees in each of the lakes (Hile et al. 1951; Christie 1973; Hartman 1973; Coble et al. 1990, 1992; Eshenroder 1992; Eshenroder et al. 1995a). Hatchery supplementation in Lake Superior, begun in earnest in 1958 to coincide with the start of chemical control of sea lampreys, continues through the present. At least three local populations of lean lake trout are currently considered to be self-sustaining (Swanson and Swedberg 1980; Curtis 1991; Curtis et al. 1995), and the numbers of naturally-produced lake trout in spawning aggregations are increasing (Hansen et al. 1994).

Stocking for purposes of rehabilitation was begun in Lake Michigan in 1965 (Wells and McLain 1973), Lake Huron in 1969 (Eshenroder et al. 1995a), Lake Erie in 1969 (Cornelius et al. 1995), and Lake Ontario in 1974 (Elrod et al. 1995). The source populations for the planted lake trout included broodstock of Lake Superior and Lake Michigan origin, as well as broodstock developed from Seneca Lake, New York, and from Canadian inland lakes. Successful natural reproduction has been documented from both Lake Michigan (Jude et al. 1981; Marsden 1994) and Lake Huron (Nester and Poe 1984; Anderson and Collins 1995; Johnson and Van Amberg 1995) though sustainable recruitment is not evidenced.

The burbot is found throughout the Great Lakes, but is restricted to cooler waters during thermal stratification (Hopkins and Ritchie 1943; Day 1983). It has a circumpolar distribution, and its range lies within the limits of glaciation (McPhail and Lindsey 1970; Scott and Crossman 1973). Burbot were generally abundant in the early 1900s, and were a conspicuous part of the commercial by-catch in nets set for the more desirable salmonid species (Dymond et al. 1929; Van Oosten and Deason 1938; Kolbe 1944). In general, burbot, like lake trout, were adversely affected by sea lampreys and declined when lake trout declined (Smith 1972). They have recovered strongly in the upper Great Lakes (Lawrie 1978; Eck and Wells 1987; Ebener 1995).

Coho salmon from the Columbia River were successfully introduced into Lakes Superior and Michigan in 1966. Between 1967 and 1988, smolts from the Cascade River (Oregon), the Toutle River (Washington), and from Alaska were planted in Lake Michigan tributaries (Keller et al. 1990). Strong runs in 1967 ushered in the modern period of salmon and trout enhancement in the Great Lakes. Chinook salmon from a Columbia River tributary were reintroduced into Lake Michigan in 1967 and a Puget Sound strain was added in 1969 (Keller et al. 1990). Between 1966 and 1988, a total of 450 million salmon and stream-spawning trout had been stocked in the Great Lakes (Great Lakes Fishery Commission, file data). Pink salmon were inadvertently released into Lake Superior in 1956, and odd-year runs were widely established by 1969 (Lawrie and Rahrer 1973). Even-year runs were evident by the late 1970s (Kwain and Chappel 1978), but populations waned in the late 1980s (Peck et al. 1994). Pink salmon are a minor species in all of the Great Lakes. Kokanee salmon were introduced in Lake Ontario in 1950 and in Lake Huron in 1964 (Parsons 1973), but these introductions failed.

Planktivore Succession

The lake herring, a shallow-water cisco, is distributed throughout the eastcentral United States and most of Canada, from eastern Quebec to Hudson Bay (Scott and Crossman 1973). Lake herring were exceedingly abundant in the Great Lakes in the late 1800s, and were fished mainly during their spawning season. By the early 1900s, they were rare in the western basin of Lake Erie (Hartman 1973), and were commercially extinct in Lake Ontario (Koelz 1926). Lake herring sustained abundant populations in the upper lakes until the 1950s, when severe population declines occurred. In Lakes Huron and Michigan, the decline in lake herring abundance was attributed to a combination of factors, including the increase in abundance of the invading alewife and smelt, overfishing, and eutrophication (Beeton 1969; Smith 1970; Wells and McLain 1973; Christie 1974; Hartman 1988). Lake herring are abundant in Lake Superior and in extreme northern Lake Huron.

The coregonine species complex in the Great Lakes included, in addition to the lake herring, several closely related, but ecologically and morphologically diverse species collectively referred to as deepwater ciscoes. The diversity of deepwater ciscoes is unique to the Great Lakes. Two species flocks comprising six species are recognized: the *hoyi-kiyi-nigripinis* flock (bloater, kiyi, and blackfin cisco) and the *zenithicus-johannae-reighardi* flock (shortjaw cisco, deepwater cisco, and shortnose cisco; Todd and Smith 1992). While the lake herring and shortjaw cisco are known from outside the Great Lakes basin, the other species are thought to be endemic to the Great Lakes (Smith and Todd 1984). Two of the endemic species, the bloater and the blackfin cisco, occur in Lake Nipigon, which connects with Lake Superior (Scott and Crossman 1973).

Among the first species flock, the bloater and kiyi were present in all the lakes except Lake Erie (Koelz 1929; Smith and Todd 1992). The bloater, the smallest in size of the deepwater ciscoes, has a less-restricted spawning time than other Great Lakes coregonines (Jobes 1949b; Dryer and Beil 1964). The kiyi, the second-smallest species, occurred most frequently at depths exceeding 100 m (Koelz 1929; Pritchard 1931). Though kiyi could be captured in certain areas of the Great Lakes, their abundance was not great, nor was there much commercial interest in them in the early 1900s (Koelz 1929). The blackfin cisco, the third member of this flock, was present only in Lakes Michigan and Huron, where it occupied deeper waters than most other ciscoes, except for the deepwater cisco (Koelz 1929; Smith and Todd 1992).

Among the second species flock, the shortjaw cisco was found in every lake but Lake Ontario, and the deepwater cisco occurred only in Lakes Huron and Michigan. The deepwater cisco was one of the largest species among the deepwater ciscoes, and was exceeded in size only by the blackfin cisco in Lake Michigan (Koelz 1929). The shortnose cisco, the third member of this flock, was found in all lakes except Lakes Erie and Superior (table 1, Todd and Smith 1992: the authors note some typesetting errors in this table). It was probably present historically in Lake Huron, but was unknown until 1956 (Scott and Smith 1962). Shortnose ciscoes were unique among the Great Lakes coregonines because they were spring rather than fall spawners (Jobes 1943).

Measures of individual abundance of the deepwater ciscoes are scarce because they were marketed collectively as chubs, their common name in the commercial fishery. By 1900, the largest cisco species (identity uncertain) was commercially extinct in Lake Ontario (Koelz 1926; Todd 1981a). In Lake Michigan, fishing caused declines of the two largest species, the blackfin cisco and the deepwater cisco. By the 1930s, both species were sparsely represented, and were being replaced by expanding populations of the shortjaw cisco, shortnose cisco, kiyi, and bloater (Smith 1964). In central Lake Huron, four species of deepwater cisco (bloater, kiyi, shortnose cisco, and shortjaw cisco) were prominent in gill nets set in 1956 from the M/V *Cisco* (U.S. Fish and Wildlife Service, cruise reports), just a decade before these populations began a pronounced decline (Brown et al. 1987).

Deepwater ciscoes are presumed extinct in the lower lakes (Christie 1974), and are reduced to a single introgressed species (i. e., the bloater) in Lakes Michigan and Huron (Smith and Todd 1984). Intensive fishing, the increase in abundance of alewives and smelt, and predation by sea lampreys all contributed to the extinction of most of the populations (Smith 1968). In Lake Superior, the bloater and kiyi remain abundant, and the shortjaw cisco is rare (Peck 1977; Todd and Smith 1992).

Benthivore Succession

The lake whitefish is widely distributed among freshwater lakes in northern North America (Scott and Crossman 1973). Like the lake trout, the distribution of the whitefish lies within the limits of glaciation, and the species is absent from Eurasia, although close relatives are present under similar ecological conditions (Behnke 1972; Bailey and Smith 1981). Both river and lake-spawning populations of whitefish were abundant, but sharp declines were widespread by the late 1890s (Koelz 1926), due mainly to overfishing and drainage modification (Smith 1995). Whitefish populations recovered, however, and harvests began to approach historical levels by the 1920s (Fleischer 1992). More recently, in the upper lakes, fishing and sea lampreys depressed whitefish populations during the 1950s (Christie 1973). Populations in Lake Erie collapsed in the 1950s, due to environmental degradation (Hartman 1973), and those in Lake Ontario fell in the 1960s without a clear cause (Christie 1973). Whitefish abundance in the Great Lakes increased steadily after sea lamprey control measures were implemented (Smith and Tibbles 1980; Fleischer 1992).

Three species of sculpin inhabited the offshore waters of the Great Lakes (Scott and Crossman 1973). Slimy sculpins are widely distributed through northern North America and extreme northeastern Siberia (McPhail and Lindsey 1970). The spoonhead sculpin, which is less widely distributed than the slimy sculpin, is confined to Canada and the Great Lakes (McPhail and Lindsey 1970). Deepwater sculpins have a circumpolar distribution, and were abundant in the deepest areas of all the Great Lakes.

During the 1950s, deepwater sculpins disappeared from Lake Ontario but remained abundant in all of the upper Great Lakes (Wells 1969; Christie 1973). This event was remarkable in that it only happened in Lake Ontario, and only after the loss of the sculpin's major predators, lake trout and burbot (Brandt 1986a). Deepwater and slimy sculpins show overlap in habitat use, especially in the summer, and Brandt (1986a) hypothesized that competition between these species, which increased in the absence of the keystone predators, caused the extinction of the deepwater sculpin.

Structural Succession

Compositional changes in the offshore fish community, especially since the 1940s, have produced marked successional changes in community structure. Most affected were the spatial distribution of biomass, size distribution of planktivores, and mean age of piscivores. These structural changes affected the flow of energy through and between trophic levels. In this section we focus on the major structural changes, and how they affect trophic connectedness.

Community Structure at the Turn of the Century

Historically, the offshore community was structurally simple. The pelagic waters contained eggs and fry of deepwater ciscoes, deepwater sculpins, and burbot (Crowder 1980), juvenile deepwater ciscoes, and lake herring (Smith 1970). Adult lake herring were common in pelagic waters in the vicinity of the thermocline, but were more abundant in inshore waters. The Atlantic salmon provided a pelagic structural element only in Lake Ontario. The benthic waters were structurally more complex than pelagic waters because of the abundance of adult deepwater ciscoes, juvenile and adult sculpins, lake trout, and burbot (Smith 1970). Adult-sized lake trout also inhabited surface waters in the upper lakes, as evidenced by catches made in the float-hook fishery (Eshenroder 1992). Sculpins and deepwater ciscoes were the major prey species for lake trout and burbot in benthic waters (Van Oosten and Deason 1938).

The division of the offshore assemblage into pelagic and benthic components is illustrative, but makes the structure appear simpler than it was because adult planktivores appear to be scarce at mid-depths. The benthic community provided structure at mid-depths, however, because deepwater ciscoes were migrating vertically at night. The significance of vertical migration as a major structural property of the benthic community and as a mechanism that trophically linked both communities has not been generally recognized.

Relative Size of Great Lakes Crustaceans

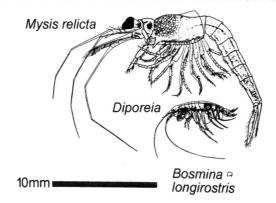

FIG. 2. *Illustration of size differences among* Mysis, Diporeia, *and* Bosmina, *a common zooplankter. Reprinted from Balcer et al. (1984) by permission of The University of Wisconsin Press.*

Importance of Vertical Migration

Coregonine planktivores colonizing the benthic waters of the Great Lakes were selected for reduced specific gravity because the amount of buoyancy regulation provided by their swim bladders inhibited their ability to feed on *Mysis relicta*, an important prey species that vertically migrates at night. *Mysis* is common in glacially scoured lakes in northeastern North America, and makes the most extensive and rapid vertical migration of all the freshwater crustaceans (Pennak 1953). Beeton (1960) observed migrations of 96m in Lakes Michigan and Huron, and he measured the rate of ascent at between 0. 5 to 0. 8m/min. *Mysis* is widely distributed in deep waters in Lakes Michigan and Ontario and is more abundant beyond 100m depths (Reynolds and DeGraeve 1972; Johannsson 1992). In contrast, *Diporeia*, the other macroinvertebrate important for planktivores, is less abundant beyond 50m (Robertson and Alley 1966; Nalepa 1989), and a much smaller fraction of the population vertically migrates (Marzolf 1965; Wells 1968). Wells and Beeton (1963) demonstrated a positive relation between water depth and increased consumption of *Mysis* by bloaters in Lake Michigan. Also, when adult bloaters began to proliferate in Lake Michigan after 1980, *Mysis* declined most severely at depths greater than 80 m (McDonald et al. 1990).

Mysis is nutritionally important for fish because of its large size and energy density. Adult *Mysis* achieve body lengths of 12–22mm (Sell 1982), whereas most zooplankters have lengths less than 1mm (fig. 2), and its energy value on a relative-weight basis exceeds that of cladocerans by a factor of 2.8 (Stewart and Binkowski 1986). Accounting for dry-weight (Morsell and Norden 1968) and energy-density differences between *Mysis* and zooplankters, ingestion of one *Mysis* by a planktivore is energetically equivalent to ingestion of two thousand to three thousand small cladocerans.

To feed on *Mysis* at night, deepwater ciscoes need to migrate vertically with them, and if salmonines want to feed on migrating coregonines at night, they must also migrate. However, the amount of swim-bladder gas that a physostomous fish needs to release to avoid bloating on an extensive diel migration cannot be replaced. Alexander (1993) noted that no fish has been shown to secrete gas fast enough to compensate for depth changes of more than 2. 5m/hr. Without other means of regulating buoyancy, a fish with a swim bladder will not be able to maintain neutral buoyancy at both ends of a vertical migration. The energetic penalty (increased swimming) for loss of buoyancy can amount up to 60% of the energy requirements for a fish swimming at one body length per second (Alexander 1993).

A freshwater fish without a swim bladder will be neutrally buoyant if it is about 50% triglycerides wet weight (Alexander 1993). Vertically migrating fish possessing a swim bladder typically regulate the quantity of swim-bladder gas so that they will be neutrally buoyant when near the top of an ascent at night. High levels of lipids help compensate for a swim bladder that is greatly compressed when a migrating fish, like a deepwater cisco, is on the bottom during the day. Thus, lipid levels are one indication of how efficiently a fish can migrate vertically, and those individuals with the highest levels should be found in the deepest waters of the Great Lakes.

The bloater has the highest lipid level of any prey-fish species in the offshore community of Lake Michigan. Bloaters collected in the fall averaged 21% lipids by weight from 1969 to 1976, after which lipid levels declined as population abundance increased (Hesselberg et al. 1990). In contrast, slimy sculpins from Lake Ontario had lipid levels of only 4% to 5% (Borgmann and Whittle 1992), and deepwater and slimy sculpins from Lake Michigan had a mean lipid level of 5% (Rottiers and Tucker 1982). Adult alewives collected from Lake Michigan from 1979 to 1981 were 16% lipids by weight in October and 5% to 7% in summer (Flath and Diana 1985). Smelt collected from throughout the open-water season in Lake Ontario from 1977 to 1988 were only 4% to 5% lipids by weight (Borgmann and Whittle 1992).

The extinct species of deepwater ciscoes, which lived at greater depths and were larger bodied than the bloater (Smith 1964, 1968; Dryer 1966), may have had even higher levels of lipids. This assumption is based on an allometric relation between size and lipid level for bloater (Hesselberg et al. 1990). Bloaters in Lake Huron are thought to have hybridized with lake herring, a less-oily pelagic cisco (Todd and Stedman 1989). Severe disruption of the cisco population of Lakes Michigan and Huron beginning in the 1940s (Smith 1968; Brown et al. 1987) favored hybridization in both lakes (Todd and Smith 1992). Consequently, recent lipid levels in Lake Michigan bloaters may be atypically low for other deepwater ciscoes of similar size.

The extraordinary level of lipids in siscowet lake trout make them the best adapted of the Great Lakes fish for vertical (i. e., diel) migration. Eschmeyer and Phillips (1965) found that the flesh of Lake Superior siscowets was 38% to 48% lipid (wet weight) for fish greater than 635mm total length whereas lean (inshore) lake trout of the same length range were 10% to 17% lipid. The largest siscowets had a lipid composition in their flesh identical to the whole-body level that Alexander (1993) estimated would achieve neutral buoyancy in a freshwater fish without other buoyancy aids (such as a swim bladder). Of interest, Crawford (1966) reported that swim bladders from large siscowets did not contain any gas (i. e., they were not functional).

Siscowet were generally considered to be a bottom fish because they are caught in the profundal zones of Lake Superior, but a rationale for the high lipid levels was always missing. Eschmeyer and Phillips (1965) showed that the difference in fat content between lean lake trout and siscowet was heritable, and not the result of a siscowet diet richer in oily deepwater ciscoes. An association between high lipid content and the ability to vertically migrate is so well accepted across many fish taxa (Alexander 1993), that a similar association for siscowet should be suspected. Siscowets

Distribution of Offshore Great Lakes Fishes

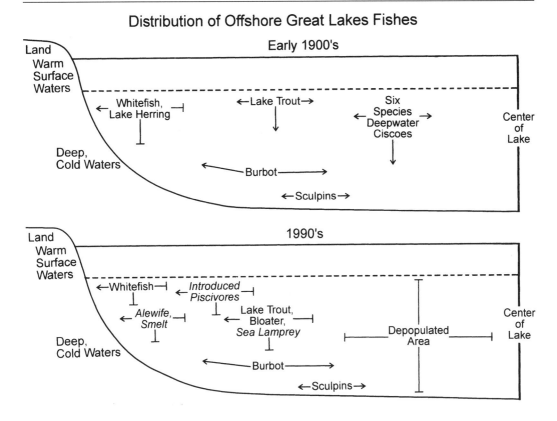

FIG. 3. *Stylized distribution of offshore fish in Lake Michigan in the early 1900s and in the 1990s. Introduced species in italics (introduced piscivores: rainbow trout, brown trout, coho salmon, and chinook salmon). Arrows indicate continuous distribution in the given direction.*

inhabit deeper water than lean lake trout, and may have had to migrate further to prey on deepwater ciscoes at night.

This discussion of the offshore fish community assumes, based on indirect evidence, that vertical migration was a key structural feature of the benthic community, and a mechanism for linking the benthic and pelagic communities. Acoustic assessment provides direct evidence that bloaters leave the bottom at night (Argyle 1992), but how high they migrate has not been reported. That all the endemics in the offshore community, five species of deepwater cisco and the siscowet, had the same morphological adaptation suggests common causation. The hypothesis advocated here is that the cause was a vertically migrating food web of exceptional scale for freshwater systems.

Recent Changes in Community Structure

The structure of the offshore fish community in the Great Lakes has been modified by extinction and depletion of native species, as well as the introduction of non-native fish (fig. 3). Among planktivores, one of the most profound structural changes has been the replacement of lake herring and deepwater ciscoes with smelt and alewife (Smith 1968, 1970). These changes have been most dramatic in Lake Ontario, and least severe in Lake Superior. In 1930, the species of deepwater cisco collected from experimental gill nets in Lake Michigan had average lengths ranging from 203 to 333mm (Smith 1964), and lake herring collected from pound nets fished from 1948–51 averaged from 236 to 285mm in total length (Smith 1956). In comparison, mean lengths (mm) of planktivores collected from Lake Michigan by mid-water trawl in 1987 were: alewife 66, smelt 109, and bloater 151 (Argyle 1992). Clearly, sizes of planktivores have declined in the Great Lakes.

Only minor changes in age structure have occurred with compositional changes among planktivores. Four-year-olds were the modal group in studies on bloater (Jobes 1949a), kiyi (Deason and Hile 1947), alpenae (now the shortjaw cisco) (Jobes 1949b), and the shortnose cisco (Jobes 1943) from Lake Michigan. Few ciscoes lived past age 6 in these fished populations, but small numbers lived to ages seven and eight. O'Gorman et al. (1987) assessed ageing techniques for alewife collected from Lakes Michigan, Huron, and Ontario, and found that scale-aged fish reported to be in age groups three and four (the modal groups) were as much as seven years older. In contrast, the partial replacement of deepwater ciscoes by rainbow smelt, a shorter-lived species in the Great Lakes, constitutes a reduction in planktivore age.

Compositional changes among offshore planktivores have resulted in structural changes in horizontal distribution. Koelz (1929) reported that blackfin ciscoes and shortjaw ciscoes were abundant out to 100 fathoms (183m) and that bloaters, deepwater ciscoes, and shortnose ciscoes were abundant out to 90 fathoms (164m). Smith (1964) showed that the Lake Michigan bloater became markedly more abundant at greater depths (91–110m) in the mid 1950s, following population reductions in the other ciscoes. In contrast to the diversity and abundance of ciscoes in the 1920s, recent acoustic surveys show that alewives and smelt inhabit much shallower waters. In Lake Michigan in 1987, the bulk of the alewife biomass was at depths less than 64m in spring and 55m in late summer (Argyle 1982). In Lake Huron from 1974 to 1976, the bulk of the biomass of both species was at depths less than 75m (Argyle 1982), and in Lake Superior smelt abundance from 1978 to 1980 was extremely low at depths greater than 50m (Heist and Swenson 1983). Hatch et al. (1981) gave a correction factor of only 10% to account for the alewife biomass beyond the 55-m depth contour in Lake Michigan. Likewise, since their recovery in Lakes Michigan and Huron in the late 1970s, bloaters were concentrated inside the 55-m depth contour (Brown et al. 1985, 1987), although they were moderately abundant in Lake Michigan out to depths of 73m in spring (Argyle 1992). The difference between the historical and contemporary distributions of bloater in Lake Michigan may be explained by introgression of bloater and lake herring when both species experienced population bottlenecks during the 1960s (Todd and Stedman 1989). Hybridization would have resulted in a leaner species less suited for vertical migration in deep water. Conversely, the more recent concentration of ciscoes in shallow water may be, in part, an artifact of sampling bias as the recent data were acquired from acoustic and trawl surveys, whereas the historical data were collected from gill nets.

Hypsographic curves indicate that about 50% of Lake Huron's main basin and 60% of Lakes Michigan's and Ontario's surface areas are at depths in excess of 55m (Christie and Regier 1988). The recent distribution of planktivores in the offshore community suggests that the deepest areas of these lakes lack connections between trophic levels. *Mysis*, which appears to be the key food source for vertically migrating planktivores, is most abundant at depths beyond 55m (Reynolds and DeGraeve 1972; Johannsson 1992), depths where planktivore abundance is now reduced, except in Lake Superior. A considerable area of Lake Ontario appears to be even more impoverished than Lakes Michigan and Huron because of the extinction of the deepwater sculpin during the 1950s (Wells 1969).

Sizes of introduced salmonines are within the range of the historically dominant native piscivore, the lake trout, but they are shorter-lived species (Stewart et al. 1981) with higher P/B ratios, a factor associated with reduced sustainability. Coho and chinook salmon, steelhead, and brown trout, all intensively stocked, seldom exceed an age of five years, whereas lake trout commonly exceed an age of twenty years (Martin and Olver 1980). Model projections for Lake Michigan indicated that the stocking rates in the 1980s would result in chinook being three times as abundant as lake trout, and coho and lake trout being of equal abundance (Stewart and Ibarra 1991). This stocking strategy, which is also common to Lakes Ontario and Huron, produced a dominant piscivore (chinook) very vulnerable to short-term fluctuations in recruitment. Two years of poor recruitment would essentially eliminate any of the introduced piscivores as a dominant species. Such a loss of dominance began in Lake Michigan in 1987, when chinook began to die off in association with a proliferation of bacterial kidney disease (Nelson and Hnath 1990).

The current management focus on introduced salmonines is intended to produce a fish community dominated by pelagic piscivores whereas benthic piscivores were historically dominant. For instance, the projections of Stewart and Ibarra (1991) indicate that the stocking strategy for Lake Michigan would result in a salmonine community with four to five times as many pelagic piscivores (chinook, coho, brown trout, steelhead) as benthic piscivores (lake trout). However, the recovery of burbot in Lakes Michigan and Huron (Eck and Wells 1987; Bowen et al. 1994) is reducing the structural disparity between benthic and pelagic piscivores.

Functional Succession

Compositional and structural characteristics of fish-community succession in the Great Lakes are ultimately shaped by the functional ability of the species to persist. Compositional changes, including extinctions and introductions, result in functional changes because the replacements have different anatomical, physiological, and behavioral traits than the originals. In this section, we examine functional characteristics of the major offshore fish and relate them to changes in community composition and structure. The effect of functional change on trophic connectedness is important because of the relationship between trophic connectedness and sustainability.

Functional Aspects of Planktivory

Offshore planktivores in the Great Lakes, whether introduced or native, exhibit similar behavior—they migrate from the lake bottom at night. Trawl data from Lake Michigan demonstrate the reduced biomass of planktivores on the bottom at night: relative to daytime estimates, bloater were reduced by 88%, alewife 99%, and smelt 96% (Argyle 1992). Brandt et al. (1991) reported reductions of a similar magnitude for all species combined. Although the distribution of planktivores

PLANKTIVORE SIZE VERSUS DEPTH IN LAKE MICHIGAN

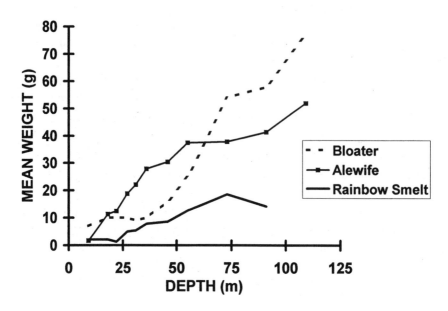

FIG. 4. *Mean weight of bloater, rainbow smelt, and alewife collected in bottom trawls in Lake Michigan versus depth during the summer of 1987 (unpublished data from Argyle 1992).*

varies with depth and season, recent efforts to enumerate planktivores acoustically provide evidence that vertical migration is a salient feature of this trophic level (Brandt et al. 1991; Sprules et al. 1991).

Diet studies suggest a relationship between planktivore size, and planktivory on *Mysis*, implying vertical migration. Wells and Beeton (1963) found that bloaters greater than 179mm in total length in Lake Michigan consumed an increasing proportion of *Mysis* with depth. The largest bloaters, alewives, and smelt in Lake Michigan tend to occur in deep water (fig. 4) where *Mysis* is more abundant. In shallow water in Lake Michigan, very large smelt (greater than 180mm in length) fed mostly on age-0 alewife, and medium-sized smelt fed predominately on *Mysis*, but in deeper water (74m) smelt of all sizes fed almost exclusively on *Mysis* (Foltz and Norden 1977).

Extensive planktivory on *Mysis* by adult alewife is less clear than for bloater and smelt. Energetic models for bloater and smelt indicate high utilization of *Mysis* (Lantry and Stewart 1993; Rudstam et al. 1994), but Stewart and Binkowski (1986), in their synthesis of alewife energetics in Lake Michigan, employed only a 10% to 15% proportion of *Mysis* in the diets of age 2+ alewife. Wells (1980), who assessed alewife diet in southeastern Lake Michigan, suggested that *Mysis* may be of importance in their total diet, since it was common in stomachs of larger alewives sampled at night at middepths. Janssen and Brandt (1980) reported that the largest alewives at their sampling site in Lake Michigan (located at the 50-m contour) migrated with, and selectively fed on the largest *Mysis*, and smaller alewives migrated higher in the water column, apparently to feed on microcrustaceans. Alewives in Lake Ontario fed primarily on microcrustacean zooplankters (Mills et al. 1992), as Wells observed in Lake Michigan. The relatively low utilization of *Mysis* by alewife has functional implications because, unlike bloater and smelt, alewife exhibit pronounced seasonal declines in lipid levels (Flath and Diana 1985; Rand et al. 1994; Rudstam et al. 1994), undergo large die-offs (Colby 1971; O'Gorman and Schneider 1986; Bergstedt and O'Gorman 1989), and adults are extremely inefficient at converting available food to growth (Stewart and Binkowski 1986).

Alewives are considered to be very efficient at feeding on microcrustacean zooplankton and larval fish, because they use three feeding behaviors (Wells 1970; Janssen 1978; Crowder 1980; Brandt et al. 1987; Johannsson et al. 1991; O'Gorman et al. 1991). Yet, in the Great Lakes, their energy content is highly variable seasonably and annually, and their annual growth is faster the more they feed on *Mysis* (Rand et al. 1994). Hewett and Stewart (1989) assumed in their models that *Mysis* and *Diporeia* in Lake Michigan make up 25% to 45% of the adult alewife diet (by weight) in summer and 0% to 25% of the yearling alewife diet, but in Lake Ontario consumption of high-energy macroinvertebrates was only 10% of their diet (Mills et al. 1992).

This difference between the two lakes likely accounts for the higher energy density of adult Lake Michigan alewives (Rand et al. 1994). Alewives in both lakes lost weight, or did not grow on a summer diet of microcrustaceans but grew best in the fall, when they did consume more *Mysis* (Stewart and Binkowski 1986; Mills et al. 1992; Rand et al. 1994). The density and/or size of microcrustaceans in the epilimnion may be marginal even for an efficient filter-feeder like alewife. Particle-size spectra indicate that marine zooplankters (the natural food of alewife) are typically larger than those in Lake Michigan (fig. 5 in Doudreau and Dickie 1992), even allowing for the fact that the Lake Michigan data was from 1987 when cladocerans were unusually large in that lake (Evans 1992).

An ability to forage on *Mysis* appears to be nutritionally important for Great Lakes planktivores, even though the temperature in the hypolimnion where *Mysis* live is well below their physiological optima. Adult alewives prefer 16°C (Stewart and Binkowski 1986), but some portion of the population in Lake Michigan exists in the hypolimnion in less than 8°C during the day (Janssen and Brandt 1980). Crowder and Magnuson (1982) theorized that adult alewives in Lake Michigan shifted from metalimnetic to hypolimnetic waters to avoid competition with juvenile bloater that became very abundant between 1977 and 1979. The bimodal distribution of adult alewife in Lake Michigan, concurrent occupation of the hypolimnion and surface waters, may represent a trade-off among individuals that is less favored in Lake Ontario, where alewives are predominately epilimnetic (Mills et al. 1992). In contrast, adult bloaters in Lake Michigan occupy only the hypolimnion. They are physiologically adapted to grow well at low temperatures, even though their optimum temperature for consumption is surprisingly high at 17°C (Binkowski and Rudstam 1994). The functional ability of bloaters to grow over a wide-range of temperatures explains how the age at which pelagic juveniles recruit to the adult (benthic) population can quickly shift in response to changes in interspecies competition (Crowder and Crawford 1984).

Laboratory estimates of temperature preference for smelt are lacking, but adults inhabit the area below the thermocline in thermally stratified lakes (Lantry and Stewart 1993). Age 2+ smelt in their energetics model consumed mostly *Mysis* and fish. Adult smelt, like bloater, probably occupy temperatures below their optima during the day. They appear to be well adapted to feed on *Mysis* even at age one, and at age 2+ forego *Mysis* only when other large nutritious prey such as *Diporeia* and fish are available (Lantry and Stewart 1993). This behavior was thought by Rand et al. (1994) to account for the less variable energy density of smelt relative to alewife.

Planktivory on *Mysis* may be enhanced when *Mysis* are off the bottom at night. Janssen and Brandt (1980) suggested that *Mysis* were more available to alewives when they were silhouetted against down-welling light. Janssen's (1978) laboratory experiments with alewives and bloaters indicated that alewives were not effective predators on *Mysis* when both were on the bottom. *Mysis* were fast swimmers, and easily evaded alewife. Photophobia in *Mysis* (Beeton 1960) may make the species less vulnerable to planktivores during the day. Mysids tolerate much higher light intensities than found in the profundal waters of the Great Lakes (Beeton 1960; Degraeve and Reynolds 1975). Although the bloater appeared to be an effective predator on *Mysis* when both were on the bottom in experimental tanks (Janssen 1978), *Mysis* may typically be more vulnerable to bloater and other planktivores when it vertically migrates, a behavior that places a premium on functional adaptations in planktivores for vertical migration.

The low lipid-content of alewives, especially during summer, may detract from their ability to vertically migrate and prey on *Mysis*. Adult alewives studied by Janssen and Brandt (1980) experienced pressure changes of from three to four atmospheres during migration, and would have had greatly compressed swim bladders when on the bottom during the day and presumably would not be neutrally buoyant. This functional limitation may explain why a substantial proportion of adult alewife populations apparently foregoes benthic foraging altogether, and why others that do inhabit the bottom remain predominately at depths less than 55m.

The winter biology of alewives may be an important determinant in their functional ability to prey on *Mysis* during the summer and fall. In Lake Michigan, lipid levels in adult alewives drop from about 16% in fall to 10% to 12% in April, when lipids are located mainly in the gonads (Flath and Diana 1985). Bergstedt and O'Gorman (1989) expected that alewives would overwinter at greater depths in Lake Ontario (a maximum of 110 m) than they did. Colby (1971) demonstrated that alewives were severely stressed by temperatures lower than 3°C. Temperatures of 3. 7°C were available in Lake Ontario in February at depths greater than 130m, but alewives did not occupy them. Maintaining neutral buoyancy may be a problem for alewives at greater depths because of increased gas diffusion from the swim bladder to the blood and gills (see Alexander 1993). In winter, alewives may be faced with selecting a depth that reduces exposure to stressful temperatures, but does not require excessive swimming to compensate for a reduced swim-bladder volume. Regardless of what causes the depletion of energy reserves in overwintering alewives (the buoyancy hypothesis is very speculative), reduced loss of lipids could improve the efficiency of alewives at preying on *Mysis* during summer and break the functional bottleneck that diminishes their fitness for the Great Lakes.

Functional Aspects of Piscivory

Functional aspects of salmonine piscivory in the Great Lakes are of interest because of continuing compositional and structural changes among planktivores. A successional process, at various stages among the lakes, has resulted in partial recovery of ciscoes in the upper lakes, decline of alewives in Lakes Michigan, Huron, and Ontario, and decline of smelt in Lakes Superior and Erie (Brown et al. 1987; Henderson and Nepszy 1989; Jones et al. 1993; Selgeby et al. 1994). Where ciscoes are in recovery, the planktivore community is comprised of larger-bodied individuals, and occupies deeper water. Planktivore succession appears to be driven by piscivores (in Lake Erie, also by a commercial fishery for smelt). Alewife and smelt predominate in the diet of salmonines, even when other planktivores are more abundant (Brandt 1986b; Jude et al. 1987; Henderson and Nepszy 1989; Diana 1990; Conner et al. 1993).

Structurally, the Great Lakes salmonine community is predominately pelagic. The question probed here is whether a pelagic species such as the chinook, the salmonine most desired by anglers, can be a dominant piscivore when ciscoes are dominant planktivores.

Stewart and Ibarra (1991) assumed a thermal preference of 11°C for adult chinook in Lake Michigan, and suggested, based on energetic models, that chinook would lose weight if they fed on adult bloater because they reside in water below 8°C. Chinook stomachs from Lake Michigan collected in spring, when water temperatures remain cold, were almost always empty, whereas lake trout fed throughout the winter (Elliott 1993). Energetic costs of planktivory in colder water during stratification could be reduced if chinook consumed adult bloaters at night when they are higher in the water column. This feeding strategy would lessen compression of the swim bladder caused by a descent all the way to the bottom. Also, if chinook returned to warmer water in the morning after feeding on bloater during the night, digestion could continue at warmer temperatures nearer their thermal preference, which should improve assimilation. Such a feeding migration is essentially the reverse of that exhibited by planktivores.

A downward migration at night by chinook to prey on adult bloaters may be thwarted, however, because they cannot visually locate bloaters at night. Other *Oncorhynchus* species, rainbow trout and kokanee, were not successful in feeding on *Mysis* introduced for them in western North American lakes (Martinez and Bergersen 1989). In this study, both the absence of coevolution with *Mysis* and dependance on sight feeding were hypothesized to account for the inability of these species to locate *Mysis* in pelagic waters at night. If these species could not see a large invertebrate like *Mysis* at night, they probably could not see prey fish were they available under the same conditions.

Chinook are considered to be visual predators, with limited ability to forage under conditions of very low light intensity. In the Pacific Ocean, chinook have been taken at depths down to 110m (Healey 1991), but Orsi and Wertheimer (1995) suggested that they were only deeper when conditions were favorable, such as in early spring and fall, for deeper transmission of light. Also, chinook rose in the water column at night, in association with declining light levels. Observations of chinook diets in Lake Michigan are consistent with the marine data. Juvenile bloaters, which are pelagic, are found in the chinook diet, but adult bloaters are rarely found (Elliott 1993), even though they are the dominant planktivore (Argyle 1992). Jude et al. (1987) concluded that the gape size of the largest salmonines (>70cm total length) in Lake Michigan could accommodate larger prey if the prey was available and susceptible. Abundance declines in large alewife, a preferred prey of chinook, were associated with lower weights-at-age for chinook (Stewart and Ibarra 1991). Ecological (i. e., prey abundance), physiological (i. e., reduced weight), and anatomical (i. e., gape size) incentives exist for chinook to prey on adult bloaters in Lake Michigan. Chinook visual acuity under conditions of low light may explain why such predation does not occur.

Lake trout, like chinook, did not readily switch to ciscoes when ciscoes became more abundant in the upper lakes. Ciscoes comprised 68% of the spring forage biomass in Lake Superior, but the frequency of occurrence of ciscoes (mostly lake herring) in the diets of inshore lake trout and chinook was only 15% and 19%, respectively (Conner et al. 1993). In Lakes Michigan and Huron, inshore lake trout rarely have adult bloater in their diet (Jude et al. 1987; Diana 1990; Elliott 1993), but the diet of offshore lake trout in Lake Michigan is dominated by bloater (Miller and Holey 1992). The low use of abundant adult ciscoes by chinook in Lakes Superior and Michigan is not readily understood (Conner et al. 1993). However, lake trout continued to consume mostly adult-sized alewife and maintained their condition in Lake Michigan, as chinook underwent size and abundance declines (Stewart and Ibarra 1991; Rand et al. 1994). When smelt abundance declined in Lake Superior, chinook populations plummeted, increasing the proportion (dominance) of lake trout in the salmonine community (Peck et al. 1994). The failure of lake trout and chinook to readily switch to ciscoes, when they became more abundant in both lakes, appears to have been more detrimental to the chinook because only its numbers declined.

Lipid levels in chinook and lake trout are of interest because they provide a measure of the energetic efficiency of vertical migration. In Lake Michigan, lipid levels in skin-on fillets were only 2% to 6% (wet weight) in fresh-run spawning chinook in 1980, when alewives, their preferred prey, were still abundant (Rohrer et al. 1982). Data from state of Wisconsin pesticide monitoring programs indicate that lipid levels in chinook muscle tissue varied from 3% to 5% from 1979 to1985, and declined to less than 2% during the first years (1986 to 1987) of the population collapse in Lake Michigan (M. Toneys, Wisconsin Department of Natural Resources, personal communication). In contrast, Pacific Ocean chinook have from 7% to 16% lipids in muscle tissue (cited in Stewart and Ibarra 1991). Lipid levels in the muscle tissue of lean varieties of large lake trout from Lakes Superior, Michigan, and Ontario averaged from 9% to 25% (Eschmeyer and Phillips 1965, DeVault et al. 1985, Borgmann and Whittle 1991, Miller et al. 1992), four times higher than in Lake Michigan chinook. The low levels of lipids in chinook as compared to lake trout suggest that the energetic costs of diel vertical migration for chinook would be higher than for lake trout, as well as predisposing chinook to disease-induced mortality (Kabré 1993).

Functional aspects of piscivory by burbot are important because this little-understood species preys on the same planktivores as do lake trout and chinook. Diet assessments of lake trout and burbot captured on offshore banks in Lake Huron from 1991 to 1993 showed that alewife were the dominant prey of both species (Bowen et al. 1994). In Lake Superior during the 1960s, burbot ate mostly bloater and smelt (Bailey 1972), and in Green Bay, Lake Michigan, burbot consumed mostly alewife and smelt (Fratt 1991). Burbot have a wide depth distribution in the Great Lakes. Van Oosten and Deason (1938) netted juvenile and subadult burbot in Lake Michigan at depths to 185m, and Boyer et al. (1989) observed burbot at depths as deep as 366m in Lake Superior.

Whether burbot vertically migrate with the diel movement of planktivores in the Great Lakes is unknown. Fish oils in the exceptionally large livers of burbot should facilitate vertical migration by reducing their specific gravity. Burbot observed resting on reefs in Lake Michigan during the day (Edsall et al. 1993) must have a specific gravity greater than 1. 0, or they would have had to swim to maintain their position in water currents amplified by the reefs (contour currents). These burbot could ascend in the water column and become neutrally buoyant. Being physoclists, burbot cannot release swim-bladder gas to allow for rapid ascent, and a decompression of only 15m may be fatal (Bernard et al. 1993). Van Oosten and Deason (1938) inferred, based on diet overlap, that lake trout and burbot were competitors in Lake Michigan. A capacity for vertical migration in burbot would add to the versatility of their feeding behavior in the Great Lakes.

Successional Setback and Sustainability

Since the late 1800s, compositional change among offshore fish in each of the Great Lakes clearly resulted in major structural change in the planktivore communities. Extinction, introgression, and/or population decline among the ciscoes and introduction of smelt and alewife generally resulted in a reduction in planktivore size and age, a redistribution of adult planktivore biomass upward in the water column, and a severe reduction in planktivore biomass in deeper water. Despite profound changes that occurred to various degrees and at various times in each of the lakes, similarities between the original and new structures indicate that the ecological properties that gave rise to the original structure continue to shape the new. The similarities between the old and new structures include (1) the positive relation between planktivore size and depth, and (2) the prominence of juvenile planktivores in the epilimnion for both native and introduced species. We have theorized that the ecological properties responsible for these similarities are (1) a marginal density/size distribution of crustaceans in surface waters, (2) a high density of very large crustaceans (*Mysis*) in deep water, and 3) the great depth of the lakes that allows *Mysis* to vertically migrate distances that favor buoyancy regulation adaptations (such as high lipid levels) in planktivores and piscivores.

The diel migration of *Mysis* can be viewed as a fundamental feature of the Great Lakes that affects species succession. Adult bloaters are physiologically better adapted than are adult alewives and smelt for planktivory on *Mysis* because they can better assimilate food at hypolimnetic temperatures (Binkowski and Rudstam 1994), and they possess higher levels of lipids, which facilitate buoyancy regulation. These physiological differences are evidenced by the recent dominance of bloaters in Lake Michigan (Brown et al. 1987), and the wider range of depths that they inhabit (Argyle 1992). Other ecological properties, besides adaptations for *Mysis* planktivory, must affect succession of offshore planktivores in the Great Lakes—otherwise the native planktivores, like the bloater, would quickly replenish themselves following population declines.

Alewife predation on ichthyoplankton also appears to affect planktivore succession. The ability of alewives to structure zooplankton communities in the Great Lakes is well established (Wells 1970; Evans and Jude 1986; Johannsson and O'Gorman 1991), and their suppression of larger zooplankters inshore can be especially severe (O'Gorman et al. 1991; Strus and Hurley 1992). The ability of alewives, when abundant, to selectively feed on and suppress ichthyoplankton is not unexpected. Researchers have attributed alewife predation to recruitment declines in a number of Great Lakes fish species that have pelagic eggs or fry: deepwater ciscoes, lake herring, emerald shiner, sculpins, and yellow perch (Smith 1970; Crowder 1980; Brandt et al. 1987; Eck and Wells 1987). Although adult alewives appear to be at a competitive disadvantage with adult deepwater ciscoes, once abundant, they can dominate planktivore biomass, as they have in Lakes Michigan,

Huron, and Ontario, by preying on the eggs and fry of potential competitors. When alewives are abundant, they inhibit the natural successional process that favors recovery of the native planktivores.

Our interpretation of species succession in the Great Lakes explains why successional setback is maximized when alewives are dominant. Alewives achieved dominance in Lakes Michigan, Huron, and Ontario when piscivores were depleted (Smith 1970; Christie 1974). In Lake Michigan, they remained dominant until piscivores and/or unfavorable environmental conditions reduced their abundance to a level where piscivory was an effective population-regulation mechanism (Kitchell and Crowder 1986; Eck and Wells 1987). Christie (1974) argued that alewives generally do not adversely affect the abundance of ciscoes except, perhaps, when they are very abundant, whereas Smith (1970), Crowder (1980), Eck and Wells (1987), and Brown et al. (1987) argued that they did adversely affect ciscoes. Christie may have been influenced by Lake Huron data indicating that alewives were not abundant in Lake Huron until the 1960s, after lake herring and deepwater ciscoes began to decline (Henderson and Brown 1985). Alewife reproduction, however, was substantial in Saginaw Bay and Lake Huron earlier, at least by 1956 (Carr 1962), when lake herring, which concentrated in Saginaw Bay for spawning (Van Oosten 1929), were declining (Christie 1974). In the main basin of Lake Huron, deepwater cisco populations proved to be very fragile after alewives became abundant (Brown et al. 1987), and they persisted longest in Georgian Bay, where alewives were less numerous (Spangler and Collins 1992) and fishing pressure was relatively light (Brown et al. 1987).

In the upper lakes, planktivore succession appears to respond to opposing processes: 1) the physiological advantages that adult deepwater ciscoes have over adult alewives for foraging and assimilation in deep, cold water and 2) the capacity of the alewife, when abundant, to suppress cisco recruitment by preying on cisco fry. These processes are mediated by piscivore abundance as postulated by Smith (1972), Ryder et al. (1981), Evans et al. (1987), and Christie et al. (1987). In Lakes Michigan and Huron, deepwater ciscoes were never abundant when the alewife was the dominant planktivore (Brown et al. 1987). In Lake Superior, where alewives never were abundant (Bronte et al. 1991), deepwater ciscoes declined (but remain numerous) in association with population increases of their main predator, the siscowet (MacCallum and Selgeby 1987; Selgeby et al. 1994). The existence of similar opposing processes between smelt and lake herring is unclear. Selgeby et al. (1978, 1994) discounted predation and competition between adult smelt and lake herring fry as factors in the decline of lake herring in Lake Superior. Therefore, smelt may not cause successional setback, but an abundance of a short-lived, small-bodied species like the smelt is indicative of it.

Sustainability and Trophic Connectedness

Earlier, we argued that higher states of system maturity would be associated with increased sustainability. Here, we consider alternative states from the standpoint of connectedness and their relation to sustainability. The concept of system maturity for the Great Lakes is somewhat ambiguous because the fish fauna varied among the lakes and an introduction may, over time, increase maturity. For instance, Atlantic salmon coexisted with ciscoes and lake trout in Lake Ontario. From a structural and functional perspective, the presence of Atlantic salmon in the upper lakes would not indicate a loss of system maturity. Therefore, we assume that a fully mature fish community would contain pelagic piscivores.

Recent declines in yields of individual salmonines conceivably related to connectedness (i. e., trophic structuring) have occurred in Lakes Superior and Michigan. The fish community in Lake Superior is structurally more mature than in the other lakes, and succession there may be reaching an end point where planktivore composition and abundance are controlled by piscivores, and piscivores are controlled mainly by the natural fertility of the lake. Smelt populations declined, and lake herring populations recovered at the same time that piscivores, especially lake trout, increased in abundance (Selgeby et al. 1994). Pink salmon in Lake Superior may have been a casualty of the decline of smelt (MacCallum and Selgeby 1987), a species extensively consumed by pink salmon in Lake Huron (Kocik and Taylor 1987).

Returns of hatchery and wild chinook also declined in Lake Superior in the last decade, but yield from other introduced salmonines appears to be stable (Peck et al. 1994). With the replacement of smelt by lake herring, planktivore sizes have increased. Although the vertical distribution of smelt and lake herring appear similar (Dryer 1966), the greater proportion of large smelt in lake trout diet, compared to chinook diet, suggests that smelt occur in deeper waters, and that planktivore biomass shifted upward in the water column when smelt populations declined. Lake Superior is the only lake where planktivore biomass and body sizes are increasing in waters above the hypolimnion. These structural changes, caused by a recovering lake herring population, should favor piscivores like chinook. Lake herring are a component of the chinook diet, but the data from Lake Superior are too limited to draw any conclusions (Conner et al. 1993).

Before chinook and coho were introduced in Lake Michigan during the 1960s, successional setback was extreme (Smith 1968, 1970). The collapse of chinook populations that began in the late 1980s occurred as planktivore structure was rapidly shifting from smaller to larger body size, and from epilimnetic to hypolimnetic waters. Thus, as the amount of successional setback was being reduced (evidenced by the recovery of yellow perch, bloater, and burbot), chinook ceased to be the dominant piscivore. The association between succession and the chinook collapse is circumstantial, but the circumstances are ecologically coherent. At the time of the collapse, chinook size in the sport catch and chinook lipid levels were declining, and chinook diet was shifting to smaller prey. The notion of a dominant piscivore that does not feed on the most abundant large prey, when the prey is below a size readily ingested (Jude et al. 1987), is inconsistent with ecological concepts of succession and dominance among fish species in lakes (Christie et al. 1987; Evans et al. 1987; Johnson 1994). The structural and functional data are consistent with a hypothesis that chinook dominance in the Great Lakes can occur only when successional setback is substantial.

Successional setback remains severe in Lake Ontario, despite extensive piscivory on alewives. Unlike Lake Michigan, chinook persist in Lake Ontario with lake trout as ecologically important piscivores. The natural successional process favoring deepwater ciscoes, when alewife piscivory is intense, cannot occur in Lake Ontario because these species are now extinct. Alewife populations have recently declined in this lake, but alewife body condition remains poor, apparently because of the effect of reduced phosphorous loads on invertebrate production (Jones et al. 1993; Rand et al. 1994). Chinook energetics may, however, be favored by the poor condition of adult alewives, which are pelagic in this lake (Mills et al. 1992). If adult alewives in Lake Ontario became hypolimnetic like the higher-energy-density alewife of Lake Michigan, lake trout would likely outcompete chinook for large alewife, and chinook diets would shift to smaller prey. Thus, inhibition of planktivore succession in Lake Ontario, brought about by extinction results in a structural arrangement conducive to a dominant role for chinook.

A chinook-alewife community is weakly connected and has low potential for stability and for high yield from all trophic levels combined. Chinook yields can nonetheless be high because of their exceptional ability to grow. In Lake Michigan, production to biomass ratios (P/B) were 1. 6 for chinook, and only 0. 6 for lake trout (Stewart and Ibarra 1991). Rapid population turnover in chinook, at the expense of storage, results in inherent instability as compared with long-lived species like lake trout that can better withstand short-term recruitment failures (Evans et al. 1987). In general, the prospect for sustainability in the Great Lakes should improve with longevity of the dominant piscivore (Christie et al. 1987; Johnson 1994).

Alewives, like chinook, have features that diminish sustainability when they are abundant. First, by preying extensively on ichthyoplankton, alewives suppress the recruitment of other species that are ecologically and economically valuable: emerald shiner, ciscoes, yellow perch, and sculpins. Second, they are inefficient at converting invertebrates to body biomass. Feeding primarily on zooplankton, adult alewife in Lake Michigan convert only 1. 3% to 2. 8% of food consumed (Stewart and Binkowski 1986). This anomaly supports our inference that the density and/ or size of epilimnetic zooplankters in the Great Lakes are only marginal for adult planktivores. Third, alewives are not fully adapted for planktivory on *Mysis* in deep water because of their low fat content and thermal physiology. Our empirical analysis suggests that *Mysis* is energetically important for all adult Great Lakes planktivores. Boudreau and Dickie (1992), using size-spectra models, demonstrate theoretically, that fish yields are higher in communities where benthic and pelagic production are coupled. An alewife-dominated planktivore community is structurally suited for fast-growing pelagic salmonines, but this successional state lacks diversity at intermediate trophic levels. We conclude that a pelagically dominated, alewife-Pacific salmon community is weakly connected trophically, and offers diminished prospects for sustainability in comparison to a benthically dominated, lake trout-cisco community.

Sustainability and Unintended Dominant Species

Ideas of balancing salmonine stocking levels with available prey have been focused on which salmonines should be dominant (Stewart and Ibarra 1991; Jones et al. 1993), but the burbot, a native piscivore, has been recovering on its own following the implementation of sea lamprey control programs, and it has potential to become dominant at the expense of the pelagic salmonines. The economics of Great Lakes fisheries favor anglers (Talhelm 1988), and this preference resulted in a vertical restructuring of piscivore biomass from benthic waters, where it was concentrated historically, to pelagic waters in recent years. Also, in large areas of the upper lakes, lake trout were not reintroduced because of potentially high losses in gill nets set for whitefish. For instance, Brown et al. (1981), using rough approximations, estimated that the number of lake trout stocked in Lake Michigan was only 30% of the number historically recruited, and that hatchery fish, spared intensive early selection, would presumably be less fit for survival than were the natives. The stage has been set for a grand experiment on whether restructuring is sustainable, and how one self-sustaining piscivore (i. e., burbot), fished lightly, responds to the reduction or absence of its closest ecological counterpart (i. e., lake trout).

The limited data available from Lake Michigan indicate that the historical piscivore structure is returning. Burbot populations expanded rapidly in Lake Michigan in the late 1980s. By 1995, adult burbot were more abundant than juvenile and adult lake trout combined (ratio of 1. 4:1) in an area of the lake (R. Rybicki, Michigan DNR, pers. comm.), now designated as a lake trout refuge (fig. 1). In the early 1930s, however, juvenile lake trout, by themselves, were 1. 2 times as abundant as adult burbot in this area of the lake, according to the scientific logs of the U.S. Fisheries Vessel, Fulmar. The recent proliferation of burbot in the lake trout refuge (and elsewhere in Lake Michigan) suggests that the combined biomass of lake trout and burbot in 1995 may well be comparable to what it was in the early 1930s, although now burbot are the dominant species.

The rise of burbot to dominance in northern Lake Michigan (possibly also in southern waters) is consistent with ecological theory outlined by Johnson (1994). Under his concept, each species in a community tends, under a given set of conditions, towards a state of maximum biomass. The *Fulmar* logs indicate that in Lake Michigan, during the 1930s, the biomass of lake trout exceeded that of burbot. Factors contributing to a dominance of lake trout over burbot then were: lake trout could achieve larger sizes allowing for ingestion of larger prey, lake trout were more pelagic (better at vertical migration), and evolution of deepwater-adapted lake trout extended the range of depths where lake trout could compete with burbot. The natural advantage that lake trout had over burbot was only modest in Lake Michigan in the 1930s, before the sea lamprey invaded the upper lakes, but the reduced abundance and diversity of lake trout since then have negated it altogether.

Our inferences about the structure and function of planktivores in the Great Lakes explain why burbot have the potential to become a dominant piscivore. Large piscivores prefer larger prey items, and the larger-sized alewife, smelt, and bloater in Lake Michigan are generally found in deeper water. Once burbot become adults, larger food items are more available to them than to the pelagic piscivores. As noted earlier, large chinook in three of the lakes ingest smaller prey than they are capable of ingesting, and than appear to be available.

Pelagic piscivores could dominate Lake Michigan, as they did until the late 1980s, as long as recruitment of burbot was suppressed. Alewives, when abundant, may have constrained burbot recruitment by preying on their fry. Burbot recruitment in Lake Michigan surged after the early 1980s, when alewife populations declined. Crowder (1980) listed burbot as one of two naturally reproducing species in Lake Michigan that have pelagic eggs and fry, and Spangler and Collins (1992) reported a negative relationship between alewife and burbot among the three basins of Lake Huron. Carl (1992) suggested that another planktivore, the lake herring, limited recruitment of burbot in Lake Opeongo at the larval stage by competition or predation. Although burbot declined in Lake Michigan because of sea lamprey predation, they recovered 20 years later than did the whitefish, which began a recovery shortly after sea lamprey control was initiated (Wells and McLain 1973). The rapidity of the burbot resurgence in Lake Michigan during the 1980s parallels that of the whitefish in the 1960s (Wells and McLain 1973), and shows how quickly early maturing, fecund species can rebound.

Burbot were neither intended nor expected to be a beneficiary of the salmonine stocking program in Lake Michigan. In hindsight, were salmonine stocking levels lower, alewives may have continued their dominance of planktivores and held burbot populations in check. A burbot resurgence in Lake Huron was apparently halted because sea lamprey and commercial fishing increased substantially in the early 1980s (Ebener 1995). Burbot are just beginning to recover in the lower lakes. In western Lake Superior, where lake trout, including siscowet, are dominant, the ratio of lake trout to burbot from 1976–93 was 4:1 (S. Schram, Wisconsin DNR, personal communication), which makes burbot a relatively minor piscivore. The future status of the burbot, particularly in Lakes Michigan, Huron, and Erie, is uncertain because it did not occur in these lakes under conditions where lake trout were a minor species and commercial fisheries, which historically killed large numbers of burbot as by-catch, were much reduced. Thus, the current management emphasis on pelagic salmonines is ecologically and economically perilous. If burbot, a species with very low fishery value becomes as abundant throughout these lakes as it is in the lake trout refuge in Lake Michigan, the potential yield from top piscivores will be greatly diminished, although a dominance of burbot should increase the sustainability of the fish community, because ecological setback will be reduced.

Sustainability and Genetic Diversity

Genetic diversity among whitefish, deepwater ciscoes, and lake trout from the Great Lakes is low or only moderate because of population constrictions during glaciation, or the short time for evolution since deglaciation (Todd 1981b; Krueger et al. 1989; Bernatchez et al. 1991). Human-induced fluctuations in the abundance of economically important native species led to even lower levels of genetic diversity, as populations were severely reduced or went extinct. Hatchery propagation of native and naturalized non-native salmonine species, which became extensive in the late 1970s, has potential to reduce genetic diversity of wild populations by diluting gene pools (Allendorf and Leary 1988), and by disrupting coadapted gene complexes (Philipp et al. 1983). Hatchery propagation is vital, however, for those species like Atlantic salmon that no longer reproduce in the Great Lakes, or those species like lake trout that do not have sustainable levels of reproduction in most of the lakes. Direct genetic effects of continued stocking include introgression, hybridization, and inbreeding, and indirect effects include shifts in rates of genetic drift, geographic isolation caused by range reductions, and alteration in selective pressures (Krueger and May 1991; Waples 1991).

The definition of fish populations as functional units, or stocks, is recommended as a basis for sustainability (Ihssen et al. 1981a). Ideally, a population is identified as a functional unit based on known genetic divergence that has phenotypic expression. Stock definition based on morphological characteristics is not reliable because morphological variation is influenced by environmental changes (Eschmeyer and Phillips 1965). Some fish populations are identified as stocks on the basis of spawning locality, if spawning-site fidelity is known to be high (Allendorf et al. 1986). Other fish populations are identified on the basis of life-history variation (Skulason et al. 1989), sometimes with no evidence of corresponding genetic variation (Magnusson and Ferguson 1987). Genetic variation in some native species can never be tested because most populations are extinct. Some divergent physiological characteristics among conspecific populations have been shown to be heritable (Eschmeyer and Phillips 1965; Ihssen and Tait 1974; Todd et al. 1981). The lack of genetic markers corresponding to phenotypic differences is more an artifact of the small number of genetic charac-teristics that have been examined than the overall lack of genetic diversity.

Among economically important Great Lakes fish, the ciscoes showed the widest array of mor-phological differences (Koelz 1926; Smith and Todd 1984), but the genetic variation detected among Great Lakes ciscoes is low (Todd 1981b; Shields et al. 1990; Snyder et al. 1992). The morphological differences used to identify cisco populations are influenced by environmental variation, hybrid-ization, and introgression (Hile 1937; Svärdson 1970; Smith 1964). Genetic studies have not found discriminating evidence for divergence among ciscoes (Todd 1981b; Smith and Todd 1984; Shields et al. 1990; Bernatchez et al. 1991), though some characteristics have been shown to be heritable (Todd et al. 1981).

Lewontin (1965) hypothesized that phenotypic plasticity in a colonizing species may be ad-vantageous because it would allow the colonizing species to take advantage of unexploited niches. Phenotypic differences among Great Lakes ciscoes could have occurred without detectable changes in genetic characteristics. This scenario is a plausible explanation for the apparent lack of genetic discontinuities among extant populations. The habitats occupied by various ciscoes have not been completely recolonized, suggesting that local adaptations played a role in the ecology of now ex-tinct populations.

Whitefish in the Great Lakes are thought to exist in discrete populations, some of which are recognized by color variation or spatial segregation (Imhof et al. 1980; Casselman et al. 1981; Ihssen et al. 1981b). The morphological variation among whitefish populations is not as distinct as that among cisco populations, and none have been designated as separate species or subspecies (Scott and Crossman 1973). The genetic variation among Great Lakes whitefish populations corresponds more closely with geographic variation than to morphological variation consistent with hypoth-esized patterns of post-glacial dispersal and colonization (Casselman et al. 1981; Ihssen et al. 1981b; Bernatchez and Dodson 1990). Depletions of whitefish caused by environmental degradation, overfishing, and sea lampreys may have resulted in population bottlenecks and losses of genetic diversity. The fact that whitefish abundance increased as conditions improved suggests that whitefish populations retained resilience, although river-spawning populations have not recovered.

The role of genetics in achieving sustainability is just as important with lake trout as with whitefish. Concerns about levels of genetic diversity have been addressed in lake trout rehabilita-tion plans in attempts to prevent the interbreeding of wild fish with hatchery-raised fish (e. g., Krueger et al. 1983; Evans and Willox 1991). The original broodstock developed for lake trout rein-troduction programs used a small number of spawning adults, prompting questions about the effect that low genetic diversity may have on the survival and growth of planted fish (Burnham-Curtis et al. 1995; Krueger and Ihssen 1995). Though stocking programs in Lake Superior used mostly native fish for broodstock, genetic diversity of the endemic populations may have been reduced in areas of high stocking intensity (Burnham-Curtis, unpublished data). Stocking efforts employing approximately twenty-five strains have increased the abundance of adult lake trout in the Great Lakes. The sustainability of the species historically benefited from a multitude of locally adapted, and likely genetically diverse populations. Some of this diversity still exists in deepwater populations in Lake Superior, but it is not used in reintroduction programs in the other lakes.

The stocking of non-native trout and salmon species in the Great Lakes is a controversial sub-ject because of direct ecological and genetic concerns (Krueger and May 1991). Stocking programs have resulted in popular sport fisheries for Pacific salmon in the Great Lakes, even though salmon likely compete for food with native piscivores (Hansen 1994). Interbreeding between closely re-lated species may lead to the production of interspecific hybrids, and effectively dilute genetic characteristics transmitted across generations (Witzel 1983; Hammar et al. 1989). Hybrids of chinook and pink salmon occur in the Great Lakes (G. Smith, University of Michigan, personal communication). Pacific salmon have become naturalized (Carl 1982), and the naturalized and source populations exhibit different characteristics: spring spawning in chinook (Kwain and Tho-mas 1984), shoal spawning in chinook (Powell and Miller 1990), and three-year-old pink salmon (Nicolette 1984). Naturalized populations of species such as brown trout (Krueger and May 1987a), rainbow trout (Krueger and May 1987b; Krueger et al. 1994), and pink salmon (Gharrett and Thomason 1987) have diverged such that genetic discontinuities are recognized.

Interspecific introgression as a consequence of stocking would, at best, increase the available genetic diversity, but at worst would disrupt coadapted gene complexes that likely enhance the sustainability of wild populations, whether native or introduced. The potential genetic impact of stocking a hatchery-raised or domesticated population on top of an existing conspecific wild popu-lation is not as clearly defined as the genetic consequences of interspecific interactions. The ge-netic characteristics of a hatchery-raised population are usually only a subset of that available in the wild (Allendorf et al. 1987). As hatchery populations are stocked onto wild populations, the wild gene pool becomes homogenized, and the potential for lowered fitness from outbreeding depression increases (Allendorf and Leary 1988; Gharrett and Smoker 1991). Reintroduction of exotic salmon and trout from the west coast could result in outbreeding depression of the existing naturalized and domesticated populations (Krueger and May 1991).

Introduced species may indirectly affect the genetic characteristics of native species. Reduced effective population sizes of native species may arise from ecological interactions, such as com-petition (for food and breeding habitat) and predation. These interactions, in effect, reduce the ability of the native population to maintain natural levels of genetic diversity and to adapt to a changing environment. To increase the prospects of achieving sustainability, the direct and indi-rect genetic consequences of any activity that could alter effective population size needs to be considered. Changes in growth and survival, environmental quality, exploitation, and competi-tion with or predation by non-native species have the potential to reduce the size of native popu-lations, their genetic diversity, and their ability to adapt to change.

Conclusions

Colonization of the offshore waters of the Great Lakes by planktivores was accompanied by buoyancy-regulation adaptations that facilitated vertical migration. Alternative explanations for the high lipid levels in deepwater ciscoes and in the siscowet are lacking. Trophic structuring of the adult planktivore community appears to be inordinately influenced by a single species—*Mysis relicta*. The original fish community may have been more evolved than believed by Ryder et al. (1981) and Christie et al. (1987). Ryder et al. (1981) proposed that the different forms of ciscoes and lake trout in the Great Lakes represented only phenotypic variation because of the short time for evolution after deglaciation. Genetic changes and significant differentiation in phenotypic responses can occur, however, on less than an evolutionary time scale (Carvalho 1993). The restricted capacity of the introduced planktivores, alewife and smelt, and of the bloater, the most shallow water of the deepwater cisco species, to fully occupy the depths inhabited by the extinct species of ciscoes in Lake Michigan, suggests that the original community, though simple, was highly evolved structurally, despite the short period since deglaciation. Isolation of individuals with different lipid levels at different depths may have occurred soon following deglaciation. The bloater and the closely related lake herring have similar upper thermal tolerances (Edsall et al. 1970) and temperature preferences (reviewed by Wismer and Christie 1987). Intrapopulation differences among individuals for a trait such as lipid content could allow expression of functional differences in a new, deepwater environment, and result in rapid separation of phenotypes.

Our findings provide an explanation for why a consistent pattern of fish-species succession occurred in the Great Lakes (Smith 1968; Evans et al. 1987): the adult native planktivores were very specialized. Smith (1968, 1972), Ryder et al. (1981), Christie et al. (1987), and Evans et al. (1987) recognized that the abundance of piscivores was important in a successional process that occurred in each of the lakes. But, succession can only be reversed if the native planktivores have advantages over introduced planktivores under certain ecological conditions. These advantages relate to buoyancy regulation, and an ability of ciscoes to assimilate nutrients at widely varying temperatures. The ideas of species succession and dominance among deepwater fish are connected. These are based on an interaction between the size and depth of the Great Lakes, and the behavior of invertebrates that are large and plentiful enough to support adult planktivores. Ryder and Kerr's (1978) concept of a harmonic community, originally applied to associations of warmwater fish, can probably be applied as well to the cisco-lake trout community of the Great Lakes. Reestablishment of a trophically efficient structure with introduced species will be difficult because so few freshwater or estuarine species of fish are likely to possess a physiology efficient for planktivory and assimilation in the deepest waters of the Great Lakes.

We doubt that intermediate successional states of the Great Lakes fish community can be sustained. Only two states appear to be structurally persistent. In one, alewives are dominant and piscivores are not abundant. Alewives maintain their dominance by suppressing, through predation, the recruitment of potential competitors (i. e., ciscoes) or predators (i. e., lake trout and/or burbot). This condition existed in Lake Ontario for over fifty years. In the other state, benthic piscivores, freed from sea lamprey predation on adults and alewife predation on fry, dominate the fish community. This condition appears near in Lake Michigan, and only a spectacular abundance of sea lampreys (Eshenroder et al. 1995a) prevents its occurrence in the main basin of Lake Huron. If smelt are simply a manifestation of successional setback, and are not a cause of it as proposed by Selgeby et al. (1978, 1994), a return to a dominance of benthic piscivores was inevitable in Lake Superior once sea lampreys were suppressed.

An alewife-salmonine community is a poor choice in terms of societal benefits. Salmonine abundance needs to be limited to ensure that alewives remain dominant. Natural reproduction of salmonines threatens the sustainability of this community, so artificial propagation, which can be quickly curtailed if needed, is logically preferred as the source of recruitment. Problems with maintaining fitness in hatchery fish actually provide positive feedback for alewife dominance. Periodic population crashes of the salmonines relieve predation pressure on the alewife. Suppression of native fish species by alewives, high control costs (Christie et al. 1987), unsteady yields of salmonines, and the riparian problems associated with mass mortalities of the alewife, which is essentially a pest species, are negative features of this community. This community appeared more productive than it was because salmonines were overstocked in Lakes Michigan and Ontario, which supported the best examples of this species complex. The high yields were temporary, and jeopardized alewife dominance (Stewart and Ibarra 1991; Jones et al. 1993). Nonetheless, the idea persists that this community can be maintained under conditions of high salmonine biomass. Because succession is already well advanced in the upper lakes, our analysis suggests that, barring extraordinary events such as a resurgence of sea lampreys, a sustainable alewife-salmonine community is possible only in Lake Ontario, where deepwater ciscoes are extinct and burbot remain scarce.

A lake trout–cisco community represents a higher level of system maturity, and greater potential sustainability; the evolutionary processes that shaped the historical community favor its reestablishment. This community is more connected, has a richer species complex (including pelagic salmonines), and demonstrates some resistance to proliferation of marine planktivores (Smith 1968; Christie 1974). Resistance to domination by introduced planktivores is probable because few invaders are likely to be as well adapted for planktivory in this system as are the ciscoes. The deepwater ciscoes and deepwater-adapted lake trout, such as siscowet, are clearly key species because of their functional abilities, and the food web of the Great Lakes cannot be fully reconnected without them.

Our analysis of structure and function clearly indicates that a more mature community, with far less ecological setback than an alewife-salmonine species complex, is more sustainable, and with sea lamprey suppression, probably inevitable. The major question is whether the dominant piscivore will be the burbot or the lake trout. A substantial majority of fishery and environmental managers favor rehabilitation of the native lake trout (Knuth et al. 1995).

Management is strongly influenced, however, by clients who demand Pacific salmon. The most challenging problems in moving the Great Lakes fish community to a more sustainable configuration are social as they are elsewhere (Ludwig et al. 1993). The Pacific salmon enhancement programs lack a strong ecological and ethical foundation (Eshenroder et al. 1995b). For example, substantial artificial propagation is necessary for their continuance, and the availability of introduced salmonines has resulted in a loss of respect for native species. Before any progress can be made in placing these programs on a stronger ecological footing, managers need to sort out their role in resource conservation, identify their clients beyond consumptive users, and establish the role of clients in policy development. Managers have a responsibility to sustain the resource, applying the best available science. The ecological framework for management is too often determined by resource users, and application of ecological principles and insights becomes restrained by social preference to a tiny range of options. This approach is not appropriate for systems over which humans have limited control.

Social preference cannot alter the ecological properties that continue to shape species succession in the Great Lakes. If social preference remains unrestrained, it will continue to pose considerable risks to achievement of a sustainable fish community, and of the benefits that it can provide.

The overall context for fishery programs should be bounded by ecological and ethical principles. Within these bounds, latitude exists for socioeconomic tradeoffs. Until responsibilities to the resource, roles of the users, and the basis for management are sorted out, we doubt that a sustainable fish community can be achieved in the Great Lakes.

Acknowledgments

We are indebted to Ray Argyle for data on the relationship between weights of alewife, smelt, and bloater and water depth in Lake Michigan. We thank Ed Brown for providing access to the logbooks of the U.S. Fisheries Vessel *Fulmar*, and acknowledge Camille Ward for compiling the associated data on lake trout and burbot catches from 1931–2. We thank Steven Schram for data on the ratio of burbot to lake trout in western Lake Superior, and Ronald Rybicki for similar data for Lake Michigan. Bryan Henderson is credited for identifying the possibility that lipids in lake trout would facilitate vertical migration. A partial review by Charles Krueger and full reviews by Gerald Smith and Thomas Todd are very appreciated. This paper is Contribution 924 of the Great Lakes Science Center, National Biological Service.

Literature Cited

Agassiz, L. 1850. Lake Superior: its physical character, vegetation, and animals, compared with those of other and similar regions. Gould, Kendall, and Lincoln, Boston, Mass.

Alexander, R. M. 1993. Buoyancy. Pages 75–98 *in* D. H. Evans, editor. The physiology of fish. CRC Press, Boca Raton, Fla.

Allendorf, F. W., and R. F. Leary. 1988. Conservation and distribution of genetic variation in a polytypic species, the cutthroat trout. Conservation Biology 2:170–184.

———, N. Ryman, and F. Utter. 1986. Genetics and fishery management: Past, present, and future. Pages 1–20 *in* N. Ryman and F. Utter, editors. Population genetics and fishery management. University of Washington Press, Seattle.

Anderson, D. M., and J. J. Collins. 1995. Natural reproduction by stocked lake trout (*Salvelinus namaycush*) and hybrid (backcross) lake trout in South Bay, Lake Huron. Journal of Great Lakes Research 21(Suppl. 1): 260–266.

Applegate, V. C., J. H. Howell, J. W. Moffett, B. G. H. Johnson, and M. A. Smith. 1961. Use of 3-triflourmethyl-4-nitrophenol as a selective sea lamprey larvicide. Great Lakes Fishery Commission Technical Report 1.

Argyle, R. L. 1982. Alewives and rainbow smelt in Lake Huron: midwater and bottom aggregations and estimates of standing stocks. Transactions of the American Fisheries Society 111:267–285.

———, 1992. Acoustics as a tool for the assessment of Great Lakes forage fish. Fisheries Research 14:179–196.

Ashworth, W. 1987. The late Great Lakes: an environmental history. Wayne State University Press, Detroit, Mich.

Bailey, M. M. 1972. Age, growth, reproduction, and food of the burbot, *Lota lota* (Linnaeus), in southwestern Lake Superior. Transactions of the American Fisheries Society 101:667–674.

Bailey, R. M., and G. R. Smith. 1981. Origin and geography of the fish fauna of the Laurentian Great Lakes basin. Canadian Journal of Fisheries and Aquatic Sciences 38:1539–1561.

Balcer, M. D., N. L. Korda, and S. I. Dodson. 1984. Zooplankton of the Great Lakes: a guide to the identification and ecology of the common crustacean species. University of Wisconsin Press, Madison, Wisc.

Baldwin, N. S., R. W. Saalfeld, M. A. Ross, and H. J. Buettner. 1979. Commercial fish production in the Great Lakes 1867–1977. Great Lakes Fishery Commission Technical Report No. 3.

Bean, T. H. 1884. On the occurrence of branch alewife in certain lakes of New York. Pages 588–593 *in* G. B. Goode, editor. Fisheries and Fishery Industry of the United States, Report to the U.S. Commission on Fish and Fisheries 1884.

Beeton, A. M. 1960. The vertical migration of *Mysis relicta* in Lakes Huron and Michigan. Journal of the Fisheries Research Board of Canada 17:517–539.

———, 1969. Changes in the environment and biota of the Great Lakes. Pages 150–187 *in* Eutrophication: causes, consequences, correctives. National Academy of Sciences, Washington, D.C.

Behnke, R. J. 1972. The systematics of salmonid fish of recently glaciated lakes. Journal of the Fisheries Research Board of Canada 29:639–671.

Bergstedt, R. A. 1983. Origins of rainbow smelt in Lake Ontario. Journal of Great Lakes Research 9:582–583.

————, R. O'Gorman. 1989. Distribution of alewives in southeastern Lake Ontario in autumn and winter: a clue to winter mortalities. Transactions of the American Fisheries Society 118:687–692.

Bernard, D. R., J. F. Parker, and R. Lafferty. 1993. Stock assessment of burbot populations in small and moderate-size lakes. North American Journal of Fisheries Management 13:657–675.

Bernatchez, L., and J. J. Dodson. 1990. Allopatric origin of sympatric populations of lake whitefish (*Coregonus clupeaformis*) as revealed by mitochondrial DNA restriction site analysis. Evolution 44:1263–1271.

————, F. Colombani, and J. J. Dodson. 1991. Phylogenetic relationships among the subfamily Coregoninae as revealed by mitochondrial DNA restriction analysis. Journal of Fish Biology 39 (Supplement A):283–290.

Berst, A. H., and G. R. Spangler. 1973. Lake Huron: the ecology of the fish community and man's effects on it. Great Lakes Fishery Commission Technical Report No. 21.

Binkowski, F. P., and L. G. Rudstam. 1994. Maximum daily ration of Great Lakes bloater. Transactions of the American Fisheries Society 123:335–343.

Borgmann, U., and D. M. Whittle. 1991. Contaminant concentration trends in Lake Ontario lake trout (*Salvelinus namaycush*): 1977–1988. Journal of Great Lakes Research 17:368–381.

Borgmann, U., and D. M. Whittle. 1992. DDE, PCB, and mercury concentration trends in Lake Ontario rainbow smelt (*Osmerus mordax*) and slimy sculpin (*Cottus cognatus*): 1977 to 1988. Journal of Great Lakes Research 18:298–308.

Boudreau, P. R., and L. M. Dickie. 1992. Biomass spectra of aquatic ecosystems in relation to fisheries yield. Canadian Journal of Fisheries and Aquatic Sciences 49:1528–1538.

Bowen, C., P. Hudson, and R. Argyle. 1994. Lake trout rehabilitation in Lake Huron—1993 progress report on coded wire tag returns. Report to Lake Huron Committee (Meeting), Great Lakes Fishery Commission, Ann Arbor, Michigan. (Cited with permission).

Boyer, L. F., R. A. Cooper, D. T. Long, and T. M. Askew. 1989. Burbot (*Lota lota*) biogenic sedimentary structures in Lake Superior. Journal of Great Lakes Research 15:174–185.

Brandt, S. B. 1986a. Disappearance of the deepwater sculpin (*Myoxocephalus thompsoni*) from Lake Ontario: the keystone predator hypothesis. Journal of Great Lakes Research 12:18–24.

————, 1986b. Food of trout and salmon in Lake Ontario. Journal of Great Lakes Research 12:200–205.

————, D. M. Mason, D. B. MacNeill, T. Coates, and J. E. Gannon. 1987. Predation by alewives on larvae of yellow perch in Lake Ontario. Transactions of the American Fisheries Society 116:641–645.

————, D. M. Mason, E. V. Patrick, R. L. Argyle, L. Wells, P. A. Unger, and D. J. Stewart. 1991. Acoustic measures of the abundance and size of pelagic planktivores in Lake Michigan. Canadian Journal of Fisheries and Aquatic Sciences 48:894–908.

Bronte, C. R., J. H. Selgeby, and G. L. Curtis. 1991. Distribution, abundance, and biology of the alewife in U.S. waters of Lake Superior. Journal of Great Lakes Research 17:304–313.

Brown, E. H., Jr. 1972. Population biology of alewives, *Alosa pseudoharengus*, in Lake Michigan, 1949–1970. Journal of the Fisheries Research Board Canada 29:477–500.

————, G. W. Eck, N. R. Foster, R. M. Horrall, and C. E. Coberly. 1981. Historical evidence for discrete stocks of lake trout (*Salvelinus namaycush*) in Lake Michigan. Canadian Journal of Fisheries and Aquatic Sciences 38:1747–1758.

————, R. W. Rybicki, and R. J. Poff. 1985. Population dynamics and interagency management of the bloater (*Coregonus hoyi*) in Lake Michigan, 1967–1982. Great Lakes Fishery Commission Technical Report No. 44.

————, R. L. Argyle, N. R. Payne, and M. E. Holey. 1987. Yield and dynamics of destabilized chub (*Coregonus spp.*) populations in Lakes Michigan and Huron. Canadian Journal of Fisheries and Aquatic Sciences 44(Suppl. 2):371–383.

Brussard, P. F., M. C. Hall, and J. Wright. 1981. Structure and affinities of freshwater sea lamprey (*Petromyzon marinus*) populations. Canadian Journal of Fisheries and Aquatic Sciences 38:1708–1714.

Burnham-Curtis, M. K. 1993. Intralacustrine speciation of Salvelinus namaycush in Lake Superior and investigation of genetic and morphological variation and evolution of lake trout in the Laurentian Great Lakes. Doctoral thesis. University of Michigan, Ann Arbor, Mich.

————, G. R. Smith. 1994. Osteological evidence of genetic divergence of lake trout (*Salvelinus namaycush*) in Lake Superior. Copeia 1994(4):843–850.

————, C. C. Krueger, D. N. Schreiner, J. E. Johnson, T. J. Stewart, R. M. Horrall, W. R. MacCallum, R. Kenyon, and R. E. Lange. 1995. Genetic strategies for lake trout rehabilitation: A synthesis. Journal of Great Lakes Research 21 (Suppl 1):477–486.

Carl, L. M. 1982. Natural reproduction of coho salmon and chinook salmon in some Michigan streams. North American Journal of Fisheries Management 4:375–380.

————, 1992. The response of burbot (*Lota lota*) to change in lake trout (*Salvelinus namaycush*) abundance in Lake Opeongo, Ontario. Hydrobiologia 243/244:229–235.

Carr, I. A. 1962. Distribution and seasonal movements of Saginaw Bay fish. United States Fish and Wildlife Service, Special Scientific Report—Fisheries No. 417, Washington, D. C.

Carvalho, G. R. 1993. Evolutionary aspects of fish distribution: genetic variability and adaptation. Journal of Fish Biology 43(Suppl. A):53–73.

Casselman, J. M., J. J. Collins, E. J. Crossman, P. E. Ihssen, and G. R. Spangler. 1981. Lake whitefish (*Coregonus clupeaformis*) stocks of the Ontario waters of Lake Huron. Canadian Journal of Fisheries and Aquatic Sciences 38:1772–1789.

Christie, G. C., and H. A. Regier. 1988. Measures of optimal thermal habitat and their relationship to yields for four commercial fish species. Canadian Journal of Fisheries and Aquatic Sciences 45:301–314.

Christie, W. J. 1973. A review of the changes in the fish species composition of Lake Ontario. Great Lakes Fishery Commission Technical Report No. 23.

———, 1974. Changes in the fish species composition of the Great Lakes. Journal of the Fisheries Research Board of Canada 31:827–854.

———, G. R. Spangler, K. H. Loftus, W. L. Hartman, P. J. Colby, M. A. Ross, and D. R. Talhelm. 1987. A perspective on Great Lakes fish community rehabilitation. Canadian Journal of Fisheries and Aquatic Sciences 44:486–499.

Coble, D. W., R. E. Bruesewitz, T. W. Fratt, and J. W. Scheirer. 1990. Lake trout, sea lampreys, and overfishing in the upper Great Lakes: a review and reanalysis. Transactions of the American Fisheries Society 119:985–995.

———, T. W. Fratt, and J. W. Scheirer. 1992. Decline of lake trout in Lake Huron. Transactions of the American Fisheries Society 121:550–554.

Colby, P. J. 1971. Alewife dieoffs: why do they occur? Limnos 4(2):18–27.

Conner, D. J., C. R. Bronte, J. H. Selgeby, and H. L. Collins. 1993. Food of salmonine predators in Lake Superior, 1981–1987. Great Lakes Fishery Commission Technical Report No. 59.

Cornelius, F. C., K. M. Muth, and R. Kenyon. 1995. Lake trout rehabilitation in Lake Erie: a case history. Journal of Great Lakes Research 21(Suppl. 1):65–82.

Crawford, R. H. 1966. Buoyancy regulation in lake trout. Doctoral dissertation. University of Toronto.

Crowder, L. B. 1980. Alewife, rainbow smelt, and native fish in Lake Michigan: competition or predation? Environmental Biology of Fish 5:225–233.

———, J. J. Magnuson. 1982. Thermal habitat shifts by fish at the thermocline in Lake Michigan. Canadian Journal of Fisheries and Aquatic Sciences 39:1046–1050.

———, H. L. Crawford. 1984. Ecological shifts in resource use by bloaters in Lake Michigan. Transactions of the American Fisheries Society 113:694–700.

Curtis, G. L. 1991. Recovery of an offshore lake trout (*Salvelinus namaycush*) population in eastern Lake Superior. Journal of Great Lakes Research 16:279–287.

———, J. H. Selgeby, and R. G. Schorfhaar. In press. Decline and recovery of lake trout populations near Isle Royale, Lake Superior 1929–1990. Transactions of the American Fisheries Society.

Day, A. C. 1983. Biological and population characteristics of and interactions between an unexploited burbot (*Lota lota*) population and an exploited lake trout (*Salvelinus namaycush*) population from Lake Athapapuskow, Manitoba. Master's thesis. University of Manitoba, Winnipeg, Man.

Deason, H. J., and R. Hile. 1947. Age and growth of the kiyi, *Leucichthys kiyi* Koelz, in Lake Michigan. Transactions of the American Fisheries Society 74:87–142.

DeGraeve, G. M., and J. B. Reynolds. 1975. Feeding behavior and temperature and light tolerance of *Mysis relicta* in the laboratory. Transactions of the American Fisheries Society 104:394–397.

Dehring, T. R., A. F. Brown, C. H. Daugherty, and S. R. Phelps. 1981. Survey of the genetic variation among Eastern Lake Superior lake trout (*Salvelinus namaycush*). Canadian Journal of Fisheries and Aquatic Sciences 38:1738–1746.

DeVault, D. S., W. A. Willford, and R. J. Hesselberg. 1985. Contaminant trends in lake trout (*Salvelinus namaycush*) from the upper Great Lakes. U.S. Environmental Protection Agency, Great Lakes National Program Office, Report 905/3–85–001, Chicago, Illinois.

Diana, J. S. 1990. Food habits of angler-caught salmonines in western Lake Huron. Journal of Great Lakes Research 16:271–278.

Dryer, W. R. 1966. Bathymetric distribution of fish in the Apostle Islands region, Lake Superior. Transactions of the American Fisheries Society 95:248–259.

———, J. Beil. 1964. Life history of lake herring in Lake Superior. Fishery Bulletin 63:493–530.

Dymond, J. R. 1932. Records of the alewife and steelhead (rainbow trout) from Lake Erie. Copeia 1932:32–33.

———, J. L. Hart, and A. L. Pritchard. 1929. The fish of the Canadian waters of Lake Ontario. University of Toronto Studies, Publications of the Ontario Fisheries Research Laboratory, No. 37.

Ebener, M. P., editor. 1995. The state of the Lake Huron fish community in 1992. Great Lakes Fishery Commission Special Publication 95–2.

Eck, G. W., and L. Wells. 1987. Recent changes in Lake Michigan's fish community and their probable causes, with emphasis on the role of the alewife (*Alosa pseudoharengus*). Canadian Journal of Fisheries and Aquatic Sciences 44:53–60.

Edsall, T. A., D. V. Rottiers, and E. H. Brown. 1970. Temperature tolerance of bloater (*Coregonus hoyi*). Journal of the Fisheries Research Board of Canada 27:2047–2052.

———, G. W. Kennedy, and W. H. Horns. 1993. Distribution, abundance, and resting microhabitat of burbot on Julian's Reef, southwestern Lake Michigan. Transactions of the American Fisheries Society 122:560–574.

Elliott, R. F. 1993. Feeding habits of chinook salmon in eastern Lake Michigan. Master's dissertation. Michigan State University, Lansing, Michigan.

Elrod, J. H., R. O'Gorman, C. P. Schneider, T. H. Eckert, T. Schaner, J. N. Bowlby, and L. P. Schleen. 1995. Lake trout rehabilitation in Lake Ontario. Journal of Great Lakes Research 21(Suppl. 1):83–107.

Emery, L. 1985. Review of fish species introduced into the Great Lakes, 1819–1974. Great Lakes Fishery Commission Technical Report No. 45.

Eschmeyer, P. H., and A. M. Phillips, Jr. 1965. Fat content of the flesh of siscowets and lake trout from Lake Superior. Transactions of the American Fisheries Society 94:62–74.

Eshenroder, R. L. 1992. Decline of lake trout in Lake Huron. Transactions of the American Fisheries Society 121:548–554.

———, N. R. Payne, J. E. Johnson, C. Bowen II, and M. P. Ebener. 1995a. Lake trout rehabilitation in Lake Huron. Journal of Great Lakes Research 21(Suppl. 1):108–127.

———, E. J. Crossman, G. K. Meffe, C. H. Olver, and E. P. Pister. 1995b. Lake trout rehabilitation in the Great Lakes: an evolutionary, ecological, and ethical perspective. Journal of Great Lakes Research 21(Suppl. 1):518–529.

Evans, D. O., and C. C. Wilcox. 1991. Loss of exploited, indigenous populations of lake trout, Salvelinus namaycush, by stocking of non-native stocks. Canadian Journal of Fisheries and Aquatic Sciences 48 (Supplement 1):134–147.

———, B. A. Henderson, N. J. Bax, T. R. Marshall, R. T. Oglesby, and W. J. Christie. 1987. Concepts and methods of community ecology applied to freshwater fisheries management. Canadian Journal of Fisheries and Aquatic Sciences 44(S2):448–470.

Evans, M. S. 1992. Historic changes in Lake Michigan zooplankton community structure: the 1960s revisited with implications for top-down control. Canadian Journal of Fisheries and Aquatic Science 49:1734–1749.

———, D. J. Jude. 1986. Recent shifts in Daphnia community structure in southeastern Lake Michigan: a comparison of the inshore and offshore regions. Limnology and Oceanography 31:56–67.

Flath, L. E., and J. S. Diana. 1985. Seasonal energy dynamics of the alewife in southeastern Lake Michigan. Transactions of the American Fisheries Society 114:328–337.

Fleischer, G. W. 1992. Status of coregonine fish in the Laurentian Great Lakes. Polish Archives of Hydrobiology 39:247–259.

Foltz, J. W., and C. R. Norden. 1977. Food habits and feeding chronology of rainbow smelt, Osmerus mordax, in Lake Michigan. U.S. National Marine Fisheries Service Fishery Bulletin 75:637–640.

Fox, W. S. 1930. The literature of Salmo salar in Lake Ontario and tributary streams. Transactions of the Royal Society of Canada, Section 2, Series 3 24:45–55.

Fratt, T. W. 1991. Trophic ecology of burbot and lake trout in Green Bay and Lake Michigan with information for burbot on fecundity, gonadosomatic index, sea lamprey marks, and intestinal acanthocephalans. Master's thesis. University of Wisconsin, Stevens Point, Wisc.

Gharrett, A. J., and M. A. Thomason. 1987. Genetic changes in pink salmon (Oncorhynchus gorbuscha) following their introduction into the Great Lakes. Canadian Journal of Fisheries and Aquatic Sciences 43:787–792.

———, W. W. Smoker. 1991. Two generations of hybrids between even- and odd-year pink salmon (Oncorhynchus gorbuscha): a test for outbreeding depression? Canadian Journal of Fisheries and Aquatic Sciences 48:1744–1749.

Goode, G. B. 1884. The fisheries and fishery industries of the United States, Section I: natural history of useful aquatic animals. Report of the U.S. Commission for Fish and Fisheries, 1884.

Goodier, J. L. 1981. Native lake trout (Salvelinus namaycush) stocks in the Canadian waters of Lake Superior prior to 1955. Canadian Journal of Fisheries and Aquatic Sciences 38:1724–1737.

Grewe, P. M., and P. D. N. Hebert. 1988. Mitochondrial DNA diversity among brood stocks of the lake trout Salvelinus namaycush. Canadian Journal of Fisheries and Aquatic Sciences 45:2114–2122.

Hammar, J., J. B. Dempson, and E. Skold. 1989. Natural hybridization between arctic char (Salvelinus alpinus) and lake char (S. namaycush): Evidence from northern Labrador. Nordic Journal of Freshwater Research 65:54–70.

Hansen, M. J., editor. 1994. The state of Lake Superior in 1992. Great Lakes Fishery Commission, Special Publication 94–1, Ann Arbor, Michigan.

———, M. P. Ebener, J. D. Shively, and B. L. Swanson. 1994. Lake trout. Pages 35–52 in M. J. Hansen, editor. The state of Lake Superior in 1992. Great Lakes Fishery Commission Special Publication 94–1.

Hartman, W. L. 1973. Effects of exploitation, environmental changes, and new species on the fish habitats and resources of Lake Erie. Great Lakes Fishery Commission Technical Report 22.

———, 1988. Historical changes in the major fish resources of the Great Lakes. Pages 103–131 in M. S. Evans, editor. Toxic contaminants and ecosystem health: a Great Lakes focus. John Wiley and Sons, N. Y.

Hatch, R. W., P. M. Haack, and E. H. Brown, Jr. 1981. Estimation of alewife biomass in Lake Michigan, 1967–1978. Transactions of the American Fisheries Society 110:575–584.

Healey, M. C. 1991. Life history of chinook salmon (Oncorhynchus tshawytscha). Pages 311–394 in C. Groot and L. Margolis, editors. Pacific salmon life histories. University of British Columbia Press, Vancouver.

Heist, B. G., and W. A. Swenson. 1983. Distribution and abundance of rainbow smelt in western Lake Superior as determined from acoustic sampling. Journal of Great Lakes Research 9:343–353.

Henderson, B. A., and E. H. Brown, Jr. 1985. Effects of abundance and water temperature on recruitment and growth of alewife (*Alosa pseudoharengus*) near South Bay, Lake Huron, 1954–82. Canadian Journal of Fisheries and Aquatic Sciences 42:1608–1613.

Henderson, B. A., and S. J. Nepszy. 1989. Factors affecting recruitment and mortality rates of rainbow smelt (*Osmerus mordax*) in Lake Erie, 1963–85. Journal of Great Lakes Research 15:357–366.

Hesselberg, R. J., J. P. Hickey, D. A. Northrup, and W. A. Willford. 1990. Contaminant residues in the bloater (*Coregonus hoyi*) of Lake Michigan, 1969–1986. Journal of Great Lakes Research 16:121–129.

Hewett, S. W., and D. J. Stewart. 1989. Zooplanktivory by alewives in Lake Michigan: Ontogenetic, seasonal, and historical patterns. Transactions of the American Fisheries Society 118:581–596.

Hile, R. 1937. Morphometry of the cisco, *Leucichthys artedi* (LeSeur), in the lakes of the northeastern highlands, Wisconsin. Sonderdruck aus; Int. Rev. d. ges. Hydr. Leipzig 36(1/2):57–130.

———, 1949. Trends in the lake trout fishery of Lake Huron through 1946. Transactions of the American Fisheries Society 76:121–147.

———, P. H. Eschmeyer, and G. F. Lunger. 1951. Status of the lake trout fishery in Lake Superior. Transactions of the American Fisheries Society 80:278–312.

Holey, M. E., R. W. Rybicki, G. W. Eck, E. H. Brown Jr., J. E. Marsden, D. S. Lavis, M. L. Toneys, T. N. Trudeau, and R. M. Horrall. 1995. Progress toward lake trout restoration in Lake Michigan. Journal of Great Lakes Research 21 (Suppl. 1):128–151.

Hopkins, E. E., and C. M. Ritchie. 1943. The burbot. U.S. Fish and Wildlife Service, Fishery Leaflet 21, Washington, D. C.

Hubbs, C. L., and T. E. B. Pope. 1937. The spread of the sea lamprey through the Great Lakes. Transactions of the American Fisheries Society 66:172–176.

Huntsman, A. G. 1931. The maritime salmon of Canada. Bulletin of the Biological Board of Canada 21:1–99.

Ihssen, P. E., and J. S. Tait. 1974. Genetic differences in relation of swim bladder gas between two populations of lake trout (*Salvelinus namaycush*). Journal of the Fisheries Research Board of Canada 31:1351–1354.

———, H. E. Booke, J. M. Casselman, J. M. McGlade, N. R. Payne, and F. M. Utter. 1981a. Stock identification: material and methods. Canadian Journal of Fisheries and Aquatic Sciences 38:1838–1855.

———, D. O. Evans, W. J. Christie, J. A. Reckahn, and R. L. DesJardine. 1981b. Life history, morphology, and electrophoretic characteristics of five allopatric stocks of lake whitefish (*Coregonus clupeaformis*) in the Great Lakes region. Journal of the Fisheries Research Board Canada 38: 1790–1807.

———, G. W. Martin, and D. W. Rodgers. 1992. Allozyme variation of Great Lakes alewife, *Alosa pseudoharengus:* genetic differentiation and affinities of a recent invader. Canadian Journal of Fisheries and Aquatic Sciences 49:1770–1777.

Imhof, M. R. Leary, and H. E. Booke. 1980. Population or stock structure of lake whitefish, *Coregonus clupeaformis*, in northern Lake Michigan as assessed by isozyme electrophoresis. Canadian Journal of Fisheries and Aquatic Sciences 37:783–793.

Janssen, J. 1978. Feeding-behavior repertoire of the alewife, *Alosa pseudoharengus*, and the ciscoes *Coregonus hoyi* and *C. artedii*. Journal of the Fisheries Research Board of Canada 35:249–253.

———, S. B. Brandt. 1980. Feeding ecology and vertical migration of adult alewives (*Alosa pseudoharengus*) in Lake Michigan. Canadian Journal of Fisheries and Aquatic Sciences 37:177–184.

Jobes, F. W. 1943. The age, growth, and bathymetric distribution of Reighard's chub, *Leucichthys reighardi* Koelz, in Lake Michigan. Transactions of the American Fisheries Society 72(1942):108–135.

———, 1949a. The age, growth, and distribution of the longjaw cisco, *Leucichthys alpenae* Koelz, in Lake Michigan. Transactions of the American Fisheries Society 76(1946):215–247.

———, 1949b. The age, growth, and bathymetric distribution of the bloater, *Leucichthys hoyi* (Gill) in Lake Michigan. Papers of the Michigan Academy of Science, Arts and Letters 33(1947):135–172

Johannsson, O. 1992. Life history and productivity of *Mysis relicta* in Lake Ontario. Journal of Great Lakes Research 18:154–168.

———, R. O'Gorman. 1991. Roles of predation, food, and temperature in structuring the epilimnetic zooplankton populations in Lake Ontario, 1981–1986. Transactions of the American Fisheries Society 120:193–208.

———, E. L. Mills, and R. O'Gorman. 1991. Changes in the near shore and offshore zooplankton communities in Lake Ontario: 1981–88. Canadian Journal of Fisheries and Aquatic Sciences 48:1546–1557.

Johnson, J. E., and J. VanAmberg. (1995). Evidence of natural reproduction in western Lake Huron. Journal of Great Lakes Research 21(Suppl. 1): 253–259.

Johnson, L. 1994. Pattern and process in ecological systems: a step in the development of a general ecological theory. Canadian Journal of Fisheries and Aquatic Sciences 51:226–246.

Jones, M. L., J. F. Koonce, and R. O'Gorman. 1993. Sustainability of hatchery-dependent salmonine fisheries in Lake Ontario: the conflict between predator demand and prey supply. Transactions of the American Fisheries Society 122:1002–1018.

Jude, D. J., S. A. Klinger, and M. D. Enk. 1981. Evidence of natural reproduction by planted lake trout in Lake Michigan. Journal of Great Lakes Research 7:57–61.

———, F. J. Tesar, S. F. Deboe, and T. J. Miller. 1987. Diet and selection of major prey species by Lake Michigan salmonines, 1973–1982. Transactions of the American Fisheries Society 116:677–691.

Kabré, J. A. T. 1993. Impact of *Renibacterium salmonirarum* on non-lipid energy, lipid and water contents in liver of infected salmon during fall and spring in Lake Michigan, 1990–1992. Ph. D. Thesis, Michigan State University.

Kay, J. J. 1990. A non-equilibrium thermodynamic framework for discussing ecosystem integrity. Pages 209–238 in C. J. Edwards and H. A. Regier, editors. An ecosystem approach to the integrity of the Great Lakes in turbulent times. Great Lakes Fishery Commission Special Publication 90–4.

Keller, M., K. D. Smith, and R. W. Rybicki, editors. 1990. Review of salmon and trout management in Lake Michigan. Michigan Department of Natural Resources, Fisheries Special Report No. 14, Lansing, Michigan.

Khan, N. Y., and S. U. Qadri. 1971. Intraspecific variation and postglacial distribution of lake char (*Salvelinus namaycush*). Journal of the Fisheries Research Board of Canada 28:465–476.

Kitchell, J. F., and L. B. Crowder. 1986. Predator–prey interactions in Lake Michigan: model predictions and recent dynamics. Environmental Biology of Fish 16:205–211.

Knuth, B. A., S. Lerner, N. A. Connelly, and L. Gigliotti. 1995. Fishery and environmental managers' attitudes about and support for lake trout rehabilitation in the Great Lakes. Journal of Great Lakes Research 21(Suppl. 1):185–197.

Kocik, J. F., and W. W. Taylor. 1987. Diet and movements of age-1+ pink salmon in western Lake Huron. Transactions of the American Fisheries Society 116:628–633.

Koelz, W. 1926. Fishing industry of the Great Lakes. United States Department of Commerce, Bureau of Fisheries Document 1001:553–617.

———, 1929. Coregonid fish of the Great Lakes. Bulletin of the U.S. Bureau of Fisheries 43(1927):297–643.

Kolbe, C. F. 1944. Food: the controlling factor of fish populations and other related factors influencing survival of valuable species. Port Dover, Ontario.

Krueger, C. C., and B. May. 1987a. Stock identification of naturalized brown trout in Lake Superior tributaries: differentiation based on allozyme data. Transactions of the American Fisheries Society 116:785–794.

———, B. May. 1987b. Genetic comparison of naturalized rainbow trout populations among Lake Superior tributaries: differentiation based on allozyme data. Transactions of the American Fisheries Society 116:795–806.

———, B. May. 1991. Ecological and genetic effects of salmonid introductions in North America. Canadian Journal of Fisheries and Aquatic Sciences 48 (Suppl. 1):66–77.

———, P. E. Ihssen. 1995. Review of genetics of lake trout in the Great Lakes: History, molecular genetics, physiology, strain comparisons, and restoration management. Journal of Great Lakes Research 21(Suppl. 1):348–363.

———, R. M. Horrall, and H. Greunthal. 1983. Strategy for the use of lake trout strains in Lake Michigan. Wisconsin Department of Natural Resources Administrative Report 17.

———, J. E. Marsden, H. L. Kincaid, and B. May. 1989. Genetic differentiation among lake trout strains stocked into Lake Ontario. Transactions of the American Fisheries Society 118:317–330.

———, D. L. Perkins, R. J. Everett, D. R. Schreiner, and B. May. 1994. Genetic variation in naturalized rainbow trout (*Oncorhynchus mykiss*) from Minnesota tributaries to Lake Superior. Journal of Great Lakes Research 20:299–316.

Kwain, W., and J. A. Chappel. 1978. First evidence for even-year spawning pink salmon, *Oncorhynchus gorbuscha,* in Lake Superior. Journal of the Fisheries Research Board of Canada 29:765–776.

———, E. Thomas. 1984. The first evidence of spring spawning by chinook salmon in Lake Superior. North American Journal of Fisheries Management 4:227.

Lantry, B. F., and D. J. Stewart. 1993. Ecological energetics of rainbow smelt in the Laurentian Great Lakes: an interlake comparison. Transactions of the American Fisheries Society 122:951–976.

Lark, J. G. I. 1973. An early record of the sea lamprey (*Petromyzon marinus*) from Lake Ontario. Journal of the Fisheries Research Board of Canada 30:131–133.

Larkin, P. A. 1977. An epitaph for the concept of maximum sustained yield. Transactions of the American Fisheries Society 106:1–11.

Lawrie, A. H. 1978. The fish community of Lake Superior. Journal of Great Lakes Research 4:513–549.

———, J. A. Rahrer. 1973. Lake Superior: a case history of the lake and its fisheries. Great Lakes Fishery Commission Technical Report 19.

Leach, J. H., and S. J. Nepszy. 1976. The fish community in Lake Erie. Journal of the Fisheries Research Board of Canada 33:622–638.

Leim, A. H., and W. B. Scott. 1966. Fish of the Atlantic coast of Canada. Fisheries Research Board of Canada Bulletin 155.

Levin, S. A. 1993. Science and sustainability. Ecological Applications 3:545–546.

Lewontin, R. C. 1965. Selection for colonizing ability. Pages 77–94 in H. G. Baker and G. L. Stebbins, editors. The genetics of colonizing species. Academic Press, New York.

Lindsey, C. C. 1964. Problems in the zoogeography of lake trout, *Salvelinus namaycush*. Journal of the Fisheries Research Board of Canada 21:977–994.

Loftus, K. H., and H. A. Regier. 1972. Introduction to the proceedings of the 1971 symposium on salmonid communities in oligotrophic lakes. Journal of the Fisheries Research Board of Canada 29:613–616.

Ludwig, D., R. Hilborn, and C. Walters. 1993. Uncertainty, resource exploitation, and conservation: lessons from history. Science 260:17–36.

MacCallum, W. R., and J. H. Selgeby. 1987. Lake Superior revisited 1984. Canadian Journal of Fisheries and Aquatic Sciences 44:23–36.

MacKay, H. H. 1934. Record of the alewife from Lake Huron. Copeia (1934)2:97.

Magnusson, K. P., and M. F. Ferguson. 1987. Genetic analysis of four sympatric morphs of Arctic charr, *Salvelinus alpinus*, from Thingvallavatn, Iceland. Environmental Biology of Fish 20(1):67–73.

Mandrak, N. E., and E. J. Crossman. 1992. Postglacial dispersal of freshwater fish in Ontario. Canadian Journal of Zoology 70:2247–2259.

Marsden, J. E. 1994. Spawning by stocked lake trout on shallow, near–shore reefs in southwestern Lake Michigan. Journal of Great Lakes Research 20(2):377–384.

Martin, N. V., and C. H. Olver. 1980. The lake charr, *Salvelinus namaycush*. Pages 205–277 in E. K. Balon, editor. Charrs: salmonid fish of the genus Salvelinus. The Hague, Netherlands, Dr. W. Junk.

Martinez, P. J., and E. P. Bergerson. 1989. Proposed biological management of *Mysis relicta* in Colorado lakes and reservoirs. North American Journal of Fish Management 9:1–11.

Marzolf, G. R. 1965. Vertical migration of *Pontoporeia affinis* (Amphipoda) in Lake Michigan. Pages 133–140 in Great Lakes Research Division, University of Michigan Publication 13.

McDonald, M. E., L. B. Crowder, and S. B. Brandt. 1990. Changes in *Mysis* and *Pontoporeia* populations in southeastern Lake Michigan: a response to shifts in the fish community. Limnology and Oceanography 35:220–227.

McPhail, J. D., and C. C. Lindsey. 1970. Freshwater fish of northwestern Canada and Alaska. Fisheries Research Board of Canada 173.

Meyer, J. L., and G. S. Helfman. 1993. The ecological basis of sustainability. Ecological Applications 3:569–571.

Miller, M. A., and M. E. Holey. 1992. Diets of lake trout inhabiting near shore and offshore Lake Michigan environments. Journal of Great Lakes Research 18:51–60.

———, C. P. Madenjian, and R. G. Masnado. 1992. Patterns of organochlorine contamination in lake trout from Wisconsin waters of the Great Lakes. Journal of Great Lakes Research 18:742–754.

Miller, R. R. 1957. Origin and dispersal of the alewife, *Alosa pseudoharengus*, and the gizzard shad, *Dorosoma cepedianum*, in the Great Lakes. Transactions of the American Fisheries Society 86:97–111.

Mills, E. L., R. O'Gorman, J. DeGisi, R. F. Heberger, and R. A. House. 1992. Food of the alewife (*Alosa pseudoharengus*) in Lake Ontario before and after the establishment of *Bythotrephes cederstroemi*. Canadian Journal of Fisheries and Aquatic Sciences 49:2009–2019.

Mitchell, S. L. 1815. The fish of New York, described and arranged. Transactions of the Literary and Philosophical Society of New York 1:355–501.

Morsell, J. W., and C. R. Norden. 1968. Food habits of the alewife, *Alosa pseudoharengus* (Wilson), in Lake Michigan. Proceedings of the Eleventh Conference of Great Lakes Research International Association for Great Lakes Research, Ann Arbor, MI.

Nalepa, T. F. 1989. Estimates of macroinvertebrate biomass in Lake Michigan. Journal of Great Lakes Research 15:437–443.

Nelson, D. D., and J. G. Hnath. 1990. Lake Michigan chinook salmon mortality-1989. Michigan Department of Natural Resources, Fisheries Division Technical Report 90–4, Lansing.

Nester, R. T., and T. P. Poe. 1984. First evidence of successful natural reproduction of planted lake trout in Lake Huron. North American Journal of Fisheries Management 4:126–128.

Nicolette, J. P. 1984. A three-year-old pink salmon in an odd-year run in Lake Superior. North American Journal of Fisheries Management 4:130–132.

O'Gorman, R., D. H. Barwick, and C. A. Bowen. 1987. Discrepancies between ages determined from scales and otoliths for alewives from the Great Lakes. Pages 203–210 in R. C. Summerfelt and G. E. Hall, editors. The age and growth of fish. The Iowa State University Press, Ames, Iowa.

———, E. L. Mills, and J. S. DeGisi. 1991. Use of zooplankton to assess the movement and distribution of alewife (*Alosa pseudoharengus*) in south-central Lake Ontario in spring. Canadian Journal of Fisheries and Aquatic Sciences 48:2250–2257.

———, C. P. Schneider. 1986. Dynamics of alewives in Lake Ontario following a mass mortality. Transactions of the American Fisheries Society 115:1–14.

Orsi, I. A., and A. C. Wertheimer. 1995. Marine vertical distribution of juvenile chinook and coho salmon in southeastern Alaska. Transactions of the American Fisheries Society 124:159–169.

Parsons, J. A. 1973. History of salmon in the Great Lakes, 1850–1970. United States Department of the Interior, Fish and Wildlife Service, Technical Paper 68, Washington, D. C.

Pearce, W. A., R. A. Braem, S. M. Dustin, and J. J. Tibbles. 1980. Sea lamprey (*Petromyzon marinus*) in the lower Great Lakes. Canadian Journal of Fisheries and Aquatic Sciences 37:1802–1810.

Peck, J. W. 1977. Species composition of deep-water ciscoes (chubs) in commercial catches from Michigan waters of Lake Superior. Michigan Department of Natural Resources, Fisheries Division, Fisheries Research Report No. 1849, Lansing.

———, W. R. MacCallum, S. T. Schram, D. R. Schreiner, and J. D. Shively. 1994. Other salmonines. Pages 35–52 in M. J. Hansen, editor. The state of Lake Superior in 1992. Great Lakes Fishery Commission Special Publication 94–1.

Pennak, R. W. 1953. Fresh-water invertebrates of the United States. The Ronald Press Company, N.Y.

Philipp, D. P., W. F. Childers, and G. S. Whitt. 1983. A biochemical genetic evaluation of the northern and Florida species of largemouth bass. Transactions of the American Fisheries Society 112:1–20.

Powell, M. J., and M. Miller. 1990. Shoal spawning by chinook salmon in Lake Huron. North American Journal of Fisheries Management 10:242–244.

Pritchard, A. L. 1931. Taxonomic and life history studies of the ciscoes of Lake Ontario. University of Toronto Studies in Biology Series 35, Publications of the Ontario Fisheries Laboratory 41.

Rand, P. S., B. F. Lantry, R. O'Gorman, R. W. Owens, and D. J. Stewart. 1994. Energy density and size of pelagic prey fish in Lake Ontario, 1978–1990: Implications for salmonine energetics. Transactions of the American Fisheries Society 123:519–534.

Reynolds, J. B., and G. M. DeGraeve. 1972. Seasonal population characteristics of the opossum shrimp, *Mysis relicta,* in southeastern Lake Michigan, 1970–71. Pages 117–131 *in* Proceedings of the 15th Conference on Great Lakes Research, International Association for Great Lakes Research.

Robertson, A., and W. P. Alley. 1966. A comparative study of Lake Michigan macrobenthos. Limnology and Oceanography 11:576–583.

Rohrer, T. K., J. C. Forney, and J. H. Hartig. 1982. Organochlorine and heavy metal residues in standard fillets of coho and chinook salmon of the Great Lakes—1980. Journal of Great Lakes Research 8:623–634.

Rottiers, D. V., and R. M. Tucker. 1982. Proximate composition and caloric content of eight Lake Michigan fish. United States Department of the Interior, Fish and Wildlife Service, Technical Paper No. 108, Washington, D.C.

Rudstam, L. G., F. P. Binkowski, and M. A. Miller. 1994. A bioenergetics model for analysis of food consumption patterns of bloater in Lake Michigan. Transactions of the American Fisheries Society 123:344–357.

Ryder, R. A., and S. R. Kerr. 1978. The adult walleye in the percid community—a niche definition based on feeding behavior and food specificity. American Fisheries Society Special Publication 11:39–51.

———, W. W. Taylor, and P. A. Larkin. 1981. Community consequences of fish stock diversity. Canadian Journal of Fisheries and Aquatic Sciences 38:1856–1866.

Scott, W. B., and S. H. Smith. 1962. The occurrence of the longjaw cisco, *Leucichthys alpenae,* in Lake Erie. Journal of the Fisheries Research Board of Canada 19:1013–1023.

———, E. J. Crossman. 1973. Freshwater fish of Canada. Fisheries Research Board of Canada, Bulletin 184.

Selgeby, J. H., W. R. MacCallum, and D. V. Swedberg. 1978. Predation by rainbow smelt (*Osmerus mordax*) on lake herring (*Coregonus artedii*) in Western Lake Superior. Journal of the Fisheries Research Board of Canada 35:1457–1463.

———, C. R. Bronte, and J. W. Slade. 1994. Forage species. Pages 53–62 in M. J. Hansen, editor. The state of Lake Superior in 1992. Great Lakes Fishery Commission Special Publication 94–1.

———, Selgeby, J. H., W. R. MacCallum, and M. H. Hoff. 1994. Rainbow smelt-larval lake herring interactions: competitors or casual acquaintances? U.S. Department of the Interior, National Biological Survey, Biological Report 25, Washington, D. C.

Sell, D. W. 1982. Size-frequency estimates of secondary production by Mysis relicta in Lakes Michigan and Huron. Hydrobiologia 93:69–78.

Shields, B. A., K. S. Guise, and J. C. Underhill. 1990. Chromosomal and mitochondrial DNA characterization of a population of dwarf cisco (*Coregonus artedii*) in Minnesota. Canadaian Journal of Fisheries and Aquatic Sciences 47:1562–1569.

Skulason, S., S. S. Snorrason, D. L. G. Noakes, M. M. Ferguson, and H. L. Malmquist. 1989. Segregation in spawning and early life history among polymorphic Arctic charr, *Salvelinus alpinus,* in Thingvallavatn, Iceland. Journal of Fish Biology 35:225–232.

Smith, B. R., and J. J. Tibbles. 1980. Sea lamprey (*Petromyzon marinus*) in Lakes Huron, Michigan, and Superior: history of invasion and control, 1936–78. Canadian Journal of Fisheries and Aquatic Science 37:1780–1801.

Smith, G. R., and T. N. Todd. 1984. Evolution of species flocks of fish in north temperate lakes. Pages 45–64 in A. A. Echelle and I. Kornfield, editors. Evolution of fish species flocks. University of Maine at Orono Press, Orono, Maine.

———, 1992. Morphological cladistic study of coregonine fish. Polskie Archiwum Hydrobiologii 39:479–490.

Smith, H. M. 1892. Report on an investigation of the fisheries of Lake Ontario. Bulletin of the United States Fishery Commission 10:195–202.

Smith, S. H. 1956. Life history of lake herring of Green Bay, Lake Michigan. United States Department of the Interior, Fish and Wildlife Service, Fishery Bulletin 57:109.

———, 1964. Status of the deepwater cisco population of Lake Michigan. Transactions of the American Fisheries Society 93:155–163.

———, 1968. Species succession and fishery exploitation in the Great Lakes. Journal of the Fisheries Research Board of Canada 25:667–693.

———, 1970. Species interactions of the alewife in the Great Lakes. Transactions of the American Fisheries Society 99:754–765.

———, 1972. Factors of ecologic succession in oligotrophic fish communities of the Laurentian Great Lakes. Journal of the Fisheries Research Board of Canada 29:717–730.

———, 1995. Early ecologic changes in the fish community of Lake Ontario. Great Lakes Fishery Commission Technical Report 60.

Snyder, T. P., R. D. Larsen, and S. H. Bowen. 1992. Mitochondrial DNA diversity among Lake Superior and inland lake ciscoes (*Coregonus artedi* and *C. hoyi*). Canadian Journal of Fisheries and Aquatic Sciences 49:1902–1907.

Spangler, G. R., and J. J. Collins. 1992. Lake Huron fish community structure based on gill net catches corrected for selectivity and encounter probability. North American Journal of Fisheries Management 12:585–597.

Sprules, W. G., S. B. Brandt, D. J. Stewart, M. Munawar, E. H. Jin, and J. Love. 1991. Biomass size spectrum of the Lake Michigan pelagic food web. Canadian Journal of Fisheries and Aquatic Sciences 48:105–115.

Stewart, D. J., and F. P. Binkowski. 1986. Dynamics of consumption and food conversion by Lake Michigan alewives: an energetics-modeling synthesis. Transactions of the American Fisheries Society 115:643–661.

———, M. Ibarra. 1991. Predation and production by salmonine fish in Lake Michigan, 1978–88. Canadian Journal of Fisheries and Aquatic Science 48:909–922.

———, J. F. Kitchell, and L. B. Crowder. 1981. Forage fish and their salmonid predators in Lake Michigan. Transactions of the American Fisheries Society 110:751–763.

Strus, R. H., and D. A. Hurley. 1992. Interactions between alewife (*Alosa pseudoharengus*), their food, and phytoplankton biomass in the Bay of Quinte, Lake Ontario. Journal of Great Lakes Research 18:709–723.

Svärdson, G. 1970. Significance of introgression in coregonid evolution. Pages 33–59 in C. C. Lindsey and C. S. Woods, editors. Biology of coregonid fish. University of Manitoba Press, Winnipeg.

Swanson, B. L., and D. V. Swedberg. 1980. Decline and recovery of the Lake Superior Gull Island Reef lake trout (*Salvelinus namaycush*) population and the role of sea lamprey (*Petromyzon marinus*) predation. Canadian Journal of Fisheries and Aquatic Sciences 37:2074–2080.

Talhelm, D. R. 1988. Economics of Great Lakes fisheries: a 1985 assessment. Great Lakes Fishery Commission, Technical Report No. 54, Ann Arbor, Mich.

Todd, T. N. 1981a. *Coregonus prognathus* Smith: A nomen dubium. Copeia (2):489–490.

———, 1981b. Allelic variability in species and stocks of Lake Superior ciscoes (*Coregoninae*). Canadian Journal of Fisheries and Aquatic Sciences 38:1808–1813.

———, L. E. Cable. 1981. Environmental and genetic contributions to morphological differentiation in ciscoes (Coregoninae) of the Great Lakes. Canadian Journal of Fisheries and Aquatic Sciences 38:59–67.

———, R. M. Stedman. 1989. Hybridization of ciscoes (*Coregonus* spp.) in Lake Huron. Canadian Journal of Zoology 67:1679–1685.

———, G. R. Smith. 1992. A review of differentiation in Great Lakes ciscoes. Polskie Archiwum Hydrobiologii 39:261–267.

Van Oosten, J. 1929. Life history of the lake herring (*Leucichthys artedi* Le Sueur) of Lake Huron as revealed by its scales, with a critique of the scale method. Fisheries Bulletin 44:265–428.

———, 1937. The dispersal of smelt, *Osmerus mordax* (Mitchill), in the Great Lakes region. Transactions of the American Fisheries Society 66(1936):160–161.

———, 1947. Mortality of smelt, *Osmerus mordax* (Mitchill), in Lakes Huron and Michigan during the fall and winter of 1942–1943. Transactions of the American Fisheries Society 74(1944):310–377.

———, H. J. Deason. 1938. The food of the lake trout (*Cristivomer namaycush namaycush*) and of the lawyer (*Lota maculosa*) of Lake Michigan. Transactions of the American Fisheries Society 67(1937):155–177.

Waples, R. S. 1991. Genetic interactions between hatchery and wild salmonids: lessons from the Pacific Northwest. Canadian Journal of Fisheries and Aquatic Sciences 48 (Supplement 1):124–133.

Wells, L. 1968. Daytime distribution of *Pontoporeia affinis* off bottom in Lake Michigan. Limnology and Oceanography 13:703–705.

———, 1969. Fishery survey of U.S. waters of Lake Ontario. Pages 51–58 in Limnological survey of Lake Ontario, 1964. Great Lakes Fishery Commission Technical Report No. 14.

———, 1970. Effects of alewife predation on zooplankton populations in Lake Michigan. Limnology and Oceanography 15:556–565.

———, 1980. Food of alewives, yellow perch, spottail shiners, trout-perch, and slimy and fourhorn sculpins in southeastern Lake Michigan. U.S. Fish and Wildlife Service, Technical papers of the U.S. Fish and Wildlife Service 98, Washington, D. C.

———, A. M. Beeton. 1963. Food of the bloater, *Coregonus hoyi,* in Lakes Michigan. Transactions of the American Fisheries Society 92:245–255.

———, A. L. McLain. 1973. Lake Michigan: man's effects on native fish stocks and other biota. Great Lakes Fishery Commission Technical Report No. 20.

Wismer, D. A., and A. E. Christie. 1987. Temperature relationships of Great Lakes fish: a data compilation. Great Lakes Fishery Commission, Special Publication No. 87–3, Ann Arbor, Mich.

Witzel, L. D. 1983. The occurrance and origin of tiger trout, *Salmo trutta* × *Salvelinus fontinalis,* in Ontario streams. Canadian Field-Naturalist 97:99–102.

Wright, J., C. C. Krueger, P. F. Brussard, and M. C. Hall. 1985. Sea lamprey (*Petromyzon marinus*) populations in northeastern North America: genetic differentiation and affinities. Canadian Journal of Fisheries and Aquatic Sciences 42:776–784.

CHAPTER 7

Non-Indigenous Species in the Great Lakes: Ecosystem Impacts, Binational Policies, and Management

Joseph H. Leach, Edward L. Mills, and Margaret R. Dochoda

Introduction

The introduction of non-native species and their impacts on indigenous biota have accelerated recently due to human activities. Increases in transoceanic ship traffic, the spread of aquaculture, and the intentional and accidental introduction of living organisms and genetic materials have all contributed to a global homogenization of plants, animals and diseases. For example, more than 4,500 non-indigenous species (NIS) have become successfully established in the United States, including more than two-thousand plant species, more than two thousand insect species, 239 plant pathogens, 142 terrestrial vertebrate species, 91 species of freshwater mollusks and 70 fish species (Office of Technology Assessment 1993). Since 1980, 205 species have invaded successfully, or have been introduced into the United States, 59 of which are causing ecological and economic damage.

The Laurentian Great Lakes drainage system has been an important corridor for settlement and development of agriculture, industry, and commerce in central North America, and therefore, has been vulnerable to invasions by, and introductions of, NIS for centuries. The number of invasions is not known but about 139 species have become successfully established in the Great Lakes watershed since the early 1800s (Mills et al. 1993a).

In this chapter, we review the history of invasions and introductions in the Great Lakes, the principal means of entry and dispersal, and impacts on ecosystem health (the concept of Great Lakes ecosystem health has been defined and discussed extensively by Ryder (1990) and Evans et al. (1990) and is not explored further here). Predictability of successful colonization by NIS and their potential to do economic or ecological harm is considered briefly, in light of ecological theory. Binational policies and the development of legislation and management strategies for prevention, containment and control of NIS in the Great Lakes are discussed.

We define NIS as successfully reproducing organisms transported by humans into the Great Lakes, where they did not previously exist. Included are organisms intentionally introduced to create established populations.

The most comprehensive sources of information on NIS in the Great Lakes are found in Mills et al. (1993a, 1993b, 1994) and Leach (1995). We have drawn extensively from these sources in compiling this review.

History of Invasions and Introductions

Exploration of the Great Lakes basin by Europeans commenced about four centuries ago, and was followed by clearing of forests, settlement, and commercial development. The Great Lakes waterway was the principal transportation route for colonization, and consequently, early settlement occurred on lake shores and tributaries (Hatcher 1944). Early commercial development was focused on fur, forests, and fish (Ashworth 1986). As forests were harvested, land was cleared for agriculture, minerals were located, and networks of roads, railroads, and canals were established to facilitate extraction and development of natural resources.

Commerce and urban centers grew, especially on the lower Great Lakes, southern Lake Michigan, and connecting waterways. The Erie Canal (1825), the Welland Canal (1829), and later, the St. Lawrence Seaway (1959), were opened to facilitate trade from and to the basin. These activities also facilitated the entry of non-indigenous species of plants and animals into the Great Lakes.

From a search of the literature, Mills et al. (1993a) have listed 139 species of NIS that have become established in the Great Lakes basin since 1810 (see table 1 for a list of fish species). No doubt exotic species entered the basin before 1800 through range expansion following the retreat of glaciation or transport by aboriginals, but clear records of such movements do not exist (see Bailey and Smith 1981, for review of origin and geography of the fish fauna). Plants and algae account for 60% of new species entering the basin since 1810, followed by invertebrates 20%, and fish 18% (table 2). The rate of entry has increased with time with 40% of all NIS becoming established since 1950.

Mills et al. (1993a) described the entry and dispersal mechanisms responsible for the establishment of the exotic species in the Great Lakes (table 3). Unintentional releases (including escape from cultivation, aquaculture and aquaria, and accidental releases due to fish stocking and from unused bait) accounted for entry of forty-seven of the exotic species. Thirteen species gained entry by unintentional release, along with one or more other vectors and are included under multiple mechanisms in table 3. Almost one-half of the exotic plant species in the basin became established through unintentional release (largely by escape from cultivation). Significant numbers of invertebrates (nine) and fish (five) were also unintentionally released.

Shipping activities directly permitted forty-three species of NIS to become established, and nine additional species gained entry through a combination of shipping and other vectors. Within the shipping vector, 62% of invasive species gained entry through ballast water, 32% from solid ballast and 6% from hull fouling. Ballast water was responsible for the entry of twenty-one species of algae, sixteen of which arrived after the opening of the St. Lawrence Seaway in 1959. Nine species of invertebrates and four species of fish were also transported to the Great Lakes in ballast water. Recent entrants of ecological significance in this group were the ruffe (*Gymnocephalus cernuus*), two species of gobies, two dreissenid species, and the cladoceran, *Bythotrephes cederstroemi*. Improved water quality in donor waterways, which interface with shipping, may

TABLE 1

Origin, date of entry and distribution of non-indigenous fish species in the Great Lakes

(Data from Mills et al. 1993a; Emery 1985)

SCIENTIFIC NAME	COMMON NAME	ORIGIN	DATE	DISTRIBUTION[a]
Petromyzon marinus	sea lamprey	Atlantic	1830s	A
Alosa pseudoharengus	alewife	Atlantic	1873	A
Carassius auratus	goldfish	Asia	<1878	A
Cyprinus carpio	common carp	Asia	1879	A
Notropis buchanani	ghost shiner	Mississippi[b]	1979	E
Phenacobius mirabilis	suckermouth minnow	Mississippi[b]	1950	E
Scardinius erythrophthalmus	rudd	Europe	1989	O
Misgurnus anguillicaudatus	oriental weatherfish	Asia	1939	H
Noturus insignis	margined madtom	Atlantic	1928	O
Osmerus mordax	rainbow smelt	Atlantic	1912	A
Oncorhynchus gorbuscha	pink salmon	Pacific	1956	A
Oncorhynchus kisutch	coho salmon	Pacific	1933	A
Oncorhynchus nerka	kokanee	Pacific	1950	H
Oncorhynchus tshawytscha	chinook salmon	Pacific	1873	A
Oncorhynchus mykiss	rainbow trout	Pacific	1876	A
Salmo trutta	brown trout	Europe	1883	A
Gambusia affinis	western mosquitofish	Mississippi[b]	1923	M, E
Apeltes quadracus	fourspine stickleback	Atlantic	1986	S
Morone americana	white perch	Atlantic	1950	A
Enneacanthus gloriosus	bluespotted sunfish	Atlantic	1971	O
Lepomis humilis	orangespotted sunfish	Mississippi[b]	1929	E
Lepomis microlophus	redear sunfish	Southern U.S.	1928	M
Gymnocephalus cernuus	ruffe	Europe	1986	S, H
Neogobius melanostomus	round goby	Eurasia	1990	A
Proterorhinus marmoratus	tubenose goby	Eurasia	1990	H, E

[a] Ontario (O), Erie (E), Huron (H), Michigan (M), Superior (S), all Great Lakes (A)   [b] Mississippi drainage system

TABLE 2

Numbers of invasive and introduced species established in the Great Lakes by taxonomic group and time period

(Data from Mills et al. 1993a)

PERIOD	FISHES	INVERTEBRATES	DISEASE PATHOGENS, PARASITES	ALGAE	PLANTS	TOTAL
1810–1849	1	—	—	—	9	10
1850–1899	6	4	—	—	23	33
1900–1949	7	8	1	6	18	40
1950–1994	11	16	2	18	9	56
Total	25	28	3	24	59	139
Percent	18	20	2	17	43	100

TABLE 3

Numbers of invasive and introduced species established in the Great Lakes by entry mechanism, time period and taxonomic group

(Data from Mills et al. 1993a)

	SHIPS[1]	UNINTENTIONAL[2] RELEASES	DELIBERATE[3] RELEASES	CANALS[4] RAILROADS HIGHWAYS	MULTIPLE MECHANISMS	UNKNOWN
TIME PERIOD						
1810–1849	—	5	—	—	3	2
1850–1899	6	11	3	2	9	2
1900–1949	9	16	7	2	3	3
1950–1994	28	15	1	2	3	7
Total	43	47	11	6	18	14
Percent	31	34	8	4	13	10
TAXONOMIC GROUP						
Fishes	4	5	8	2	6	—
Invertebrates	11	9	1	3	2	2
Disease Pathogens, Parasites	—	3	—	—	—	—
Algae	21	2	—	—	—	1
Plants	7	28	2	1	10	11
Total	43	47	11	6	18	14

[1]Includes entry of NIS through ballast water, solid ballast and by fouling.
[2]Releases of NIS without intention of creating established populations including releases from aquaria, cultivation, unused bait and disease pathogens and plankton associated with fish stocking.
[3]The planting of NIS with the intention of creating established populations.
[4]Man-made transport corridors which aid the entry and dispersal of NIS.

also have contributed to the recent increase in ballast water NIS. Solid ballast was directly responsible for the entry, mostly in the 1800s, of two species of invertebrates and at least seven species of plants.

Deliberate releases of fish commenced in the 1800s with plantings of common carp (*Cyprinus carpio*), goldfish (*Carassius auratus*), rainbow trout (*Oncorhynchus mykiss*), brown trout (*Salmo trutta*), and chinook salmon (*O. tshawytscha*). Successful plantings in the Great Lakes basin of rainbow smelt (*Osmerus mordax*), kokanee (*O. nerka*), western mosquito fish (*Gambusia affinis*) and red-ear sunfish (*Lepomis microlophus*), occurred in the first half of the twentieth century (Emery 1985).

Canals, railroads and highways directly allowed six NIS to reach the Great Lakes basin, and an additional seven entered in association with other vectors (table 3). Some very ecologically significant NIS entered the Great Lakes via canals, including the sea lamprey (*Petromyzon marinus*), alewife (*Alosa pseudoharengus*), white perch (*Morone americana*), and purple loosestrife (*Lythrum salicaria*).

The rate of colonization of the Great Lakes by NIS of fish and invertebrates increased dramatically since the mid 1800s (table 2). The majority of plant species (70%) invaded from 1850–1950, mostly from solid ballast, accidental release, or escape from cultivation. All of the disease pathogens and algae invaded the system since 1900; however, we recognize that earlier invasions may not have been recorded.

The trend in entry mechanisms has been towards more invasions from shipping activities, especially due to ballast water releases (table 3). With the opening of the St. Lawrence Seaway in 1959, the amount of ballast water released into the Great Lakes and the risk of new invasions increased greatly. In fact, nearly 30% of all NIS in the Great Lakes became established since 1959. Unintentional releases also increased with time, and have contributed forty-two NIS since the mid nineteenth century. Most deliberate releases have been plantings of fish species which occurred mainly in the period from 1850–1950. Chinook salmon and coho salmon (*Oncorhynchus kisutch*), were stocked in this period but later (1960s) plantings are credited with establishment of these species in the Great Lakes (Emery 1985).

The main donor areas for NIS in the Great Lakes are Europe and Asia with eighty-nine species to date which is a reflection of immigration and trade patterns (Mills et al. 1993a). Within North America, the Atlantic coast contributed eighteen species, the Mississippi drainage seven species, and five species (salmonid fishes), were introduced from the Pacific coast. The source of many algae species is unknown because they are indigenous to many parts of the world.

Introductions, Invasions and Ecological Theory

We know that approximately 139 NIS have colonized the Great Lakes basin (Mills et al. 1993a). We also know that a number of species (especially fish, Emery 1985) have been unsuccessfully introduced; but we do not know how many invasive species failed to become established. In a study of natural invasions and deliberate introductions into the United Kingdom over the past century, Williamson and Brown (1986) found that about 10% of the exotic species had become established. Similarly, Groves and Burdon (1986) estimated that about 10% of plant introductions to Australia became established. If the 10% rule also applies in North America, then perhaps some 1400 NIS have attempted colonization in the Great Lakes basin.

Why do some invasions and introductions succeed while others fail? Why are some systems more vulnerable to colonization than others? Why are some colonists more disruptive than others? Are there generalizations which will help us to predict what species will colonize the Great Lakes system, and to what extent a colonizing species will degrade ecosystem health?

Many ecological concepts, particularly those concerning communities, have been explored to explain success or failure of introductions and invasions (Pimm 1991). Two general hypotheses have been advanced (Simberloff 1986; Baltz and Moyle 1993) to help explain invasion resistance: environmental resistance hypothesis (e.g., inability of a species to adapt to physical and/or chemical characteristics of the environment); and biotic resistance hypothesis (e.g., competition and/or predation interaction with native species).

Elton (1958) referred to these together as "ecological resistance" to invasion. In the invasions that he reviewed, he found that most occurred in disturbed habitats, and he hypothesized that ecological resistance is lowered in a disturbed system. This hypothesis is particularly useful in understanding plant invasions, and was clearly evident in the successful invasions of plants from

TABLE 4

Attributes of habitats, invading species and communities which may contribute to success of colonization by non-indigenous species

(From Leach 1995)

HABITATS	INVADING SPECIES	COMMUNITIES
climate similar in donor and recipient habitats	high fecundity	species-poor
	short generation time	low complexity
disturbed	rapid dispersal	stressed native populations
suitable physical and chemical characteristics	eurybiont	trophic opportunities available
available substrata for attachment	polyphagous	reduced competition and predation from native species
	high genetic variability	
	high inoculation rate	
	affiliated with *Homo sapiens*	

Europe and Asia into North America. Of 124 common weeds of Canada, 74% are NIS, almost all from Europe and Asia (Mulligan 1987). Many of the colonists became established when the land was being cleared for agriculture, which represented a highly disturbed system with reduced native flora.

A summary of habitat, organism, and community characteristics that may influence the success or failure of an invader to overcome habitat or biotic resistance is listed in table 4. In general, none of the attributes may be sufficient or essential by itself to permit colonization.

Similar climate in donor and recipient habitats is an important ecological requirement for establishment of an invasive species. For example, dreissenid mussels occur between latitudes 30° and 60° N in Europe, and will likely occupy a similar latitudinal range in North America. However, other habitat requirements, such as calcium concentration and pH of the water, might preclude establishment of dreissenid mussels within a broad climate scale (Neary and Leach 1992). The environment (i.e., climate and habitat) was considered more important for the colonization of dreissenid mussels in Ontario than the existence of other biota (Neary and Leach 1992). This observation agrees with a hypothesis suggested by Simberloff (1986), who stated that the probability of successful colonization by an invader depends mainly on its habitat requirements and availability of suitable habitat, and less on the presence of native biota.

Vulnerability of an ecosystem to colonization was considered by Groves and Burdon (1986) to be determined by the rate at which NIS inoculate a system. For example, about 30% of NIS in the Great Lakes have been discovered since the opening of the St. Lawrence Seaway in 1959 (Mills et al. 1993a). Consistent with the ecosystem vulnerability concept developed by Groves and Burdon (1986), the increase in the rate of colonization of Great Lakes NIS appears to be directly related to the increase in the inoculation rate of ballast water.

TABLE 5

Number of native fish species and number and percent
of non-indigenous fish species in the Great Lakes

(Data from Ryder 1972; Mills et al. 1993a).

LAKE	NUMBER OF NATIVE SPECIES	NUMBER OF NON-INDIGENOUS SPECIES	PERCENT INTRODUCED
Superior	67	14	17
Michigan	114	14	11
Huron	99	16	14
Erie	113	17	13
Ontario	112	15	12

Opportunities for successful invasion decrease with increased species richness and connectance (i.e., the actual, divided by the possible number of inter-specific interactions) in a system (Pimm 1991). The rationale is that competition for resource use increases directly with connectance in a community and with more competition there is less opportunity for invasion and consequently, species-rich communities will have a smaller proportion of colonists than species-poor communities. For example, in the Great Lakes, Lake Superior has the fewest native species of fish and the highest percentage of non-indigenous fish species (table 5). However, Pimm's theory does not hold for undisturbed streams in California, where Baltz and Moyle (1993) found that species-poor fish communities continuously resisted invasion by introduced species. Moreover, the endemic fish community of Lake Victoria, in Africa, with over three-hundred species was easily invaded by Nile perch (*Lates sp.*), with subsequent loss of approximately two-thirds of the native cichlid species (Lowe-McConnell 1994).

The "empty" niche concept has been proposed as a reason for the success of natural invasions and an excuse for the intentional introduction of new species (Herbold and Moyle 1986; Li and Moyle 1993). The problem has been aggravated by confusion over niche definition (Hutchinson 1957; Kerr and Ryder 1977), which has led to intentional introductions of species to fill niches considered vacant. Pimm (1991) considered that the concept of empty niche was not particularly useful for prediction of success or failure of colonization because a niche must be occupied to be recognized. Richard A. Ryder, Ontario Ministry of Natural Resources, (personal communication) has proposed the use of "trophic opportunity" as more useful and realistic than "trophic niche."

Another principal concern about NIS is our ability or inability to predict if a colonizing species will reduce the health of a receiving ecosystem. Will habitat and/or native communities be degraded, and if so, how severely? Will indigenous species ultimately be displaced or driven to extinction?

The overall percentage of successful colonists considered to be economically and ecologically harmful in the United States, the United Kingdom, and the Great Lakes basin is surprisingly uniform, and ranges from 10% to 15% (Office of Technology Assessment 1993; Williamson and Brown 1986; Mills et al. 1993a). Of the 139 NIS established in the Great Lakes system, 20% of fish, 21% of

TABLE 6

Degree of impact on ecosystem health by taxa of non-indigenous species in the Great Lakes

(Data from Mills et al. 1993a).

DEGREE OF IMPACT	FISHES	INVERTEBRATES	DISEASE PATHOGENS, PARASITES	ALGAE	PLANTS
Very Harmful	2	2	2	—	1
Harmful	3	2	—	1	2
Potentially Harmful	3	2	1	—	9
Harmful and Beneficial	7	—	—	—	—
Unknown	10	23	—	23	47

invertebrates and disease pathogens, and only 5% of algae and aquatic plants are considered harmful to ecosystem health. As a general rule, animal and pathogen (including parasites) invaders are more likely to be damaging than plants. However, impacts of non-indigenous flora at the population or community levels in the Great Lakes have neither been well observed nor documented, and could be much higher than indicated.

Extirpation of indigenous species and extinction of endemic species are probably the ultimate measures of ecosystem damage by NIS. In a review of ten studies involving 850 invasions and introductions of plants and animals, Simberloff (1981) found that fewer than 10% of NIS caused extinctions of native species. Of seventy-one extinctions, predation was the main cause with 51, followed by habitat change with 11 and competition with 3.

Impacts of Non-Indigenous Species on Ecosystem Health

Mills et al. (1993a) found that only about 10% of NIS in the Great Lakes significantly degraded the health of the ecosystem. This proportion is similar to that found in other areas. However, ecological effects of NIS worldwide are not well understood (Pimm 1991), and usually only highly visible and disruptive impingements are observed and recorded. In the Great Lakes, impacts from more than one hundred NIS (including seventy species of algae and aquatic plants) are virtually unknown (table 6).

Mechanisms for damage to ecosystems by colonizing species include habitat modification, competition for food and habitat, predation, associated pathogens and parasites, and genetic effects (Ferguson 1990; Krueger and May 1991; Li and Moyle 1993). All have been implicated in impacts on the health of the Great Lakes. Some NIS, particularly fishes, mussels, and plants have severely perturbed habitats and native species (tables 7 and 8).

For the taxa reviewed in tables 7 and 8, particularly introduced salmonids, alewife, white perch, ruffe, dreissenid mussels, and purple loosestrife, competition has implicitly been a major factor in damage to native species. Common carp, through its feeding activity, and purple loosestrife and Eurasian watermilfoil (*Myriophyllum spicatum*), through displacement of native plants, have inflicted major changes to habitats. Many of the taxa caused adverse predator-prey interactions with the most serious resulting from the sea lamprey.

TABLE 7

Time of colonization and mechanism of damage to ecosystem health (destabilization of community or habitat alteration) by selected NIS in the Great Lakes

(Data from Mills et al. 1993a: Leach 1995)   XX = major damage; X = minor damage; ? = uncertain

NON-INDIGENOUS SPECIES	TIME OF COLONIZATION	ASPECTS OF ECOSYSTEM HEALTH DAMAGE			
		HABITAT MODIFICATION	COMPETITION	PREDATION	DISEASES & PARASITES
Sea lamprey	1830s	X		XX	
Purple loosestrife	1869	XX	XX		
Alewife	1873		XX	X	
Chinook salmon	1873	X	XX	X	
Rainbow trout	1876	X	XX	X	
Common carp	1879	XX	X		
Brown trout	1883	X	XX	X	
Aeromonas salmonicida	1902				XX
Rainbow smelt	1912		X	X	
Coho salmon	1933	X	XX	X	
White perch	1950s		XX	X	
Eurasion watermilfoil	1952	XX	X		
Glugea hertwigi	1960				XX
Eurasian ruffe	1986		XX	X	
Dreissena spp.	1988	X	XX	X	
Round goby	1990		X	?	

TABLE 8

Major perturbations of selected non-indigenous species in the Great Lakes

(Data from Mills et al. 1994; Leach 1995).

NIS	PERTURBATIONS
Sea lamprey	Decline and/or extirpation of native lake trout and burbot populations.
Alewife	Decline and suppression of several native fish populations.
Common carp	Destruction of habitat for favoured fish and waterfowl species.
White perch	Competition with native fish species for food and space.
Introduced salmonids (coho and chinook salmon, rainbow and brown trout)	Competition with and predation on native fish species (became valuable sport species).
Rainbow smelt	Competition with and predation on native fish species (became a valuable commercial species in Lake Erie).
Eurasian ruffe	Competition with native fish species.
Aeromonas salmonicida (furunculosis)	Infection of salmonid species.
Glugea hertwigi	Parasitism of rainbow smelt.
Zebra mussel	Alteration of habitat through biofouling and filtering, competition with other filter feeders, predation on small zooplankton, decline and/or extirpation of native bivalve populations.
Purple loosestrife	Competition with native plants, loss of habitat for waterfowl, mammals and fish.
Eurasian watermilfoil	Competition with native plants, loss of habitat and recreational opportunities due to luxuriant growths.

Native species have been reduced in abundance or eliminated locally in the Great Lakes by perturbations from at least seven taxa of NIS. Direct causes of local extinction of native species are difficult to delineate because of the wide array of cultural stresses on the Great Lakes ecosystem (Loftus and Regier 1972; Francis et al. 1979). However, predation by the sea lamprey has been implicated in local extermination of native lake trout (Lawrie 1970), and competition from purple loosestrife has led to losses of cattails (*Typha latifolia*) (Skinner et al. 1994). There is evidence (Nalepa 1994; Schloesser and Nalepa 1994) that populations of some unionid species have been extirpated from Lake St. Clair and western Lake Erie due to extensive biofouling and competition for food by zebra mussels (*Dreissena polymorpha*).

Management Strategies

Current management strategies for future invasions and introductions in the Great Lakes basin include prevention, containment and control (Mills et al. 1993b). The long history of invasions into the basin is a good indication that virtually no attention was directed towards prevention in the past. Until the entry of the sea lamprey into the upper Great Lakes, the recent invasions of the zebra mussels and the ruffe, very little effort was made to contain an organism after it became established in the system. The rapid dispersal of the zebra mussel in North America (New York Sea Grant 1993) points to the difficulty of containing an invader. Despite efforts to contain the ruffe in western Lake Superior, where it was first established in Duluth harbor and the St. Louis River estuary, it has spread eastward about 150km along the south shore (Anonymous 1992a), and leap-frogged to Thunder Bay, Ontario on the north shore (Busiahn and McClain 1995), and then to the Alpena area on Lake Huron. Established introduced and invasive species are virtually impossible to eradicate. Control measures to minimize impacts of NIS are usually difficult and expensive (e.g., chemical control of sea lamprey larvae (Anonymous 1992b); prevention and control of biofouling by zebra mussels (Claudi and Mackie 1994)).

Aquaculture, live bait, and the aquarium trade have all been responsible for the release of unwanted NIS in the Great Lakes, particularly fishes, invertebrates, plants, and disease organisms. This has occurred despite the existence of preventative legislation in both the United States and Canada (Stanley et al. 1991; Wingate 1991; Leach and Lewis 1991). In the United States, the list of introduced fish species continues to grow (Crossman 1991; Robins et al. 1991). In Canada, the efficacy of the legislation in preventing and controlling the introduction of undesirable species has been disappointing, due to insufficient commitment by jurisdictions to enforce regulations and obtain compliance (Leach and Lewis 1991).

The use of live bait has led to the introduction of unwanted species in the Great Lakes and elsewhere (Mills et al. 1993a; Crossman 1991; Litvak and Mandrak 1993). Damage to ecosystem health from introductions due to release of live bait includes alterations of habitats and food webs, displacement of valuable native species, and introduction of diseases (Litvak and Mandrak 1993). As a general rule, the transportation and use of live bait is not adequately managed in North America, despite regulations in most jurisdictions. Clearly, there is a need for review of the live bait industry and development of specific legislation and educational programs to minimize the distribution of unwanted species from this source.

With a worldwide decline in oceanic and freshwater stocks of fishes (Larkin 1992), there is increasing interest in aquaculture as a source of protein, including the production of fin fishes, shellfishes, and crustaceans. Favored organisms include salmon, shrimp and bivalve mollusks. Much of North America's commercial aquaculture is located in coastal areas. However, freshwater culture, especially of salmonid species has been increasing recently. For example, the Province of Ontario amended The Game and Fish Act in 1994 to support the growing aquaculture industry by broadening the number of fish species that can be commercially produced. The number of allowable species for culture increased from four to thirty, plus baitfish and crayfish. Included are cichlid species of the genera *Tilapia, Sarotheradon* and *Oreochromis*, nine baitfish species and five species of crayfish. Development of the industry will likely result in more proposals to import non-native species.

In addition to risk of escape to the wild, there is the risk of importation of disease pathogens and parasites with stocks for aquaculture. In 1994, for example, an outbreak of whirling disease forced three New York state hatcheries to destroy and bury 570,000 rainbow trout. In this instance, stocks of infected fish from the probable site of initial infestation were transferred to other hatcheries, where the disease organisms infected the resident fish (New York Sea Grant 1994). Sinderman (1993) reviewed case histories of diseases and parasites associated with importation of non-indigenous organisms, and concluded that "severe disease outbreaks can result from inadequately controlled or uncontrolled movements of marine animals." Sinderman's (1993) recommendations to reduce risks associated with NIS include reduction in the need to import non-native species through development of native stocks, comprehensive inspection, diagnostic and quarantine systems, and extensive evaluation of introduction proposals on the basis of ecological, genetic, behavioral, and pathological considerations.

Several non-indigenous species of fishes, mollusks, and plants established in the Great Lakes have been released from aquariums (Mills et al. 1993a). Currently, the aquarium industry is focused on tropical fishes with rather rigorous thermal requirements which, if released into the Great Lakes, would not likely survive. However, the industry is virtually unregulated, and is a potential source of inoculation of unwanted organisms, particularly if fashion favors temperate water species (Mills et al. 1994). The recent popularity of garden ponds and demands for hardy plants and animals may constitute such a worrisome development.

Well-intentioned stocking programs, with introduction of species perceived to be beneficial (e.g., Pacific salmons), have also contributed to ecosystem health damage and loss of ecosystem integrity (Ryder and Kerr 1984). Protocol and guidelines for evaluating proposed fish introduction have been suggested (Kohler and Stanley 1984; Li and Moyle 1993). Strategies for reducing risk from introduction of aquatic organisms have been put forward by several authors (Kohler 1986; Aquatic Nuisance Species Task Force 1994). The protocols for evaluating fish introductions are quite explicit, and if implemented carefully, managers could minimize problems arising from intentional introductions. For example, to reduce risks to oligotrophic systems, Ryder and Kerr (1984) recommended consideration of ecological principles, including niche theory, interactive segregation, dominance-subordinance and resource partitioning in selecting candidate fish species for introduction.

The increase in the rate of invasions of NIS in the Great Lakes in recent decades from ballast water has motivated control efforts. In May 1989, the Canadian Coast Guard implemented voluntary guidelines that requested exchange of original ballast water in the open ocean before a ship entered the St. Lawrence Seaway. Most freshwater organisms are unable to live in sea water. Although compliance has been about 90% (Mills et al. 1994), opportunities are still available for entry of alien species. The United States has legislated exchange of ballast under the Non-indigenous Aquatic Nuisance Species Prevention and Control Act of 1990. Mandatory ballast water exchange, under this legislation, did not take effect until regulations were in place with the 1993 shipping season. The ballast water exchange requirement was extended to ships entering the Hudson River (which is connected to the Great Lakes through the Erie Canal) in the Water Resources Development Act of 1992. Slowing or preventing the spread of NIS already established through voluntary management of ballast water movement among lakes was attempted for ruffe (Great Lakes Maritime Industry Voluntary Ballast Water Management Plan for the Control of Ruffe in Lake Superior Ports 1994).

Ballast water exchange reduces the risk but does not totally eliminate the possibility of invasion by NIS. While full salinity sea water (35ppt) may kill freshwater organisms, estuarine or brackish water species may survive the exchange treatment. Brackish water species (e.g., crustaceans and algae) have become established in Great Lakes waters. Ballast exchange by Great Lakes confined vessels relies on flushing only (and not salinity) to remove organisms. A five inch gap between the end of the pipe (through which ballast water enters and leaves), and the bottom of the ballast chamber prevents complete flushing (Weathers and Reeves 1995). Inefficiencies in flushing, salt tolerance of brackish water organisms, and lapses in compliance have all likely played a role in recent failures, (i.e., the 1994 reports of young flounder (*Platichthys flesus*) in Lake Superior, Chinese mitten crabs (*Eriocheir sinensis*) in Lake Erie, and movement of ruffe from Duluth to Thunder Bay, Ontario, and Alpena, Michigan).

Consequently, ballast water management should consider alternative strategies (such as ship design, thermal alteration, microfiltration, ultraviolet treatment, ozonation, or other chemical additions) to remove living organisms from water destined to be discharged into the Great Lakes. Studies of alternative methodologies are underway in Canada and in the United States, where they are mandated under the Non-indigenous Aquatic Nuisance Species Act. The development of new technologies is the key to success in overcoming the ballast water problem.

Binational Policies and Legislation

Jurisdictional management of resources in the Great Lakes drainage basin is complex. In addition to two federal governments, bureaucracies from two provinces, eight states, and Native Americans are involved in the administration of the basin. Further policy and management perspectives are provided by the International Joint Commission and the Great Lakes Fishery Commission with binational treaties and agreements. Many of these agencies have specific policies and legislation which address NIS (Leach and Lewis 1991; Stanley et al. 1991; Wingate 1991, Appendix 1).

Many of the older legal instruments did not firmly address vector management of invasive species, but dealt broadly with issues of aquaculture and introducing species within jurisdictional boundaries (Appendix 1). However, Mills et al. (1993a) have shown that almost two-thirds of NIS established in the Great Lakes arrived via two major vectors: unintentional releases, 34%, and shipping activities, 31% (table 3). Throughout the period from 1960–91, these two entry mechanisms were responsible for all forty-one new introductions and invasions. Reporting on a 1991 workshop sponsored by the Great Lakes Fishery Commission, Mills et al. (1993b) stated that "legislative and regulatory efforts must consider transport mechanisms used by exotic species to gain access to the Great Lakes basin" and warned that "vector management is critical in order to prevent more non-indigenous species from entering the Great Lakes basin." Shipping activities and unintentional releases, increasingly through actions of the private sector, should be the focus of policy makers and law makers wishing to preserve the health and integrity of, and social benefits from, the Great Lakes ecosystem.

As indicated in the previous section, sound management strategies to preserve Great Lakes integrity are being undermined by unintentional releases of NIS, increasingly through private sector activities including aquaculture, bait, horticulture, and aquarium industries. Recent examples include infection of New York state fish hatcheries with whirling disease (New York Sea Grant 1994), striped bass and striped bass hybrids in Lake Erie (Dochoda and Koonce 1994), and grass carp in the Great Lakes (Crossman et al. 1987).

These examples suggest that the ecological integrity of the Great Lakes cannot be guaranteed, in spite of recent actions on ship ballast discharge, until North American authorities address unintentional introductions as a byproduct of private sector activities. National governments must control the import of live organisms into North America by the private sector. Similarly, states and provinces, supported by their respective national governments, must assume control of private aquaculture's movement of fish and disease organisms between watersheds (Aquatic Nuisance Species Task Force 1994).

The North American Commission for Environmental Cooperation, newly established under the North American Free Trade Agreement, may provide an opportunity to minimize continental risk of introductions as a byproduct of private sector activity. Certainly, Canada, Mexico, and the United States all have an interest in ensuring that they have collectively achieved North American security from unintentional introductions associated with private sector activity. Once introduced, species migrate naturally across national borders; for example, thirty-eight species have moved into some part of Canada from the United States since 1800. And, for fishes, the southern boundary for North America extends into Mexico to 20° N on the Atlantic slope, and just below 17° on the Pacific slope (which corresponds to the southern limit of Cyprinidae) (Crossman 1991).

Since both resource management and agriculture agencies are now involved with private aquaculture, uncertainties concerning regulation exist. Therefore, cooperation across disciplines as well as borders will be an essential factor in determining whether the growth industry of private aquaculture will decide the future aquatic communities and management options for the Great Lakes.

Shipping was the other active vector which injected new species into the Great Lakes over the last thirty years. Ballast management initiatives (primarily discouraging discharge of ballast water originating in other fresh or brackish waters) collectively provide the greatest degree of protection to the Great Lakes of any water body in the world subject to ocean shipping. Risk of introductions with discharge of ships' ballast began to decline in May, 1989, with implementation of the Canadian Coast Guard's voluntary "Great Lakes Ballast Water Control Guidelines," and later, with implementation in 1993 of regulations in the United States under the Non-indigenous Aquatic Nuisance

Prevention and Control Act of 1990. Preventive measures for the Hudson River in the Water Resources Act of 1992 further reduced risk, as the Hudson River and Great Lakes are connected through the Erie Canal.

However, reports in 1994 of specimens of Chinese mitten crabs and European flounder being taken again from the Great Lakes (these species do not reproduce in fresh water) suggest that ballast management guidelines and legislation as implemented only reduce risk. More rigorous enforcement, conversion of Canadian guidelines to regulation, and extension of ballast management exercises to contiguous coastal water would further reduce risk. However, risk reduction, although welcome in itself, will most benefit the Great Lakes ecosystem if it buys time to develop techniques to eliminate risk entirely. The best long-term hope for the Great Lakes and global aquatic resources lies in research and development of ship designs and mechanisms that, upon adoption by classification societies, will one day produce fleets which are incapable of transporting viable aquatic organisms in ballast.

Although deliberate release constitutes a relatively small vector-source of new Great Lakes introductions, 8%, prior to the 1960s, and none since, state authority in these matters has been the focus of two reports to the United States Congress (Office of Technology Assessment 1993; Aquatic Nuisance Species Task Force 1994). However, in 1980, Great Lakes fishery agencies signed the Joint Strategic Plan for Management of Great Lakes Fisheries (GLFC 1994), in which they agreed to consult with one another on management which would significantly influence the interests of more than one jurisdiction (as is the potential of a new introduction).

In support of the Strategic Plan, agencies have developed formal consultation procedures on intentional introductions (GLFC 1992), as well as a fish disease model program (Hnath 1993) and quarantine protocol (Horner and Eshenroder 1993). Thus, the Great Lakes states, Ontario, and Tribal Management Authorities (using the committee forum of the Great Lakes Fishery Commission) have prevented two serious salmonine fish diseases (infectious hematopoietic necrosis virus, IHNV, and viral hemorrhagic septicemia, VHS) from being introduced into the Great Lakes (Dochoda 1991), and have decided against planned introductions of striped bass by Pennsylvania and Ohio, grass carp by Michigan, and arctic char by Ontario (Dochoda and Koonce 1994). In addition, Great Lakes fishery agencies have virtually eliminated from their hatcheries endemic diseases such as infectious pancreatic necrosis virus, IPNV, and epizootic epitheliotropic disease virus, EEDV, and until recently, the established protozoan *Myxobolus cerebralis* which causes whirling disease.

Although builders of new inter-watershed highways, railroads, and canals have been less active during the last thirty years (and thus, have not introduced new species during that period), continued vigilance is warranted. Within the last decade, newspapers have documented proposals for diversion of Great Lakes water to the United States southwest, as well as diversion of fresh water from James Bay tributaries to the Great Lakes. Society is unlikely to rule out such projects *a priori*. Thus, environmental impact studies are an important tool to ensure that future decision-making on proposed new construction takes into account the permanent consequences of potential exotic introductions. Under the Boundary Waters Treaty of 1909, the International Joint Commission may also play an important role, as in its 1977 report to Governments on the transboundary implications of the proposed Garrison Dam (I.J.C. 1977).

Also worrisome is evidence that pollution abatement can expose new watersheds to established exotics. For example, cleanup in Wisconsin's Fox River would have exposed Lake Winnebago to Great Lakes sea lamprey, had not the Wisconsin Department of Natural Resources, in cooperation with the Great Lakes Fishery Commission and the United States Corps of Engineers, requested the sealing of a lock on that waterway in 1988 (GLFC 1988). Even in its presently degraded state, the Chicago Sanitary and Ship Canal was suitably hospitable in 1992 to serve as a vector for the exotic zebra mussel to spread from the Great Lakes to the Mississippi River. The Erie Canal allowed zebra mussels from the Great Lakes to reach the Hudson River and Lake Champlain. Conversely, Mississippi River fishes such as the skipjack herring (*Alosa chrysochloris*), have accessed the Great Lakes, presumably via the Chicago Sanitary and Ship Canal (Fago 1993). In the future, preventing exotic invasions via newly hospitable waterways will depend on the awareness of those participating in pollution cleanup initiatives of any attendant invasion risks, and on their ability to alert those most concerned of the need for remedial measures.

The Future

Very clearly, the ability to predict potential invaders to the Great Lakes, and their possible impacts on ecosystem health is desirable. However, the Great Lakes experience to date indicates that questions about vulnerability of the ecosystem to invasion, predictions of success or failure of colonization, and predictions of ecosystem damage are difficult (Mills et al. 1993b; Pimm 1991; Leach 1995).

Prediction of colonization by future invaders is sometimes possible from consideration of similarities in donor and recipient habitats, characteristics of the invading species which favor inoculation rate, rapid dispersal, reproduction rate, etc., and vulnerability of the recipient community (table 4). However, because of high variability associated with habitats and communities and diversity of invading species, uniform ecological concepts are not evident (Leach 1995). For example, Leach (1995) found that colonization of the Great Lakes was predictable for some NIS (e.g., ruffe) but not others (e.g., gobies). Cultural disturbance, which Elton (1958) found to be an important factor in the success of most of the invasions he reviewed, was considered by Leach (1995) to be a factor in the success of some of the Great Lakes invasions, but was not an overriding ecological requirement. Organisms such as the sea lamprey and the zebra mussel can successfully colonize pristine and undisturbed environments, providing other habitat criteria are met, presumably because ecological analogues did not exist, and accordingly, trophic opportunities were present.

In general, there is a match between climate and other basic habitat criteria of donor and recipient systems. However, Mills et al. (1993b) found no correlation between an invader's distribution range in donor and recipient regions. For example, *Bythotrephes cederstroemi* has spread widely in the Great Lakes basin, but has limited distribution in Europe.

Prediction of damage to ecosystem health by a colonizing NIS is also difficult. In a review of damage by Great Lakes NIS, Leach (1995) found that perturbations were not readily predictable from knowledge about the ecology of invaders or native communities. Impact by early colonists, such as sea lamprey, purple loosestrife, and alewife were not generally envisaged. Managers stocking salmonids concerned themselves primarily with providing recreational opportunities.

In stocking salmon species, however, they were responding more to an already disrupted fish community than considering the implications of planting NIS. According to Tody and Tanner (1966):

> Invasion of the upper lakes by the sea lamprey in the mid-1930s upset the prey-predator relationship. The sea lamprey population increased rapidly. The lake trout and the burbot were virtually eliminated. The sea lamprey preyed upon the other fish to a lesser extent. Thus, with the removal of their normal predators, the populations of chubs and other prey species expanded enormously. The loss of predatory species also set the stage for invasion of the upper lakes by other species of small fish. One particular invading species, the alewife, from Lake Ontario and the Atlantic Ocean via the Welland Canal has exploded to superabundance in lakes Michigan and Huron since 1955.

> The well-known sea lamprey control program of Canada and the United States coordinated by the Great Lakes Fishery Commission has brought the sea lamprey under a relatively high degree of control in Lake Superior. The control program is well under way in Lake Michigan and is scheduled to begin on Lake Huron in 1966.

> The alewife is building to an almost unbelievably dominant role in the fish populations of lakes Michigan and Huron. Recent estimates by the Bureau of Commercial Fisheries indicate that alewife may comprise 90% or more of the total fish population by weight in Lake Michigan. The alewife is present, but not particularly abundant as yet in Lake Superior. It is an extremely efficient competitor and appears capable of eliminating these species with which it competes for food and space. It may also eliminate or decimate all of the fish species that spawn directly in the Great Lakes by consuming their eggs or young. The spawning success of even the predatory lake trout may be seriously affected. Alewives are so abundant that swimming and boating are degraded by periodic accumulation of dead fish on the beaches and in harbor areas. Even city water supply intakes are clogged at times with hordes of alewife. Ecosystem health damage by some recent ballast water invaders (e.g., Dreissenid species, ruffe) has been predictable.

From a review of fauna in the European donor region, Mills et al. (1993b) predicted that the Antipodean snail (*Potamopyrus antipodarum*), and the Caspian amphipod (*Corophium curvispinum*) will invade North America and the Great Lakes. These organisms have an extensive invasion history in Europe, interface with shipping to North America, are eurybionts, and are highly fecund. Subsequent to the prediction, it was discovered that the Antipodean snail had been found in Idaho in the 1980s (Mills et al. 1994). Notwithstanding the difficulties surrounding prediction of invasions, the accumulation of advance knowledge of potential invaders provides a basic step towards preventing their entry to the Great Lakes. Agricultural quarantines, for example, have been useful in preventing the introduction of well-known pests with potential for economic and ecologic effects on North American livestock and crops (Mills et al. 1993b).

Global warming, associated with increased levels of carbon dioxide and other gases, has been predicted (Kellogg and Schware 1981) and measured (Gates 1993), and will influence the distribution of organisms in the future. Resulting from this trend, thermal structure in the Great Lakes is expected to change (Meisner et al. 1987), leading to range expansion by southern North American fish species. Mandrak (1989) has identified nineteen species of fish from the Mississippi and

Atlantic Coastal basins as potential invaders of the lower Great Lakes, and has predicted that the invading species would greatly alter existing fish communities.

It is likely that the Great Lakes system will continue to receive NIS, particularly from Europe and Asia, although the rate of inoculation may decrease (Mills et al. 1993b). Mills et al. (1993b) have suggested that global strategies be developed for the prevention of spread and the control of invasive species. Included would be an inventory of potential invaders from other continents and a clearinghouse network among agencies in North America, and donor regions worldwide to exchange information on ecological requirements and control methodologies. Legislative and regulatory policies designed to prevent unplanned introductions must consider management of entry vectors and be broad enough geographically, to protect not only the Great Lakes region, but also the entire North American continent.

Acknowledgments

We acknowledge, with appreciation, contributions to Appendix 1 from Bruce Bandurski, Tim Carey, Dick Hassinger, Rod Horner, Julian Hynes, Bill James, Lee Kernen, Bob Lange, Wayne MacCallum, Bill Mattes, Steve Nepszy, Ken Paxton, John Robertson, Andrew Shiels and Jay Troxel. Helpful review comments were provided by Thomas Busiahn, Steve Nepszy, Richard Ryder and William Taylor. We are grateful to Debbie Warner for processing the manuscript. Contribution No. 95–11 of the Ontario Ministry of Natural Resources, Aquatic Ecosystems Science Section, P.O. Box 7000, Peterborough, Ontario, Canada, K9J 8M5. Contribution No. 176 of the Cornell University Biological Field Station.

Literature Cited

Anonymous. 1992a. Ruffe in the Great Lakes: A Threat to North American Fisheries. Report of the Ruffe Task Force. Great Lakes Fishery Commission. Ann Arbor, Michigan.

———, 1992b. Strategic vision for the Great Lakes Fishery Commission for the decade of the 1990s. Great Lakes Fishery Commission, Ann Arbor, Michigan.

Aquatic Nuisance Species Task Force. 1994. Report to Congress: Findings, conclusions and recommendations of the intentional introductions policy review. U.S. Congress.

Asworth, W. 1986. The Late, Great Lakes. New York: Alfred A. Knopf, Inc.

Bailey, R. M., and G. R. Smith. 1981. Origin and geography of the fish fauna of the Laurentian Great Lakes basin. Canadian Journal of Fisheries and Aquatic Sciences 38:1539–1561.

Baltz, D. M., and P. B. Moyle. 1993. Invasion resistance to introduced species by a native assemblage of California stream fishes. Ecological Applications 3:246–255.

Busiahn, T. R., and J. R. McClain. 1995. Status and control of ruffe (*Gymnocephalus cernuus*) in Lake Superior and potential for range expansion. Pages 461–470 *in* M. Munawar, T. Edsall, and J. H. Leach, editors. The Lake Huron ecosystem: ecology, fisheries and management. Ecovision World Monograph Series, SPB Academic Publishing, Amsterdam, The Netherlands.

Claudi, R., and G. L. Mackie. 1994. Practical Manual for Zebra Mussel Monitoring and Control. Lewis Publishers, Boca Raton, Fla.

Crossman, E. J. 1991. Introduced freshwater fish: A review of the North American perspective with emphasis on Canada. Canadian Journal of Fisheries and Aquatic Sciences 48(Suppl. 1): 46–57.

———, S. J. Nepszy, and P. Krause. 1987. The first record of grass carp, *Ctenopharyngodon idella*, in Canadian waters. Canadian Field-Naturalist 101:584–586.

Dochoda, M. R. 1991. Meeting the challenge of exotics in the Great Lakes: the role of an international commission. Canadian Journal of Fisheries and Aquatic Sciences 48 (Suppl. 1): 171–176.

———— and J. F. Koonce. 1994. A perspective on progress and challenges under A Joint Strategic Plan for Management of Great Lakes Fisheries. University of Toledo Law Review 25:425–442.

Elton, C. S. 1958. The Ecology of Invasions by Animals and Plants. Chapman and Hall. London. 181p.

Emery, L. 1985. Review of fish species introduced into the Great Lakes, 1819–1974. Great Lakes Fishery Commission Technical Report No. 45. Great Lakes Fishery Commission, Ann Arbor, Michigan.

Evans, D. O., G. J. Warren, and V. W. Cairns. 1990. Assessment and management of fish community health in the Great Lakes: synthesis and recommendations. Journal of Great Lakes Research 16:639–669.

Fago, D. 1993. Skipjack herring, *Alosa chrysochloris*, expanding its range into the Great Lakes. Canadian Field-Naturalist 107: 352–353.

Ferguson, M. M. 1990. The genetic impact of introduced fishes on native species. Canadian Journal of Zoology 68: 1053–1057.

Francis, G. R., J. J. Magnuson, H. A. Regier, and D. R. Talhelm (Eds.). 1979. Rehabilitating Great Lakes ecosystems. Great Lakes Fishery Commission Technical Report No. 37. Great Lakes Fishery Commission, Ann Arbor, Michigan.

Gates, D. M. 1993. Climate change and its biological consequences. Sinauer Associates, Inc., Sunderland, Massachusetts.

GLFC (Great Lakes Fishery Commission). 1988. Minutes of the 1988 Annual Meeting. Great Lakes Fishery Commission Ann Arbor, Michigan.

GLFC (Great Lakes Fishery Commission). 1992. Minutes of the Council of Lake Committees' 1992 Annual Meeting. Great Lakes Fishery Commission Ann Arbor, Michigan.

GLFC (Great Lakes Fishery Commission). 1994. A joint strategic plan for management of Great Lakes fisheries. Great Lakes Fishery Commission Ann Arbor, Michigan.

Groves, R. H., and J. J. Burdon. 1986. Ecology of Biological Invasions. Cambridge University Press, Cambridge, UK.

Hatcher, H. 1944. The Great Lakes. London: Oxford University Press.

Herbold, B., and P. B. Moyle. 1986. Introduced species and vacant niches. American Naturalist 128:751–760.

Hnath, J. G. (Ed.). 1993. Great Lakes fish disease control policy and model program. Great Lakes Fishery Commission Special Publication 93–1:1–38.

Horner, R. W., and R. L. Eshenroder (Eds.). 1993. Protocol to minimize the risk of introducing emergency disease agents with importation of salmonid fishes from enzootic areas. Great Lakes Fishery Commission Special Publication 93–1: 39–54.

Hutchinson, G. E. 1957. Concluding remarks. Cold Spring Harbor Symposia on Quantitative Biology 22:415–427.

I.J.C. (International Joint Commission). 1977. Transboundary implications of the Garrison Diversion Units: An I.J.C Report to the Governments of Canada and the United States. International Joint Commission. Ottawa.

Kellogg, W. W., and R. Schware. 1981. Climate change and society, consequences of increasing atmospheric carbon dioxide. Westview Press Boulder, Colorado.

Kerr, S. R., and R. A. Ryder. 1977. Niche theory and percid community structure. Journal of the Fisheries Research Board of Canada 34:1952–1958.

Kohler, C. C. 1986. Strategies for reducing risks from introductions of aquatic organisms. Fisheries 11:2–3.

———— and J. G. Stanley. 1984. A suggested protocol for evaluating proposed exotic fish introductions in the United States. Pages 387–406 *in*: W. R. Courtenay, Jr. and J. R. Stauffer, (Eds.), Distribution, biology and management of exotic fishes. John Hopkins University Press, Baltimore.

Krueger, C. C., and B. May. 1991. Ecological and genetic effects of salmonid introductions in North America. Canadian Journal of Fisheries and Aquatic Sciences 48(Suppl. 1):66–77.

Larkin, P. A. 1992. Future prospects and their implications for research on the ecology of freshwater fish. Ecology of Freshwater Fish 1:1–4.

Lawrie, A. H. 1970. The sea lamprey in the Great Lakes. Transactions of the American Fisheries Society 99:766–775.

Leach, J. H. 1995. Non-indigenous species in the Great Lakes: were colonization and damage to ecosystem health predictable? Journal of Aquatic Ecosystem Health 4:117–128.

———— and C. A. Lewis. 1991. Fish introductions in Canada: provincial views and regulations. Canadian Journal of Fisheries and Aquatic Sciences 48(Suppl. 1):156–161.

Li, H. W., and P. B. Moyle. 1993. Management of introduced fishes. Pages 287–307 *in* C. C. Kohler and W. A. Hubert, (eds.), Inland Fisheries Management in North America. American Fisheries Society, Bethesda, Md.

Litvak, M. K., and N. E. Mandrak. 1993. Ecology of freshwater baitfish use in Canada and the United States. Fisheries 18(12):6–13.

Loftus, K. H., and H. A. Regier. 1972. Introduction to the proceedings of the 1971 Symposium on Salmonid Communities in Oligotrophic Lakes. Journal of the Great Lakes Fisheries Research Board of Canada 29:613–986.

Lowe-McConnell, R. 1994. The changing ecosystem of Lake Victoria, East Africa. Freshwater Forum 4:76–89.

Mandrak, N. E. 1989. Potential invasion of the Great Lakes by fish species associated with climatic warming. Journal of Great Lakes Research 15:306–316.

Meisner, J. D., J. L. Goodier, H. A. Regier, B. J. Shuter, and W. J. Christie. 1987. An assessment of the effects of climate warming on Great Lakes basin fishes. Journal of Great Lakes Research 13:340–352.

Mills, E. L., J. H. Leach, J. T. Carlton, and C. L. Secor. 1993a. Exotic species in the Great Lakes: a history of biotic crises and anthropogenic introductions. J. Great Lakes Res. 19:1–54.

———, J. H. Leach, C. L. Secor, and J. T. Carlton. 1993b. What's next? The prediction and management of exotic species in the Great Lakes (report of the 1991 workshop). Great Lakes Fishery Commission. Ann Arbor, Michigan.

———, J. H. Leach, J. T. Carlton, and C. L. Secor. 1994. Exotic species and the integrity of the Great Lakes: Lessons from the past. BioScience 44:666–676.

Mulligan, G. A. 1987. Common Weeds of Canada. Minister of Supply and Services Canada. 140 pp.

Nalepa, T. F. 1994. Decline of native unionid bivalves in Lake St. Clair after infestation by the zebra mussel, *Dreissena polymorpha*. Canadian Journal of Fisheries and Aquatic Sciences 51:2227–2233.

Neary, B. P., and J. H. Leach. 1992. Mapping the potential spread of the zebra mussel (*Dreissena polymorpha*) in Ontario. Canadian Journal of Fisheries and Aquatic Sciences 49:406–415.

New York Sea Grant. 1993. *Dreissena polymorpha* information review. Vol. 4, No. 2. New York Sea Grant, Brockport, N.Y.

New York Sea Grant. 1994. Charterlines. New York Sea Grant Extension Program. Issue 43.

Office of Technology Assessment. 1993. Harmful non-indigenous species in the United States. Office of Technology Assessment F–565, U.S. Government Printing Office, Washington, D.C. pp. 391.

Pimm, S. L. 1991. The Balance of Nature? The University of Chicago Press, Chicago.

Robins, C. R., R. M. Bailey, C. E. Bond, J. R. Brooker, E. A. Lachner, R. N. Lea, and W. B. Scott. 1991. Common and scientific names of fishes from the United States and Canada, Fifth Edition. American Fisheries Society Special Publication 20:1–183.

Ryder, R. A. 1972. The limnology and fishes of oligotrophic glacial lakes in North America (about 1800 A. D.). Journal of the Fisheries Research Board of Canada 29:617–628.

———, 1990. Ecosystem health, a human perception: definition, detection, and the dichotomous key. Journal of Great Lakes Research 16:619–624.

——— and S. R. Kerr. 1984. Reducing the risk of fish introductions: a rational approach to the management of integrated cold-water communities. Pages 510–533 *in* Introductions and Transplantations. EIFAC Tech. Paper 42. Suppl. Vol. 2. FAO Rome.

Schloesser, D. W., and T. F. Nalepa. 1994. Dramatic decline of unionid bivalves in offshore waters of western Lake Erie after infestation by the zebra mussel, *Dreissena polymorpha*. Canadian Journal of Fisheries and Aquatic Sciences 51:2234–2242.

Simberloff, D. 1981. Community effects of introduced species. Pages 53–81 *in* II. Nitecki (ed.) Biotic Crises in Ecological and Evolutionary Time. Academic Press, New York.

———, 1986. Introduced insects: a biogeographic and systematic perspective. Pages 3–26 *in* H. Mooney and J. A. Drake, (eds.), Ecology of Biological Invasions of North America and Hawaii. Springer-Verlag, Berlin.

Sinderman, C. J. 1993. Disease risks associated with importation of non-indigenous marine animals. Marine Fisheries Review 54:1–10.

Skinner, L. C., W. J. Rendall, and E. L. Fuge. 1994. Minnesota's purple loosestrife program: history, findings and management recommendations. Minnesota Department of Natural Resources Special Publication 145.

Stanley, J. G., R. A. Peoples Jr., and J. A. McCann. 1991. U.S. federal policies, legislation, and responsibilities related to importation of exotic fishes and other aquatic organisms. Canadian Journal of Fisherise and Aquatic Sciences 48 (Suppl 1):162–166.

Thomas, S. K., R. M. Sullivan, R. L. Vertrus, and D. W. Floyd. 1992. Aquaculture law in the north central states: a digest of state status pertaining to the production and marketing of aquacultural products. Technical Bulletin Series #101. North Central Regional Aquaculture Center, Michigan State University, 13 Natural Resources Building, East Lansing, MI 48824–1222.

Tody, W. H., and H. A. Tanner. 1966. Coho salmon for the Great Lakes. Fish Management Report No. 1. Department of Conservation. Lansing, Michigan.

Weathers, K., and E. Reeves. 1995. The defense of the Great Lakes against the invasion of non-indigenous species in ballast water. Appendix B of 1995 Joint Report of the Canadian and United States Coast Guard on Progress Towards Achievement of the Objectives of the Great Lakes Water Quality Agreement of 1978 (as amended by its protocol of 1987).

Williamson, M. H., and K. C. Brown. 1986. The analysis and modelling of British invasions. Philosophical Transactions of the Royal Society of London, Series B 314:505–522.

Wingate, P. J. 1991. U.S. State's view and regulations on fish introductions. Canadian Journal of Fisheries and Aquatic Sciences 48 (Suppl. 1):167–170.

Appendix I

Legal and policy instruments (with geographic areas and vectors addressed) useful in preventing the introduction and spread of non-indigenous species in the Great Lakes basin (Leach and Lewis 1991; Stanley et al. 1991; Thomas et al. 1992). (Geographic areas: global, North America, United States, Canada, state or province, Great Lakes watershed; vectors: shipping, unintentional release, deliberate release, canals, highways, railroads).

United States
Lacey Act of 1900 with subsequent amendments (18 U.S.C. 42)
- authorizes Secretary of Interior to enforce any law, treaty, or regulation of the United States, any Indian tribal law, or any interstate transfer violating any state or foreign law relative to import, transport, sale, receipt, acquisition, or purchase of any fish, wildlife, or plant (United States, state; deliberate release, unintentional release)

Endangered Species Act of 1973 (16 U.S.C. 1531–1543)
- provides for regulation of imports to protect endangered species (United States, Great Lakes watershed; deliberate release, unintentional release)

Executive Order 11987, 1977
- prohibits federal agencies from engaging in introductions or imports of exotic species, or exports of native species (Global, United States, Great Lakes watershed; deliberate release)

National Aquaculture Improvement Act of 1985
- establishes national aquaculture policy, establishes and implements a national aquaculture plan, and encourages aquaculture activities and programs through interagency task force under leadership of Secretary of Agriculture (United States; unintentional release)

Non-indigenous Aquatic Nuisance Prevention and Control Act of 1990 (16 U.S.C. § 4711 (Supp. II 1990))
- authorizes Secretary of Transportation to regulate ballast discharge of vessels (Great Lakes watershed; shipping)
- authorizes a task force to prevent, monitor, control, and study introductions (United States; unintentional release)
- directs task force to report to Congress on approaches for reducing risks associated with intentional introductions (United States; intentional release)
- authorizes a Great Lakes Panel to advise on priorities, coordinate, and provide control advice (Great Lakes watershed; all vectors)
- invites Governors to advise task force on technical and financial assistance needed to address harmful effects of exotics (states; all vectors)
- encourages the Department of State to initiate negotiations with foreign governments on preventing, controlling and studying aquatic nuisance species infesting shared waters (Global, United States; all vectors)

Water Resources Development Act of 1992 (33 U.S.C. § 2201 (Supp. IV 1992))
- authorizes the Secretary of Transportation to regulate deballasting by ocean vessels in the Hudson River, to prevent introduction and spread of aquatic nuisance species in the Great Lakes (Great Lakes watershed; shipping)

Great Lakes Maritime Industry Voluntary Ballast Water Management Plan for the Control of Ruffe in Lake Superior Ports, 1994
- encourages ships to implement specific ballasting procedures to prevent the further spread of ruffe (Great Lakes watershed; shipping)

Minnesota
Commissioner's Order No. 2239 — Regulations for the importation, transportation, and stocking of fish sperm, fish eggs, and live fish; supersedes Commissioner's Order No. 2076
- requires importation-transportation-stocking permit, except for Minnesota minnows, Minnesota caught food fish, appropriately documented live fish in transit through Minnesota, aquarium fish, processed fish, appropriately documented live Minnesota hatchery fish for consumption, appropriately documented live fish for export, sucker or fathead minnow eggs or fry purchased for bait propagation, or 14-day exhibits of Minnesota fish not destined for stocking (state; deliberate release, unintentional release)

Commissioner's Order No. 2290—prescribes regulations for the maintenance and operation of private fish hatcheries and fish farms; supersedes Commissioner's Order No. 2237
- regulates maintenance and operation of private fish hatcheries and fish farms, in particular license requirements, inspection fees, operation requirements (e.g. not connected with public waters), vehicle identification requirements, and specific requirements for minnow, game fish, and other species (state; unintentional release)

Wisconsin
Chapter 29, Fish and Game Act 1917—In 1989, rule amendments were proposed declaring non-indigenous fish detrimental, except where their presence is specifically permitted under §.29.535, Stats. Harvesting proposals relating to bait fishes, frogs and crayfish were not subsequently enacted.
- requires a permit for any introduction or importation of fish. Fish imported under permit are subject to inspection (state; deliberate release)

- establishes provision that permits will only be issued if introduction is not detrimental to the natural resources of the state (state; deliberate release)
- declares all fish species, strains or hybrids that are not indigenous to the waters of the state to be detrimental except where their presence is permitted under §29.535, Stats. The department may remove them or cause them to be removed (state; deliberate release; unintentional release)

Great Lakes Indian Fish and Wildlife Commission—Voigt Model Code (Sections 9.17 and 9.22)

- restricts release of (bait) fish, except for trout, in designated off-reservation waters within the 1837 and 1842 Wisconsin-ceded waters (state; deliberate release; unintentional release)
- prohibits individual tribe members from releasing any fish or spawn into any water body except for returns from which it came (state; unintentional release)

Illinois

Title 17: Conservation Chapter 1: Department of Conservation Subchapter b: Fish and Wildlife. Part 870. Aquaculture, Transportation, Stocking Importation and/or Possession of Aquatic Life (1988)

- regulates importation, transportation, culture management and utilization of aquatic life resources; establishes aquaculture permit and requirements for facility, operations, and release; conditionally exempts the aquarium industry, the state, and universities; establishes an aquatic life approved species list; species not appearing on this list are illegal unless a letter of authorization to import/possess is granted by the state (state; deliberate release, unintentional release)

Aquaculture Development Act. Public Act 85–856, Senate Bill No. 762 of the 85th General Assembly (1987)

- establishes Department of Agriculture as regulatory agency of aquaculture; DOA shall consult with Department of Conservation to ensure protection of aquatic life in the native environment; defines aquaculture as agriculture; requires permit for raising aquatic life, requires permission to import non-indigenous aquatic life (state; deliberate release, unintentional release)

Indiana

Title 14. Fish and Wildlife Act

- regulates sale, hauling, importation (permit required), and stocking of fish (permit required except for private pond) (state; deliberate release, unintentional release)

Title 15. Agriculture and Animals

- governor may prohibit animal imports from disease-endemic areas (state; deliberate release; unintentional release)

DNR rule 310 IAC 3.1–6–7 Grass carp and exotic catfish

- prohibits importation, possession, or release of grass carp or exotic catfish (state; deliberate release, unintentional release)

Michigan

Fish Breeders Act 196, 1957

- regulates the propagation, importation, use, transportation, possession, and sale of game fish in private waters (state; deliberate release, unintentional release)

DNR Regulation 299.1051. Importation, transportation and sale of live fishes or viable eggs of the salmonidae family, 1969

- regulates by permit the importation, transport and sale of salmonids, with special attention to disease (state; deliberate release, unintentional release)
- regulates by permit the importation of Japanese weatherfish, grass carp, ide, rudd, bitterling, and tench (state; deliberate release, unintentional release)

Michigan Sports Fishing Law, Act 165, 1929

- regulates by license, the taking, possession and sale of minnows. Prevents transportation, outside of the state, of minnows taken in the public waters of the state (not enforceable). Prevents the importation of non-indigenous minnows into the state. Requires a permit to stock spawn, fry or fish of any kind into the public waters of the state (state; deliberate release, unintentional release)

Ohio

Revised Code, Chapter 1531: Wildlife, 1953

- authorizes temporary written orders pertaining to protection, preservation, propagation, possession, and management of wild animals (state; deliberate release, unintentional release)
- prohibits possession of wildlife contrary to division orders, e.g., unlawful to possess, import or sell species of fish or hybrids not naturally found in Ohio thereof for introduction into any body of water that is connected to or otherwise drains into any body of water . . . that would allow . . . egress of the fish into public waters . . . without permission (state; deliberate release, unintentional release)

Revised Code, Chapter 1533: Hunting; Fishing, 1953
- authorizes regulation of import, export, or transport of live wild animals except for interstate commerce which does not originate or terminate in the state (state; deliberate release, unintentional release)
- regulates by permit collecting or dealing in bait (state; deliberate release, unintentional release)

Pennsylvania
Title 30, Pennsylvania Consolidated Statutes
- requires permits for taking, catching, killing, possession, introduction, removal, importing, transporting, exporting or distributing any fish in Pennsylvania waters (state; deliberate release, unintentional release)
- requires licenses or permits for live fish dealers and for transportation of live fish in Pennsylvania (state; deliberate release, unintentional release)

Chapter 71, Introduction of Fish into Commonwealth Waters
- provides for the issuing of limited propagation licenses to permit non-indigenous fish species to closed system aquaculture (state; unintentional release)
- permits (as a limited exception to the prohibition of grass carp in the commonwealth) triploid grass carp, certified by the United States Fish and Wildlife Service (USFWS) and procured from a producer participating in the USFWS certification program, to be introduced into Commonwealth waters (state; deliberate release)

Chapter 73 Transportation of Live Fish into the Commonwealth
- prohibits transport and release into Pennsylvania, or into new watersheds within the State, or the transport of grass carp, or the import of potentially problematic tropical fish, or unlisted species without written permission (state; deliberate release, unintentional release)

New York
Environmental Conservation Law
- requires permit to place fish or fish eggs in any waters of the state (state; deliberate release, unintentional release)
- prohibits use or sale of carp, goldfish, and lamprey larvae for bait (state; unintentional release)
- authorizes regulation of importation, and transportation of any live fish or eggs of Salmonidae (state; unintentional release)
- authorizes hatchery permits for private trout and black bass hatcheries (state; deliberate release, unintentional release)
- prohibits liberation of zebra mussels into any water of the state; requires a permit to buy, sell, possess or transport zebra mussels (state; deliberate release, unintentional release)

Canada
Fisheries Act (R.S.C., 1985, c. F–14)
- provides for regulation of 1) catching, loading, landing, handling, transporting, possession, and disposal of fish; 2) taking or carrying of fish of any part thereof from one province to another; and 3) conservation and protection of fish (Canada, provinces, Great Lakes watershed; all vectors)
- Fish Health Protection Regulations which prohibit importation of cultured fish, eggs of wild fish, or dead cultured salmonids into Canada or between provinces without permit and fish health inspection for certain disease agents (Canada, provinces; deliberate release, unintentional release)

Great Lakes Ballast Water Control Guidelines (Canadian Coast Guard, revised 1993)
- requests that oceangoing vessels exchange ballast for ocean water before entering the St. Lawrence River (Great Lakes watershed; shipping)

A Wildlife Policy for Canada (Department of Environment, 1990)
- provides a framework for federal, provincial, territorial, and nongovernmental policies and programs that affect wildlife (including fish) (Canada, provinces; all vectors)

Federal Aquaculture Development Strategy (Department of Fisheries and Oceans, 1995)
- outlines the federal role in developing the aquaculture industry in Canada (Canada, provinces; unintentional release)

Ontario
Ontario Fishery Regulations
- prohibits transport of live fish (other than bait fish) and no depositing of live fish from one body of water to another without a license (province, Great Lakes watershed; deliberate release)

Environmental Assessment Act
- requires class environmental assessment for fish stocking in new waters, and individual environmental assessment for stocking of exotic species (province, Great Lakes watershed; deliberate release)

Ontario Game and Fish Act (Amended 1994 in Omnibus Bill 175)
- authorizes the Ministry of Natural Resources to regulate private aquaculture (province, Great Lakes watershed; deliberate release, unintentional release)

Quebec

Loi Sur La Conservation De La Faune C–61 (before 1983) and Loi Sur La Conservation Et La Mise En Valeur De La Faune C–61–1 (since 1983) (an act respecting the conservation and development of wildlife)
- provides for conditions, including disease certification, on the importation of fish (province; deliberate release, unintentional release)

International

Boundary Waters Treaty of 1909 (Canada–United States)
- created International Joint Commission whose approval is required for projects in boundary waters and, in some cases, for projects in transboundary rivers which affect the water levels or flows on the other side of the boundary; as well, when questions or matters of difference arise between the United States and Canada involving the rights, obligations, or interests on either side in relation to the other or to the inhabitants of the other, they are to be referred to the International Joint Commission for examination and report. These assignments are called "references." (Canada, United States; canals, shipping)

Convention on Great Lakes Fisheries (Canada–United States 1955)
- created and charged Great Lakes Fishery Commission to eradicate or minimize sea lamprey in the Great Lakes watershed—includes containment actions (Great Lakes watershed; canals)
- charged Commission to investigate and advise on issues affecting sustainable benefits from fish stocks of common concern (Great Lakes watershed; all vectors)

Joint Strategic Plan for Management of Great Lakes Fisheries (Great Lakes Fishery Agencies, 1980)
- "Consensus must be achieved when management will significantly influence the interests of more than one jurisdiction" and supporting consultation procedures (Great Lakes watershed; deliberate release)
- Great Lakes fish disease policy and model program (Great Lakes watershed; unintentional release)
- Protocol to minimize the risk of introducing emergency disease agents with importation of salmonid fishes from enzootic areas (Great Lakes watershed; unintentional release)

Great Lakes Water Quality Agreement 1972 (Canada–United States, amended in 1987)
- in the context of the Agreement's purpose (to restore and maintain the chemical, physical, and biological integrity of the waters of the Great Lakes Basin Ecosystem), commits *inter alia* to review of pollution (including biological pollution) from shipping sources; and, where review identifies additional areas for improvement, commits to a study to establish improved procedures for the abatement and control of pollution from shipping sources (Great Lakes watershed; shipping)

International Guidelines for Preventing the Introduction of Unwanted Aquatic Organisms and Pathogens from Ships' ballast water and Sediment Discharges (Marine Environment Protection Committee of the UN's International Maritime Organization, 4 July 1991)
- provides administrations with guidance on short-term (e.g. ballast exchange) and long-term (e.g. structural or equipment modifications on ships) strategies for minimizing the risk from introduction of unwanted aquatic organisms and pathogens from ships' ballast water and sediment (global; shipping)

North American Free Trade Agreement, 1994
- creates mechanism to address environmental issues related to commerce (North America; deliberate release, unintentional release)

American Fisheries Society—Position Statement on Introduced Aquatic Species (1986)
- urges prohibition of all introductions unless formal review of a species' probable impact etc. finds the introduction to be desirable (North America, Great Lakes watershed; deliberate release)
- urges that all governmental authorities refrain from introductions which may contaminate waters in other jurisdictions without agreement of exposed jurisdiction (North America, Great Lakes watershed; deliberate release)
- encourages fish importers, farmers, dealers, and hobbyists to prevent and discourage the accidental or purposeful introduction into local ecosystems (North America, Great Lakes watershed; deliberate release; unintentional release)

American Fisheries Society—Position Statement on Ballast Water Introductions (1990)
- encourages studies of ballast water introductions and methodologies to find ways to halt their occurrence (North America, Great Lakes watershed; shipping)

CHAPTER 8

Issues Affecting Fish Habitat in the Great Lakes Basin

Daniel B. Hayes

Introduction

One of the primary goals of ecology is to understand the abundance and distribution of organisms. At its core, the study of fish habitat is an attempt to accomplish this goal. Fish and other aquatic organisms need habitat to survive, and the productive capacity of the environment depends upon how well their needs are met. The themes I will develop in this chapter focus on how human activities in the Great Lakes basin have altered the quality and quantity of aquatic habitats, and the resulting effects on fish populations. The issues faced by fishery managers and researchers in this region will then be examined in this context.

What Is Fish Habitat?

Fish habitat has been defined in various ways. Historically, most definitions considered only the physical structure (e.g., water depth, overhead cover, substrate) of the environment to be fish habitat (e.g., Orth and White 1993). The focus on physical environmental characteristics stems from the traditional partitioning of limiting factors into abiotic (i.e., nonliving) and biotic (i.e., living) factors, with habitat generally including abiotic factors and community ecology generally including biotic factors. An advantage of this definition is that fishery managers and researchers can quantify and compare habitat at different locations and times relatively easily, because physical and chemical characteristics of the environment are often readily measured. A difficulty with views focusing solely on physical and chemical habitat variables is that other (i.e., biological) features of the environment also affect fish growth, mortality, reproductive success and habitat choice. For example, Werner et al. (1983) have shown that the abundance of predators affects fish habitat choice in addition to directly affecting mortality. The abundance of prey is also an important determinant of fish habitat choice. Hayes et al. (1992), for example, found that yellow perch in an inland lake were distributed closer to the bottom following a whole-lake treatment that increased the abundance of benthic invertebrates, even though the physical and chemical characteristics of the lake remained unchanged. As these examples illustrate, models of habitat choice focusing purely on the physical characteristics of the environment are not always adequate for predicting habitat choice. Because of this, I, (and others) have argued that we need to expand our concept of fish habitat to include physical, chemical, and biological attributes of the environment. As such, I feel that the following definition (based on Hudson et al. 1992) best captures the way fish relate to their environment:

> Fish habitat is the set of places where a fish (or group of fish) could potentially live. The needs of the fish are determined by its (their) biology; the physical, chemical, and biological characteristics of the environment then delineate the places where they can live.

One of the key features of this definition is that habitat is a place or a location in space. Accordingly, habitat is an attribute or property of the environment rather than a property of an organism. This is an important distinction, because it implies that the characteristics of the environment can be described without reference to any specific organism, and habitats can be classified based on the similarity of different locations. For example, pools, riffles, and runs are three stream habitats that can be distinguished based on a site's depth, water velocity, and substrate characteristics. An important implication of the definition I use is that descriptions of fish habitat can, and often should, include a description of the biological characteristics of a site.

Many habitat classification schemes have been developed based on various physical, chemical, or biological variables. For example, Hutchinson (1957a) presents lake classification systems based on lake origin, lake morphometry, and mixing properties, among others. Leach and Herron (1992) provide a current review of lake classification schemes. Although habitats can be characterized or classified without reference to a specific organism, in practice these habitat classifications are often used as explanatory variables in analyses of fish distribution or abundance (e.g., Fausch et al. 1988). One difficulty in applying various habitat classification methods is that of scale: some schemes are intended to describe macrohabitat characteristics, whereas others are focused on mesohabitat or microhabitat descriptions. For example, even though rapids in rivers are characterized by fast, turbulent water flows, there are places within the rapids (e.g., behind boulders) that have low current velocities. Thus, our description of the habitat (in this example our description of water velocity) depends on our measurement scale.

To overcome some of these difficulties in habitat classification, Sly and Busch (1992) describe a hierarchical system for classifying aquatic habitats in the Great Lakes basin. A hierarchical approach to habitat classification simplifies the discussion of fish habitat, as it begins with broad categories that can be described in general terms, but allows the incorporation of finer site-level details when needed. At the top level of Sly and Busch's classification, they recognize several major Great Lakes "systems," including: lakes, connecting channels, estuaries, and rivers (fig. 1). These encompass the largest divisions among aquatic habitats in the basin.

The lake systems comprise the main bodies of the Great Lakes (i.e., Lake Michigan, Lake Erie, etc.). Connecting channels are the major rivers linking the Great Lakes, and include the St. Mary's, St. Clair, Detroit, Niagara and St. Lawrence Rivers. Although estuaries are traditionally recognized where freshwater rivers empty into saltwater systems, Sly and Busch (1992) use this term for habitats in the Great Lakes where rivers empty into the Great Lakes. Great Lakes estuaries thus repre-

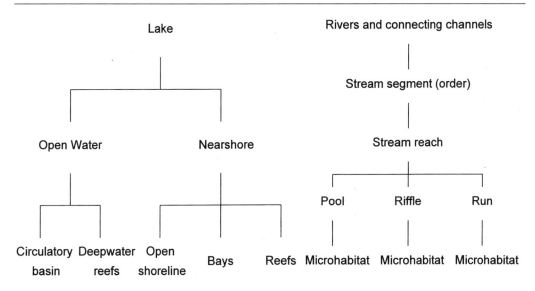

FIG. 1. *Examples of finer levels of classification for lakes and rivers and connecting channels based on Sly and Busch (1992).*

sent zones where lotic environments meet the lentic environment of the lakes. Rivers are the final top level aquatic habitat in the classification scheme, and provide the major linkage between the drainage basin and the lakes themselves. Although Sly and Busch (1992) include wetlands as a finer level habitat category, I would argue that they represent a major system as wetlands and the shoreline represent the interface between terrestrial and aquatic habitats. The bridge between terrestrial and aquatic habitats created by wetlands is important because it determines the pathway for the flow of nutrients and energy from terrestrial to aquatic ecosystems. Furthermore, many fish species take advantage of the productivity of wetlands that result from nutrient and energy flows from terrestrial systems.

How Do Fish Populations Depend on Habitat?

The idea of an organism's niche is an essential concept that is complementary to the idea of habitat. The most commonly cited definition of niche is probably that of Hutchinson (1957b, 416), where he describes the niche as "the sum of all factors acting on an organisms; the niche thus defined is a region of n-dimensional hyper-space." Although the term n-dimensional hyper-space (or hypervolume) is commonly used, it is a difficult concept to envision. Perhaps a simpler way of expressing this is that a fish's niche encompasses the combination of biological, physical, and chemical factors (the n-dimensions) it can tolerate. It is important to note that the niche is a property of an organism, and is determined by its phenotype, which is a function of the fish's genotype and environmental influences on development. Although other definitions of niche have been proposed in the literature (Hutchinson 1957a; Kerr 1976; Kerr and Werner 1980), they all share the underlying concept that an organism's well-being (i.e., its fitness) is determined by how well its needs are met by environmental conditions. As such, the niche acts to define the utility (i.e., the contribution to the organism's fitness) of possible habitats based on the conditions present. Thus, a fish's niche determines what is usable habitat (i.e., what it can tolerate), preferred habitat (i.e., what habitat it selects), and optimal habitat (i.e., the habitat where the fish's fitness is highest), from the set of all habitats available to it.

When linking habitat quality to fish population dynamics (i.e., the abundance and rate of change in abundance), a complication in using a fish's niche to determine the utility of a habitat is that a fish's needs typically change during its life. The term "ontogenetic niche" (Werner and Gilliam 1984) has been used to capture the notion that a fish's requirements change over the course of its lifetime. Thus, the habitat needs over a fish's entire life cycle must be considered in order to link habitat conditions with fish population dynamics.

Some of the broad stages fish show during their ontogeny are egg, juvenile, and adult. As such, a fish population needs habitat (although not necessarily separate habitats) for spawning of eggs, for juveniles to feed, rest and find cover (refugia), and for adults to feed, rest and find cover in order to be self-propagating. Furthermore, each of these habitats must be present at the same time the fish is passing through the corresponding life stage. The temporal linkage between a fish's life stages has important consequences for their habitat needs. Not only do habitats for each life stage need to be present at the proper time, but these habitats must be physically and tempo-rally linked themselves. For example, chinook salmon that live in the open lake as adults and use gravel beds in tributary rivers for spawning require a pathway connecting these gravel beds with the open lake. Thus, the connecting habitats are important components of the total habitat needs of a species as well as those habitats where fish reside during the primary stages of their life. The concept that a species' entire lifetime habitats need to be connected is particularly important when considering the impact of dams or alterations in wetlands on Great Lakes fish production.

Several patterns in ontogenetic habitat use are evident among Great Lakes fish. Some of the most common patterns and representative species include:
1. offshore habitat use throughout most or all of life (lake trout);
2. early life stages in near shore habitats, and later life stages offshore (lake whitefish);
3. near shore habitat use throughout most, or all of life (largemouth bass);
4. early life stages in river habitats, and adult life stages in near shore habitats (white sucker);
5. early life stages in river habitats, and later life stages in offshore habitats (chinook salmon);
6. early life stages found in rivers or near shore habitats and later life stages found in both near shore and offshore habitats (alewife, walleye).

The presence of these life-history patterns among the Great Lakes fish results in our percep-tion of discrete fish communities. For example, we recognize an open-water community in the upper Great Lakes composed of lake trout, the Pacific salmonids, bloaters, alewives, smelt, and lake herring, among other species. In the context of fish habitat, it is important to recognize, how-ever, that the open-water fish community is supported not only by offshore habitats, but also by the near shore and river habitats needed during the ontogeny of several of the major fish species in this community.

For many species, spawning habitat and nursery grounds for juveniles are the habitats that have the greatest influence on fish population abundance. These habitats are particularly important for at least two reasons. First, change or variation in the survival of young fish has a much greater impact on fish populations than similar variability at later life stages (e.g., Caswell 1989). Secondly, the habitat requirements for spawning and juvenile survival are often narrower than the habitat needs of adult fish. Eggs and young fish are often more sensitive to environmental conditions, and have smaller energy stores to withstand prolonged periods of adversity. Also, early life stages are much less mobile than adult fish, and less capable of seeking out more suitable conditions (Shuter 1990).

Habitat Requirements of Fish: A Guild Approach

Habitat needs are often evaluated in the context of a single species, since the needs of a single species are relatively easier to define that those of a fish community. The single-species approach has often been criticized, however, because of the need to evaluate the community or ecosystem response to habitat changes. One solution to multispecies habitat definitions is to consider the overlapping needs of co-occurring fish, and to group fish with similar habitat needs into guilds. The primary utility of grouping species by their ontogenetic habitat needs is that human alterations to habitat generally affect the entire community using that habitat, rather than a single species. The above grouping of fish into similar ontogenetic habitat needs is a coarse-level description of the habitat needs of a fish community. Such a coarse-level description may be useful, however, in evaluating the implication of change in the water quality of the open lake (for example) where all species using the open lake would potentially be affected.

A more detailed example of the guild approach is the concept of reproductive guilds as elaborated by Balon (1975, 1981). One component of reproductive guild designation is a delineation of groups based on habitat choice (i.e., where progeny are deposited), and another component is the fishs' behavior (i.e., if parents guard young) in regard to their eggs or progeny (table 1). An example of two species with different spawning habitat needs are the lake herring and rainbow trout in Lake Huron (fig. 2). The lake herring is representative of the lake-spawning rock and gravel spawner with pelagic larvae guild (lithopelagophil), which have relatively broad environmental tolerances for spawning and nursery areas, and have broad distribution of spawning grounds. Rainbow trout, on the other hand, are typical of the rock and gravel river spawning guild (lithophil), with more specific habitat requirements. As such, rainbow trout and other species in this guild have a more limited distribution of spawning grounds. Based on this classification system, Great Lakes fish are obviously not uniformly distributed across all reproductive guilds (table 2). The majority of species prefer rock and gravel substrates for spawning, with plants being the next most preferred substrate. There are roughly equal numbers of species that choose lakes, rivers, or both for their spawning location, however. It is important to note that the ability of some species to successfully spawn in either the Great Lakes or their tributaries has been important to their survival. Lake trout, for example, are often viewed as spawning primarily on the reefs of the open lakes. Stocks of river spawning lake trout are present in some areas of the Great Lakes (Loftus 1958; Lawrie and Rahrer 1973; Peck 1992), and possibly were more important historically, before access to their spawning grounds was cut off by dams, and the spawning grounds themselves were degraded through changes in land use.

Another example of fish guilds based on habitat preference is temperature guilds. In the Great Lakes region, there are three commonly recognized temperature guilds: cold-water fish, cool-

TABLE 1

Reproductive guild membership of common Great Lakes fish

Adapted from Balon (1975, 1981), and from information on spawning in Goodyear et al.(1982b) and Scott and Crossman (1985).
Guilds with no representatives in the Great Lakes region are not shown. Primary spawning sites are indicated by R (river), L (lake), B (both), O (ocean).

NONGUARDERS

PELAGIC SPAWNERS (PELAGOPHILS)

American Eel	*Anguilla rostrata*	O
Longjaw Cisco	*Coregonus alpenae*	L
Blackfin Cisco	*Coregonus nigripinnis*	L
Shortnose Cisco	*Coregonus reighardi*	L
Shortjaw Cisco	*Coregonus zenithicus*	L
Emerald Shiner	*Notropis atherinoides*	L
Freshwater Drum	*Aplodinotus grunniens*	L

ROCK AND GRAVEL SPAWNERS WITH PELAGIC LARVAE (LITHOPELAGOPHILS)

Lake Sturgeon	*Acipenser fulvescens*	B
Gizzard Shad	*Dorosoma cepedianum*	B
Lake Herring	*Coregonus artedi*	L
Bloater	*Coregonus hoyi*	L
Deepwater Cisco	*Coregonus johannae*	L
Kiyi	*Coregonus kiyi*	L
Mooneye	*Hiodon tergisus*	R
Burbot	*Lota lota*	B
Walleye	*Stizostedion vitreum*	B

ROCK AND GRAVEL SPAWNERS WITH BENTHIC LARVAE (LITHOPHILS)

Sea Lamprey	*Petromyzon marinus*	R
Lake Whitefish	*Coregonus clupeaformis*	L
Pygmy Whitefish	*Prosopium coulteri*	L
Round Whitefish	*Prosopium cylindraceum*	B
Rainbow Smelt	*Osmerus mordax*	B
Lake Chub	*Couesius plumbeus*	B
Silverjaw Minnow	*Ericymba buccata*	R
Pugnose Shiner	*Notropis anogenus*	L
Rosyface Shiner	*Notropis rubellus*	R
Blacknose Dace	*Rhinichthys atratulus*	R
Longnose Dace	*Rhinichthys cataractae*	B
River Carpsucker	*Carpoides carpio*	R
Longnose Sucker	*Catostomus catostomus*	B
White Sucker	*Catostomus commersoni*	B
Creek Chubsucker	*Erimyzon oblongus*	R
Northern Hog Sucker	*Hypentelium nigricans*	R
Silver Redhorse	*Moxostoma anisurum*	R
Black Redhorse	*Moxostoma duquesnei*	R
Golden Redhorse	*Moxostoma erythrurum*	R
Shorthead Redhorse	*Moxostoma macrolepidotum*	R
Greater Redhorse	*Moxostoma valenciennesi*	R
Trout-perch	*Percopsis omiscomaycus*	B

Sauger	*Stizostedion canadense*	B

NONOBLIGATORY PLANT SPAWNERS (PHYTOLITHOPHILS)

Alewife	*Alosa pseudoharengus*	L
Silver Chub	*Hybopsis storeriana*	B
Spotfin Shiner	*Notropis spilopterus*	R
Mimic Shiner	*Notropis volucellus*	L
Brook Silverside	*Labidesthes sicculus*	B
White Perch	*Morone americana*	B
White Bass	*Morone chrysops*	R
Iowa Darter	*Etheostoma exile*	B
Yellow Perch	*Perca flavescens*	B

OBLIGATORY PLANT SPAWNERS (PHYTOPHILS)

Spotted Gar	*Lepisosteus oculatus*	L
Longnose Gar	*Lepisosteus osseus*	L
Goldfish	*Carassius auratus*	R
Carp	*Cyprinus carpio*	B
Central Mudminnow	*Umbra limi*	R
Grass Pickerel	*Esox americanus vermiculatus*	R
Northern Pike	*Esox lucius*	B
Muskellunge	*Esox masquinongy*	B
Golden Shiner	*Notemigonus crysoleucas*	L
Bridle Shiner	*Notropis bifrenatus*	B
Pugnose Minnow	*Notropis emiliae*	L
Blackchin Shiner	*Notropis heterodon*	L(?)
Northern Redbelly Dace	*Phoxinus eos*	R
Lake Chubsucker	*Erimyzon sucetta*	B
Bigmouth Buffalo	*Ictiobus cyprinellus*	B
Banded Killifish	*Fundulus diaphanus*	B
Greenside Darter	*Etheostoma blennioides*	R

SAND SPAWNERS (PSAMMOPHILS)

Blacknose Shiner	*Notropis heterolepis*	B
Spottail Shiner	*Notropis hudsonius*	L
Sand Shiner	*Notropis stramineus*	L
Quillback	*Carpiodes cyprinus*	B
Logperch	*Percina caprodes*	B

BROOD HIDERS

ROCK AND GRAVEL SPAWNERS (LITHOPHILS)

Pink Salmon	*Oncorhynchus gorbuscha*	R
Coho Salmon	*Oncorhynchus kisutch*	R
Rainbow Trout	*Oncorhynchus mykiss*	R

Kokanee	*Oncorhynchus nerka*	R
Chinook Salmon	*Oncorhynchus tshawytscha*	R
Atlantic Salmon	*Salmo salar*	R
Brown Trout	*Salmo trutta*	R
Brook Trout	*Salvelinus fontinalis*	R
Lake Trout	*Salvelinus namaycush*	B
River Chub	*Nocomis micropogon*	R
Creek Chub	*Semotilus atromaculatus*	B
Fallfish	*Semotilus corporalis*	B
Rainbow Darter	*Etheostoma caeruleum*	B
Channel Darter	*Percina copelandi*	B

GUARDERS

SUBSTRATE CHOOSERS

PLANT SPAWNERS (PHYTOPHILS)

White Crappie	*Pomoxis annularis*	B

NEST SPAWNERS

MISCELLANEOUS SUBSTRATE AND MATERIAL NESTERS (POLYPHILS)

Pumpkinseed	*Lepomis gibbosus*	L

ROCK AND GRAVEL NESTERS (LITHOPHILS)

Stoneroller	*Campostoma anomalum*	R
Common Shiner	*Notropis cornutus*	B
Black Bullhead	*Ictalurus melas*	L
Flathead Catfish	*Pylodictus olivaris*	R
Rock Bass	*Ambloplites rupestris*	B
Green Sunfish	*Lepomis cyanellus*	B
Orangespotted Sunfish	*Lepomis humilis*	B

Bluegill	*Lepomis macrochirus*	L
Smallmouth Bass	*Micropterus dolomieui*	B
Fourhorn Sculpin	*Myoxocephalus quadricornis*	L

GLUEMAKING NESTERS (ARIADNOPHILS)

Brook Stickleback	*Culea inconstans*	B
Threespine Stickleback	*Gasterosteus aculeatus*	L
Ninespine Stickleback	*Pungitius pungitius*	B

PLANT MATERIAL NESTERS (PHYTOPHILS)

Bowfin	*Amia calva*	L
Largemouth Bass	*Micropterus salmoides*	L
Black Crappie	*Pomoxis nigromaculatus*	B

HOLE NESTERS (SPELEOPHILS)

Bluntnose Minnow	*Pimephales notatus*	B
Fathead Minnow	*Pimephales promelas*	L
Yellow Bullhead	*Ictalurus natalis*	L
Brown Bullhead	*Ictalurus nebulosus*	B
Channel Catfish	*Ictalurus punctatus*	B
Stonecat	*Noturus flavus*	B
Tadpole Madtom	*Noturus gyrinus*	B
Bridled Madtom	*Noturus miurus*	B
Fantail Darter	*Etheostoma flabellare*	R
Johnny Darter	*Etheostoma nigrum*	B
Mottled Sculpin	*Cottus bairdi*	B
Slimy Sculpin	*Cottus cognatus*	B
Spoonhead Sculpin	*Cottus ricei*	L

TABLE 2

Number of species using specific habitats and substrates for spawning.

HABITAT	PELAGIC	SAND	ROCK & GRAVEL	PLANTS	HOLE NESTERS	TOTAL
Lake	6	2	10	9	3	30
River	0	0	24	7	1	32
Both	0	2	22	16	9	49
Total	6	4	56	32	13	111

water fish, and warm-water fish. Species from these guilds can co-occur in the Great Lakes, due to the thermal diversity occurring through lake stratification and localized areas of warming (e.g., shallow bays). Lake whitefish are an example of the cold-water fish guild, and are found across much of the northern part of Lake Michigan at depths of 50–200 ft during the summer (fig. 3). In contrast, yellow perch are a member of the cool-water fish community, and are more prevalent in the southern part of Lake Michigan at depths less than 75 ft (fig. 3). Yellow perch are common,

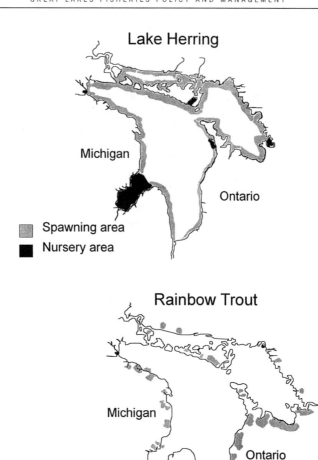

FIG. 2. *Spawning and nursery grounds of lake herring (A) and rainbow trout (B) in Lake Huron (from Goodyear et al. 1982a).*

however, in Green Bay, northern Lake Michigan, where water temperatures are higher than in the open lake nearby.

Issues Affecting Great Lakes Fish Habitat

Water Quality

Although water quality is typically viewed as a limnological issue (i.e., having an impact mostly on lower trophic levels) and not a fish habitat issue, the chemical characteristics of the Great Lakes have direct bearing on the lakes' ability to support productive fisheries. Although this topic is covered in more detail by Beeton et al. (this volume), I will focus on the impact of changes in water quality from a fish habitat perspective.

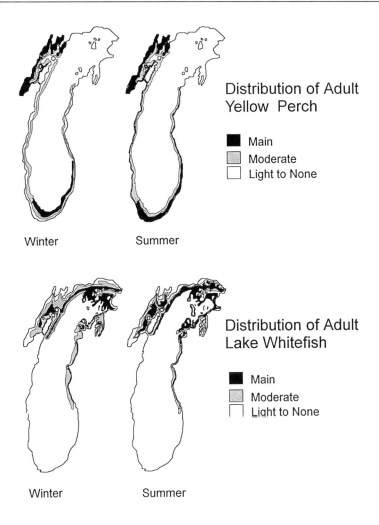

FIG. 3. *Distribution of yellow perch (A) and lake whitefish (B) in Lake Michigan, illustrating the habitat utilization of a cool-water and cold-water species, respectively (from Sommers et al. 1981).*

One of the major effects of degraded water quality is the limitation that low oxygen levels has on the quality of fish habitat. Most of the cold-water and cool-water fish species require relatively high (e.g., greater than 5ppm O_2) levels of dissolved oxygen. When oxygen levels fall below the threshold for a given species, it can limit their use of extensive areas which are otherwise suitable. Some of the initial impacts on water quality began with the early sawmills that discharged saw-dust and bark into the Great Lakes. In addition to directly covering the natural sediments, these materials have a high biological oxygen demand. As such, localized areas of low oxygen concentration occurred near sawmill discharges. With the growth of the human population in the Great Lakes basin, the discharge of nutrient-rich sewage and other effluents increased, resulting in cultural eutrophication across much of the Great Lakes (Beeton et al., this volume). Phosphorus

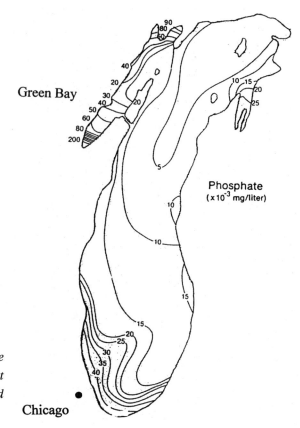

FIG. 4. *Phosphorus concentrations in Lake Michigan illustrating elevated nutrient loading near urbanized areas and enclosed embayments (from Sommers et al. 1981).*

concentrations, for example, are elevated across the Great Lakes, and show "hot-spots" in areas where there is extensive human development (e.g., the south end of Lake Michigan, Green Bay; fig. 4). Heightened phosphorus loading has serious implications for water quality and the limnology of the lakes, as well as for fish habitat. Of greatest concern for fish habitat quality is perhaps the reduction or loss of hypolimnetic oxygen that occurs with increasing nutrient inputs. In Lake Erie, for example, eutrophication and the resulting anoxia of the hypolimnion were observed as early as the 1920s. By the 1960s, oxygen levels were consistently low in the hypolimnion in the Western and Central Basins. Loss of hypolimnetic oxygen directly reduces the suitability of hypolimnetic habitats for fish such as walleye, and for some species of benthic invertebrates. In Lake Erie and Saginaw Bay, loss of hypolimnetic oxygen has been implicated as a reason for the decline of the burrowing mayfly (*Hexagenia*), resulting in a lower quality habitat for yellow perch and walleye feeding.

Fishery Management and Water Quality

Since the signing of the Great Lakes Water Quality Agreement in 1972, great strides have been made in reducing the input of nutrients, and in improving water quality. There remain, however, localized areas where water quality and oxygen concentrations limit fish habitat. These problems are most common in bays and other near shore areas close to anthropogenic sources of nutrients.

In addition to eutrophication, another water quality concern is that of toxic contaminants in the Great Lakes. In high enough concentrations, anthropogenic contaminants can cause direct mortality of fish and other aquatic organisms. At sublethal concentrations, toxic substances can also limit the abundance of fish. Although the basinwide impact of contaminants on fish populations may never be known, they have certainly had a major effect on the ability of the Great Lakes to produce fish safe for human consumption (Hesse 1997).

The problem of contaminants and degraded water quality is perhaps most critical in the forty-three Areas of Concern identified by the International Joint Commission. The Areas of Concern are locations that have the most degraded near shore habitats in the Great Lakes basin. Virtually all of the Areas of Concern have problems with elevated levels of toxic substances in the sediments due to recent or past industrial chemical discharges. In addition to problems with high levels of contaminants, most of the Areas of Concern have additional sources of environmental degradation. Hartig (1993) recently summarized fish community and habitat concerns identified by managers associated with the Areas of Concern. Some of the most common problems identified were: wetland loss, dredging, low levels of dissolved oxygen, shoreline development, and alterations to sediments (e.g., sludge from mill effluents). The fact that most of the Areas of Concern suffer from several types of fish habitat degradation makes it more difficult to restore the habitat's productive capacity.

Dams

Dams have at least three major effects on fish populations and their habitats. First, dams can impede fish migration into or within rivers. Secondly, dams convert stream habitats into lakes (impoundments). Finally, dams alter the physical characteristics of the streambed downstream, and the physical, chemical, and biological composition of the water flowing from the dam. These impacts will be reviewed in more depth below.

Dams have been constructed in the Great Lakes basin since at least the late 1700s. Trautman (1981) relates that the first dam on an Ohio stream was completed in 1790 on Wolfe Creek, near Marietta. These early dams were often constructed to provide power to run gristmills or sawmills. Many of these dams were constructed at the first rapids upstream from the Great Lakes because these sites provided easy access to the lakes for transporting lumber or other products by ship. The fact that early dams were built low in the watershed is important to fish production, since they cut off access to the entire upstream reaches of the river and its tributaries for migratory fish. As such, impact on potamodromous fish (i.e., fish that migrate for spawning between lakes and streams or within streams) in the Great Lakes began shortly after these early dams were built. Unfortunately, these impacts began long before information on fish abundance was available to evaluate the effects of dams.

Over time, additional dams were built further upstream and in more remote areas. Dams have been constructed for a variety of purposes, including power generation, water level control, flood control, and lake creation. Presently, many, if not most of the major watersheds in the Great Lakes basin have impoundments. The number of dams on some Great Lakes tributaries is surprising. For example, the Huron River in southeastern Michigan has nearly one hundred dams in its watershed (fig. 5), few of which have fish ladders or other fish passage devices. Such a large number

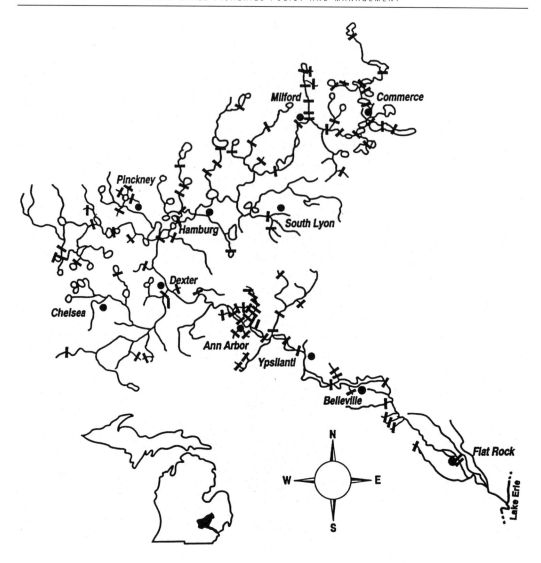

FIG. 5. *Approximate location of dams in the Huron River, MI, watershed (from Hay-Chmielewski et al. 1995).*

of dams creates a fragmented river habitat interspersed with reservoirs. Thus, the utility of rivers for fish spawning is affected by physical changes in the streams, as well as the barrier to migration that many dams create. The impact of preventing fish passages is serious, since this disconnects habitats that previously were connected. The fragmentation of habitats is particularly important for fish that require stream habitats for spawning. The lake sturgeon is a good example of a species whose population has been drastically reduced by inundation of its former spawning habitats as well as blocking access to upstream habitats (Auer, this volume).

Even for dams equipped with fish ladders or other fish passage devices, fish passage continues to be an issue. Fish ladders are often designed for species such as trout and salmon that are swift swimmers, and are able to jump over low barriers. Common ladder designs do not accommodate many other species (e.g., suckers) that are not able to jump as well as the salmonids. Lake sturgeon also have problems passing most fish ladders, because they are too big for common ladder designs. In addition to creating a barrier to upstream movement, survival of fish moving downstream is also a concern for many dams. Fish often experience mortality caused directly by turbine operation, or indirectly through increased predation when fish become disoriented after passing below the dam (Garcia 1989).

Dams have the obvious effect of changing stream habitat into reservoir habitat, thus favoring fish adapted to lake environments. Because the resulting reservoir occupies a much greater area and contains more water, fish production in that river reach may be much greater than the original river, but with a different species composition. The localized increase in fish production is often offset, however, by the impact of the dam on downstream ecosystems, and decreased production of fish species whose migration is impaired by dam operation (Balon 1978). For example, many rivers contain resident populations of trout, as well as containing suitable spawning and nursery grounds for migratory salmon and steelhead trout. When a river is dammed, an impoundment is created which may favor lake species such as walleye and largemouth bass. Because the impoundment is larger than the original river reach, production of walleye and largemouth bass may be greater than resident trout production prior to dam construction. Fishery managers must thus balance the possible benefits of impoundment (e.g., walleye and largemouth bass fisheries) with losses of localized fish production (e.g., resident stream trout fishery), loss of migratory fish spawning habitat, and the subsequent loss of Great Lakes fishing opportunities.

In addition to converting stream habitats into lakes, dams have other effects on fish habitats above and below the dam. Some of these impacts directly affect stream habitat, and its utility for fish production. Changes in stream water temperature below impoundments is an example where the quality of the habitat is altered directly. Other impacts indirectly affect fish production by altering the transport of nutrients and materials by rivers to wetlands, estuaries, and the Great Lakes themselves.

Altered sediment dynamics is an example of a stream process that is strongly affected by dam construction that has both direct and indirect impacts on fish habitat. Before a dam is built, there are typically localized areas of deposition and erosion in a given stream reach that are nearly in equilibrium. As the stream slowly erodes downward in its channel, there is generally a small net export of sediment and nutrients to the lake (Taylor 1978). After dam construction, the sequence of depositional and erosional zones is disrupted. Because current velocities are greatly decreased within the reservoir, sediments settle to the bottom, and the stream bed aggrades (i.e., increases in height and thickness; fig. 6). As such, sediments are deposited in upstream reaches of the reservoir, and stream reaches below the dam are deprived of their usual sediment inputs. Reduced sediment transport directly affects the quality of fish habitat by changing the amount of sediment in the water column. A secondary consequence of disrupting the normal sediment transport process is to eventually fill in the reservoir with fine sediments, and cause down cutting (i.e., degradation) of the channel below the dam (fig. 6; Taylor 1978). As the stream bed below the dam degrades, the substrate typically becomes coarser. As such, the habitat for gravel-spawning fish (i.e., lithophils) may improve in habitats below dams where finer sediments previously existed. The

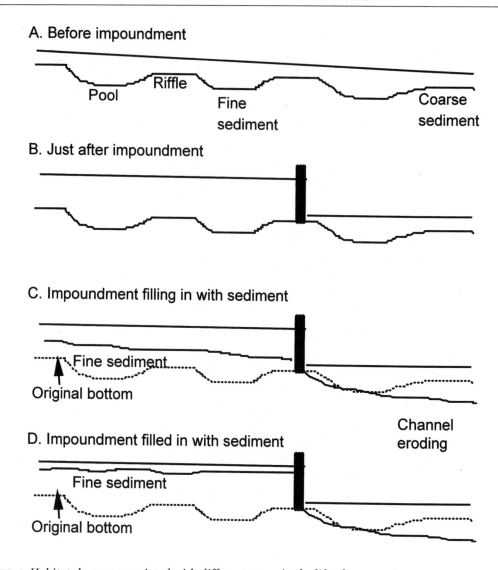

FIG. 6. *Habitat changes associated with different stages in the life of a reservoir.*

extent of down cutting depends on local conditions, and is difficult to predict in advance (Taylor 1978), but bed cutting of several feet has been observed in some western U.S. streams (Stanley 1951). Although the impact of change in streambed level on fish production was not determined in these studies, it is apparent that the habitat itself underwent substantial change.

Another direct habitat impact of dam construction is to change the temperature regime of the impoundment above, and the river below the dam. Generally, the surface water in the reservoir is warmer (due to solar heating) than the river would typically be. If the reservoir is deep enough, the water column will stratify, and the bottom water will be colder than the river would normally be. Thus, surface spillways that pass surficial water from the reservoir to the river below will result in warming of the river downstream. In contrast, if water is drawn from the hypolimnion of the reservoir, it will be colder than the river would normally be. In either case, the normal daily and seasonal fluctuations in temperature are typically moderated below dams. The larger water volume of the reservoir and longer residence time provide thermal inertia that reduces temperature fluctuation. In addition to changing the thermal habitat characteristics, hypolimnetic water releases can be anoxic due to oxygen depletion in the hypolimnion of the reservoir (e.g., Grizzle 1981). Below high dams with surface spillways, dissolved nitrogen may become supersaturated in the plunge pool below the dam, with supersaturation extending for considerable distances downstream (e.g., Legg 1978; Crunkilton et al. 1980). If supersaturation exceeds 110%, gas bubble disease may occur; above 125% this condition is often lethal (Weitkamp and Katz 1980).

Dams also alter the natural sequence of biological production in rivers, thus altering the export of nutrients and organic matter from the river to Great Lakes estuaries, and the Lakes themselves. Free-flowing rivers tend to show a gradient of conditions from headwaters to large rivers. The river continuum concept (Vannote et al. 1980) suggests that in headwater streams, primary production is low, due to shading of the stream by riparian vegetation. Energy input into headwater streams is primarily through allochthonous materials (e.g., leaves and woody debris). This material is classified as coarse particulate organic matter (CPOM). As a stream gets progressively larger, autochthonous primary production becomes more important, as shading of the stream bed decreases. In these stream reaches, organic material is composed of smaller pieces (fine particulate organic matter, FPOM). In the lowest sections of large rivers, autochthonous production often declines because of increased water turbidity. In these rivers, allochthonous material transported from upstream sections provides the basis for secondary production. Because material transported from upstream sections has already been processed, the particles are very small (ultra fine particulate organic matter, UPOM). Although few fish (e.g., catostomids) in the Great Lakes region feed extensively on particulate organic matter, this material is a critical food source for benthic invertebrates, which are an important food source for many fish species. Thus, changes in the composition of transported organic matter may have a large impact on the food resources for many fish.

After a dam is constructed, the usual longitudinal gradient of environmental conditions is often disrupted. Reservoir construction replaces the lotic conditions of the stream with conditions closer to that of a large river. Reservoirs act as a sink for coarser particles, because they settle to the bottom in the impoundment. Downstream of the dam, particulate organic matter is often dominated by planktonic algae swept from the reservoir. Thus, the characteristics and amount of organic matter and nutrients delivered to the Great Lakes from impounded watersheds differs from free-flowing rivers. The consequences of these changes are unknown. However, in marine estuaries and their associated marshes, primary production and fish production have been shown to be strongly affected by the input of organic material and sediments from upstream sources (Wiegert and Pomeroy 1981). By analogy, decline in nutrient, organic material, and sediment inputs below dams may have a negative impact on the productivity of Great Lakes estuaries.

Fishery Management and Dams
Most dams are created for reasons other than fishery interests. As such, fishery managers typically have only partial control over dam construction or operations. Several pieces of legislation in the

United States and Canada require, however, that fishery interests be taken into consideration as a component of the decision process regarding dam operations. In Canada, the Fisheries Act covers all activities (including dams) that have an effect on fish production. In the United States, the Federal Energy Regulatory Commission has jurisdiction over dams used for energy production.

At a basic level, the decision of whether to construct a dam is an important consideration that fundamentally affects the connection between fish habitats within a watershed. The next issues are how the dam will be built, and what provisions will be made for fish passage. As emphasized earlier, one of the major effects of dams is to disconnect habitats that would typically be accessible to fish through migration. Because of this, fish passage facilities should be included in the dam design phase, taking into consideration the species using the river as a migratory route. Another design criterion that is important to fish habitats below the dam is where the water intake is located for water flow through the dam, since this can affect the temperature of water discharge. Generally, water intakes near the bottom of the dam are better if the goal is to maintain cold-water fish habitat downstream of the dam.

Once the dam design criteria are set, another concern is how the water releases from the dam are managed. At issue here is the pattern and volume of water release on a within-day, between-day, seasonal, and annual basis. Unimpounded rivers typically show relatively small within-day fluctuations in flow, but have pronounced between-day, seasonal, and annual variability. In contrast, within-day flow variability is often increased in impounded rivers, while between-day and seasonal variability is often decreased. The clearest impact on fish habitat occurs when dams work in a peaking operation. In a peaking operation, flows within a day are scheduled to meet water or power needs. This results in widely varying flows within a day, a situation that few fish or their prey are adapted to. This causes a reduction in the abundance of fish and other aquatic organisms. Because of the negative impact peaking operations have on fisheries, managers have tried to impose "run-of-the-river" (i.e., flows below a dam have the same flow pattern as above the dam) operations on many hydroelectric dams. In contrast to natural streams, reaches below dams may have less variable flows within a year or between years. This is often the case for dams built as flood control structures. The short-term effect of flood control structures is often to benefit fish reproduction by reducing catastrophic loss of young fish. The long-term effects on fish populations and their habitat is less clear, as the variability in flow strongly affects the recurrence time of channel-shaping (flood) flows.

Altered patterns of flow certainly affect the quality and quantity of fish habitat below dams. The net effect of these changes, however, are difficult to assess. The Instream Flow Incremental Methodology (Bovee 1986) is an example of an assessment technique designed for such situations, and illustrates the complexity of the assessment needs. Briefly, the Instream Flow Incremental Methodology models the amount and quality of fish habitat as a function of stream flow. At the same time, fish habitat usage and preference is determined. The estimated impact on fish populations is then modeled as a function of dam management options. One of the key questions and difficulties, however, is to determine how fish populations respond to varying amounts of habitat present over time.

The final issue with dams is what to do with them when they can no longer function. The life span of all reservoirs is limited because of the deposition of sediments. Because of sediment deposition, the environmental conditions of reservoirs show a progression from lake back to river habitats. The rate that the reservoir fills in depends on a number of factors including climate, soils, geology, watershed slope, land use, and reservoir size. In temperate climates, Dendy et al. (1973) found sedimentation rates of small reservoirs (less than 10 acre-feet) averaged 3.5% loss of capacity per year. Relative sedimentation rates in larger reservoirs was lower, ranging from 2.7% per year for medium sized reservoirs (roughly 100 acre-feet) to 0.16% for very large reservoirs (>1,000,000 acre-feet). Based on these averages, the life span of small to medium reservoirs is roughly fifty years. Larger reservoirs have longer life spans, but still face the same fate. As the reservoir fills in, choices must be made on what to do with the dam. Sometimes it is feasible to dredge sediments out of the reservoir, restoring its water storage capacity. This is expensive, and can damage fish habitats below the dam by quickly flushing sediment downstream. Often dams are left standing, with minimal maintenance to keep them structurally sound. This option is not expensive on a year-to-year basis, but total costs continue to accumulate over time. Also, if the dam is not maintained properly, it is subject to catastrophic failure during floods. A final option is to remove the dam entirely. As with dredging, this is an expensive process, and has the potential to release excessive sediment to down-stream reaches. Because of these concerns, dam removal has been rarely done in a controlled man-ner. This option, however, is the only one with the potential for restoring impounded rivers back to a free-flowing state and removing impediments to fish migration.

Wetlands

The climate, soil types, and topography of the Great Lakes after the last glacial period left a profu-sion of wetlands in our region. Numerous types of wetlands occur in our region, ranging from continuously wetted areas to sites that are wet only seasonally. Wetlands provide habitat for many fish, including spawning habitat for phytophyllic reproductive guild species (table 1), and adult habitat for many cool-water and warm-water species. Juvenile fish often use wetlands as nursery grounds (Goodyear et al. 1982a) because of the cover from predators that aquatic macrophytes provide, and the high production rates in Great Lakes coastal wetlands.

Some of the defining characteristics of wetlands are that: (1) hydric soils predominate, (2) hy-drophytic (water loving) plants predominate, and (3) the water table is at or above the soil surface for some part of the year (Cowardin et al. 1979). Where wetlands are adjacent to open-water areas, the lower limit of the wetland is defined as the boundary of emergent, tree or shrub vegetation. Because of their dependence on the level of the water table, wetlands are very sensitive to changes in lake levels (e.g., Burton 1985), land use practices that alter the hydrologic cycle (e.g., conversion of forestland to urban areas), and human construction that alters water flow patterns (e.g., drain-age systems).

Cowardin et al. (1979) present a classification system for wetlands that illustrates some of the different wetland types found in the Great Lakes region. Some of the major ecological systems of wetlands found here are: riverine systems, lacustrine systems, and palustrine systems. Riverine systems are characterized by water flowing through an identifiable channel, and can be further subdivided into habitats with similar characteristics (fig. 7). Lacustrine systems include lakes or ponds which have some deep water area devoid of persistent emergent vegetation, and the shore-line surrounding these bodies of water (fig. 7). Cowardin et al. (1979) define palustrine systems as

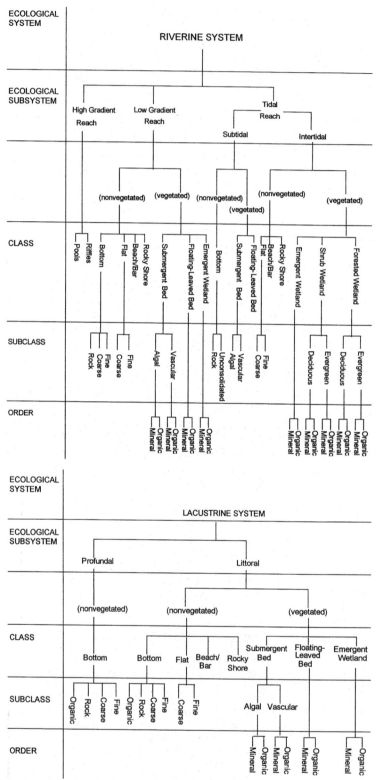

FIG. 7. *Classification of wetlands (from Cowardin et al. 1979).*

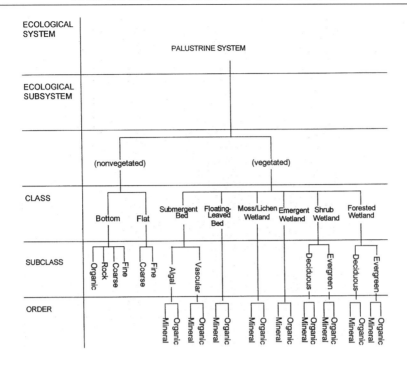

TABLE 3

Common classes of vegetated wetlands as defined by Cowardin et al. (1979)

CLASS	DOMINANT VEGETATION
Forested wetlands	Trees (e.g., spruce, white cedar, red maple, elm)
Shrub wetlands	Shrubs (e.g., leatherleaf, bog laurel, alder, willow)
Emergent wetlands	Herbaceous hydrophytes (e.g., cattails, bulrushes, arrowheads, sedges, grasses)
Moss/lichen wetlands	Nonvascular plants (e.g., mosses, lichens, liverworts)
Floating-leaved beds	Rooted herbaceous hydrophytes with floating leaves (e.g., water lilies, smartweed, water shield, floating-leaved pondweed)
Submergent beds	Rooted herbaceous hydrophytes without floating leaves (e.g., *Potamogeton*, *Ceratophyllum*, *Myriophyllum*)

areas where surface waters are not restricted to a definable habitat, and do not include wave-formed or bedrock shoreline features. An example of a palustrine system is forested wetlands (e.g., cedar swamps) that are not contained within the channel of a stream. Since lacustrine and riverine wetlands are directly connected to open water areas, they provide usable habitat for fish. In contrast, palustrine systems often do not provide usable fish habitat because of the lack of access by fish to them. Palustrine systems, however, have an indirect effect on downstream habitats by moderating water flows (i.e., reducing high flows and increasing minimum flows).

Within each of these broad ecological settings for wetlands, there are a number of different types (classes in Cowardin et al.'s (1979) terminology) of wetlands (table 3). These wetland classes are based primarily on the vegetation or substrate present. Although similar classes of wetlands

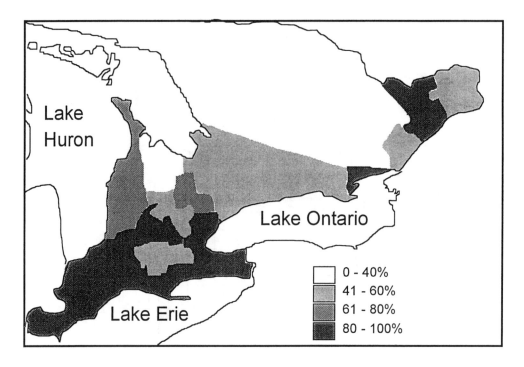

FIG. 8. *Losses of wetlands in southern Ontario (from Glooschenko 1985).*

may occur among the various ecological systems, the hydrological conditions vary between the higher levels of the classification. For example, submergent beds of vegetation may occur in riverine, lacustrine or palustrine systems. Although similar species of plants may occur in these habitats, they would have different effects on the habitat they provide. Although macrophytes would provide cover in either a river or protected bay, their effect on water velocity would be less in the protected bay where current velocity would be lower than in the river.

Wetland loss in the Great Lakes Basin began with early settlement when wetlands were drained to provide reclaimed land for agricultural use (Trautman 1981). The loss of wetlands has continued, with conversion to suburban, urban, and agricultural uses contributing to this decline. The rate of loss has been high during the past several decades, due in large part to the greatly expanding human population in the Great Lakes basin (Beeton et al., this volume). Presently, many areas in the Great Lakes basin have already lost the majority of their original wetlands (fig. 8; Glooschenko 1985).

As with dams, human impact on wetlands began long before estimates of fish abundance were available. As such, it is difficult to directly determine the historical importance of wetlands to the overall fish production of the Great Lakes. Historical accounts of fish abundance in Ohio rivers related by Trautman (1981) suggest that large numbers of fish (e.g., walleye and pike) migrated upstream to spawn in wetlands and riverine habitats. Even presently, it is difficult to assess the total contribution of wetlands to fishery production in the Great Lakes (Chubb and Liston 1986). Many of the difficulties are methodological; it is very difficult to obtain accurate estimates of larval and juvenile fish density in wetlands because of the difficulty in sampling these areas. Furthermore, it is prohibitively expensive to obtain synoptic sampling of all of the wetlands across substantial lengths of shoreline. Despite these difficulties in determining the actual fish production sustained by wetlands, it is apparent that they make a significant contribution to the total. Nearly one-third (32 of 111 species, Table 2) of fish species in the Great Lakes require or prefer plants for spawning habitat. Furthermore, many fish also utilize vegetated areas for rearing of young and as adult habitat.

Fishery Management and Wetlands

As with dam building, the issues surrounding wetland losses do not center on fisheries. Draining and conversion of wetlands are motivated by economic pressure for development of underutilized land. As such, fishery managers rarely, if ever, are the primary decision makers concerning land and water-use issues affecting wetlands.

Another factor that affects wetlands and their value for fisheries is Great Lakes water levels. The functioning of wetlands that are hydrologically connected to the Great Lakes (fig. 9) is greatly affected by water levels. Long-term cycles and trends in water level affect the location and extent of coastal wetlands, but shorter-term variation can also affect the fish and plant communities (Burton 1985). Seiches, caused by winds or water displacement by passing ships, can cause sudden, short-term changes in water level. These temporary water level fluctuations can strand fish on land, or transport fish or eggs out of their preferred habitat (Liston and Chubb 1985). Although natural fluctuations in lake levels cannot, at present, be controlled, proposed water diversion plans (Smith 1986) have the potential for causing extensive loss of coastal wetlands.

The difference in the hydrologic connectedness of palustrine, lacustrine, and riverine ecological systems is particularly important when considering the issue of water levels in the Great Lakes. Riverine and palustrine wetlands that are not directly dependent on the Great Lakes water level are more insulated from the effects of lower lake levels through water diversion than lacustrine systems are. Thus, fishery production in these wetlands would be less affected by water diversion or natural lake level fluctuation. The extent of riverine and palustrine coastal wetlands along the Great Lakes is substantial. Patterson and Whillans (1985) report that 18,225 of the 60,750 hectares of wetlands along the lower Great Lakes are mophometrically isolated from lake levels. Some of the isolation of the wetlands is natural, although much of it is attributable to dyking and other human influences. The benefits of more stable wetland water levels in dyked areas may be offset, however, by barriers to fish migration into the wetlands.

Land Use Practices

Some of the effects of land use practices have already been discussed in the context of wetland losses. Human development has many other effects on fish habitat beyond wetland loss. Originally, much of the Great Lakes basin was forested. When forestland is converted to agricultural use, water runoff and soil erosion both tend to increase. This alters the flow regime in streams and increases their sediment and nutrient loading. Up to a point, increases in both of these factors can be tolerated, but stream fish populations often decline in the face of extensive agricultural development because of heightened siltation of gravel spawning areas, and lower oxygen

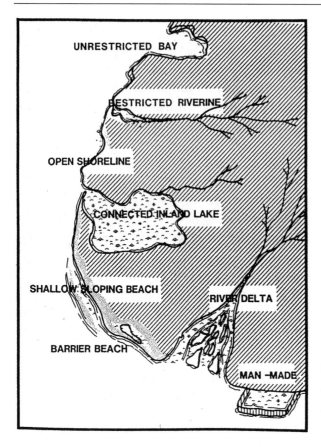

FIG. 9. *Types of wetlands in the Great Lakes region that are hydrologically dependent on Great Lakes water level (from Chubb and Liston 1985).*

levels brought about by increased primary production and biological oxygen demand. Declines are often most evident in cold-water species (e.g., brook trout) which are less tolerant than many of the warm-water fish species (e.g., carp).

The effects of increased soil and nutrient runoff are felt not only in the stream being directly affected, but this material is transported downstream, eventually to the Great Lakes themselves. Changes in land use patterns, in combination with point source pollution have greatly altered the trophic conditions across much of the Great Lakes (see previous section on water quality). Although the effects of increased nutrient input on primary production are quite clear, the overall impact on fish production is unknown. Near shore habitats show mixed responses to upland development. Although increased nutrient input can increase macrophyte growth, soil erosion and the resulting turbidity can limit macrophyte growth. Although the species richness (i.e., the number of species present) of the near shore fish community in the Great Lakes tends to increase with increasing macrophyte density, the response may be nonlinear (Randall 1996). Thus, it is difficult to predict the general response of fish habitat and the near shore fish community to upland development. Furthermore, the effects of increased nutrient inputs and sediment transport on the offshore fish community is less well known.

In addition to nutrient dynamics, urbanization also affects stream flow regimes. Nearly all urban and suburban development results in an increase in runoff rates by decreasing the amount of vegetation (thereby decreasing evapotranspiration) and by decreasing soil permeability. Increased runoff due to urbanization affects fish habitat in the short term by changing the temporal pattern of water flows, as well as changing channel morphology by increasing the volume and frequency of flood flows. By increasing runoff, stream flows tend to be more "flashy" (i.e., high flows are higher, and occur more quickly following rains, and low flows are lower). Flashier streams tend to support lower fish populations because high flows can transport young fish out of the river, and low flows can restrict adult and juvenile habitats. Production of potamodromous fish may also be reduced in streams where flow regimes have been altered.

Although it is difficult to directly determine the effect of changing land use practices on fish production, this is an important issue in the long-term management of the Great Lakes and their fish communities. Many of the habitat changes resulting from land use practices are cumulative and difficult to reverse. It is difficult to imagine removing sediments accumulated in the Great Lakes because of agriculture and urbanization. Even in tributary streams, stream beds degraded from past sedimentation may take hundreds of years to recover (Trimble 1983).

Fishery Management and Land Use

As with dams and wetlands, fishery issues are only one component of land use decisions and regulations. Some progress has been made to reduce the impact of human development by improvements in land use planning (e.g., green belts, protecting riparian corridors) and bridge and roadway construction techniques. Despite this progress, it seems that our population has put an ever increasing demand on the Great Lakes watershed. In addition to the effects on fish habitat considered here, contaminant loads are also elevated in urbanized watersheds. In some watersheds (e.g., in the Areas of Concern) the habitat has been rendered unfit for fish and many other aquatic organisms.

Examples of Species-Specific Effects of Habitat Changes on Fish Populations

Habitat alterations have played a part in the decline observed for many fish species in the Great Lakes basin. A major difficulty in determining the ultimate effect of habitat changes is that few, if any cases are available where habitat alterations are the sole factor affecting fish population dynamics. Fishing pressure and introduced species are examples of other factors that have impacted most Great Lakes fish. Our lack of knowledge on the impact of habitat changes on fish populations is exacerbated by the fact that the best known species (in terms of abundance and distribution) are those targeted by commercial and recreational fisheries. Thus, the effects of habitat changes are confounded by the effects of fishery removals.

Despite these difficulties, the fact that changes in fish habitat have reduced fish production is clear for some species. The lake sturgeon is an example of a potamodromous species whose population has declined because dams have cut off access to its natal spawning grounds, as well as degradation of its spawning areas by logging and changes in land use practices (Auer, this volume). These habitat changes have reduced the survival of eggs and young sturgeon, as well as limited the ability of adult sturgeon to deposit their eggs. As indicated earlier, fish population dynamics are generally most sensitive to changes in spawning success and early survival; this is also the case with lake sturgeon. As adults, lake sturgeon feed and live along the bottom of the lakes. At present, there is little evidence indicating whether adult habitats have changed enough to impact their growth or survival, but body burdens of contaminants in lake sturgeon have increased to the point where there is concern over their suitability for human consumption.

In contrast to lake sturgeon, the Pacific salmonids are an example of potamodromous fish where the impact of changes in stream habitats are less clear. Naturalized populations of these fish have developed relatively recently, after many of the basin's rivers were already dammed. These naturalized stocks depend on access to streams and rivers with suitable spawning gravel for reproduction. As such, dams and degradation of stream habitats do have an effect on their abundance. It is difficult, however, to predict what levels of salmonid production could be attained if fish habitats were in their "pristine" state, since we have no historical basis for comparison. Even though it is difficult to make such predictions, the potential benefits of more spawning habitat for these species should be considered when evaluating proposed removal of dams or improved fish passage facilities.

Another consideration for predicting the response of the Pacific salmonids to habitat changes is the difference in these species' specific life-histories that create somewhat different habitat needs. While all these species generally need a gravel substrate for spawning, the amount of time that fry spend in their natal river varies among species. After emerging from the gravel, pink salmon fry generally outmigrate within days (Higgs et al. 1995). Likewise, most young chinook salmon also migrate to the lakes relatively soon after emergence (days to weeks). In contrast, coho salmon juveniles typically spend several months to 1.5 years, and steelhead trout juveniles may spend up to three years rearing in their natal stream (Kocik and Jones, this volume). As such, coho salmon and steelhead trout depend more upon cold-water tributaries for juvenile rearing than do pink or chinook salmon.

Habitat changes have also impacted fish inhabiting the open lakes; the walleye stocks in Lake Erie have already been used as an example of a species that has suffered declines due to water quality degradation. In this case, habitat losses affected all life stages. Poor water quality decreased the success of spawning and rearing of young walleye, as well as limiting the distribution and quality of feeding habitats for adult walleye. In addition to habitat degradation, fishing pressure was a major factor in the decline of Lake Erie walleye. After the walleye fishery in Lake Erie was closed due to contaminants in their tissue, the population started to rebuild (Nepszy et al. 1991). Thus, it appears that at the time of the fishery closure, fishing pressure limited the walleye population more than habitat degradation.

The lake trout is another species that has possibly been impacted by changes in the open water habitat of the Great Lakes. It is particularly difficult to ascertain the degree to which habitat limitations have played in the decline of lake trout across the basin. Overfishing, sea lamprey predation, contaminants, and loss of genetic diversity are some of the other factors that have influenced lake trout population dynamics. The supposition that some aspect of habitat quality is having an impact is clear from the fact that lake trout have not shown any substantial reproductive success in Lakes Michigan and Huron during the last three decades of attempts to rehabilitate their populations (Sly 1984), even though fish stocked as juveniles have survived to maturity. The mechanism causing poor reproductive success is not clear, and may be confounded with other factors such as differences in reproductive behavior by stocked fish. In any case, efforts to determine problems with and methods for improving reproductive habitat will be needed to reach the goal of self-sustaining stocks of lake trout in Lakes Michigan and Huron.

Fish Habitat Policy

At the international level, several agreements set the policy for fisheries habitat and environmental quality. A particularly important agreement is the 1955 Convention on Great Lakes Fisheries between Canada and the United States. This agreement created the Great Lakes Fishery Commission, which is charged with developing programs of research and recommending management actions to maximize sustained production by Great Lakes fisheries. The Great Lakes Fishery Commission has set forth a policy of net gains in the quality of aquatic habitats (Anonymous 1992a), with particular emphasis on improved stream habitats. The Commission's focus on improving stream habitats is a result of their goal for improving conditions for natural reproduction of important sport and commercial fish (particularly salmonids), in order to minimize the need for stocking, and eventually, achieve self-regulating fish communities. In addition, the Great Lakes Fishery Commission has identified the need to work in partnership with Environment Canada, the U.S. Environmental Protection Agency, and the International Joint Commission to develop quantifiable environmental objectives for each of the Great Lakes that would enhance fishery production. Although the individual lake committees have been charged to develop habitat objectives compatible with their fish community objectives, none have been completed to date. Development of these habitat objectives is important to identify and remove substantial habitat limitations, and to provide guidelines for the maintenance of "good" habitats, where they already exist.

Water quality mandates are set at the international level by the Great Lakes Water Quality Agreement. As discussed previously in this chapter, this agreement has resulted in dramatic reductions in nutrient loadings across the basin, thereby improving the environmental conditions for many desirable fish species (e.g., walleye in Lake Erie). Furthermore, current directives of this agreement emphasize the need to reduce or eliminate toxic discharges, and to remediate, where feasible, past contamination sites. As part of the planning process, the Great Lakes Water Quality Agreement requires that Lakewide Management Plans (LaMPs) be developed for each of the Great Lakes. The intent of the LaMPs is to restore beneficial uses in the open waters of each of the lakes. In the near shore area, the International Joint Commission has identified 43 Areas of Concern where environmental conditions are severely degraded. For each of these Areas of Concern, both countries have agreed to develop a Remedial Action Plan (RAP) to identify pollution problems and restore and protect the site's uses.

At the national level, Canada and the United States differ sharply in their national policies on fish habitat. In Canada, fish habitat management, conservation, and restoration at a federal level are authorized under the federal Fisheries Act, with primary responsibility for these functions being vested in the Department of Fisheries and Oceans. Essentially, this Act prohibits the destruction or degradation of fish habitat without the approval of the Department of Fisheries and Oceans. In the Fisheries Act, fish habitat is explicitly defined as "spawning grounds and nursery, rearing, food supply, and migration areas on which fish depend directly or indirectly in order to carry out their life processes" (Fisheries Act 1986, 13). In 1986, the Department of Fisheries and Oceans published an explicit statement outlining their policy of no net loss of the productive capacity of fish habitats nationwide. In contrast, U.S. policies concerning fish habitat are incorporated into numerous acts, such as the Endangered Species Act, the Clean Water Act, and the Federal Power Act. Thus, no single agency has authority over fish habitats in the United States, and there is no analogous policy concerning net loss or gain of fish habitat.

The management of fish habitat in Canada is also facilitated by the fact that Ontario is the only province bordering the Great Lakes basin above the St. Lawrence River. Although the Canadian federal government is charged with fish habitat management through the Fisheries Act, Ontario also has a number of laws regulating impacts on aquatic habitats. Some examples include the Lakes and Rivers Improvement Act and the Planning Act. These acts, among others, serve to strengthen the ability of fishery managers in Ontario to protect and conserve fish habitat by giving them an opportunity to comment or intercede in nearly all human activities that impact fish habitat.

The complexity of legal issues, and the lack of a single policy concerning fish habitat in the United States is illustrated in the process of licensing dam operations. The agency that oversees dam licensing is the Federal Energy Regulatory Commission. In addition to engineering concerns and issues of dam safety, the Federal Energy Regulatory Commission must take into account whether the proposed dam (or dam being considered for relicensing):

- meets the requirements of the Clean Water Act by complying with applicable effluent limitations;
- complies with the state's programs for coastal development, as authorized in the Coastal Zone Management Act;
- affects the habitat of endangered or threatened species as outlined in the Endangered Species Act;
- provides for adequate fish passage as recommended by the Fish and Wildlife Service;
- will have a "significant environmental impact" as defined in the National Environmental Policy Act;
- will "unreasonably diminish" the scenic, recreational, or fish and wildlife values of rivers identified in the Wild and Scenic Rivers Act (Anonymous 1992b).

Although the impact of dams or other actions on fish habitat must be taken into account, fishery habitat concerns are not the primary concern of many of these pieces of federal legislation. As such, fishery habitat policy varies from project to project, depending on which legislative actions are applicable for the project in question. For example, the Wild and Scenic Rivers Act only applies to rivers designated under the act. Likewise, the Endangered Species Act becomes a consideration only when an endangered or threatened species is present in the area affected by a project.

Summary

Human actions often have a detrimental effect on fish habitats. Sadly, the true extent of our losses may never be known, because our impact of fish populations and habitats started soon after settlement. Historical accounts of fish migration in Great Lakes rivers (in Trautman 1981, for example) speak of tremendous runs of fish in rivers that are now dammed, drained, dredged, channelized, and otherwise detached from the production they once achieved. Likewise, our wetland resources are only a fraction of what was originally present in the region. Many of the factors degrading the quality of fish habitat are the result of issues beyond the direct control of fishery managers. Fishery issues do, however, play a role in the decision process, and can make a difference in the final resolution of possible threats to fish habitat. National and international policy initiatives, such as the Great Lakes Water Quality Agreement, have been successful in improving the quality of open lake habitats for many species. There remain, however, numerous threats to fish habitat that have not been comprehensively considered in such broad policy agreements.

Literature Cited

Anonymous. 1992a. Strategic vision of the Great Lakes Fishery Commission for the decade of the 1990s. Great Lakes Fishery Commission, Ann Arbor, MI.

———, 1992b. A quick'n'dirty guide to hydro licensing at the FERC. Unpublished manuscript, Federal Energy Regulatory Commission, Washington, D.C.

Auer, N. 1999. Lake sturgeon: a unique and imperilled species in the Great Lakes. Pages 515–535 *in* W. W. Taylor and C. P. Ferreri (eds.) Great Lakes Fisheries Policy and Management: A Binational Prespective. Michigan State University Press, East Lansing, MI.

Balon, E. K. 1975. Reproductive guilds of fish: A proposal and definition. Journal of the Fisheries Research Board of Canada 32:821–864.

———, 1978. Kariba: the dubious benefits. Ambio 7:40–748.

———, 1981. Additions and amendments to the classification of reproductive styles in fish. Environmental Biology of Fish 6:377–389.

Beeton, A. M., C. E. Sellinger, and D. F. Reid. 1998. An introduction to the Laurentian Great Lakes ecosystem. Pages 1–51 *in* W. W. Taylor and C. P. Ferreri (eds.). Great Lakes Fisheries Policy and Management: A Binational Perspective. Michigan State University Press, East Lansing, MI.

Bovee, K. D. 1986. Development and evaluation of habitat suitability criteria for use in the instream flow incremental methodology. Instream Flow Information Paper No. 21. U.S. Fish and Wildlife Service Biological Report 86(7). Fort Collins, CO.

Burton, T. M. 1985. The effects of water level fluctuations on Great Lakes coastal marshes. Pages 3–13 *in* H. H. Prince, and F. M. D'Itri, editors. Coastal wetlands. Lewis Publishers, Inc., Chelsea, MI.

Caswell, H. 1989. Matrix population models. Sinauer Associates, Sunderland, Massachusetts. 328 p.

Chubb, S. L., and C. R. Liston. 1986. Density and distribution of larval fish in Pentwater Marsh, a coastal wetland on Lake Michigan. Journal of Great Lakes Research 12:332–343.

Cowardin, L. M., V. Carter, F. C. Golet, and E. T. LaRoe. 1979. Classification of wetlands and deepwater habitats of the United States. FWS/OBS-79/31. U.S. Fish and Wildlife Service, Washington, D. C.

Crunkilton, R. L., J. M. Czarnezki, and L. Trial. 1980. Severe gas bubble disease in a warm-water fishery in the midwestern United States. Transactions of the American Fisheries Society 109:72–33.

Dendy, F. E., W. A. Champin, and R. B. Wilson. 1973. Reservoir sedimentation surveys in the United States. Pages 349–357 *in* Ackermann, W. C., G. I. White, E. B. Worthington, and I. J. Loreena, editors. Man-made lakes: their problems and environmental effects. American Geophysical Union, Washington, D. C.

Fausch, K. D., C. L. Hawkes, and M. G. Parsons. 1988. Models that predict standing crop of stream fish from habitat variables: 1950–1985. United States Forest Service Gen. Tech. Rep. PNW-GTR-213. Pacific Northwest Research Station, Portland, OR.

Garcia, A. 1989. The impacts of squawfish predation on juvenile chinook salmon at Red Bluff Diversion Dam and other locations in the Sacramento River. U.S. Fish and Wildlife Service Report No. AFF/FAO-8-5.

Glooschenko, V. 1985. Characteristics of provincially significant wetlands as assessed by the Ontario wetland evaluation system. Pages 187–199 *in* H. H. Prince, and F. M. D'Itri, editors. Coastal wetlands. Lewis Publishers, Inc., Chelsea, MI.

Goodyear, C. D., T. A. Edsall, D. M. Ormsby Dempsey, G. D. Moss, and P. E. Polanski. 1982a. Atlas of the spawning and nursery areas of Great Lakes fish. Volume 5: Lake Huron. U.S. Fish and Wildlife Service, Washington, DC. FWS/OBS-82/52.

———, 1982b. Atlas of the spawning and nursery areas of Great Lakes fish. Volume 13: Reproductive characteristics of Great Lakes fish. U.S. Fish and Wildlife Service, Washington, DC. FWS/OBS-82/52.

Grizzle, J. M. 1981. Effects of hypolimnetic discharge on fish health below a reservoir. Transactions of the American Fisheries Society 110:2–3.

Hartig, J. H. 1993. A survey of fish-community and habitat goals/objectives/targets and status in Great Lakes Areas of Concern. Great Lakes Fishery Commission, Ann Arbor, MI.

Hay-Chmielewski, E. M., P. W. Seelbach, G. E. Whelan, and D. B. Jester, Jr. 1995. Huron River Watershed Assessment. Fisheries Division, Michigan Department of Natural Resources, Lansing, MI.

Hayes, D. B., W. W. Taylor, and J. C. Schneider. 1992. Response of yellow perch and the benthic invertebrate community to a reduction in the abundance of white suckers. Transactions of the American Fisheries Society 121:36–53.

Hesse, J. L. 1997. Case study: sport fish consumption advisories in the Great Lakes region. Pages 151–170 *in* M. A. Kamrin (ed.) Environmental risk harmonization: Federal and state approaches to environmental hazards in the U.S.A. John Wiley and Sons, Chichester.

Higgs, D. A., J. S. MacDonald, C. D. Levings, and B. S. Dosanjh. 1995. Nutrition and feeding habitats in relation to life history stage. Pages 161–315 *in* Groot, C. L. Margolis, and W. C. Clarke, editors. Physiological ecology of Pacific salmonids. University of British Columbia Press, Vancouver.

Hudson, P. L., R. W. Griffiths, and T. J. Wheaton. 1992. Review of habitat classification schemes appropriate to streams, rivers, and connecting channels in the Great Lakes drainage basin. Pages 73–107 *in* W. D. N. Busch and P. G. Sly, editors, The development of an aquatic habitat classification system for lakes. CRC Press, Ann Arbor, MI.

Hutchinson, G. E. 1957a. Treatise on Limnology, Vol. 1. John Wiley and Sons, New York.

———, 1957b. Concluding remarks. Cold Spring Harbor Symposia on Quantitative Biology 22:415–427.

Kerr, S. R. 1976. Ecological analysis and the Fry paradigm. Journal of the Fisheries Research Board of Canada 33:329–332.

———, E. E. Werner. 1980. Niche theory in fisheries ecology. Transactions of the American Fisheries Society 109:254–260.

Kocik, J. F. and M. L. Jones. 1999. Pacific salmonines in the Great Lakes Basin. Pages 455–87 *in* W. W. Taylor and C. P. Ferreri (eds.) Great Lakes Fisheries Policy and Management: A Binational Perspective. Michigan State University Press, East Lansing, MI.

Lawrie, A. H., and J. F. Rahrer. 1973. Lake Superior: a case history of the lake and its fisheries. Great Lakes Fishery Commission Technical Report 19, Ann Arbor, MI.

Leach, J. H., and R. C. Herron. 1992. A review of lake habitat classification. Pages 27–57 *in* W. D. N. Busch and P. G. Sly, editors, The development of an aquatic habitat classification system for lakes. CRC Press, Ann Arbor, MI.

Legg, D. L. 1978. Gas supersaturation problem in the Columbia River basin. Pages 149–164 *in* Binger, W. V., J. P. Buehler, F. J. Clarke, E. R. DeLuccia, R. D. Harza, R. B. Jansen, J. C. Peters, M. F. Thomas, and W. L. Chadwick, editors. Environmental effects of large dams. American Society of Civil Engineers, New York, New York.

Liston, C. R., and S. Chubb. 1985. Relationships of water level fluctuations and fish. . Pages 121–140 *in* H. H. Prince, and F. M. D'Itri, editors. Coastal wetlands. Lewis Publishers, Inc., Chelsea, MI.

Loftus, K. H. 1958. Studies on river-spawning populations of lake trout in Eastern Lake Superior. Transactions of the American Fisheries Society 87:259–279.

Nepszy, S. J., D. H. Davies, D. Einhouse, R. W. Hatch, G. Isbell, D. MacLennan, and K. M. Muth. 1991. Walleye in Lake Erie and Lake St. Clair. Pages 145–168 *in* P. J. Colby, C. A. Lewis, and R. L. Eshenroder, editors. Status of walleye in the Great Lakes: case studies prepared for the 1989 workshop. Great Lakes Fishery Commission Special Publication 91–1.

Orth, D. J., and R. J. White. 1993. Stream habitat management. Pages 205–230 *in* C. C. Kohler and W. A. Hubert, editors. Inland fisheries management in North America. American Fisheries Society, Bethesda, Maryland.

Patterson, N. J., and T. H. Whillans. 1985. Human interference with natural water level regimes in the context of other cultural stresses on Great Lakes wetlands. Pages 209–251. *in* H. H. Prince, and F. M. D'Itri, editors. Coastal wetlands. Lewis Publishers, Inc., Chelsea, MI.

Peck, J. W. 1992. The sport fishery and contribution of hatchery trout and salmon in Lake superior and tributaries at Marquette, Michigan, 1984–1987. Michigan Department of Natural Resources, Fisheries Research Report 1956, Ann Arbor, MI.

Randall, R. G., C. K. Minns, V. W. Cairns, and J. E. Moore. 1996. The relationship between an index of fish production and submerged macrophytes and other habitat features at three littoral areas in the Great Lakes. Canadian Journal of Fisheries and Aquatic Science 53(Suppl. 1): 35–44.

Scott, W. B., and E. J. Crossman. 1985. Freshwater fish of Canada. Fisheries Research Board of Canada Bulletin 184. Ottawa, Canada.

Shuter, B. J. 1990. Population-level indicators of stress. American Fisheries Society Symposium 8:14–66.

Sly, P. G. 1984. Habitat. Pages 35–39 *in* R. L. Eshenroder, T. P. Poe, and C. H. Olver, editors. Strategies for rehabilitation of lake trout in the Great Lakes: proceedings of a conference on lake trout research, August 1983, Great Lakes Fishery Commission, Technical Report No. 40.

———, W.-D. N. Busch. 1992. Introduction to the process, procedure, and concepts used in the development of an aquatic habitat classification system for lakes. Pages 1–26 *in* W. D. N. busch and P. G. Sly, editors. The development of an aquatic habitat classification system for lakes. CRC Press, Ann Arbor, MI.

Smith, M. F. 1986. Great Lakes water diversion: protecting Michigan's interests. Monograph No. 9, Public Sector Consultants, Inc., Lansing, MI.

Sommers, L. M., C. Thompson, S. Tainter, L. Lin, T. W. Colucci, and J. M. Lipsey. 1981. Fish in Lake Michigan: distribution of selected species. Michigan Sea Grant Advisory Program.

Stanley, J. W. 1951. Retrogression of the Lower Colorado River after 1935. Transactions of the American Society of Civil Engineers 116: 943–957.

Taylor, K. V. 1978. Erosion downstream of dams. pages 165–186 *in* Binger, W. V., J. P. Buehler, F. J. Clarke, E. R. DeLuccia, R. D. Harza, R. B. Jansen, J. C. Peters, M. F. Thomas, and W. L. Chadwick, editors. Environmental effects of large dams. American Society of Civil Engineers, New York, New York.

Trautman, M. B. 1981. The fish of Ohio, with illustrated keys, revised edition. Ohio State University Press, in collaboration with the Ohio Sea Grant Program Center for Lake Erie Area Research, Columbus, Ohio.

Trimble, S. W. 1983. A sediment budget for Coon Creek Basin in the Driftless Area, Wisconsin, 1853–1977. American Journal of Science 283:445–474.

Vannote, R. L., G. W. Minshall, K. W. Cummins, J. R. Sedell, and C. E. Cushing. 1980. The river continuum concept. Canadian Journal of Fisheries and Aquatic Science 37:130–137.

Weitkamp, D. E., and M. Katz. 1980. A review of dissolved gas supersaturation literature. Transactions of the American Fisheries Society 109:659–702.

Werner, E. E., and J. F. Gilliam. 1984. The ontogenetic niche and species interactions in sizestructured populations. Annual Review of Ecology and Systematics 15: 393–425.

———, D. J. Hall, and G. G. Mittelbach. 1983. An experimental test of the effects of predation risk on habitat use in fish. Ecology 64:1540–1548.

Wiegert, R. G., and L. R. Pomeroy. 1981. The salt-marsh ecosystem: a synthesis. Pages 219–230 *in* Pomeroy, L. R., and R. G. Wiegert, editors. The ecology of a salt marsh. Springer Verlag, New York.

CHAPTER 9

Impact of Airborne Contaminants on the Great Lakes

Raymond M. Hoff

No whitefish or trout here, we leave them alone
The inspectors raise hell if we take any home
What kind of fisherman can't eat his catch
Or call what he's taken his own?

In the Norfolk Hotel over far too much beer
The old guys remember when the water ran clear
No poisons with names that we can't understand
And no tiny fish for Japan

> —Stan Rogers, "Tiny Fish for Japan" (1984), © Fogarty's Cove Music Inc.,
> Dundas, Ontario (used by permission of Mrs. Ariel Rogers)

Toxic Chemicals and the Great Lakes Fishery

When the Canadian folk singer Stan Rogers wrote those words, he was chronicling the demise of the Lake Erie fishery. Phosphorus, 2,3,7,8-dibenzodioxin, gamma-hexachlorocyclohexane (lindane), methyoxychlor, atrazine, 2,4,5-trichlorophenoxyacetic acid (2,4,5-T), and 1,1,1-trichloro-2–2-bis (parachlorophenyl) ethane (DDT) are hardly the stuff to be placed in a song about the fishery, let alone to be placed in the lake.

Many of these chemicals are airborne contaminants, and may have their sources far from the Great Lakes Basin. Chemicals which are bioaccumulative and persistent have been identified under legislation (Clean Air Act Amendments-CAAA, 1990; Toxic Substances Control Act-TSCA, 1985; Canadian Environmental Protection Act-CEPA, 1990; the Great Lakes Water Quality Agreement-GLWQA, 1987). These compounds are variously referred to as hazardous air pollutants (HAPs). The governments of Canada and the United States have sought to control the sources of many of these HAPs through legislation. While some of these controls have been successful, some HAPs (polychlorinated biphenyls and DDT, for example) are still observed in the Great Lakes some twenty

TABLE 1

Persistent Substances of Concern Identified in the Great Lakes

(GLWQA, 1987)

1	Polychlorinated Biphenyls	10	γ-HCH
2	Benzo(apyrene)	11	Chlordane
3	Dieldrin	12	nonachlor
4	Hexachlorobenzene	13	Lead
5	Tetrachlorodibenzo(p)dioxin	14	Cadmium
6	Tetrachlorodibenzofuran	15	Mercury
7	DDT	16	Arsenic
8	Toxaphene	17	Copper
9	α-HCH	18	Zinc

or more years after their use was restricted (Environment Canada and U.S. Environmental Protection Agency, 1995). The reason for such a long residual lifetime in the lakes has to do both with a slow removal process within the lakes themselves, but also continual addition of these chemicals from the atmosphere.

The GLWQA agreement (International Joint Commission 1988) defines a toxic substance as:

a substance which can cause death, disease, behaviour abnormalities, cancer, genetic mutations, physiological or reproductive malfunctions or physical deformities in any organism or its offspring, or which can become poisonous after concentration in the food chain or in combination with other substances.

This definition implicitly focuses on bioaccumulative substances, which can become increasingly concentrated in lipids at higher levels in the food chain. A second criteria, persistence, is defined in the GLWQA for substances which have a half-life (the time by which the concentration decreases by 50%) in water of greater than eight weeks (GLWQA Annex 12, 1987).

Using a combination of criteria on toxicity, pathways for exposure, concentration in the environment, and bioaccumulative potential, the 1987 Amendments to the GLWQA identified 221 chemicals as present and toxic in water (this is Appendix 1 of Annex 10 of the GLWQA). More recent reviews of the persistence of these chemicals have prompted this list to be narrowed to eighteen chemicals, or chemical classes, which are defined to be of concern (table 1). This means that these compounds do not meet the objectives of the GLWQA, one of which states "the discharge of any or all persistent toxic substances be virtually eliminated." Chemical classes 1–12 are persistent organochlorine pollutants (POPs), and chemicals 13–18 are elements that are largely found on airborne particulates (and are called airborne trace elements).

Processes Affecting Atmospheric Concentrations, Transport and Deposition

Figure 1, adapted from Schroeder and Lane (1988), shows schematically the processes involved in the emission of chemicals to the air, their transport, and their redeposition to the Earth's surface.

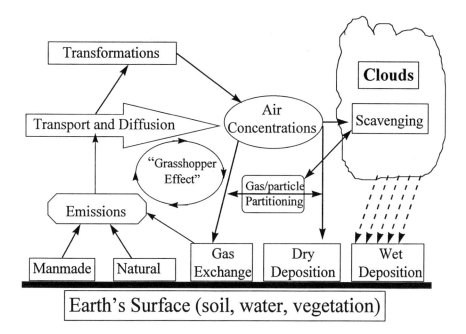

FIG. 1. *The processes of environmental cycling of HAPS through the atmosphere and their removal to the Earth's surface (adapted from Schroeder and Lane (1988))*

There are three direct routes by which chemicals can be deposited to the lakes from the atmosphere and one indirect route. The first three are wet deposition (rain, snow, and fog inputs to the water surface), dry deposition (the sedimentation of particles in the atmosphere to the water surface), and gas exchange (the movement of chemicals across the air/water interface by gaseous absorption and volatilization). The indirect route involves the same three processes impacting the land and vegetative surfaces around the lakes, and subsequently running off into the water through riverine inputs, municipal outflows, groundwater seepage, etc.

Emission Sources to the Atmosphere

Sources of HAPs to the atmosphere are both natural and manmade. For the trace elements in table 1, there are significant natural sources for the emission of these chemicals from weathering of the earth surface and biological production. An excellent review of these sources to the atmosphere can be found in Nriagu (1989), and Nriagu and Pacyna (1988). Primary sources of these elements are industrial (i.e., primary and secondary nonferrous metal production, coal and oil electric generation), refuse incineration, and from mobile sources (i.e., gasoline additives). For lead, the predominant atmospheric source is (in Europe) and used to be (in North America) leaded gasoline. This practice was discontinued in the late 1980s in much of the world, and atmospheric lead concentrations have dropped dramatically. For the POPs in table 1, some are used as herbicides and pesticides (α- and γ-HCH, toxaphene, DDT, dieldrin, chlordane, and nonachlor), while others are industrial products and by-products. Polychlorinated biphenyls were produced, and

TABLE 2

1983 natural versus anthropogenic emissions of trace elements

(From Nriagu, 1989) All values in 10^6 kg/y.

ELEMENT	ANTHROPOGENIC	NATURAL	TOTAL	% NATURAL
Lead	332	12	344	4
Cadmium	7.6	1.3	8.9	15
Mercury	3.6	2.5	6.1	41
Arsenic	19	12	31	39
Copper	35	28	63	44
Zinc	132	45	177	34

used primarily as transformer and capacitor oils. Hexachlorobenzene was produced until the early 1970s, but it continues to be found as a contaminant in the production of pesticides and chlorinated organic compounds. Benzo(a)pyrene is only one member of a class of multi-ring aromatic compounds, known as polycyclic aromatic compounds (PACs, also known as polynuclear aromatic hydrocarbons or PAHs). Many of the PACs, and their oxidation products are carcinogenic. This class of compounds, as well as the chlorinated dibenzo(p)dioxins and dibenzofurans, are formed from low temperature combustion processes. Dioxins and PACs have been detected in emissions from coal, oil, and wood combustion, from industrial sources (such as coke ovens in steelmaking), and even from cigarette smoking. Recently, cigarette smoke has been estimated to comprise over one percent of the total organic loading to the air in the Los Angeles basin (Hildemann et al. 1994).

There are natural sources for the combustion-related POPs in table 1 (forest fires, as one source), however, most are emitted from anthropogenic activities. For the trace metals in table 1, however, Nriagu (1988) estimated that in 1983, between 4% and 44% of emissions to the atmosphere are from natural sources, such as wind-borne soil, sea salt, volcanic emissions, wildfires, and biogenic emissions (see table 2).

In figure 1, a third source of emissions is noted. This source is from previously deposited material which revolatilizes from the Earth's surface. This is part of the process called the "grasshopper effect" which will be discussed below.

Transport, Diffusion, and Transformation in the Atmosphere

After being emitted from the surface by manmade or natural processes, pollutants travel through the air with the prevailing winds. This process of transport, and the spreading of the material vertically in the atmosphere via diffusion is crucial to the degree to which a pollutant moves. The pollutants are found in the atmosphere in two forms: as gases or as part of particulate matter. The form in which the chemical is found is important in determining its atmospheric lifetime. Large particles formed by windblown erosion or as fly ash during combustion processes will not travel far since they rapidly settle out to the surface. Smaller particles (generally less than one micrometer in size) can remain aloft much longer, often for weeks or more.

Particles may also be more efficiently scavenged via cloud and rain processes than gases. Thus, the gas and particle partitioning in the atmosphere is important in determining the chemical's fate. Similarly, if the chemical transforms via photochemical action in the air, it will not be transported far. Many persistent organic chemicals in water (lindane, dioxins, benzo(a)pyrene) have short half-lives in air, due to their reactivity with the hydroxyl radical, OH (Arey et al.,1986). These transformation processes may remove the toxic chemical from air by processing them to another species. However, for some species, such as nitrogenated-PACs, the secondary product can be more toxic than the primary pollutant (Ramdahl et al. 1986).

Wet Deposition

Chemicals are absorbed into rain through gaseous adsorption onto cloud droplets, inclusion of particulate chemicals into cloud water through nucleation processes and scavenging, and scavenging of gases and particles into rain or snow falling through the atmospheric column under clouds (Seinfeld 1986). From what is known about particulate inorganic species (such as sulphates involved in the acid rain issue), inclusion of the chemical into cloud water with subsequent droplet growth and deposition appears to be the predominant process.

For gaseous species, such as the PCBs or lighter organochlorine pesticides, wet removal processes occur primarily below cloud. The partitioning of gases into liquid water is controlled by an equilibrium constant (the Henry's law constant). When a gas is in equilibrium with a water surface, the concentration of the species in air ($C_{a,eq}$) and water ($C_{w,eq}$) are related at a given temperature by their Henry's Law constant; $H = C_{a,eq} / C_{w,eq}$. This equilibration should take only a matter of milliseconds in the atmosphere, and falling rain drops would have time to equilibrate with the concentration of organic chemicals near the Earth's surface. Studies of gas/rainwater concentrations have shown that, on average, a Henry's law equilibration is achieved (Leister and Baker 1994).

Measurements of rain and snow have been made for many years around the Great Lakes (Chan and Perkins 1989; Strachan 1985; Strachan and Huneault 1995). In these measurements, it is important to exclude the dry fall of chemical into the collector during periods when it is not raining or snowing. For this reason, wet-only collectors are used (Strachan and Huneault 1984). A standard for organic chemical measurements is the Meteorological Instruments Company (MIC) collector, which has an aperture of $0.2m^2$, and a cover which opens only when a sensor is activated by falling rain or snow. The precipitation is collected by a slightly heated funnel (to melt falling snow) and passes through a funnel with an organic resin to trap the toxic chemical. Integrated samples over a two-week or monthly period are accumulated, the resin cartridges extracted, and analysed by gas chromatography. The wet atmospheric flux of a chemical to a water body is the precipitation volume multiplied by the concentration of the chemical in the collected precipitation. To get the total mass entering the lake, one multiplies this loading by the area of the lake (Strachan and Eisenreich 1988).

Another component of wet deposition, however, which may need to be studied, is the inclusion of organics in fog water and subsequent removal by fog deposition or drizzle. Studies of pesticide inclusion in fogwater in California (Glotfelty et al. 1990) and Switzerland (Capel et al. 1991) have shown remarkable enhancements in droplet concentrations over that in rainwater. It is possible that fog scavenging may be an important removal process for organics from the atmosphere over lakes. The magnitude of this input currently is unquantified.

Dry Deposition

Particles are known to settle out of the atmosphere at a rate which is strongly dependent on their particle size. The flux of such particles to the surface is the product of the particulate-borne portion of the air concentration times a deposition velocity of particles of a given diameter (Strachan and Eisenreich 1988; Slinn and Slinn 1980). Conventional understanding of the mass-size distribution of organic and trace metal particles in air puts their median diameter at 0.2–0.7mm (Dodd et al. 1991; Poster et al. 1995). Over water, these particles are believed to absorb water rapidly and grow, which speeds up their deposition to the surface. For a 0.4mm diameter aerosol over water, the deposition velocity would be about 0.2cm s^{-1}.

Although there are artifacts which can be found in sampling for atmospheric particulates, the measurement of the air particulate concentration is done by putting a glass fibre (for organics) or cellulose (for trace elements) filter in series with a gaseous absorbent trap. Air is pulled through these trapping media at a rate of 0.1–2m^3 min^{-1} for periods of between twenty-four and ninety-six hours, gaining large air volumes of several hundreds to thousands of cubic meters of air sampled. The filters are extracted with an organic solvent for the POP analysis. For the trace element analysis, the filters are either extracted in an acid and subjected to inductively coupled plasma emission spectroscopy (ICP), or analysed directly with proton-induced x-ray emission (PIXE), nuclear activation analysis (NAA), or energy dispersive x-ray fluorescence analysis (EDXRF). Typically, for most organochlorine compounds, their particulate-borne fraction is small, and dry deposition is relatively unimportant. This is not the case for the trace elements or PACs, which can have substantial dry deposition, especially around urban areas.

Gas Transfer

Gas transfer across the water surface has generally been understood by the two-film resistance model developed by Liss and Slater (1974). Generally, the concentration in air over the lakes is not exactly in equilibrium with the water concentration given by the Henry's law constant. The imbalance of the concentrations gives a piston pressure which either forces gas out of the water column or into it. The air-water flux is governed by a mass transfer coefficient, K_{oL}, which has units of distance time^{-1} (analogous to a deposition velocity or precipitation rate). The mass flux can be calculated from an equation which relates the air-water imbalance to the mass transfer rate through the water surface (Mackay and Yeun, 1983; Baker and Eisenreich, 1990). In the past, many of the organochlorine chemicals were found at higher concentrations in air, and were absorbed by the water at the lake surfaces. At present, the situation has reversed, and the lakes are now sources of many of the chemicals entering the atmosphere (Hoff et al., 1996).

Tributary Component Which May be Atmospheric in Origin

In addition to these direct inputs, the indirect input of chemicals of atmospheric origin can occur via deposition and subsequent runoff to rivers and the lakes. This component is difficult to quantify, as the retention of organic chemicals and trace metals in soils, vegetation, and improved surfaces all have different time constants. In a recent study of the inputs to Lake Superior, Dolan and his coworkers (1993) concluded that about 10% of the material which fell into the watershed ultimately evolved to end up in Lake Superior by migration through the water column.

Historical Evidence of an Atmospheric-Input Source to the Great Lakes

With the mounting evidence of damage to fish stocks and wildlife which feed off the fishery (literature reviewed in Swain et al. (1994)), a search intensified in the 1980s for a continuing source of input of toxic chemicals other than direct discharges from urban areas and industry ringing the Great Lakes. Several studies had remarkable impact in making the case for the significance of atmospheric inputs to the basin.

Siskiwit Lake

In 1983 and 1984, Ron Hites, a professor at Indiana University's School of Public and Environmental Affairs, and two of his graduate students, Deborah Swackhamer and Bruce McVeety, decided to monitor an isolated lake for polychlorinated biphenyls (Swackhamer et al. 1988). The results provided a compelling case for the importance of atmospheric pathway to water bodies in the Great Lakes. Siskiwit Lake is in the middle of Isle Royale on Lake Superior. The site has been a national park since 1940, and has had minimal anthropogenic activity near the lake. Sampling air on an island in Siskiwit Lake (which is on a bigger island (Isle Royale) in a bigger lake(Lake Superior)) provided isolation from what might be local ground-based sources of pollution, such as rivers, runoff, or groundwater.

What Hites and his coworkers found was that a mass balance could be found between the deposition of airborne PCBs and the water and sediment concentrations of PCBs. The water concentration of PCBs in Siskiwit Lake was 2.5 times higher than that in Lake Superior itself, and implied that the dynamics and sedimentation rates in this smaller lake were less efficient in removing the continuing input of PCBs. The conclusion was that atmospheric inputs from rainfall and dry deposition were the most likely source of the PCBs to this isolated lake. Since that work was published, we have learned that gas transfer to these lakes was also very important.

Ombrotropic Peat Bogs

Another strong piece of evidence in the case for atmospheric deposition of toxic chemicals to the Great Lakes basin comes from another receptor which is unlikely to be affected by local polluting sources. From 1986 to 1988, Professor Steven Eisenreich, of the University of Minnesota, and his student Rob Rapaport sought to find a historical record which would document the use and deposition of toxic chemicals. Such climatological markers have been found in tree rings, indicating yearly growth of trees or in ice cores which indicate the growth of glaciers. They found that evidence within the Sphagnum peat bogs in eastern North America (Rapaport and Eisenreich 1988). Peat bogs are ombrotrophic, meaning that they derive their nutrients from atmospheric sources, not from the groundwater and surface water flow. Since they are highly organic (over 90% of their mass by weight is organic carbon), they are an excellent trapping medium for airborne gaseous organics. They dated the cores using inputs of ^{210}Pb and ^{137}Cs which are known to be atmospheric in origin (^{137}Cs is from atom-bomb testing) and have a strong temporal signal. Rapaport and Eisenreich constructed PCB, HCB, toxaphene, HCB, and DDT, Pb and Zn profiles for seven remote peat bogs across eastern North America, from Minnesota, Ontario, Quebec, North Carolina, Maine, and Nova Scotia. The atmospheric deposition profile of PCBs (fig. 2) follows the production and

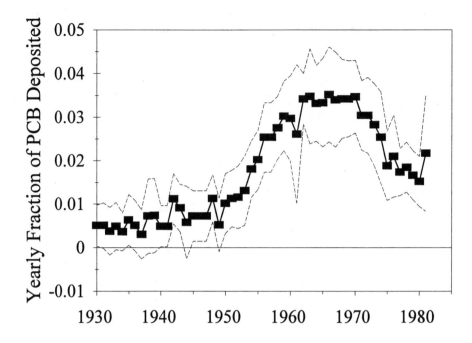

FIG. 2. *Sedimentation rate of PCBs in peat bogs in the northern US and Canada (Rapaport and Eisenreich, 1988)*

use cycle of PCBs in North America, showing the immediate decline in PCB deposition after they were banned from sale in the early 1970s (the 1983 peak is due to current-year input to the bog, some of which was subsequently lost back to the atmosphere before the surface layer becomes buried). PCBs were first introduced into commerce in 1929 as high-density insulating oils for electrical transformers. Their use increased steadily until the 1960s and decreased abruptly in North America after the ban. The cores also show that PCB inputs had not gone to zero, even fifteen years after they were regulated.

Sediment Patterns

A similar type of evidence can be found from the profiling of lake bottom sediment cores. While atmospheric input is obviously not as conclusively shown, due to the other fluvial and direct inputs to the lakes, the time history of accumulation of a chemical in sediments can be a good marker of its removal from the water column. Again, the use of ^{210}Pb and ^{137}Cs allows the dating of the airborne inputs to these cores, as these two metals are known to have a predominant atmospherically derived input to the surface. Swain et al. (1992) examined historical sediment cores for mercury Hg from seven headwater lakes in Minnesota and Wisconsin, and Oliver and his co-workers (1989) examined Lake Ontario sediments for PCBs, DDT, mirex, and HCB. In the former study, Swain showed that Hg inputs to the lakes have a relatively large natural background, but the anthropogenic inputs could be discerned by removing a constant natural input. In Oliver's work, the core profiles mimic the historical production and use patterns of the chemicals well. Particularly important in the work by Oliver was the observation that sediment recycling (the resuspension of

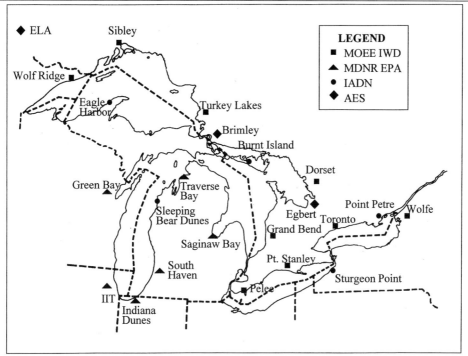

FIG. 3. *Sites of stations in the Integrated Atmospheric Deposition Network.*

particles from the lake bottom during storm events or rapid mixing in the water column) is also sustaining the levels of organic chemicals in the lake.

Estimating the Inputs to the Lakes

Since the late 1980s, several reviews have attempted to determine the mass balance of the transfer of toxic chemicals from the air to the water column (Strachan and Eisenreich 1988; Eisenreich and Strachan 1992; Hoff et al. 1996). In addition, a larger scale effort is required to determine the relative magnitude of those inputs to other sources of chemical inputs to the lakes (riverine, direct discharge, etc.). At least two major field studies have been carried out which attempt to do the latter: the Green Bay Mass Balance Study in 1989 (Achman et al. 1993; Hornbuckle et al.,1992) and the Lake Michigan Mass Balance Study (1994–96). The 1987 Amendments to the Great Lakes Water Quality Agreement required the governments of Canada and the United States to institute an Integrated Atmospheric Deposition Network (IADN) on the Great Lakes (Egar and Adamkus 1990). In operation since 1990, stations have been collecting air and precipitation data on each of the Great Lakes (fig. 3). Coupled with short-term and intensive monitoring sites (also shown in the fig. 3), IADN has provided the basis to improve our understanding of the deposition of organics and trace metals to each of the Great Lakes. A recent publication (Hoff et al. 1996) has used these IADN results to determine the atmospheric fluxes of HAPs to each of the five Great Lakes.

In 1988, an International Joint Commission study (Strachan and Eisenreich 1988) strongly implicated the atmosphere as a major source of toxic chemicals to the Great Lakes (table 3). For the

TABLE 3

1988 Estimates of the Percentage Atmospheric Contribution to each Lake

SPECIES	L. SUPERIOR	L. MICHIGAN	L. HURON	L. ERIE	L. ONTARIO
PCBs	90	58	78	13	7
t-DDT	97	98	97	22	31
Pb	97	99	98	46	73

upper Great Lakes (Superior, Michigan and Huron), atmospheric deposition dominates the current inputs to the water bodies with between 58% and 96% of the total input. Further downstream, industrial sources are believed to make larger contributions, so that by the time Lake Ontario accepts waters from the upper Great Lakes, atmospheric deposition can be a less major component of the total loadings to the lakes (between 7% and 73%).

In 1992, a second detailed estimate of the wet-and-dry deposition to the lakes was made (Eisenreich and Strachan 1992). This update included the effects of seasonality in the air concentration of a greater number of organochlorine species, some current use pesticides and herbicides (alachlor, atrazine, endosulfan, and trifluralin, for example). Table 4 shows the annual loading of each of the lakes in kilograms per year. The study concluded that the largest lake (Superior) had the largest overall inputs. This is because the difference in rates of deposition do not vary widely across the five Great Lakes, and the surface area is a major factor. There were few east-west differences seen in deposition across the lakes. This fact, too, points to more distant rather than local sources being the source of many of the POPs.

In 1994, a third review of the deposition of toxins to the Great Lakes was made based on more recent data from IADN (Hoff et al. 1996). In this latter study, the impact of gas transfer across the air-water interface has proven to be the most important process in delivering chemicals to and removing them from the Great Lakes. Additional estimates of the deposition to the Great Lakes can be found in reports by Baker et al. (1993), Cohen et al. (1995), and in a report to the U.S. Congress (U.S. Environmental Protection Agency, 1994). In each of these studies, a consistent finding has been that atmospheric inputs of many restricted toxic chemicals are larger than other current inputs from rivers, discharges, or spills. In Cohen et al. (1995), for example, sources as far away as Colorado and New Mexico are shown to have impact on the Great Lakes. Clearly, long-range transport of airborne toxic chemicals is a major component of the deposition process to the Great Lakes.

Current Sources of Airborne-Toxic Chemicals

Urban Areas

The Great Lakes region, with over 40 million inhabitants, contains significant local anthropogenic pollution impacts. Combustion sources such as automobiles, home heating, and industry (hydroelectric generation, steel making), create tons of polynuclear aromatic compounds on a regional scale. That impact has been seen in air sampling around the Great Lakes. At Sturgeon Point, near Buffalo, the polynuclear aromatic compound concentrations in air are more than a factor of two

TABLE 4

1992 Estimates of Atmospheric Loading of Toxic Chemicals to the Great Lakes (kg/y)

(Eisenreich and W. M. J. Strachan, 1992)

SPECIES	L. SUPERIOR	L. MICHIGAN	L. HURON	L. ERIE	L. ONTARIO
Σ-PCB	160	110	110	53	42
α-HCH	310	230	230	110	87
γ-HCH	160	110	110	54	43
HCB	4	3	3	1	1.1
Dieldrin	5	4	4	2	1.4
Σ-DDT	34	25	25	12	9.5
Heptachlor	1	1	1	0.46	0.37
HepEpoxide	6	5	5	2	2
Chlordane	13	10	10	4	4
Toxaphene	17	13	13	6	5
Σ-Endosulphan	98	72	71	34	27
Atrazine	11000	8100	8100	3800	3000
Alachlor	35000	25000	25000	12000	9700
Trifluralin	1000	710	730	312	240
Fluorene	160	110	110	55	43
Phenanthrene	340	250	250	120	93
Fluoranthene	460	330	330	160	130
Pyrene	330	240	240	110	90
Benzanthracene	83	61	61	28	22
Chrysene	160	110	140	53	41
B(k)F	190	140	140	63	50
B(b)F	220	160	160	74	59
B(a)P	120	84	84	39	31
B(e)P	150	110	110	50	40
B(ghi)P	210	150	150	71	56
Acenaphene	62	46	46	22	17
Indeno(c,d)Pyr	210	150	150	68	54
Acenaphylene	31	23	23	11	9
TCDD	0.02	0.02	0.02	0.01	0.01
PeCDD	0.03	0.02	0.02	0.01	0.01
HxCDD	0.10	0.07	0.07	0.04	0.03
HpCDD	1.5	1.1	1.1	0.52	0.42
OCDD	3.4	2.5	2.5	1.2	0.94
TCDF	0.37	0.27	0.27	0.13	0.10
PeCDF	0.19	0.17	0.17	0.07	0.05
HxCDF	0.69	0.50	0.50	0.24	0.19
HeCDF	0.15	0.11	0.11	0.05	0.04
OCDF	0.04	0.03	0.03	0.01	0.01
Hg	2200	1600	1600	720	570
Pb	67000	26000	10000	97000	48000
Cd	12000	8800	8800	4100	3300
As	17000	12000	12000	5800	4600

higher than at other IADN sites, which are more removed from urban areas (Hoff and Brice, 1994). For lead in 1992, air concentrations in urban Chicago were 28ng m^{-3}, while at more remote IADN sites on Lakes Superior, Huron, and Ontario, the concentrations were 2.7, 3.1, and 3.2 ng m^{-3}, respectively (a ten-fold difference between urban and rural areas). Until the late 1980s, alkyl-lead was widely used in automobile fuel as an additive to increase octane and protect engine valves. After its elimination from gasoline, air concentrations have dropped markedly (Reid et al. 1993), but in urban areas there are still sources of lead from the residual combustion of recycled oil from automobiles.

Chlordane (a pesticide used in ant and termite control) and DDT (a pesticide banned in North America in the 1970s) also show increased concentrations in urban areas. In 1992, levels of these pesticides in Chicago were five and four times higher than the regional IADN sites for these pesticides, respectively. This indicates that, in the past, residential use of these pesticides increased the burden in soils and buildings, and that volatilization from past use is still occurring.

The impact of pollution from urban areas on the lakes is poorly understood. The low deposition velocities of most particulate toxic chemicals and the low mass transfer velocity of organic gases would lead to the conclusion that pollution emitted from the cities will not significantly deposit during the few hours of travel time that the air takes to pass over the lakes and move on. There is evidence, though, that regionally high deposition of PCBs occurs around Chicago (Holsen et al. 1991), and the interpretation is that these PCBs are bound to very large particles, which have rapid deposition to the lakes in the immediate vicinity of the city. This conclusion has yet to be confirmed in other cities surrounding the Great Lakes.

Industrial Sources and Fugitive Emissions

For most of the persistent toxic chemicals in the Great Lakes (those with a lifetime in air of several days or more), current use has been regulated or discontinued. PCBs, DDT, toxaphene, chlordane, Pb, and As all have had stringent emission controls placed on their release to the environment. Combustion activity produces polynuclear aromatic compounds and dioxins, many of which are toxic. Controls on particulate emissions, however, have reduced particulate loadings in the air from the high values experienced in the 1960s. Without a doubt, the decreases in airborne concentration of these chemicals in the Great Lakes region is one of the major successes of the U.S. Clean Air Act, the Canadian Environmental Protection Act, and the Canada/U.S. Great Lakes Water Quality Agreement.

Nevertheless, many discontinued chemicals continue to be found in the lakes. Their removal to sediments and by revolatilization to the atmosphere is slow. There are continued ambient air concentrations of these chemicals, which help sustain the water concentrations. For most of the POPs, current inputs from spills and runoff are minor in comparison to the equilibrium between the air and water through gas exchange (Dolan et al. 1993).

Surveillance of fugitive emissions of these species continues to be necessary, however. Until a few years ago, road oil used to suppress dust on Ontario unpaved roads could contain up to 50ppm PCBs, and this was, potentially, a significant source of PCBs to the atmosphere. Anecdotal stories have been heard for many years about the contamination of current-use pesticides with a small

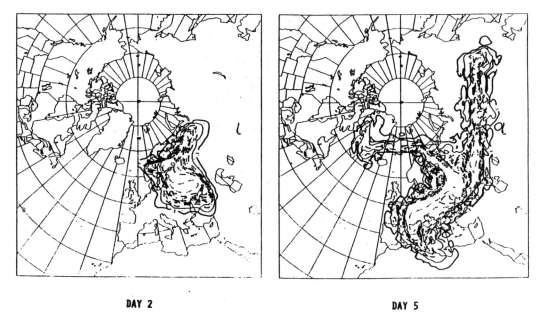

DAY 2　　　　　　　　　　　　**DAY 5**

FIG. 4. *Movement of the radioactive cloud from Chernobyl (a) 2 days after the release and (b) 5 days after the release (Pudykiewicz, 1989)*

amount (a few percent by weight) of restricted pesticide. These tales are difficult to verify and police. Clearly, if such residual emission practices are tolerated, the toxin problem in the Great Lakes will be further aggravated.

Agricultural Practice

Pesticides and herbicides are widespread in their use in agriculture. Current-use chemicals are regulated, and their usage is not allowed if they are persistent, bioaccumulate through the food chain, or are mobile through ground water transport. Unfortunately, non-persistence in soils can be demonstrated in a number of ways. If a pesticide disappears from soil it may be because it breaks down microbially to a more (or less) benign component (alachlor, for example, is believed to be degraded in three to six weeks by soil fungi (McEwen and Stephenson, 1979)). It may volatilize from soil. Recent studies of trifluralin and atrazine have indicated that significant fractions of the chemicals are emitted to the atmosphere with every rainfall. This is believed to be due to a wicking action of the water in the top layers of the soil (Scholtz et al. 1994). Thus, chemicals can be mistakenly stated to be nonpersistent because they disappear rapidly from soils, while in reality they have only moved on to other media.

Past-use persistent chemicals, however, are also a legacy of agriculture. Studies in England have shown that tilling of fields which had large deposition of PCBs decades ago, will release these PCBs as they are exposed to the surface (Alcock et al. 1993). These persistent chemicals provide a continuing source of chemicals to the air and to long-range transport.

Long-Range Transport (Ecosystem Recycling)

The Chernobyl Nuclear Reactor accident brought home to many the true size of our planet. Figure 4 shows the size of the radiation cloud from Chernobyl two and five days after the accident (Pudykiewicz 1989). Within two weeks, Canadians and Americans were exposed to excess radiation which was not of their making.

Recently, it has been discovered that Inuit in the Canadian, U.S. and Greenland Arctic are now exposed to toxaphene through their food supply (Kinloch et al. 1992). Toxaphene was widely used in the United States as a pesticide on cotton crops. Andy Hamilton of the International Joint Commission has commented that the levels of toxaphene found in Inuvik (a high Arctic settlement) are clearly out of proportion to the number of cotton fields in the area.

There is little doubt that long-range transport of air pollution is significant, and provides enough material to bioaccumulate through the food chain, even several thousand miles from the source of the chemical. This, plus the observation that chemicals can be reemitted by volatilization from soil, many years after their supposed elimination, leads to a pessimistic view about the ability of the ecosystem to purge itself of unwanted chemicals.

The volatilization-condensation process has been compared to a grasshopper, moving from one area to another throughout the year. Another analogy is a global still, evaporating chemicals from warm climes, and condensing them in cold (Wania and Mackay 1993). Yet the latter cold finger analogy is not perfect. PCBs, which have been measured in Arctic glacier snowpacks during one year are not there when the scientists return the next year (Gregor and Gummer 1989). The conclusion is that even in the Arctic, a significant fraction of these volatile chemicals evaporate from snow during summer months, and move on to recirculate throughout the Northern Hemisphere.

The evidence for long-range transport is powerful. Figure 5 shows the air movement corresponding to the five highest concentrations of toxaphene measured at Egbert, Ontario, during a one year study in 1988 (Hoff et al. 1993a). Each line on this diagram represents the path that air took for the five days prior to the toxaphene measurement. These trajectories pass over states in the United States which, in 1980, used significant quantities of toxaphene (Voldner and Schroeder 1989). Similar air trajectories have been seen for chlordane and DDT. In each case, high inputs of air concentrations of these banned chemicals are being tracked by the use of meteorological models thousands of miles back to areas where the known use was high.

Potential Controls

In North America, toxic chemicals are controlled via legislation. All chemicals that can be shown to be toxic in nature are subject to a variety of legislative controls, including production, sale, use, and disposal. It is clear, however, from the results shown earlier, that once a chemical is released into the environment, man has essentially no control over its movement. It is also clear that many chemicals currently in use may be found to have deleterious impacts in the future. This emphasizes the importance of understanding the negative impacts of chemicals before they are widely released into the environment.

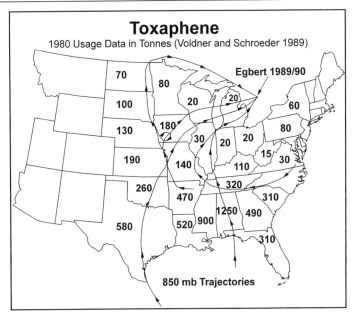

FIG. 5. *Air parcel trajectories arriving at Egbert, Ontario, Canada, on the days with the five highest toxaphene air concentrations during a study in 1988–89 (Hoff et al., 1993b). The toxaphene usage in 1980 is shown by state (Voldner and Schroeder, 1989).*

Virtual Elimination of Toxins

The Governments of Canada and the United States have agreed, through the Great Lakes Water Quality Agreement, to principles of zero discharge and virtual elimination of toxins. In practice, this means that toxic substances, which have a controllable source, should be controlled. What atmospheric studies have shown us is that after the chemical is out of the can, you can watch it all you want but you cannot control it. Movement of a chemical in a landfill might be measured in meters per year, in a lake at meters per hour, and in the air at meters per second. Clearly, common sense would indicate that you do not want to put a chemical into its most mobile compartment if you want to control it.

Similarly, one must be realistic about controls and the response of the system to their behavior. The response of the atmosphere to the elimination of nuclear testing, as reflected in the ^{137}Cs levels in the atmosphere, and the response of lead levels in eastern Canada to its elimination in gasoline (Reid et al. 1993) have taken over a decade to fully be seen. These responses to policy decisions to stop releasing a chemical to the atmosphere show that after you turn off the source, the air concentrations decrease slowly. The time constant for their removal from the atmosphere may be years to decades. Figure 6 shows the projected removal of PCBs from Lake Superior, given two scenarios of the removal by sediments at the bottom of the water column (Mackay et al. 1992). Toxic chemicals can be virtually eliminated, but politicians and government officials who are responsible for such changes should be aware that the changes take time. This means that management decisions must be made even though current information is incomplete or inconclusive. As an example, the Montreal Protocol to reduce the emissions of chlorofluorocarbons (CFCs) to

Time Response of PCBs in Lake Superior
Mackay et al. (1992)

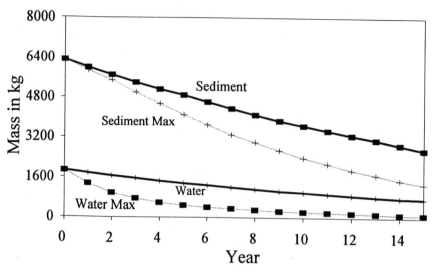

FIG. 6. *Predictions of the water concentration levels in Lake Superior with "virtual elimination" and two scenarios for the removal rate of material at the lake sediments (Mackay et al., 1992).*

protect the Earth's ozone layer was signed in 1987, without unanimity of the scientific community as to the eventual outcome. We have seen, however, that the level of CFCs in the atmosphere has now reached its maximum concentration and is starting to decrease. Recovery of the ozone layer may take decades. In the intervening years since the Protocol was signed, we have stronger evidence that ozone depletion is occurring, even over southern Canada (Kerr and McElroy 1993), reaffirming that the decision taken in the Montreal Protocol was a wise one.

Global Solutions to a Global Problem

Global transport of persistent pollutants has been demonstrated for over a decade. The observation of toxaphene in mother's milk of Inuit in the Arctic should be no more surprising than finding PCBs in the milk of Michigan mothers (Jacobson and Jacobson 1988).

Global solutions are required for a global problem. Reducing DDT in North America by restricting its use in North America is a positive achievement. However, DDT will not be eliminated until nearly a century after the last country stops using it. Third world use of pesticides is increasing, not decreasing.

The United Nations Economic Commission for Europe (UNECE) Executive Body approved the Aarhus protocol for persistent organic pollutants (POPs) in June 1998. By mid-1999, only Canada had ratified the protocol. Like the 1987 Montreal Protocol, which committed nations to reduce the use of fluorocarbons to protect the ozone layer, a protocol will be required to regulate the use of POPs internationally before local pollutant levels can be reduced to acceptable levels. This internationally coordinated effort is not a prerequisite to change which will improve the water quality in the Great Lakes. The reduction in PCB levels in fish is an indication that improvements occur when we can act locally. But without accelerated, concerted action on the global scale, we should expect that toxic chemicals will continue to persist in the Great Lakes region well into the next century.

Literature Cited

Achman, D. R., K. C. Hornbuckle, and S. J. Eisenreich. 1993. Volatilization of Polychlorinated Biphenyls from Green Bay, Lake Michigan. Environmental Science and Technology 27: 75–87.

Alcock, R. E., A. E. Johnston, S. P. McGrath, M. L. Berrow, and K. C. Jones. 1993. LongTerm Changes in the Polychlorinated Biphenyl (PCB) Content of UK Soils. Environmental Science and Technology 27: 1918–1923.

Arey, J., B. Zielinska, R. Atkinson, A. M. Winer, T. Ramdahl, and J. N. Pitts, Jr. 1986. The Formation of NitroPAH from the GasPhase Reactions of Fluoranthene and Pyrene with the OH Radical in the Presence of NOx. Atmospheric Environment 20: 2339–2345.

Baker, J. E., T. M. Church, S. J. Eisenreich, W. F. Fitzgerald, and J. R. Scudlark. 1993. Relative atmospheric loading of toxic contaminants and nitrogen to the Great Waters. Office of Air Quality Planning and Standards, US Environmental Protection Agency. Research Triangle Park, N.C.

——, S. J. Eisenreich. 1990. Concentrations and Fluxes of Polycyclic Aromatic Hydrocarbons and Polychlorinated Biphenyls across the Air Water Interface of Lake Superior. Environmental Science and Technology 24: 342–352.

Capel, P. D., C. Leuenberger, and W. Giger. 1991. Hydrophobic Organic Chemicals in Urban Fog. Atmospheric Environment 25: 1335–1346.

Chan, C. H. and L. H. Perkins. 1989. Monitoring of Trace Organic Contaminants in Atmospheric Precipitation. Journal of Great Lakes Research 15: 465–475.

Cohen, M. and 7 co-authors. 1995. Quantitative Estimation of the Entry of Dioxins, Furans and Hexachlorobenzene into the Great Lakes from Airborne and Waterborne Sources. Center for the Biology of Natural Systems, Queens College, City University of New York. Flushing, NY.

Dodd, J. A., J. M. Ondov, G. Tuncel, T. G. Dzubay, and R. K. Stevens. 1991. Multimodal Size Spectra of Submicrometer Particles Bearing Various Elements in Rural Air. Environmental Science and Technology 25: 890–903.

Dolan, D. M., K. P. McGunagle, S. Perry, and E. Voldner. 1993. Source Investigation for Lake Superior. International Joint Commission. Windsor, Ontario, Canada.

Egar, D. and V. Adamkus. 1990. Integrated Atmospheric Deposition Implementation Plan. Environment Canada and U.S. Environmental Protection Agency, 25 St. Clair Ave., Toronto, Ontario, Canada.

Eisenreich, S. J. and W. M. J. Strachan. 1992. Estimating Atmospheric Deposition of Toxic Substances to the Great Lakes An Update. Gray Freshwater Biological Institute, University of Minnesota, P. O. Box 100, Navarre, MN 55392.

Environment Canada and US Environmental Protection Agency. 1995. State of the Great Lakes. Environment Canada. En40–11/35–1995. 867 Lakeshore Road, Burlington, Ontario.

Glotfelty, D. E., M. S. Majewski, and J. N. Seiber. 1990. Distribution of Several Organophosphorus Insecticides and their Oxygen Analogues in a Foggy Atmosphere. Environmental Science and Technology 24: 353–357.

Gregor, D. J. and W. Gummer. 1989. Evidence of atmospheric transport and deposition of organochlorine pesticides and PCB in Canadian arctic snow. Environmental Science and Technology 23: 561–565.

Hildemann, L. M., D. B. Klinedinst, G. A. Klouda, L. A. Currie, and G. R. Cass. 1994. Sources of urban contemporary carbon aerosol. Environmental Science and Technology 28: 1565–1576.

Hoff, R. M., T. F. Bidleman, and S. J. Eisenreich. 1993a. Estimation of PCC loadings from the atmosphere to the Great Lakes. Chemosphere 27:2047–2055.

——, D. C. G. Muir, N. P. Grift, and K. A. Brice. 1993b. Measurement of PCCs in air in Southern Ontario. Chemosphere 27:2057–2062.

——, 11 co-authors. 1996. Atmospheric deposition of toxic chemicals to the Great Lakes: A review of data through 1994. Atmospheric Environment 30:3505–3528.

——, K. A. Brice. 1994. Atmospheric Dry Deposition of PAHs and Trace Metals to Lake Ontario and Lake Huron. 94RA110.04. Annual Meeting. Air Waste Management Association, Pittsburgh, PA.

Holsen, T. M., K. E. Noll, S. P. Liu, and W. J. Lee. 1991. Dry Deposition of Polychlorinated Biphenyls in Urban Areas. Environmental Science and Technology 25: 1075–1081.

Hornbuckle, K. C., D. R. Achman, and S. J. Eisenreich. 1992. Over Water and Over Land Polychlorinated Biphenyls in Green Bay, Lake Michigan. Environmental Science and Technology 27: 87–98.

International Joint Commission. 1988. Revised Great Lakes Water Quality Agreement of 1978. International Joint Commission, Windsor, Ontario, Canada.

Jacobson J. L. and S. W. Jacobson. 1988. New methodologies for assessing the effects of prenatal toxic exposure on cognitive functioning in humans. *In* M. Evans, ed. Pages 373–388 Toxic Contaminants and Ecosystem Health: A Great Lakes Focus. Wiley Interscience Series, New York.

Kerr, J. B. and C. T. McElroy. 1993. Evidence for large upward trends of ultravioletB radiation linked to ozone depletion. Science 262: 1032–1034.

Kinloch, D., H. Kuhnlein, and D. C. Muir. 1992. Inuit food and diet: a preliminary assessment of benefits and risks. Science of the Total Environment. 122:245–276.

Leister, D. L. and J. E. Baker. 1994. Atmospheric Deposition of Organic Contaminants to the Chesapeake Bay. Atmospheric Environment 28: 1499–1520.

Liss, P. S. and P. G. Slater. 1974. Flux of Gases across the Air Sea Interface. Nature 247: 181–184.

Mackay, D., S. Sang, M. Diamond, P. Vlahos, E. Voldner, and D. Dolan. 1992. Virtual elimination of toxic and persistent chemicals from Lake Superior: the role of simple mass balance models. International Joint Commission. Windsor, Ontario, Canada.

———, A. T. K. Yeun. 1983. Mass Transfer Coefficient Correlations for Volatilization of Organic Solutes from Water. Environmental Science and Technology 17: 211–217.

McEwen F. L. and G. R. Stephenson. 1979. The Use and Significance of Pesticides in the Environment. John Wiley & Sons, New York.

Nriagu, J. O. 1989. A Global Assessment of Natural Sources of Atmospheric Trace Metals. Nature 338: 47–49.

———, J. M. Pacyna. 1988. Quantitative assessment of worldwide contamination of air, water and soils by trace metals. Nature 333:134–139.

Oliver, B. G., M. N. Charlton, and R. W. Durham. 1989. Distribution, Redistribution and Geochronology of Polychlorinated Biphenyl Congeners and Other Chlorinated Hydrocarbons in Lake Ontario Sediments. Environmental Science and Technology 23:200–208.

Poster, D. L., R. M. Hoff, and J. E. Baker. 1995. Measurement of the particlesize distributions of semivolatile organic contaminants in the atmosphere. Environmental Science and Technology 29:1990–1997.

Pudykiewicz, J. 1989. Simulation of Chernobyl dispersion with a 3D hemispheric tracer model. Tellus 41B:391–412.

Ramdahl, T., B. Zielinska, J. Arey, R. Atkinson, A. M. Winer, and J. N. Pitts, Jr. 1986. Ubiquitous occurrence of 2nitrofluoranthene and 2nitropyrene in Air. Nature 321:425–427.

Rapaport, R. A. and S. J. Eisenreich. 1988. Historical atmospheric inputs of high molecular weight chlorinated hydrocarbons to eastern North America. Environmental Science and Technology 22:931–941.

Reid N. W., P. Kiely, and W. Gizn. 1993. Time trends in ambient lead in Ontario. *In* Heavy Metals in the Environment. R. J. Allan, J. O. Nriagu, editors. CEP Consultants, Edinburgh, Scotland.

Scholtz, M. T., E. C. Voldner, and E. Pattey. 1994. Pesticide volatilization model comparison with field measurements. 94MP 5. 03. Air and Waste Management Association, Pittsburgh, PA.

Schroeder, W. H. and D. A. Lane. 1988. The fate of toxic airborne pollutants. Environmental Science and Technology. 22:240–246.

Seinfeld J. H. 1986. Atmospheric Chemistry and Physics of Air Pollution. John Wiley and Sons, New York, N. Y.

Slinn, S. A. and W. G. N. Slinn. 1980. Predictions for Particle Deposition on Natural Waters. Atmospheric Environment. 14:1014–1016.

Strachan, W. M. J. 1985. Organic substances in the rainfall of Lake Superior: 1983. Environmental Toxicology and Chemistry 4:677–683.

———, S. J. Eisenreich. 1988. Mass Balancing of Toxic Chemicals in the Great Lakes: The Role of Atmospheric Deposition. International Joint Commission, Windsor, Ontario. Windsor, ON Canada.

———, H. Huneault. 1984. Automated rain sampler for trace organic substances. Environmental Science and Technology 18: 127–130.

———, 1995. Polychlorinated Biphenyls and Organochlorine Pesticides in Great Lakes Precipitation. Journal of Great Lakes Research 5: 61–68.

Swackhamer, D., B. D. McVeety, and R. A. Hites. 1988. Deposition and Evaporation of Polychlorobiphenyl Congeners to and from Siskiwit Lake, Isle Royale, Lake Superior. Environmental Science and Technology 22: 664–672.

Swain, E. B., D. R. Engstrom, M. E. Brigham, T. A. Henning, and P. L. Brezonik. 1992. Increasing rates of atmospheric mercury deposition in midcontinental North America. Science 257: 784–787.

Swain, W. and 6 co-authors. 1994. Exposure and Effects of Airborne Contamination. Office of Air Quality Planning and Standards, U.S. Environmental Protection Agency. Research Triangle Park, NC. .

U.S. Environmental Protection Agency. 1994. Deposition of Air Pollutants to the Great Waters: First Report to Congress. Office of Air Quality Planning and Standards, US Environmental Protection Agency. EPA–453/R–93–055. Research Triangle Park, NC.

Voldner, E. C. and W. H. Schroeder. 1989. Modeling of atmospheric transport and deposition of toxaphene into the Great Lakes ecosystem. Atmospheric Environment 23:1949–1961.

Wania, F. and D. Mackay. 1993. Global fractionation and cold condensation of low volatility organochlorine compounds in polar regions. Ambio. 22:10–18.

Allocation of Fishery Resources

CHAPTER 10

An Overview of Recreational Fisheries of the Great Lakes

James R. Bence and Kelley D. Smith

Introduction

Recreational fishing is an activity where anglers attempt to catch fish primarily for the enjoyment of the activity involved, although harvest and personal consumption of the catch is often an important motivation. Commercial sale of fish caught while recreational fishing is usually prohibited by law. Angling using rod-and-reel gear is the most common method of recreational fishing on the Great Lakes, although specialized fishing using dip nets and other gear types does occur.

By the early to mid 1960s, recreational fishing on the Great Lakes was insignificant because many target species, such as walleye, yellow perch, rainbow trout[1] and lake trout were either in scarce supply or regarded as unfit for human consumption, due to high contaminant concentrations (Hartman 1973; Wells and McLain 1973; Hatch et al. 1987; Keller et al. 1987; Keller et al. 1990; Lange et al. 1995). The sport fishing that did occur was generally based on or near shore, and there was little investment in access points and harbors, or in vessels that could safely venture far from shore. Specific causes for the collapse of fisheries in the Great Lakes are discussed in other chapters in this volume, and are summarized lake-by-lake later in this chapter.

Starting in the 1960s, successful control of sea lamprey, restrictions on commercial harvest, and massive stocking programs for lake trout and introduced salmonines initiated the modern period of recreational fishing in the Great Lakes. The resulting presence of large and attractive game fish led to substantial investment in vessels by individual anglers and charter operators, and in access sites by state and local governments (Dawson et al. 1989; Dann 1994). Other factors that helped fuel growth of these recreational fisheries included declining concentrations of toxic substances in fishes, resurgence of some yellow perch populations, and in some areas, restoration efforts for walleye.

Allocation of harvest to recreational fishing is generally believed to provide larger economic benefits than allocation to commercial fishing on the Great Lakes (Talhelm 1988). Great Lakes recreational fisheries have often been emphasized in areas near population centers, especially when species vigorously sought after by sport anglers are present. Recreational anglers exert

TABLE 1

Summary of estimated recreational fishing effort and expenditures in the Great Lakes system based on federal mail and phone surveys

For comparison, fishing effort and expenditures from other U.S. freshwater and U.S. saltwater fishing are included. The Great Lakes system as defined in these surveys included Lakes Michigan, Huron, Superior, Erie, St. Clair, and Ontario, their connecting waters and the St. Lawrence River. Estimates are based on results of a 1991 quarterly phone survey of U.S. residents (USFWS and Bureau of the Census 1993), and a 1990 annual mail survey of the Canadian fishery (DFO 1994).

CATEGORY	DAYS FISHING (MILLIONS)	EXPENDITURES (BILLIONS)
US residents – Great Lakes	25.3	1.34 (1991 $US)
Ontario residents – Canadian Great Lakes	10.5	0.26 (1990 $CAN)
US residents – other freshwater	430.9	15.15 (1991 $US)
US residents – saltwater	74.7	4.99 (1991 $US)

substantial effort fishing on the Great Lakes, although the exact magnitude of this effort remains uncertain. Estimates based on phone and mail surveys indicate that more than thirty-five million angler days and about $1.5 billion (U.S.) were spent fishing on the Great Lakes by United States and Ontario residents from 1990 to 1991 (Table 1).

Fishery and environmental managers on the Great Lakes are concerned about more than economic return from fishing and satisfaction of anglers. For example, some stakeholders are interested in restoration of populations of native species. Others desire habitat and environmental changes, such as increased water clarity, that may reduce fishery productivity. This broader set of concerns influences recreational fisheries, because there are limits to both lake productivity and resources that can be invested in management. Investments are being made in habitat restoration, stocking of introduced and native species, and control efforts for sea lamprey. Furthermore, harvest is being distributed among recreational, commercial, and tribal fisheries. To some extent, conflict between recreational anglers and commercial harvesters has been avoided by allocating to commercial fisheries those species which are not targeted by many sport anglers. As productivity limits of the Great Lakes have been pushed, however, some of these commercial species are becoming valued as forage for important sport species (O'Gorman and Stewart, 1999). Thus, new tradeoffs are continually arising, and debate is ongoing in the management community about allocation of resources and appropriate goals for the ecosystem.

Management of the Great Lakes ecosystem, including allocation of resources, should be based, in part, on understanding the status, magnitude, and trends in the recreational fishery. In this chapter we describe the recreational fisheries that operate on each of the five Great Lakes and their tributaries. Fishing on these five lakes constitutes most of the angling in the Great Lakes system (Table 2). Our synthesis of existing information is intended to provide a backdrop for ongoing discussions and decisions regarding management of Great Lakes' ecosystems.

Much of the material on recent fishery attributes which we present here is taken from agency reports, or from unpublished tables provided directly to us by scientists and resource managers monitoring these fisheries. This information is the best available for describing the current state of the fisheries of the Great Lakes, and is the same information that managers are using to make decisions about the fisheries. We caution, however, that not all the survey and estimation procedures used by these agencies has received the same degree of scientific peer review. Given the

TABLE 2

Estimates of fishing effort by water body/type within the Great Lakes by Ontario and United States residents based on federal mail and phone surveys

WATER BODY	ONTARIO RESIDENTS[1]		U.S. RESIDENTS[2]	
	DAYS FISHING (THOUSANDS)	PERCENT	DAYS FISHING (THOUSANDS)	PERCENT
Lake Michigan	—	—	5,090	21.1
Lake Superior	352	3.4	883	3.7
Lake Huron	4,579	43.6	2.113	8.8
Lake St. Clair	419	4.0	1,658	6.9
Lake Erie	1,497	14.3	7,084	29.4
Lake Ontario	3,397	32.4	2,394	9.9
St. Lawrence River	250	2.4	218	0.9
Great Lakes tributaries	—	—	1,616	6.7
Connecting waters	—	—	3,021	12.5

[1]Data from DFO (1994) based on a 1990 annual mail survey. Lake totals for Ontario residents are for the Canadian Great Lakes and include tributary fishing and fishing in associated connecting waters.

[2]Data from USFWS and Bureau of the Census (1993) based on a 1991 quarterly phone survey. Totals are for all fishing on the Great Lake, but tributary fishing and fishing in connecting waters are not included in lake totals. Effort does not sum to total Great Lakes effort because of "no response" and multiple responses. Percent values calculated as percent of the sum of days fished over all water bodies listed.

complexities of surveying recreational fisheries and analyzing the resulting data (Guthrie et al. 1991; Pollock et al. 1994), some estimated quantities such as harvest and effort might be revised as the agencies further evaluate their methods and data.

Because we use recent unpublished information, our description of these fisheries is sometimes markedly different than those based on historical, more widely available information. This reflects the dynamic nature of the Great Lakes recreational fisheries. We think our description of the current situation will provide an important reference point for future evaluations of how the fishery has changed.

In the remainder of this chapter, we first provide separate descriptions of recreational fisheries for each lake, followed by a section describing tools available to managers for protecting and enhancing fishery resources, and their use. We end with a discussion of emerging issues and future directions in the management of these sport fisheries.

Lake Superior

A significant recreational fishery exists on Lake Superior, but because fewer people live near its shores, recreational angling effort is less than on the other Great Lakes (Table 2). Ontario's shoreline makes up the northern and most of the eastern margin of the lake; Michigan's shoreline makes up a majority of the southern margin of the lake; and Wisconsin's and Minnesota's shorelines form the southwestern and northwestern shores (fig. 1). This lake is largely oligotrophic, with a predominantly cold-water fish community.

Lake trout are more important to anglers fishing on Lake Superior than to those fishing on the other Great Lakes. Nearly all anglers interviewed in Michigan Department of Natural Resources

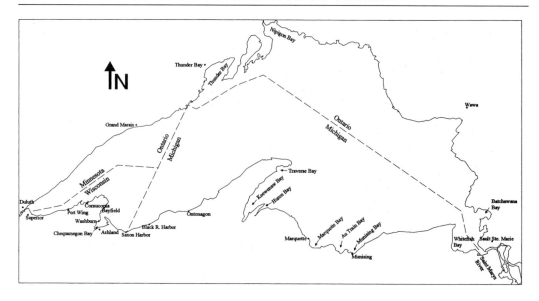

FIG. 1. *Schematic map of Lake Superior with major landmarks and jurisdictional boundaries.*

(Michigan DNR) contact creel surveys indicate that they are fishing for salmonines (Peck 1992), and lake trout constitute nearly 70% of the reported recreational harvest of major salmonines in U.S. waters of Lake Superior (fig. 2), and an even larger percentage of the charter harvest (Halpern 1993; Rakoczy and Svoboda 1993; Hulse 1994). This species is also the most commonly harvested by Ontario anglers on Lake Superior (W. MacCallum, Ontario Ministry of Natural Resources (OMNR), personal communication). Introduced salmonines, walleye, and lake and round whitefish also contribute to recreational harvests. Walleye continues to be a valued recreational species, but many stocks are at low levels (Schram et al. 1991; MacCallum et al. 1994). Exceptions are the Saint Louis River and Chequamegon Bay stocks in Wisconsin's waters. Although angling for other species such as yellow perch occurs, these species are less important on a lakewide basis.

Historical Background

Lake trout, rainbow trout, coaster brook trout, and to a lesser extent, walleye and yellow perch contributed to the early recreational fishery. All these species are native to Lake Superior, except rainbow trout, which became naturalized following introductions in 1895. There are numerous early reports of successful angling for lake trout (Herbert 1851; Roosevelt 1865; Brown 1876), and a substantial sport fishery developed beginning in 1926 off Munising, Michigan, following successful introduction of west coast trolling techniques (Lawrie and Rahrer 1973). Anecdotal reports indicate that lake trout dominated the recreational fishery until stocks collapsed during the 1950s. Even after collapse of the lake trout fishery had begun, 2,000 lbs. of lake trout were caught in one day by anglers at the Rossport Derby in 1953[2].

With the severe reduction in lake trout abundance during the 1950s, and depletion of many brook trout and walleye populations, recreational fishing activity declined substantially. Recreational fishing was described as insignificant near Thunder Bay, Ontario by the late 1950s[2]. Inter-

Recreational Harvest of Major Salmonines
from Lake Superior during 1993

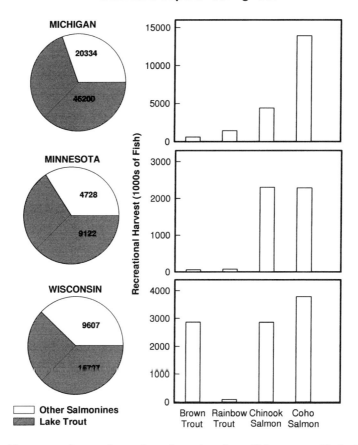

FIG. 2. *Recreational harvest estimates for major salmonines from U.S. waters of Lake Superior during 1993. Michigan estimates from unpublished table compiled by J. Peck (Michigan Department of Natural Resources) for Lake Superior Committee Meeting (March, 1994) agenda item 5. Wisconsin estimates from Hulse (1994) and Minnesota estimates from Halpern et al. (1994). Tributary fishing is not included because it was incompletely surveyed. Figures on pie charts are estimates of number of fish harvested.*

views by Michigan conservation officers indicate that yellow perch became dominant in the sport fishery during this time (Eschmeyer 1959).

Lake trout stocking rates were increased substantially in the early 1960s, subsequent to successful control of sea lamprey (Hansen et al. 1994a), and have continued at high rates since that time. Recreational fishing effort on Lake Superior increased in the 1960s and early 1970s in response to the stocking programs for trout and salmon in Lake Superior (Peck and Schorfhaar 1991). By 1962, improved lake trout fishing was attracting increasing numbers of anglers to western Lake Superior, and in particular, to the Keweenaw Bay area (Stauffer 1966). At that time nearly all harvested lake trout were stocked fish, most of which were caught by ice anglers. A year-round fishery for lake trout developed on Keweenaw Bay by 1964, and concerns were raised that this increased angling would jeopardize lake trout restoration efforts. Stauffer (1966) concluded, however, that recreational fishing on Lake Superior at that time was an insignificant source of mortality, based on a creel survey harvest estimate for Keweenaw Bay, and his expansion of this estimate to approximate lakewide harvest. This increase in angling on Keweenaw Bay was, however, only the beginning of a resurgence in recreational fishing on Lake Superior.

More recently, lake trout abundance has fallen from lakewide highs experienced during the 1970s, although timing of changes in abundance have differed among areas of Lake Superior (Hansen et al. 1994a, 1995). Declines appear to have resulted from poorer performance of hatchery fish and increased fishing mortality. In many areas, however, abundance of wild lake trout has increased since 1970 and these wild fish now dominate angler harvests.

In addition to wild populations of lake trout, and long naturalized populations of rainbow trout and brown trout, chinook salmon and coho salmon have become naturalized in Lake Superior over the past three decades. Most rainbow trout caught in Lake Superior are wild fish, although this species has also been stocked in recent years (Peck et al. 1994). Most harvested coho salmon are also wild. Michigan DNR was the first agency to plant coho salmon in Lake Superior in 1966, and although they plan to stock some coho salmon in one river in 1996, no agency still regularly stocks this species. Stocking of chinook salmon in Lake Superior began in 1967, also in Michigan's waters, and this species was subsequently stocked by Minnesota, Wisconsin and Ontario. In recent years, annual stocking has been at about 1.5 million spring fingerlings, but most harvested fish are of wild origin (J. Peck, Michigan DNR, personal communication). Stocking of lake trout and other salmonines is no longer as effective as it was originally, because survival of hatchery-released fish appears to have declined in areas where wild fish are abundant (Hansen et al. 1994b; Peck et al. 1994). Generally, wild fish dominate recreational harvests in the lake. Stocked anadromous fishes constitute most of the harvest in some streams where they are stocked, and stocked lake trout also tend to dominate harvest of lake trout in the eastern most part of Lake Superior (W. MacCallum, OMNR, personal communication), and in Minnesota's waters. The relatively low proportion of wild lake trout in Minnesota's waters (Hansen et al. 1995) may reflect the later start to the stocking program in that state and high mortality rates caused by sea lamprey (Hansen et al. 1995; D. Schreiner, Minnesota DNR, personal communication). The low proportion of wild lake trout at the eastern end of the lake may reflect high mortality caused by a tribal whitefish fishery in Michigan's waters (Hansen et al. 1995).

Current Fishery and Recent Trends

The sport fishery in U.S. waters is monitored through contact creel surveys and charter boat reporting systems. Monitoring of the recreational fishery in this way began in 1969 in Wisconsin, and expanded to other areas over time. Even today, harvest and effort are not surveyed for all portions of the fishery on an annual basis. For example, not all jurisdictions have surveyed tributary waters and ice fisheries each year. In addition, Michigan does not survey east of Munising, an area thought to support little recreational fishing, and Wisconsin has discontinued sampling certain cool-water sites to allow greater concentration on cold-water species. Faced with a long coast, many access points, and a generally low population density, Ontario uses creel surveys only in

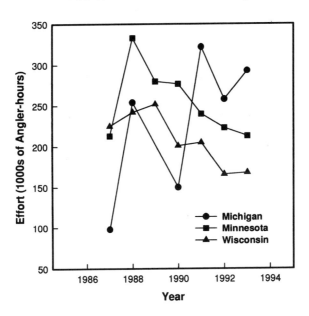

Recreational Effort on Lake Superior

FIG. 3. *Recreational fishing effort for U.S. waters of Lake Superior. Estimates do not include tributary fishing, Wisconsin cool-water fishing, or fishing in Michigan east of Munising. The only ice fishery included is that at Keweenaw Bay. Sources of Wisconsin and Minnesota estimates as in fig. 2. Michigan estimates from Rakoczy and Svoboda 1993.*

areas of special concern and at sport derbies. They also monitor the recreational fishery through evaluation of sport logbooks kept by volunteer anglers.

Each state contributes substantially to lakewide fishing effort in U.S. waters (fig. 3), and comparison of federal surveys (but see The Future of Recreational Fisheries) suggests that Ontario effort may be similar to that in other jurisdictions (table 2). In U.S. waters, fishing from boats is the dominant method reported to be used by anglers, although ice and shore angling are significant. The significance of the ice fisheries may be underestimated because not all ice fisheries are surveyed (J. Peck, Michigan DNR, personal communication). Ice fishing has generally been more important along the Canadian shore, but boat fishing effort has increased as greater investment in larger vessels has been made. Charter boats are an important part of the overall recreational fishery, contributing from 10% to 20% of all angler-hours during the open water season in U.S. waters and are also believed to contribute a significant, but unknown proportion of the total in Ontario (W. MacCallum, OMNR, personal communication).

In Wisconsin, open water angler-hours have declined substantially since 1989, and in Minnesota, effort peaked in 1988 and has declined since (fig. 3). Estimated effort in Michigan's waters has fluctuated, but trends are difficult to discern (fig. 3). Although substantial angling effort is targeted at anadromous fish in tributaries (Peck et al. 1994), only in Minnesota are data sufficient to evaluate recent trends, where effort increased during the late 1980s and has declined from this peak. The causes for these changes in effort are not well understood. The decline in stream fishing

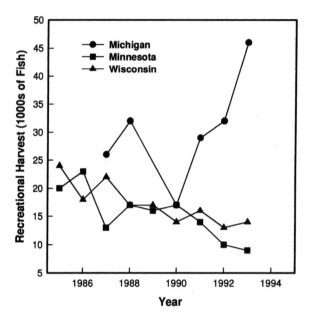

Recreational Harvest of Lake Trout from Lake Superior (U.S. Waters)

FIG. 4. *Recreational harvest of lake trout from U.S. waters of Lake Superior. Harvest includes all surveyed ice fisheries. Sources of estimated values as in figure 2.*

effort in Minnesota may reflect angler responses to the combination of more restrictive fishing regulations and low fish abundance (D. Schreiner, Minnesota DNR, personal communication). The open water decline in fishing effort in Wisconsin and Minnesota are not clearly tied to changes in fish stocks. Modest decline in harvest per unit effort (see below), which are partly regulation driven, could play a role, but larger unknown social causes may also be involved.

Each jurisdiction estimates total recreational yield of lake trout in spite of sampling difficulties described earlier. The 1991 lakewide recreational yield reported to the Great Lakes Fishery Commission was 152,000 kg dressed weight (Swanson et al. 1992), which constituted just over a third of the total lake trout yield, and is roughly equivalent to 108,000 fish. The largest reported yield was from Ontario, but this estimate required extensive extrapolation to unsampled areas (W. MacCallum, OMNR, personal communication). The majority of recreational harvest of lake trout from the U.S. side of Lake Superior comes from Michigan's waters (fig. 2). Annual number of lake trout harvested from surveyed areas in Michigan's waters have fluctuated nearly three-fold during the 1987 through 1993 period (fig. 4), but have been at high levels in recent years. In Minnesota's waters, lake trout harvests have declined since the mid 1980s (fig. 4). Annual harvest reported by Wisconsin has not fluctuated as much over time, but numbers harvested in recent years are lower than during the 1980s (fig. 4), probably resulting from more conservative regulations after 1985.

Harvest of other species of salmonines reported by Michigan, Minnesota, and Wisconsin has fluctuated over the past decade in response to changes in stocking rate, effort, and catchability.

The most important species were coho salmon, chinook salmon, brown trout, splake (brook trout crossed with lake trout), and rainbow trout (fig. 2). Recreational harvest of rainbow trout has declined to low levels in recent years, as a result of restrictions on harvest imposed by Minnesota's and Wisconsin's DNRs (Peck et al. 1994).

Harvest per unit effort (fish per angler-hour) of lake trout by Michigan's charter fishery has not shown substantial variation since it was first monitored in 1989 (Rakoczy and Svoboda 1993). Because this fishery concentrates on lake trout, it provides a good index of fishing success for lake trout on the southern shore west of Munising. There are reports of declining harvest per unit effort from some locations along the Ontario shore[3]. In Minnesota waters, charter harvest of lake trout per angler-hour is somewhat lower than in Michigan, possibly because the fishery targets a broader suite of species. Harvest per angler-hour declined during the late 1980s and early 1990s (Halpern 1993). In Wisconsin, harvest per hour is intermediate between Minnesota and Michigan. It peaked from 1982 through 1985, and declined to lower levels by 1988, possibly due to more restrictive regulations imposed after 1985. Recent harvest per angler-hour in Wisconsin has been about 70% of that seen from 1982 to 1985 (Hulse 1994).

Important areas of sport fishing in Ontario's waters include Thunder Bay, Nipigon Bay, and near Sault Ste. Marie in Whitefish Bay (including Batchawana Bay). Significant recreational fishing also occurs off WaWa in the northeast part of the lake. The ice fishery in Nipigon Bay targets lake trout, and to an increasing degree, lake whitefish[3]. During the open water season, this area is the most popular in Ontario for sport fishing because of high success rates for lake trout[3,4]. The eastern basin of Batchawana Bay has a history of high fishing pressure during the winter sport fishery, with yellow perch dominating the harvest. Recently, lake trout has also become a more important species there, although catch rates have been quite low[4]. Fishing in tributaries and off river mouths for anadromous salmonines is an important part of the recreational fishery in some regions within Ontario. Rainbow trout are important in spring, while chinook and coho salmon are important in fall.

One of the most active areas for sport fishing on Michigan's coast is off Marquette, due to the population center there, the presence of developed launch sites, and promotion of fishing by a local sport fishing organization (J. Peck, Michigan DNR, personal communication). Most fishing effort near Marquette is directed at lake trout and salmon, with an emphasis on coho salmon in spring, and lake trout in summer (Peck 1992). Substantial numbers of round whitefish are also caught in this area in October and November. A majority of in-lake effort is by boat anglers, but shore (approximately 16% to 20%) and ice fishing (1% to 14%) are also important (Peck 1992). The winter ice fishery is primarily for coho salmon. Fishing in tributaries near Marquette is most active in spring for rainbow trout, and fall for coho and chinook salmon. From 1984 to 1987, nearly a quarter of all angler-hours were spent on the tributaries (Peck 1992). Perhaps the largest concentration of shore fishing effort on Lake Superior exists at Marquette Bay.

Other areas with substantial sport fishing effort in Michigan include: Munising Bay, Au Train, Huron Bay, Keweenaw Bay, Traverse Bay, Ontonagon, and Black River Harbor. During the open water season (April-October), lake trout dominate sport harvests at most locations. Coho salmon are harvested in large numbers in April at many ports. The largest ice fishery occurs at Keweenaw Bay, which represents a majority of recreational fishing at this location. Excluding rainbow smelt, lake trout is the most abundant species in this fishery's harvest. There is a substantial ice fishery at Munising also, and lake whitefish and coho salmon are important in the harvest there.

The most important fishing ports on Wisconsin's shore of Lake Superior include: Saxon Harbor near the Michigan border, Ashland and Washburn in Chequamegon Bay, Bayfield, Cornucopia, Port Wing, and Superior near the Minnesota border. Somewhat surprisingly, approximately 15% of the harvest from the boat fishery operating from cold-water ports were walleyes during 1993, ranking this species third most abundant in the harvest. Walleye, yellow perch, northern pike, and smallmouth bass are important species in the cool-water fishery of Chequamegon Bay. There is also a significant ice fishery for yellow perch, splake, and brown trout in this bay.

Fishing activity in Minnesota centers on the southern shore near Duluth. Another area used by anglers is to the north near Grand Marias. As in other jurisdictions, the sport fishery in Minnesota is based on salmonines, with lake trout dominating harvests and coho salmon, chinook salmon, and rainbow trout also contributing. In the spring, stream fishing is almost exclusively for rainbow trout, while in the fall chinook salmon are of some importance. Lesser numbers of Atlantic salmon and pink salmon also contribute to the fall harvest. During the open water season, most fishing on the lake is from boats, but about one quarter of the effort is shore-based. Winter creel surveys indicate that harvests of rainbow trout in the ice fishery can sometimes be as large or larger than those in the open water fishery (Morse and Schreiner 1991; Peck et al. 1994).

Lake Michigan

Approximately 21% of all recreational fishing in the Great Lakes system by U.S. residents is on Lake Michigan, ranking it second in usage by U.S. residents among the five Great Lakes (Table 2). Lake Michigan receives substantial fishing pressure because of the concentration of urban centers (e.g., Gary, Chicago, Milwaukee), and because of investments in harbors and boat ramps (Dawson et al. 1989; Keller et al. 1990). Lake Michigan consists of Green Bay, Big Bay de Noc and Little Bay de Noc in the north, and the main body of the lake consisting of a distinct northern and southern basin (fig. 5). Grand Traverse Bay extends inland from the eastern edge of the northern basin. Michigan borders the northern portion of Green Bay, Big Bay de Noc, Little Bay de Noc, and the entire eastern edge of the lake. Indiana borders the southern tip, and Illinois the southwest edge of the southern basin. Wisconsin borders on the western edges of both the southern and northern basins, as well as the southern portion of Green Bay. In all jurisdictions, most fish harvested by anglers are yellow perch, but lakewide salmonines draw a larger portion of angling effort. Coho salmon, chinook salmon, lake trout, rainbow trout, and brown trout are all important to the salmonine sport fishery. Lake Michigan's sport fishery is largely supported by an extensive stocking program for salmonines; significant production of wild lake trout has not been reported, and significant natural reproduction of introduced salmonines is limited largely to the east side of the lake where migrating salmon have access to suitable stretches of streams and rivers.

In recent years, Lake Michigan's sport fishery has undergone substantial changes. Lakewide fishing effort has fallen in excess of 60% from peak levels seen in 1987, and salmonine fishing effort has shifted away from chinook salmon toward other species. These changes are, in part, a response to substantial declines in fishing success for chinook salmon.

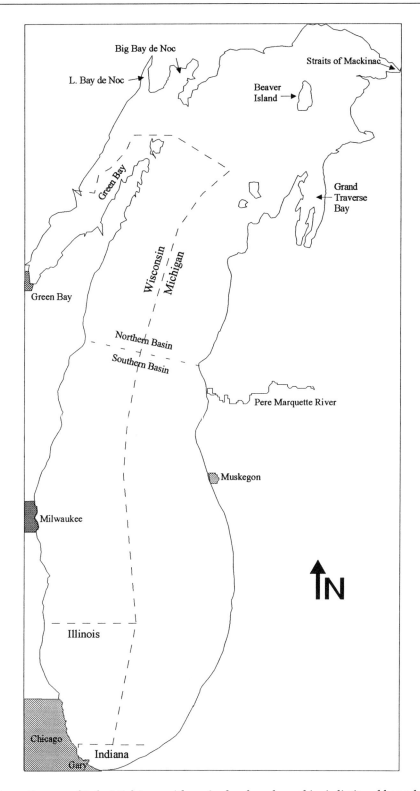

FIG. 5. *Schematic map of Lake Michigan with major landmarks and jurisdictional boundaries.*

Historical Background

Yellow perch were probably the most important species in the early sport fishery of Lake Michigan. Smith and Snell (1891) report extensive pleasure fishing in the Chicago area in 1885, and yellow perch were caught from shore as well as from small vessels near shore. This species sustained the sport fishery through the early 1960s. Trolling fisheries for lake trout developed in some areas of the northern basin of Lake Michigan prior to 1950 (Wells and McLain 1973). There was also fishing for brook trout in tributary streams, and the Pere Marquette was considered one of the better trout streams in the country before 1900 (Keller et al. 1990). Rainbow trout were stocked as fry in streams to enhance sport fishing opportunities, and by 1898 naturalized populations were attracting attention from sport fishers (Foster 1963). Brown trout were also introduced in streams before the turn of the century as a sport fishing alternative, although they were not initially as popular a target as brook trout. Walleye has been a prized sport fish in Lake Michigan; however, populations were of modest abundance over most of this century, and have supported only localized fisheries. Several very strong year classes in the late 1940s supported a pulse of concentrated sport fishing for this species in Green Bay during the 1950s (Wells and McLain 1973). Sport fishing for rainbow smelt in Lake Michigan occurs near-shore and in streams, mostly by dip netting of spawning runs in the spring, and by hook-and-line fishing through the ice in the winter. This fishery was probably at its peak in 1942, prior to a massive die-off that greatly reduced abundance of this species (Wells and McLain 1973).

During the 1950s and 1960s, changes in Lake Michigan's fish community, which appear to have been driven by the combined effects of sea lamprey parasitism and commercial fishing, had substantial effects on the sport fishery (Wells and McLain 1973). Burbot and lake whitefish declined in abundance, and wild lake trout were extirpated from the lake. All but the smallest species of deep water cisco disappeared from the system during this period. The resulting scarcity of native predators was a contributing factor to a substantial increase in abundance of the exotic alewife during the 1960s. This increased abundance eventually led to massive die-offs of alewives, with dead fish littering beaches in 1967. Yellow perch catch rates began declining during the early 1960s, probably in response to adverse effects of alewife (hypothesized to be competition for food and egg predation) as well as harvest by an intense commercial fishery. By the mid 1960s, the lake was supporting little sport fishing.

Management agencies bordering Lake Michigan responded to this ecological crisis with a massive control program for sea lamprey. Following the initiation of this program, agencies began a program to stock substantial numbers of salmonine predators into the lake during the mid 1960s. These stocking programs were motivated by multiple goals, including reestablishment of reproducing lake trout populations, development of quality sport fisheries for introduced salmonines, and control of alewife populations (Stewart et al. 1981; Hansen et al. 1990; Keller et al. 1990). Total numbers of salmonines stocked each year increased through the 1960s and 1970s (Stewart et al. 1981; Hansen et al. 1990; Keller et al. 1990), with most of the increased number after 1970 being chinook salmon. Since the early 1980s, stocking rates have remained relatively constant, although modest increases and decreases in lakewide stocking have occurred in response to concerns about salmonine survival and overexploitation of the forage by piscivorous fish (Stewart et al. 1981; Keller et al. 1990; Stewart and Ibarra 1991).

The rise of the modern Lake Michigan salmonine fishery was reviewed by Hansen et al. (1990) and Keller et al. (1990). From 1969 through 1985, salmonine catch rates in Wisconsin waters doubled, and angler effort increased an order of magnitude leading to a twenty-fold increase in salmonine harvest (Hansen et al. 1990). Similar, but less well documented changes were seen throughout the lake. Furthermore, yellow perch were recruited in large numbers during the early 1980s, and again supported an important recreational fishery. These changes resulted in development of a multimillion dollar sport fishery, with substantial investments in access points, and both charter and private fishing boats (Keller et al. 1990; Dawson et al. 1989). As investments in fishing vessels increased, trolling became an increasingly common fishing method. The contribution of lake trout to the harvest declined during the early 1980s as anglers increasingly targeted on chinook salmon (Hansen et al. 1990; Keller et al. 1990). Lake trout continued to be an important component of the sport harvest, however, especially when the Pacific salmonines were scarce.

A dramatic change in Lake Michigan's salmonine fishery first became apparent in Michigan's waters in 1987. Lakewide fishing effort, and chinook salmon harvest and catch per unit effort began declining during the late 1980s (Keller et al. 1990; Stewart and Ibarra 1991, Rakoczy 1992). Spring die-offs of large chinook salmon were observed in southern Lake Michigan in 1988 and 1989 (Nelson and Hnath 1990). The cause of death was from bacterial kidney disease (BKD), and this increased mortality led to declines in harvest and harvest per unit effort of chinook salmon. Fishing opportunities have been further impaired by a decline in yellow perch abundance due to weak year classes during the 1990s, and the resulting imposition of more restrictive regulations on yellow perch harvest (Francis et al. 1996).

The Current Fishery and Recent Trends

Contact creel surveys, which measure effort as it occurs, indicate that total angling effort on Lake Michigan exceeds eight million hours and that lakewide angling effort has declined since 1987 (fig. 6). A decline in effort has occurred in the waters of each of the four states. Substantial uncertainty remains about the size of the recreational fishery. One indication of this uncertainty is that effort estimates from the creel surveys do not agree with results from phone surveys run by federal agencies. The creel survey estimate of effort for 1991, equivalent to approximately 2.7 million angler days of effort, is little more than half the effort for that year estimated by the phone survey (Table 2). Our view is that the creel surveys provide a more accurate accounting of the magnitude of Lake Michigan's recreational fishery. While these contact creel surveys do not provide complete coverage of some components of the fishery, it seems unlikely that they miss half the effort.

The Lake Michigan sport fishery of the 1990s has principally focussed on yellow perch and a variety of salmonines. Lakewide annual sport harvest of yellow perch was near 7.5 million fish in 1993. This species is the most common in angler harvests in all jurisdictions. Yellow perch harvest became increasingly dominated by older fish, following a series of weak year classes during the 1990s (Francis et al. 1996), but declines in lakewide sport harvest of yellow perch during the early 1990s were modest (fig. 7).

The most dramatic trend in the salmonine fishery has been a decline in sport harvest of chinook salmon (fig. 7). Associated with this decline in harvest has been substantial decreases in harvest per angler-hour. It is likely that this poorer success in fishing for chinook salmon has led to some

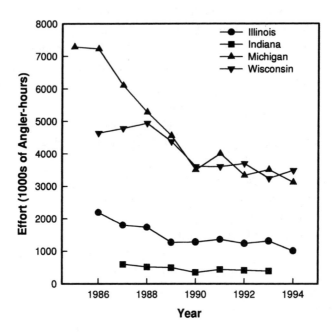

Recreational Effort on Lake Michigan

FIG. 6. *Recreational effort on Lake Michigan. Estimates for Michigan are from unpublished Michigan Department of Natural Resource tables based on estimates in published reports on charter and creel harvest (e.g., Rakoczy and Svoboda 1995). Charter values from Illinois are from unpublished Illinois Department of Conservation tables and non-charter Illinois estimates are from Marsden and Brofka (1994). Estimates for Indiana are from Francis (1994). Wisconsin estimates from unpublished Wisconsin Department of Natural Resources tables. Indiana in-lake estimate is for four major access sites and is not total for all Indiana waters. Michigan estimates do not include effort in tributaries.*

of the decline in angling effort on Lake Michigan, and may have also acted to redirect effort toward other salmonines, such as lake trout and rainbow trout. Perhaps as a result, harvest per angler-hour for these latter two species increased during the late 1980s and early 1990s. Improved angling success for rainbow trout is associated with increased participation in an offshore fishery, which targets fish that have aggregated to feed on terrestrial insects concentrated along a temperature break. The end result of these trends is a sport salmonine fishery that is now more diverse than that of the mid 1980s, but it is also substantially smaller (fig. 8).

Michigan has the longest Lake Michigan shoreline. Sport angling in Michigan's waters of the lake is dominated by boat fishing, which comprises 85% of all lake angler-hours (Rakoczy 1992; Rakoczy and Svoboda 1994, 1995). A number of important tributaries extend quite far inland, and provide additional fishing for anadromous salmonines. There is a substantial ice fishery in Little Bay de Noc for yellow perch and walleye, and a smaller one in Little Traverse Bay for lake whitefish.

The greatest harvests and harvest per angler-hour of yellow perch in Michigan's waters have been reported for anglers fishing in the southern basin, and from Little and Big Bay de Noc in the

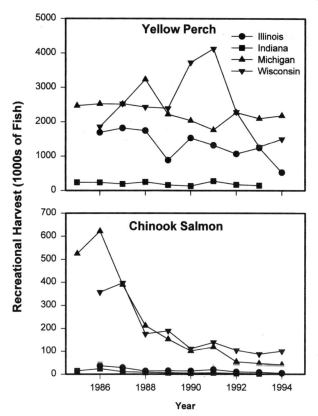

FIG. 7. *Recreational harvest of yellow perch and chinook salmon from Lake Michigan. Sources of estimates as for Figure 6. Indiana in-lake estimate is for four major access sites and is not total for all Indiana waters. Michigan estimates do not include effort in tributaries.*

north. There is also a walleye fishery in Little Bay de Noc, and most of Michigan's harvest of this species occurs there. Smallmouth bass are important in a few locations in Michigan's waters of the northern basin.

Nearly twice as many anglers target salmon and trout compared to yellow perch in Michigan, even though larger numbers of yellow perch are harvested (Rakoczy 1992; Rakoczy and Svoboda 1994, 1995). More than 90% of the resulting in-lake salmonine harvest comes from boat fisheries, and five salmonine species contribute to the harvest (fig. 8). The coho salmon fishery is most important in Michigan's southern basin in the early spring, after which this species migrates to the south and west. Coho salmon become important again in late summer and fall as they return to spawning streams and stocking locations. Other salmonines are important components of fisheries in both the northern and southern basins. Significant chinook salmon harvest first occurs during late spring along Michigan's southern coast, and significant harvests begin at progressively later

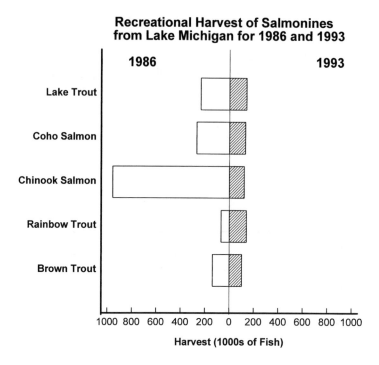

FIG. 8. *Comparison of 1986 and 1993 in-lake harvest of salmonines by the recreational fishery from Michigan's and Wisconsin's waters of Lake Michigan combined. Sources of Michigan and Wisconsin estimates as for fig. 6.*

dates to the north, following the northward movement of fish. Lake trout and rainbow trout attracted more attention by anglers in recent years as chinook salmon became less abundant. The offshore fishery for rainbow trout, in particular, has increased dramatically as anglers have learned to locate and target this species. The resulting fishing pressure has led the Michigan DNR to consider new regulations for rainbow trout.

An extensive system of tributaries in Michigan provides opportunities for recreational fishing for salmonines in streams. This spatially-extensive fishery has, however, not been surveyed regularly. A survey of the most important tributaries in 1985 indicated that fishing effort on streams and rivers with spawning runs of anadromous species equaled approximately 15% of the in-lake effort, and significant late summer-fall stream fisheries took 20% of Michigan's chinook salmon and 10% of the coho salmon sport harvests (Rakoczy and Lockwood 1988). Recent calculations indicate that more rainbow trout are still harvested from Michigan's tributaries than from Michigan's waters of the lake (K. Smith, unpublished). A fall-winter stream fishery for rainbow trout begins in September and extends through March and April, and summer fisheries for stocked Skamania strain occur during July and August.

Indiana has the smallest share of Lake Michigan's shoreline, and much of this area is relatively urban. Boat fishing is the primary mode of angling in Indiana, but more than 10% of all effort is attributed to shore fisheries, and more than a quarter is executed by stream anglers (Francis 1994).

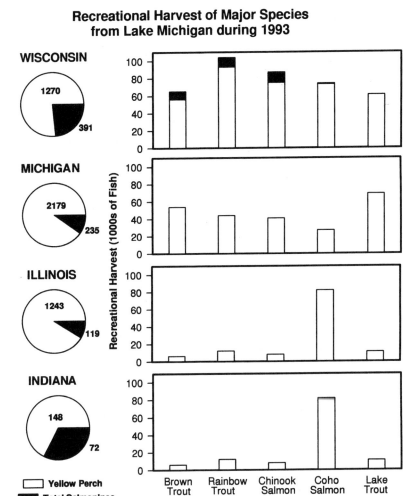

Recreational Harvest of Major Species from Lake Michigan during 1993

FIG. 9. *Recreational harvest for major species from Lake Michigan by jurisdiction during 1993. Sources of estimates as for fig. 6. Figures on pie charts are numbers of fish harvested in thousands. There is no significant tributary fishing in Illinois and harvest is presumed to be zero. Indiana in-lake estimate is for four major access sites and is not total for all Indiana waters. No estimates of stream harvest are available for Michigan.*

Harvest rates are substantially higher from boats, and more than 90% of the harvest is taken using this method. The salmonine fishery is dominated by coho salmon (fig. 9), with most fish harvested in the spring. Lake trout and rainbow trout also contribute to Indiana's sport harvest, and rainbow trout is the dominant species in stream harvests.

Chicago occupies one-third of Illinois' shoreline, which is highly urbanized (Brofka and Marsden 1994). Shore angling, primarily from piers and in harbors, is the dominant fishing mode, and accounts for about 60% of the angler-hours. Yellow perch make up nearly 99% of the harvest by shore anglers, and 80% of the harvest by boat anglers. Salmonines make up nearly all the remaining harvest, with coho salmon dominating (fig. 9), and being harvested in significant numbers from April through July.

Boat anglers contribute 70% of all angler-hours in Wisconsin and most of this effort is from private launched vessels (Wisconsin DNR, unpublished tables). Shore fishing and stream fishing each contribute about 15% of the effort. Most tributaries are blocked by dams near their outflows into the lake, but concentrated fisheries on stocked anadromous salmonines still occur. There is significant harvest of yellow perch, and five species of salmonines contribute to Wisconsin's sport harvest (fig. 9).

Wisconsin's waters of Green Bay support a sport fishery for yellow perch, and a significant sport fishery for walleye has developed as a result of stocking efforts. Within Green Bay, salmonine fisheries are largely restricted to cooler waters of the northern portion of the bay, with brown trout being most numerous in the harvest.

Although the majority of yellow perch harvested by sport fishers in Wisconsin's waters comes from Green Bay, significant numbers are also harvested in the southern basin. Much of the sport fishing effort in Wisconsin's waters of the northern basin is directed at salmon and trout, and significant fishing effort for salmonines also occurs in the southern basin. Brown trout is especially important to sport anglers fishing near Milwaukee, and sport fishing for coho salmon is centered in the southern basin. The chinook salmon fishery shows a seasonal pattern like that in Michigan, with substantial harvests beginning later in the season at northern sites.

Lake Huron

Recreational fishing on Lake Huron plays an important role in the overall sport fishery of the Great Lakes, especially in Canada, where 44% of all days fishing the Great Lakes are spent on this lake (Table 2). The high fraction of the Canadian effort on Lake Huron probably results from a combination of high-quality fishing, low contaminant concentrations (in contrast with Lake Ontario), and a high-density of temporary cottage residents on Georgian Bay, many of whom come from the high-population density area bordering the west side of Lake Ontario (D. Reid, OMNR, personal communication). Lake Huron receives inflow from Lake Superior and Lake Michigan, and is generally regarded as consisting of a main basin, Saginaw Bay, a north channel, and Georgian Bay (fig. 10). Michigan forms the western border of Lake Huron including Saginaw Bay. Ontario borders the eastern and northern shores, including most of the north channel and all of Georgian Bay.

In Michigan, the sport fishery for yellow perch and walleye is important in Saginaw Bay, while most effort is directed at salmonines outside of Saginaw Bay. Chinook salmon make up most of the salmonine harvest, with lake trout, rainbow trout, and brown trout also contributing substantial numbers to the harvest. There is some fishing for northern pike, smallmouth bass, and channel catfish, particularly in Saginaw Bay. Rockbass, pumpkinseed sunfish, brown bullhead, and freshwater drum are also caught in moderate numbers in some areas.

In Ontario's waters, habitat for smallmouth bass is relatively plentiful, and this species has attracted significant sport effort (DFO 1994). Fishery managers believe this species probably sup-

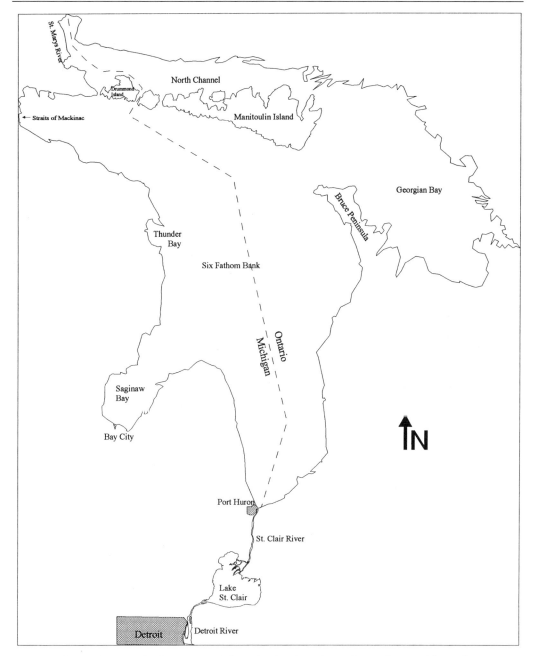

FIG. 10. *Schematic map of Lake Huron with major landmarks and jurisdictional boundaries.*

ported much of Ontario's sport fishery effort (LHMU 1994). Walleyes, yellow perch, and salmonines also attract significant effort. Among salmonines, chinook salmon is the most heavily harvested by the sport fishery, although rainbow trout and lake trout also contribute significantly to the harvest.

Historical Background

Because vessels large enough to be safe on the open lake were uncommon, much of the early recreational fishing activity (before 1950) was restricted to areas close to shore, and to Saginaw Bay and Georgian Bay (Keller et al. 1987). Warm and cool-water fishes were important in the sport fishery in the major bays and in other shallow near shore locations. These species included yellow perch, smallmouth bass, largemouth bass, and northern pike. Lake herring were an important sport species also. Walleye were of some importance in tributary streams, but tended to be less important in the lake because vessels needed to safely fish this species were not generally available (Keller et al. 1987). Lake trout were relatively unimportant to the sport fishery for the same reason.

The fish community of Lake Huron has undergone dramatic changes during the twentieth century, due to habitat alterations, fishing, introductions of exotic species, and subsequent restoration efforts. These changes have had a profound influence on development of the recreational fishery. Habitat impairment (due to pollution and siltation), combined with intense commercial fishing, caused collapse of walleye populations in Saginaw Bay during the late 1940s and early 1950s (Schneider and Leach 1977; Keller et al. 1987). These factors, along with sea lamprey parasitism, caused lake herring populations to collapse. Fishing and sea lamprey parasitism have been cited as causes for the collapse of burbot and lake trout populations, leading to greatly reduced predation pressure on exotic alewives and rainbow smelt. By 1966, lake trout were driven to extinction in Lake Huron except for two small populations (Berst and Spangler 1973). Resulting increases in abundance of alewives and rainbow smelt may in turn have had negative influences on yellow perch and other species through predation or competition (Smith 1970; O'Gorman and Stewart 1999).

Responses to deterioration of Lake Huron's ecosystem included sea lamprey control on a lakewide basis beginning in 1968, restriction of commercial fisheries, and stocking of native and introduced predator species. Substantial numbers of coho and chinook salmon were stocked beginning in 1968 (Ebener 1995). For reasons that are not clear, coho salmon did not do well in the lake, and stocking was terminated in 1989. Stocking of chinook salmon, and subsequent natural reproduction (especially in Canadian waters) continues to support a vigorous recreational fishery. Stocking of lake trout in U.S. waters began in 1973, and by the mid 1980s, lake trout seemed to be on the verge of a recovery similar to that seen in Lake Superior. Limited reproduction by lake trout has been documented at Thunder Bay, Six-fathom Bank, and South Bay (Manitoulin Island, Ontario) (Johnson and VanAmberg 1995; Eshenroder et al. 1995). Since 1986, however, lake trout populations have suffered a reversal, especially in the northern part of the lake, which contains the best lake trout spawning habitat (Eshenroder et al. 1995). An increase in sea lamprey abundance and tribal fishing activity appear to be responsible (S. Sitar, Michigan State University, unpublished). The increase in sea lamprey abundance has occurred because of large-scale reproduction of this species in the St. Mary's River.

Other salmonines stocked and harvested in significant numbers include brown trout and rainbow trout. Pink salmon are not stocked, and were first noted in the lake in 1969 (Parsons 1973), entering through the St. Mary's River from Lake Superior. This species contributed substantially to recreational fisheries during some odd-numbered years, but recent harvests have been low.

Stocking of walleye fingerlings has helped rebuild the Saginaw Bay walleye stock. Stocking began in 1978 and increased through 1982, resulting in the development of a substantial sport fishery. The sport fishery for yellow perch in Saginaw Bay has historically shown large fluctuations (Keller et al. 1987).

Current Fishery and Recent Trends

Most of the quantitative information on harvest and fishing effort comes from Michigan's waters. Creel surveys are done regularly on southern Georgian Bay, where much of the Ontario sport fishing activity is thought to occur, and estimates of Ontario's main basin recreational harvest were made from creel surveys done in 1989 and 1990. Most fishing on Lake Huron is done from private boats, with this mode contributing 85% of the effort during the open water season in Michigan's waters (Rakoczy and Svoboda 1992, 1994, 1995). Charter fishing contributes a relatively small amount of the total effort on Lake Huron, on the order of 2% to 3% in Michigan's waters during recent years, and walleyes, lake trout, and chinook salmon are primary targets. Ice fishing, primarily on Saginaw Bay, contributes nearly 20% of the total fishing effort in Michigan's waters (Rakoczy and Svoboda 1995).

A majority of the human population bordering Lake Huron is in the Saginaw Bay vicinity, and thus, a substantial sport fishery occurs in this area. Much of the sport fishing effort on Saginaw Bay is directed at walleyes and yellow perch. The abundance of walleyes in Saginaw Bay increased substantially during the 1980s, as large numbers of fingerlings were stocked. Although large numbers of hatchery-reared walleyes are stocked, contribution of natural reproduction to the Saginaw Bay walleye population was confirmed in 1993, when a small year class was produced in the absence of stocking. The proportion of the Michigan sport effort that occurs in the Bay increased substantially with the resurgence of walleyes. From 1969 to 1982, mail surveys indicated that 24% to 43% of the effort in Michigan waters was on Saginaw Bay (Keller et al. 1987). By 1986, a contact creel survey indicated that this increased to 60%, and the Saginaw Bay percentage was about this level in 1993.

In 1983 and 1984, 98% of sport harvest from Saginaw Bay was yellow perch (Ryckman 1986). Since 1983, effort on Saginaw Bay has varied between 1.0 million and 2.6 million angler-hours, and walleyes now make up a larger portion of the harvest (fig. 11). Harvest of yellow perch in Saginaw Bay has varied greatly, falling to a recent low in 1993 (fig. 12), then recovering in 1994 (Michigan DNR, unpublished data). Harvest rate of yellow perch declined precipitously between 1983 and 1986, from 2.0 to less than 1.0 per angler-hour, probably because effort was redirected toward walleyes. Harvest rate declined further to about 0.55 from 1991 through 1993, and this probably reflects a combination of declining abundance due to weak year classes and redirection of effort toward other species. Harvest rates for yellow perch increased in 1994 for reasons that are not yet clear.

Most of the sport fishing in the main basin is aimed at salmonines, and most (92%) of Michigan's harvest of salmon and trout from Lake Huron comes from this area. Total effort and harvest have not been sampled in many years, but trends in Michigan's waters can be inferred from data collected at index ports. These data indicate an overall decline in effort in the main basin since 1986 (fig. 11). There is some fishing for walleyes in the main basin, but this is relatively limited. In

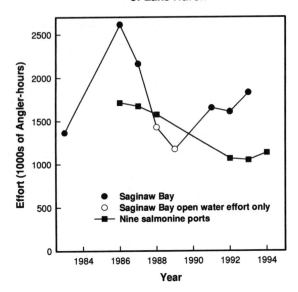

FIG. 11. *Recreational fishing effort for Saginaw Bay and at nine regularly sampled salmonine ports on Michigan's waters of Lake Huron. Saginaw Bay data include the ice fishery when possible. Estimates from Rakoczy and Svoboda (1995). Charter harvest is for calender year, while non-charter estimates are for a April 1–March 31 year.*

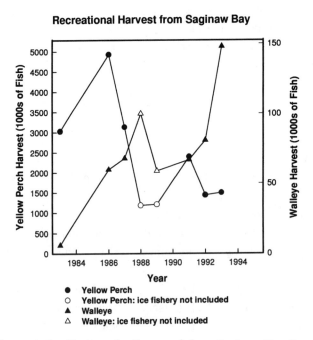

FIG. 12. *Recreational harvest of walleyes and yellow perch from Saginaw Bay. Sources of estimates as in fig. 11.*

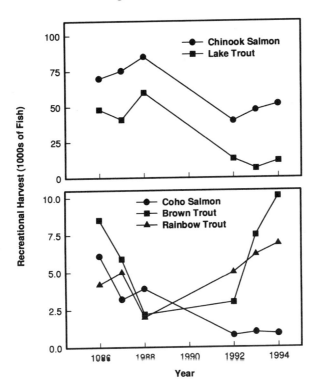

Recreational Harvest of Salmonines from Michigan's Waters of Lake Huron

FIG. 13. *Recreational harvest of salmonines from nine index ports in Michigan's waters of Lake Huron based on charter and non-charter anglers during April–October. Table after Rakoczy and Svoboda (1995).*

contrast with the increased abundance of Saginaw Bay walleyes, walleye stocks in Canadian waters of the main basin (and Georgian Bay) have declined in abundance (D. Reid, OMNR, personal communication).

Chinook salmon were the most abundant salmonine in Michigan's harvest every year in which sampling occurred (fig. 13). Harvest of chinook salmon per angler-hour varied without trend, and harvest declined in response to lower effort. There have been some reports of declines in fishing success for chinook salmon in the mid lake, but available data are equivocal. There appears to be a north-south migration of chinook salmon, so that the fishery is strongest in late spring in the southern part of the main basin, and fishing activity moves north as the season progresses. Fishing for chinook salmon occurs in Ontario's waters of the main basin, especially in the Manitoulin Island and Bruce Peninsula areas. In 1989 and 1990, chinook salmon dominated Ontario's recreational harvest from the main basin, to an even greater extent than is the case in Michigan (OMNR unpublished results). Harvest in those years suggests that Ontario harvest of chinook salmon from the main basin is about 10% of that in Michigan's waters. There is evidence of reproduction by chinook salmon in many Ontario streams, and also in some Michigan streams (Johnson et al., 1995). Returns of stocked fish to the fishery tend to be highest for fish stocked in northern areas (Michigan DNR, unpublished data). In addition to the sport fishery, chinook salmon are harvested commercially at harvest weirs and by tribal gill netters. Michigan's sport fishery contributed about half the total Michigan harvest of stocked chinook salmon in the mid 1980s, but its contribution has declined to 26%. This change reflects an increase in Native American commercial harvest, resulting from increased stocking in northern waters where this fishery is located.

Significant numbers of brown trout, rainbow trout, and lake trout are also harvested by sport anglers from Michigan's waters of the main basin (fig. 13), and pink salmon harvest was generally significant in odd-numbered years, but has been lower in recent years (Rakoczy and Svoboda 1995). The lake trout fishery continues throughout the open water fishing season. Nearly all lake trout stocked into the main basin are stocked on the U.S. side. Because of the tendency of lake trout to remain in the general vicinity where they are stocked, sport fishing for lake trout in Ontario's waters of the main basin is limited. Fishing for rainbow trout is concentrated in the spring. A significant portion of fishing for rainbow trout occurs in tributaries during spawning runs. Since these rivers are not surveyed, Michigan's creel survey underestimates the total harvest of rainbow trout from Lake Huron.

A notable decline in lake trout and coho salmon harvest, and harvest rate from Michigan's waters of the main basin has occurred since 1988 (fig. 12) (Rakoczy and Svoboda 1995), and may, in part, explain the decline in recreational fishing effort seen on Lake Huron. The decline in lake trout harvest may partly reflect population declines associated with resurgence of sea lamprey populations and changes in stocking locations. The decline in coho salmon reflects cessation of stocking of this species. Rainbow trout harvest in Michigan's waters of the main basin increased from 1988 to 1992, in response to increased stocking. In Canadian waters on the Bruce Peninsula and the south shore of Manitoulin Island, the sport fishery is increasingly supported by stocked rainbow trout, and the abundance of wild fish has declined (LHMU 1994). Brown trout harvest and harvest rate in Michigan's waters have increased substantially in the past few years, after an extended period of declining survival. The improved brown trout fishery probably is a result of stocking more successful large-lake strains (e.g., Seeforellen and Wild Rose), and of moving stocking windows from April and May to mid June, when attacks by predators may be buffered by the presence of alewives as alternative prey (J. Johnson, Michigan DNR, personal communication). Sport fishing for salmonines also occurs in the north channel. Although there is little quantitative information on the recreational fishery there, lake trout, chinook salmon, and brown trout are known to be important (D. Reid, OMNR, personal communication).

Georgian Bay is a mix of deep colder water and near shore warmer areas, and supports a diverse recreational fishery. Lakewide mail surveys suggest that fishery harvest was dominated by yellow perch, walleyes, smallmouth and largemouth bass, and northern pike in 1990 (DFO 1994). Smallmouth bass was argued to be critically important in supporting recreational fishing activities (LHMU 1994), but harvest and catch per unit effort have been at low levels during the 1990s (D. Reid, OMNR, personal communication). Chinook salmon dominates the cold-water salmonine harvest from Georgian Bay, while rainbow trout and lake trout are also important (LHMU 1994).

A contact creel survey estimate of effort for southern Georgian Bay in 1993 was over 700,000

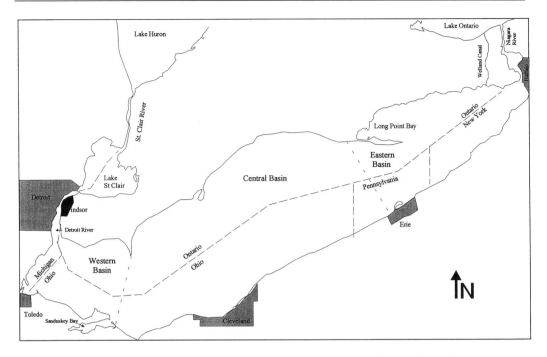

FIG. 14. *Schematic map of Lake Erie with major landmarks and jurisdictional boundaries.*

rod-hours, with chinook salmon being the primary target, and boat fishing the primary fishing method. Twenty-nine thousand fish were harvested, comprising by weight 48,000 kg of chinook salmon, 11,000 kg of lake trout, five thousand kg of rainbow trout, and smaller harvests of brown trout and lake whitefish (LHMU 1994).

Lake Erie

Ontario makes up most of the northern shore of Lake Erie, Michigan borders the western edge of the lake, while Ohio, Pennsylvania, and New York form the lake's southern shore (fig.14). An estimated 29% of all angling by U.S. citizens on the Great Lakes was on Lake Erie, and 14% of total recreational effort in Canadian waters occurred on this lake (Table 2). Most of the fishing effort on Lake Erie occurs in U.S. waters, and Lake Erie receives more recreational fishing effort by U.S. residents than any of the other Great Lakes. The high level of effort on the U.S. side reflects the higher human population density there.

Lake Erie consists of a western, central, and eastern basin, with the eastern basin being deepest, and the western basin being shallowest. The western basin of Lake Erie receives inflow from the upper Great Lakes via Lake St. Clair and the Detroit River, and outflow from the eastern basin occurs through the Welland Canal and Niagara River (fig. 14). The western basin contains much of the walleye spawning habitat (Regier et al. 1969) and is also near urban centers (Detroit, Windsor, Toledo), and thus, is a center of sport fishing activity. Other large urban areas bordering Lake Erie include: Cleveland, Ohio on the south shore of the central basin, Erie, Pennsylvania at the break between the central and eastern basin, and Buffalo, New York at the eastern end of the lake.

Lake Erie is shallower than the other Great Lakes, and consequently, the fishery is dominated by cool and warm-water fishes, with cold-water salmonines playing a lesser role. In contrast with the other Great Lakes, the recreational fishery has been supported almost entirely by naturally reproducing populations of native species. Together, yellow perch and walleye dominate the harvest (excluding rainbow smelt) in all jurisdictions. Smallmouth bass is a primary target for fishing in some areas, and other centrarchids, as well as percichthyiids also contribute to the sport fishery. There is some recreational fishing for lake trout and other salmonines, particularly in the deeper eastern basin bordered by Pennsylvania, New York, and Ontario. Among introduced salmonines, rainbow trout have been the most successful in Lake Erie.

Historical Background

Yellow perch and walleye have been important to the sport and commercial fisheries of Lake Erie during most of this century (Baldwin et al. 1979; Thomas and Haas 1994). Angling success for both yellow perch and walleyes has varied greatly over time, in response to varying year-class strength, with strong year classes of both species tending to coincide (Regier et al. 1969; Hartman 1973). This temporal variation in angling success for yellow perch has continued to the present, while success for walleyes have shown longer-term trends in the past few decades. Historically, angling for walleyes was most prevalent in Ohio, but Keller (1964) estimated Ohio recreational harvest of walleyes in the late 1950s from creel survey results as substantially less than the commercial harvest. Based on periodic creel surveys from 1948 to 1980, the recreational fishery in Ontario waters was characterized as diverse, with smallmouth bass, largemouth bass, rock bass, other centrarchids and northern pike all playing a role (Sztramko and Paine 1984). Consequently, recreational fishing for walleyes was not as extensive in Ontario's waters as in Ohio's (Regier et al. 1969). Walleye fishing in Ontario waters of the western basin, however, ranked among the best in North America (Sztramko and Paine 1984).

Walleye stocks in western Lake Erie collapsed during the late 1950s and early 1960s due to habitat degradation, overfishing, and competition with exotic rainbow smelt (Hartman 1973). Catch rates for walleyes were generally below 0.10 fish per angler-hour during the late 1950s and early 1960s (Keller 1964; Sztramko and Paine 1984; Hatch et al.1987). Discovery of mercury contamination in western basin walleyes led to a ban on commercial fishing for walleyes in the western basin, and a ban on sport fishing for this species in Michigan and Ontario in 1970. The resulting reduction in fishing pressure is thought to have allowed a major resurgence of walleye stocks in the lake (Hatch et al. 1987; Nepszy et al. 1991). Closure of the commercial fishery also provided management agencies an opportunity to implement lakewide management of walleye and yellow perch stocks through an interagency agreement for quota allocation of harvest when fishing resumed (Hatch et al. 1987). Commercial fisheries were never reopened in Michigan or Ohio, and New York's commercial fishery was closed in 1986. Limited sport fishing was allowed again throughout the lake by the mid 1970s. The sport fishery recovered first in the lake's western basin, and then in the central basin, perhaps fueled by fish migrating from the western basin (Nepszy et al. 1991). Sport catch rates and harvest rose slowly during the early 1970s and then rapidly during the mid 1970s, with catch rates reaching over 0.5 walleyes per angler-hour (Hatch et al. 1987). By 1987, total walleye harvest exceeded peak levels seen during the 1950s, with angling contributing more

TABLE 3

Estimated harvest (thousands of fish) of major species or species of special interest
and fishing effort (millions of hours) by the Lake Erie recreational fishery

Effort is reported by angler-hours except for Ontario where it is by rod-hours. Effort is for boat fisheries except for Pennsylvania, which includes shore fishing effort. Data are for 1993, except Ontario where data are for 1992. Michigan data from Thomas and Haas (1994), New York data from NYDEC (1994), Ohio data from Ohio DNR (1994), Ontario data from OMNRSR (1993), and Pennsylvania data from unpublished tables provided by Pennsylvania Fish and Boat Commission.

	MICHIGAN	NEW YORK	OHIO	ONTARIO	PENNSYLVANIA
Yellow perch	452	9	2,809	41	101
Walleye	230	40	2,669	109	125
Smallmouth bass	<1	14	40	49	9
Rainbow trout	<1	<1	34	5	34
Lake trout	NA	2	NA	<<1	2
TOTAL EFFORT	0.89	0.56	7.60	0.77	1.21

of the harvest than commercial fishing. Catch per angler-hour also continued to substantially exceed values reported from the 1950s (Nepszy et al. 1991).

Current Fishery and Recent Trends

Most components of the Lake Erie recreational fishery have been surveyed annually in recent years, although Ontario's sport fishery effort has not been surveyed lakewide since 1992, and tributaries are not surveyed every year. Estimates of effort measured on the lake, or at access points by state and provincial surveys indicate that between three and four million days have been spent angling on Lake Erie in recent years. These estimates of effort (and associated harvest) are used to manage the fishery, but as was the case for Lake Michigan, the lakewide estimate of effort is about half that from federal surveys, calculated as the sum of estimates from a U.S. phone survey and a Canadian mail survey (Table 2).

Yellow perch and walleyes are harvested in the greatest numbers throughout Lake Erie (Table 3). Although many fewer smallmouth bass are harvested, significant effort is directed at this species, and many fish are released rather than harvested (OMNR Southwestern Region (OMNRSR) 1994; Ohio DNR 1994; Pennsylvania Fish and Boat Commission, unpublished tables).

There is some fishing for cold-water salmonines in the lake, but in comparison with effort for cool and warm-water species this is relatively minor. Consideration of lake fishing alone, however, appears to underestimate the overall importance of salmonines, particularly rainbow trout, to the sport fishery. Significant numbers of rainbow trout are stocked in the central and eastern basins by Michigan, Ohio, Pennsylvania, New York, and Ontario, and most fishing for this species occurs on tributary streams. For example, creel surveys in Pennsylvania show that when tributary effort is taken into account, rainbow trout is the second most sought after species in this state. Angler diaries from Ontario and New York fishers (OMNRSR 1994; New York Department of Environmental Conservation (NYDEC) 1994) indicate that nearly 80% of the effort directed at rainbow trout occurs in tributaries, but total effort from tributaries is not surveyed. These diaries show a growing dominance of rainbow trout in the salmonine fishery (NYDEC 1994). A Canadian mail survey of recreational fishing in 1990 indicates that rainbow trout are important to the sport fishery

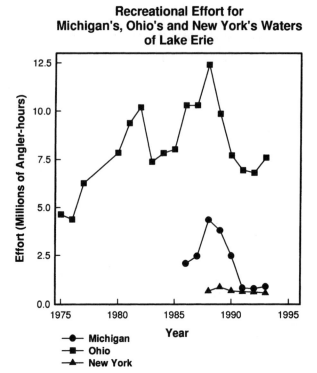

FIG. 15. *Recreational fishing effort (millions of angler-hours) in Michigan's, Ohio's, and New York's waters of Lake Erie. Only effort from boat fisheries included. Sources of estimates as for Table 3.*

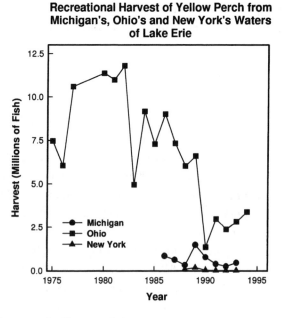

FIG. 16. *Recreational harvest of yellow perch from Michigan's, Ohio's, and New York's waters of Lake Erie. Only harvest from boat fisheries included. Sources of estimates as for Table 3.*

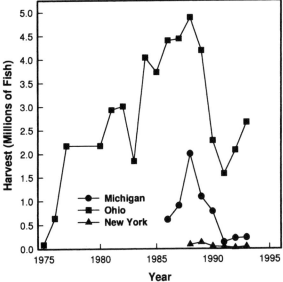

Recreational Harvest of Walleye from Michigan's, Ohio's and New York's Waters of Lake Erie

FIG. 17. *Recreational harvest of walleyes from Michigan's, Ohio's, and New York's waters of Lake Erie. Only harvest from boat fisheries included. Sources of estimates as for Table 3.*

in Ontario's waters, with a harvest equaling approximately 10% of the lakewide walleye harvest (DFO 1994).

Sport fishing effort and harvest in Ohio far surpasses that of any other jurisdiction on Lake Erie (Table 3, Figs. 15–17). Sixty percent of all sport effort and 80% of charter effort in Ohio occurs in the western basin (Ohio DNR 1994). Most of Ohio's effort is directed at walleyes, with significant effort for yellow perch and smallmouth bass (fig. 18). Fishing effort peaked in Ohio's waters in 1988, and declined from that point to a lower level in the past several years. Similar declines from peak effort levels are evident in other jurisdictions. Fishing effort by Ohio's charter fleet peaked at the same time as total effort, but increased proportionally more and declined proportionally less than non-charter effort did.

Harvests of yellow perch and walleyes in Ohio are up from those reported from 1990 to 1992, but are far below peak harvests observed in earlier years (Figs. 16 and 17). Effort directed by Ohio anglers at yellow perch has generally declined since the early 1980s (fig. 18). Effort directed at walleye has increased somewhat from its 1991 low, but is substantially below the 1988 peak (fig. 18). These changes in targeted effort are partly due to switching by anglers between yellow perch and walleye, but an explanation for the decline in total recreational fishing effort is not obvious. The decline in recreational effort is largely responsible for lower harvests in recent years, but a lower harvest of walleyes per angler-hour since the late 1980s has also contributed (Ohio DNR

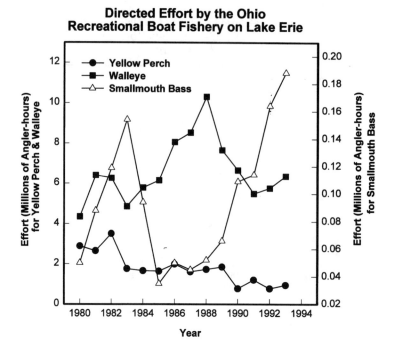

Directed Effort by the Ohio Recreational Boat Fishery on Lake Erie

FIG. 18. *Directed (i.e. at a target species) fishing effort by the Ohio recreational boat fishery on Lake Erie. Source of estimates given in Table 3.*

1994). Effort directed at smallmouth bass has increased steadily since 1987, although effort for this species remains well below levels targeted at walleyes and yellow perch (fig. 18). Harvest of smallmouth bass by Ohio's anglers has not, however, increased substantially because of an increasing tendency for anglers to catch-and-release this species (Ohio DNR 1994).

More than 80% of all fish harvested in Michigan are yellow perch or walleyes. Even so, Michigan's harvests of yellow perch and walleyes are down substantially from peak levels reported for 1988 and 1989 (Figs. 16 and 17). Although a decline in yellow perch abundance may be partly responsible for the decline in harvest of that species, lower effort has certainly played a major role (fig. 18). Decline in walleye harvest reflects both a decline in fishing effort, and a decline in harvest per unit effort from that seen at the peak in 1988 (Rakoczy and Svoboda, 1995). Although the charter fishery reported its highest number of excursions in 1993 since the reporting system started in 1989, charter effort is still much less than non-charter recreational effort.

Walleye are a dominant target in Pennsylvania, attracting two-thirds of the boat fishing effort. There is also significant directed fishing for both yellow perch and smallmouth bass (Pennsylvania Fish and Boat Commission, unpublished tables), and substantial shore angling. Shore angling amounted to 515,000 angler-hours in 1993, with most of this effort directed at rainbow trout on tributaries. Yellow perch attracts the second most attention of shore anglers, with about 10% of the total. Most yellow perch and walleyes caught in Pennsylvania are harvested, and these two species are most numerous in the harvest (Table 3). Almost as many smallmouth bass are caught as walleyes, but few are harvested. About half the rainbow trout caught are harvested, and this species ranks third in harvest. Brown trout and a variety of centrarchids, in addition to small-mouth bass also contribute to the sport harvest.

Walleye tends to dominate harvest throughout most of Ontario's waters of Lake Erie (Table 3). In Ontario, yellow perch contributes less to the sport fishery than in Michigan, Ohio, or Pennsylvania, and there is substantial directed effort toward smallmouth bass, both in the western basin and in some near shore areas of the eastern basin. Long Point Bay, a relatively small area in the eastern basin, has long been renowned for its smallmouth bass fishing, and contributes a substantial portion of total fishing effort in Ontario's waters (OMNRSR 1994). Most walleyes that were caught were harvested, while less than half the yellow perch and smallmouth bass that were caught were harvested. The high release rate for yellow perch may indicate these fish were undesirable, perhaps because they were small. High release rates for smallmouth bass probably reflect a growing ethic of catch-and-release fishing among those targeting this species.

Catch per rod-hour of effort (CPUE) has been available from Ontario's waters for over ten years from an angler diary program and from creel surveys of Long Point Bay (OMNRSR 1994). CPUE for walleyes is lower at locations in the eastern basin than in either the western or central basin. In the central basin, walleye CPUE increased from very low levels in 1985, to a peak in 1988, then underwent a modest decline through 1991, possibly indicating declining abundance. In the western basin walleye CPUE was already high by 1985, and fishing success has been somewhat lower after 1989. Smallmouth bass CPUE was generally low in the western basin, but fishing success has increased in the past few years. Smallmouth bass CPUE is not available from the central basin. Patterns seen at several different locations in the eastern basin differ from one another. The highest CPUEs for smallmouth bass are reported from Long Point Bay in the eastern basin, and this high level of fishing success has been quite stable since 1985.

In New York's waters, more than half the sport effort is directed at walleyes (NYDEC 1994). Sport harvest of walleyes in 1993 was substantially higher than in 1992, but well below the peak in 1989 (fig. 17). Harvest per angler-hour directed at walleyes (HPUE) in 1993 was substantially higher at all locations than in 1992, but still below levels seen in the late 1980s at four out of five locations. The increase in HPUE of walleyes from 1992 to 1993 may reflect increased abundance, but the sharp change suggests that increased catchability may also play a part. Smallmouth bass were the second most common fish in the New York sport harvest (Table 3). There is now little directed effort for yellow perch, and harvest of this species fell by 1993 to about 5% of that in 1989 (fig. 16), reflecting a much reduced abundance in the eastern basin. There was a modest fishery for lake trout with 1,900 fish harvested, even though the harvest limit is one fish. Harvests of less than one-thousand rainbow trout, brown trout, and coho salmon were reported from the lake, but no results were available for tributaries, where much of the harvest of these anadromous fishes is believed to occur.

Lake Ontario

New York borders most of the southern and eastern shores of Lake Ontario, with the remaining shoreline under Ontario's jurisdiction (fig. 19). The Ontario shore on the western end of the lake is

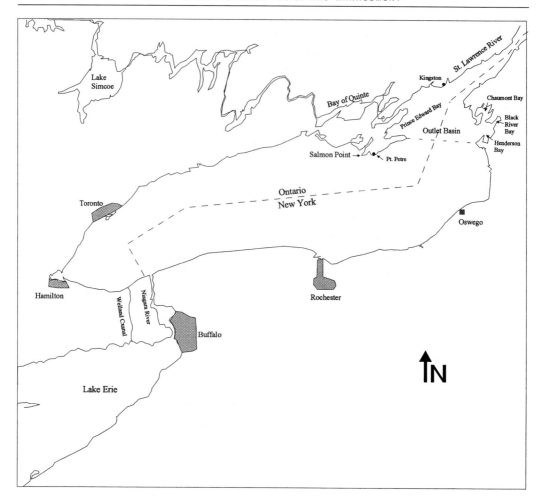

FIG. 19. *Schematic map of Lake Ontario with major landmarks and jurisdictional boundaries.*

highly populated, including the cities of Toronto, Mississauga, and Hamilton. Other major cities on or near Lake Ontario include: St. Catherine's, Ontario; Rochester, New York; Watertown, New York; Kingston, Ontario; and Belleville, Ontario. About 14% of the sport fishing effort by U.S. citizens on the Great Lakes is on Lake Ontario, and about 29% of the effort in Canadian waters of the Great Lakes occurs on this lake (table 2). The relative importance of Lake Ontario to recreational fishing on the Canadian Great Lakes stems from the large population concentrated near the western shore. Care should be taken, however, in using table 2 to gauge the relative magnitude of U.S. and Canadian effort on Lake Ontario. The Canadian mail survey estimate includes effort on tributaries and connecting waters, while the U.S. phone survey estimate does not, and the kinds of differences in methods used in the two surveys (i.e., phone versus mail, annual versus quarterly) have been shown to influence results by up to several-fold (Talhelm 1988; USFWS and Bureau of the Census 1993). Furthermore, partial creel surveys suggest, contrary to Table 2, that angler effort may be greater in New York than in Ontario. These problems (see also the Lake Michigan and Lake

Erie sections) again highlight our uncertainty about the true amount of recreational fishing effort on the Great Lakes.

The main body of Lake Ontario consists of the main lake with three subtly defined basins, and a northeastern outlet basin (fig. 19). The Niagara River enters the lake in the west, and is the major inflow. There are, however, an additional 780 tributaries (Kerr and Le Tendre 1991), some of which support important fisheries for anadromous fishes. The outlet basin (including the Bay of Quinte) is separated from the rest of the lake by a distinct sill and is relatively shallow. Warm and cool-water fisheries tend to dominate in this area, while cold-water salmonines are the most important part of the sport fishery in the deeper main lake. The sole outlet from Lake Ontario is the St. Lawrence River.

Important species in the salmonine fishery include: chinook salmon, coho salmon, lake trout, brown trout, and rainbow trout. Chinook salmon dominates the harvests of salmonines in Ontario, while salmonine harvest is more diverse in New York. Walleye and smallmouth bass contribute to the lake's sport fishery. Walleye fishing is concentrated at the eastern end of the lake and in the Bay of Quinte, and smallmouth bass fishing in concentrated on the U.S. shore.

Historical Background

The fish community of Lake Ontario has a long history of change and response to human actions (Christie 1973), and this has had profound influences on the sport fishery that developed. Atlantic salmon, a likely target for sport fishing, was extinct by the mid 1800s, largely due to habitat alterations in streams used for spawning. Habitat changes in streams, including increased soft sediment and warming caused by removal of forest canopy, also improved conditions for sea lamprey reproduction. Increased parasitism by sea lamprey contributed to the decline and eventual collapse of lake trout populations during the 1930s and 1940s, and to a great reduction in burbot abundance. Other contributing factors to the collapse of lake trout stocks include commercial fishing and predation by alewives on lake trout eggs (Krueger et al. 1995). Scarcity of open water predators, in turn, allowed a build-up of smelt and alewife populations, which may have adverse effects on native species, like yellow perch (Christie 1973; Brandt et al. 1987) and lake trout (Kreuger et al. 1995).

Historically, much of the recreational fishing on Lake Ontario was in shallow water for cool-water species such as yellow perch, walleye, smallmouth bass, and northern pike. In deeper open water, there was some fishing for lake trout, but this was considered insignificant relative to commercial fishing for this species. Even after closure of the commercial fishery in 1962, there was no sport harvest limit for lake trout in New York waters (Elrod et al. 1995). Walleye was a dominant species in the lake in the 1950s, and yellow perch were abundant even in open waters in the 1960s. Abundance of walleyes declined during the 1960s, perhaps due to effects of eutrophication and increased abundance of white perch (Hurley and Christie 1977). They began recovering in the late 1970s in the eastern portions of the lake, especially the Bay of Quinte, possibly in response to improved sewage treatment and a large winter kill of white perch (Bowlby et al. 1991). Yellow perch abundance has declined steadily, at least since 1985, and are now insignificant to the recreational fishery.

The modern salmonine fishery began after the stocking of three species of Pacific salmon (coho salmon, chinook salmon, and kokanee salmon) by New York and Ontario in the 1960s, additional stocking of lake trout, brown trout, rainbow trout, and chinook salmon starting in the mid 1970s, and a substantial reduction in sea lamprey abundance that occurred by the 1980s (Kerr and LeTendre 1991). Stocking of salmonines increased from the late 1970s through the mid 1980s, primarily through increases in numbers of chinook salmon, and remained fairly steady through 1992, at about eight million total fish lakewide. The result was the development of a substantial and economically important fishery (Lange et al. 1995).

The last several years have brought changes that could potentially have important consequences to the sport fishery in the near future. By 1990, the stocking rate of salmonines in Lake Ontario was the highest per unit area of any of the Great Lakes, while natural productivity of the forage base was thought to be decreasing due to reversal of cultural eutrophication and invasion of the zebra mussel (Jones et al. 1993; Lange et al. 1995). Simulation modeling suggested that the salmonine fishery in Lake Ontario might be sensitive to a die-off of alewives following a severe winter (Jones et al. 1993). In response to these concerns and observations of a collapse of the chinook salmon fishery in Lake Michigan, New York, and Ontario agreed to substantially reduce salmonine stocking rates starting in 1993, primarily by reducing numbers of chinook salmon and lake trout planted (Orsatti and LeTendre 1994; Lange et al. 1995). Lakewide stocking was reduced from 8 million to 5 million salmonines in 1993, and further reduced to 4.5 million in 1994.

Current Fishery and Recent Trends

Not all components of the sport fishery (e.g., shore, some tributaries, and the outlet basin) have been surveyed on an annual basis. Because of the partial coverage, estimates of lakewide harvest and effort are not currently available. Personnel at the Ontario Ministry of Natural Resources and the New York State Department of Environment and Conservation are in the process of combining information collected over different years to estimate lakewide values. Both New York and Ontario creel surveys target the boat portion of the salmonine fisheries. Annual surveys of the boat fishery in Ontario's waters of the main lake include about 25% of the total salmonine effort (Savoie and Mathers 1994). The portion of New York's Lake Ontario recreational effort (including tributaries) covered by the annual boat survey was estimated at 13% in 1984, but this portion of the fishery contributed most of the lake trout harvest (Elrod et al. 1995), and probably includes most of the non-tributary effort directed at other salmonines.

Boat fishing is the most important mode of recreational fishing on the lake, but shore and stream fishing also contribute. In 1984 (NYDEC, undated), 20% and 13% of angler trips were to shore and stream fishing sites in New York's waters, respectively. The salmonine fishery has been centered in the deeper main lake. There is significant fishing on tributaries for chinook salmon during the fall, and for rainbow trout in both fall and spring. In 1984, more than half of the tributary effort in New York was on the Salmon River (NYDEC, undated). In Ontario, the salmonine fishery was centered in the portion of the lake west of Pt. Petre through 1985, and targeted Pacific salmonines and rainbow trout. The boat fishery in the eastern part of the lake (from Salmon Point to Kingston) developed rapidly between 1985 and 1987, and now contributes most of the harvest of lake trout from Ontario waters, but relatively few other salmonines (Elrod et al. 1995).

Fishing effort in New York's waters of the outlet basin is difficult to monitor due to widely distributed access, but appears to have increased in recent years in response to resurgence of wall-

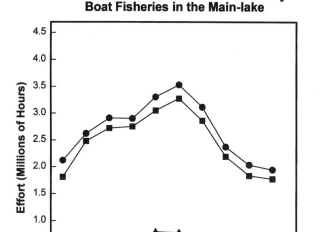

Recreational Effort on Lake Ontario by Boat Fisheries in the Main-lake

FIG. 20. *Recreational fishing effort on Lake Ontario in surveyed portions of the main lake boat fisheries. New York estimates from Eckert (1994) and unpublished NYDEC tables. Ontario estimates from Savoie and Mathers (1994).*

eye populations and good fishing for smallmouth bass. Walleye fishing in Black River Bay, Henderson Bay, Chaumont Bay, and near main-lake islands and shoals along the northeast New York shore is considered among the best in the northeast United States, yielding particularly large fish. In Ontario, walleyes are the major target of the Bay of Quinte sport fishery, and more than 10% of the effort is from an ice fishery.

Boat fishing effort in the main lake peaked from 1989 to 1990, and has decreased since then (fig. 20). The trend in effort of the charter boat component of the boat fishery in New York was similar to that of total boat fishery effort, but the increase was more dramatic, and the decrease less so (Eckert 1994). Over 80% of effort in New York's waters is directed specifically at salmon and trout, with much of the remaining effort directed at smallmouth bass. New York's Lake Ontario recreational fishery has increasingly attracted out-of-state participants; the percentage of angler trips by nonresidents increased from 21% in 1985 to 45% in 1994.

Harvest of salmonines by boat fisheries in the main lake has largely tracked fluctuations in effort during recent years. Total salmonine harvest in New York has declined from a peak in 1989 (fig. 21). In Ontario, total harvest of salmonines in the surveyed boat fishery fell to lower levels after 1991, in comparison with harvests during 1988 through 1991 (fig. 21). In New York, five salmonine species and smallmouth bass contribute substantially to the fishery harvest, while in Ontario most of the harvest has been chinook salmon (fig. 22).

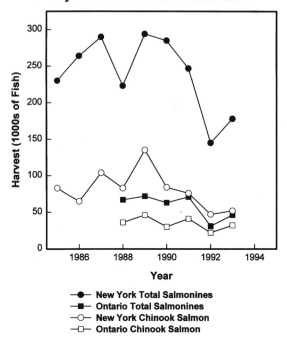

FIG. 21. *Recreational harvest from Lake Ontario of all salmonines combined and of chinook salmon in surveyed portions of the main lake boat fisheries. Sources of estimates as in fig. 20.*

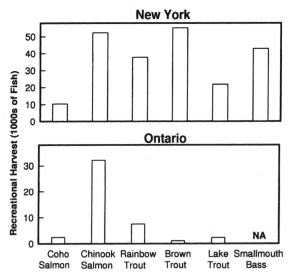

FIG. 22. *Recreational harvest from Lake Ontario during 1993 of major species by the surveyed portions of the main lake boat fisheries. Sources of estimates as for fig. 20.*

In addition to harvested fish, many fish are also caught and released. The percentage of caught fish that have been harvested has varied through time and among species. These percentages are influenced by factors such as conservation ethics, size and creel limit regulations, palatability, and concerns regarding potentially unhealthy concentrations of toxic substances. For the combined catch of salmonines by the New York boat anglers, percentage harvested has varied between 59% and 71%. For chinook salmon, coho salmon, brown trout, and rainbow trout, percentages have been relatively high (i.e., 64% to 83%) during the 1990s. Harvest percentages were much lower for lake trout (i.e., 28% to 50%) and Atlantic salmon (i.e., 21% to 40%). The percentage for lake trout had pronounced variations in response to changes in size limits (Elrod et al. 1995). The percentage harvested was generally lower in Ontario's fishery than in New York's, except for lake trout and Atlantic salmon. For other salmonines combined (chinook salmon, coho salmon, rainbow trout, and brown trout) this percentage was 49% in Ontario.

Ontario's boat fishery has experienced a downward trend in harvest per unit effort for salmon and trout over the past decade, declining from 0.15 fish per rod-hour in 1984, and leveling off at less than 0.10 fish per rod-hour in recent years (Savoie and Mathers 1994). This decline has been interpreted along with other evidence as suggesting a decline in survival of stocked fish (Savoie and Mathers 1994). In contrast, harvest per angler-hour among New York's anglers targeting salmon and trout does not indicate a similar downward trend.

The important Bay of Quinte walleye fishery is also surveyed on an annual basis by the Ontario Ministry of Natural Resources (Savoie and Mathers 1994). About one-third of the annual effort there (957 thousand rod hours) was by ice fishing. The ice fishery contributed much less than a third of the walleye catch and harvest since the catch per unit effort by ice fishing is lower than for boat fishing. Fishing effort has increased since the early 1980s. Total harvest and harvest per unit effort were quite high when the fishery started in the late 1970s. Harvest per unit effort stabilized around a lower level by the early 1980s, leading to lower harvests that subsequently increased as effort increased. 58% (coming to 166,000 fish) of captured walleyes were harvested in 1993.

Management Tools

Various methods have been used to manage the sport fisheries of each Great Lake, and these have also varied temporally and spatially within lakes. There is generally little historical information available about the rationale behind adoption of specific regulations, nor about effects of the regulations on sport anglers. Thus, rather than detailing temporal changes and spatial differences in regulations, we present a general discussion of tools available to fishery managers, and how such tools may be used to enhance fish stocks and achieve other goals.

Historically, a major impetus of fishery management has been to pursue the elusive maximum sustainable yield by restricting mortality from fishing (e.g., Beverton and Holt 1957). While the concept of maximum sustained yield has been challenged (Larkin 1977; Sissenwine 1978; Hilborn and Walters 1994), most alternatives implicitly or explicitly acknowledge that at some point increases in fishing mortality can lower sustainable yields. This can occur because yield-per-recruit declines when fish are harvested at too young an age, because recruitment declines substantially if fishing causes too great a reduction in spawning stock size, or through some combination of

TABLE 4

Daily harvest limits of salmon and trout in the Great Lakes

These harvest limits were in effect for in-lake fishing in 1994 or 1995. More restrictive limits applied in some local areas.

	SALMON AND TROUT IN COMBINATION	LAKE TROUT	CHINOOK SALMON, COHO SALMON, OR BROWN TROUT	RAINBOW TROUT (STEELHEAD)	PINK SALMON
Lake Michigan					
Illinois	5	2	3	3	3
Indiana	5	2	5	5	5
Michigan	5	2	3	3	5
Wisconsin	5	2	5	5	5
Lake Superior					
Michigan	5	3	3	3	5
Minnesota	21	3	10	3[1]	10
Ontario	5	3	5	5	5
Wisconsin	10	3	5	1	5
Lake Huron					
Michigan	5	3	3	3	5
Ontario	5	3	5	5	5
Lake Erie					
Michigan	5	3	5	5	5
Ohio	5	5	5	5	5
Ontario	5	3	5	5	5
Pennsylvania					
Jan.–March	8	2	8	8	8
April–Dec.	3	2	3	3	3
New York	5	1	5	5	5
Lake Ontario					
New York	5	3	5	5	5
Ontario	5	2	5	5	5

[1]Only one may be unclipped (wild).

such growth overfishing or recruitment overfishing (e.g., Hilborn and Walters 1994). The idea of maintaining sufficient spawning stock is usually applied to naturally reproducing populations, but is also important for fisheries being maintained by hatchery releases, which require enough mature fish for egg-take operations.

While the idea of a target stock size (at least on average) is well accepted, in practice, fishery managers usually react to perceived changes in fish abundance and fishing quality. Fishery management agencies on the Great Lakes attempt to control mortality due to sport fishing through harvest (creel) limits, size limits, closed areas and seasons, and gear restrictions. Harvest limits restrict the number of fish an angler can take in a day, and these are usually enforced with complementary limits on the number of fish that can be possessed at one time. The concept behind this type of regulation is that once anglers reach their limit, they will cease angling, and thus, not kill fish they would have without the regulation. Creel limits for salmon and trout on the Great Lakes vary among lakes and jurisdictions, and sometimes on smaller scales (Table 4). Harvest limits for these species and other premier (large and sought after) species, such as walleye, are often on the order of five fish or less, but often only a small proportion of anglers approach these limits. Thus, managers have little room to maneuver, because reductions in creel limits can lead to strong negative responses from user groups, yet, resulting reduction in harvest may not be large. In addition to expressing vocal opposition, some anglers might respond to a reduced creel limit by reducing the number of fishing trips they take, with significant economic consequences. Similar problems with the use of creel limits to reduce fishing mortality have been expressed for other systems (Noble and Jones 1993).

Creel limits do have potential to limit fishing mortality to some degree, when the catch of a larger proportion of fishers approach them. This appeared to be the case for yellow perch in Lake Michigan. Individual trip data from South Haven, Michigan collected in 1993 and 1994, showed that 10% of the harvest was contributed by anglers' harvest in excess of fifty fish on a trip (P. Schneeberger, Michigan DNR, personal communication). For 1995, the creel limit in Michigan was reduced from one hundred to fifty yellow perch. Many yellow perch anglers on Lake Michigan seemed to prefer a reduction in creel limits over other restrictions on harvest, such as closed seasons (P. Schneeberger, Michigan DNR, personal communication).

Managers use regulations to restrict which sizes of fish are harvested by sport anglers. The simplest regulation of this type is a minimum size below which no fish can be retained. More complex size restrictions include slot limits, where only fish within a specified length-range can be retained, reverse slot limits where fish within a specified range cannot be kept, and other variants (Noble and Jones 1993). Size limits in place for lake trout in the Great Lakes illustrate the range of regulations of this type being used by management agencies (Table 5). The concept underlying size limits is that fishing mortality on certain sizes is reduced (e.g., Beverton and Holt 1957). Size limits are often designed to protect young small fish, whose expected growth and increases in reproductive capacity outweigh the risk they will die before they can be harvested (Ricker 1975). Other considerations (such as protecting spawning stock or providing fishing opportunities) also play a role in setting size limits, and sometimes lead to more complex regulations. An example of this type is the imposition of reverse slot limits to allow a fishery on smaller fish, while also providing motivated anglers a chance to obtain large trophy size fish.

Fishing mortality is also controlled by restricting harvest on a spatial or temporal scale. Closed areas are often used to protect stocks that are thought to be in serious trouble, or are being rehabilitated. Examples include lake trout refuges which have been established in a number of locations within the Great Lakes, and closures of walleye fishing in local areas where this species is not doing well. Closed areas have also been suggested as part of a sustainable management strategy, rather than just as restoration (Hill and Shell 1975). For example, fish moving out of a protected area sometimes provide significant recreational harvest in nearby waters, but provided the closed area is large enough it still may protect the target population. Some lake trout refuges on the Great Lakes, such as those in northern Lake Michigan and in Wisconsin's waters of Lake Superior (WSTTC 1995), may be operating in this fashion. Hilborn and Walters (1994) suggest that some fish populations could be managed with a rotational harvest scheme, as is common in forestry. Modeling by MacCall (1990) showed that closing fishing in productive areas can sometimes actually increase total sustainable yield, as offspring move into other areas and are harvested.

TABLE 5

Length restrictions on harvest of lake trout in the Great Lakes

These restrictions were in effect for in-lake fishing as published in 1994 or 1995. More restrictive regulations are applied in some local areas.

Lake Michigan
    Illinois   10" minimum

Lake / region	Restriction
Lake Michigan	
Illinois	10" minimum
Indiana	16" minimum
Michigan	
Northern Lake Michigan and Grand Traverse Bay	24" minimum
Southern and central Lake Michigan	10" minimum
Wisconsin	10" minimum
Lake Superior	
Michigan	10" minimum
Minnesota	none
Wisconsin	15" minimum, only one fish over 25"
Ontario	none
Lake Huron	
Michigan	10" minimum
Ontario	none
Lake Erie	
Michigan	10" minimum
New York	12" minimum
Ohio	none
Ontario	none
Pennsylvania	9" minimum
Lake Ontario	
New York	<25" or >30" only
Ontario	none

Closed seasons are another device used to reduce fishing mortality. An example is the prohibition of yellow perch harvest during June that was implemented lakewide on Lake Michigan for 1995 only. This regulation was proposed by management agencies on Lake Michigan because of reduced abundance and weak year classes of yellow perch in recent years (Francis et al. 1996). Often, the period when harvesting is prohibited coincides with the spawning season when many species are especially vulnerable to angling, either because they tend to aggregate or be more aggressive at that time (Noble and Jones 1993). Sometimes these closures are designed to keep spawning fish from being disturbed, but a more general goal of closing seasons is simply to reduce fishing mortality.

Gear restrictions can limit fishing power, and thus, lower overall fishing mortality (Noble and Jones 1993). For most species, recreational anglers can only take fish by hook-and-line. A second restriction widely applied throughout the Great Lakes limits the number of fishing rods that can be used simultaneously by an angler. Gear restrictions may also be designed for special purposes, including reducing mortality of released fish (e.g., restrictions on the use of bait with certain types of lures), or to prevent unwanted or illegal angler behavior (e.g., restrictions on hook size to prevent snagging of fish).

Sport fisheries in the Great Lakes are managed by altering the production of fish, as well as by controlling fishing mortality. One tool that fishery managers employ is stocking hatchery reared fish (e.g., Heidinger 1993; McGurrin et al. 1995). Stocking programs sometimes have an explicit goal of rehabilitating or creating self-sustaining populations. For example, this is the ultimate goal of stocking programs for lake trout in the Great Lakes (Eshenroder and Burnham-Curtis 1999), and also of many of the walleye stocking programs. Another goal of stocking programs is to support recreational fisheries on a put-grow-take basis (e.g., Heard et al. 1995; Kinman 1995; Lange et al. 1995). This has been a goal of many stocking programs for Pacific salmon in the Great Lakes. In some cases, stocking programs have both these goals (e.g., Perry 1995).

An assumption underlying many stocking programs is that eventual recruitment to the fishery is, at least on average, positively related to numbers stocked, and implies that some facet of spawning production is a primary factor limiting fishery productivity. During the 1970s and early 1980s, this appeared to be generally true for stocked salmonines. For example, Hansen (1989) showed that returns to Wisconsin's sport fishery in Lake Michigan through 1985 were proportional to numbers stocked. As the Great Lakes have become more saturated, supplies of forage fish and other in-lake influences appear to have limited fish populations. Some of the best evidence for this comes from lake trout populations inhabiting Lake Superior (Hansen et al. 1994b), but concerns about such limits have been raised for other species and locations (Stewart and Ibarra 1991; Peck et al. 1994; Savoie and Mathers 1994; Ebener 1995). As a result of these concerns, fishery managers have become reluctant to increase stocking levels (Ebener 1995), and in some cases have suggested or implemented cuts in stocking levels (Hansen 1986; Jones et al. 1993; Hansen et al. 1994b; Peck et al. 1994).

Habitat or other environmental changes can also influence fishery productivity by altering recruitment rates, individual growth rates or rates of non-fishing mortality (Hayes 1999). Although sport fishery managers can sometimes manipulate certain environmental factors, most remain outside their direct control. For example, it is likely that phosphorous loadings from watersheds into the Great Lakes influences production of sport fishes, but regulation of these loadings is directed by a broader management group which is concerned about water quality, economic consequences to industry, and other factors (Hartig et al. 1991; Lange 1995). In some cases, fishery managers may take the lead in decisions to make alterations to a watershed. For example, they might advocate the removal of dams, with a goal of reconnecting fragmented reaches with a tributary (Hayes 1999). Even within the fishery management arena, such an action can involve tradeoffs. For example, a dam removal might increase spawning habitat for sea lamprey, or lead to a reduced abundance of resident trout if runs of anadromous salmon were to have adverse effects on them.

The Future of Recreational Fisheries

Recreational fishing is one of many factors which influence the complex ecosystems of the Great Lakes. These ecosystems are managed by fishery and environmental managers from a variety of state, provincial, tribal, and federal agencies. Authority and information are fragmented, and goals differ among these management groups. Furthermore, management decisions made in other parts of the watershed can influence recreational fisheries of the Great Lakes, but these fisheries may not be considered when the decisions are made. Given these considerations, we believe that management of Great Lakes sport fisheries needs to work within a broader framework which employs a multidisciplinary approach (see also Hartig et al. 1991; Lange et al. 1995). Portions of this ecosystem vision are already incorporated in many strategic planning documents for the Great Lakes (Knuth et al. 1994). To make ecosystem management a reality, however, total fish production from both natural and hatchery sources, mortality due to sea lamprey parasitism and other natural sources, and sport, commercial, and tribal fishing all need to be considered together rather than singly. In addition, interactions within and between species, and changes in habitat and environmental characteristics need to be considered. Information on how anglers and other stakeholders respond to management decisions is needed, as is information on how stakeholders will value their changed access to resources, if managers are to become more proactive in the future. Although progress in being made toward this goal, much is left to be done.

One lesson we have learned in writing this chapter is how short, fragmentary, and uncertain the time-series of recreational fishery data are. For example, the only comprehensive information on total angling effort on the Great Lakes comes from mail and phone surveys done at five year intervals in both the United States and Canada. Effort estimates resulting from these surveys are widely used to determine the economic value of the Great Lakes recreational fisheries. How these kind of surveys are done can have large influences on the resulting estimates (Talhelm 1988; USFWS and Bureau of the Census 1993); methods used in these surveys have changed over time, and differ between the United States and Canada. Furthermore, measures of sport fishing effort made at the time it occurs, as part of creel surveys on Lakes Michigan and Erie have been substantially below estimates from these national surveys, even though these creel surveys provide nearly complete coverage of sport fishing. This uncertainty about the amount of recreational fishing effort (and hence, harvest) contrasts with the long time-series of relatively reliable harvest and effort information available for many commercial fisheries. This reflects, in part, the relatively recent recognition that recreational fishing is an important influence on the ecology and economy of the Great Lakes system. It also demonstrates the diverse and dynamic nature of recreational fisheries. Programs for monitoring Great Lakes recreational fisheries have grown to be more comprehensive over the past twenty-five years, and agencies are working toward greater standardization and more complete sampling coverage of fisheries. Currently, the most comprehensive creel survey data come from Lakes Michigan and Erie, which also support the largest amount of angling effort by sport fishers. We believe that this is not a coincidence. Yet, we also think that even more comprehensive information about sport fisheries on the Great Lakes is needed. If the move toward ecosystem management is to be successful, lakewide and basinwide quantitative information on recreational fishery catch and removals, their composition, and the associated effort is needed. We think that information on sport effort targeted at a specific species or species group, and estimates of both catch (including released fish) and harvest are critical information to help evaluate trends in fish populations, and the sport fisheries they support. These data are not collected or reported by all agencies, but could be added to existing surveys. Creel survey methods and results for the Great Lakes are only occasionally published in peer-reviewed literature, and we believe such review would enhance quality of estimates, and encourage standard approaches.

Understanding angler behavior and human population trends may be as important as obtaining reliable information on current status of recreational fisheries and fish stocks. Developing new predictive tools based on this information should allow managers to make more realistic projections about what a given management action is likely to achieve. There are examples where changes in fishing effort and angler behavior are explained by relating them to management actions (e.g., Elrod et al. 1995), publicity regarding toxins (e.g., Brofka and Marsden 1994), and declines in fishing success (Keller et al. 1990), but more can be learned through designed studies and additional analyses of existing data. A pressing need is to better understand the cause of the general decline in sport fishing effort throughout most of the Great Lakes described in this chapter (see also USFWS and U.S. Bureau of the Census 1993), as well as reasons for differences in effort trends among the lakes.

Better information about recreational fisheries will certainly influence stakeholder attitudes. The future of recreational fisheries in the Great Lakes will still be molded, however, by competing visions and different attitudes among stakeholders, even if managers achieve a firm grasp on ecosystem status and underlying dynamics of the Great Lakes. From a recreational fishing standpoint, lake trout rehabilitation versus maintenance of sport fisheries for introduced salmonines is one key area where some disagreement exists among stakeholders. Successful rehabilitation of lake trout populations is a major goal of fishery and environmental management programs (Knuth et al. 1994). Although Knuth et al. (1994) document differences among groups of managers, we were struck by the qualitative similarity of views. The majority of managers in all groups viewed rehabilitation of self-sustaining, native fish populations as the single most important priority for Great Lakes fishery management, and over three quarters supported rehabilitation of lake trout populations that can support some harvest (Knuth et al. 1994).

The views of these groups of managers, as reported by Knuth et al. (1994), generally seem to be in closer agreement with one another than with those we would expect from the majority of sport anglers. Managers placed more importance on reestablishing native species and protecting ecological diversity than on satisfying anglers or obtaining economic return. Many differences among management groups may be related to the degree of direct exposure and interaction of their agency with sport anglers. For example, those managers working in jurisdictions with larger sport angler populations (e.g., United States versus Canada), or having greater responsibility for managing the sport fishery (e.g., fishery managers versus environmental managers or state/provincial versus federal managers) gave greater weight to angling and resource use than managers with less exposure to anglers. It seems likely to us that the opinions of managers reflect an internal and informal balancing of the desires of multiple stakeholder groups to which they are responsible.

We believe that attempts to rehabilitate lake trout and other native species should continue, with a goal of establishing self-sustaining populations that can support recreational, commercial, and tribal harvest. We also believe that this management goal should continue to recognize that recreational fisheries supported by introduced salmonines are an important part of the system (see also Lange et al. 1995). Obviously, tradeoffs and allocation of resources will be necessary components of such a broad goal. The challenge will be to determine what is ecologically possible, identify costs and benefits involved, and finally to build consensus among stakeholders about what they desire (e.g., Eshenroder 1987; Peyton 1987; Milliman et al. 1987; Dawson et al. 1993).

Acknowledgments

Numerous colleagues provided us with information, data, and reports that allowed us to write this chapter. We wish to thank Rick Clark, Jim Peck, Jim Johnson, Dick Schorfhaar, Phil Schneeberger, Bob Haas, Mike Thomas and Jerry Rakoczy (Michigan DNR), Brad Eggold, Stephen Schram and Scott Hulse (Wisconsin DNR), Jim Francis and Janel Palla (Indiana DNR), Ellen Marsden and Wayne Brofka (Illinois Natural History Survey), Rich Hess (Illinois Department of Conservation), Dave Reid, Paul Savoie, Les Sztramko, Wayne MacCallum, Mike Jones, Tom Stewart, Douglass Legg and Jerry Payne (Ontario Ministry of Natural Resources), Tom Eckert and Don Einhouse (New York Department of Environmental Conservation), Chuck Murray (Pennsylvania Fish and Boat Commission), Don Schreiner, Thomas Jones and Ted Halpern (Minnesota DNR), Roger Knight and Doug Johnson (Ohio DNR), Mike Hansen and Joe Elrod (Great Lakes Science Center), Carol Swinehart (Michigan Sea Grant), Chad Dawson (State University of New York), Mark Ebener (COTFMA), and Gavin Christie (Great Lakes Fishery Commission). We appreciate reviews of earlier drafts by Paola Ferreri, Mike Hansen, Roger Knight, and Bill Taylor and technical assistance by S. Bence and J. Clevenger.

Literature Cited

Baldwin, N. S., R. W. Saalfeld, M. A. Ross, and H. J. Buettner. 1979. Commercial fish production in the Great Lakes 1867–1977. Great Lakes Fishery Commission, Technical Report No. 3, Ann Arbor, Michigan.

Berst, A. H., and G. R. Spangler. 1973. Lake Huron: the ecology of the fish community and man's effects on it. Great Lakes Fishery Commission, Technical Report No. 21, Ann Arbor, Michigan.

Beverton, R. J. H. and S. J. Holt. 1957. On the dynamics of exploited fish populations. Chapman and Hall, London. 533 pp.

Bowlby, J. N., A. Mathers, D. A. Hurley, and T. H. Eckert. 1991. The resurgence of walleye in

Lake Ontario. Pages 169–206 in P. J. Colby, C. A. Lewis, R. L. Eshenroder, editors. Status of walleye in the Great Lakes: case studies prepared for the 1989 workshop. Great Lakes Fishery Commission, Special Publication 91–1, Ann Arbor, Michigan.

Brandt, S. B., D. M. Mason, D. B. MacNeill, T. Coates, and J. E. Gannon. 1987. Predation by alewives on larvae of yellow perch in Lake Ontario. Transactions of the American Fisheries Society 116:641–645.

Brofka, W. A., and J. E. Marsden. 1994. A survey of sport fishing in the Illinois portion of Lake Michigan. Illinois Natural History Survey, Aquatic Ecology Technical Report 94/4.

Brown, J. J. 1876. The American angler's guide: or, complete fisher's manual for the United States: containing the opinions and practices of experienced anglers of both hemispheres. 5th edition, D. Appleton and Co., New York. 421 pp.

Christie, W. J. 1973. A review of the changes in the fish species composition of Lake Ontario. Great Lakes Fishery Commission Technical Report No. 23, Ann Arbor, Michigan.

Dann, S. L. 1994. The life of the lakes: A guide to the Great Lakes fishery. Michigan Sea Grant Extension Bulletin #E-2440, East Lansing, Michigan.

Dawson, C. P., F. R. Lichtkoppler, and C. Pistis. 1989. The charter fishing industry in the Great Lakes. North American Journal of Fisheries Management 9:493–499.

———, N. A. Connelly, and T. L. Brown. 1993. Salmon snagging controversy: New York's salmon river. Fisheries 18:6–10.

DFO (Department of Fisheries and Oceans). 1994. 1990 survey of recreational fishing in Canada: Selected results for the Great Lakes fishery. Economic and Commercial Analysis Report No. 142.

Ebener, M. P. (editor). 1995. State of the Lake Huron fish community in 1992. Great Lakes Fishery Commission Special Publication 95–2, Ann Arbor, Michigan.

Eckert, T. H. 1994. New York's 1993 Lake Ontario fishing boat census. Pages 59–117 in New York State Department of Environmental Conservation, 1994 Annual Report: Bureau of Fisheries Lake Ontario Unit to the Lake Ontario Committee and the Great Lakes Fishery Commission.

Elrod, J. H., R. O'Gorman, C. P. Schneider, T. H. Eckert, T. Schaner, J. N. Bowlby, and L. P. Schleen. 1995. Lake trout rehabilitation in Lake Ontario. Journal of Great Lakes Research 21 (Suppl. 1):83–107.

Eschmeyer, P. H. 1959. Great Lakes fishery research by the Michigan Department of Conservation, 1958. Michigan Department of Natural Resources, Research Report 1959.

Eshenroder, R. L. 1987. Socioeconomic aspects of lake trout rehabilitation in the Great Lakes. Transactions of the American Fisheries Society 116:309–313.

——, M. K. Burnham-Curtis. 1999. Species succession and sustainability of the Great Lakes fish community. Pages 145–184 *in* W. W. Taylor and C. P. Ferreri, eds. Great Lakes Fisheries Policy and Management: A Binational Perspective. Michigan State University Press, East Lansing, MI.

——, N. R. Payne, J. E. Johnson, C. Bower, III, and M. P. Ebener. 1995. Lake trout rehabilitation in Lake Huron. Journal of Great Lakes Research 21 (suppl. 1):108–127

Foster, H. 1963. Twenty-five years of conservation in Michigan, 1921–1946. Michigan Department of Conservation, Fisheries Division Pamphlet 37, Lansing, Michigan.

Francis, J. 1994. Indiana's Lake Michigan creel survey results, Indiana Department of Natural Resources, 1993.

——, S. R. Robillard, and J. E. Marsden. 1996. Yellow perch management in Lake Michigan: a multi-jurisdictional challenge. Fisheries 21:18–20.

Guthrie, D., J. M. Hoenig, M. Holliday, C. M. Jones, M. J. Mills, S. A. Moberly, K. H. Pollock, and D. R. Talhelm, editors. 1991. Creel and angler surveys in fisheries management. American Fisheries Society Symposium 12.

Halpern, T. N., T. S. Jones, and D. R. Schreiner. 1993. Completion report: Lake Superior and anadromous tributary creel surveys. Minnesota Department of Natural Resources, Division of Fisheries and Wildlife Report.

——, 1993. Charter fishing summary: Minnesota waters of Lake Superior. Minnesota Department of Natural Resources, Division of Fisheries and Wildlife Report.

Hansen, M. J., P. T. Schultz, and B. A. Lasee. 1990. Changes in Wisconsin's Lake Michigan salmonid sport fishery, 1969–1985. North American Journal of Fisheries Management 10:442–457.

——, 11 coauthors. 1995. Lake trout (*Salvelinus namaycush*) populations and their restoration in Lake Superior, 1959–1993. Journal of Great Lakes Research 21 (Suppl. 1):152–175.

——, 1986. Size and condition of trout and salmon from the Wisconsin waters of lake Michigan, 1964–84. Wisconsin Department of Natural Resources, Fish Management Report 126, Madison, Wisconsin.

——, 1989. Chinook salmon and Wisconsin's Lake Michigan sport fishery 1969–1985. Proceeding of the 1988 Northeast Pacific Chinook and Coho Salmon Workshop, Bellingham, Washington.

——, M. P. Ebener, J. D. Shively, and B. L. Swanson. 1994a. Lake trout. Pages 13–34 *in* M. J. Hansen, editor. The State of Lake Superior in 1992, Great Lakes Fishery Commission Special Publication 94–1, Ann Arbor, Michigan.

——, R. Schorfhaar, S. Schram, D. Schreiner, and J. Selgeby. 1994b. Declining survival of lake trout stocked in U.S. waters of Lake Superior during 1963–1986. North American Journal of Fisheries Management 14:395–402.

Hartig, J. H., J. F. Kitchell, D. Scavia, and S. B. Brandt. 1991. Rehabilitation of Lake Ontario: the role of nutrient reduction and food web dynamics. Canadian Journal of Fisheries and Aquatic Sciences 48:1574–1580.

Hartman, W. L. 1973. Effects of exploitation, environmental changes, and new species on the fish habitats and resources of Lake Erie. Great Lakes Fishery Commission Technical Report 22, Ann Arbor, Michigan. 43 pp.

Hatch, R. W., S. J. Nepszy, K. M. Muth, and C. T. Baker. 1987. Dynamics of the recovery of the western Lake Erie walleye (*Stizostedion vitreum vitreum*) stock. Canadian Journal of Fisheries and Aquatic Sciences 44(Suppl. 2):15–22.

Hayes, D. B. 1999. Issues affecting fish habitat in the Great Lakes basin. Pages 209–237 *in* W. W. Taylor and C. P. Ferreri, eds. Great Lakes Fisheries Policy and Management: A Binational Perspective. Michigan State University Press, East Lansing, MI.

Heard, W., R. Burkett, F. Thrower, and S. McGee. 1995. A review of chinook salmon resources in Southeast Alaska and development of an enhancement program designed for minimal hatchery-wild stock interaction. American Fisheries Society Symposium 15:21–37.

Heidinger, R. C. 1993. Stocking for sport fisheries enhancement. Pages 309–334 *in* C. C. Kohler and W. A. Hubert, editors. Inland fisheries management in North America. American Fisheries Society, Bethesda, Maryland.

Herbert, H. W. 1851. Frank Forester's fish and fishing of the United States and British Provinces of North America. 3rd edition, Stringer and Townsend, New York. 359 pp.

Hilborn, R. and C. J. Walters. 1994. Quantitative fisheries stock assessment: Choice, dynamics and uncertainty. Chapman and Hall, New York. 570 pp.

Hill, T. K., and E. W. Shell. 1975. Some effects of a sanctuary on an exploited fish population. Transactions of the American Fisheries Society 104:441–445.

Hulse, S. R. 1994. 1993 Lake Superior creel census. Wisconsin Department of Natural Resources. 36 pp.

Hurley, D. A, and W. J. Christie. 1977. Depreciation of the warm-water fish community in the Bay of Quinte, Lake Ontario. Journal of the Fisheries Research Board of Canada 34:1849–1860.

Johnson, J. E., G. M. Wright, D. M. Reid, C. A. Bowen, and N. R. A. Payne. 1995. Status of the cold-water fish community in 1992. Pages 21–72 *in* M. P. Ebener, editor. State of the Lake Huron fish community in 1992. Great Lakes Fishery Commission Special Publication 95–2, Ann Arbor, Michigan.

——, J. VanAmberg. 1995. Evidence of natural reproduction of lake trout in western Lake Huron. Journal of Great Lakes Research 21 (Suppl. 1):253–259.

Jones, M. L., J. F. Koonce, and R. O'Gorman. 1993. Sustainability of hatchery-dependent salmonine fisheries in Lake Ontario: The conflict between predator demand and prey fish supply. Transactions of the American Fisheries Society 122:1002–1018.

Keller, M., K. D. Smith, and R. W. Rybicki. 1990. Review of salmon and trout management in Lake Michigan. Michigan Department of Natural Resources, Fisheries Special Report No. 14, Lansing, Michigan.

———, 1964. Lake Erie sport fishing survey. Ohio Department of Natural Resources, Division of Wildlife Publication W-316.

———, J. C. Schneider, L. E. Mrozinski, R. C. Haas, and J. R. Weber. 1987. History, status, and management of fishes in Saginaw Bay, Lake Huron, 1891–1986. Michigan Department of Natural Resources, Fisheries Technical Report 87–2, Lansing, Michigan.

Kerr, S. J. and G. C. LeTendre. 1991. The state of the Lake Ontario fish community in 1989. Great Lakes Fishery Commission, Special Report No. 91–3, Ann Arbor, Michigan. 38 pp.

Kinman, B. 1995. Use of cultured fish for put-grow-and-take fisheries in Kentucky impoundments. American Fisheries Society Symposium 15:518–526.

Knuth, B. A., S. Lerner, N. A. Connelly, and L. Gigliotti. 1994. Value systems and attitudes of fishery and environmental managers related to lake trout rehabilitation. Great Lakes Fishery Commission, Research Completion Report.

Krueger, C. C., D. L. Perkins, E. L. Mills, and J. E. Marsden. 1995. Predation by alewife on lake trout fry in Lake Ontario: role of an exotic species in preventing restoration of a native species. Journal of Great Lakes Research 21(Suppl. 1):458–469.

Lange, R. E., G. C. LeTendre, T. H. Eckert, and C. P. Schneider. 1995. Enhancement of sportfishing in New York waters of Lake Ontario with hatchery-reared salmonines. American Fisheries Society Symposium 15:7–11.

Larkin, P. A. 1977. An epitaph for the concept of maximum sustained yield. Transactions of the American Fisheries Society 106:1–11.

Lawrie, A. H., and J. F. Rahrer. 1973. Lake Superior: A case history of the lake and its fisheries. Great Lakes Fishery Commission, Technical Report 19, Ann Arbor, Michigan.

LHMU (Lake Huron Management Unit), Ontario Ministry of Natural Resources. 1994. 1993 annual report. Ontario Ministry of Natural Resources.

McGurrin, J., C. Ubert, and D. Duff. 1995. Use of cultured salmonids in the federal aid in sport fish restoration program. American Fisheries Society Symposium 15:12–15.

MacCall, A. D. 1990. Dynamic geography of marine fish populations. University of Washington Press.

MacCallum, W. R., S. T. Schram, and R. G. Schorfhaar. 1994. Other species. Pages 63–76 in M. J. Hansen, editor. The State of Lake Superior in 1992. Great Lakes Fishery Commission, Special Publication 94–1, Ann Arbor, Michigan.

Milliman, S. R., A. P. Grima, and C. J. Walters. 1987. Policy making within an adaptive management framework, with an application to lake trout (*Salvelinus namaycush*) management. Canadian Journal of Fisheries and Aquatic Sciences 44(Suppl. 2):425–430.

Morse, S. D. and D. R. Schreiner. 1991. Completion report: winter creel survey, Minnesota waters of Lake Superior. Minnesota Department of Natural Resources, Study 4, Job 166.

Nelson, D. D., and J. G. Hnath. 1990. Lake Michigan chinook salmon mortality-1989. Michigan Department of Natural Resources, Fisheries Technical Report 90–4, Ann Arbor, Michigan.

Nepszy, S. J., D. H. Davies, D. Einhouse, R. W. Hatch, G. Isbell, D. MacLennan, and K. M. Muth. 1991. Walleye in Lake Erie and Lake St. Clair. In P. J. Colby, C. A. Lewis and R. L. Eshenroder, editors. Status of walleye in the Great Lakes: Case studies prepared for the 1989 workshop. Great Lakes Fishery Commission, Special Publication 91–1, Ann Arbor, Michigan.

NYDEC (New York State Department of Environmental Conservation). 1994. 1994 Annual Report: Bureau of Fisheries Lake Erie Unit to the Lake Erie Committee and the Great Lakes Fishery Commission.

———, Undated. 1984 New York State Great Lakes angler survey, Volume I.

Noble, R. L. and T. W. Jones. 1993. Managing fisheries with regulations. Pages 383–404 in C. C. Kohler and W. A. Hubert, editors. Inland fisheries management in North America. American Fisheries Society, Bethesda, Maryland.

O'Gorman, R., and T. J. Stewart. 1999. Ascent, dominance, and decline of the alewife in the Great Lakes: food web interactions and management strategies. Pages 489–513 in W. W. Taylor and C. P. Ferreri, eds. Great Lakes Fisheries Policy and Management: A Binational Perspective. Michigan State University Press, East Lansing, MI.

Ohio DNR (Ohio Department of Natural Resources). 1994. Status and trend highlights Ohio's Lake Erie fish and fisheries.

OMNRSR (Ontario Ministry of Natural Resources, Southwestern Region). 1994. Lake Erie Fisheries Report 1993.

———, 1993. Lake Erie Fisheries Report 1992.

Orsatti, S. D., and G. C. LeTendre. 1994. Lake Ontario stocking and marking program 1993. Pages 14–58 in New York State Department of Environmental Conservation, 1994 Annual Report: Bureau of Fisheries Lake Ontario Unit to the Lake Ontario Committee and the Great Lakes Fishery Commission.

Parsons, J. W. 1973. History of salmon in the Great Lakes, 1850–1970. U.S. Department of the Interior, Fish and Wildlife Service Technical Paper 68.

Peck, J. W. and R. G. Schorfhaar. 1991. Assessment and management of lake trout stocks in Michigan waters of Lake Superior, 1970–1987. Michigan Department of Natural Resources, Fisheries Research Report 1956, Ann Arbor, Michigan.

———, W. R. MacCallum, S. T. Schram, D. R. Schreiner, and J. D. Shively. 1994. Other salmonines. Pages 35–51 in M. J. Hansen, editor. The State of Lake Superior in 1992, Great Lakes Fishery Commission, Special Publication 94–1, Ann Arbor, Michigan.

———, 1992. The sport fishery and contribution of hatchery trout and salmon in Lake Superior and tributaries at Marquette, Michigan, 1984–87. Michigan Department of Natural Resources, Fisheries Research Report 1975, 62 p.

Peyton, R. B. 1987. Mechanisms affecting public acceptance of resource management policies and strategies. Canadian Journal of Fisheries and Aquatic Sciences 44(Suppl. 2):306–312.

Pollock, K. H., C. M. Jones, and T. L. Brown. 1994. Angler survey methods and their applications in fisheries management. American Fisheries Society Special Publication 25.

Rakoczy, G. P., and R. F. Svoboda. 1993. Charter boat catch and effort from the Michigan waters of the Great Lakes, 1992. Michigan Department of Natural Resources, Fisheries Technical Report 93–2, Lansing, Michigan.

———, 1992. Sportfishing catch and effort from the Michigan waters of Lakes Michigan, Huron, Erie, and Superior, and their important tributary streams, April 1, 1990–March 31, 1991. Michigan Department of Natural Resources Technical Report 92–8.

———, R. F. Svoboda. 1994. Sportfishing catch and effort from the Michigan waters of Lakes Michigan, Huron, Erie, and Superior, April 1, 1992–March 31, 1993. Michigan Department of Natural Resources Fisheries Technical Report 94–6.

———, 1995. Sportfishing catch and effort from the Michigan waters of Lakes Michigan, Huron, Erie, and Superior, April 1, 1993–March 31, 1994. Michigan Department of Natural Resources Fisheries Technical Report 95–1.

———, R. N. Lockwood. 1988. Sportfishing catch and effort from the Michigan waters of Lake Michigan and their important tributary streams, January 1, 1985–March 341, 1986. Michigan Department of Natural Resources Fisheries Division Technical Report 88–11a.

Regier, H. A., V. C. Applegate, and R. A. Ryder. 1969. The ecology and management of walleye in western Lake Erie. Great Lakes Fishery Commission Technical Report No. 15, Ann Arbor, Michigan.

Ricker, W. E. 1975. Computation and interpretation of biological statistics of fish populations. Bulletin of the Fisheries Research Board of Canada, Bulletin 191. Ottawa, Ontario.

Roosevelt, R. B. 1865. Superior fishing; or the striped bass, trout and black bass of the northern states. Carleton, New York. 304 pp.

Ryckman, J. R. 1986. A creel survey of sportfishing in Saginaw Bay, Lake Huron, 1983–1984. Michigan Department of Natural Resources Technical Report 86–4.

Savoie, P. J., and A. Mathers. 1994. Recreational fisheries. *In* Ontario Ministry of Natural Resources 1993 Annual Report.

Schneider, J. C., and J. H. Leach. 1977. Walleye (*Stizostedion vitreum vetreum*) fluctuations in the Great Lakes and possible causes, 1800–1975. Journal of Fisheries Research Board of Canada 34:1878–1889.

Schram, S. T., J. R. Atkinson, and D. L. Pereira. 1991. Lake Superior walleye stocks: Status and management. Pages 1–22 *in* P. J. Colby, C. A. Lewis, and R. L. Eshenroder, editors. Status of Walleye in the Great Lakes: Case Studies Prepared for the 1989 Workshop. Great Lakes Fishery Commission Special Publication 91–1, Ann Arbor, Michigan.

Sissenwine, M. P. 1978. Is MSY an adequate foundation for optimum yield? Fisheries 3:22–24, 37–42.

Smith, S. H. 1970. Species interactions of the alewife in the Great Lakes. Transactions of the American Fisheries Society 99:754–765.

Smith, H. M., and M. Snell. 1891. Review of the fishery of the Great Lakes in 1885, with introduction and description of fishing vessels and boats by J. W. Collins. Report of the U.S. Commission of Fish and Fisheries 1887:1–333.

Stauffer, T. M. 1966. Lake trout angling on Keweenaw Bay in 1964. Michigan Department of Natural Resources, Research and Development Division Report 71.

Stewart, D. J., J. F. Kitchell, and L. B. Crowder. 1981. Forage fishes and their salmonid predators in Lake Michigan. Transactions of the American Fisheries Society 110:751–763.

Stewart, D. J., and M. Ibarra. 1991. Predation and production by salmonine fishes in Lake Michigan, 1978–88. Canadian Journal of Fisheries and Aquatic Sciences 48(5):909–922.

Swanson, B. L., J. D. Shively, and J. W. Heinrich. 1992. 1991 Lake Superior lake trout extraction. *In* W. MacCallum and L. Bird, editors. Lake Superior Committee 1992 Annual Meeting Minutes, Great Lakes Fishery Commission, Ann Arbor, Michigan.

Sztramko, L. K., and J. R. Paine. 1984. Sport fisheries in the Canadian portion of Lake Erie and connecting waters, 1948–1980. Ontario Ministry of Natural Resources, Technical Report Series No. 13.

Talhelm, D. R. 1988. Economics of Great Lakes fisheries: a 1985 assessment. Great Lakes Fishery Commission, Technical Report No. 54, Ann Arbor, Michigan.

Thomas, M. V., and R. C. Haas. 1994. Status of selected fish species in Michigan waters of Lake Erie and Lake St. Clair - 1993. Michigan Department of Natural Resources Fisheries Division, Lake St. Clair Fisheries Station, Mt. Clemens, Michigan. Prepared for the Great Lakes Fishery Commission Lake Erie Committee Meeting, Niagara Falls, Ontario, March 30, 1994.

———, 1994. Status of yellow perch and walleye in Michigan waters of Lake Erie, 1989–93. Michigan Department of Natural Resources Fisheries Division, Fisheries Research Report 2011.

USFWS and Bureau of Census. 1993. 1991 National survey of fishing, hunting, and wildlife-associated recreation. U.S. Department of the Interior, Fish and Wildlife Service and U.S. Department of Commerce, Bureau of the Census. U.S. Government Printing Office, Washington, D. C.

Wells, L., and A. L. McLain. 1973. Lake Michigan: Man's effects on native fish stocks and other biota. Great Lakes Fishery Commission, Technical Report No. 20, Ann Arbor, Michigan.

WSTTC (Wisconsin State/Tribal Technical Committee). 1995. Recommended maximum lake trout harvest for the Apostle Islands region of Lake Superior for the 1996–2000 fishing years (November 28, 1995-September 30, 2000).

Notes

1. Throughout we use American Fisheries Society common names. Thus various varieties of rainbow trout such as domestic rainbow trout and steelhead are included as rainbow trout.
2. Internal report, Lake Superior Fisheries Unit, Ontario Ministry of Natural Resources, provided by Wayne MacCallum, OMNR: Cullis, K. 1987. Inner Thunder Bay winter creel survey 1983–1986.
3. Internal report, Lake Superior Management Unit, Ontario Ministry of Natural Resources, provided by W. MacCallum, OMNR: Sport fish report 1993, dated 1994. Quik Report Number 94-2.
4. Internal report, Lake Superior Fisheries Unit, Ontario Ministry of Natural Resources, provided by W. MacCallum, OMNR: dated 1989, Co-operative angler program report for 1987–1988. Quik Report Number 89-4.

CHAPTER 11

Great Lakes Commercial Fisheries: Historical Overview and Prognosis for the Future

Russell W. Brown, Mark Ebener, and Tom Gorenflo

Introduction

Commercial fisheries on the Great Lakes played an important role in the settlement of the Great Lakes basin. Abundant fishery resources were a key factor in the establishment of early settlements in many areas of the Great Lakes. Along with timber, trapping, and mining, commercial fishing represented one of the key natural resource extraction industries that generated economic wealth to stimulate settlement and development of many Great Lakes ports.

The commercial fisheries have changed markedly over the past 150 years due to changes in the Great Lakes ecosystem, continual technological development of the commercial fishing industry, and recent shifts in fisheries management focus. Although the dynamic history of these fisheries are partially explained by examination of catch statistics, one cannot gain a full appreciation without understanding many factors influencing the fisheries including, technological changes in gear, changes in processing and distribution, marketing and economics, and regulation and management of the fisheries. In this chapter, we provide an overview of these factors, along with a brief history of the commercial fisheries in each of the five Great Lakes. More detailed analyses in the form of case studies for Lake Erie percids (Koonce et al. 1999), lake trout (Hansen 1999), and lake sturgeon (Auer et al. 1999) are contained elsewhere in this volume.

Technological Development

The continual technological development of fishing equipment, processing methods, and transportation/marketing has had profound impacts on the efficiency of commercial harvesters and the status of fish stocks in the Great Lakes (table 1). Early fisheries were relatively primitive, near shore operations that harvested fish to be sold locally because transportation to distant markets was lacking. As fishing gear developed, commercial harvesters were able to harvest fish from deeper waters, extending both the length of the fishing season and the number of species that could be harvested. The development of boats and vessels allowed harvesters to spend more days on the

TABLE 1

Technological innovations in gear development, transportation, processing, and preservation that resulted in improved efficiency of the Great Lakes commercial fishery

DATE	GEAR	INNOVATION	IMPACT	CURRENT USE
early 1800s	haul seines	—	allowed active harvest of near shore stocks	limited
early 1800s	gill nets	twine mesh	allowed effective harvest of offshore stocks	no longer used
early 1800s	set hooks	—	allowed deepwater harvest of predators (mainly lake trout)	limited
1830s	transportation	rail transport	allowed export of fisheries products outside of region	limited
1850s	pound nets	—	allowed efficient harvest of migrating fish	still used
1860s	gill nets	cork floats	improved net positioning	no longer used
1868	preservation	pan freezing	allowed preservation of fish for transport outside of region	no longer used
late 1800s	gill nets	cedar floats	improved floatation of net	no longer used
late 1800s	trap nets	—	allowed fishing in deeper waters and increased mobility compared with pound nets	still widely used
late 1800s	preservation	commercial freezers	allowed rapid freezing of fish	still widely used
1889	gill nets	steam driven lifters	increased the amount of gill net that could be fished	no longer used
1890s	vessels	steam powered tugs; gasoline launches	ability to fish more gear at greater distances	no longer used
1892	preservation	chilled ammonia freezing	improved ability to freeze large quantities of fish	no longer used
1905	gill nets	bull net (very deep gill nets)	greatly improved capture efficiency for lake herring	still used
1909	vessels	outboard engines	allowed for continuation of small boat fishery	still used
1913	gill nets	Crossley lifter	decreased lifting time allowing for more net to be fished	widely used
1920s	processing	filleting	reduced shipping weight of fish	still widely used
1925	gill nets	aluminum floats	allowed fishing at greater depths for deepwater chubs	still used
late 1920s	trap nets	deep trap nets (greatly increased depth of net)	greatly increased capture efficiency for lake whitefish and lake trout	still used
late 1920s/'30s	vessels	diesel power and steel hauled vessels	increased fleet durability and ability to fish in marginal weather	widely used
1930s	gill nets	cotton mesh	improved capture efficiency over linen and twine	no longer used
1932	electronics	fathometers	allowed for rapid determination of fishing depth	still widely used
late 1930s/'40s	processing	filleting and scaling machines	greatly reduced processing time and costs	widely used
late 1940s/'50s	gill nets	multifiliment mesh	improved capture efficiency over cotton; could be fished with removal for drying and treating	limited use
1950s	gill nets	plastic floats, floam core line	improved floatation and durability	widely used
1950s	processing	marketing rough fish as "fish sticks"	improved market for carp and other rough fish	still used
1953	electronics	echosounders	ability to locate concentrations of fish	still used
mid 1950s	electronics	radiotelephones	improved safety and marketing ability	widely used
late 1950s	trawls	—	allowed for development of alewife and rainbow smelt fisheries	widely used in Canadian Lake Erie fishery; limited elsewhere

DATE	GEAR	INNOVATION	IMPACT	CURRENT USE
early 1960s	processing	smelt processing equip.	allowed for the development of the Lake Erie smelt fishery	still used
1960s	gill nets	monofilament mesh	improved capture efficiency over cotton due to visibility and flexibility	widely used
1970s	electronics	loran-C navigation	improved ability to locate fishing areas and gear	widely used, but being phased out
1970s/'80s	gill nets	bull nets reintroduced to upper lakes whitefish fishery	capture efficiency for lake whitefish improved 1.5- to 2.5-fold	used in some areas
late 1980s/'90s	gill nets	reduced diameter of monofiliment twine	increased capture efficiency (1.2 fold) due to reduced visibility and increased elasticity	still used
early 1990s	electronics	Global Positioning Navigation	improved ability to locate fishing areas and gear	widely used

water, fish with more and larger gear, and transport their catch to more distant markets. These technological developments have resulted in overexploitation of fish stocks, and a general reduction in the number of commercial fishers that can be supported by each fishery.

Nets and Fishing Gear

Historically, commercial harvesters on the Great Lakes used a large variety of nets and fishing gear in commercial operations. Gear use changed as fishing effort shifted among species, and from inshore to offshore waters. Technological innovations including net materials, net configurations, and vessel improvements have resulted in greatly increased capture efficiency over the past 150 years (table 1).

Many fishers used set hooks to catch lake trout in the upper Great Lakes from the mid 1800s until about the 1940s. Set hooks consisted of a strong main line with many dropper lines to attach hooks. Routinely, fishers deployed 2,000 to 3,000 hooks per set baited with bloater, lake herring, and rainbow smelt that floated off the bottom. The entire gang of hooks was floated at any desired depth through the use of float and anchor lines. Hook-and-line gear was prevalent in the U.S. fisheries, but was less common in Canadian waters, except for southern Georgian Bay and the Bruce Peninsula (McCullough 1989). Commercial trollers fished for lake trout in Georgian Bay and the Bruce Peninsula until the 1920s, when most of the fishery was converted to recreational charter boats (Hile et al. 1951). Set lines are still used in the 1990s to harvest channel catfish, eels, and other less desirable species.

Haul seines were used in areas where fish congregated near shore during migrations or spawning (fig. 1). Haul seines consist of vertical walls of netting fastened on either side of a bag of netting. During operation, a section of shore waters is surrounded by the net and the wings of the net are drawn to shore, reducing the area of enclosure, and forcing fish into the bag of the net. Haul seines were large, often exceeding 700m in length, and were retrieved by hand or with the assistance of horses. Seines were extensively fished in the 1800s to fish spawning aggregations of lake whitefish, but their use declined as near shore stocks of fish were depleted. Seines are still used in some areas of the Great Lakes, although use has declined substantially since the late 1800s.

PANEL A.

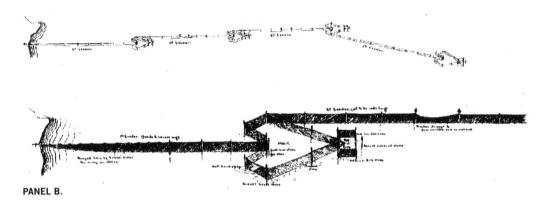

PANEL B.

FIG. 1. *Haul seine (Panel A) and pound net (Panel B) commercial fishing gear used to harvest fish from the Great Lakes. (Figures from Smith and Snell (1891), Appendix 1, Plates 26 and 29).*

Pound nets were a passive type of entrapment gear (i.e., fish encounter the net during normal movements) fished along the shoreline in many areas of the Great Lakes (fig. 1). Pound nets consist of vertical walls of netting maintained in position by a series of stakes driven into the bottom. The netting is deployed to form four parts: the pound, the lead, the heart, and the pot (fig. 1). Fish traveling along the shoreline encounter the lead, follow along the lead into deeper water, and are eventually funneled into the pot of the net. The mesh size of the net gradually decreases from the lead down into the pot to prevent fish from being gilled. Pound nets were normally lifted daily when possible, and fish were alive when removed from the pot.

Although pound nets were normally more efficient at catching fish than seining operations, their use was limited by several factors. Nets had to be set in shallow water (50ft or less) because of the length of stakes required. Nets were vulnerable to storm and ice damage, and were normally removed each fall, and set again in the spring. Because of the amount of effort required to install stakes, nets were fairly immobile. Pound nets were in use on Lakes Ontario and Erie by 1850, and were extensively used on Lake Erie, Saginaw Bay, southern Lake Michigan, and Green Bay (Van Oosten 1938). As of 1994, pound nets were fished in Wisconsin waters of Lakes Superior and

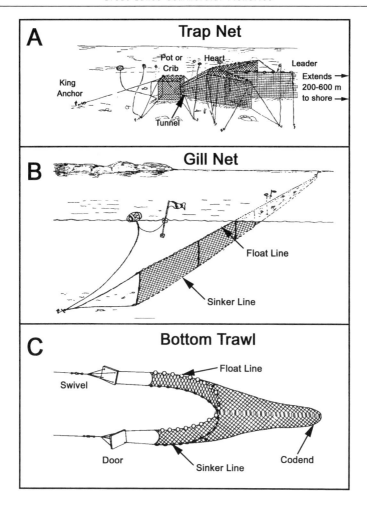

FIG. 2. *Trap net (Panel A), gill net (Panel B), and bottom trawl (Panel C) gear used to harvest fish from the Great Lakes. (Figures A and B from Rounsefell and Everhart (1953); Figure C from National Research Council (1988)).*

Michigan for lake whitefish, and in selected areas of the Great Lakes for rainbow smelt and several other species.

Gill nets were among the first gears used by commercial fishing operations, and their use continues to the present time. Native American bands used gill nets in the Great Lakes basin as early as 400 B.C. Gill nets consist of two lines, a float, and a lead (sinker) line with a vertical wall of netting between the lines (fig. 2). Nets are normally set in a straight line, anchored at each end, and fished passively. Fish are captured when they become wedged, gilled, or entangled in the netting. Gill nets gained popularity in the late 1800s because they required less labor and capital investment than pound nets, were easily moved among locations, and could be fished in deeper waters where fish stocks were less depleted. Because many fish captured in gill nets die, fish captured with this gear can have lower flesh quality and bring a lower market price if nets are not lifted regularly. Also, deployment of gill nets can result in mortality of nontarget species, which has emerged as an important management issue since the 1970s.

Historically, gill nets were made from crude twine, but improvements in net material resulted in large increases in fishing efficiency. Commercial fishers switched from traditional linen and cotton netting, to multifilament, and finally, monofilament nylon materials. During the 1930s, softer, more elastic cotton thread replaced linen in most major fisheries, resulting in greatly improved capture efficiency. The conversion from cotton to multifiliment gill nets occurred between 1949 and 1952 for the lake trout fishery (Pycha 1962) and between 1951 and 1961 for the lake herring fishery (Selgeby 1982). Multifilament nylon net materials were superior to linen and cotton twine because of durability, elasticity, and reduced visibility to fish (Jester 1977). Published information on the relative efficiency of nylon and cotton nets suggests that nylon nets are two to three times more efficient than cotton (Atton 1955; Hewson 1951; Lawler 1950; Pycha 1962). Unlike cotton and linen nets, multifilament nylon nets did not rot, and could be reset multiple times without removing the nets completely for drying and treating. These factors resulted in increases in the length of net that could be fished per vessel and the total number of days on the water (Christie 1978). In the 1960s, commercial fishers began to switch from multifilament to monofilament nylon webbing in gill nets, and this change resulted in nearly a two-fold increase in efficiency toward lake whitefish (Collins 1979). The conversion from multifilament to monofilament webbing was complete by the late 1970s.

The positioning of gill nets in the water column has important implications for their fishing efficiency. Nets may be fished on the bottom, floated, suspended at mid-depths, or fished from surface to bottom (deep nets). The practice of canning nets, or the use of floats (i.e., cans) to fish gill nets at intermediate depths, greatly improved the capture efficiency of the lake herring fishery about 1900 (Koelz 1926). The introduction of the bull net in 1905, a very deep gill net that fished effectively in mid-depths, resulted in drastically increased fishing efficiency for lake herring in Lake Erie. Bull nets were outlawed by Ohio in 1929, and by New York and Pennsylvania in 1934 (Regier et al. 1969).

Mesh size is perhaps the most important feature determining capture efficiency and species selection by gill nets. Bait nets (1½ in to 1⅝ in stretched mesh) are used to catch bait-sized bloaters and rainbow smelt, while small mesh nets (2 in to 3 in stretched mesh) for bluepike, deepwater chubs, lake herring, round whitefish, and yellow perch (Hile 1962). Larger mesh gill nets (4 in to 12 in stretched mesh) are used to target lake trout, lake whitefish, suckers, walleye, carp, and sturgeon (Hile 1962). Mesh size is the most frequently regulated aspect of gill net fisheries.

Changes in depth and twine diameter of gill nets occurred in the Great Lakes basin during the decades of the 1970s and 1980s, and have probably introduced some bias into the current commercial fisheries catch statistics. In the late 1970s, commercial gill net fisheries for lake whitefish began changing from nets that were 28 and 36 meshes deep to nets that were 50 meshes deep. Comparison of catches from the different depth of nets in Lake Huron show that the 50 mesh deep nets were 1.7 times more efficient at capturing lake whitefish than the 36 mesh deep nets (Collins 1987). During the early 1990s, Native American commercial harvesters in Michigan waters began converting from 50 mesh deep gill nets to nets of 60 and 75 meshes deep. Initial comparisons of lake whitefish catches found that the 75 mesh deep nets were 1.4 times more efficient than the 50 mesh deep nets in Lakes Superior, Huron, and Michigan during 1992–94 (unpublished data, Chippewa/Ottawa Treaty Fishery Management Authority, Sault Ste. Marie, Michigan).

In the 1800s, gill nets were lifted by hand over a roller, until steam-powered lifters were introduced on Lake Erie in 1889 (Keyes 1894). Early lifters consisted of revolving drum bearings with rows of rubber teeth to catch and pull the net (Koelz 1926). The introduction of the Crossley gill net lifter in 1913 allowed gill net operations to fish greater lengths of net (Prothero and Prothero 1990).

Trap and fyke nets (fig. 2) were introduced on the Great Lakes in the late 1800s. These nets operated on the same principles as a pound net, except that the heart and pot were enclosed on all sides so that the net could be submerged beneath the water, and the net was held in position by anchors instead of stakes. Trap nets were widely used in U.S. waters to target lake whitefish, lake trout, and walleye. Trap nets were prohibited in Michigan and Ontario waters in the early 1900s, because of fear that their high capture efficiency would rapidly deplete fish stocks. Deep trap nets, extremely tall trap nets fished in deeper waters of the upper Great Lakes, were introduced in the late 1920s (Van Oosten et al. 1946). The deep trap net was quickly prohibited by many management agencies because it was believed these nets caused depletion of lake whitefish stocks and excessive mortality of undersized fish in Lake Huron (Van Oosten et al. 1946). In the 1990s, trap nets are preferred by many fisheries managers because of the potential to release nontarget species. However, because the capital outlay necessary to fish trap nets is high (>$5000 (U.S.) per net) and the volume of fish that can be caught by trap nets is large, the number of trap net operations that can exist in an area is much less than with a gill net fishery. The areas that can be fished by trap nets are limited, as it is difficult to set and maintain the nets over rocky bottom, steep banks, in strong currents, and in depths greater than about 40m.

In the Great Lakes, recent uses of trawls have focused on the harvest of alewife and rainbow smelt in Lakes Michigan and Erie, and to a more limited extent, lake whitefish and lake herring in other areas. Trawls are bag-shaped nets dragged along the bottom or through the water column to strain fish from the water (fig. 2). The net opening is maintained by outward forces generated by water pressure and bottom friction against wooden or metal doors, towed at a specific angle to the net. The headrope is usually suspended with floats, and the bottom line is weighted to maintain contact with the bottom. Development of trawling in Great Lakes waters has been limited because many areas lack the clean bottom needed to fish this gear, the required capital investment and operating costs are high, and management agencies have discouraged or regulated against use of trawls.

Boats and Electronic Equipment

Although early fishery operations were shore-based operations using haul seines and pound nets, the development of boats and vessels allowed for the establishment of offshore fisheries. Fishing boats have progressed from row and sail powered vessels, through steam powered ships, and finally to gas and diesel powered boats.

In the early 1870s, Mackinaw, Huron or square stern, Norwegian, and pound-net were four types of sailing vessels used in U.S. waters (Milner 1874). Milner (1874) describes the Mackinaw as:

bow and stern sharp, a great deal of sheer, the greatest beam forward of amidships and taper with little curves to the stern. She is either schooner-rig, or with a lugsail forward, is fairly fast, the greatest surf-boat known, and with an experienced boatman will ride out any storm, or, if necessary, beach with greater safety than any other boat. . . . They have been longer and more extensively used on the upper lakes than any other boats, and with less loss of life or accident. . . . They are employed entirely with the light-rig gill-net stocks, and are usually from twenty-two to twenty-six feet in length.

The Huron or square stern were usually 9m to 12m in length and were preferred by large gill net operators fishing far from shore. The Norwegian is described as a "huge, unwieldy thing, with flaring bows, great sheer, high sides, and is sloop rigged" (Milner 1874). This vessel was preferred by Scandinavian fishers, but was known to be slow and difficult to row in calm winds. Pound net boats generally had flat bottoms, wide beams, and were well suited for the task of driving pound net stakes.

Fishery operations began converting from these early sailboats to steam-powered tugs and gasoline-powered launches by the late 1880s (Koelz 1926; Kennedy 1970). This conversion was gradual, and by 1919, there were 119 tugs, 626 gasoline launches, and 984 sail and rowboats fishing on the Great Lakes (McCullough 1989). Most gasoline powered vessels were small launches, 25 to 50 feet long, used by near shore pound and gill net fishing operations. Most early steam tugs in use were less than 65 feet long, so that they could be operated by unlicensed pilots and engineers (Koelz 1926). The introduction of steam tugs resulted in increased fishing intensity by allowing operations to fish up to five times as much net as they formerly fished using sails and oars (Toner 1939).

The first steel ship in the Great Lakes was built in Cleveland, Ohio in 1886, while diesel marine engines were manufactured at Two Rivers, Wisconsin by 1895. The first outboard motors were introduced in 1909. During the 1920s and 1930s, the fishing fleets began converting to diesel power and steel hulled vessels.

Steel-hulled gill net boats evolved toward enclosing the entire deck with a high box-like super-structure, known as the turtle deck, extending the entire length of the vessel (fig. 3). Initially developed by Lake Erie fishers, these enclosures were gradually upgraded from canvas to wood structures, and spread to other areas of the Great Lakes (Thompson 1978). Deployment and lifting of gill nets in the large steel-hulled boats are accomplished through large, sliding doors. Trap net vessels normally have enclosed pilot houses on the bow, and large open decks to provide room for lifting of nets (fig. 3). In the 1990s, the majority of Great Lakes fishery operations use 10–25m steel-hulled tugs rigged specifically for trap net, gill net, or trawl fisheries. Small aluminum boats (<7 m in length) equipped with outboards are still commonly used by some commercial fishers, primarily in small-scale gill net operations (fig. 3).

Electronic sounders were introduced on the Great Lakes in 1932 (Applegate et al. 1970), and by 1939 fathometers were in use on commercial fishing vessels in Lake Superior. This technology allowed fishers to locate specific depths needed for fishing deepwater species. By the mid 1950s, radio telephones were standard equipment on many larger Great Lakes vessels. Echo sounders were first utilized by Lake Erie fishers in 1953 to locate concentrations of fish. Loran-C navigation units were used on commercial fishing vessels in the 1970s, and Global Positioning Systems were

FIG. 3. *Fishing vessels commonly used by commercial fishers on the Great Lakes including a gill net tug (top), a trap net tug (center), and a small, open gill net boat (bottom).*

adopted around 1990. Use of other technologies such as sidescan sonar and satellite imagery have not been widely adopted by Great Lakes fishers.

Processing, Marketing and Distribution

The initial development of the commercial fisheries before 1850 was limited by the capability to preserve and transport landed fish. Fresh fish were sold in local markets, but could not be transported outside the local area without spoiling. Most fish transported outside local markets were gutted, beheaded, and packed in barrels of salt brine. Large sailing vessels were used to transport salt and barrels into Great Lakes waters from the East Coast, and returned with brined fish from Great Lakes ports to eastern markets. Salted fish were shipped as early as 1807, and shipments increased greatly with the opening of the Erie Canal in 1825 (Ashworth 1987) and the Ohio Canal in 1832 (Mansfield 1899). Shipments of barrels of salted fish from Cleveland, Ohio increased substantially following the completion of the Ohio Canal (Klippart 1877). The development of a salt-mining operation at Goderich, Ontario in the 1870s established a local supply of salt for fish preservation (Belden 1877). Fish were shipped by rail as early as the 1830s in the United States, and 1856 in Canada.

When adequate transportation became available, fish were transported on ice or in frozen form. Fish caught in the upper lakes during the late fall and winter could be frozen in the open air, bagged, and transported by ship or rail to distant markets. In the winter, ice was cut from ponds and lakes, and stored for transport of fish during warmer weather. S. H. Davis of Detroit introduced pan freezing to the Great Lakes in 1868. In his patented system, fish were placed in covered metal trays, packed in ice and salt, and frozen (Standsby 1963). Artificially-produced ice was available by 1870 with the invention of the Lowe Compression Ice Machine. The first freezers were installed by fish wholesalers on the U.S. side of Lake Erie in the late 1800s. By 1885, Sandusky, Ohio processed 4,100mt of fresh fish, 2,700mt of salted fish, 1,500mt of frozen fish, and 1,050mt of smoked fish (Smith and Snell 1891). In 1892, an alternative method of freezing using chilled ammonia was introduced, and eventually, freezing with chilled brine systems became popular (Stansby 1963). Gradually, shipments of fresh and frozen fish exceeded those of salted fish, and by 1900 salted fish were relatively rare (McCullough 1989).

Before 1900, almost all fish were dressed by removing the head and entrails. Filleting of fish did not become common in the Great Lakes until the 1920s (Anonymous 1929). Filleting had several advantages, including rapid freezing of fish and reduction of shipping weight. In 1937, Grow Brothers Fishery (Painesville, Ohio) patented a machine capable of scaling and washing 1,200 fish per hour. In 1942, Kishman Fish Company (Huron, Ohio) developed a new scaler capable of processing 45kg of fish every nine minutes. Previous electrical hand scalers took thirty minutes to process the same weight of fish. Marketing of coarse fish including carp improved with the production of fish sticks in the United States in 1952, and Canada in 1954 (Stansby 1963). The invention of a mechanized machine for cleaning smelt by Omstead Fishery (Wheatly, Ontario) was the impetus for development of a major trawl fishery for smelt in Lake Erie. Shrimp graders were used as early as 1960 to sort rainbow smelt during processing. During the 1980s, marketing of lake whitefish, deepwater chub, and chinook salmon roe for caviar increased the demand and price for female fish of each species caught during fall.

Once a Great Lakes fish is landed, it traverses a complex distribution network before finally reaching the consumer. Historically, the distribution was relatively simple: fish were landed; sold to a wholesale dealer; shipped to large metropolitan wholesale dealers; then distributed to smaller retail markets for final sale. That distribution system still exists today, but on a much broader scale. For example, in the Upper Peninsula of Michigan and nearby northern Michigan, there are ten large wholesale buyers and many more small buyers now dealing in fish harvested in the Great Lakes. Before the 1980s, the number of wholesale buyers in the area was five or less. Commercially-caught fish are bought and sold among these wholesale buyers. These buyers also distribute fish to buyers in larger metropolitan areas like New York, Chicago, and Detroit, restaurants, grocery stores, large food processing companies, and directly to the consumer.

Since the 1980s, there has been a growing trend for fishers to market their catch directly. Cooperative ventures, in which groups of commercial fishers band together to market their product, are common. The basic purpose of these cooperatives is to help commercial fishers receive a higher price per pound for their fish than they could get by simply selling their catch to a wholesale buyer. There are commercial fishery cooperatives in the Door County and Bayfield Peninsula areas of Wisconsin. The Ontario Fish Producers and Lake Erie Fish Packers and Processors Associations have operated much like fishery cooperatives, providing significant marketing advantages to fishers in Ontario. The Freshwater Fish Marketing Board of Canada operates as a cooperative and serves as a conduit for the sale of Canadian-caught fish into the United States. The Great Lakes Fisheries Development Foundation, Inc. currently serves to introduce Great Lakes fish to wholesale buyers and consumers in countries outside North America. Many commercial fishers in both the United States and Canada now market part of their catch from small retail shops located near their commercial fishing bases in much the same way the wholesale buyers do by selling fish directly to the consumer, restaurants, and grocery stores.

Employment and Economics

Commercial fishing was an important industry during the settlement and development of the Great Lakes watershed. At the height of the commercial fishery, over ten thousand people were directly employed in the industry as fishers, processors, or marketing personnel. The importance of the fishing industry has declined from historical levels, resulting in a decline in the percentage of the population employed in the fishing industry. The economic development of recreational fisheries since the late 1960s has greatly exceeded the value of the current commercial fisheries (Talhelm 1988).

In the late 1800s, Milner (1874) estimated investment in commercial fishery operations on Lake Michigan alone at approximately $580,000. In 1872, an incomplete record estimated a total harvest of 14,650mt of fish with a value of $1.6 million (U.S.) (Milner 1874). True (1887) reported that commercial operations in U.S. Great Lakes waters employed 5,050 fishers, had $1.3 million dollars in capital equipment investment, and produced $1.8 million in fisheries products annually. Koelz (1926) reported that from 1875 to 1925, annual production averaged over 45,000mt, and 1922 landings exceeded 63,500mt valued at over $9 million dollars from U.S. waters. By 1926, the Great Lakes fishing industry employed over 12,000 people, including fishers, processors, and marketing employees.

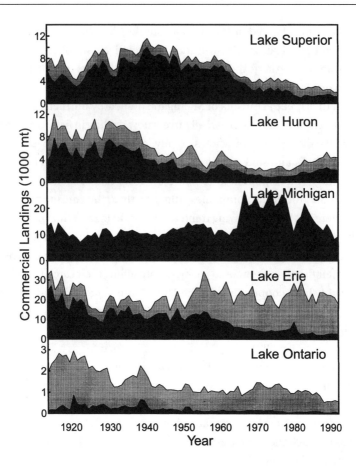

FIG. 4. *The relative contribution of U.S. (dark gray), Canadian (light gray), and tribal (black) fisheries to overall landings on each of the five Great Lakes from 1914–1992.*

The McKinley tariff, enacted by the United States in 1890, greatly affected the development of commercial fisheries in the United States and Canada. This legislation placed a tariff of three-quarters of one cent on fresh and frozen fish imported from other countries, including Canada. Fish captured by U.S. owned vessels or gear in Canadian waters were exempt from the tariff. The tariff encouraged American ownership of Canadian fisheries operations, and poaching by U.S. fishers in Canadian waters. During this period, poaching by U.S. fishers was common with hundreds of American gill nets seized annually by Canadian enforcement vessels. The McKinley tariffs were gradually reduced and canceled in 1914, providing some relief to Canadian fishers. Following cancellation of the tariff, the Canadian fisheries on Lakes Erie and Ontario increased in size, and eventually, Canadian landings surpassed those of U.S. fisheries on the lower Great Lakes (fig. 4).

Dockside value of landings from the Great Lakes remained relatively stable from 1930s to the 1970s (fig. 5), with a small increase in the late 1940s and early 1950s associated with increased landings of lake whitefish, herring, walleye, and blue pike resulting from strong year-classes produced in the mid 1940s. Dockside value has increased steadily since 1970 (fig. 5), with the

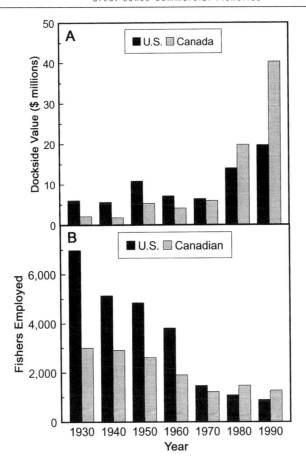

FIG. 5. *Dockside value of commercial landings (Panel A) and the number of commercial fishers employed (Panel B) in U.S. and Canadian commercial fisheries on the Great Lakes from 1930 to 1990. U.S. Data from Fishery Statistics of the United States, published by the U.S. Fish and Wildlife Service (1931–1955), U.S. Bureau of Commercial Fisheries (1956–1967), and the National Marine Fisheries Service (1968–1994). Canadian data from Fisheries Statistics (title and imprint vary), published by the Canadian Department of Trade and Commerce, Dominion Bureau of Statistics (1930–1964), Canada Dominion Bureau of Statistics (1965–1970), Statistics Canada (1971–1980), and Ontario Ministry of Natural Resources (1981–1990).*

recovery of lake whitefish, walleye, and bloater stocks, development of fisheries for exotic species (alewife and rainbow smelt), and worldwide increases in demand and prices for fishery products. Historically, the value of Canadian landings was less than U.S. landings, but the value of Canadian landings exceeded those of U.S. landings by 1979 (fig. 5). In 1979, the total dockside value of commercial landings was approximately $25 million (U.S.) with a regional economic impact of about $160 million (Fetteroff 1980).

Although the value of commercial fisheries has increased markedly since 1960, it has been dwarfed by tremendous increases in the value of recreational fisheries. In 1991, the dockside value of commercial landings was approximately $52.1 million (U.S.) with a regional economic impact of approximately $200 million. However, these levels paled in comparison with the established recreational fisheries, with trip and equipment expenditures of $1.3 billion in 1991 in the United States alone (U.S. Dept. of the Interior, Fish and Wildlife Service and U.S. Department of Commerce, Bureau of the Census 1991), and a total economic impact between $2 and $4 billion (U.S.) during the late 1970s and 1980s (Fetterolff 1980; Talhelm 1988).

The total number of U.S. commercial fishers has declined steadily from nearly 7,000 in 1930, to approximately 1,500 in the early 1970s (fig. 5). By the early 1990s, the U.S. fisheries employed less than 700 full-time fishers, with an additional 300 fishers licensed through Native American management agencies. The number of Canadian fishers declined from almost 3,300 fishers in 1920 to approximately 1,200 in 1970 (fig. 5). Canadian fisheries employment has stabilized with the number of licensed fishers fluctuating between 1,000 and 1,500 since 1970 (fig. 5). Several factors have contributed to this decline, including fluctuating abundance of fish, high operating costs, marketing problems, contaminant issues, and restrictive management policies favoring recreational fisheries. Technological advances and associated operating costs have reduced the economic carrying capacity of the Great Lakes ecosystem to support a commercial fishing industry. The landings needed to economically support the average commercial fisher have increased significantly over the past two centuries due to increased capital investment and higher standards of living. Restrictive management practices including limited entry, quota systems, area restrictions, and gear restrictions have also contributed to the decline in the number of fishers. The recent conversion from gill nets to impoundment gear in U.S. fisheries has also limited the number of participants in the fisheries due to the high level of investment required in netting gear and vessels.

Management and Regulation of Commercial Fisheries

Management of the commercial fisheries in the Great Lakes has always been confounded by conflicts between state, provincial, federal, and tribal organizations with management authority over various regions of the Great Lakes. In Canada, the federal government assumed primary management authority for the Canadian waters of the Great Lakes from 1867 to 1899. The Canadian federal government established the Department of Marine and Fisheries under the Fisheries Act of 1868 (McCullough 1989). In 1885, the Province of Ontario passed a fisheries act similar to the federal legislation in an effort to exert provincial control of the management of the fisheries. Enforcement responsibility, including licensure of commercial fisheries, was transferred from federal to provincial control in 1899. Legislative responsibility for establishing regulations was disputed until the 1910s, when the federal government agreed to increased provincial input into the formulation of legislation. Currently, the federal Fisheries Act of Canada still provides legislation to protect and conserve fish stocks, but the provincial Game and Fish Act provides for licensure of the commercial fisheries. Although U.S. federal agencies, including the U.S. Fish Commission, the U.S. Bureau of Commercial Fisheries, the U.S. Fish and Wildlife Service, and the U.S. Geological Survey have historically provided research and management advice, fisheries management in the U.S. waters of the Great Lakes has largely remained the jurisdiction of individual states.

International attempts to coordinate fisheries management were initiated in 1892, when Canada and the United States established an international fishery commission to investigate overfishing and pollution, the establishment of closed seasons for the fisheries, and the stocking of fish in border waters. This committee recommended that a joint commission be established with the authority to regulate the international fisheries by establishing uniform regulations for both U.S. and Canadian waters. Although a draft treaty was signed in 1908, the U.S. House of Representatives refused to approve the treaty, and the treaty was abrogated in 1914 (Piper 1967). In 1909, the International Boundary Waters Treaty established the International Joint Commission to arbitrate questions concerning the use and pollution of boundary waters (Piper 1967).

In the early 1940s, the United States and Canada convened a Board of Inquiry to investigate the current state of fisheries management on the lakes (Gallagher and Van Oosten 1943). Recommendations from this board produced a treaty signed in 1946 that established an international commission to formulate common fishery regulations for the Great Lakes (Great Lakes Fisheries 1946). The states of Wisconsin and Ohio passed resolutions in objection to the treaty, and the agreement was never ratified (Anonymous 1947). In response to the collapse of lake trout populations due to sea lamprey depredation, Canada and the United States established the Great Lakes Fishery Commission in 1954 to eradicate sea lamprey, and coordinate fishery research programs. In the 1990s, the commission produced a Joint Strategic Plan for the Management of Great Lakes Fisheries (Great Lakes Fishery Commission 1994) that establishes a set of fish community objectives for each lake.

Since the 1970s, a number of groups of native peoples in the United States and Canada have sought to reaffirm fishing and management authority granted by historic treaties. In the United States, several of these groups have won legal cases, establishing their right to fish, and in some cases, to manage areas of the Great Lakes. Although several groups are exercising subsistence and commercial fisheries in Canadian waters, the issue of treaty fishing is still emerging among native peoples in Canada. Management jurisdiction of commercial fisheries on the Great Lakes is currently shared by eight states, one province, two federal, and several Native American management agencies.

Management of State and Provincial Licensed Fisheries

Regulatory efforts by Canadian management agencies were initiated earlier, and were generally more restrictive than regulatory efforts by U.S. agencies. The Ontario legislature passed legislation providing for fish passage over mill dams, regulating fishing methods, seasons, and locations for Atlantic salmon in Lake Ontario tributaries as early as 1828. Additional legislation between 1823 and 1843 regulated lake herring fisheries in Burlington Bay, lake whitefish in the Niagara, St. Clair, and Detroit Rivers, and lake trout in Kent and Essex Counties in Lake Erie (McCullough 1989). The passage of the Fisheries Inspection Act of 1840 regulated the quality of fish products packed in Canada. These acts also established an enforcement network consisting of a superintendent, fishery overseers, and guardians to enforce fishing regulations.

Management regulations in U.S. waters were not initiated until the mid 1800s. Ohio instituted regulations in 1857 to prevent the disruption of natural fish movements in rivers (Woner 1961), while Michigan enacted initial commercial fishing laws in 1865. Initial efforts to coordinate management efforts between U.S. states were made at a meeting of the U.S. Fish and Game Commission meeting in 1883 (True 1887). Although the commission meeting resulted in thirteen specific recommendations, none of the state legislatures adopted the resulting recommendations. Between 1883 and 1941, twenty-six additional interstate and international conferences were largely unsuccessful at achieving interjurisdictional coordination of management efforts.

Currently, there is better coordination of interstate and international management activities facilitated by individual lake committees of the Great Lakes Fishery Commission. Lake committees are composed of one fishery manager from each of the political jurisdictions bordering each lake. During the late 1980s and early 1990s, lake committees developed objectives for managing fish communities of each lake (Busiahn 1990; Kerr and LeTendre 1991; DesJardine et al. 1995; Eshenroder et al. 1995). Lake committee objectives address habitat management, sea lamprey control, introduced species, stocking of hatchery-reared fish, and establish goals for the structure of future fish communities. Goals for each species, or groups of species, typically focus on the harvesting of fish and incorporate some expectation of future yields based on historic harvests over some specified stable time period (Busiahn 1990; DesJardine et al. 1995; Eshenroder et al. 1995).

State and provincial management has focused on several objectives, including protection of exploited stocks, regulation of unintended by-catch, protection of recreationally-sought species, segregation of sport and commercial fisheries, and reduction of interference with navigation and other activities. Management agencies have used several regulatory measures to achieve management objectives, including limitations on the number of licenses, gear restrictions, minimum size limits, quotas, protected species, closed seasons, and closed areas (refuges).

Licensing and permitting have been used historically to regulate the amount and distribution of commercial fishing activity. The Province of Ontario passed fishery legislation in 1885 that formally instituted a licensing system in Great Lakes waters (Province of Ontario 1885). Before the 1900s, pound net sites were assigned and leased on both the Canadian and U.S. shores of Lake Erie. The state of Michigan began licensing nonresident commercial fishers in 1865 and resident fishers in 1907 (Brege and Kevern 1978), while Ohio initiated licensing in 1906 (Woner 1961). Licensing allowed managers to collect information about commercial landings through mandatory reporting programs and to enforce fishing regulations.

Management agencies have enacted size limits to prevent exploitation until recruiting fish could reach market size or sexual maturity. Michigan enacted minimum size limits on a regional basis for yellow perch and white suckers as early as 1893, and on a statewide basis in 1897 (Brege and Kevern 1978). In 1922, Ontario instituted minimum size limits on lake herring, sturgeon, blue pike, yellow perch, white bass, sheepshead, lake whitefish, and lake trout. With the recent shift from gill nets to impoundment gear in many U.S. fisheries, size limits continue to be an important regulatory tool because of the opportunity for release of undersized fish. Size limits are of little utility in regulating fisheries conducted with gear causing direct mortality of captured fish (i.e., gill nets and trawls), unless they are accompanied by corresponding gear restrictions and adequate enforcement.

Gear restrictions have been enacted by management agencies to reduce mortality of undersize or unintended species, to reduce competition between fishers and gear types, to reduce the efficiency of the fishery, or to reduce physical damage to habitat by active gear (e.g., seines and trawls). Mesh size was regulated to prevent mortality to undersized fish in seine, pound, gill, and trap net fisheries as early as 1889 in Michigan (Brege and Kevern 1978), and by 1906 in Ohio (Woner 1961). On Lake Erie, where fishing intensity was greatest, shoreline areas were assigned to specific pound net operations, and regulations limited the distance gill nets could be set from shore (Keyes 1894). Several types of gear, including trap nets, deep trap nets, and bull nets have been prohibited for periods of time because of fears that their efficiency would result in stock depletion. The state of Michigan banned trap nets in 1905, but reinstated their use in the 1920s (Brege and Kevern 1978). The use of deep trap nets was outlawed in Michigan and Wisconsin waters of Lake Michigan in 1935, Indiana waters of Lake Michigan and Michigan waters of Lake Superior in 1936 (Van Oosten et al. 1946). Trap net fishing was not permitted in Canadian waters in the early years, but illegal use of trap nets in Georgian Bay was widespread (McCullough 1989). Trap nets were finally legalized in Canadian waters in 1950, and have since replaced most pound nets on Great Lakes waters (Kennedy 1970).

Gill nets were banned completely as a commercial gear by the states of New York, Ohio, and Indiana as recently as 1994, and the state of Michigan banned the use of large-mesh gill nets (greater than 4½ in stretched mesh) in the 1970s. Although small-mesh gill nets are still used in some fisheries (e.g., deepwater chub and yellow perch), only Wisconsin and Ontario currently allow the use of large-mesh gill nets. In the 1990s, trap nets are a preferred fishing gear in U.S. fisheries because the unintended by-catch can be released alive, and the market value of fish landed with trap nets is greater than those taken by gill nets. Gill nets are still widely used by Canadian commercial fishers in many areas of the Great Lakes. Large-mesh gill nets are also used extensively by Native American fishers, because of the low capital investment required to participate in the fishery. In 1994, there were only fourteen Native American trap net operations out of the nearly three hundred active Native American commercial licenses in U.S. waters of the Great Lakes.

As recreational fisheries developed on the Great Lakes, management agencies enacted regulations to protect recreationally-sought species from commercial harvest. Michigan prohibited commercial harvest of recreational species, including black bass in Saginaw Bay (1895); black bass statewide, crappie, brook, brown and rainbow trout (1909); rock bass, sunfish, and bluegill (1921); muskellunge and sturgeon (1927); coho salmon and northern pike (1966); chinook salmon, lake trout, and splake (1967); blue pike, walleye, and sauger (1970); and Atlantic salmon in 1974 (Brege and Kevern 1978). In 1922, Ontario banned the commercial harvest of largemouth bass, smallmouth bass, muskellunge, salmon, brook trout, brown trout, and rainbow trout. Recent regulation in U.S. waters has focused on elimination of by-catch of recreationally-sought species and on interference with recreational fishing activity.

Management agencies also instituted closed seasons and refuges for many species to protect spawning aggregations of fish from exploitation. The Canadian government established closed seasons for lake whitefish and lake trout in 1868, although early enforcement of these regulations was questionable. Michigan enacted closed seasons during the winter for lake whitefish in the Detroit and St. Clair Rivers in 1875, and statewide closures by 1897 (Brege and Kevern 1978). In the late 1800s, areas of the St. Clair and Detroit Rivers were closed to fishing to protect spawning populations of lake whitefish. Currently, closed fishing seasons center on the spawning seasons for lake trout and lake whitefish. More recently, a system of refuges has been established in Lakes Michigan, Huron, and Superior to protect lake trout from commercial and recreational fishing.

Quota management systems, which allocate a fixed number or weight of fish within areas of the Great Lakes to management jurisdictions (states and provinces), fisheries (commercial or recreational), or individual commercial operators (individual quotas), have been recently implemented in some areas of the Great Lakes. Canadian fish processors established informal quota systems for rainbow smelt (1960) and yellow perch (1979) as a mechanism for stabilizing prices. Lake Erie agencies implemented quota management systems in Lake Erie for walleye in 1976 and yellow perch in 1986 (Berkes and Pocock 1987; Hatch et al. 1990). These quota systems allocate an agreed-upon level of harvest to each management jurisdiction (Ontario, New York, Pennsylvania, Ohio, Michigan). Each management jurisdiction is then responsible for allocating its share of the overall quota between fishery types (commercial versus recreational), and adopting regulation and enforcement practices to ensure that its quota is not exceeded. Following negotiations between the Ontario Ministry of Natural Resources and the Ontario Council of Commercial Fisheries, Ontario implemented a system of individual quotas to allocate its share of the lakewide quota to individual commercial operations (Berkes and Pocock 1987). Michigan and Ohio have each allocated their entire quotas of walleye to the recreational fishery, but Ohio still allocates part of its yellow perch quota to the commercial fishery. Tribal fisheries for lake whitefish, lake trout, and bloater in treaty-ceded waters of the upper Great Lakes are currently managed under a quota system that determines the Total Allowable Catch (TAC) for each management area (TRFC 1992). Individual transferable quota systems are used to manage many of the remaining important state and provincial licensed fisheries, including fisheries in Wisconsin and Ontario.

In the early 1970s, many U.S. State management agencies began to carry out management actions that favored recreational fishery interests over commercial fishing interests. In many cases, the reallocation of fish resources toward recreational fisheries was driven by larger economic returns of recreational fisheries. The states with the two largest commercial fisheries, Michigan and Ohio, enacted highly restrictive regulations that resulted in considerable downsizing of the commercial fisheries. Wisconsin pursued a similar course of action, until the mid 1970s, when management was redirected to maintain viable commercial fisheries (University of Wisconsin Sea Grant Institute 1988). Michigan and Wisconsin also reduced the areas in which commercial fishing could be pursued (Brege and Kevern 1978). During this period, the Province of Ontario continued to manage resources to promote commercial fisheries in Canadian waters.

Development of Native American Fishing Rights

Native American bands of the Great Lakes basin were involved in some form of commercial fishing as early as the 1700s. Kinietz (1940) noted that local bands would travel to the Straits of Mackinaw to sell or trade freshly caught fish with European traders. At the time of European settlement in North America, native fishers were using gill nets, spears, weirs, and hook-and-line fishing methods to harvest Great Lakes fish (Kinietz 1940). Fish species harvested for commercial or subsistence use included lake whitefish, lake trout, lake sturgeon, lake herring, walleye and several species of suckers.

Between 1781 and 1854, Native American bands signed ten treaties or agreements that ceded lands and Great Lakes waters to the United States, British, and Canadian governments

(Minnesota Historical Society 1973). The treaties of 1836, 1842, and 1854 ceded lands and waters of the Great Lakes region to the U.S. federal government, while establishing native fishing rights in large areas of the U.S. Great Lakes waters. Many additional treaties were signed between the British and Canadian governments and native bands in what is now the Province of Ontario. These treaties or agreements essentially sold the land and water of the Great Lakes basin to the U.S. and Canadian governments. However, germane to most of the treaties was the guarantee that the signatory bands would be permitted to continue to hunt, fish, trap, and gather resources on lands and waters ceded to the various governments (World Wildlife Fund 1993) until the land was required for settlement. The Treaty of Ghent (1814) between the U.S. and British governments laid the foundation for U.S. federal policy in dealing with Native American bands by recognizing the sovereignty of the tribal governments.

The end result of the treaty process was that the Great Lakes bands and their associated fishing activities were restricted to reservations created by the treaties. Settlement of the Great Lakes basin after signing of the treaties also severely restricted resource use by the bands. Non-native commercial fishing operations quickly developed during the mid to late 1800s, and out-competed tribal subsistence fisheries for available near-shore populations of fish. The Canada Fisheries Act of 1857 was created to control the impact of the expanding non-native commercial fishing industry on Great Lakes fish by establishing regulations to protect fish populations. However, the Act did not recognize or accommodate native fishing rights, subjected native fishing activity to licensing, implemented closed seasons and other regulations, and restricted native fishing activities to domestic consumption. Native fishing rights were also restricted in the United States, but there were no federal laws that specifically mandated controls over native fishing activities. Instead, regulations adopted by state agencies were imposed on native fisheries. Although some Native Americans were fishing commercially under state licenses by the mid 1960s, the use of Great Lakes fish by native bands was limited.

Fishery management policies of state agencies began changing in the 1960s, providing the impetus for reasserting of treaty-protected fishing rights in the Great Lakes. Many state agencies promoted recreational fishing over commercial fishing through policies that restricted commercial fisheries, and allocated historically important commercial fish species solely to the recreational fishery. Large areas of the Great Lakes were closed to commercial fishing, the number of licenses was reduced, and the use of gill nets was banned by several state agencies. At the same time, there was an increased awareness among Native American bands of their inherent sovereign rights, and many used the judicial process to reaffirm treaty-reserved fishing rights that were being severely restricted by changing state management policies. Of key importance to their efforts was the basic principle of construction established by the U.S. Supreme Court in cases dating back to the 1800s, which stated that treaties must be liberally interpreted in favor of the Indians (Tulee v. Washington 1942). In 1979, the U.S. Supreme Court ruled that treaties must be interpreted as the Indians would have understood them, and ambiguities resolved in favor of Native Americans (United States v. Washington 1979).

There were three major court decisions which presented sweeping implications for Native American commercial and subsistence fishing in the Great Lakes: United States v. Michigan (1979), Lac Court Oreilles Band v. Voight (1983), and Sparrow v. The Queen (1990). These three court decisions facilitated implementation of native fishing by recognizing aboriginal rights to fish free of

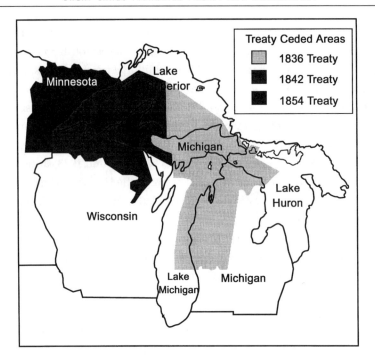

FIG. 6. *Treaty-ceded waters of Lakes Michigan, Huron, and Superior where tribal fishing rights have been reaffirmed based on the treaties in 1836 and 1842 between Native Americans and the U.S. Federal Government.*

state or provincial regulation throughout areas ceded to the U.S. and Canadian governments. In United States v. Michigan (1979), called the Fox Decision, the federal district court ruled the Bay Mills, Sault Ste. Marie, and Grand Traverse Band of Ottawa and Chippewa Indians retained rights to fish free of state regulation in waters of Lakes Superior, Huron, and Michigan ceded in the Treaty of 1836 (fig. 6). After the Fox Decision the 6th Circuit Court of Appeals further stipulated that the State of Michigan retained the right to regulate the tribal fishery in the event the State could prove native fishing activities were depleting the resource. In Lac Court Oreilles Band v. Voight (1983), called the Voight Decision, the federal district court ruled that Native bands signatory to the treaties of 1837 and 1842 retained the right to hunt, fish, trap, and gather resources outside of reservation boundaries in areas which now encompass parts of the states of Minnesota, Wisconsin, and Michigan. In Sparrow v. the Queen (1990), called the Sparrow Decision, the Canadian Supreme Court: recognized aboriginal rights to fish for food, ceremonial purposes, and barter throughout Canada; reminded governments that they are held to a high standard in their dealings with native peoples and must be liberal and generous; and provided a two-part test for ensuring that laws have due regard for aboriginal rights. The two-part test of Sparrow asked, "Does the legislation infringe on an aboriginal right, and is the infringement as reasonable and limited as possible?"

Prior to the Fox and Voight decisions, there were lower court rulings in the United States that also reaffirmed the existence of treaty-reserved fishing rights in the Great Lakes. However, these decisions addressed treaty rights within or adjacent to reservation boundaries, and not the much broader off-reservation rights addressed in Fox and Voight. The Jondreau Decision, in 1971, stated that treaty fishing rights did exist on Lake Superior waters within the boundaries of the Keweenaw Bay Indian Community Reservation. In the State v. Gurnoe (1972), called the Gurnoe Decision, the Wisconsin Supreme Court found that the Red Cliff Band of Lake Superior Chippewas retained the right to fish commercially and for subsistence in Lake Superior waters adjacent to the reservation. People v. LeBlanc (1976), again, supported the existence of treaty fishing rights, this time in Michigan waters of Lake Superior near the Bay Mills Indian Community reservation.

The fundamental difference between native fishing rights in Canada and the United States is that treaty rights are specifically protected under the Constitution in Canada, but not in the United States. Section 35(1) of the Canadian Constitution Act (1982) specifically recognizes and affirms the existence of aboriginal and treaty rights. There is no comparable section in the U.S. Constitution; instead, treaty rights are protected in the Supremacy Clause of the constitution which provides that federal law and treaties are the "supreme law of the land" (Jannetta 1991). It must be remembered that a "treaty was not a grant of right to the Indians, but a grant of rights from them" (United States v. Winans 1905). Therefore, treaty rights belong to each native government; the rights are not individual rights of members.

Management of Native American Fisheries in the United States

Central to the exercise of treaty rights by Native American bands is the premise that with the right to harvest the resource also comes the responsibility to protect and manage the resource. Courts in both Canada and the United States have recognized the authority of state, provincial, and federal governments to regulate the exercise of treaty fishing rights, and have generally required the bands to show that commercial fishing activities were taking place at the time of the treaty in order for there to be a commercial right (Jannetta 1991). Regulation of treaty fishing rights by non-native governments is permitted if the regulations do not discriminate against the native fisheries, and are in the best interest of protecting the fishery resource as outlined in the two-part test of Sparrow. Since the basic philosophy of using Great Lakes fishery resources is very different between Native American Bands and non-native governments, it is in the best interest of the bands to maintain regulation of native fishing under native control.

Native American governments view commercial and subsistence fishing as traditional ways for band members to support themselves and family members, whereas non-native governments in the United States tend to view commercial and subsistence fishing as an entity with lesser social value. A large proportion of band members in the Great Lakes region depend upon fishery resources for food, income and a connection to their spiritual and traditional past. Most current reservations (and many historical encampments) in the Great Lakes are near historical fish spawning grounds. Conversely, most of the non-native population does not derive its income or food from Great Lakes fish. Currently, U.S. state governments promote recreational uses of Great Lakes fishery over commercial uses because they believe that the economic return from recreational fishing is much greater than from commercial fishing, and because more people can participate in recreational fishing than commercial fishing. Essentially, state governments use Great Lakes fish resources based on the greatest economic return to the state, whereas Native American governments use of Great Lakes resources is based on tradition, food, and livelihood.

Management of Native American commercial and subsistence fisheries in U.S. waters of the Great Lakes has been promulgated through inter-tribal agreements, state and tribal governments, federal regulations, and court-sanctioned agreements or decrees. After the Fox Decision in 1979, regulations governing the native fisheries in the 1836 ceded waters of Lakes Superior, Huron and Michigan (fig. 6) were adopted by the U.S. Department of Interior. In 1981, the Secretary of the Interior allowed the federal rules to expire. In response to the expired federal regulations, the three native bands created the Chippewa/Ottawa Treaty Fishery Management Authority (COTFMA) to govern fishing activities by band members. COTFMA regulations were approved in 1981 by the same federal District Court, which two years earlier, had reaffirmed the existence of the treaty right (U.S. v. Michigan 1979). Allocation of fish resources between COTFMA member tribes and the State of Michigan had not been addressed in any of the court decisions up to this point. In 1985, COTFMA member tribes, the State of Michigan, and the U.S. government negotiated a fifteen-year agreement that addressed the allocation of the fishery resources within the 1836 ceded waters. This agreement provided for tribal-managed commercial fisheries within portions of the treaty-ceded waters of Lakes Michigan, Huron, and Superior. The negotiations resulted in the implementation of a comprehensive Consent of Decree in 1985, which was instituted by the federal District Court, despite objections by one of the bands (U.S. v. Michigan 1985). The District Court continues to have jurisdiction to settle disputes between the parties involved in the U.S. v. Michigan case.

Management of native fisheries in the 1842 waters of Lake Superior (fig. 6) has been less organized than management in the 1836 waters. Following the Voight Decision in 1983, two native bands from Wisconsin and one Michigan-based native band developed annual inter-tribal agreements which were designed to govern commercial fishing activities by members of all three bands in Michigan waters of the 1842 ceded territory of Lake Superior. The State of Michigan was not included in the development of the inter-tribal agreements in the 1842 ceded waters, and there has been no specific judicial resolution or allocation of resources between native and non-native fisheries in the 1842 waters. However, the State of Michigan unsuccessfully attempted to gain management control over native fisheries in the 1842 waters during the mid 1980s by petitioning the U.S. Federal government to impose state promulgated regulations over treaty fisheries. The federal government supported the contention of the three bands that the inter-tribal agreement contained sufficient regulations to protect the fishery resources.

Management of native commercial and subsistence fisheries within Wisconsin waters of Lake Superior has been established through native initiatives and negotiated settlements. Subsequent to the Gurnoe Decision in 1972, the Red Cliff Band held informal discussions with staff members of the Wisconsin Department of Natural Resources regarding band commercial and subsistence fishing activities on Lake Superior waters adjacent to the reservation. Red Cliff was generally adhering to fishing regulations already governing the state-licensed commercial fishery during that time. The informal discussions broke down in the late 1970s, after a Red Cliff Band member was arrested for harvesting walleyes in a tributary to Lake Superior. As a result, in 1979 the Red Cliff Band developed the first fishery regulations and court system to govern Native American fishing rights in the Great Lakes. The band subsequently negotiated a five-year fishery management agreement with the Wisconsin Department of Natural Resources in 1981. In 1986, Red Cliff, along with the Bad River Band, negotiated a ten-year fishery management agreement with the State of Wisconsin which allocated lake trout populations and fishing areas in the Apostles Island area of Lake Superior. The management agreement was modified in 1993 to address issues related to limiting the amount of gill net set, and the allowable harvest of lake trout.

As of 1994, there has been no exercise of off-reservation treaty fishing rights within the Minnesota waters of Lake Superior. During the late 1980s, the Grand Portage Band did negotiate a fishing agreement with the State of Minnesota. The agreement defined the on-reservation fishing zone in Lake Superior, allocated 12.25mt of lake trout annually to the band, created a season when the harvest of lake trout was prohibited, and called for mandatory reporting of catches and licensing of fishers. In exchange, non-native commercial fishing was prohibited in the Grand Portage zone.

Management of Native American commercial and subsistence fisheries within U.S. waters of the Great Lakes is currently facilitated through two independent inter-tribal organizations; COTFMA and the Great Lakes Indian Fish and Wildlife Commission (GLIFWC). As discussed earlier, COTFMA directly manages the native fishery in the 1836 ceded waters of Lakes Superior, Michigan, and Huron. GLIFWC is an organization of eleven bands, but has no direct management control over Great Lakes fisheries. Instead, GLIFWC serves as an advisory organization providing technical, legal and enforcement capabilities to bands involved in fishing issues. Each organization has an administrative, biological, enforcement, and public information branch. COTFMA also supports a judicial system for trying violations of the regulations.

Despite the formation of inter-tribal regulatory bodies, nearly all fishery management decisions are made by the individual native bands. Every native band involved with fishing on the Great Lakes possesses an internal structure for managing the fishery. The Grand Portage, Red Cliff, Bad River, Keweenaw Bay, Bay Mills, Sault Ste. Marie, and Grand Traverse bands have all formed conservation committees where fishing-related issues are first discussed and generally approved or denied. Ultimate regulatory authority, however, lies with the individual native governments. Fishery management recommendations are typically passed on from the conservation committees to the respective tribal government who possesses the ultimate regulatory authority over its members. In the case of COTFMA, issues pass from the conservation committee to the tribal government, and finally to COTFMA, who may or may not adopt a joint regulation or management strategy. It is important to note that each of the COTFMA member bands may adopt more stringent regulations that govern only the particular band's members.

The treaty-based commercial fishery in U.S. waters of the upper Great Lakes has developed rapidly. Although some Native Americans who fished prior to the affirmation of tribal fishing rights under state licenses were simply relicensed under the tribal management system, many Native Americans entered the fishery in response to court decisions, and at least three hundred tribal fishers were actively fishing by the early 1990s. In the upper Great Lakes, U.S. tribal harvest accounted for 19% of the total (U.S., U.S. Tribal, and Canadian) landings, and 16% of the dockside value in 1992.

The evolution of native management of fishery resources in Canada has not progressed as rapidly as in the United States. The existence of treaty-fishing rights in Canadian waters of the Great Lakes was only recently acknowledged with the Sparrow Decision in 1990. Since then, native bands in Canada have been attempting to gain recognition of treaty rights by the Province of

TABLE 2

Definitions used to standardize fishing effort of commercial fishers on the Great Lakes

As defined by Hile (1962) for the Great Lakes Fishery Commission.

GEAR TYPE	EFFORT DEFINITION	COMMENTS
Gill Nets	Lift of 1000 linear feet of net	Depth of net, mesh size, and number of nights fishing are not considered.

Impoundment Nets
  Pound Nets
  Shallow Trap Nets
  Deep Trap Nets
  Fyke Nets
	Lift of one net	Size of net, mesh size, length of set, and number of nights fishing are not considered.

Set Hook	Lift of 1000 hooks	Hook spacing, length of line set, and number of nights fishing are not considered.
Haul Seines	One haul of a 100-rod (503-m) seine	Seine depth or length of haul are not considered.
Trawls	One hour of actual dragging	Trawl size or configuration are not consider.

Ontario, and have begun exploring means to develop fishery management capabilities. As of 1994, numerous negotiations between Ontario and native bands over commercial and subsistence fishing were ongoing. Two native bands in the Bruce Peninsula area of Lake Huron were establishing regulations and the biological expertise necessary to govern fishing activities of their members. The treaty-fishing situation in Canada is potentially more complex than in the United States because there are many native bands in Canada, including twenty-three native bands located in close proximity to Lake Huron alone.

Native American bands have been elevated to the status of co-managers of fishery resources in the Great Lakes basin as a consequence of the reaffirmation of treaty fishing rights. That is, native bands now have equal status in planning, organizing, and implementing fishery management strategies in the Great Lakes (Smith and Burt 1987). The negotiated agreement in U.S. waters of the 1836 treaty area not only addressed allocation and lake trout rehabilitation, but provided for various cooperative fishery enhancement programs, and created the institutional mechanism for resolving disputes and furthering cooperation (Eger 1987). Joint technical, law enforcement, and public information programs were established between COTFMA and the Michigan Department of Natural Resources as a result of the 1985 Consent Decree. Technical committees composed of fishery biologists from the bands, state and federal agencies meet regularly to review the status of fish populations, develop assessment strategies, and produce estimates of the allowable harvests of important commercial fish species (PRFC 1992). These technical committees in both Wisconsin and Michigan have produced the most cooperative working relationships among native bands and state agencies.

Overview of Trends in Commercial Fisheries Yield

Information on commercial fish production in the Great Lakes was collected as early as 1867 in Canada, and 1879 in the United States (Baldwin et al. 1979), although written records of landings

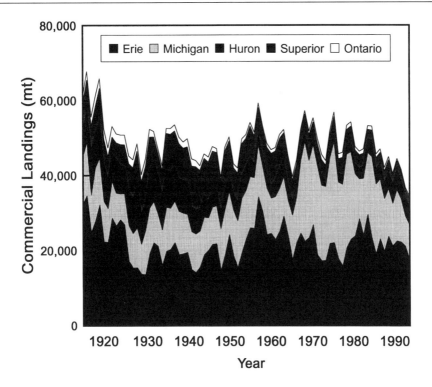

FIG. 7. *Total commercial landings from each of the Great Lakes from 1914 to 1992.*

date from the early 1800s. Based on the descriptive writings of early authors (Milner 1874; Klippart 1877; True 1887; Keyes 1894), there is considerable evidence to suggest that yields of many species had already begun to decline before the initial collection of commercial landings. In this chapter, commercial landings are reported back to 1914, the first year that landings were tabulated from all management jurisdictions. Licensed commercial fishers in Michigan were not required to submit daily reports detailing their catches until 1929, and mandatory catch reporting by commercial fishers was not instituted in Ontario until the 1940s. Therefore, both historical and modern landings data should be viewed as minimum estimates of actual harvests due to incomplete reporting.

Currently, mandatory landings reports from fishers are compiled by state, provincial, and tribal management agencies. Summary landings data are compiled by the U.S. Geological Survey (U.S. landings) and the Ontario Ministry of Natural Resources (Canadian landings). Fishers are currently required to report the location of their landings and effort data. Harvest locations are recorded based on a grid system that divides each of the Great Lakes into larger statistical districts and smaller scale grids (Smith et al. 1961). Catch per unit of standardized effort (CPUE) is often used as a rough index of fish population abundance. Fishing effort is reported in standardized units by gear type as outlined in table 2. A complete description of statistical districts and fishing effort standardization can be found in Hile (1962).

The five Great Lakes vary greatly in their ability to support commercial fishery yields. Lake Erie, the shallowest and warmest lake, has historically supported the highest total yield (fig. 7) and has sustained production per area from three to five times greater than any of the other Great Lakes (Loftus et al. 1987). Lake Michigan has sustained the next highest total yield and production per area. Loftus et al. (1987) compared yield per area from an early period (1920 to 1950) to that from a more recent period (1979 to 1983) and found that yield had declined in Lakes Superior, Huron, and Ontario due to declines in native stocks, but increased in Lakes Michigan and Erie due to high yields of exotic alewife and rainbow smelt, respectively. Although many historically important native fish stocks have declined, collapsed, or been extirpated, total yields have remained at a relatively constant level during the past century (fig. 7).

Lake Superior

All of the Great Lakes supported subsistence fisheries by native peoples long before colonial times. Commercial activity on Lake Superior was initiated in the seventeenth and eighteenth centuries, when two fur trading companies established commercial fishing stations (Nute 1944). The first pound net operations were established at Whitefish Point in 1850 (Koelz 1926), and commercial pound net operations were established in Canadian waters less than ten years later. Early fisheries targeted lake whitefish and lake trout, which were brined and shipped in barrels to urban markets, primarily in the United States. Total commercial landings were approximately 1,800mt in the early 1870s, rose to almost 11,600mt in 1941 (Lawrie and Rahrer 1972), and have declined to approximately 2,000mt in 1992.

During the 1900s, lake herring supported larger commercial harvests than all other species combined (fig. 8). Lake herring fisheries developed in the 1890s and exceeded the combined catch of lake trout and lake whitefish by 1908 in U.S. waters (Baldwin et al. 1979). Development of the lake herring fishery was slower in Canadian waters, and herring catches did not exceed other species until the mid 1910s (Baldwin et al. 1979). Lakewide commercial landings of lake herring averaged 5,500mt from 1936 to 1962, and declined from a peak harvest of 8,600mt in 1941, to less than 1,000mt by the early 1980s (Hansen 1990; fig. 8). Selgeby (1982) identified sequential overexploitation of discrete stocks as the cause of fishery declines in the Wisconsin waters of Lake Superior, although competition from exotic rainbow smelt has also been implicated. Fish community objectives established by the Lake Superior Lake Committee of the Great Lakes Fishery Commission focus on the rehabilitation of lake herring stocks to levels of abundance maintained from 1916 to 1940 for the purpose of lake trout rehabilitation, production of other predators, and fishery harvest (Busiahn 1990).

Commercial landings of lake whitefish in U.S. waters declined from 2,300mt in 1885 to 172mt in 1922 (Koelz 1929), almost resulting in commercial extinction along the U.S. shore. Catches of lake whitefish in Canadian waters also declined from 454mt in 1895 to 136mt in 1922 (Baldwin et al. 1979). There was a steady recovery of lake whitefish landings from 1925 to 1950 (fig. 8), but populations declined again in response to sea lamprey depredation (Lawrie and Rahrer 1972). Lake trout sustained an average annual yield of 1,800mt from 1929 to 1943, but a combination of fishery overharvest and sea lamprey depredation (Jensen 1978, 1994; Coble et al. 1990) resulted in drastic declines in harvest during the 1950s and 1960s (fig. 8). Lake trout and lake whitefish landings have rebounded since the initiation of sea lamprey control programs in the 1960s. By the late 1980s, lake whitefish landings were at the highest levels observed this century (fig. 8). Fishery

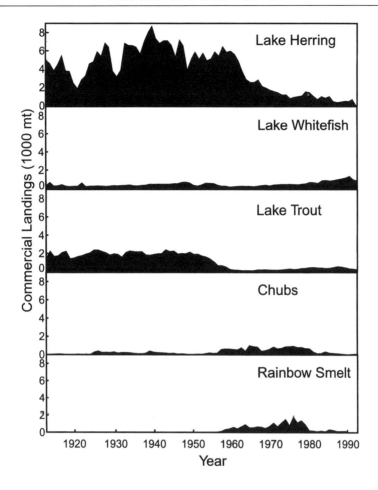

FIG. 8. *Commercial landings (mt) of five important species in the Lake Superior commercial fishery from 1914 to 1992.*

agencies expect to achieve a sustained annual harvest of 1,800mt of lake trout from the future fish community of Lake Superior (Busiahn 1990).

The deepwater chub species complex in Lake Superior was comprised of seven species (deepwater cisco, blackfin cisco, shortjaw cisco, kiyi, shortnose cisco, longjaw cisco, and bloater; Todd and Smith 1992) that were harvested and reported as a single-stock complex. This stock complex was fished with small-mesh gill nets in the deeper waters of Lake Superior (MacCallum and Selgeby 1987). The deepwater chub fishery was affected by market demand and the sequential collapse of deepwater chub fisheries on Lakes Michigan, Huron, and Erie (Lawrie and Rahrer 1972). By the late 1880s, stocks of blackfin cisco were depleted in Lake Michigan, resulting in a shift in fishing effort to Lake Superior. Blackfin ciscoes were intensively fished in Lake Superior for a decade, and were reported to be commercially extinct by 1907 (Koelz 1929). The collapse of the Lake Erie herring fishery in the 1920s resulted in increased fishing effort and elevated yields of lake herring and shortjaw ciscoes from Lake Superior (Lawrie and Rahrer 1973). Harvests of shortjaw ciscoes sustained the deepwater chub fishery through the 1930s, but there was a gradual shift in species composition from shortjaw ciscoes to bloaters (Koelz 1929; Dryer and Beil 1968). Reduced predation following the collapse of lake trout resulted in increased deepwater chub abundance during the 1950s and early 1960s (Hansen 1990). Deepwater chub fisheries did not develop beyond a by-catch fishery in Canadian waters until experimental fisheries were initiated in 1971. Redirected fishing effort toward deepwater chubs and increased efficiency of nylon gill nets resulted in intense exploitation of deepwater chub stocks. Deepwater chub landings averaged 700mt from the late 1950s until 1980, but subsequently declined to 35mt by 1990 (fig. 8).

Rainbow smelt invaded Lake Superior in the early 1930s, and reached commercially harvestable levels by the early 1950s (Lawrie and Rahrer 1972). Rainbow smelt are commercially harvested with small-mesh gill nets, trap nets, and trawls, although pound nets (United States) and dip nets (Ontario) are used to target spawning aggregations (MacCallum and Selgeby 1987). Commercial harvest increased in the 1950s and 1960s, with peak landings exceeding 1800mt occurring in 1976 (fig. 8). The commercial fishery for rainbow smelt was concentrated in western Lake Superior from Thunder Bay, Ontario to Ashland, Wisconsin. After 1976, there were major declines in abundance and landings (Hansen 1990; Selgeby et al. 1994), and by 1992, landings had declined to less than 30mt (fig. 8).

The reaffirmation of Native American fishing rights in Michigan waters of Lake Superior in the mid 1980s has resulted in the establishment of a Native American managed commercial fishery. In Lake Superior, Native American fisheries target lake whitefish and lake trout, which comprised 80% and 14% of the 1992 tribal landings, respectively. U.S. tribal fisheries accounted for 34% of the total landings, and almost 41% of the total value of landings from Lake Superior in 1992.

In the early 1990s, total annual commercial landings of 2,200mt represent only 37% of the long-term (1914 to 1992) average landings from Lake Superior. Current landings from Lake Superior are composed of historically-important native species including lake herring, lake trout, lake whitefish, and bloater. Although lake whitefish currently support yields approximately equal to historical average yields, lake trout and lake herring landings remain substantially below historical levels. Populations of lake sturgeon, blackfin cisco, shortjaw cisco, kiyi, and sauger are either commercially extinct or extirpated from Lake Superior (Table 3). Over the past eighty years, U.S. landings comprised about 75% of the total landings, but by the early 1990s, the U.S. share of landings had declined to 63% (fig. 4).

Lake Michigan

Commercial fishing operations were established in Lake Michigan in the 1820s, and were dominated by near shore fisheries using haul seines in the early years (Milner 1874). Gill nets were first utilized by non-native commercial fishers in 1846, and pound nets were introduced in the 1850s. Early fisheries targeted near shore populations of lake whitefish, and landings had already begun to decline before the first available landings records in 1879 (Milner 1874; Smith and Snell 1891). Early declines were attributed to commercial overfishing and sawdust pollution by sawmills, which covered important feeding and spawning areas in the lake and its tributaries (Milner 1874; Smith and Snell 1891).

TABLE 3

Future prospects for commercial exploitation of historically important species

SPECIES	LAKE SUPERIOR	LAKE MICHIGAN	LAKE HURON	LAKE ST. CLAIR	LAKE ERIE	LAKE ONTARIO
Lake Sturgeon	Poor	Poor	Fair	Fair	Poor	Poor
Lake Trout	Good	Fair	Fair	None	Poor	Poor
Atlantic Salmon	Introduced for Rec. Fishery	Not Present	Introduced for Rec. Fishery	Not Present	Not Present	Reintroduced for Rec. Fishery
Lake Herring	Good	Poor	Fair	Poor	Poor	Poor
Blackfin Cisco	Extirpated	Extirpated	Extirpated	Not Present	Not Present	Not Present
Deepwater Cisco	Not Present	Extirpated	Extirpated	Not Present	Not Present	Not Present
Shortjaw Cisco	Poor	Extirpated	Extirpated	Not Present	Extirpated	Extirpated
Shortnose Cisco	Not Present	Extirpated	Poor	Not Present	Not Present	Extirpated
Kiyi	Fair	Extirpated	Extirpated	Not Present	Not Present	Extirpated
Bloater	Fair	Excellent	Excellent	Not Present	Not Present	Extirpated
Lake Whitefish	Excellent	Excellent	Excellent	Poor	Fair	Good
Round Whitefish	Good	Good	Good	Poor	Poor	Poor
Walleye	U.S.: Allocated to Rec. Fishery Canada: Fair	Allocated to Rec. Fishery	U.S.: Allocated to Rec. Fishery Canada: Excellent	Allocated to Rec. Fishery	U.S.: Allocated to Rec. Fishery Canada: Excellent	U.S.: Allocated to Rec. Fishery Canada: Fair
Blue Pike	Not Present	Not Present	Not Present	Extirpated	Extirpated	Extirpated
Sauger	Poor	Poor	Unknown	Unknown	Poor	Unknown
Yellow Perch	Poor	Primary Allocation to Rec. Fishery	Good	Allocated to Rec. Fishery	U.S.: Primary Allocation to Rec. Fishery Canada: Excellent	Good
Northern Pike	Fair	Poor	Fair	Fair	Poor	Poor
Burbot	Good	Excellent	Excellent	Poor	Poor	Poor
Suckers	Good	Excellent	Excellent	Excellent	Excellent	Excellent
Alewife	Poor	Primary Allocation for Forage	Poor	Poor	Poor	Poor
Rainbow Smelt	Fair	Good	Fair	Poor	Excellent	Fair
Carp	Fair	Good	Excellent	Excellent	Excellent	Excellent
Channel Catfish	Poor	Fair	Good	Excellent	Excellent	Good
Bullheads	Poor	Poor	Poor	Good	Good	Excellent
White Perch	Not Established	Poor	Fair	Fair	Good	Excellent
White Bass	Poor	Poor	Poor	Fair	Good	Poor
Rock Bass	Poor	Poor	Fair	Good	Good	Good
Sunfish	Poor	Poor	Poor	Fair	Poor	Good
Sheepshead	Poor	Good	Good	Good	Good	Good
American Eel	Poor	Poor	Poor	Poor	Poor	Excellent

Lake whitefish, lake trout, lake herring, and lake sturgeon were the most important commercial species landed in the late 1800s. By 1890, lake whitefish landings had declined, but landings of all species from Lake Michigan averaged 18,600mt from 1893 to 1908 due to increased landings of lake herring (Baldwin et al. 1979). Although the separation of lake herring landings from those of other coregonines were not reliable, landings from the 1890s through 1908 are believed to have averaged approximately 9,000mt annually (Wells and McLain 1972, 1973).

The sturgeon fishery collapsed before 1900, and the lake herring fishery declined sharply between 1908 and 1911 (Baldwin et al. 1979; Wells and McLain 1972, 1973). By the late 1930s, lake trout, deepwater chubs, and yellow perch dominated commercial landings and value (Gallagher and Van Oosten 1943). Lake trout populations supported remarkably stable landings averaging 3,100mt between 1890 and 1945 (fig. 9).

Commercial landings of deepwater chubs were combined with those of lake herring or not recorded at all before 1926 (Baldwin et al. 1979). Since 1926, landings of this species complex including blackfin cisco, deepwater cisco, shortjaw cisco, shortnose cisco, longjaw cisco, kiyi, and bloater (Todd and Smith 1992) have been recorded together. Landings of deepwater chubs increased markedly from less than 1,000 mt in the late 1930s to approximately 5,000mt by the early 1950s (fig. 9). Shifts from a large-mesh (4½") gill net fishery in the late 1800s (Smith and Snell 1891) to sequentially smaller mesh (2½") by 1950 (Smith 1964) resulted in the early collapse of the largest species (blackfin and deepwater ciscos) and severe depletion of medium size species (shortjaw, long jaw, and shortnose ciscos, and kiyi). Gill net fishing intensity for deepwater chubs reached a record high level in 1953, following a substantial transfer of fishing effort from the lake trout fishery (Brown et al. 1987). Deepwater chub landings peaked in 1960, but declined temporarily after 1963, following the botulism deaths of several people from consuming smoked chubs (Baldwin et al. 1979). By the mid 1960s, only the bloater, the smallest of the deepwater chub species, contributed significantly to commercial landings (Brown et al. 1987). Bloater landings declined from 4,600mt in 1968 to less than 500mt in 1975 despite regulatory restrictions to stem the decline (fig. 9). The states of Michigan, Wisconsin, Illinois, and Indiana agreed to close the deepwater chub fishery in 1976 (Brown et al. 1985). A strong 1977 year-class resulted in the reopening of the fishery in 1979, and landings rebounded to almost 1,400mt by 1985 (fig. 9).

From 1940 to 1960, there were dramatic shifts in the species composition of commercial landings due to several factors (Smith 1972). Following the establishment of rainbow smelt, commercial landings of rainbow smelt peaked at almost 2,200mt before massive die-offs occurred in the winter of 1942 and 1943 (Van Oosten 1947). Rainbow smelt populations recovered quickly, reaching a peak yield of 4,100mt in 1958 (Baldwin et al. 1979). The exceptionally large 1943 year-classes of lake whitefish and lake herring supported elevated yields of these species in the late 1940s and early 1950s (fig. 9). From 1945 to 1952, landings of lake herring increased markedly to 4,400mt. Strong year-classes of walleye in Green Bay in 1943, 1950, 1951, and 1952 resulted in peak landings from 1949 to 1950 and from 1955 to 1956 (Pycha 1961). Landings of lake trout collapsed from almost 3,200mt in 1945 to only 15kg by 1954 (Baldwin et al. 1979), due to a combination of sea lamprey depredation and commercial overfishing (Eschmeyer 1957) although sea lamprey depredation alone could have caused the extinction of lake trout (Hile et al. 1951; Jensen 1994). By the late 1950s, the lake herring fishery also collapsed due to overfishing, and possibly competition or

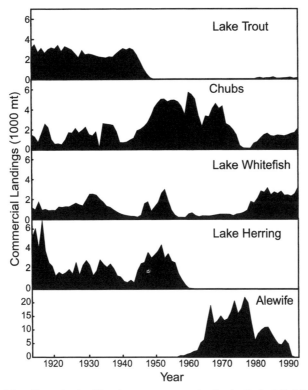

FIG. 9. *Commercial landings (mt) of five important species in the Lake Michigan commercial fishery from 1914 to 1992. Alewife landings are scaled differently than the other four graphs.*

predation on their larvae by expanding populations of alewife (Smith 1968, 1970). Lake whitefish populations collapsed by 1960, due to sea lamprey predation, but a long-term recovery of lake whitefish populations has continued through the early 1990s (fig. 9). Lake whitefish landings in the mid-1990s were at or near all-time record levels.

The alewife, first reported in Lake Michigan in 1949, reached commercially harvestable levels in the late 1950s, when a trawl fishery was initiated (Smith 1968). Although not used for human consumption in the Great Lakes, alewives were marketed for use in fish meal, fertilizer, and pet food (Smith 1968; Brown 1972). Commercial landings exploded from less than 45mt in 1956 to close to 19,000mt by 1967 (fig. 9). The trawl fishery, located mainly in Wisconsin waters since the 1970s, maintained landings of more than 4,500mt annually, until restrictive regulations by the Wisconsin Department of Natural Resources were enacted in 1991 to conserve alewife stocks as forage for salmonines.

Contaminant issues have had significant effects on commercial harvesting of several species of fish since the 1960s. Although commercial fishers began to harvest abundant coho salmon in southern Lake Michigan in the late 1960s, the U.S. Food and Drug Administration shut down the commercial fishery in 1969 because of high levels of DDT in canned salmon products. By the time that contaminant levels had dropped in salmon, these chinook and coho salmon were being allocated to the recreational fishery by management agencies. Landings of deepwater chubs declined following a botulism scare in 1963, high levels of DDT in 1971, and unacceptable levels of dieldrin in 1977 (Baldwin et al. 1979). Contaminant issues also resulted in declines in effort and landings of carp. Contaminant levels in nearly all Lake Michigan fish declined significantly from the 1960s to the 1990s, yet levels in some species, including lake trout, are still high enough to warrant health advisories restricting the consumption of some species. The goal of fishery management agencies is to reduce and eliminate toxic chemicals, where possible, to enhance fish survival and allow for promotion of fish that are safe for human consumption (Eshenroder et al. 1995).

The reaffirmation of Native American fishing rights in Michigan waters of northern Lake Michigan in the late 1970s has resulted in the establishment of a Native American managed commercial fishery. In Lake Michigan, Native American fisheries target four primary species, lake whitefish, bloater, lake trout, and round whitefish, which comprised nearly 97% of tribal landings in 1992. Lake whitefish are harvested by large boat trap net operations and large and small boat gill net fisheries (fig. 3). U.S. tribal fisheries accounted for 19% of the total landings, and almost 17% of the value of landings from Lake Michigan in 1992.

Lake whitefish, rainbow smelt, yellow perch, bloater, and white sucker currently dominate the commercial fishery landings in Lake Michigan. In the early 1990s, total landings averaged 9,600mt, or approximately 71% of the 1914 to 1992 long-term yield. Below average yields are largely the result of reduced landings from historically important lake herring and lake trout fisheries. Lake sturgeon and five species of the deepwater chub complex (blackfin, deepwater, shortjawed, and shortnose ciscoes, and kiyi) are commercially extinct or were extirpated from Lake Michigan (Table 3).

Lake Huron

Gill net operations were initially established on Lake Huron at or near Alpena, Michigan, and in Georgian Bay, Ontario around 1835 (Koelz 1926). Early fisheries targeted lake whitefish in shallow water, gradually shifting to fishing lake trout in deeper waters, as the inshore stocks of lake whitefish were depleted. Lake whitefish landings declined from 1,225mt in 1880 to 270mt by 1899 in U.S. waters, and from 544mt in 1875 to 12mt by 1900 in Canadian waters (Koelz 1929). By 1890, lake trout superceded lake whitefish as the primary fishery on Lake Huron in both U.S. and Canadian waters. Lake trout landings peaked in U.S. waters in the early 1890s, and in Canadian waters in the late 1890s (Baldwin et al. 1979). In the early 1900s, these fisheries gradually shifted from targeting lake whitefish and lake trout, to lake herring and deepwater chubs (Koelz 1929). By 1917, the combined harvest of herring and deepwater chubs rose to approximately 2,300mt (Baldwin et al. 1979). Following their introduction into Lake Huron, a concentrated fishery for carp in Saginaw Bay landed of over 500mt annually by 1917 (Berst and Spangler 1972).

Lake whitefish landings remained relatively stable from 1910 to 1930, with average lakewide landings of 1,100–1,400mt (Baldwin et al. 1979). The introduction of deep trap nets, the production of an exceptionally large 1929 year-class, and lamprey depredation all resulted in the destabilization of lake whitefish yield (Van Oosten et al. 1946). Deep trap nets were introduced off Alpena, Michigan in 1925 (Koelz 1926), and were widely used throughout U.S. waters of Lake Huron by 1933 (Van Oosten et al. 1946). Recruitment of the 1929 year-class and increased efficiency associated with deep trap nets resulted in peak yields approaching 2,600mt in the early 1930s (Van Oosten

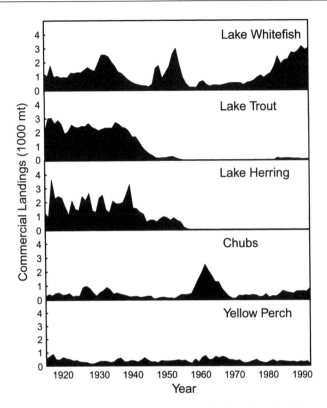

FIG. 10. *Commercial landings (mt) of five important species in the Lake Huron commercial fishery from 1914 to 1992.*

et al. 1946). Increased fishing effort attracted by recruitment of the 1929 year-class is thought to have resulted in overfishing that resulted in a partial collapse of stocks by the late 1930s (Van Oosten et al. 1946). By 1945, lake whitefish landings had declined to 250mt, approximately 10% of the level observed only twelve years earlier (fig. 10). Recruitment of an exceptionally strong 1943 year-class resulted in separate peaks in U.S. and Canadian commercial landings due to delayed recruitment of this year-class in Georgian Bay (Cucin and Regier 1966). U.S. landings peaked at over 1,400mt in 1947, while Canadian landings peaked at close to 3,000mt in 1953 (Baldwin et al. 1979). Lake whitefish landings declined again in the late 1950s after sea lamprey became numerous in the lake, but have recovered following sea lamprey control to current record levels in the early 1990s (fig. 10). Commercial harvest of lake whitefish during the 1990s met or exceeded the harvest goals set by fishery agencies (DesJardine et al. 1995).

The lake trout fishery produced landings in the 2,000 to 3,000mt range between 1914 and 1940 (fig. 10), but the fishery collapsed in U.S. waters by 1946, and in Canadian waters by 1955 (Berst and Spangler 1972). The lake trout collapse was caused by a combination of sea lamprey depredation and commercial exploitation, although the relative contribution of each of these factors is still debated (Hile 1949; Coble et al. 1990; Eshenroder 1992). Commercial landings of lake trout during the early 1990s represented only 10% of the average historic harvest from Lake Huron (Johnson et al. 1995).

Harvest of lake herring was concentrated in U.S. waters, with approximately 80% of the historical landings coming from Saginaw Bay. Lake herring landings peaked in the late 1930s, declined sharply from 1940 to 1945, and completely collapsed during the late 1950s (fig. 10). Although commercial fishing contributed to the collapse of lake herring, habitat destruction and pollution in Saginaw Bay, and the lakewide invasion of exotic rainbow smelt and alewife are probably the dominant causes for declines in lake herring stocks (Berst and Spangler 1972).

Following the collapse of lake trout and lake herring, fishing operations began to target deepwater chubs. Information on the distribution of deepwater chubs provided by the U.S. Bureau of Commercial Fisheries resulted in increased fishing effort in the late 1950s (Berst and Spangler 1972). Commercial landings peaked in 1961 and declined sharply by 1970 (fig. 10). Bloater fisheries recovered quickly during the 1980s, beginning with the 1977 year-class, and landings in the mid-1990s were approaching long-term averages.

The yellow perch fishery was initiated later than fisheries for lake whitefish, lake trout, and lake herring, although landings exceeded 1,700mt by 1900 (Baldwin et al. 1979). Eshenroder (1977) describes three landings stanzas for the Saginaw Bay fishery: a period from 1891 to 1916 when landings averaged 550mt; a period of low abundance from 1917 to 1963 when landings averaged 210mt; and a period from 1964 to 1977 when landings increased substantially to 640mt, and declined to 125mt. Hile and Jobes (1941) attribute the 1920s decline in landings to overfishing. Since 1900, Canadian landings have been irregular due to recruitment variability, but have shown an increasing trend, averaging 440mt during the 1960s (Spangler et al. 1977). On a lakewide basis, yellow perch landings have remained relatively stable during the twentieth century (fig. 10). Fish community objectives for Lake Huron are to maintain yellow perch as the dominant near shore omnivore, while sustaining a harvestable annual surplus of 500mt (DesJardine et al. 1995).

The reaffirmation of Native American fishing rights in U.S. waters of northern Lake Huron in the early 1980s has resulted in a shift in commercial fishery exploitation patterns. Native American fisheries target three primary species, lake whitefish, chinook salmon, and bloater, which comprised 93% of tribal landings in 1992. Lake whitefish are harvested by large boat trap net operations and small boat gill net fisheries. Commercial fisheries for chinook salmon resemble salmon ranching operations. Chinook salmon are stocked as age 0+ smolts and are harvested by intercept gill net fisheries as they return to stocking areas as mature spawners. Following declines in the recreational harvest, commercial harvest of chinook salmon exceeded recreational harvests in 1992. U.S. tribal fisheries accounted for 20% of the total landings and almost 30% of the value of landings from Lake Huron in 1992.

The annual average landings of 5,100mt in the early 1990s represents 83% of the long-term average from 1914 to 1992. Although these species are approximately equal to the long-term historical average, lake trout and lake herring landings remain substantially below historical levels. Sauger, and five species of deepwater chubs (blackfin, deepwater, short jaw and shortnose cisco, and kiyi) are either commercially extinct or extirpated from Lake Huron (table 3). Over the past eighty years, U.S. landings comprised about 58% of the total landings, but in the early 1990s, the U.S share declined to 43% (fig. 4).

Lake St. Clair and Connecting Waters

Historically, Lake St. Clair and its connecting waters supported small, but productive commercial fisheries for lake sturgeon, lake herring, lake whitefish, walleye, and carp. In the 1800s, productive fisheries for spawning lake whitefish and other species occurred in the Detroit River system. Commercial production peaked in the Lake St. Clair system in 1891, with total recorded landings of 1,700mt (Baldwin et al. 1979). Michigan has prohibited fishing by commercially licensed fishers since 1909 (Brege and Kevern 1978), but special permitted fisheries for the harvest of noxious fish have resulted in limited production of carp and channel catfish (Baldwin et al. 1979).

Several significant species in the early fisheries were rapidly depleted, and landings were reduced before 1930. Lake sturgeon landings approached 500mt in 1879, the first year that landings data was available (Baldwin et al. 1979). Landings quickly dwindled in the 1880s, and never exceeded 50mt after 1893. Lake herring landings were sporadically reported during the late 1800s, and probably exceeded 700mt in 1885, although Canadian landings are not available for that year (Baldwin et al. 1885). Landings declined quickly to 25mt by 1895, and never exceeded 1mt after 1918. Lake whitefish landings exceeded 130mt in the late 1880s, but declined gradually and did not exceed 1mt after the mid 1920s.

Several warm and cool-water species supported relatively stable yields until the fishery was closed in 1970. Northern pike were harvested primarily in Canadian waters, and landings remained relatively stable, averaging 12mt from 1876 to 1969 (Baldwin et al. 1979). Walleye landings showed three distinctive stanzas of production: averaging 186mt from 1889 to 1909, 24mt from 1910 to 1959 after U.S. commercial harvest was prohibited, and 117mt in the 1960s when the collapse of other stocks resulted in increased fishing pressure (Johnson 1977). With the exception of high landings in the late 1910s, Canadian yellow perch landings averaged approximately 20mt annually after the U.S. commercial harvest of perch was prohibited in 1909 (Johnson 1977). Carp landings averaged 156mt annually from 1940 to 1969, representing 41% of total landings during this period.

The discovery of mercury contamination in 1970 resulted in the closure of the Ontario commercial fishery. The Ontario fishery reopened a live capture fishery (i.e., trap nets, seines and hooks) in 1980 once contaminant levels had declined, with the issuance of ten limited commercial licenses, with a total commercial quota of 339mt (OMNR 1981). Commercial licensees were limited to harvesting yellow perch, northern pike, rock bass, crappie, lake sturgeon, and coarse fish species including carp, channel catfish, and suckers. In the mid 1980s, six licensees relinquished their licenses due to an inability to fish profitably without harvesting walleye and yellow perch (OMNR 1986). By 1994, only two of the remaining four licensees were still fishing and the total harvest of less than 6mt of channel catfish, bowfin, northern pike, sheepshead, crappie, rock bass, bullhead, and lake sturgeon was valued at approximately $6,000 (U.S.) (OMNR 1995).

Lake Erie

Although the second smallest of the Great Lakes in area and smallest in volume, Lake Erie has supported the largest commercial fisheries on the Great Lakes. Commercial production peaked at more than 34,000mt in 1915 and 1956, and reached all time lows of 13,000mt in 1929 and 1941

(Baldwin et al. 1979). Commercial fishery production in Lake Erie has often exceeded the production from the other four lakes combined (Baldwin et al. 1979).

Although the earliest net fishery in the Detroit River was initiated in 1774 (Swan 1977), commercial fishery operations were not initiated on Lake Erie until the 1790s, using seines in rivers and bays (Ashworth 1987). By 1824, catches of lake whitefish in the Detroit River approached five thousand barrels (approximately 225mt) annually (Swan 1977). The first exportation of brined fish to eastern markets occurred in 1826 (Hatcher 1945), and by 1837, export of brined fish from the Michigan territory approached 13,500 barrels, or approximately 600mt (Whitaker 1892). By the late 1830s, total fish sales of brined fish from Cleveland, Ohio averaged ten thousand barrels (approximately 450mt) annually (Klippart 1877). In the 1850s, lake whitefish and lake herring in the U.S. waters of Lake Erie were targeted with gill nets in the east end and pound nets in the west end. Pound nets were introduced in Canadian waters near Point Pelee around 1869 (Regier et al. 1969). By 1875, at least five-hundred pound nets were fished between the Detroit and Cuyahoga Rivers, harvesting 5,400mt of lake whitefish (Klippart 1877). The combined landings of pound, trap, and gill nets fished intensively along the shoreline exceeded that of offshore gill net catches by larger boats. Total landings in Lake Erie started to decline by 1877 (Baldwin et al. 1979).

Fishery development along the south shore of Lake Erie (Ohio and Michigan) proceeded more rapidly compared to development of the north-shore fishery (Ontario). In fact, more than 90% of the commercial catch during the 1800s was landed in U.S. ports, although U.S. operators also fished extensively in Canadian waters. Because the McKinley tariff discouraged importation of Canadian caught fish, American interests invested heavily in Canadian fishing operations to avoid the tariff. Sandusky, Ohio emerged as the major Lake Erie port for fisheries processing and marketing, and was referred to as the "freshwater fishing capital of the world" (Regier et al. 1969). Canadian fisheries expanded gradually until 1914 and then rapidly between 1914 and 1924. By 1925, Canadian landings exceeded U.S. landings for the first time (fig. 4).

In the early days of the Lake Erie fishery, lake sturgeon were regarded as a nuisance species that often damaged or destroyed fishing gear when captured. Fishers captured sturgeon with large mesh gill nets to remove them from fishing areas, and often piled sturgeon along the shoreline to be burned. Lakewide sturgeon landings peaked in 1885 at almost 2,400mt, with greater than 90% of the landings occurring in U.S. ports. By 1900, sturgeon landings had declined to less than 500mt. Additional details regarding lake sturgeon can be found in the chapter by Auer (1999).

Lake trout yields from Lake Erie exceeded 50mt in the late 1800s, but by the early 1900s, lake trout populations were largely depleted. Early anadromous runs of lake whitefish and lake herring in the Detroit and Maumee Rivers were destroyed by overfishing and pollution (Trautman 1957). By 1893, the lake whitefish stocks in the western basin of Lake Erie were substantially depleted, and lake herring and blue pike had become the dominant species in both landings and value (Wakeham and Rathbun 1897).

Lake herring most likely dominated commercial landings in Lake Erie from the early 1880s until the early 1920s, but landings fluctuated erratically even in the late 1800s (Scott 1951; Hartman 1972). Annual landings of lake herring exceeded 8,000mt from 1914 to 1924 (fig. 11). The lake herring fishery collapsed in the late 1920s, with landings dropping to 14,600mt in 1924, and below 225mt by 1929 (fig. 11). Regier et al. (1969) attributed this collapse to increased fishing intensity in

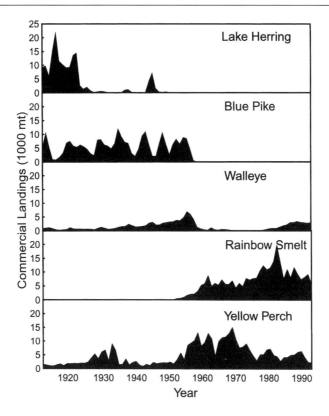

FIG. 11. *Commercial landings (mt) of five important species in the Lake Erie commercial fishery from 1914 to 1992.*

Canadian waters, coupled with continued intense fishing with bull nets in U.S. waters (Regier et al. 1969). Van Oosten (1930) reported high exploitation of heavily concentrated schools of lake herring in 1923 and 1924 off of Long Point, Ontario. An exceptionally strong 1944 year-class resulted in a resurgence of the fishery from 1945 to 1947, but commercial landings have been insignificant since 1960 (fig. 11). Introductions of exotic rainbow smelt and alewife, coupled with elevated nutrient and silt loading contributed to the final collapse of lake herring (Smith 1968).

Landings of blue pike fluctuated markedly between 1900 and 1960, with peak landings resulting from the recruitment of one or two strong year-classes per decade (fig. 11). During this period, approximately 75% of the total harvest of blue pike was landed by U.S. fishers. Strong year-classes in 1939, 1940, 1944, and 1949 sustained the fishery through the mid 1950s (Parsons 1967), but recruitment failure occurred after 1954, resulting in the collapse of the fishery from 1958 to 1960 (fig. 11). Although blue pike were fished intensively, habitat alteration and water quality deterioration may have contributed to their collapse. A few specimens of blue pike were captured throughout the 1960s and 1970s, but the subspecies is now thought to be extirpated from the lake.

Strong year-classes of lake whitefish in 1926, 1936, and 1944 resulted in short-term peaks in total landings, but landings were generally insignificant from the 1950s to the mid 1980s (Lawler 1965). Leach and Nepszy (1976) attributed declines in lake whitefish populations to siltation of spawning habitats, declining oxygen levels in the hypolimnion of the central basin, and increased water temperatures. Lawler (1965) reported that successful reproduction of lake whitefish is linked to a specific water temperature pattern occurring during spawning and incubation. Lake whitefish abundance increased sufficiently in Lake Erie during the 1980s to support a small fishery, but 1992 landings of 300mt were only about 10% of the peak landings of the late 1940s.

Annual walleye landings remained relatively stable, averaging 825mt between 1915 and 1934 (Leach and Nepszy 1976), but increased steadily to 2,400mt between 1934 and 1939 (fig. 11). A series of strong year-classes from 1942 to 1953 resulted in peak landings that approached 7,000mt in 1956 (fig. 11). Landings declined rapidly as a series of weak year-classes recruited to the fishery from 1954 to 1961 (Parsons 1970). The decline in landings has been attributed to overfishing (Regier et al. 1969), environmental degradation leading to oxygen depletion in the central and western basins (Regier et al 1969; Parsons 1970), declines in mayflies and other benthos (Beeton 1966; Parsons 1970), and competition and predation with expanding rainbow smelt populations (Regier et al. 1969).

The walleye fishery was closed in 1971 because of mercury contamination, and Ohio instituted a five-year ban on the commercial harvest. Reductions in exploitation, coupled with improved water quality conditions, allowed walleye populations to recover. As walleye populations increased through the 1970s and 1980s, recreational angler interest in walleyes expanded rapidly. During the 1970s, U.S. states shifted quota allocations for walleye from commercial to recreational fisheries, and by 1977, Ohio recreational anglers were catching more than twice their allocated quota. The four U.S. state agencies instituted regulations to restrict commercial harvest of walleyes, including outright harvest bans and restrictions on gill net fishing (Hatch et al. 1990). In contrast, the Ontario Ministry of Natural Resources continued to allocate walleye harvest to the commercial fishery.

Although walleye populations exhibited a tremendous recovery, commercial landings rebounded modestly due to restrictive management practices. Current allocation of walleye for recreational anglers by U.S. management agencies prevents commercial exploitation of the resurgent populations. The majority of commercial walleye harvest occurs in Ontario waters, where the fishery is managed under an individual transferable quota system (Berkes and Pocock 1987). In 1992, Canadian walleye landings, with an approximate value of $9.4 million (U.S.), represented 42% of the total value of Canadian landings from Lake Erie. By the early 1990s, combined commercial and recreational landings exceeded previous historical high levels of the mid 1950s (Hatch et al. 1990).

Rainbow smelt were first documented in Lake Erie in 1935 (Van Oosten 1937), and by 1948, some fishers began to deploy small mesh nets to target rainbow smelt in Lake Erie. In the mid to late 1950s, Ontario commercial fishers experimented with trawls in an attempt to harvest more rainbow smelt, and by 1959, large catches were taken. By 1965, Ontario instituted a trawling license system with a daily quota of 20mt, and the fishery was worth $2.7 million (U.S.) by 1967. During the late 1970s and 1980s, rainbow smelt were marketed to Japan, where they were served as an expensive delicacy and considered an aphrodisiac. An industry-initiated quota system was used as early as 1960 to stabilize prices for rainbow smelt, and harvesting is currently regulated under an individual transferable quota system by the Ontario Ministry of Natural Resources. Rainbow smelt landings peaked in 1982 at nearly 20,000mt, and have averaged 9,000mt from 1983 to 1992 (fig. 11). The rainbow smelt fishery has always been dominated by Canadian landings. The U.S. share of rainbow smelt landings from Lake Erie has never exceeded 4% since the inception of the fishery.

Yellow perch were a relatively unimportant commercial species before 1950. Landings remained relatively stable between 1910 and 1950 except for three peaks with yields exceeding 5,000mt during the late 1920s and 1930s (fig. 11). Shifts in fishing effort from collapsed stocks of blue pike and walleye to perch resulted in increased landings to 11,300mt in the late 1950s (Hartman 1972). A strong 1965 year-class resulted in peak yellow perch landings of 15,000mt in 1969 with almost 13,600mt landed in Canadian ports (Baldwin et al. 1979). Since 1970, yellow perch landings have averaged 5,500mt with the majority of fish landed by the Canadian fishery. In 1992, Canadian landings were valued at $4.8 million (U.S.), representing 37% of the value of the Canadian Lake Erie landings.

Historically, commercial harvest in Lake Erie has shifted from production by native coregonines (lake herring and lake whitefish), through native percids (blue pike, walleye, sauger), to its current state where commercial yield is dominated by walleye, rainbow smelt, and yellow and white perch (Smith 1968; Regier and Hartman 1973). Several ecological stresses including eutrophication (Beeton et. al. 1999), habitat alteration (Hayes 1999), species introductions (Leach et. al. 1999), and overfishing have contributed to shifts in commercial exploitation patterns. Current landings of 21,900mt are approximately 11% greater than the average from the previous eighty years, primarily due to large landings of exotic rainbow smelt. In 1992, walleye, rainbow smelt, and yellow and white perch accounted for 78% of total landings, and almost 91% of the total value of the Lake Erie fishery. Many of the historically important native species including lake trout, lake herring, sauger, and blue pike have failed to recover following population collapses (table 3), although recent population increases of lake whitefish are promising. Over the past eighty years, the Canadian fishery has accounted for about 54% of the total landings, but in the early 1990s, the Canadian share of landings has risen to 87% (fig. 4).

Lake Ontario

The magnitude of commercial fishery development and landings from Lake Ontario has been minor compared to the other Great Lakes. Historically, most of the commercial landings have come from Canadian waters (fig. 4), primarily from shallow areas in the eastern end and the Bay of Quinte. As the Lake Ontario basin was settled earlier than other regions of the Great Lakes, much of the early development of its fisheries occurred before accurate accounts of landings and fishery development were recorded.

By 1890, New York fishers used three vessels and 373 boats to fish over 330km of gill net, 42km of set lines, 288 pound and trap nets, 288 fyke nets, and 27 seines (Smith 1892). Lake herring dominated landings in 1890, but perch-pike (probably walleye and blue pike) were the most valuable species for these fisheries (Smith 1892). At least three species (lake sturgeon, Atlantic salmon, and blackfin cisco) were substantially reduced or extirpated before 1900 (Smith 1892). Atlantic sturgeon and Atlantic salmon populations were impacted by a combination of habitat alteration due to damming of tributaries used for spawning, and overfishing (Smith 1995). Blackfin cisco

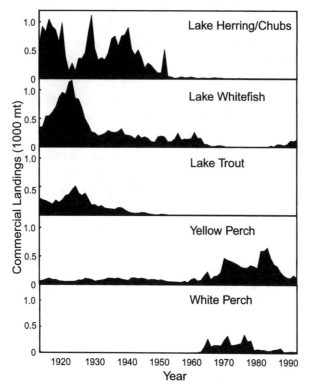

FIG. 12. *Commercial landings (mt) of five important species in the Lake Ontario commercial fishery from 1914 to 1992.*

populations collapsed in response to overfishing, in similar fashion to the collapses on Lakes Huron, Michigan, and Superior.

The species comprising commercial landings since 1900 are related to regional differences in habitat within Lake Ontario. The most productive fisheries in eastern Lake Ontario and the Bay of Quinte harvested a variety of species, including both cold-water (lake trout, ciscoes, lake herring, and whitefish) and warm-water (walleye, American eel, yellow perch, northern pike, carp, and sunfish) species. In western Lake Ontario and U.S. waters, the fishery was primarily dependent on cold-water species, including lake herring, deepwater chubs, lake whitefish, lake trout, and burbot, although blue pike supported a productive fishery in the extreme western end of the lake.

The primary cold-water species in Lake Ontario, including lake herring, deepwater chubs, lake whitefish, lake trout, and burbot, all collapsed between 1930 and 1960, and none currently produce landings at or near historical levels (fig. 12; Baldwin et al. 1979). Lake trout landings peaked in 1925 at 500mt and declined steadily to less than 45mt by 1947 (fig. 12). By the early 1960s, lake trout were extirpated from the lake, but have been subsequently reintroduced during rehabilitation efforts. Burbot populations were substantially reduced during the same time period, due to by-catch in the lake trout gill net fishery and sea lamprey depredation.

Landings of lake herring and deepwater chubs were not recorded separately before 1952. From 1914 to 1944, combined landings were erratic, due to variable year-class strength, occasionally reaching 900mt to 1,000mt (fig. 12). Severe reductions in the abundance of kiyi occurred in the early 1930s, leaving bloater as the only significant species in the deepwater chub fishery after 1935. Lake herring and deepwater chub populations collapsed in Lake Ontario in the mid 1950s, probably due to overexploitation resulting from the introduction of nylon gill nets and competition with expanding rainbow smelt populations (Christie 1972).

Lake whitefish in eastern Lake Ontario produced exceptionally high yields, exceeding 1,100mt in the 1920s. By 1930, yields declined although the fishery still produced stable yields, averaging 227mt annually from 1930 to the late 1950s (fig. 12). Landings of lake whitefish declined to negligible levels between 1961 and 1970, but recent landings suggest some recovery has occurred. Lake whitefish populations recovered slowly during the 1970s and 1980s, and by 1990 lake whitefish were the most landed species in the Ontario fishery.

The Bay of Quinte and shallow areas in the northeastern portion of Lake Ontario historically produced high levels of commercial fisheries landings. By 1910, approximately 50% of the total commercial landings in Lake Ontario was shipped from Picton, Ontario (Koelz 1926). Early landings were dominated by lake whitefish, lake herring, and northern pike, but these populations declined and yields dwindled by 1940. Hurley and Christie (1977) attribute declines in northern pike populations to overfishing and low water levels during the 1930s. Christie (1972, 1974) theorized that lake herring populations suffered from competition or predation with expanding rainbow smelt populations. Christie (1963) attributed reduced production of lake whitefish to destabilization of the stock due to commercial exploitation, unfavorable weather conditions for reproduction, and cultural eutrophication. Bullheads, channel catfish, and carp dominated landings in the Bay of Quinte during the 1940s and 1950s, accounting for 50% to 70% of total landings during this period (Baldwin et al. 1979). White perch colonized the Bay of Quinte around 1950 (Scott and Christie 1963), and commercial yields increased rapidly to over 250mt by 1965 (Hurley and Christie 1977).

The Lake Ontario fishery compensated for the loss of landings from cold-water species by increasing fishing effort on warm water species, including yellow perch, white perch, carp, bullheads, American eel, rainbow smelt, and sunfish (fig. 12). In the 1990s, lake whitefish, eel, yellow perch, and bullhead represent approximately 80% of landings, and 90% of the value of landings by the Ontario fishery. In the U.S. fishery (New York), brown bullheads, white perch, and yellow perch comprise 80% of U.S landings, and 95% of the value of U.S. landings. Nine historically important commercial species are commercially extinct or were extirpated from Lake Ontario (Table 3). Atlantic salmon and lake trout have been reintroduced to Lake Ontario, but harvest is currently allocated to the recreational fishery. In the early 1990s, commercial landings of 611mt annually represent 42% of the long-term average landings from 1914 to 1992. Over the past eighty years, Canadian landings comprised about 86% of the total landings, and have remained at this level during the early 1990s (fig. 4).

Historical Perspective and Prognosis for the Future

Historical Summary

High levels of exploitation and other stresses (habitat destruction, water quality degradation, species introductions) have resulted in the partial or complete collapse of fisheries for most populations of commercially important fish species at some point in the past two hundred years (table 3). In many cases, commercial exploitation has resulted in destabilization of fish communities, leaving them more vulnerable to other disruptive processes, including species introductions and changes in environmental conditions. Although investigators have been quick to attribute population collapses solely to overharvesting by commercial fisheries, the cumulative effects of habitat alteration/destruction and water quality degradation have often been unrecognized or underestimated (Egerton 1985).

Several species, including Atlantic salmon (Lake Ontario), lake trout (Lakes Michigan, Erie, and Ontario), blue pike (Lakes Erie and Ontario), and several species of deepwater chubs (all lakes) have been extirpated from individual lakes or the entire system (table 3). Restoration of species with long life spans and late maturity (lake sturgeon and lake trout) has not been achieved, and these species may never support commercial fisheries at historical levels. Lake whitefish and bloater in Lakes Michigan and Huron have recovered, and now support high yields for remaining state and provincial licensed commercial fishers, and the reestablished Native American fisheries. The failure of lake herring to recover in any of the lakes except Superior is most likely related to the establishment of exotic alewife and rainbow smelt (Smith 1970; Crowder 1980).

Although historically important, currently, commercial fisheries production in the Great Lakes makes a relatively minor contribution to the overall commercial fishery landings in the United States and Canada due to increased production of marine fisheries in the Atlantic and Pacific oceans. In 1994, commercial landings from U.S. waters of the Great Lakes were 13,400mt or about 0.3 % of the U.S. domestic landings (NMFS-NOAA 1995). These landings were worth $19.3 million (U.S.) or less than 0.5 % of the value of U.S. Domestic landings in 1994 (NMFS-NOAA 1994). The U.S. share of the Great Lakes commercial fishery landings has declined in three of the four binational lakes due to increased emphasis of U.S. management agencies to promote recreational fisheries (fig. 4). In Canadian waters, fishers landed 19,100mt of fish worth approximately $29 million (U.S.) in 1992, which is less than 2% of the total value of Canadian commercial landings. Currently, total Great Lakes landings are worth at least $50 million (U.S.) annually, and have a regional economic impact of exceeding $200 million (U.S.).

Current Allocation Issues

There are now several contentious issues related to commercial and recreational fishing in the Great Lakes including; direct allocation of fish, allocation of exotic and native forage species, by-catch, and gear interference concerns. Commercial harvesters and recreational anglers currently compete for yellow perch in Lake Michigan, walleye and yellow perch in Lake Erie, lake trout in the upper three Great Lakes, and chinook salmon in Lake Huron. For yellow perch and walleye, allocation issues have generally been addressed by allocation of harvest through fishery-specific quota systems. There are conflicts over the allocation of alewife (particularly in Lake Michigan) and rainbow smelt between commercial fishing harvest and recreationally-sought piscivores, as recreational anglers recognize the importance of these forage species for the growth and survival of highly prized salmonine species. Recreational fishing interests are also concerned about by-catch of salmonine species in commercial gill nets, which are still extensively used in Canadian waters, and by Native American and some state licensed fishers in U.S. waters. Deployment of passive gear that interferes with navigation and recreational activities is a locally important issue in some areas.

Interjurisdictional conflicts between states, provinces, and countries have been minimal in relation to other transboundary fisheries, primarily because most fishes in the Great Lakes are not highly migratory. Although the history of the fisheries is rich with instances of cross border fishing by both U.S. and Canadian fishers, current enforcement limits this type of activity. Conflicts over allocation of migratory walleye and yellow perch in Lake Erie between Ontario commercial fishers and Michigan, Ohio, Pennsylvania, and New York recreational anglers are addressed through an interjurisdictional quota management system.

Allocation of fishery resources typically forms the basis of agreements between native bands and state fishery agencies. The 1985 Consent Decree in 1836-ceded waters was essentially an allocation agreement, whereby native bands suspended commercial fishing activities in certain waters for exclusive commercial fishing activities in northern Lakes Michigan and Huron, and eastern Lake Superior. The fishery management agreements in Wisconsin waters of Lake Superior created recreational fishing areas where native and state-licensed commercial fishers are prohibited from fishing, and other areas where state-licensed commercial fishers can fish, but native fishers cannot. Issues and conflicts related to inter-tribal allocation of Great Lakes fishery resources have also arisen in the 1980s and 1990s, and may arise again in the future.

The Future

We foresee a continued shift of management focus from commercial to recreational fisheries management by state and provincial management agencies. The number of non-native commercial fishers may continue to decline, as the recreational fishery continues to seek a larger allocation of Great Lakes fish (yellow perch, walleye). State and provincial-licensed commercial fisheries will be allowed to continue as long as interference with recreational fishing and other recreational activities is minimized. Remaining non-native fisheries will continue to target healthy populations of traditional commercial species, including lake whitefish and bloater in the upper Great Lakes, rainbow smelt and percids in Lake Erie, and percids, white perch, bullhead, and eel in Lake Ontario. Remaining fishers will strive to develop or expand markets for underutilized species including lake herring (Lake Superior), suckers, carp, sheepshead, and burbot.

There will continue to be an increase in the importance of Native American commercial fisheries as Native Americans increasingly exercise their treaty fishing rights. The advent of treaty fishing rights in Canada will probably result in an increase in the number of commercial fishers, primarily in Lakes Huron and Superior. However, increased technological buildup among tribal fishers, and increased cost of operating a fishing operation will gradually result in a reduction in the number of fishers over a broader timeframe. It is conceivable that within the next several decades, native managed fisheries will dominate the commercial fishing industry on the Great Lakes.

Although the improvement of water quality, restoration of habitat, and effective management programs has fostered the recovery of some traditional commercial species, the continued influence of exotic species (e.g., sea lamprey, alewife, rainbow smelt, and salmon) and the threat of new introductions (e.g., white perch, ruffe, zebra mussels, and gobies) cloud the future of commercial fishing. Societal forces, including management actions and legal decisions will play an important role in determining the future of commercial fishing in the Great Lakes.

Acknowledgements

We thank Ed Brown, Steve Clark, Steve Nepszy, Bob Haas, and Mike Thomas for their thoughtful reviews of this chapter. John Tilt (Ontario Ministry of Natural Resources) and Scott Nelson (U.S. Geological Survey) provided recent commercial landings data. We thank Brenda Figuerido for her assistance in the production of graphics.

Literature Cited

Anonymous. 1929. History of filleting. The Canadian Fisherman 16:34.

———, 1947. Wisconsin and Ohio pass resolutions against U.S.-Canada pact. The Fisherman 15(5):9.

Applegate, V. C. and H. D. Van Meter. 1970. A brief history of commercial fishing in Lake Erie. U.S. Fish and Wildlife Service Fishery Leaflet 630:28 p.

Ashworth, W. 1987. The Late Great Lakes, an Environmental History. Wayne State University Press, Detroit. 274p.

Atton, F. M. 1955. The relative effectiveness of nylon and cotton gill nets. Canadian Fish Culture 17:18–26.

Auer, N. A. 1999. Lake Sturgeon: a unique and imperiled species in the Great Lakes. Pages 515–536 in W. W. Taylor and C. P. Ferreri eds. Great Lakes Fisheries Policy and Management: a Binational Perspective. Michigan State University Press, East Lansing, Michigan.

Baldwin. N. S., R. W. Saalfeld, M. A. Ross, and H. J. Buettner. 1979. Commercial fish production in the Great Lakes 1867–1977. Great Lakes Fishery Commission Technical Report 3:187p.

Beeton, A. M. 1966. Indices of Great Lakes eutrophication. Great Lakes Research Division, University of Michigan Publication 15:1–8.

———, C. E. Sellinger, and D. F. Reid. 1999. An introducyion to the Laurentian Great Lakes ecosystem. Pages 3–54 in W. W. Taylor and C. P. Ferreri eds. Great Lakes Fisheries Policy and Management: a Binational Perspective. Michigan State University Press, East Lansing, Michigan.

Belden, H. 1877. Historical atlas of Huron County. H. Beldon & Company, Toronto. Reprint by Richardson, Bond & Wright Ltd. Owen Sound, Ontario. 96 p.

Berkes, F. and D. Pocock. 1987. Quota management and people problems: a case history of Canadian Lake Erie fisheries. Transactions of the American Fisheries Society 116:494–502.

Berst, A. H. and G. R. Spangler. 1972. Lake Huron: effects of exploitation, introductions and eutrophication on the salmonid community. Journal of the Fisheries Research Board of Canada 29:877–887.

Brege, D. A. and N. R. Kevern. 1978. Michigan commercial fishing regulations: a summary of Public Acts and Conservation Commission Orders, 1865 through 1975. Michigan Sea Grant Program Reference Report. 62 p.

Brown, E. H., Jr. 1972. Population biology of alewives, *Alosa pseudoharengus*, in Lake Michigan, 1949–1970. Journal of the Fisheries Research Board of Canada 29:477–500.

———, R. W. Rybicki, and R. J. Poff. 1985. Population dynamics and interagency management of the bloater (*Coregonus hoyi*) in Lake Michigan, 1967–1982. Great Lakes Fishery Commission, Technical Report 44:34p.

———, R. L. Argyle, N. R. Payne, and M. E. Holey. 1987. Yield and dynamics of destabilized chub (*Coregonus* spp.) populations in Lakes Michigan and Huron, 1950–1984. Journal of Fisheries and Aquatic Sciences 44(Suppl. 2):371–383.

Busiahn, T. R. 1990. Fish community objectives for Lake Superior. Great Lakes Fishery Commission Special Publication 90–1. 23 p.

Christie, W. J. 1963. Effects of artificial propagation and the weather on recruitment in the Lake Ontario whitefish fishery. Journal of the Fisheries Research Board of Canada 20:597–646.

———, 1972. Lake Ontario: effects of exploitation, introductions, and eutrophication on the salmonid community. Journal of the Fisheries Research Board of Canada 29:913–929.

———, 1974. Changes in the fish species composition of the Great Lakes. Journal of the Fisheries Research Board of Canada 31:827–854.

———, 1978. A study on freshwater fishery regulation based on North American experience. FAO Fisheries Technical Paper 180:46p.

Coble, D. W., R. E. Bruesewitz, T. W. Fratt, and J. W. Scheirer. 1990. Lake trout, sea lampreys and overfishing in the upper Great Lakes: a review and reanalysis. Transactions of the American Fisheries Society 119:985–995.

Collins, J. J. 1979. Relative efficiency of multifilament and monofilament nylon gill net toward lake whitefish (*Coregonus clupeaformis*) in Lake Huron. Journal of the Fisheries Research Board of Canada 36:1180–1185.

———, 1987. Increased catchability of the deep monofilament nylon gillnet and its expression in a simulated fishery. Canadian Journal of Fisheries and Aquatic Sciences 44:129–135.

Crowder, L. B. 1980. Alewife, rainbow smelt and native fishes in Lake Michigan: competition or predation. Environmental Biology of Fishes 5:225–233.

Cucin, D. and H. A. Regier. 1966. Dynamics and exploitation of lake whtiefish in southern Georgian Bay. Journal of the Fisheries Research Board of Canada 23:221–274.

DesJardine, R. L., T. K. Gorenflo, R. N. Payne, and J. D. Schrouder. 1995. Fish community objectives for Lake Huron. Great lakes Fishery Commission Special Publication 95–1. 37 p.

Dryer, W. R. and J. Beil. 1968. Growth changes of the bloater, *Coregonus hoyi*, of the Apostle Islands region of Lake Superior. Transactions of the American Fisheries Society 97:146–158.

Eger, W. H. 1987. Government, agency and citizen co-operation in management of the Michigan 1836 treaty area Great Lakes fisheries (abstract only). Pages 46–47 *in* B. L. Smith and P. Burt eds. Fisheries Co-Management; A Response to Legal, Social and Fiscal Imperatives. Symposium Sponsored by the Native Peoples' Fisheries Committee of the American Fisheries Society at the 117th Annual Meeting, Winston-Salem, North Carolina.

Egerton, F. N. 1985. Overfishing or pollution? Case history of a controversy on the Great Lakes. Great Lakes Fishery Commission Technical Report 41:28 p.

Eschmeyer, P. H. 1957. The near extinction of the lake trout in Lake Michigan. Transactions of the American Fisheries Society 85:102–119.

Eshenroder, R. L. 1977. Effects of intensified fishing, species changes, and spring water temperatures on yellow perch, *Perca flavescens*, in Saginaw Bay, Lake Huron. Journal of the Fisheries Research Board of Canada 24:1830–1838.

———, 1992. Decline of lake trout in Lake Huron. Transactions of the American Fisheries Society 121:548–554.

———, M. E. Holey, T. K. Gorenflo, and R. D. Clark Jr. 1995. Fish community objectives for Lake Michigan. Great Lakes Fishery Commission Special Publication 95–3. 56 p.

Fetterolff, C. M., Jr. 1980. Why a Great Lakes Fishery Commission and why a Sea Lamprey International Symposium. Canadian Journal of Fisheries and Aquatic Sciences 37:1588–1593.

Gallagher, H. R. and J. Van Oosten. 1943. Supplemental report of the United States members of the International Board of Inquiry for the Great Lakes Fisheries. International Board of Inquiry for the Great Lakes Fisheries Report. Supplement 25–213.

GLFC (Great Lakes Fishery Commission). 1994. A joint strategic plan for management of Great Lakes fisheries. Great Lakes Fishery Commission. Ann Arbor, MI. 25 p.

Great Lakes Fisheries. 1946. Great Lakes Fisheries. Convention between Canada and the United States of America signed at Washington, April 2, 1946. Edmond Cloutier, King's Printer: Ottawa, Ontario.

Hansen, M. J. 1999. Lake trout in the Great Lakes: Basinwide stock collapse and binational restoration. Pages 417–453 *in* W. W. Taylor and C. P. Ferreri eds. Great Lakes Fisheries Policy and Management: a Binational Perspective. Michigan State University Press, East Lansing, Michigan.

———, (ed.) 1990. Lake Superior: The state of the lake in 1989. Great Lakes Fishery Commission Special Publication 90–93.

Hartman, W. L. 1972. Lake Erie: effects of exploitation, environmental changes and new species on the fishery resources. Journal of the Fisheries Research Board of Canada 29:899–912.

Hatch, R. W., S. J. Nepszy, and M. R. Rawson. 1990. Management of percids in Lake Erie, North America. Pages 624–636 *in* W. L. T. van Densen, B. Steinmetz, and R. H. Hughes eds. Management of freshwater fisheries. Proceedings of a symposium organized by the European Inland Fisheries Advisory Commission, Göteborg, Sweden.

Hatcher H. 1945. Lake Erie. Bobbs-Merill Co., New York. 416p.

Hayes, D. B. 1999. Issues affecting fish habitat in the Great Lakes basin. Pages 209–237 *in* W. W. Taylor and C. P. Ferreri eds. Great Lakes Fisheries Policy and Management: a Binational Perspective. Michigan State University Press, East Lansing, Michigan.

Hewson, L. C. 1951. A comparison of nylon and cotton gill nets used in the Lake Winnipeg winter fishery. Canadian Fish Culture 11:1–3.

Hile, R. 1949. Trends in the lake trout fishery of Lake Huron through 1946. Transactions of the American Fisheries Society 76:121–147.

———, 1962. Collection and analysis of commercial fishery statistics in the Great Lakes. Great Lakes Fishery Commission Technical Report 5:31p.

———, F. W. Jobes. 1941. Age, growth, and reproduction of yellow perch, *Perca flavescens* (Mitchill), of Saginaw Bay. Transactions of the American Fisheries Society 70:102–122.

———, P. H. Eschmeyer and G. F. Lunger. 1951. Decline of the lake trout fishery in Lake Michigan. U.S. Fish and Wildlife Service Fishery Bulletin 52:77–95.

Hurley, D. A. and W. J. Christie. 1977. Depreciation of the warm-water fish community in the Bay of Quinte, Lake Ontario. Journal of the Fisheries Research Board of Canada 34:1849–1860.

Jannetta, J. M. 1991. The constitutional status of native hunting and fishing rights in Canada and the United States: a comparative analysis. Native American Studies Conference, Lake Superior State University, Sault Ste. Marie, MI.

Jensen, A. L. 1978. Assessment of the lake trout fishery in Lake Superior: 1929–1950. Transactions of the American Fishery Society 107:543–549.

———, 1994. Larkin's predation model of lake trout (*Salvelinus namaycush*) extinction with harvesting and sea lamprey (*Petromyzon marinus*) predation: A qualitative analysis. Canadian Journal of Fisheries and Aquatic Sciences 51: 942–945.

Jester, D. B. 1977. Effects of color, mesh size, fishing in seasonal concentrations, and baiting on catch rates of fishes in gill nets. Transactions of the American Fisheries Society 106:43–56.

Johnson, D. A. 1977. Population dynamics of walleye (*Stizostedion vitreum vitreum*) and yellow perch (*Perca flavescens*) in Lake St. Clair, especially during 1970–1976. Journal of the Fisheries Research Board of Canada 34:1869–1877.

Johnson, J. E., G. M. Wright, D. M. Reid, C. A. Bowen II, and N. R. Payne. 1995. Status of the cold-water fish community in 1992, pages 21–72. *In* M. P. Ebener [ed.], The State of Lake Huron in 1992. Great Lakes Fishery Commission Special Publication 95–2. 140 p.

Kennedy, W. A. 1970. A history of commercial fishing in Inland Canada. Fisheries Research Board of Canada, Manuscript Report Series (Biological), 871.

Kerr, S. J. and G. C. LeTendre. 1991. The state of the Lake Ontario fish community in 1989. Great Lakes Fishery Commission Special Publication 93–3. 38p.

Keyes, C. M. 1894. The fishing industry of Lake Erie, past and present. U.S. Fishery Commission Bulletin 13(1893):349–353.

Kinietz, W. V. 1940. The Indians of the Western Great Lakes, 1615–1760. Ann Arbor Paperbacks, The University of Michigan Press. 427 p.

Klippart, J. H. 1877. History of Toledo and Sandusky fisheries. Ohio State Fish Commission Annual Report 1:31–42.

Koelz, W. 1926. Fishery industry of the Great Lakes. Pages 554–617, *In* Report of the U.S. Commissioner of Fisheries for 1925.

———, 1929. Coregonid fishes of the Great Lakes. Bulletin of the U.S. Bureau of Fisheries 43(Part 2):297–643.

Koonce, J. F., A. B. Locci, and R. L. Knight. 1999. Contributions of fishery management to changes in walleye and yellow perch populations in Lakes Erie. Pages 397–416 *in* W. W. Taylor and C. P. Ferreri eds. Great Lakes Fisheries Policy and Management: a Binational Perspective. Michigan State University Press, East Lansing, Michigan.

Lac Courte Oreilles Band v. Voight (1983), U.S. Circuit Court of Appeals, Seventh Circuit, Federal Reporter (Second series), 700:341.

Lawler, G. H. 1950. The use of nylon netting in the gill net fishery of Lake Erie whitefish. Canadian Fish Culture 7:22–24.

———, 1965. Fluctuations in the success of year-classes of whitefish populations with special reference to Lake Erie. Journal of the Fisheries Research Board of Canada 22:1197–1227.

Lawrie, A. H. and J. F. Rahrer. 1972. Lake Superior: effects of exploitation and introductions on the salmonid community. Journal of the Fisheries Research Board of Canada 29:765–776.

———, 1973. Lake Superior: A case history of the lake and its fisheries. Great Lakes Fisheries Commission Technical Report 19: 69 p.

Leach, J. H. and S. J. Nepszy. 1976. The fish community in Lake Erie. Journal of the Fisheries Research Board of Canada 33:622–638.

Leach, J. H., E. L. Mills, and M. A. Dochoda. 1999. Non-indigenous species in the Great Lakes: ecosystem impacts, binational policies, and management. Pages 185–207 *in* W. W. Taylor and C. P. Ferreri eds. Great Lakes Fisheries Policy and Management: a Binational Perspective. Michigan State University Press, East Lansing, Michigan.

Loftus, D. H., C. H. Olver, E. H. Brown, P. J. Colby, W. L. Hartman, and D. H. Schupp. 1987. Partitioning potential fish yields from the Great Lakes. Canadian Journal of Fisheries and Aquatic Sciences 44(Suppl. 2):417–424.

MacCallum, W. R. and J. H. Selgeby. 1987. Lake Superior revisited: 1984. Canadian Journal of Fisheries and Aquatic Sciences 44(Suppl. 2):23–36.

Mansfield, J. 1899. History of the Great Lakes. J. H. Beers & Co., Chicago. Reprinted by Freshwater Press Inc. Cleveland, Ohio. Vol. 1, 926p., Vol 2., 1,108 p.

McCullough, A. B. 1989. The Commercial Fishery of the Canadian Great Lakes. Canadian Parks Service, Environment Canada. 153 p.

Milner, J. W. 1874. Report on the fisheries of the Great Lakes: the result of inquiries prosecuted in 1871 and 1872. Pages 1–78, *In* Report of the Commissioner for 1872 and 1873. U.S. Commission of Fish and Fisheries.

Minnesota Historical Society. 1973. The Land of the Ojibwe. Ojibwe Curriculum Committee, American Indian Studies Department, University of Minnesota, and the Educational Services Division, Minnesota Historical Society. 48 p.

National Research Council. 1988. Fisheries Technologies for Developing Countries. National Academy Press, Washington, D. C.

NMFS-NOAA 1994. Fisheries of the United States, 1993. U.S. Department of Commerce, National Oceanic and Atmospheric Administration, National Marine Fisheries Service. 121 p.

Nute, G. L. 1944. Lake Superior. The American Lakes Series. Bobbs-Merrill, Indianapolis, Indiana. 376 p.

OMNR (Ontario Ministry of Natural Resources). 1981. Lake St. Clair Fisheries Report 1980. Lake St. Clair Fisheries Assessment Unit, Ontario Ministry of Natural Resources. 16 p.

———, 1986. Lake St. Clair Fisheries Report 1985. Lake St. Clair Fisheries Assessment Unit, Ontario Ministry of Natural Resources. 32 p.

———, 1995. Lake St. Clair Fisheries Report 1994. Lake St. Clair Fisheries Assessment Unit, Ontario Ministry of Natural Resources. 13 p.

Parsons, J. W. 1967. Contributions of year-classes of blue pike to the commercial fishery of Lake Erie, 1943–1959. Journal of the Fisheries Research Board of Canada 24:1035–1066.

———, 1970. Walleye fishery of Lake Erie in 1943–1962 with emphasis on contributions of the 1942–1961 year classes. Journal of the Fisheries Research Board of Canada 27:1475–1489.

People v. LeBlanc (1976), Michigan Supreme Court, Michigan Reports, 399:31.

Piper, D. C. 1967. The International Law of the Great Lakes: A Study of Canadian-United States Cooperation. Commonwealth Studies Center, Duke University Press, Durham, N.C.

Prothero, F. and N. Prothero. 1990. The lone survivor: the Katherine V of Rogers City. Nan-Sea Publications Limited., Port Stanley, Ontario. 32 p.

Province of Ontario. 1885. An act to regulate the fisheries of this province. Statutes of the Province of Ontario. John Notman, Queen's Printer, Toronto.

Pycha, R. L. 1961. Recent changes in the walleye fishery of northern Green Bay and history of the 1943 year class. Transactions of the American Fisheries Society 90:475–488.

———, 1962. The relative efficiency of nylon and cotton gill nets for taking lake trout in Lake Superior. Journal of the Fisheries Research Board of Canada 19:1085–1094.

Regier, H. A., V. C. Applegate and R. A. Ryder. 1969. The ecology and management of the walleye in western Lake Erie. Great Lakes Fishery Commission Technical Report 15:101p.

———, W. L. Hartman. 1973. Lake Erie's fish community: 150 years of cultural stresses. Science (Washington) 180:1248–1255.

Rounsefell, G. A. and W. H. Everhart. 1953. Fishery Science: Its Methods and Applications. Wiley & Sons, Inc., New York.

Scott, W. B. 1951. Fluctuations in abundance of the Lake Erie cisco (*Leucichthys artedi*) population. Royal Ontario Museum, Contribution 32: 41 p.

———, W. J. Christie. 1963. The invasion of the lower Great Lakes by the white perch, *Roccus americanus* (Gmelin). Journal of the Fisheries Research Board of Canada 20:1189–1195.

Šelgeby, J. H. 1982. Decline of lake herring (*Coregonus artedii*) in Lake Superior: an analysis of the Wisconsin herring fishery, 1936–1978. Canadian Journal of Fisheries and Aquatic Sciences 39:554–563.

———, C. R. Bronte, and J. W. Slade. 1994. Forage species. Pages 53–62 *in* M. J. Hansen, ed. The state of Lake Superior in 1992. Great Lakes Fishery Commission Special Publication 94–1.

Smith, B. L., and P. Burt, ed. 1987. Fisheries Co-Management: A Response to Legal, Social and Fiscal Imperatives. Symposium Sponsored by the Native Peoples' Fisheries Committee of the American Fisheries Society at the 117th Annual Meeting, Winston-Salem, North Carolina.

Smith, H. M. 1892. Report on the fisheries of Lake Ontario. Bulletin of the U.S. Fish Commission 10:177–215.

———, M. M. Snell. 1891. Review of the Fisheries of the Great Lakes in 1885. Pages 1–333, *in* Report of the Commissioner for 1887. U.S. Government Printing Office, Washington, D. C.

Smith, S. H. 1964. Status of the deepwater cisco populations of Lake Michigan. Transactions of the American Fisheries Society 93:155–163.

———, 1968. Species succession and fishery exploitation in the Great Lakes. Journal of the Fisheries Research Board of Canada 25:667–693.

———, 1970. Species interactions of the alewife in the Great Lakes. Transactions of the American Fisheries Society 99:754–765.

———, 1972. Factors in ecologic succession in oligotrophic fish communities of the Laurentian Great Lakes. Journal of the Fishery Research Board of Canada 29:717–730.

———, 1995. Early changes in the fish community of Lake Ontario. Great Lakes Fishery Commission Technical Report 60:38 p.

———, H. J Buettner, and R. Hile. 1961. Fishery statistical districts of the Great Lakes. Great Lakes Fishery Commission Technical Report 2:24 p.

Spangler, G. R., N. R. Payne, and G. K. Winterton. 1977. Percids in the Canadian waters of Lake Huron. Journal of the Fisheries Research Board of Canada 34:1839–1848.

Sparrow v. The Queen (1990), Supreme Court of Canada, slip opinion, May 31, 1990.

Stansby, M. E. 1963. Industrial Fishery Technology; A Survey of Methods for Domestic Harvesting, Preservation, and Processing of Fish Used for Food and for Industrial Products. Reinhold Press, New York.

State v. Gurnoe (1972), Wisconsin Supreme Court, 53 Wis. 2D 390.

Swan, I. 1977. The Deep Roots, Grosse Isle 1776–1876. Isabel Swan Grosse Isle, Michigan. 445p.

Talhelm, D. R. 1988. The international Great Lakes sport fishery of 1980. Great Lakes Fishery Commission Special Publication 88–4. 70 p.

TFRC (Technical Fisheries Review Committee). 1992. Status of the fishery resource: 1991. A report of the Technical Fisheries Review Committee on the assessment of lake trout and whitefish in waters of the upper Great Lakes ceded in the Treaty of 1836. 87 p.

Thompson, R. 1978. Fishing ports on Lake Erie. Historical Planning and Research Branch, Ontario Ministry of Culture and Recreation, Toronto. 61 p.

Todd, T. N., and G. R. Smith. 1992. A review of differentiation in Great Lakes ciscoes. Polish Archives of Hydrobiology 39:261–267.

Toner, G. C. 1939. The Great Lakes fisheries: unheeded depletion. Canadian Forum 19(224):178–180.

Trautman, M. B. 1957. The Fishes of Ohio. University of Ohio Press, Columbus, Ohio. 683p.

True, F. W. 1887. The fisheries of the Great Lakes. Pages 631–673, in G. B. Goode, ed. The Fisheries and Fishing Industries of the United States. U.S. Commission on Fish and Fisheries. Washington D. C.

Tulee v. Washington (1942), United States Supreme Court, United States reports, 315:61. United States v. Michigan (1979), U.S. District Court, Western District of Michigan, Federal Supplement, 471:192.

United States v. Michigan (1979) U.S. District Court, Western District of Michigan, Federal Supplement, 520:207.

United States v. Michigan (1985), U.S. District Court, Western District of Michigan, Case No. M–26–73CA.

United States v. Winans (1905), United States Supreme Court, United States Reports, 198:371.

United States v. Washington (1979), U.S. District Court, Western District of Washington, Federal Supplement, 384:312.

U.S. Department of Interior, Fish and Wildlife Service and U.S. Department of Commerce, Bureau of the Census. 1993. 1991 National Survey of Fishing, Hunting, and Wildlife-Associated Recreation. U.S. Government Printing Office, Washington, D.C.

University of Wisconsin Sea Grant Institute. 1988. The fisheries of the Great Lakes. University of Wisconsin Sea Grant Institute, Madison, WI. 19 p.

Van Oosten, J. 1930. The disappearance of the Lake Erie cisco: a preliminary report. Transactions of the American Fisheries Society 60:20

———, 1937. Dispersal of smelt, Osmerus mordax (Mitchill), in the Great Lakes region. Transactions of the American Fisheries Society 66:160–171.

———, 1938. Michigan's commercial fisheries of the Great Lakes. Michigan History Magazine 22(1):13.

———, 1947. Mortality of smelt, Osmerus mordax (Mitchill), in Lakes Huron and Michigan during the fall and winter of 1942–1943. Transactions of the American Fisheries Society 74:310–337.

———, R. Hile and F. W. Jobes. 1946. The whitefish fishery of Lakes Huron and Michigan with special reference to the deep trapnet fishery. U.S. Fish and Wildlife Service Fishery Bulletin 50:297–394.

Wakeham, W. and R. Rathbun. 1897. Report of the joint commission relative to the preservation of the fisheries in waters contiguous to Canada and the United States. Queen's Printer, Ottawa. 146 p.

Wells, L. and A. L. McLain. 1972. Lake Michigan: effects of exploitation, introductions and eutrophication on the salmonid community. Journal of the Fisheries Research Board of Canada 29:889–898.

———, 1973. Lake Michigan: man's effects on native fish stocks and other biota. Great Lakes Fishery Commission Technical Report 20:55p.

Whitaker, H. 1892. Early history of the fisheries on the Great Lakes. Transactions of the American Fisheries Society 21:163–179.

Woner, P. 1961. Ohio fisheries. Unpublished manuscript. Ohio Division of Wildlife.

World Wildlife Fund. 1993. Protected areas and aboriginal interests in Canada. A World Wildlife Fund Canada Discussion Paper. 37 p.

CHAPTER 12

Allocating Great Lakes Forage Bases in Response to Multiple Demand

Edward H. Brown, Jr., Thomas R. Busiahn, Michael L. Jones and Ray L. Argyle

Introduction

Throughout most of the period from the 1950s to the 1990s, the predominant trend in the biota of the offshore waters of each of the Great Lakes was one of dramatically increasing abundance of piscivorous fishes. In Lake Erie, the trend was largely due to improving habitat quality and relaxation of exploitation pressure on native walleye (*Stizostedion vitreum vitreum)* populations (Hatch et al. 1987). In the other lakes, however, the build up of piscivores was primarily due to simultaneous reductions in abundance of parasitic sea lamprey (*Petromyzon marinus)* and major increases in plantings of both native and non-native salmonine species (Smith and Tibbles 1980; Pearce et al. 1980; Kocik and Jones 1999). The survival of these stocked fish was exceptional, and the result was the development of major sport fisheries for salmonines on each of the lakes. Collectively, the Great Lakes sport fisheries of the 1980s were estimated to contribute billions of dollars to the Great Lakes basin economy (Talhelm 1988), of which the hatchery-based salmonine fishery was the single largest component. Public interest in the availability of forage to sustain these fisheries is, therefore, substantial and has been ever since their economic and recreational potential became apparent.

Forage base allocation, which has become an important issue because of these major changes in the fish communities and fisheries of the Great Lakes since the 1950s is examined and documented in this chapter (Paine and Lange 1985; Krueger and Paine 1985). Management initiatives that were used to address the issue, and supporting research and development that provided new or improved methods of field sampling and analysis are also highlighted.

The term forage base allocation, as used here, implies a deliberate channeling of production to various uses, including competing fisheries and restoration or maintenance of the living resource itself. Yet the tools or methods available to fishery managers for accomplishing that complex task are limited. In the case of Lake Michigan and other Great Lakes, managers can regulate access to and harvesting of predator and prey species, and/or they can control to some extent the numbers and species mix of hatchery-reared salmonine predators that are stocked.

TABLE 1

Categorical abundance in the 1980s and early 1990s, of the major prey species in the offshore areas of the Great Lakes

3 = dominant, 2 = secondary, 1 = scarce, and 0 = locally extinct.
Based mainly on surveys by research vessels of the U. S. Geological Survey–Biological Resources Division.

DISTRUBUTION AND SPECIES	LAKES				
	SUPERIOR	HURON	MICHIGAN	ONTARIO	ERIE*
Pelagic					
Alewife	1	2	2	3	1
Rainbow Smelt	2	2	2	2	3
Bloater	2	3	3	0	—
Lake Herring	3	1	1	1	1
Benthic					
Deepwater sculpin	3	3	3	0	—
Slimy sculpin	2	2	2	3	1

*Young gizzard shad *Dorosoma cepedianum* and emerald shiners *Notropis atherinoides* are important prey fish in some areas of Lake Erie during certain periods (Wolfert and Bur 1992).

The availability, however, of both lake trout (*Salvelinus namaycush)* and Pacific salmon (*Oncorhynchus spp.*) from federal and state hatcheries has been unpredictable in recent years because of disease epidemics and other rearing problems (Holey et al. 1995; Keller et al. 1989).

Forage Bases of the Great Lakes

The principal forage bases in the offshore areas of Lakes Superior, Michigan, Huron, and Ontario proper and of Lake Erie's deeper central and eastern basins are composed mainly of two to four pelagic species, which occupy midwater, and one or two benthic species, which occupy the bottom substrate (table 1). Since the 1950s, the pelagic component has been dominated either by the non-native alewife (*Alosa pseudoharengus)* or rainbow smelt (*Osmerus mordax*) or by the native bloater (*Coregonus hoyi)* or lake herring (*Coregonus artedii).* The benthic component is composed of the deepwater sculpin (*Myoxocephalus thompsoni),* a glacial relic, and/or the slimy sculpin (*Cottus cognatus).*

Alewife

This small Clupeid, which seldom exceeds a length of 225mm in the Great Lakes (Brown 1972; Argyle 1982), has been a mixed blessing for fishery managers since its establishment throughout the Great Lakes. On one hand, fishery managers value alewives highly as the preferred forage of lake trout and salmon in Lakes Michigan, Huron and Ontario; on the other hand, alewives have negatively impacted fish communities of those lakes and possibly impeded efforts to rehabilitate lake trout by preying on fish larvae (Krueger et al. 1995). Those species with pelagic larvae were the most vulnerable to alewives (Crowder 1980; Eck and Wells 1987). Indeed, the alewife was implicated in the extirpation of the bloater and deepwater sculpin in Lake Ontario, and in the severe depression of bloater populations in Lakes Michigan and Huron in the 1960s and 1970s (Smith 1970; Brown et al. 1987). It was also implicated in the sudden disappearance of the emerald shiner (*Notropis atherinoides*) and the depression of the deepwater sculpin, yellow perch (*Perca flavescens*) and burbot (*Lota lota*) in Lake Michigan (Smith 1970; Crowder 1980).

Alewives may also have contributed to the extirpation of several species of deepwater ciscoes (*Coregonus* spp.) closely related to the bloater in one or more of Lakes Michigan, Huron and Ontario along with size-selective commercial fishing and predation by the sea lamprey (Smith 1964, 1968, and 1970). The affected species include the kiyi (*C kiyi*), the shortnose cisco (*C. reighardi*), the shortjaw ciscoe (*C. zenithicus*), and the blackfin cisco (*C. nigripinis*) (Smith 1964, 1968, and 1970).

Recent laboratory and field research on Lake Ontario confirmed that alewives also prey on lake trout larvae, and therefore, may help to prevent the rehabilitation of self-sustaining lake trout populations in Lakes Ontario, Michigan and Huron (Krueger et al. 1995). The adverse effects of alewives on the program to rehabilitate lake trout populations destroyed by sea lampreys in the Great Lakes may be more complex, however, than alewife predation on lake trout larvae alone. Recent laboratory research in Canada suggested that thiaminase in alewives consumed by lake trout may lead to swim-up syndrome in larval lake trout because vitamin B administered to the larvae reduced the symptoms (Fitzsimons 1995).

In Lake Superior, water temperature is too low in most years for the alewife to reproduce (Bronte et al. 1991). Therefore, it has not had an impact on the fish community of that lake. Alewives were abundant in the central basin of Lake Erie in the 1960s, but they have been scarce there in recent years (Wolfert and Bur 1992; Bowman 1974). The numerous predators in Lake Erie, the shallowest and most productive of the Great Lakes (per surface hectare), and the lack of winter refugia, which could have been critical during the extremely cold winters of the late 1970s, may have helped to constrain alewife populations (Smith 1968; Eck and Brown 1985; Colby 1973).

Bloater

The bloater, like the alewife, is highly pelagic in the first and early part of the second year of life, but occupies the cold depths of the hypolimnion as an adult, where it cannot be preyed upon efficiently by salmon, which prefer higher water temperature (Sprules et al. 1991; Stewart and Ibarra 1991). Bloaters also attain larger sizes as adults in the Great Lakes than do alewives and rainbow smelt, but only a few taken in gill net surveys in Lakes Michigan and Huron in the early 1980s were 300m or longer (Brown et al. 1987). The growth rate of the bloater has fluctuated mark-edly in Lake Michigan and probably other Great Lakes, in response to drastic changes in popula-tion density that have occurred since the 1950s (Brown et al. 1987). When population density was high in Lake Michigan in the early 1960s, and again in the late 1980s, periods of low alewife abun-dance, the bloater growth rate fell and consequently only a moderate percentage of those in the commercial catch were marketable as a smoked product for human food (i.e. those fish about 229mm and greater in length; Brown et Eck 1992). As a result, the percentage of No. 1 chubs, the largest and choicest weight grade in the Wisconsin commercial fishery (fig. 1), also decreased (i.e., bloaters and other deepwater ciscoes are known and marketed as chubs by commercial fishermen). In contrast, proportionally more of the slow-growing bloaters were vulnerable in size to salmonine predators (Brown et al. 1987), although switching by predators from alewives to bloaters as their principal prey was first observed mainly on the Mid-Lake Reef and other offshore areas where alewives were scarce (Stewart and Ibarra 1991).

The Lake Michigan commercial chub catch, which was composed mainly of bloaters, peaked at an all-time high of 5,400 metric tons in 1960, when trawls, targeted on numerous under-sized bloaters and newly established alewives, were used in addition to the conventional gill net. Forage base allocation had not evolved as a management tool at that time, and was largely the result of chance and natural forces. For example, destruction of lake trout and burbot by sea lampreys greatly reduced predation pressure on bloaters, while the gradual decline of the other deepwater ciscoes, noted above, reduced competition for available food resources (Smith 1970, 1968, and 1964). Bloaters became overabundant and stunted, hence, trawls were introduced to utilize them together with alewives as animal food (Smith 1968). In sharp contrast, the commercial chub catch averaged only about 1,400 metric tons from 1979 to 1993, when various forms of limited entry and catch quotas were in effect (Brown et al. 1985, 1987; U.S. Geological Survey–Biological Resources Division, COMCAT data base). The commercial fishery for chubs on Lake Michigan was actually closed from 1976 to 1978, except for limited assessment fishing, because bloaters themselves appeared to be endangered (Brown et al. 1985).

Lake Herring

Historically, the pelagic lake herring or shallow-water cisco, which commonly exceeds a length of 300mm (Smith 1956; Dryer and Beil 1964), produced greater commercial harvests for human food in Lakes Erie (peak of 22,142 metric tons in 1918) and Superior (peak of 8,740mt in 1941) than did any other species. The lake herring also produced substantial harvests in Lakes Michigan (peak of 4,395 metric tons in 1952) and Huron (peak of 3,667 metric tons in 1916), even while most of the populations of lake herring in the Great Lakes and their bays collapsed from the late 1920s into the 1960s (Van Oosten 1930; Smith 1968; Baldwin et al. 1979). The collapses in Lakes Erie and Superior have been mainly attributed to overexploitation by the commercial fisheries (Van Oosten 1930; Selgeby 1982); whereas predation by invading alewives on lake herring larvae was, most likely, the main cause of the collapses in lakes Huron and Michigan, although commercial fishing and sea lamprey predation may have been secondary contributing factors (Smith 1968, 1970). Lake herring and chub landings in Lake Ontario were combined prior to 1952; hence they cannot be evaluated (Baldwin et al. 1979). The lake herring has made a strong recovery from a generally depressed state in much of Lake Superior since the early 1980s, but still ranks second to rainbow smelt as the principal prey of lake trout and salmon, as it has since rainbow smelt became abundant in the 1950s (Hoff and Selgeby 1994).

Rainbow Smelt

In addition to being an important prey of salmonines in all the Great Lakes, the rainbow smelt, which seldom exceeds a length of 225mm (Argyle 1982; Conner et al. 1993), has also supported some commercial fishing in each lake. The largest smelt fishery by far, and one of the largest operations of its kind on the Great Lakes, is the government-subsidized trawl fishery that has operated in Ontario waters of the deeper central and eastern basins of Lake Erie since about the 1950s (Brown et al. 1999). Rainbow smelt from this fishery are sold as frozen products for human food in

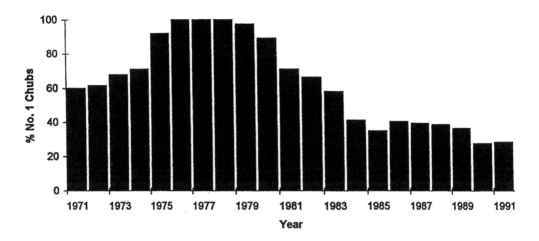

FIG. 1. *Percentage of No. 1 chubs (i.e., deep water ciscoes, mostly bloaters, about 22 per kg dressed) in the commercial catch from Wisconsin waters of Lake Michigan, 1971 to 1991; COMCAT data base, U. S. Geological Survey–Biological Resources Division.*

North America, Europe, and Asia (Baldwin et al. 1979). Rainbow smelt landings from Ontario waters of Lake Erie peaked at 10,428 metric tons in 1977, but the fishery was temporarily closed in late 1994, after the catch dropped to a few thousand metric tons in the early 1990s (U.S. Geological Survey–Biological Resources Division, COMCAT data base; R. O'Gorman, U.S. Geological Survey–Biological Resources Division, personal communication). Predation, resulting from a strong walleye recovery that had extended into the Eastern Basin of Lake Erie in the 1980s and from subsequent plantings of lake trout and salmon, lower primary production, from point-source phosphorus reduction and possibly from oligotrophic-like effects on water clarity from filter-feeding by recently introduced Dreissenid mussels, and exploitation by the commercial fishery, may all have contributed to the decline of the rainbow smelt population in Lake Erie (Nepzy et al. 1991; Wolfert and Bur 1992; Einhouse et al. 1993; Herbert et al. 1989 and 1991; P. Ryan, Ontario Ministry of Natural Resources, personal communication).

Slimy and Deepwater Sculpins

The slimy sculpin, which is present in all of the Great Lakes, and the deepwater sculpin, which is found today only in the three upper lakes (Table 1), are important prey of the lake trout and burbot. Both the slimy sculpin, which seldom exceeds a length of about 120mm (Owens and Weber 1995), and the deepwater sculpin, which reaches a length of 159mm or more (Wojcik et al. 1986), are found in close proximity to the substrate where they are not readily available to salmon and other pelagic salmonines. The deepwater sculpin is found at depths from about 64m out to midlake in Lake Michigan, where they were an important prey of a fat, deepwater form of native lake trout, before the sea lamprey invasion (Van Oosten and Deason 1938; Brown et al. 1981). Like other native species that were impacted by alewives, the deepwater sculpin has pelagic larvae, which may have been vulnerable to predation by alewives (Crowder 1980). Low abundance of slimy

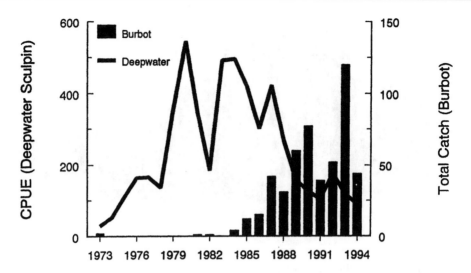

FIG. 2. *Number of deepwater sculpins per 10-min bottom trawl tow (CPUE) and total number of burbot in lakewide forage base surveys at 7 transects in Lake Michigan, fall 1973 to 1994; RVCAT data base, U. S. Geological Survey–Biological Resources Division.*

sculpin in Lakes Michigan and Ontario in the 1970s and 1980s were associated with high stocking rates of lake trout in inshore waters, where juvenile trout prey on sculpins (Eck and Wells 1987; Owens and Bergstedt 1994). However, an overlapping decline in abundance of deepwater sculpins in Lake Michigan since the mid 1980s was attributed to a lakewide increase in size of the predatory burbot population (fig. 2). Neither sculpin species has commercial value in the Great Lakes.

The Players Affecting the Forage Bases

The Province of Ontario and the eight states bordering the Great Lakes have primary responsibility for prey fish allocation and for regulating the fisheries that rely on that allocation in their nine jurisdictions. However, many of the management decisions that provincial and state natural resource agencies must make involve transboundary fish populations and predator-prey systems that are parts of regional and lakewide ecosystems. A forum through which forage-base allocations of this scope and complexity can be addressed is provided by the international Great Lakes Fishery Commission (established by Canada and the United States in 1955) and its infrastructure of Lake Committees and interagency technical committees which it sponsors (Fetterolf 1980). In addition, the United States Fish and Wildlife Service and, more recently, the United States Geological Survey–Biological Resources Division have supported the states and province by conducting long-term forage base assessment and research, using five research vessels, one of which is stationed on each lake. The Canadian Department of Fisheries and Oceans has provided additional support. Finally, the network of University Sea Grant Institutes and unaffiliated universities and colleges have conducted important contractual research on the status and functions of forage bases in each of the lakes. Some of this research has been funded by the National Sea Grant College Program, the Federal Aid in Sport Fish Restoration Act, the Natural Sciences and

Engineering Research Council of Canada, the Great Lakes Fishery Commission and other sources (Stewart et al. 1981; Stewart and Ibarra 1991; Sprules et al. 1991; Koonce and Jones 1994).

Stewardship of the forage base and other fishery resources in the Great Lakes is entrusted to the government players above in behalf of private-sector players who are the stakeholders of these natural resources (Koonce and Jones 1994). The main players in the private sector are representatives of, and participants in: commercial fisheries, which developed the capacity to overexploit some fish stocks in the Great Lakes by the late 1800s and early 1900s, sport fisheries which have surpassed commercial fisheries in economic importance since the 1950s, Native American fisheries, environmental interest groups, and raw water users (Holey et al. 1995; Talhelm 1988; Brown et al. 1999; Koonce and Jones 1994). Less visible players are private individuals and groups who value the natural beauty and other intrinsic values of fishery resources and their environment and want them preserved.

Conflicting Interests in Forage Allocation

Conflicts and shifts in the allocation of fishery resources, and prey fish in particular, among user groups have been intensified throughout the Great Lakes, mainly because of the rapid growth of sport fishing since the 1950s (Eshenroder 1987). The development of sport fisheries was emphasized more by some agencies than others, and, as noted earlier, economic returns to regional economies were often much greater than those from commercial fishing (Francis et al. 1979; Talhelm 1988).

The State of Michigan, which pioneered the development of hatchery-based sport fisheries for Pacific salmon and trout in the 1960s, offered some compensation through the legislature to reduce the number of commercial fishing licensees for chubs in the 1970s and later (M. Keller, Michigan Department of Natural Resources, personal communication). Since the lakewide ban on chub fishing in Lake Michigan in the late 1970s (Brown et al. 1985), Michigan has licensed five or less commercial fishing operators for chubs in that lake, compared to about forty-one that were licensed in the mid 1960s and twenty-three in the early 1970s (D. Nelson, Michigan Department of Natural Resources, personal communication). However, some fishermen left the chub fishery because of economic conditions before the closure in 1976. Wisconsin, in contrast, is licensing about fifty-four commercial operators for chubs (A. Blizel, Wisconsin Department of Natural Resources, personal communication) and from 1989 to 1992 accounted for about 77% of the annual lakewide chub harvest(United States Geological Survey–Biological Resources Division, COMCAT data base).

Although the walleye is a piscivore and not a prey fish, its almost complete turnaround in Lake Erie, from a heavily exploited commercial species and gamefish to solely a gamefish in U.S. waters in the 1970s, represents an indirect reallocation of its forage base. That is because Ohio and Michigan have elected to allocate all of their walleye quotas determined by the Interagency Walleye Scientific Protocol Committee to sport fisheries since the early 1970s. Ontario, in contrast, has chosen to divide its walleye quotas between both commercial and sport fisheries (Nepszy 1985; Hatch et al. 1987). Commercial walleye fisheries were closed in Ontario, Ohio and Michigan waters of Lake Erie in 1970 because of mercury contamination (Nepszy 1985).

Fundamentally different values placed on fishery resources by commercial operators and by sport anglers and charter-boat operators, who serve the latter, as well as the differences in management philosophies and policies of the agencies indicated above, have precluded unified, lakewide forage base allocation in the past (Dochoda et al. 1987). The series of drastic changes undergone by the fish communities of the Great Lakes, which are discussed in more detail later, and the interactive ecological, economic and social consequences of those changes have further complicated the process (Eshenroder 1987). Indeed, before allocations of forage can be implemented, fishery managers must make difficult decisions on a higher level and address questions, which have not all been answered. Do you allocate forage for lake trout rehabilitation, although stocked lake trout are not reproducing successfully, except in Lake Superior and several sites in Lake Huron (Selgeby et al. 1995)? Should efforts to rehabilitate lake trout be abandoned in favor of put-and-take planting of salmon? Do you allocate forage for salmonines, and hence, sport fisheries at the expense of commercial fisheries? How do you avoid the overstocking of chinook salmon (*Oncorhynchus tshawytscha*) that may have depressed the alewife forage bases in Lakes Michigan and Ontario in the 1980s (O'Gorman and Stewart 1999)?

Additional insight into some of these forage base issues and conflicts are provided in a historical context in the two case histories that follow: Lake Michigan, where the greatest ecological changes and conflicts among user groups have occurred, and Lake Superior, where the ecological changes and conflicts have not been as great, and forage base allocation has not been implemented on a large scale.

Recent application of a dynamic management-driven modeling approach to the predator-prey systems of chinook salmon and alewives in Lakes Ontario and Michigan, aimed at understanding the demand on the forage bases, is covered next. This is followed by a review and evaluation of past and present sampling gear and methods used to assess the forage base in Lake Michigan and provide information for making decisions on forage base allocation.

Lake Michigan Case Study

Lake Michigan is a classic example of the complex interplay between ecological and socioeconomic conditions and the ways that agencies have attempted to respond to change, and hence, have influenced forage base allocation and fishery yields in general. From about 1945 to 1965, when it was brought under control with a chemical lampricide, the sea lamprey was the driving force, destroying or severely depleting stocks of the most valuable commercial species, including lake trout and lake whitefish (*Coregonus clupeaformis*) (Eschmeyer 1955; Smith et al. 1980; Smith 1968). An explosive increase in population size of the exotic alewife adversely affected various native species in the 1960s and 1970s, and the introduction of Pacific salmon by the State of Michigan from 1966 to 1967 not only helped to control alewives, but stimulated the growth of an intensive recreational fishery for large, mostly hatchery-reared salmonines in the four states bordering Lake Michigan (Smith 1970; Borgeson 1981).

As stocking programs for lake trout restoration, alewife control, and sport fisheries developed rapidly in the 1960s, the states quickly banned commercial fishing for the salmonines (Holey et al. 1995). This diversion of forage into sport fishing channels has been offset, to some extent, since the late 1970s by Tribal commercial and subsistence fishing for lake trout in Treaty-ceded waters in the State of Michigan. However, in 1987, the Chippewa-Ottawa Treaty Fishery Management Authority enacted zoning regulations and other fishing restrictions intended to control both the

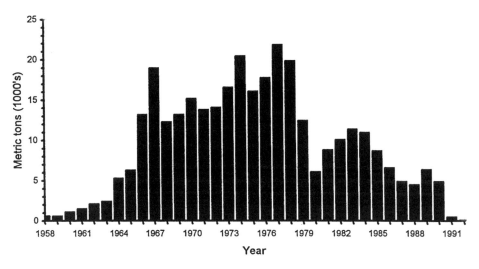

FIG. 3. *Annual commercial catch (metric tons) of alewives from Lake Michigan, 1958 to 1992; COMCAT data base, U. S. Geological Survey–Biological Resources Division.*

targeted catch and by-catch of lake trout in accordance with the Consent Decree of 1985, that settled the Tribal fishing controversy.

The salmon, lake trout, and other trout foraged mainly on alewives in Lake Michigan, in competition with a commercial fishery for alewives that reported some of the largest landings ever recorded for the Great Lakes (fig. 3). Commercial trawlers converted from gill net tugs and trap net boats had been introduced to Lake Michigan in the late 1950s, with technical assistance from the U.S. Bureau of Commercial Fisheries (Gordon and Brouillard 1961), to utilize large stocks of small bloaters and recently introduced alewives. The developing trawl fisheries in Michigan and Wisconsin waters shifted their main effort to the expanding alewife population in the 1960s (Smith 1968; Brown 1972), when fish meal plants to process alewives were in operation on Green Bay (Greenwood 1970; Borchardt and Pohland 1970; Stuiber and Quigley 1973). After recovering rapidly from a mass die-off in 1966–1967 which resulted from overpopulation, and cost the tourist industry millions of dollars from alewife-littered beaches (Wells 1970; Brown 1972; Time 1967), the alewife declined to lower levels again in the late 1970s and 1980s (Brown et al. 1987).

Consumption of alewives by salmonines surpassed commercial landings every year beginning in 1970, and averaged about five to ten times greater than the diminishing commercial landings of the 1980s (fig. 3; O'Gorman and Stewart 1999). As a result, conflicts of interest intensified between sport and commercial fishermen over the question of forage base allocation. The State of Wisconsin finally banned trawling for alewives in 1991, after several years of intense pressure from sport fishing organizations, which alleged that trawling caused the alewife to decline (J. Moore, Wisconsin Bureau of Fish Management, Madison, personal communication). Near the peak of the trawl fishery for alewives in 1969, about eleven trawling vessels operated in Michigan and Wisconsin waters of Lake Michigan proper and Green Bay (Brown 1972). In addition, commercial pound net fisheries for alewives, mainly in Green Bay, landed roughly the same tonnage of alewives as the trawlers (Brown 1972). At the time of its closure in 1991, the Wisconsin trawl fishery for alewives in Lake Michigan proper was composed of five vessels. These fishing vessels were capable of depleting local alewife concentrations but certainly not the lakewide population of the 1980s. That feat has been attributed to predation by the salmonines, and by the chinook salmon in particular, although the generally much colder than average winters from 1976 to 1981 were also probably involved (Kitchell and Crowder 1986; Eck and Brown 1985; O'Gorman and Stewart 1999).

Need For Holistic Management

The creation of hatchery-reared salmonine sport fisheries to help control and utilize alewives in Lake Michigan, where they attained world class status in the 1970s (Wells and McLain 1973), and later in the other Great Lakes, did more to focus interagency attention on the importance of forage base allocation than did any other single factor. As a result, fishery managers in the Great Lakes Region began to recognize the need for holistic management of predator-prey systems rather than for piecemeal management of the uncoupled species, which was a common practice in the past (Stewart et al. 1981; Kitchell 1985; Crowder 1985; Busiahn 1985a; Krueger and Paine 1985). The severe decline in alewife abundance in Lake Michigan in the late 1970s and early 1980s, attributed partly to effects on survival of a series of cold winters (Eck and Brown 1985) and to increasing predation from salmonines (Kitchell and Crowder 1986), emphasized the need for holistic management, as did a subsequent decline in survival of chinook salmon in the late 1980s concomitant with outbreaks of bacterial kidney disease (BKD) caused by *Renibacterium salmoninarum*, which is ubiquitous to salmon in the Pacific Northwest (Withler and Evelyn 1990). Roughly 40% to 50% fewer chinook salmon may have returned to some Michigan streams to spawn in the late 1980s as compared to the early 1980s, and BKD was the apparent final cause of spring die-offs of large chinook salmon in southern Lake Michigan from 1989 to 1990, according to Michigan and Indiana fishery managers (Stewart and Ibarra 1991). Because of the alewife decline and its possible implication in the salmon die-offs, fishery managers in Michigan and Wisconsin lowered salmonine stocking rates by about 8% in the late 1980s (Keller et al. 1989).

At least two hypotheses were initially proposed to explain the outbreaks of BKD and the salmon die-offs: (1) stress from depletion of alewives, the preferred forage of salmon (Stewart and Ibarra 1991), as indicated above; and (2) a more virulent disease organism or lower resistance to the disease organism because of salmon propagation in hatcheries (R. Rybicki, Michigan Department of Natural Resources, personal communication). Evidence was also found of a genetic component in resistance of coho salmon to the bacterium (Withler and Evelyn 1990).

In January 1996, a study group of state and academic researchers in Michigan reported that high stocking rates of chinook salmon in Lake Michigan may be perpetuating BKD (R. Clark, Michigan Department of Natural Resources, personal communication). The study group concluded that BKD is transmitted between fish in the lake and that infection and mortality rates are probably a function of fish density at early life stages. Catch rates of one-year-old chinook were quite stable from 1985 through 1993, in spite of the collapse of the sport fishery for chinook. Therefore, the study group hypothesized that BKD infection occurs among the high-density young in the lake, and they die from it as adults if it becomes virulent from poor nutrition or some other stressor. To test this hypothesis, the study group calculated that a reduction of three million chinook smolts annually, or about 46% of the current number would achieve fish-community goals for Lake Michigan (Eshenroder et al. 1995) if BKD were eliminated. These calculations took into consideration an estimate that about two million natural smolts are produced in the lake each year.

Forage Requirements of Predators

As the salmonine population continued to build in Lake Michigan in the 1970s, fishery managers were becoming increasingly aware of the need for information on the forage requirements of the various species being planted. Stewart et al. (1981) provided some of that information by using energetics-based population models (Stewart et al. 1983; Hansen et al. 1993) to estimate total annual food consumption in the lake by lake trout, coho salmon (*Oncorhynchus kisutch)* and chinook salmon. Their results showed that a cohort of one million chinook salmon during its life in the lake consumed almost twice as much forage as a shorter-lived cohort of one million coho salmon, while a single cohort of lake trout, the longest lived of the salmonines, consumed 1.5 times as much as the coho. They also reported that consumption of adult alewives by chinook salmon was almost three times greater than that by coho salmon. Yet the states at that time were planning to cut back on coho plantings and apparently increase planting rates of chinook, the least expensive of the salmon to raise.

Stewart et al. (1981) used the concept of predation inertia (i.e. the time from stocking until most of the predation impact has occurred) as an indicator of the potential of a species for short-term manipulation by managers in response to changes in the forage base. This research prompted fisheries managers to consider forage availability when setting stocking rates for hatchery-reared salmonines (Kitchell et al. 1977; Stewart et al. 1981). The State of Wisconsin began using mean weight of salmon in their sport fisheries as an index of forage availability in a computerized system of allocating planted fish to zoned fishing waters in Lake Michigan (Dehring and Krueger 1985). Other input to the Wisconsin system included a catch objective, species preferences of anglers, and number of lake trout available for planting.

The State of Michigan reviewed its management program for salmon and trout a few years later and recognized the need for more timely information on the status of the forage base as well as for research on salmon biology and dynamics (Keller et al. 1989). As a result, it diverted some of its stock assessment work on lake trout and other activities into field and laboratory research on salmon. Part of this research was an attempt to explain the high mortalities of chinook salmon in Michigan waters. The fact that BKD symptoms and high mortality in chinook salmon were traced to the first years of life in the lake, as well as among the adults, led Michigan biologists to lean more toward the more virulent disease or lower resistance hypothesis than that espousing a shortage of forage (i.e. alewives) for adults (R. Rybicki, Michigan Department of Natural Resources, personal communication). As a result, hatchery managers in Michigan obtained chinook salmon eggs for a few years from New York state where the incidence of BKD was extremely low, compared to that of feral salmon returning to weirs on Lake Michigan tributaries. However, if the current hypothesis of the chinook study group in Michigan were accepted, it would clear hatchery practices as causes of the BKD epidemic, and instead implicate higher infection rates at higher stocking densities of young chinook in the lake, resulting in their death as adults, which was indicated earlier in the chapter.

Forage Base Allocation in the 1980s and 1990s

Research on forage base allocation accelerated on Lake Michigan in the 1980s and 1990s, as it did on other Great Lakes (Stewart et al. 1981; Eck and Brown 1985; Kitchell and Crowder 1986; O'Gorman et al. 1987; Einhouse et al. 1993). The central theme of much of this research was to maintain the hatchery-dependent suite of large salmonine predators, while not endangering the forage base (Ney 1990; Jones et al. 1993). At the same time, management agencies continued to support the goal of a rehabilitated lake trout population in Lake Michigan to bring lake trout into natural synchrony with the forage base. Although, as noted by Addis (1988), reaching that goal ". . . has proven to be far more difficult, both biologically and socially than originally anticipated."

To provide managers with a more comprehensive and current basis for making informed decisions on forage base allocation, stocking rates of salmonines, and related issues, Stewart and Ibarra (1991) remodeled (after Stewart et al. 1981) the bioenergetics of lake trout and salmon and integrated the results with population estimates of the salmonines from 1978 to 1988 (a period of peak abundance of chinook salmon and total predators, and of low alewife abundance). Their population estimates were based on stocking records and survival rates provided by state and federal agencies. Stewart and Ibarra (1991) found diet shifts in both chinook and coho salmon from feeding primarily on large alewives to feeding proportionately more on juveniles and other prey, when alewives declined markedly from 1981 to 1983. As a result the chinook salmon suffered a 20% decline in gross conversion efficiency of biomass and a 25% decline in average weight of fish caught by anglers from 1981 to 1983. Total predation by all the salmonines, including brown trout (*Salmo trutta)* and rainbow trout (*Onchorynchus mykiss)*, peaked at 71,000 metric tons in 1983, and 76,000 metric tons in 1987. Although the bloater population was far larger than the alewife population of Lake Michigan from 1987 to 1988, Stewart and Ibarra (1991) estimated that alewives comprised 70% of all prey eaten by the salmonines in that period. Moreover by the 1990s the salmonines and commercial fisheries together were consuming and landing an estimated 60% of the annual production of alewives (Stewart and Ibarra 1991; Brandt et al. 1991). Those extraction rates led Stewart and Ibarra (1991) to express the concern that, according to their research, greater densities of Pacific salmon "would affect alewife first and foremost." Hence they emphasized that increases in stocking densities by managers to utilize the available bloater production, a possibility suggested by Keller et al. (1989), was incompatible with the goal of maintaining alewife as an important prey in the system. The controversy and consequences of adding more predators to the predator-prey system of Lake Michigan are evaluated in a management-driven modeling context later in this chapter.

Lake Superior Case Study

Although Lake Superior is the largest, coldest, least productive, and still most pristine of the Great Lakes, its fish community and forage base in particular have undergone substantial changes due to exploitation, species introductions, and habitat degradation (Lawrie and Rahrer 1972 and 1973; Waters 1987; Hartman 1988). Until the last decade or so, most of the changes were negative—that is, they resulted in fish populations that were less stable and less productive for human use than was previously the case. Lake Superior, however, was the last of the Great Lakes invaded by the sea lamprey, and the first in which that predator was brought under control with a chemical lampricide (Smith et al. 1974; Smith and Tibbles 1980). Remnant stocks of lake trout survived and provided a possible nucleus for lake trout rehabilitation. Also, as noted earlier in this chapter, the cold environment of Lake Superior prevented the buildup of a large alewife population such as that which supported the world class sport fishery for Pacific salmon in Lake Michigan from the 1960s into the early 1980s. Because of the less productive forage base and lower fishing pressure from a sparser human population in its watershed, conflicts among competing fisheries on Lake Superior have been much less intense and less focused than on the other Great Lakes.

Lake Herring Decline as the Preeminent Prey and Commercial Species

The lake herring was historically the dominant prey fish in Lake Superior for native lake trout and burbot (Dryer and al. 1965) and also the largest component in the commercial catch (Dryer et al. 1965). From 1916 to 1940, the period used as a benchmark by Lake Superior fishery managers, harvests of lake herring by the commercial fishery averaged more than 4.5 thousand metric tons per year—or 64% of all commercial landings (Baldwin et al. 1979). Lake herring stocks subsequently declined from west to east, beginning in Minnesota, where the yield fell below the historical average in 1941. Landings dropped below the historical averages in Wisconsin in 1963 and in Michigan in 1970. By the 1970s, lake herring were scarce in most of the U.S. waters of Lake Superior. In contrast, commercial landings in Ontario remained near the historical average until 1988, when the harvest fell sharply, partly at least, because of a decline in abundance of the intensively fished Black Bay Stock (Hansen 1990).

Predation, Forage-base Interactions, and Fishing.

Causes for the lake herring's progressive decline in U.S. waters of Lake Superior from about mid-century have been debated by researchers, who knew that lake trout and burbot preyed on juvenile and adult herring and suspected that smelt preyed on the larvae. (The non-indigenous rainbow smelt, which was first recorded in Lake Superior in 1930 and became abundant during the 1930s, is the only major addition to the native forage base; Van Oosten 1937). Neither Anderson and Smith (1971), who used correlation of relative abundance of rainbow smelt and lake herring, nor Selgeby (1978), who evaluated the stomach contents of rainbow smelt, found predation by rainbow smelt on lake herring to be a major factor in the lake herring decline. Anderson and Smith (1971) also mentioned the fact that lake trout stocks declined in the 1950s, while lake herring were declining, which might rule out predation by the trout as a major factor. They proposed instead that increasing populations of rainbow smelt and bloaters affected lake herring survival through competition for copepods during larval stages. However, Selgeby et al. (1994) found that competition between larval rainbow smelt and lake herring was almost nonexistent. Although Anderson and Smith (1971) also argued that overfishing was not correlated with the decline of lake herring, Selgeby (1982) showed that lake herring in the Apostle Islands area of Wisconsin were sequentially fished up while segregated into discrete stocks during the fall spawning season. Thus, commercial harvest and catch rates remained deceptively high, as the fishery depleted the nearer stocks first and then moved sequentially to the next closest ones that were still productive.

Other Coregonines in the Forage Base

In addition to the lake herring, the Lake Superior forage base contains populations of bloaters and two other species of deepwater ciscoes, the kiyi and the shortjaw. All three species are preyed on by burbot and lake trout, particularly by the siscowet strain of lake trout which lives mainly in deep water. Since the 1970s, both commercial harvest and abundance of the deepwater ciscoes have declined markedly (MacCallum and Selgeby 1987; Hansen 1994). Increasing predation from siscowets and other predators may be the reason for the decline (Hansen 1994). The deepwater ciscoes, and their close relative the lake herring, are difficult or impossible to identify when they are partially digested. For that reason, they have been reported only as coregonines in published food habits studies of lake trout in Lake Superior, which masks their individual contributions to the predator's diet (Dryer et al. 1965; Conner et al. 1993).

Rehabilitating and Newly-naturalized Salmonine Predators

The dominant native predators in the open waters of Lake Superior were the burbot and several distinct morphological and/or behavioral variants of the lake trout (Goodier 1981). The latter included the lean lake trout, which supported a relatively stable commercial fishery with yearly landings that averaged about 1.8 million kg from 1920 to 1950, the fat, or siscowet, half-breeds, and humpers. By the time sea lamprey control became effective in Lake Superior in 1961, the abundant stocks of lean lake trout had been reduced to one or more remnant local populations. However, widespread stocking of hatchery-reared lean lake trout produced rapid increases in abundance in some areas of the lake in the 1960s and 1970s (Pycha and King 1975), and surviving wild stocks have slowly recovered since 1961 (Hansen 1990, 1999).

The above array of native predators that subsisted mainly on lake herring, deepwater ciscoes, sculpins (i.e., slimy, deepwater, and spoonhead (*Cottus ricei*) and, more recently, smelt, has been broadened considerably by introductions of non-native salmonines, mostly since the mid 1960s (Hansen 1990). These introductions, which include only one that was not deliberate (i.e., that of the pink salmon (*Oncorhynchus gorbuscha*)), have affected predator-prey relationships and, hence, the forage base in possibly important but poorly understood ways. They have also had mixed success. Pink, coho, and chinook salmon have all established reproducing populations since the 1950s or 1960s and are now considered to be naturalized (like the rainbow and brown trout which preceded them in about 1900). The average numbers of coho salmon harvested in Michigan, Wisconsin, and Minnesota waters of Lake Superior from 1984 to 1988 ranked either first or second behind those of lake trout. Moreover, the 1988 coho harvest in Wisconsin waters exceeded the lake trout harvest for the first time (Hansen 1990). Pink salmon, in contrast, peaked in Michigan waters in 1979 and by 1980 were spawning in fifty-six Michigan streams tributary to the lake. However, most of the major spawning streams among the fifty-six had few, if any, spawning runs of pink salmon by 1989, according to fishery managers (Hansen 1990).

Partial Recovery by the Forage Base Since the Early 1980s

Rainbow smelt and lake herring have fluctuated considerably in abundance in Lake Superior in recent years (fig. 4; MacCallum and Selgeby 1987; Hoff and Selgeby 1994). Rainbow smelt, which were very abundant in the 1960s and early 1970s, declined sharply in abundance from 1978 to

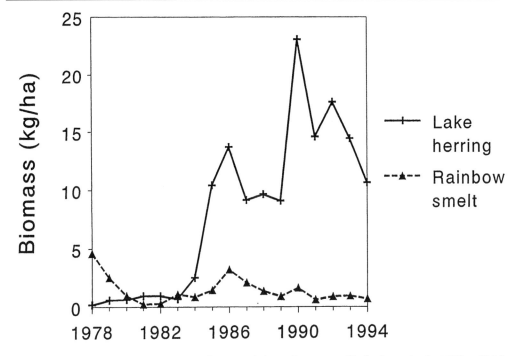

FIG. 4. *Biomass of lake herring and rainbow smelt in U. S. waters of Lake Superior in 1978 to 1994, as measured by area swept with a 12m bottom trawl; data were provided by the Lake Superior Biological Station of the Great Lakes Science Center, U. S. Geological Survey–Biological Resources Division.*

1981, but have partially recovered since then. Lake herring have made an even stronger recovery, in some areas, since 1983, as reflected in a more than ten-fold increase in biomass, but they have not yet reached historic levels of abundance (Hansen 1994). Fishery managers view the recovery of the lake herring from its formerly depressed state as one of the most important and positive ecological occurrences in Lake Superior in recent decades because of the herring's unique ability to transform biological production in the large offshore pelagic zone to fish biomass needed to support predator fishes.

Despite the strong recovery of the lake herring, the rainbow smelt continues to be the dominant food item of inshore stocks of lake trout and other salmonine predators, just as it was when it was more abundant in the 1960s (Dryer et al. 1965; Conner et al. 1993). The importance of rainbow smelt as the preferred prey of salmonines in Lake Superior is comparable to that of the alewife as their preferred prey in Lake Michigan. Indeed, most salmonines consistently selected rainbow smelt during the 1980s, even though coregonine biomass was more than double that of rainbow smelt (Conner et al. 1993). The lake herring is the most pelagic of the coregonines and perhaps less available in the lower water column than are rainbow smelt. Only offshore were the diets of lake trout dominated by coregonines, apparently because fewer rainbow smelt are available over the greater depths than they are closer to shore. More recently, however, lake trout have been eating proportionally more coregonines than they did in past decades, and wild lake trout have been consuming fewer rainbow smelt than have hatchery-produced fish (Hansen 1994). This suggests that the wild lake trout are better adapted to preying on the native fraction of the forage base and/or occupy deeper water on average than do planted lake trout.

Because of concern about changes in abundance of prey fish and their effects on lake trout, fishery management agencies established the objective to maintain a predator-prey balance that would allow for normal lake trout growth (Busiahn 1990), apparently based on the range of growth observed in past years. For example, indices of growth declined in nearly all areas of Lake Superior during the early 1980s, most likely due to the decline in abundance of rainbow smelt that preceded the sharp increase in abundance of lake herring (Busiahn 1985b; Hansen 1994). Increased predator abundance, decreased prey abundance, and changes in species composition and energy density of prey may have affected lake trout growth rates (Busiahn 1985b; Rand et al. 1994).

A System for Allocating the Forage Base Needed

Fishery agencies on Lake Superior are aware of the need for developing a system of allocating the forage base in harmony with both production of predators and fishery harvest (Busiahn 1990). However, the scientific basis for forage base allocation is poorly developed at present. Negus (1992) used a bioenergetics model to relate predation to available prey and found large, unexplained discrepancies. For example, estimates of available rainbow smelt and coregonines were far lower than estimates of consumed and harvested rainbow smelt and coregonines. She then identified and prioritized the information needed for effective bioenergetics modeling, the most critical being improved estimates of prey fish biomass at all depths and positions in the water column.

Lake Superior fishery managers have recommended that lake herring be protected through appropriate commercial fishing regulations so that the herring stocks can further recover, while supporting increased predation by lake trout and salmon (Hansen 1994). However, the harvest of lake herring is not regulated in some jurisdictions. This is a sensitive point, but apparently some jurisdictions do not have the political support to "stand-up" to user groups. In such cases, low market prices have provided the only real protection from the commercial fishery.

Lake Superior fishery management agencies, unlike those on the other Great Lakes, have also agreed that the non-indigenous rainbow smelt is an undesirable species that should not be protected from harvest. Based on the history of rainbow smelt in the Great Lakes, it is unlikely, however, that any economically feasible level of harvesting could greatly reduce the size of the smelt population. If it did, then any suspected, but yet unproven, negative impacts of rainbow smelt on the fish community would be reduced to the possible benefit of the more desirable lake herring and its salmonine predators. Although stocking of predators provides some control over the allocation of the forage base, provincial and state management agencies have also recognized that, eventually, stocks of lake trout and other salmonine predators can most likely be totally sustained by natural reproduction. This implies the reemergence of natural predator-prey relations and their modulating effects on the forage base. In addition, the agencies recognized that stocking may still be necessary to provide harvest opportunities in some geographic areas (Busiahn 1990). More recently, however, fishery managers jointly recommended through the Lake Superior Technical Committee of the Great Lakes Fishery Commission that stocking programs and fisheries should not compromise any naturally reproducing population of lake trout or other salmonine (Hansen 1994).

In summary, the lake herring stocks and the lake trout stocks that prey on them in Lake Superior are well along in their recoveries from severe depletions. The completion of their restorations to fully self-sustaining and/or productive members of the fish community will require the following: (1) coordinated management actions by agencies; (2) use of advanced assessment techniques, such as those described later in this chapter for Lake Michigan; (3) use of sophisticated bioenergetics and ecosystem modeling, such as that described later for Lake Ontario and Lake Michigan; and (4) establishment of a system of allocating forage stocks between predacious fish and competing fisheries.

Understanding Multiple Demands on Great Lakes Forage Bases

Earlier in this chapter, it was noted that, as the salmonine biomass in the Great Lakes continued to increase during the 1970s and early 1980s, scientists began to question whether the forage base would be able to sustain the predatory demand imposed by the expanding piscivore populations (Stewart et al. 1981). Much of the early work focused on Lake Michigan, where the Pacific salmon stocking programs began, and where piscivore biomass built up to very high levels relatively early. Calculations based on bioenergetics modeling (Stewart et al. 1981; Stewart and Ibarra 1991) and whole-lake acoustic estimates of alewife abundance (Brandt et al. 1991) suggested that predators could have been consuming as much as 53% of the alewife biomass in Lake Michigan annually. This suggestion led investigators to call for a reexamination of current stocking levels, due to concerns that the population of the primary prey fish, the alewife, would collapse under the heavy predation imposed by the stocked piscivores. On the other hand, management agencies were pressured to maintain stocking levels by special interest groups, such as operators of sport-fishing charter boats, whose livelihood had come to depend on the high catch rates enjoyed by anglers participating in these fisheries.

The Salmonine Stocking Level Controversy

Because of the growing controversy over the effects of stocking on the forage base, both Canadian and U. S. resource management agencies faced a dilemma: given the considerable uncertainty regarding the sustainability of the current standing stocks of salmonines, was it wiser to reduce stocking rates and risk undesirable economic impacts on the recreational fishing industry or maintain stocking rates and risk even greater undesirable biological and economic impacts on the piscivores and the fishing industry in the event of a forage base collapse? The decline in alewife abundance in Lake Michigan during the early 1980s, followed by heavy mortalities of chinook salmon infected with the BKD bacterium (which some believed resulted from insufficient forage) finally brought the issue to a head. Indeed, fishery managers throughout the basin responded by calling for an open, objective assessment of the situation, with the goal of arriving at a future stocking strategy based, where possible, on a consensus among the major stakeholders.

Formal Investigation of Salmonine Sustainability

The GLFC agreed to sponsor a technical investigation of the sustainability issue through its Board of Technical Experts, employing the methods of Adaptive Environmental Assessment and

SIMPLE

Sustainability of Intensively Managed Populations in Lake Ecosystems

FIG. 5. *Time line of major activities of the SIMPLE Task Group while it addressed the issue of the sustainability of salmonine fisheries on Lake Ontario and Lake Michigan; from Figure 1 of Koonce and Jones (1994).*

Management (Holling 1978). The essential features of this approach are that a resource management problem is analyzed using quantitative tools, such as computer simulation models, and that these models are developed through interactive workshop discussions with technical experts, fishery managers, and stakeholder group representatives. This mix of contributors to the analysis ensures that the product of the effort is simultaneously technically sound and relevant to the issues of greatest concern to the stakeholders. The workshop modeling process also exposes all participants to the many uncertainties associated with a complex issue of this sort, and allows explicit examination of the risks associated with assuming greater knowledge than actually exists.

Focus and Procedure of the SIMPLE Study

The resulting GLFC study, known as SIMPLE (Sustainability of Intensively Managed Populations in Lake Ecosystems), was led by Canadian and U.S. coprincipal investigators, overseen by a panel of two fishery managers (one U.S. and one Canadian) as well as a representative from the GLFC. From the outset, it was agreed that the study would focus on Lakes Michigan and Ontario, the two lakes with the highest rates of salmonine stocking per unit surface area. The study began with a technical workshop, at which participants agreed upon the scope and contents of the models that would guide the remainder of the study (fig. 5). Following the workshop, the two principal investigators conducted a series of technical sessions, aimed at compiling the necessary information to develop the models. Later they held a second workshop, that focused on socioeconomic issues. The process culminated with a pair of round-table meetings to evaluate relevant management policies, one meeting for Lake Ontario, and one for Lake Michigan. The model was used at these meetings to explore the predicted consequences of various management options and through this to identify opportunities for agreement on appropriate action. The process is described in detail in Koonce and Jones (1994).

Prey Supply Versus Predator Demand: A Key Issue

The overall issue of sustainability of hatchery-dependent fisheries encompasses much more than the interactions between stocked salmonine predators and their prey. However, the issue of prey supply versus predator demand (Ney 1990) dominated discussions during the SIMPLE task, mainly because the issue is highly relevant to important decisions managers must make in answer to the question: how many fish should be stocked? The models developed during the SIMPLE task reflect this emphasis. Dynamic computer simulations employing the models were developed for Lakes Ontario and Michigan and included all of the major species of salmonine predators and prey-fish present in each lake. Predator demand within each model is calculated using simplified representations of the bioenergetic models identified earlier in this chapter (Stewart et al. 1981; Stewart and Ibarra 1991) and assumptions about post-stocking survival of the hatchery-derived salmonines as well as about the recruitment of naturalized populations of these predator species.

To properly determine whether predator demand for forage is excessive, however, one needs to compare estimated demand levels to the rates of production of the prey fish. Prey species supply was therefore modeled using stock-recruitment relationships derived from observed trends over time in the abundance of these species (Eck and Wells 1987; O'Gorman et al. 1987; Jones et al. 1993). For a particular level of predator abundance to be sustainable, then, predator demand cannot exceed the supply of prey generated by the latter's natural recruitment processes. In Lake Michigan, the picture is complicated further by evidence of community interactions among the species of prey fish (Crowder and Binkowski 1983; Crowder et al. 1987; Eck and Wells 1987) that must be accommodated within the prey-fish recruitment model. For example, recruitment of prey-fish species which have pelagic larvae that are vulnerable to predation by alewives was computed as a function, not only of adult prey-fish biomass, but of alewife abundance as well.

Although the SIMPLE models have the advantage over the earlier bioenergetic models of including a dynamic prey supply, they remain major simplifications of reality. They treat each lake as a single spatial unit, model fish population dynamics using an annual time step, and include numerous untested assumptions about key processes, such as predator survival rates, foraging rates, and preferences for different prey types. These model simplifications are, in part, necessitated by uncertainty, but they also afford the advantage of making the models sufficiently simple for a non-technical audience of decision makers and stakeholders to understand them. Without this understanding, the impact on decision making and policy would have been greatly diminished.

Predator Forage Demands at or above Sustainable Levels

For both Lakes Ontario and Michigan, the models suggested that by the late 1980s predator demand was at or exceeded levels that were sustainable by the existing prey-fish community (Jones et al. 1993; Koonce and Jones 1994). In Lake Ontario, where fish community dynamics are somewhat less complicated (i.e., fewer interacting species), and model analysis has proceeded further (Jones et al. 1993), the model predicts that alewife populations would be extremely vulnerable to collapse under excessive predation pressure if a cold winter were to lead to extraordinary over-winter mortality of alewives (fig. 6), such as that observed from 1976 to 1977 (O'Gorman and Schneider 1986). Uncertainty analyses of the model suggest that these predictions are sensitive to assumptions about model parameters for which there is considerable uncertainty, such as

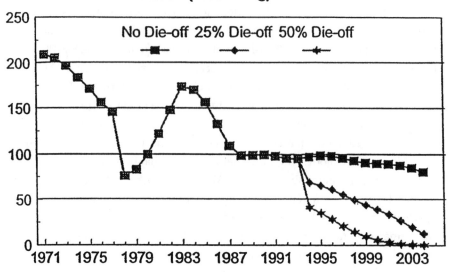

FIG. 6. *Predictions from the SIMPLE Model of trends in adult alewife biomass in Lake Ontario at 1992 salmonine stocking levels, given over-winter die-offs of varying severity; adapted from Figure 1 of Jones et al. (1993).*

predator foraging rates and variability of the alewife stock-recruitment relationship (Jones et al. 1993). Nevertheless, the risk of collapse cannot be discounted. For Lake Michigan, the uncertainty is perhaps even greater, yet the model predictions suggest, even without extraordinary alewife mortalities, that predator demand resulting from salmonine stocking during the 1980s exceeded that which could be sustained by the alewife population (fig. 7). This result was obtained despite a substantial upward adjustment to alewife biomass in the model, relative to estimates based on trawls (Eck and Brown 1985), in order to calibrate the model against the historic patterns (1968 to 1988) of alewife abundance changes over time.

Uncertainty and Risks of Management Actions Noted

The SIMPLE study exposed the profound uncertainty associated with making proactive decisions about a complex ecosystem problem, such as predator-prey management, and was very successful in fostering dialogue among the major stakeholders concerned with this issue. The use of the models at the policy evaluation round-tables allowed stakeholders to explore the dynamics of the predator and prey fish communities in each of the lakes, examine the predicted consequences of a variety of management strategies or scenarios (e.g., reductions in overall salmonine stocking rates, or shifts in relative stocking rates among predator species, such as chinook salmon and steelhead, with contrasting degrees of dependence on alewives), discuss the implication of these scenarios with representatives of other interest groups, and consider how the uncertainty, that became readily apparent as these discussions proceeded, influenced the risks associated with a particular action.

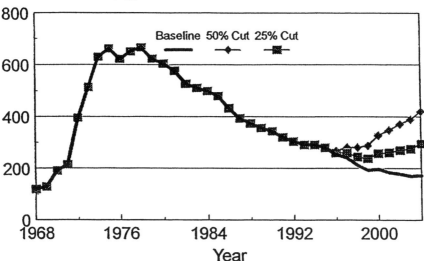

FIG. 7. *Predictions from the SIMPLE Model of trends in adult alewife biomass in Lake Michigan, given salmonine stocking level reductions (i.e., scenarios) of 0%, 25%, and 50% from 1992 levels; from Figure 2 of Koonce and Jones (1994).*

Conclusions and Follow-Up Management Actions

Lake Ontario

Soon after the Lake Ontario round-table, the two agencies responsible for stocking decisions on Lake Ontario (i.e., the New York Department of Environmental Conservation or NYDEC, and Ontario Ministry of Natural Resources or OMNR) began a process of public consultation with the objective of determining the appropriate actions to take in light of emerging evidence that predator demand may be excessive (Anonymous 1992). From this consultation emerged a widespread, although certainly not unanimous, view that action to reduce predator demand was needed. As a consequence, the two agencies implemented coordinated stocking cuts aimed at reducing predator demand by about 50%, consistent with the conclusions of an expert task group (Anonymous 1992). Many managers and stakeholder group representatives participated in the SIMPLE study, which began before the perception of a crisis on Lake Ontario was widespread, and the insights gained from their involvement almost certainly had a large influence on their thinking about the issue when the decision-making/consultation process began.

Lake Michigan

It is not as clear that the SIMPLE study has had an impact on Lake Michigan. Major stocking cuts have not been implemented on Lake Michigan, although a sharp decline in chinook salmon survival, due to BKD infection, may be achieving a similar effect on overall predator demand (Keller et al. 1989; Stewart and Ibarra 1991).

Timing is probably the most plausible of several possible reasons for the difference in management response on Lake Michigan. In contrast to Lake Ontario, the issue of excessive predator demand had been a subject of much discussion and controversy on Lake Michigan for several years before the SIMPLE study began (Stewart et al. 1981; Eck and Brown 1985; Kitchell and Crowder 1986). Agency positions differed as to the likely causes of the decline in alewife abundance in the early 1980s and in chinook salmon survival in the late 1980s. As a result, by the time the SIMPLE study began, the positions of key stakeholders were already somewhat entrenched. It is always easier to objectively discuss a crisis before it occurs, than while it is occurring.

In addition, lakewide management decisions on Lake Michigan, although only involving one federal government, require the support of five agencies, four state and one tribal. Often, differences among state agencies in management priorities can be just as great as transnational differences, and the more players there are sitting at the table, the more difficult it is to reach a consensus. Last but not least, prey-fish community dynamics are more uncertain in Lake Michigan than in Lake Ontario because of additional species and our limited understanding of the roles of weather versus predation in the decline in abundance of alewives in the early 1980s, and of disease versus disease mediated by nutritional deficiencies in the poor survival of chinook salmon in the late 1980s (Eck and Brown, 1985; Kitchell and Crowder 1986; Eck and Wells 1987; Stewart and Ibarra 1991). The greater uncertainty about community dynamics in Lake Michigan means that the relationship of the SIMPLE model predictions to reality is more tenuous there than on Lake Ontario. This uncertainty tends to allow divergent hypotheses with divergent management implications to persist.

Nevertheless, the chinook study group in Michigan concluded subsequent to the SIMPLE study, that the best stocking rate for chinook salmon in Lake Michigan, based on their mathematical modeling analysis, and the best scientific information available, is 3.5 million smolts per year (R. Clark, Michigan Department of Natural Resources, personal communication). This rate represents about a 46% reduction from the 6.5 million chinook per year that produced the present "suppressed" fishery, and is close to the 50% stocking-cut scenario of the SIMPLE study.

Providing Sound Assessment Data For Allocation Decisions

Changes in the attitudes and expectations of the fishing public, together with changes in fish stocks in the Great Lakes during the past four decades, have underscored the needs of fishery managers for more effective ways to address socioeconomic problems that confront their agencies. To solve many of these problems, managers need better tools for assessing fish stocks and to enhance their predictive capabilities. These requirements have not been fully met by conventional assessment programs (Christie et al. 1987). Indeed, Lewis et al. (1987) observed that short-term forecasting capabilities were critically needed by fisheries managers in the Great Lakes region. Although single-species models are available for predicting fish yield, those models are not satisfactory for many Great Lakes situations (Lewis et al. 1987). The changing status of the alewife, which was evaluated earlier in this chapter, is a good example. In less than two decades, it changed from a nuisance species to a useful prey species needed to sustain the high-value salmonine sport fisheries created since 1965. Demands for more sophisticated information on the biology and dynamics of alewives and other prey species have increased accordingly. Without this information to help understand the Lake Michigan predator-prey system, fishery managers were often unable to respond effectively to demands on the forage base by competing user groups.

Before this information could be provided, however, the methods in place for assessing fish stocks in Lake Michigan had to be examined, so that the best of old methods and technologies could be retained and new methods and technologies could be adapted or developed. Progress in that direction has been admittedly slow, partly because it is easier and safer to maintain programs already in place than to reallocate resources to unproven new methods or programs that may not yield better results (Lewis et al. 1987). Nevertheless, the limitations of conventional survey methods are obvious when the size and complexity of the aquatic system, and its fish communities, are contrasted with the available and affordable sampling effort. These limitations should ensure at least the gradual adoption of more advanced technologies that are now available.

The following discussion provides an overview of the conventional methods and gear that have been used by the Great Lakes Science Center and its predecessor agencies to assess prey fish populations, and then covers more recent applications of hydroacoustic methods. Although this discussion focuses on Lake Michigan, the assessment methods apply to all of the Great Lakes, and technologies developed on one lake can readily be adapted and used on the others.

Initiation and Expansion of Conventional Assessments

Surveys of the bloater and other chubs or deepwater ciscoes of Lake Michigan with graded-mesh gill nets by the U. S. Bureau of Fisheries from 1930 to 1932 were the basis for the earliest assessment of this exploited segment of the native fish community, which provided forage for lake trout and burbot (Jobes 1943 and 1949; Van Oosten and Deason 1938). These early lakewide surveys were conducted from the R/V *Fulmar*, leased specifically for that purpose, and were aimed at providing information on mesh-size selectivity for more efficient exploitation of the deepwater ciscoes and to lessen the take of small lake trout in chub gill nets. The survey operations were supported financially, or in principle, by the states of Wisconsin and Michigan and by net manufacturers. Ancillary biological data on age, growth, sex, maturity, and fecundity were collected only when the opportunity presented itself during the course of the survey operations (Hile 1957). Despite these limitations, the biological and hydrological data collected in conjunction with the primary sampling provided a major source of information on the biology and distribution of the deepwater ciscoes, as well as information on invertebrates and lake hydrology.

The Great Depression and World War II effectively halted nearly all field programs from 1933 through 1947. Except for the short-term investigation of the chub stocks in the 1930s, assessments of the fish stocks up until the early 1950s were limited primarily to those which could be gleaned from analysis of the annual catch and effort statistics from mandatory reports of licensed commercial fishermen (Hile 1957, 1962). These records provided valuable information on local and lakewide changes in relative abundance, harvest, and fishing effort. In fact, the decline and collapse of the commercial lake trout fishery were traceable through these data.

A major reason, other than insufficient funding, for not conducting stock assessments on Lake Michigan prior to the 1950s was the lack of a suitable research vessel. Finally, in 1951, construction of the R/V *Cisco* was authorized, and the vessel was accepted the same year by the U. S. Bureau of Fisheries. The R/V *Cisco* provided a much needed sampling platform (Moffett 1954), and has been used in that capacity mainly on Lake Michigan up to the present day. The chub stocks were reinvestigated on a lakewide basis from the R/V *Cisco* from 1951 to 1952 and again from 1960

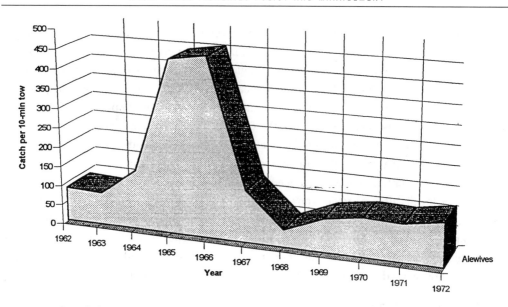

FIG. 8. *Number of alewives per 10-minute bottom trawl tow (CPUE) in surveys of the Lake Michigan forage base off Saugatuck, MI, fall 1962–1972; RVCAT data base, U. S. Geological Survey–Biological Resources Division.*

to 1961 to assess the loss of the larger species of ciscoes to the combination of size-selective fishing, predation from sea lampreys, and the impact of the alewife invasion (Smith 1964 and 1968). Limited exploratory trawling was conducted in conjunction with the gill-net surveys in the 1950s and early 1960s, but because the focus of the surveys was on chubs, systematic trawl surveys of the overall forage base did not begin in Lake Michigan until 1962.

In that year, prey-fish surveys with bottom trawls began to take on a more formal and systematic approach. From 1962 through 1966, systematic surveys with trawls were conducted in the fall along a single transect off Saugatuck, Michigan in southeastern Lake Michigan. The trawl used in those surveys has been described as a semiballoon-bottom trawl, a Yankee Standard number 25 trawl, and a North Atlantic whiting trawl (Hatch et al. 1981; Argyle 1982). Regardless of its several names, this trawl had a 12m headrope, 15.5m footrope, and 13mm mesh (nominal size) in the cod end. Its effective width when fishing, based on measurements made by divers when the trawl was fished at shallow depths (< 20m) was 6.5m, with a maximum height of 2.4m, and a mean height of 1.7m. Its frontal area, based on the same measurements, was 10.8m², and a ten-minute tow covered 0.5 Ha (Hatch et al. 1981). The 12m bottom trawl used on Lake Michigan has also been the standard trawl of the Great Lakes Science Center, and its predecessor agencies, for prey fish assessments on Lake Huron, until recently, and of the Center's field stations on the other Great Lakes.

Tows with the standard trawl at the original transect off Saugatuck were made at depths of 5, 9, 18, 27, 37, 46, 55, 64, 73, 82, and 91m along the contour during daylight hours. In 1964, Wells (1968) added depths of 13, 22, and 31m to the Saugatuck transect and collected data there for a comprehensive study of the seasonal depth distribution of prey fish with reference to water temperature. In order to extend the survey coverage to more of the lakewide forage base, three trawling transects were added to the annual fall surveys in 1967, off Waukegan, Illinois and St. Joseph (adjacent to Benton Harbor) and Ludington, Michigan (Hatch et al. 1981). The same series of depth contours, with the exception of the 82m and 91m contours, were fished along these transects. Because changes in the dynamics of the alewife population were a major focus of these surveys, the deeper stations were omitted from 1967 through 1972.

In 1973, four additional transects were added off Port Washington and Sturgeon Bay, Wisconsin, and Manistique and Frankfort, Michigan. With the addition of these transects, deeper trawling contours or fishing stations at 82m and 91m were added to all transects, a 110m station was added at Manistique, and 110m and 128m stations were added at Frankfort, Ludington, Port Washington and Sturgeon Bay. In order to complete a transect within one day and include trawling at the deeper locations, trawling at 5, 13, 22 and 31m was discontinued. Based on a statistical analysis of the trawl catch data, Hatch (1981) recommended dropping the St. Joseph transect and increasing sampling effort in the northern and western parts of Lake Michigan. Therefore transects were added near Charlevoix, Michigan, in 1988 and Two Rivers, Wisconsin, in 1990. Although the St. Joseph transect was dropped in 1990, seven of the original eight locations have formed the core transects for the bottom trawl surveys. These are Manistique, Frankfort, Ludington, and Saugatuck, Michigan, Waukegan, Illinois, and Port Washington and Sturgeon Bay, Wisconsin.

The selection of trawlable locations was based largely on exploratory trawling conducted by the U.S. Fish and Wildlife Service's Bureau of Commercial Fisheries (BCF), Branch of Exploratory Fishing (BEF). These locations are generally smooth sand, gravel, or mud bottomed areas, free of rocks and other obstructions to accommodate bottom trawls. The BEF conducted extensive trawling surveys at these and other locations in Lake Michigan, from 1962 through 1970, and assisted with the systematic forage base surveys in some years (Riegle 1969a and 1969b). It also evaluated a variety of sizes and types of trawls, catch and tow-time relationships, and trawl door and gear configurations, and conducted comparisons between day and night trawl catches. In addition, the BEF investigated the use of underwater television and experimented with electrified trawls, lampara seines, and early echo sounders.

It is arguable, however, whether the trawling gear and the sampling strategy actually used for prey fish assessment in the Great Lakes were selected by design or by default. The size and diversity of the system, along with logistic, weather, and practical constraints, such as the maximum-size catch that can be adequately processed and recorded, have dictated the final gear and design criteria. These constraints have certainly influenced the robustness of the data. However, the validity of the trends portrayed by the trawling data has been demonstrated repeatedly. Even the single transect surveys conducted off Saugatuck, Michigan from 1962 to 1968 clearly mirrored the actual buildup and crash of alewives in Lake Michigan (fig. 8). These dramatic changes in the alewife population were also evident from the commercial catch data collected off Saugatuck and from Two Rivers south to Kenosha, Wisconsin (Brown 1972).

The three locations added from 1967 through 1972, and the inclusion of data from these stations with those from Saugatuck, provided a broader perspective and probably more accurate depiction of the alewife population than that afforded solely by the Saugatuck data set (fig. 9). The data set was further expanded in 1973 and subsequent years with the addition of the more northern sampling locations identified earlier and the inclusion of some deeper sampling sites (fig. 10).

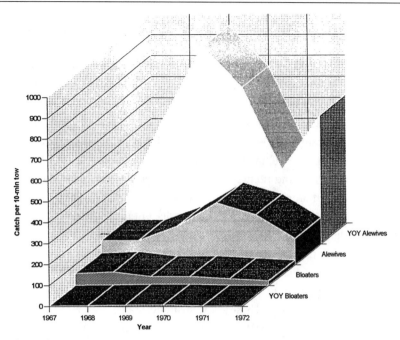

FIG. 9. *Numbers of young-of-the-year (YOY) and numbers of older alewives and bloaters per 10-minute bottom trawl tow (CPUE) off St. Joseph, Saugatuck, and Ludington, Michigan and Waukegan, Illionois in forage base surveys on Lake Michigan, fall 1967 to 1972; RVCAT data base, U. S. Geological Survey–Biological Resources Division.*

In addition to the trend or index data, the trawl surveys provided a basis for determining year class strength, age structure, mortality, growth and other population characteristics.

To provide the information needed by resource agencies to evaluate salmonid stocking strategies and make the data more responsive to the needs of the fisheries community, estimates of the weight or biomass of alewives in Lake Michigan were needed. Hatch et al. (1981) examined analytical methods for stratifying the historical trawl catch data and estimating from it the alewife biomass with sufficient precision to detect statistically significant fluctuations in the population during the period 1967 through 1978. They also examined the data set for potential biases and concluded that the biomass estimates were conservative because coefficients of availability and catchability were not incorporated in the calculations.

In view of drastic shifts in abundance among the three major prey-fish species since the early 1980s, it is unlikely that the stratification developed by Hatch et al. (1981) for alewives is still valid. Therefore, the index of forage fish biomass is presently calculated by multiplying the catch per unit area swept by the trawl (g per ha) by the area of the depth stratum, summing the stratum estimates within geographic sectors, and then summing the estimates across sectors for a lakewide estimate (Brown et al. 1987; Argyle 1992)(fig. 11). The biomass indices are believed to be preferable to simple catch per unit effort for indicating trends, because they provide a correction for spatial distribution by areal weighting.

Planktivorous species such as alewives, bloaters, and rainbow smelt, which are pelagic to varying degrees and have patchy spatial distributions which change daily, seasonally, and with age,

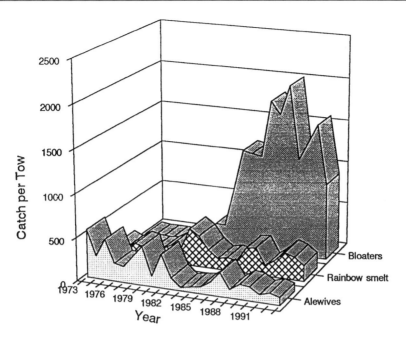

FIG. 10. *Number of alewives, rainbow smelt, and bloaters per 10-minute bottom trawl tow (CPUE) in lakewide forage base surveys at seven locations in Lake Michigan, fall 1973 to 1993; RVCAT data base, U. S. Geological Survey–Biological Resources Division.*

are inherently difficult to sample accurately with nets (Brandt et al. 1991). Acoustic monitoring of trawls while they are fishing has also shown that both configuration of the bottom trawl and the time that it actually fishes on the bottom differ at various depths, producing additional biases. It is, therefore, inappropriate to use the biomass estimates as estimates of absolute prey-fish abundance.

In spite of these various shortcomings in the bottom trawl assessments of the forage base, the resulting catch data provide the only information available on prey-fish trends. Furthermore, the continuity, and hence value, of such long-term data sets for use in constructing predictive models of the populations should not be underestimated. An example is the simulation analysis with the SIMPLE model of the Lake Michigan pelagic fish community, described in the preceding section of this chapter, which suggested that reversal of the alewife decline was unlikely without a rather large cut-back in predator stocking rates (Koonce and Jones 1994). Although new technology promises to provide alternatives to the bottom trawl surveys of the forage base, it is prudent to maintain the long-term data base until the capabilities of alternative methods are proven.

Initiation and Refinement of Acoustic Assessments

Sound has been used to locate fish in the marine environment for more than four decades, but it was not until the echo integrator was developed in the mid 1960s that echo sounders were used to quantify fish abundance (Dragesund and Olsen 1965). About this same time, the BEF was using echo sounders routinely, in conjunction with the trawl surveys conducted in Lake Michigan;

L Mich forage index

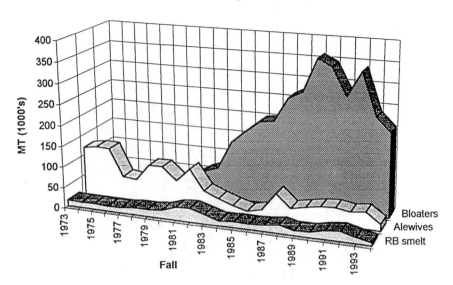

FIG. 11. *Lakewide biomass indices of age-one and older alewives, rainbow smelt, and bloaters based on bottom trawl surveys at seven to nine transects in Lake Michigan, fall 1973 to 1994; RVCAT data base, U. S. Geological Survey–Biological Resources Division.*

however, the recordings on the echograms were only used descriptively. The echogram recordings showed that fish were frequently concentrated in midwater during the night and sometimes were off bottom during the day where they were unavailable to bottom trawls. Some off-bottom trawling was attempted, but generally the identity of species in midwater was simply extrapolated from daytime catches in bottom trawls. For example, if alewives were the dominant species caught on the bottom during the day within a range of depths the assumption was that if few alewives were caught on the bottom at night within the same range of depths, alewives must be the dominant species shown in midwater on the echograms. No attempt was made to quantify species abundance from the paper records.

One of the early published applications using echosounders as a tool to acoustically estimate abundance of fishes in the Great Lakes was that of Kelso et al. (1974). The system used by Kelso et al. (1974) demonstrated the feasibility of using acoustic methods to estimate density of pelagic alewives and rainbow smelt. Brandt (1978) also made an acoustic estimate of the alewife biomass in Lake Michigan; however, some of the assumptions upon which the estimate was based were untested. Further developments in the processing of echo signals in the late 1970s and early 1980s made it possible to estimate fish size and calculate absolute abundance (Brandt et al. 1991). Brandt et al. (1991), in cooperation with the Michigan and Wisconsin departments of natural resources and the U.S. Fish and Wildlife Service's National Fisheries Research Center in Ann Arbor, used acoustics in combination with midwater and bottom trawling to estimate the biomass of pelagic species in Lake Michigan. The biomass estimates relied on a combination of Love's empirical relation between target strength (TS) and fish length (Love 1971) and mean size (weight) and relative number of fish caught in bottom trawls during the day to proportion the size and number in the water column at night. In spite of the known difficulties associated with establishing a solid link between TS and fish length or weight, hydroacoustic assessment of Lake Michigan prey species was the most promising technological advancement on the horizon. However, implementation on a routine basis required bringing together a combination of expertise, funding for acoustic and peripheral equipment, and sampling platforms.

In 1991, The National Fisheries Research Center–Great Lakes, in cooperation with the Illinois Department of Conservation and the Indiana, Michigan, and Wisconsin departments of natural resources, initiated a six-year integrated program of acoustics and midwater trawling to assess prey fish in Lake Michigan. The objectives of the program were to develop the strategy and sampling protocols for an integrated survey system that would define the spatial distributions of the pelagic prey fish communities. Inherent within the objectives were the requirements to examine and optimize trawl design, determine target strength for the dominant prey species, develop target strength estimators needed for multiple species acoustic surveys, and determine optimum survey design within realistic constraints of time, cost and logistics. In concert with accomplishing these requirements was the goal to better define the population dynamics and production of the major prey species and to develop an integrated model, based on data from the acoustic and trawl surveys, which would provide estimates of prey fish abundance.

The acoustic program was initiated in August 1991 and focussed initially on developing TS and fish size relations for the major pelagic fish species of the Great Lakes and determining the height and spread of the midwater trawl when fishing. A BioSonics® 120kHz dual-beam system with post processing software was used for collecting the hydroacoustic data. Because of the importance of determining an accurate relation between TS and fish size, it is essential to identify the stratum fished by the trawl and match only acoustic target returns from fish in that stratum to the trawl catch. The spread and height measurements, together with the temperature and depth where the trawl fished, were determined from transducers mounted on the trawl and transmitted to the vessel receiver via an acoustic link.

The hydroacoustic data collected concurrently with the midwater trawl tows were used to develop a relation between TS and weight. Although log-length is commonly used for TS and fish size regressions, staff used log-weight because biomass can be calculated directly from the echo integrator data and because backscattering cross section per unit weight is less sensitive to variation in fish size (Burczynski et al. 1987).

Several approaches can be used to calculate biomass from the hydroacoustic data. Regardless of the approach, an estimate of the average length or weight of the target species is required. In a very simple and ideal aquatic system, all targets are the same size and each target would be recorded as an echo once and only once. Using this scenario the number of targets multiplied by the mean size (weight per target) would yield an estimate of total biomass. In reality, fish are not the same size nor are they ideally spaced. Therefore, the approach is to collect the acoustic data in two parts and, although both data sets are collected simultaneously, the sets are discrete. One of these data sets is the average size or target strength which is generally derived for discrete strata for ranges. Because the echo intensity is dependent upon the location of the fish in the beam, some method must be used to correct for the reduced intensity of off-axis echoes. This can be accomplished by using single beam systems and then relying on statistical procedures to infer the TS distribution, dual-beam systems which correct for targets off the acoustic axis of the beam and discriminate between multiple targets, or split beam systems which measure phase differences, allowing the angular position to be determined. The complementary data set is that of echo integration which integrates the sums of the voltage returns squared for discrete ranges. The total energy within the strata divided by the average TS times the mean weight yields the biomass for the strata.

Acoustic surveys were conducted on Lake Michigan in various seasons to determine if there was an optimal season for assessment of prey fishes. There is little or no thermal structure in the lake in spring, when most prey fish are concentrated near the bottom. This distribution pattern biased the TS distributions and affected echo integration by shadowing the acoustic signal and disrupting the function of the bottom tracking algorithms. Also, the acoustic target distribution in the water column was composed of disproportionately high percentages of rainbow smelt, which move shoreward for spawning in spring. This combination of known biases prompted cancellation of the spring assessments until the biases could be more closely examined.

In July and August, the thermal structure is established, and although upwellings are common, adult fish are distributed more uniformly in the water column at night which is preferable for acoustic assessment. However, young-of-the-year alewives, rainbow smelt and bloaters are small then and most are in shallow water near shore where they cannot be assessed with hydroacoustic equipment. Because of the near shore distribution of young fish in summer when production is very high, the summer months were chosen for investigations on TS, diel migrations, food habits, and spatial distributions of adult fish in the deeper zones.

By late summer or early fall (i.e., early September) upwellings are less frequent, epilimnetic cooling has begun (Carr et al. 1973) and most young-of-the-year fish have moved offshore. The thermal structure of the lake, although beginning to show some signs of erosion, acts to partition the pelagic species to some extent. Alewives, bloaters and rainbow smelt occupy slightly different thermal regimes (Brandt 1980; Crowder et al. 1981). These associations with temperature were important because they provided a means to partition the midwater targets into species assemblages. This feature, coupled with more uniform distribution of fish targets throughout the water column, dictated that the lakewide acoustic surveys aimed at estimating biomass should be conducted in fall.

Because of the rapid advances that were occurring in computer technology, several changes in equipment and sampling design were made as the study progressed. In 1992, the staff began using BioSonics® dual frequency Echo Signal Processors (ESP). These systems were faster, and the processing algorithms were improved with better integration of the TS and integrator data. In 1994, staff stopped using towed bodies for housing the transducers, which constrained operations in heavy seas, and began using acoustically transparent windows built into the hull of the vessel.

The fall lakewide assessments of the prey fish were designed for two vessels. One vessel was devoted mainly to the collection of acoustic data along preselected transects that ran perpendicular to shore, whereas the other vessel was used primarily for midwater trawling at locations and depth strata that were selected from the acoustic returns along the transects. Midwater trawling locations were based on a combination of the number of targets in the strata, temperature, depth, and fish and habitat associations. In general, the trawl tows were made within an hour or two of the acoustic collections. If fish were not evident on the vessel sounder during midwater trawling at assigned strata, the station was omitted and trawling was continued at the next assigned location.

In the fall, targets were generally distributed into distinct groups throughout the water column, with smaller targets in the upper portion of the water column near or in the thermocline and larger targets in deeper areas. Most small targets, such as rainbow smelt, alewives and young bloaters, were in the shallower 30m to 50m stratum in association with the metalimnion. Most larger targets occupied the hypolimnetic areas at strata depths of 70m to 90m. Alewives, which included young-of-the-year, were concentrated in the upper 30m to 50m depth strata along the transect extending offshore over depths in excess of one hundred meters at some locations. Rainbow smelt were mostly in the upper strata associated with the thermocline, whereas bloaters were distributed throughout the deeper strata. In addition to the distributional characteristics associated with depth and temperature, differences in species composition in some regions of the lake added further complexity to the assessment (fig. 12).

Although many inherent difficulties affect the assessment of pelagic prey fish stocks in a system as large and complex as Lake Michigan, hydroacoustic technology is well suited to the task because of its unique ability to scan and electronically quantify fish throughout most of the water column. The major requirements for acoustic assessment are establishment of a reliable and accurate TS versus size relation, understanding how the major pelagic target species partition by depth, geographic and temperature preferences, and a sampling design or strategy that is statistically valid.

Summary and Conclusions

Accelerated interagency research and development since the 1970s on the greatly altered pelagic fish communities of the Great Lakes is providing fishery biologists and managers new tools and better understanding of the predator-prey systems in these lakes. The supply and demand aspects of this research have strongly influenced forage base allocation, in regard to maintaining the hatchery-dependent suite of large salmonine predators at population densities that do not endanger the forage base as it may have been in Lake Michigan and Lake Ontario in the 1980s (Ney 1990; Jones et al. 1993).

Forage base allocation, itself, has become an increasingly important management issue on the Great Lakes, mainly as a result of the phenomenal growth of sport fishing for coho and chinook salmon that were introduced by the State of Michigan to Lake Michigan and other lakes in the mid to late 1960s and the early 1970s, to control and utilize large stocks of non-indigenous alewives and rainbow smelt. Alewives are the preferred prey of native and introduced salmonine predators in Lakes Michigan, Huron, and Ontario, where they have had serious negative effects on the native fish communities. In Lake Superior, where alewives are scarce because of lower water temperature, rainbow smelt are the preferred prey of the salmonines, but whether smelt have harmed the lake herring and other native prey fishes is still open to debate.

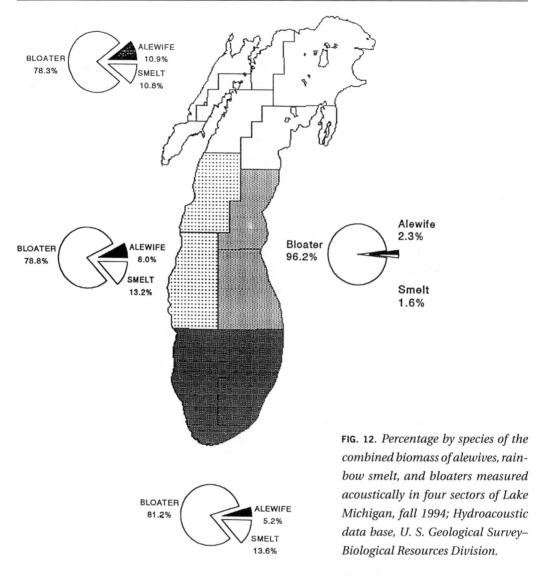

FIG. 12. *Percentage by species of the combined biomass of alewives, rainbow smelt, and bloaters measured acoustically in four sectors of Lake Michigan, fall 1994; Hydroacoustic data base, U. S. Geological Survey–Biological Resources Division.*

Conflicts over allocation of the forage base among the principal stakeholders, including anglers and charter boat operators on one hand, and commercial operators on the other, intensified as the hatchery-based salmonine populations increased to high levels in the 1970s and 1980s, first in Lake Michigan, and somewhat later in Lake Ontario. The State of Michigan compensated some of its commercial chub fishermen for nonrenewal of their licenses through the legislature beginning in the 1970s, and the State of Wisconsin finally closed its commercial trawl fishery for alewives in 1991. Fishery managers on Lake Superior recommended that lake herring be protected through appropriate commercial fishing regulations so that the stocks can further recover, while supporting increased predation by lake trout and salmon (Hansen 1994). However, the harvest of lake herring is still not regulated in every jurisdiction.

Vital information needed by fishery managers on the forage requirements of the expanding salmonine populations was provided by energetics-based population modeling of the Lake

Michigan stocks in 1981 (Stewart et al. 1981). A decade later the modeling was extended and refined on a whole-lake basis to cover the high chinook salmon and total salmonine abundance and low alewife abundance from 1978 to 1988 (Stewart and Ibarra 1991). Diet shifts to younger alewives and other prey by chinook and coho salmon were reported, as were losses in gross conversion efficiency of biomass and average weight in the sport catch by the chinook salmon. Predation by all salmonines combined peaked at 71,000 metric tons in 1983, and at 76,000 metric tons in 1987. In about 1987, high mortalities of chinook salmon infected with BKD were first reported from Lake Michigan. Some biologists and managers attributed the outbreak of BKD to poor condition of the salmon resulting from an insufficient supply of their preferred prey, the alewife. Others attributed the outbreak of BKD to a more virulent organism or lower resistance because of rearing conditions in the hatcheries.

The bioenergetics modeling and lakewide acoustic estimates of alewife abundance suggested that the salmonines could have been consuming as much as 53% of the alewife biomass in Lake Michigan annually (Stewart and Ibarra 1991; Brandt et al. 1991; Jones et al. 1993). That potentially high rate of consumption together with the high mortality of chinook salmon, resulted in further controversy. Investigators called for a reexamination of the then high stocking levels that might cause the alewife stocks to collapse, whereas some sport-fishing interests pressured management agencies to maintain stocking levels.

Because of the potential implications to the forage base, the GLFC sponsored a technical investigation conducted by one Canadian and one U. S. principal investigator (Jones et al. 1993). This investigation was based on the methods of Adaptive Environmental Assessment and Management, and was focused both on Lakes Ontario and Michigan. The SIMPLE model, which had the advantage over the bioenergetics models of a dynamic prey supply, was developed and used as the primary investigative tool (Holling 1978; Koonce and Jones 1994). Information for developing the model was compiled in a series of technical sessions at the various agency offices. The principal investigators also worked directly with groups of scientists, managers, and stakeholders in the fishery resources at technical workshops on model development and socioeconomic issues, and later, round-table meetings to evaluate management policies and make management recommendations. The central issue of prey supply versus predator demand dominated the discussions and was addressed in the various modeling simulations run for each lake (Ney 1990; Jones et al. 1993).

The modeling simulations suggested that by the late 1980s predator demand was at or exceeded levels that were sustainable by the prey-fish communities in the two lakes (Jones et al. 1993; Koonce and Jones 1994). Indeed, the model predicted that the Lake Ontario alewife population would be extremely vulnerable to collapse under excessive predation pressure if a cold winter like that of 1976–1977 were to cause a severe mortality of alewives. The model predictions for Lake Michigan also suggested that, even without severe alewife mortalities, predator demand from salmonine stocking during the 1980s exceeded that which could be sustained by the alewife population (Koonce and Jones 1994).

After the Lake Ontario round-table and a series of public consultations on the depleted status of the forage base, the NYDEC and the OMNR implemented coordinated stocking cuts aimed at reducing predator demand in Lake Ontario by about 50%, which was consistent with the conclusions of an expert task group (Anonymous 1992). In contrast, major stocking cuts were not implemented on Lake Michigan, where the sharp decline of chinook salmon survival, associated with BKD infection, may have been having a similar overall effect on predator demand (Keller et al. 1989; Stewart and Ibarra 1991). Also, the more complicated fish community dynamics in Lake Michigan make the predictions of the SIMPLE model more uncertain and tenuous. However, subsequent mathematical modeling analysis by the chinook study group in Michigan led to the conclusion that the best stocking rate for chinook in Lake Michigan is 3.5 million smolts per year. That is a 46% reduction from the 6.5 million smolts per year that produced the present suppressed fishery and is close to the 50% stocking-cut scenario of the SIMPLE study.

Because of the dynamic and complex spatial distributions of pelagic prey fish in the Great Lakes, rapidly improving hydroacoustic technology is being used in conjunction with midwater trawls, which provide ground truth, to estimate prey fish biomass, and from that, prey production available to predators and competing fisheries. A six-year cooperative investigation based on hydroacoustics, midwater trawling and remote electronic trawl mensuration by the GLSC and the four states with jurisdictions on Lake Michigan was begun in 1991 to develop the methodology and sampling protocol for future surveys on that lake. Toward this goal, the investigators have been determining more accurate relations between TS and fish size, and how major pelagic species are distributed in relation to size, age, depth, temperature, geographic location and season. Although not covered in this chapter, hydroacoustic surveys of prey fish stocks have also been conducted on Lake Ontario in recent years (Mathers et al. 1992; Schaner and Mathers 1991).

The considerable value to prey fish assessment and modeling of data bases acquired through more conventional long-term surveys with bottom trawls on Lake Michigan since the early 1960s, and on the other lakes in more recent years, were also recognized, together with some of their limitations and short comings. In fact, relative abundance of the major prey fish in these surveys (i.e., alewives, rainbow smelt, and bloaters) were used to apportion into species the prey fish remotely quantified in midwater in the first lakewide hydroacoustic assessment on Lake Michigan (Brandt 1991).

Finally, in spite of the impetus of the research and developments described above, a number of other biological, technological, and logistical problems or impediments will, in all likelihood, continue to affect forage-base allocation and resource management in general in the Great Lakes. The value of alewives as prey of hatchery-sustained salmonines will be diminished by any costs placed on their adverse ecological effects on other species in the fish communities (Smith 1970; Eck and Wells 1987; Brandt et al. 1987; Krueger et al. 1995). The possible negative effects of non-indigenous invertebrates on prey fish, including the alewife, are already foreboding. Such effects could be lengthening of the food chain, at the expense of juvenile prey fish, by the predatory cladoceran *Bythotrephes cederstromei* (Branstrator and Lehman 1996), and reduction of nutrients, evident in increased water clarity, from filtering by Dreissenid mussels (Herbert et al. 1991). Nutrient reduction by these mussels might exacerbate the present adverse effects of reduced phosphorus loading on the forage base productivity of Lake Ontario (Lean et al. 1990; Jones et al. 1993). Despite important technological advantages in hydroacoustic assessment of prey fish populations, the Great Lakes will continue to pose serious logistical problems because of their large size and adverse weather. To help reduce these constraints, substantially larger research vessels (>30 m) are needed on all of these lakes with the possible exception of shallower Lake Erie.

In retrospect, no single agency or institution can afford the personnel and material resources needed to solve major problems in forage-base allocation and related issues, many of which extend beyond individual state and provincial jurisdictions. Interagency and university cooperation in conducting and sharing the results of research and assessment activities has therefore been a key factor in the many notable successes to date. Yet much of this cooperation might never have materialized without the forum, leadership, and sponsorship provided by the Great Lakes Fishery Commission and its various advisory groups and committees (Fetterolf 1980).

Literature Cited

Addis, J. T. 1988. Implementation of the Joint Strategic Plan for Management of Great Lakes Fisheries: Lake Michigan Committee goals, issues, progress and problems. Pages 51–54 *in* M. R. Dochoda, editor. Committee of the Whole Workshop on Implementation of the Joint Strategic Plan for Management of Great Lakes Fisheries (reports and recommendations from the 18–20 February and 5–6 May 1986 meetings). Great Lakes Fishery Commission Special Publication 88–1.

Anderson, E. D., and L. L. Smith, Jr. 1971. Factors affecting abundance of lake herring (*Coregonus artedii*) in western Lake Superior. Transactions of the American Fisheries Society 100:691–707.

Anonymous. 1992. Status of the Lake Ontario offshore pelagic fish community and related ecosystem in 1992. Report to the Lake Ontario Committee of the Great Lakes Fishery Commission. Great Lakes Fishery Commission, Ann Arbor, Michigan.

Argyle, R. L. 1992. Acoustics as a tool for the assessment of Great Lakes forage fishes. Fisheries Research (Amsterdam)14:179–196.

———, 1982. Alewives and rainbow smelt in Lake Huron: Midwater and bottom aggregations and estimates of standing stocks. Transactions of the American Fisheries Society, 111:267–285.

Baldwin, N. S., R. W. Saalfeld, M. A. Ross, and H. J. Buettner. 1979. Commercial fish production in the Great Lakes 1867–1977. Great Lakes Fishery Commission, Technical Report No. 3, Ann Arbor, Michigan.

Borgeson, D. P. 1981. Changing management of Great Lakes Fish Stocks. Canadian Journal of Fisheries and Aquatic Sciences 38:1466–1468.

Bowmen, W. B. 1974. Lake Erie bottom trawl exploration, 1962–1966. National Marine Fisheries Service, Special Scientific Report-Fisheries Series 674, Seattle, Wash.

Borchardt, J. A., and F. G. Pohland. 1970. Anaerobic treatment of alewife processing wastes. Journal Water Pollution Control Federation 42:2060–2068.

Brandt, S. B. 1978. Thermal ecology and abundance of alewife *Alosa pseudoharengus* in Lake Michigan. Doctoral dissertation. University of Wisconsin-Madison, Madison, Wisconsin.

———, 1980. Spatial segregation of adult and young-of-the-year alewives across a thermocline in Lake Michigan. Transactions of the American Fisheries Society 109:469–478.

———, D. M. Mason, D. B. MacNell, T. Coates, and J. E. Gannon. 1987. Predation by alewives on larvae of yellow perch in Lake Ontario. Transactions of the American Fisheries Society 116:641–645.

———, E. V. Patrick, R. L. Argyle, L. Wells, P. A. Unger, and D. J. Stewart. 1991. Acoustic measures of the abundance and size of pelagic planktivores in Lake Michigan. Canadian Journal of Fisheries and Aquatic Sciences 48:894–908.

Branstrator, D. K., and J. T. Lehman. 1996. Evidence for predation by young-of-the-year alewife and bloater chub on *Bythotrephes cederstroemi* in Lake Michigan. Journal of Great Lakes Research 22: 917–924.

Bronte, C. R., J. H. Selgeby, and G. L. Curtis. 1991. Distribution, abundance, and biology of the alewife in U. S. waters of Lake Superior. Journal of Great Lakes Research 17:304–313.

Brown, E. H. Jr. 1972. Population biology of alewives, *Alosa pseudoharengus,* in Lake Michigan, 1949–70. Journal of the Fisheries Research Board of Canada 29:477–500.

———, G. W. Eck, N. R. Foster, R. M. Horrall, and C. E. Coberly. 1981. Historical evidence for discrete stocks of lake trout (*Salvelinus namaycush*) in Lake Michigan. Canadian Journal of Fisheries and Aquatic Sciences 38:1747–1758.

———, R. W. Rybicki, and R. J. Poff. 1985. Population dynamics and interagency management of the bloater (*Coregonus hoyi*) in Lake Michigan, 1967–1982. Great Lakes Fishery Commission, Technical Report No. 44.

———, R. L. Argyle, N. R. Payne, and M. E. Holey. 1987. Yield and dynamics of destabilized chub (*Coregonus* spp.) populations in Lakes Michigan and Huron, 1950–1984. Canadian Journal of Fisheries and Aquatic Sciences 44(Supplement 2):371–383.

————, G. W. Eck. 1992. Density-dependent recruitment of the bloater (*Coregonus hoyi*) in Lake Michigan. Polish Archives of Hydrobiolgy 39:37–45.

Brown, R. W., M. Ebener, and T. Gorenflo. 1999. Great Lakes commerical fisheries: Historical overview and prognosis for the future. Pages 307–354 in W.W. Taylor and C. P. Ferreri, eds. Great Lakes Fisheries Policy and Management: A Binational Perspective. Michigan State University Press, East Lansing, Michigan.

Burczynski, J. J., P. H. Michaletz, and G. M. Marrone. 1987. Hydroacoustic assessement of the abundance and distribution of rainbow smelt in Lake Oahe. North American Journal of Fisheries Management 7:106–116.

Busiahn, T. R. 1985a. Implications of fish community changes for fishery management on the Great Lakes. Pages 79–83 in R. L. Eshenroder, editor. Presented papers from the Council of Lake Committees Plenary Session on Great Lakes predator-prey issues, March 20, 1985. Great Lakes Fishery Commission, Special Publication 85-3, Ann Arbor, Michigan.

————, 1985b. Predator responses to fish community changes in Lake Superior. Pages 39–56 in R. L. Eshenroder, editor. Presented papers from the Council of Lake Committees Plenary Session on Great Lakes predator-prey issues, March 20, 1985. Great Lakes Fishery Commission, Special Publication 85-3, Ann Arbor, Michigan.

————, (editor). 1990. Fish community objectives for Lake Superior. Great Lakes Fishery Commission, Special Publication 90–1, Ann Arbor, Michigan.

Carr, J. F., J. W. Moffett, and J. E. Gannon. 1973. Thermal characteristics of Lake Michigan, 1954-1955. Technical Papers of the Bureau of Sport Fisheries and Wildlife. Number 69, 143 p.

Chippewa-Ottawa Treaty Fishery Management Authority. 1987. Rules and regulations governing Tribal commercial and subsistence fishing activities in the 1836 Treaty ceded waters of Lake Superior, Lake Huron and Lake Michigan. Sault Ste. Marie, Michigan.

Christie, W. J., J. J. Collins, G. W. Eck, C. I. Goddard, J. M. Hoenig, M. Holey, L. D. Jacobson, W. MacCallum, S. J. Nepszy, R. O'Gorman, and J. Selgeby. 1987. Meeting future information needs of Great Lakes fisheries management. Canadian Journal of Fisheries and Aquatic Sciences. 44(suppl. 2):439–447.

Colby, P. J. 1973. Response of the alewives, *Alosa pseudoharengus*, to environmental change. Pages 163–196 in W. Chavin, editor. Responses of fish to environmental changes. Charles C. Thomas, Springfield, Illinois.

Conner, D. J., C. R. Bronte, J. H. Selgeby, and H. L. Collins. 1993. Food of salmonine predators in Lake Superior, 1981–1987. Great Lakes Fishery Commission Technical Report 59, Ann Arbor, Michigan.

Crowder, L. B. 1980. Alewife, rainbow smelt and native fishes in Lake Michigan: Competition or predation? Environmental Biology of Fishes 5:225–233.

————, J. J. Magnuson, and S. B. Brandt. 1981. Complementarity in the use of food and thermal habitat by Lake Michigan fishes. Canadian Journal of Fisheries and Aquatic Sciences 38:662–668.

————, 1985. Indicators of the status of cold water predator-prey systems. Pages 99–105 in R. L. Eshenroder, editor. Presented papers from the Council of Lake Committees Plenary Session on Great Lakes predator-prey issues, March 20, 1985. Great Lakes Fishery Commission, Special Publication 85-93, Ann Arbor, Michigan.

————, F. P. Binkowski. 1983. Foraging behaviors and the interaction of alewife, *Alosa pseudoharengus*, and bloater, *Coregonus hoyi*. Environmental Biology of Fishes 8:105–113.

————, M. E. McDonald, and J. A. Rice. 1987. Understanding recruitment of Lake Michigan fishes: The importance of size-based interactions between fish and zooplankton. Canadian Journal of Fisheries and Aquatic Sciences 44(Supplement 2):141–147.

Dehring, T. R., and C. C. Krueger. 1985. A computer program system to allocate the annual stocking of salmonids in the Wisconsin waters of Lake Michigan. Wisconsin Department of Natural Resources, Bureau of Fish Management, Fish Management Report 124, Madison.

Dochoda, M. R., and C. M. Fetterolf, Jr. 1987. Public purpose of Great Lakes fishery management: Lessons from the management experience. Transactions of the American Fisheries Society 116:302–308.

Dragesund, O., and S. Olsen. 1965. On the possibility of estimating year-class strength by measuring echo-abundance of 0-group fish. Fiskeridirektoratets Havforskningsinstitutt 13:48–75.

Dryer, W. R., and J. Beil. 1964. The life history of lake herring in Lake Superior. Fishery Bulletin 63:493–530.

————, L. F. Erkkila, and C. L. Tetzloff. 1965. Food of lake trout in Lake Superior. Transactions of the American Fisheries Society 94:169–176.

Eck, G. W., and E. H. Brown, Jr. 1985. Lake Michigan's capacity to support lake trout (*Salvelinus namaycush*) and other salmonines: An estimate based on the status of prey populations in the 1970s. Canadian Journal of Fisheries and Aquatic Sciences 42:449–454.

————, L. Wells. 1987. Recent changes in Lake Michigan's fish community and their probable causes, with emphasis on the role of the alewife (*Alosa pseudoharengus*). Canadian Journal of Fisheries and Aquatic Sciences 44(Suppl. 2):53–60.

Einhouse, D., and eleven others. 1993. Consumption of rainbow smelt by walleye and salmonine fish in the central and eastern basins of Lake Erie. Report of Forage Task Group to Standing Technical Committee, Lake Erie Committee, Great Lakes Fishery Commission, March 1993.

Eshenroder, R. L., T. P. Poe, and C. H. Olver (editors). 1984. Strategies for rehabilitation of lake trout in the Great Lakes: Proceedings of a conference on lake trout research, August 1983. Great Lakes Fishery Commission, Technical Report No. 40, Ann Arbor, Michigan.

————, 1987. Socioeconomic aspects of lake trout rehabilitation in the Great Lakes. Transactions of the American Fisheries Society 116:309–313.

————, M. E. Holey, T. H. Gorenflo, and R. D. Clark, Jr. 1995. Fish-community objectives for Lake Michigan. Great Lakes Fishery Commission, Special Publication 95–3.

Eschmeyer, P. H. 1955. The near extinction of lake trout in Lake Michigan. Transactions of the American Fisheries Society 85:102–119.

Fetterolf, C. M. Jr. 1980. Why a Great Lakes Fishery Commission and why a Sea Lamprey International Symposium? Canadian Journal of Fisheries and Aquatic Sciences 37:1588–1593.

Fitzsimons, J. D. 1995. The effects of B-vitamins on a swim-up syndrome in Lake Ontario lake trout. Journal of Great Lakes Research 21 (Supplement 1):286–289.

Francis, G. R., J. J. Magnuson, H. A. Regier, and D. R. Talhelm. 1979. Rehabilitating Great Lakes ecosystems. Great Lakes Fishery Commission, Technical Report No. 37, Ann Arbor, Michigan.

Goodier, J. L. 1981. Native lake trout (*Salvelinus namaycush*) stocks in the Canadian waters of Lake Superior prior to 1955. Canadian Journal of Fisheries and Aquatic Sciences 38:1724–1737.

Gordon, W. G., and K. D. Brouillard. 1961. Great lakes trawler conversion. United States Department of the Interior, Fish and Wildlife Service, Bureau of Commercial Fisheries, Fishery Leaflet 510, Washington, D. C.

Greenwood, M. R. 1970. 1968 state-federal Lake Michigan alewife die-off control investigation. U. S. Department of the Interior, Bureau of Commercial Fisheries, Ann Arbor, MI.

Hansen, M. J., D. Boisclair, S. B. Brandt, S. W. Hewett, J. F. Kitchell, M. C. Lucas, and J. J. Ney. 1993. Applications of bioenergetics models to fish ecology and management: Where do we go from here? Transactions of the American Fisheries Society 122:1019–1030.

————, (editor). 1990. The state of Lake Superior in 1989. Great Lakes Fishery Commission Special Publication 90–3.

————, (editor). 1994. The state of Lake Superior in 1992. Great Lakes Fishery Commission Special Publication 94–1, Ann Arbor, Michigan.

————, 1999. Lake trout in the Great Lakes: Basinwide stock collapse and binational restoration. Pages 417–453 *in* W. W. Taylor and C. P. Ferreri, eds. Great Lakes Fisheries Policy and Management: A Binational Perspective. Michigan State University Press, East Lansing, Michigan.

Hartman, W. L. 1988. Historical changes in the major fish resources of the Great Lakes. Pages 103–131 *in* M. Evans, editor. Toxic contaminants and ecosystem health, a Great Lakes Focus. John Wiley & Sons, Inc, New York.

Hatch, R. W. 1981. Evaluation of fall forage fish surveys, Lake Michigan, 1962–1969. United States Fish and Wildlife Service, Great Lakes Fishery Laboratory, Ann Arbor, Michigan, Administrative Report No. 81–7.

————, S. J. Nepszy, K. M. Muth, and C. T. Baker. 1987. Dynamics of the recovery of the Western Lake Erie walleye (*Stizostedion vitreum vitreum*) stock. Canadian Journal of Fisheries and Aquatic Sciences 44(Suppl. 2):15–22.

————, P. M. Haack, and E. H. Brown, Jr. 1981. Estimation of alewife biomass in Lake Michigan, 1967–1978. Transactions of the American Fisheries Society 110:575–584.

Herbert, P. D. N., B. W. Muncaster, and G. L. Mackie. 1989. Ecological and genetic studies on *Dreissena polymorpha* (Pallas): A new mollusc in the Great Lakes. Canadian Journal of Fisheries and Aquatic Sciences 46:1587–1591.

————, C. C. Wilson, M. H. Murdoch, and R. Lazar. 1991. Demography and ecological impacts of the invading mollusc *Dreissena polymorpha*. Canadian Journal of Zoology 59: 405–409.

Hile, R. 1957. U. S. Federal fishery research on the Great Lakes through 1956. Fish and Wildlife Service, Special Scientific Report, Fisheries Number 226, 46p.

————, 1962. Collection and analysis of commercial fishery statistics in the Great Lakes. Great Lakes Fishery Commission, Technical Report No. 5, Ann Arbor, Michigan.

Hoff, M. J., and J. H. Selgeby. 1994. Population status and trends for Lake Superior forage fish, 1978 to 1993. Pages 17–27, *in* Minutes of Annual Meeting, Lake Superior Committee, Great Lakes Fishery Commission, Ann Arbor, Michigan, March 22, 1994.

Holey, M. E., R. W. Rybicki, G. W. Eck, E. H. Brown, Jr., J. E. Marsden, D. S. Lavis, M. L. Toneys, T. N. Trudeau, and R. M. Horrall. 1995. Progress toward lake trout restoration in Lake Michigan. Journal of Great Lakes Research 21 (Supplement 1):128–151.

Holling, C. S., editor. 1978. Adaptive environmental assessment and management. Wiley, Chichester, United Kingdom.

Jobes, F. W. 1943. The age, growth, and bathymetric distribution of Reighard's chub, *Leucichthys reighardi* Koelz, in Lake Michigan. Transactions of the American Fisheries Society 72:108–135.

————, 1949. The age, growth, and bathymetric distribution of the bloater, *Leucichthys hoyi* (Gill), in Lake Michigan. Papers Michigan Academy of Science, Arts, and Letters 33:135–172.

Jones, M. L., J. F. Koonce, and R. O'Gorman. 1993. Sustainability of hatchery-dependent salmonine fisheries in Lake Ontario: The conflict between predator demand and prey supply. Transactions of the American Fisheries Society 122:1002–1018.

Keller, M. J., K. D. Smith, and R. W. Rybicki (editor). 1989. Summary of salmon and trout management in Lake Michigan. Michigan Department of Natural Resources, Fisheries Technical Report 89–1, Lansing, Michigan.

Kelso, J. R. M., E. E. Pickett, and R. G. Dowd. 1974. A digital echo-counting system used in determining abundance of freshwater pelagic fish in relation to depth. Journal of the Fisheries Research Board of Canada 31:1101–1104.

Kitchell, J. F. 1985. An ecological rationale for managing predator-prey systems in the Great Lakes. Pages 85–97 in R. L. Eshenroder, editor. Presented papers from the Council of Lake Committees Plenary Session on Great Lakes predator-prey issues, March 20, 1985. Great Lakes Fishery Commission Special Publication 85–3.

———, L. B. Crowder. 1986. Predator-prey interactions in Lake Michigan: Model predictions and recent dynamics. Environmental Biology of Fishes 16:205–211.

———, D. J. Stewart, and D. Weininger. 1977. Applications of a bioenergetics model to yellow perch (*Perca* flavescens) and walleye (*Stizostedion vitreum vitreum*). Journal of the Fisheries Research Board of Canada 34:1922–1935.

Kocik, J. F., and M. L. Jones. 1999. Pacific salmonines in the Great Lakes. Pages 455–488 in W. W. Taylor and C. P. Ferreri, eds. Great Lakes Fisheries Policy and Management: A Binational Perspective. Michigan State University Press, East Lansing, Michigan.

Koonce, J. F., and M. L. Jones. 1994. Sustainability of the intensively managed fisheries of Lake Michigan and Lake Ontario. Great Lakes Fishery Commission, Board of Technical Experts, Final Report of the SIMPLE Task Group, Ann Arbor, Michigan.

Krueger, C. C., and J. Paine. 1985. A plenary session about predator-prey issues of the Great Lakes. Pages 127–134 in R. L. Eshenroder, editor. Presented papers from the Council of Lake Committees Plenary Session on Great Lakes predator-prey issues, March 20, 1985. Great Lakes Fishery Commission Special Publication 85–3, Ann Arbor, Michigan.

———, D. L. Perkins, E. L. Mills, and J. E. Marsden. 1995. Predation by alewives on lake trout fry in Lake Ontario: Role of an exotic species in preventing restoration of a native species. Journal of Great Lakes Research, 21(Supplement 1):458–469.

Lawrie, A. H., and J. F. Rahrer. 1972. Lake Superior: Effects of exploitation and introductions on the salmonid community. Journal of the Fisheries Research Board of Canada 29:763–776.

———, J. F. Raher. 1973. Lake Superior: A case history of the lake and its fisheries. Great Lakes Fishery Commission Technical Report 19, Ann Arbor, Michigan.

Lean, D. R. S., M. A. Neilson, J. J. Stevens, and A. Mazumder. 1990. Response of Lake Ontario to reduced phosphorus loading. Internationale Vereinigung fur theoretische und angewandte Limnologie Verhandlungen 24;420–425.

Lewis, C. A., D. H. Schupp, W.W. Taylor, J.J. Collins, and R.W. Hatch. 1987. Predicting Great Lakes fish yields: Tools and constraints. Canadian Journal of Fisheries and Aquatic Sciences. 44(Suppl. 2):411–416.

Love, R.G. 1971. Dorsal aspect target strength of an individual fish. Journal of the Acoustic Society of America 49:816–823.

MacCallum, W. R., and J. H. Selgeby. 1987. Lake Superior revisited 1984. Canadian Journal of Fisheries and Aquatic Sciences 44:23–36.

Mathers, A., T. J. Stewart, and A. Goyke. 1992. Alewife and smelt abundance estimates from midwater beam trawl and hydroacoustics surveys in Canadian waters of Lake Ontario, 1991. Section 13 in Lake Ontario Fisheries Unit 1991 annual report. Ontario Ministry of Natural Resources, Glenora.

Moffett, J. W. 1954. Fisheries knowledge increased through research vessel. The Fisherman, 22 (3):7–14.

Negus, M. T. 1992. Evaluation of bioenergetics modeling in the study of predator-prey dynamics in Minnesota waters of Lake Superior. Minnesota Department of Natural Resources Investigational Report 414.

Nepszy, S. J. 1985. Walleye in Western Lake Erie: Responses to change in predators and prey. Pages 57–76 in R. L. Eshenroder, editor. Presented papers from the Council of Lake Committees Plenary Session on Great Lakes predator-prey issues, March 20, 1985. Great Lakes Fishery Commission, Special Publication 85–3, Ann Arbor, Michigan.

———, D. H. Davis, D. Einhouse, R. W. Hatch, G. Isbell, D. MacLennan, and K. M. Muth. 1991. Walleye in Lake Erie and Lake St. Clair. Great Lakes Fishery Commission, Special Publication 91–1. Ann Arbor, Michigan.

Ney, J. J. 1990. Trophic economics in fisheries: Assessment of demand-supply relationships between predators and prey. Reviews in Aquatic Sciences, 2(1):55–81.

O'Gorman, R., R. A. Bergstedt, and T. H. Eckert. 1987. Prey fish dynamics and salmonine predator growth in Lake Ontario, 1978–84. Canadian Journal of Fisheries and Aquatic Sciences 44(Supplement 2):390–403.

———, C. P. Schneider. 1986. Dynamics of alewives in Lake Ontario following a mass mortality. Transactions of the American Fisheries Society 115:1–14.

———, T. J. Stewart. 1999. Ascent, dominance, and decline of the alewife in the Great Lakes: Food web interactions and management strategies. Pages 489–513 in W. W. Taylor and C. P. Ferreri, eds. Great Lakes Fisheries Policy and Management: A Binational Perspective. Michigan State University Press, East Lansing, Michigan.

Owens, R. W. and R. A. Bergstedt. 1994. Response of slimy sculpins to predation by juvenile lake trout in southern Lake Ontario. Transactions of the American Fisheries Society 123:28–36.

———, P. G. Weber. 1995. Predation on Mysis relicta by slimy sculpins (*Cotus cognatus*) in southern Lake Ontario. Journal of Great Lakes Research 21:275–283.

Paine, J. R., and R. E. Lange. 1985. Effecting a policy: Institutional arrangements for allocating the resource. Pages 119–123 in R. L. Eshenroder, editor. Presented papers from the Council of Lake Committees Plenary Session on Great Lakes predator-prey issues, March 20, 1985. Great Lakes Fishery Commission, Special Publication 85–3, Ann Arbor, Michigan.

Pearce, W. A., R. A. Braem, S. M. Dustin. and J. J. Tibbles. 1980. Sea lamprey (*Petromyzon marinus*) in the Lower Great Lakes. Canadian Journal of Fisheries and Aquatic Sciences 37:1802–1810.

Pycha, R. L., and G. R. King. 1975. Changes in the lake trout population of southern Lake Superior in relation to the fishery, the sea lamprey, and stocking, 1950–1970. Great Lakes Fishery Commission Technical Report 28, Ann Arbor, Michigan.

Rand, P. S., B. F. Lantry, R. O'Gorman, R. W. Owens, and D. J. Stewart. 1994. Energy density and size of pelagic prey fishes in Lake Ontario, 1978–1990: Implications for salmonine energetics. Transactions of the American Fisheries Society 123:519–534.

Reigle, N. J., Jr. 1969a. Bottom trawl explorations in southern Lake Michigan, 1962 to 1965. United States Department of the Interior, Bureau of Commercial Fisheries Circular 301.

———, 1969b. Bottom trawl explorations in northern Lake Michigan, 1963 to 1965. United States Department of the Interior, Bureau of Commercial Fisheries Circular 318.

Schaner, T., and A. Mathers. 1991. Lake Ontario hydroacoustic survey for alewife and smelt. Section 15 *in* Lake Ontario Fisheries Unit 1990 annual report. Ontario Ministry of Natural Resources, Glenora.

Selgeby, J. H. 1982. Decline of lake herring (*Coregonus artedi*) in Lake Superior: An analysis of the Wisconsin herring fishery, 1936–78. Canadian Journal of Fisheries and Aquatic Sciences 39:554–563.

———, W. R. MacCallum, and D. V. Swedberg. 1978. Predation by rainbow smelt (*Osmerus mordax*) on lake herring (*Coregonus artedii*) in western Lake Superior. Journal of the Fisheries Research Board of Canada 35:1457–1463.

———, M. H. Hoff. 1994. Rainbow smelt-larval lake herring interactions: competitors or casual acquaintances? U. S. Department of the Interior, National Biological Survey Biological Report 25.

———, C. R. Bronte, E. H. Brown, Jr., M. J. Hansen, M. E. Holey, J. P. VanAmberg, K. M. Much, D. B. Makauskas, P. McKee, D. M. Anderson, C. P. Ferreri, and S. T. Schram. 1995. Lake trout restoration in the Great Lakes: stock-size criteria for natural reproduction. Journal of Great Lakes Research 21 (Supplement 1):498–504.

Smith, B. R., and J. J. Tibbles. 1980. Sea lamprey (*Petromyzon marinus*) in Lakes Huron, Michigan, and Superior: History of invasion and control, 1936–78. Canadian Journal of Fisheries and Aquatic Sciences 37:1780–1801.

———, B. G. H. Johnson. 1974. Control of the sea lamprey (*Petromyzon marinus*) in Lake Superior, 1953 to 1970. Great Lakes Fishery Commission Technical Report 26, Ann Arbor, Michigan.

Smith, S. H. 1956. Life history of lake herring of Green Bay, Lake Michigan. U. S. Department of Interior, Fish and Wildlife Service, Fishery Bulletin 57: 109.

———, 1964. Status of the deepwater cisco population of Lake Michigan. Transactions of the American Fisheries Society 93(2):155–163.

———, 1968. Species succession and fishery exploitation in the Great Lakes. Journal of the Fisheries Research Board of Canada 25:667–693.

———, 1970. Species interactions of the alewife in the Great Lakes. Transactions of the American Fisheries Society 99:754–765.

Sprules, W. G., S. B. Brandt, D. J. Stewart, M. Munawar, E. H. Jin, and J. Love. 1991. Biomass size spectrum of the Lake Michigan pelagic food web. Canadian Journal of Fisheries and Aquatic Sciences 48:105–115.

Stewart, D. J., J. F. Kitchell, and L. B. Crowder. 1981. Forage fishes and their salmonid predators in Lake Michigan. Transactions of the American Fisheries Society 110:751–763.

———, D. Weininger, D. V. Rottiers, and T. A. Edsall. 1983. An energetics model for lake trout, *Salvelinus namaycush*: application to the Lake Michigan population. Canadian Journal of Fisheries and Aquatic Sciences 40:681–698.

———, M. Ibarra. 1991. Predation and production by salmonine fishes in Lake Michigan, 1978–88. Canadian Journal of Fisheries and Aquatic Sciences 48:909–922.

Stuiber, D. A., and J. T. Quigley. 1973. Wastewater treatment in fish processing. International Association for Great Lakes Research. Proceedings of the 16th Conference on Great Lakes Research 1973:958–966.

Talhelm, D. R. 1988. Economics of Great Lakes Fisheries: A 1985 assessment. Great Lakes Fishery Commission, Technical Report No. 54, Ann Arbor, Michigan.

Time. 1967. Ecology: alewife explosion. Time 90(1; July 7):56.

Van Oosten, J. 1930. The disappearance of the Lake Erie cisco—A preliminary report. Transactions of the American Fisheries Society 60:204–214.

———, 1937. The dispersal of smelt, *Osmerus mordax* (Mitchill), in the Great Lakes Region. Transactions of the American Fisheries Society, 66:160–171.

———, H. J. Deason. 1938. The food of the lake trout (*Christivomer namaycush*) and of the lawyer (*Lota maculosa*) of Lake Michigan. Transactions of the American Fisheries Society 67:155–177.

Waters, T. F. 1987. The Superior North Shore. University of Minnesota Press, Minneapolis, Minnesota.

Wells, L. 1968. Seasonal depth distribution of fish in southeastern Lake Michigan. United States Fish and Wildlife Service, Fishery Bulletin 67:1–15.

———, 1970. Effects of alewife predation on zooplankton populations in Lake Michigan. Limnology and Oceanography 15:556–565.

————, A. L. McLain. 1973. Lake Michigan, man's effects on native fish stocks and other biota. Great Lakes Fishery Commission, Technical Report No. 20, Ann Arbor, Michigan.

Withler, R. E., and T. P. T. Evelyn. 1990. Genetic variation in resistance to Bacterial Kidney Disease within and between two strains of coho salmon from British Columbia. Transactions of the American Fisheries Society 119:1003–1009.

Wojcik, J. A., M. S. Evans, and D. J. Jude. 1986. Food of deepwater sculpins Myoxocephalus thompsoni, from southeastern Lake Michigan. Journal of Great Lakes research 12:225–231.

Wolfert, D. R., and M. T. Bur. 1992. Selection of prey by walleyes in the Ohio waters of the Central Basin of Lake Erie, 1985 to 1987. United States Department of the Interior, Fish and Wildlife Service, Resource Publication 182.

Note

1.   Contribution 941 of the National Biological Service, Great Lakes Science Center, Ann Arbor, Michigan.

PART 4

Case Studies

CHAPTER 13

Contribution of Fishery Management in Walleye and Yellow Perch Populations of Lake Erie

Joseph F. Koonce, Ana B. Locci, and Roger L. Knight

Introduction

The Lake Erie ecosystem has changed dramatically over the last two hundred years. Regier and Hartman (1973) referred to this record of change as 150 years of cultural stress. During the 1800s, westward expansion of the American frontier led to deforestation, conversion of prairies to farmland, drainage and filling of wetlands, widespread erosion and siltation of streams, and construction of dams for mill sites (Trautman 1981). With growing industrialization of the basin through the early 1900s, and with the accompanying increase in numbers of people in the basin, pollution began to emerge as a major stress. Discharge of untreated sewage depleted oxygen in the lower reaches of large rivers like the Cuyahoga, and artificial eutrophication associated with increased phosphorus loading caused a shift in composition of benthic invertebrates to dominance of pollution tolerant oligochaetes, especially in the western basin (Britt 1955; Carr and Hiltunen 1965). By the late 1960s, the result of these physical and chemical stresses, coupled with overexploitation of Lake Erie's fish resources was an impression of a dying lake.

Prior to these changes, Lake Erie was the most productive and biologically diverse of the Laurentian Great Lakes (Leach and Nepszy 1976). Its southern location, mixture of cold-water, mesothermal, and warm-water habitats, and its fertile drainage basin sustained this productive and diverse ecosystem. For example, Van Meter and Trautman (1970) provide evidence for the historical presence of 138 fish species in Lake Erie, exclusive of the Detroit River. Further reflecting its productive status, Lake Erie yielded the greatest commercial fish harvest of all the Great Lakes for most years since 1914 (Leach and Nepszy 1976). These sustained high harvests, however, belie the stability of the fish community. Hartman (1973) noted a progressive loss of commercially important fish species. First salmonids, then coregonids, and finally, the percids declined as a result of the history of cultural stress, and non-indigenous species like smelt began to sustain Lake Erie fish harvests (Leach and Nepszy 1976).

Since the nadir of the 1960s, the walleye *(Stizostedion vitreum vitreum)* fishery of Lake Erie has shown a remarkable recovery (Hatch et al. 1987). Through implementation of a quota manage-

Harvest (t)

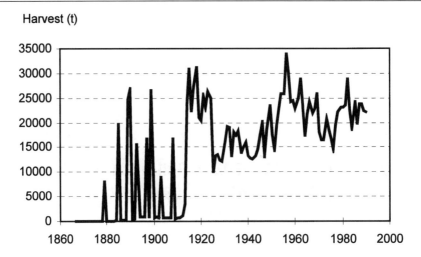

FIG. 1. *Historical harvests of commercial fisheries from Lake Erie for the period 1868 to 1990. Data from Baldwin* et. al *1979 and Margaret Dochoda, Great Lakes Fishery Commission, personal communication.*

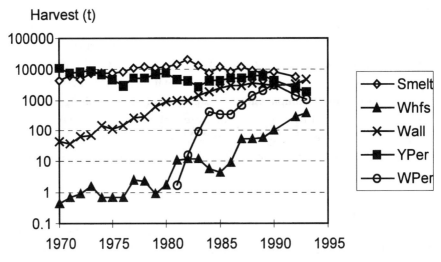

FIG. 2. *Commercial harvests of selected species from Lake Erie since 1970. Data from Baldwin* et. al *1979 and Margaret Dochoda, Great Lakes Fishery Commission, personal communication. Species codes are for smelt, whitefish (Whfs), walleye (Wall), yellow perch (YPer), and white perch (WPer).*

ment procedure (involving the States of Michigan, Ohio, Pennsylvania, and New York, and the Province of Ontario), the walleye fishery has recovered from a severely depressed status in the late 1960s, to a highly successful recreational fishery through the 1980s. Commercial fisheries declined from 1960 to the late 1970s with percids accounting for over 50% of harvest, but rainbow smelt (*Osmerus mordax*) proved to be a replacement for some of the earlier losses, and commercial harvests increased in the 1980s (figs. 1 and 2). Other species also increased in abundance in this

Harvest

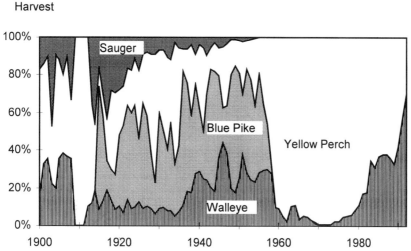

FIG. 3. *Historical trends in composition of Lake Erie percid harvest. Data from Baldwin et. al 1979 and Margaret Dochoda, Great Lakes Fishery Commission, personal communication.*

period. White perch (*Morone americana*) became very abundant and may have had a role in the decline of yellow perch, (*Perca flavescens*), (Parrish and Margraf 1990). Lake whitefish (*Coregonus clupeaformis*) also began to increase as reflected in the increase of harvests (fig. 2).

The improvement of fisheries of Lake Erie, however, did not continue in the 1990s. There is no clear explanation for these recent changes, but three lines of explanation have emerged: effectiveness of fishery management in preventing overexploitation, declining productivity due to restrictions on phosphorus loading, and the cumulative effect of invasions of non-indigenous species. Certainly the greatest change in the Lake Erie ecosystem followed the invasion of zebra and quagga mussels (*Dreissena polymorpha* and *D. burgensis*). In this paper, we will examine these recent changes in the context of the history of management of percids. The purpose of the paper is to review the interaction of fishery management and changes in walleye and yellow perch populations in Lake Erie, and to evaluate the relative contribution of changing productivity, fishery regulation, and invasion of non-indigenous species on recent declines in both species.

History

Percid harvest from Lake Erie was an important component of commercial fisheries during the period from 1900 to 1990, accounting on average for 45% of total catch. Combined harvests of walleye, sauger (*Stizostedion canadense*), blue pike (*Stizostedion vitreum glaucum*), and yellow perch averaged 3,560 metric tons per year, while total harvests averaged 8,060 metric tons per year (Baldwin et al. 1979 Great Lakes Fishery Commission, personal communication). Figure 3 shows the changing contribution of each of these species to the commercial fishery. Until recently, walleye catches were consistently smaller than yellow perch. Since the 1970s, recreational fishing has become an increasingly important contribution to walleye harvest (fig. 4). Yellow perch harvests are dominated by commercial fisheries (fig. 5), with about 75% of this harvest occurring in

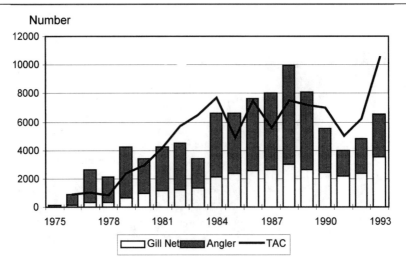

FIG. 4. *Relative contribution of gill net catch and angling catch to the walleye harvest from 1976 to 1994. Also included is the total allowable catch (TAC) approved by the Lake Erie Committee each year from 1976 to 1993. Data were derived from Walleye Task Group Reports (Knight, personal communication).*

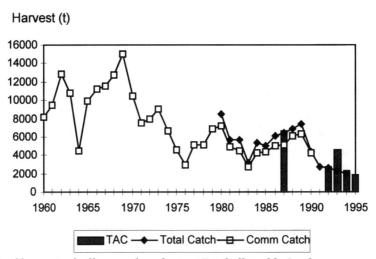

FIG. 5. *Historical harvests of yellow perch and recent Total Allowable Catch recommendations. Commercial harvests statistics are from Baldwin et al. 1979 and Margaret Dochoda, Great Lakes Fishery Commission, personal communication. Total biomass of harvest is from Yellow Perch Task Group Reports, Knight, personal communication.*

Ontario waters over the period 1985 to 1994 (Yellow Perch Task Group; Knight, Ohio Division of Wildlife, personal communication).

Using harvest statistics collected for walleye and yellow perch in Lake Erie, state and provincial agencies have performed virtual population analysis for both species. The Lake Erie Committee of the Great Lakes Fishery Commission established technical committees for walleye (Walleye

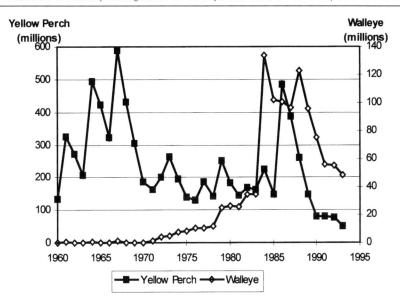

FIG. 6. *Population reconstruction of yellow perch and walleye for Lake Erie. Yellow perch estimates for the period 1960 to 1977 are from Rawson, personal communication, and for the period 1978 to 1993 are from Knight, personal communication. Both relied on virtual population with the CAGEAN analysis program. Walleye data are from CAGEAN analysis for the period 1978 to 1993 and for 1960 to 1977, from forward projection estimates based on the work of Shuter et al. 1979.*

Task Group) and yellow perch (Yellow Perch Task Group) to compile harvest data and analyze trends. Estimates of population trends for walleye and for yellow perch ages two and older are in figure 6. Walleye abundance reached a peak in 1984, and remained at high levels until beginning to decline in 1989. Yellow perch populations were high in the 1960s, and declined through the 1970s and 1980s. Following an increase in 1986, the population has since been in decline. In general, harvests follow variation in population size for both species ($r = 0.82$, $p < 0.005$ for walleye and $r = 0.53$, $p < 0.005$ for yellow perch).

In 1980, the Great Lakes States, Tribes with Treaty fishing rights in the Great Lakes, the Province of Ontario, and the U.S. and Canadian federal governments agreed to cooperate formally in the management of fisheries stocks of common concern through the Strategic Plan for the Management of Great Lakes Fisheries, SGLFMP (GLFC 1980; Dochoda and Koonce 1994). Although many elements of SGLFMP had been operating informally for many years, the agreement specifically called for the creation of individual lake committees for coordination of management. SGLFMP required these committees to develop a set of fish community objectives for each of the lakes, as well as a set of environmental objectives that would be necessary to achieve the stated fish community objectives.

Prior to the implementation of SGLFMP, Lake Erie fishery managers had already begun cooperative management for walleye in western Lake Erie. Hatch et al. (1987) reviewed the development of a quota management scheme for walleye in western Lake Erie and showed that quota management reduced the annual mortality rate from 90% for walleye older than 1.5 years in the

1960s to 30% by 1985. Except for a few years, harvests of walleye have been at or less than the total allowable catches, which the Lake Erie Committee approved each year since 1976. In U.S. waters, agencies elected to allocate the entire quota to the recreational fishery, while in Canadian waters, Ontario allocated its quota between commercial and recreational fisheries.

Because application of quota management to yellow perch started later than walleye, implementation of yellow perch quotas lagged behind walleye. In 1980, the Lake Erie Committee created a Yellow Perch Task Group (YPTG), which was modeled after the Walleye Task Group to develop a basis for setting total allowable catch of yellow perch. The YPTG collected harvest data, estimated yellow perch standing stock through virtual population analysis, and developed methodology for recommending total allowable catches (TAC).

Disagreements about methods of allocation of allowable catch among jurisdictions and fisheries, and about recommended size of harvest precluded systematic action on the recommendations by the Lake Erie Committee (LEC). The LEC, formally adopted a total allowable catch only for 1987, and annually since 1992 (fig. 5). As early as 1980, however, agencies were concerned with the continuing low levels of yellow perch harvest. YPTG population analyses showed that the yellow perch population had declined lakewide relative to the 1960s. Although general agreement on a quota management scheme was slow in developing, individual agencies were taking action within their own jurisdictions. Through the early 1980s, all agencies were independently implementing measures to reduce yellow perch harvest. In 1984 and 1985, the Ontario Ministry of Natural Resources implemented quotas for yellow perch and increased minimum gill net mesh size to 2.25 inches. New York, Ohio, and Pennsylvania also adopted increased gill net size, and restricted fishing seasons and zones as early as 1980. These three agencies eliminated gill net fisheries for yellow perch by 1986.

In addition to these changes in the fish community and its fisheries, other characteristics of the Lake Erie ecosystem have changed since the 1960s. With the Great Lakes Water Quality Agreement, Canada and the United States set targets for phosphorus loading to Lake Erie of eleven thousand metric tons per year (Vallentyne and Thomas 1978). Annual loadings of phosphorus are at this target level (Dolan 1993), but the extent of anoxia in the central basin hypolimnion has not fallen to expected levels. Achieving this target on a lakewide basis reduced phosphorus loading about 39% from the level in 1968 (Rathke and McRae 1989), and Bertram (1993) has shown a positive correlation between total phosphorus concentrations and oxygen depletion rates in the central basin of Lake Erie. Nevertheless, the influence of water level and spring weather on physical characteristics makes the hypolimnion of the central basin less responsive to nutrient reductions than near shore areas or the western basin (Charlton et al. 1993).

Regulation of other ecosystem stresses has not been as extensive as with phosphorus loading and eutrophication. Two stresses in particular seem to make large contributions to the instability of the Lake Erie ecosystem: habitat loss and invasion of non-indigenous species. Near shore and tributary habitat disruption, which mainly occurred in the late 1800s, remains a problem (Koonce et al. 1996). Despite improving water quality of many of the major rivers (Krieger and Ross 1993), few of the estuarian reaches of these rivers have been restored to minimum standards as measured by an index of biotic integrity (R. Thoma, Ohio Environmental Protection Agency, personal communication). Limitations of spawning and nursery habitat imposed by continuing stress on near shore and tributary zones may contribute to the lagging recovery of some native fish species (Koonce et al. 1996).

Invasion of non-indigenous species is both an indication of disturbance of the Lake Erie ecosystem, as well as a potential cause of instability. Mills et al. (1993) have documented invasion of 139 species of aquatic organisms into the Great Lakes basin since the 1800s. Invasion of sea lamprey (*Petromyzon marinus*) into the upper Great Lakes resulted in the catastrophic loss of native predators (Smith 1971). Because Lake Erie has less cold-water habitat than the other Great Lakes, its fish community was less affected by sea lamprey, but invading rainbow smelt and white perch have become dominant species (Leach and Nepszy 1976; Parrish and Margraf 1990). Since their introduction into Lake St. Clair in the late 1980s, *Dreissena* species have become very abundant in Lake Erie (Dermott and Munawar 1993). Associated with this invasion have been changes in water transparency, plankton abundance, and macrobenthos composition. Leach (1993) reported an 85% increase in water transparency and a 43% decrease in chlorophyll between 1988 and 1989. In contrast to an observed increase in the amphipod *Gammarus* in near shore benthic communities dominated by zebra mussels (Leach 1993), Dermott (1993) found an inverse relation for abundance of *Diporeia* and the quagga mussel in the deeper areas of Lake Erie. Invasion of zebra and quagga, therefore, have fundamentally altered energy flow and nutrient cycling pathways.

A Theoretical Analysis of the Fish Community

The complexity of these interactions is difficult to analyze empirically. To aid understanding of the history of fishery management in the context of recent changes in the Lake Erie ecosystem, we relied on a simulation model of the fish community of the western basin of Lake Erie. Koonce et al. (1983) originally developed the Lake Erie Fish Community Model. Locci (1988) modified and calibrated the model for western Lake Erie, and recently we have expanded it to include effects of nutrient loading, contaminant loading, and zebra mussel invasion. The version of the model we used for this paper is thus, a predator-prey model constrained by energy flow and nutrient availability. Following many of the conventions of Jones et al. (1993), functional characteristics of the model are based on major ecological principles, which depend upon bioenergetics and population theory, rather than fixed empirical relations.

The model includes seven fish species (walleye, yellow perch, white perch, white bass(*Morone chrysops*), drum, gizzard shad(*Dorosoma cepedianum*), and emerald shiners (*Notropis atherinoides*)). For each fish species, the model includes numbers of individuals and size by age. Annual changes in abundance of each age group are functions of instantaneous mortality rates due to nonpredatory natural mortality ($m_{i,j}$), consumption by predators ($v_{i,j}$), and fishing mortality ($f_{i,j}$) according to the following equations:

$$N_{i,j,t} = N_{i,j-1,t-1} \cdot e^{-(m_{i,j} + v_{i,j} + f_{i,j})} \tag{1}$$

where $N_{i,j,t}$ is the abundance of the jth age group of the *i*th species at time *t* and $N_{i,j-1,t-1}$ is the abundance of this cohort in the previous year, $t - 1$. Natural mortality is constant for each species and age group, and mortality due to predation follows a functional response model (Murdoch 1973) as applied to fish populations in Lake Ontario by Jones et al. (1993):

$$v_{i,j} = \sum_k \sum_l \frac{a_{i,j,k,l} \cdot N_{k,l}}{1 + \sum_i \sum_j \frac{1}{Cmax_{k,l}} \cdot \frac{W_{i,j}}{W_{k,l}} \cdot a_{i,j,k,l} \cdot N_{i,j}} \tag{2}$$

where $a_{i,j,k,l}$ is the effective search rate of predator species k, age group l, on prey species i, age group j, $N_{k,l}$ is the abundance of lth age group of the kth predator species, and $Cmax_{k,l}$ is the maximum consumption rate of prey per predator, $W_{i,j}$ is the mean weight of prey, and $W_{k,l}$ is in mean weight of predator. Fishing mortality is a product of catchability ($q_{i,j,y}$) and fishing effort (E_y) for each fishery, y (for example, commercial or recreational) by species i and age j:

$$f_{i,j} = \sum_y q_{i,j,y} \cdot E_y \tag{3}$$

Estimation of parameters in the functional response equation (equation 2) depends on a number of bioenergetic and behavioral assumptions. Maximum consumption rates are allometric functions of body size (Kitchell et al. 1977):

$$Cmax_{k,l} = a \cdot w_{k,l}^b \tag{4}$$

where a and b are constants. To allow for variation in habitat preference among species, the predator search rate ($a_{i,j,k,l}$) depends upon the product of the reactive distance (r), swimming speed (ω), prey preference (λ), and a habitat overlap coefficient (H):

$$a_{i,j,k,l} = \pi \cdot r_{k,l}^2 \cdot \omega_{k,l} \cdot \lambda_{i,j,k,l} \cdot H_{i,j,k,l} \tag{5}$$

Prey preference depends directly upon the length ratio of prey and predator:

$$\lambda_{i,j,k,l} = e^{-(L_{i,j} - p_{k,l} \cdot L_{k,l})^2 / 2 S_{k,l}^2} \tag{6}$$

where $L_{i,j}$ is the length of prey species i at age j, $L_{k,l}$ is length of predator species k at age l, $p_{k,l}$ is the maximum proportion of prey to predator length, and $S_{k,l}$ is the standard deviation of that preferred ratio. Derivation of this function assumes that prey preference of a predator depends exclusively upon the relative size of prey and predator. Equation 6 is a symmetrical, bell-shaped curve about the preferred prey length. Based on observed distributions of species and age groups among three basins and three habitat zones (near shore, pelagic, and benthic), we calculated habitat overlap coefficients for all combinations of species and age groups. Effective search rate, therefore, is zero for combinations having no overlap in distribution or for prey to predator length ratios greater than a species-specific maximum.

In the model, growth rates depend upon consumption rates. Based on results of detailed bioenergetics models (c.g., Kitchell et al. 1977), we assumed that growth rate increases to a maximum value as consumption approaches its maximum value. We adopt the assumption by Jones et al. (1993) that growth rate is zero below maintenance consumption levels. Because several predators consume each prey category (species i and age j), actual consumption depends upon a proportional allocation of annual mortality to each mortality source:

$$C_{i,j,k,l} = \frac{\frac{A_{i,j,k,l}}{Z_{i,j}} \cdot (1 - e^{-Z_{i,j}}) \cdot N_{i,j} \cdot W_{i,j}}{N_{k,l} \cdot W_{k,l}}$$

$$A_{i,j,k,l} = \frac{a_{i,j,k,l} \cdot N_{k,l}}{1 + \sum_i \sum_j \frac{1}{Cmax_{k,l}} \cdot \frac{W_{i,j}}{W_{k,l}} \cdot a_{i,j,k,l} \cdot N_{i,j}}$$

(7)

where $Z_{i,j}$ is the instantaneous mortality rate of the ith prey species of the jth age group.

We assume that annual growth rates follow a simple Walford function:

$$W_{i,j,t} = \rho \cdot W_{i,j-1,t-1} + w_k$$

(8)

where $W_{i,j,t}$ is the weight of the jth age of the ith species at time t, and w_k is a constant. Growth rate (ρ is a function of the maximum growth rate (ρmax) and a relation between maximum ($Cmax$) and predicted consumption ($C_{i,j}$):

$$\rho = \rho_{max} \frac{C_{i,j}}{g_i Cmax_{i,j} + C_{i,j}}$$

(9)

where g_1 is the proportion of maximum consumption at which growth rate is half of maximum growth rate.

We calculated recruitment of age zero (YOY) from age specific fecundity and maturity schedules for females of each species. Female fertility schedules vary with growth rate and fecundity depends upon size as shown by Ware (1982). The recruitment function also includes a time-dependent random variable to simulate the effects of environmental variability on recruitment success.

As indicated above, the age-structured model of fish growth and survival depends upon productivity of lower trophic levels. To accommodate possible effects of zebra mussel dynamics on the structure of the Lake Erie food web, we included four components to represent the energetics of lower trophic levels of the Lake Erie ecosystem: annual primary production, mean zooplankton abundance, mean zoobenthos abundance, and zebra mussel biomass. Annual primary production depended upon mean annual phosphorus loading and affected production of zooplankton and zoobenthos. We assumed that zebra mussel biomass affected the allocation of primary production between benthos and zooplankton. The result of this explicit representation of lower trophic level dynamics is that both growth and recruitment of fish depends upon nutrient loading and the biomass of zebra mussels.

Parameter Estimation and Model Calibration

Three major sources of information were used to obtain estimates of trends in abundance or relative abundance of walleye, yellow perch, white bass, drum, white perch, gizzard shad, and emerald shiners in western Lake Erie: trawling and gill net data, diet studies, and virtual population reconstruction. Trawling and gill net surveys are reported annually by Lake Erie fisheries managers from Michigan Department of Natural Resources, Ohio Department of Natural Resources, and the Ministry of Natural Resources Ontario. These data provide indices of relative abundance of the species in the model. In addition to occasional diet surveys reported by the U.S. Fish and Wildlife Service, we relied on diet studies of Price (1963), Parson (1971), and Knight et al. (1984) for data to test diet predictions of the model, principally for walleye and yellow perch. Reports of the Walleye Task Group and the Yellow Perch Task Group were the sources of data on abundance trends of walleye and yellow perch.

Model Calibration

The goal of model calibration was to represent known species trends and diets, and the effect of fisheries on the abundance of the selected fish species. We tested the model with historical data using historical fishing effort as an input variable, and we compared predicted values of species abundance with observed values. We evaluated model fit first in terms of agreement between predicted and observed abundance trends for walleye and yellow perch over the period 1960 to 1994. Secondarily, we compared predicted trends to relative abundance trends of the remaining species.

Testing the model with historical data for walleye and yellow perch showed that expected values of species abundance corresponded to observed values (figs. 7 and 8). We included the effects of environmental variability on year-class strength of walleye and yellow perch by estimating the relationship between observed and expected YOY production based on virtual population analyses. Simulations show the walleye recovery and a reasonable fit to the dynamics of yellow perch (figs. 7 and 8).

Dynamics of the remaining species represented in the model correspond with the trends derived from trawl data. These trends consisted of an increased of white perch, since 1982, and a continuous decline of white bass (fig. 9). Simulated dynamics of the forage species were also consistent with the abundance trends reflected by the trawling data in the western basin of the lake. Those trends were a general decline of emerald shiner with an accelerated decrease in the last ten years, and a pattern of initial increase and decline of the gizzard shad population after 1985 (fig. 10).

Because the model does not have fixed prey preferences, comparison of predicted and observed diet is an important test. We assume that changes in diet reflect changes in prey availability relative to preferred prey sizes, using available data for walleye and yellow perch for the periods of 1958 to 1969 and 1979 to 1981 (Parson 1971; Price 1963; Knight et al. 1984). Figure 11 shows that simulated walleye diet is almost strictly piscivorous, with preference for soft-rayed fish (shiner and gizzard shad) over the spiny-rayed fish (walleye, yellow perch, white bass, and white perch). The change in diet of walleye between 1961 to 1981 is reasonably consistent with diet changes observed by Parson (1971) and Knight et al. (1984). These changes consisted of a marked decrease

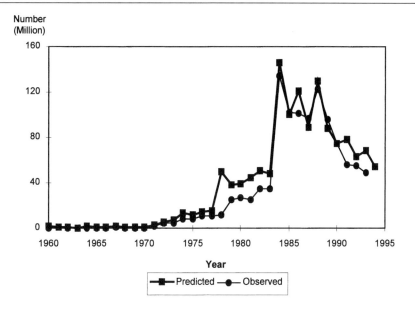

FIG. 7. *Comparisons of observed and predicted dynamics of walleye (age 2 and older) in the western basin of Lake Erie.*

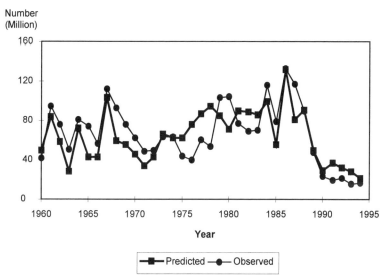

FIG. 8. *Comparisons of observed and predicted dynamics of yellow perch in the western basin of Lake Erie.*

of shiner consumption, and a strong increase of clupeids in the diet. An increase of white perch consumption was also detected in the walleye diet during early 1990s. The simulated diet of yellow perch consisted of more than 80% invertebrates, mostly zoobenthos (fig. 12). Overall, predicted and observed diet were consistent. This consistency, coupled with reasonable representation of abundance trends for those species with observed relative abundance trends, only means that the model is a reasonable representation of the historical dynamics of these seven fish species of

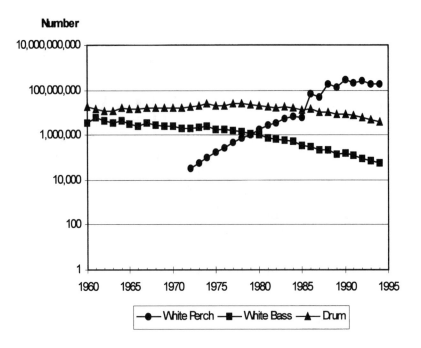

FIG. 9. *Dynamics of white perch, white bass, and drum in the western basin of Lake Erie predicted by the Lake Erie Ecological Model during period 1960–1994.*

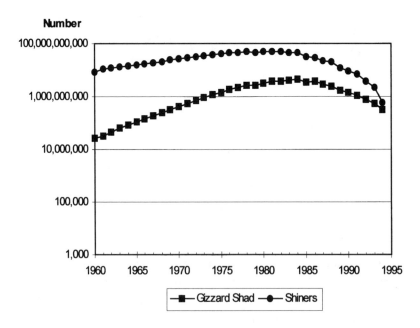

FIG. 10. *Dynamics of gizzard shad and shiners in western Basin of Lake Erie predicted by the Lake Erie Ecological Model during period 1960–1994.*

kg/y/ind

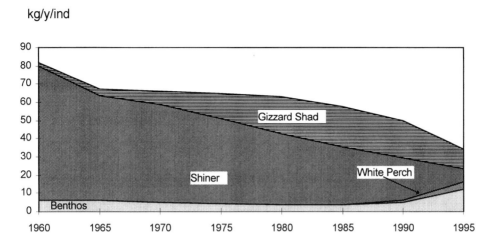

FIG. 11. *Predicted diet for 2 age and older walleye in the western basin of Lake Erie from 1960 to 1994.*

kg/y/ind

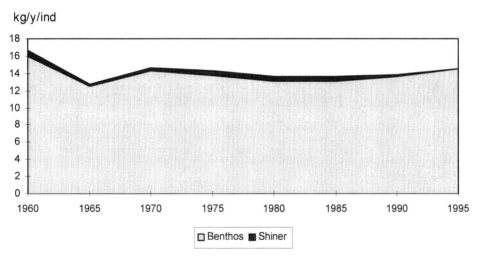

FIG. 12. *Predicted diet for 2 age and older yellow perch in the western basin of Lake Erie from 1960 to 1994.*

western Lake Erie. On the basis of this agreement, therefore, we can use the model to explore the consequences of various hypotheses for recent changes in the fish community of Lake Erie.

Fishery Management Issues

Implementation of a quota management scheme led to successful restoration of walleye in western Lake Erie (Hatch et al. 1987). Strong year-classes, especially 1982, also contributed to the recovery, but they would not have occurred without the buildup of adult stock that was the result of lowered fishing mortality. Through binational and multiagency coordination under SGLFMP, Lake Erie fishery managers were able not only to increase the abundance and harvests of walleye in the

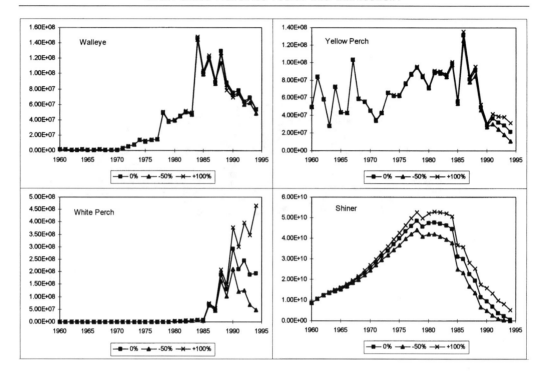

FIG. 13. *Predictions of abundance trends for four species in western Lake Erie for the period, 1960 to 1994. Simulations represent variations in phosphorus loading over the historical (0%) trends of either a 50% reduction or 100% increase of historical, annual phosphorus loads.*

western basin, but they also were able to improve the abundance of stocks of walleye in the central basin. The results of quota management of yellow perch have been less successful. Despite a brief peak in 1986, yellow perch populations have remained depressed relative to the levels in the 1960s, and they show a sustained decline since the mid 1980s. As shown in figure 2, yellow perch has not been the only species to show a recent decline of population and harvests. Walleye population (fig. 6) and harvests of smelt and white perch have also declined.

With declining nutrient loading and zebra mussel invasion, concern arises that decreased availability of lower-trophic level productivity may be an important factor causing declines in fish populations. To explore this possibility using model simulations, we varied phosphorus loading from the historical levels by a 50% reduction and a 100% increase. This range of variation in phosphorus loading affects abundance of all the species, but the magnitude of these effects varied among species. In figure 13, we show the response of a top predator, walleye, two secondary predators (yellow perch and white perch), and a planktivore (emerald shiner) to these variations in productivity. Change in productivity levels has the strongest effect on the abundance of the secondary predators (i.e., yellow perch and white perch) while the walleye abundance was only slightly affected by variation of productivity. For the secondary predators, the cumulative effect of changing phosphorous loading was apparent after twenty-five years, but these effects were apparent after only ten years for a planktivore. The implication of these results is that the decreased availability

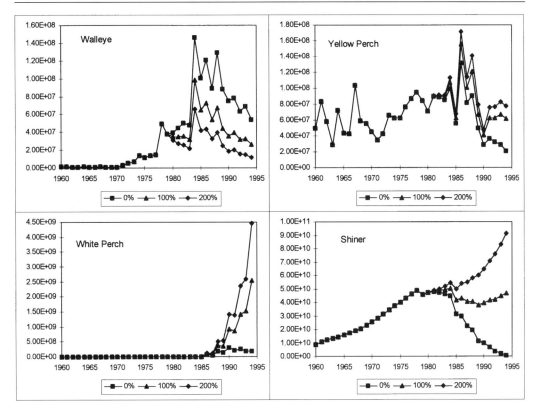

FIG. 14. *Predictions of abundance trends for four species in western Lake Erie for the period, 1960 to 1994. Simulations represent variations in fishing effort on walleye populations from 1980 to 1994. Historical trends in effort (0%) contrast with 100% and 200% increases from historical levels.*

of the production of lower trophic levels has the potential to lower abundance of fish, especially of planktivorous species, but does not lead to similarity of the decline observed in yellow perch and walleye.

Exploitation and community interactions are obvious alternative candidates for consideration. Quota management of walleye and yellow perch populations has generally followed conservative fishing mortality targets. Certainly, quota management of walleye allowed dramatic recovery of the walleye population (fig. 6) and recent harvests have been lower than total allowable catches (fig. 4). Examination of walleye population trends (fig. 6), however, raises a question about the possible effects of the rapid increases of walleye in the 1980s. To explore this effect, we used model simulations to vary fishing history of the walleye population in western Lake Erie by increasing fishing mortality from 1980 to 1994 by either 100% or 200% of historical levels (fig. 14). Results of these simulations suggest much greater sensitivity of fish community structure to walleye abundance than to productivity. Most notably, declines of yellow perch, white perch, and emerald shiner become uncoupled from the walleye decline at lower walleye abundance. At high abundance levels, the model clearly shows that walleye demand is a major source of mortality for these species, but the diet of walleye reflects the greater contribution of shiners and gizzard shad (e.g., fig. 11).

Discussion

Declining abundance and harvests from 1990 to 1994 are causes of concern for fishery managers. The success with quota management of walleye in the 1980s does not appear to be sustainable, and does not seem to transfer to yellow perch management. Quota management does minimize the problems of overexploitation as long as the assumptions for calculation of total allowable catch remain correct (Koonce and Shuter 1987). The fundamental issue confronting fishery managers, therefore, is whether they can achieve any regulatory gains through traditional fish management practices, such as harvest regulation on a species-by-species basis, or stocking programs. Given the rapid changes in the Lake Erie ecosystem since the late 1980s, for example: is the potential decline in available production and the cumulative effect of introductions of non-indigenous species overwhelming efforts to prevent overexploitation of the fishery resources?

White perch and smelt became major components of the fish community of Lake Erie during the 1980s. Increases in these non-indigenous species could have had a wide-range of consequences. Parrish and Margraf (1990), for example, showed substantial competitive interactions between yellow perch and white perch, and there were reasons to suspect that white perch was suppressing yellow perch. The invasion of zebra mussels beginning in the late 1980s further worsened this problem. Although Wu and Culver (1991) argue that zooplankton grazing still controlled edible algal density and water transparency during 1989, longer-term analyses by Nicholls and Hopkins (1993) suggest that phosphorus reduction and zebra mussel increases coincide with decreases in total phytoplankton densities in western Lake Erie. The impact of these changes is an overall reduction in pelagic production available to fish. Griffiths (1993) referred to a similar pattern in Lake St. Clair as an oligotrophication effect. Expectation of a decline of fish population abundance is not unreasonable under these circumstances; less food for a fish species should lead to decreased growth, recruitment, and population size.

Our analyses confirm the complexity of this problem, but also point to important lessons to be learned from the history of percid management in Lake Erie. Model results show (fig. 13) that species vary in their response to changes in productivity. White perch and shiners seem to be more responsive to reductions in phosphorus loading than yellow perch and walleye. Analyses of simulated diets suggest that white perch is much less piscivorous at older ages than yellow perch, and thus, more dependent upon invertebrate production, which would be more directly affected by decreases in primary productivity. Walleye, in contrast, changes little in abundance with declining productivity. Under the limited alterations of historical phosphorus loading, walleye has a much greater ability to substitute prey than secondary predators and planktivores. The simulations (fig. 14) further suggest that walleye abundance patterns dominate the population trends of other species. In an unexpected way, therefore, the success of the quota management of walleye in the 1980s led to high walleye abundance that then caused the decline of other populations in the 1990s through predation.

Percid management of Lake Erie has produced mixed results and has an uncertain future. The spectacular recovery of walleye was the direct result of coordinated management of agencies and the implementation of a quota management scheme. Although declining, the walleye abundance during the period of 1990 to 1995 is much greater than during the period of collapse in the 1960s.

From the perspective of recreational fishing, however, judgment of success is relative to the peak abundance in the 1980s. Recovery of walleye population thus generated levels of expectation that may not be attainable in the long run, owing to fundamental constraints and changes in the Lake Erie ecosystem over which fishery managers have little control. Several important lessons for fishery management seem to follow from this predicament.

The most obvious lesson from this history of percid management is that management on a species-by-species basis is futile. It is futile because fish species are not independent. Management decisions for walleye affect population dynamics of other species. Similarly, harvest of prey species, such as smelt, will affect growth of predators and the dynamics of other species. Setting harvest quotas for individual species without considering these interactions is bound to lead to unintended consequences in the long run. From theoretical studies (Beddington and May 1977; Hilborn 1979; Koonce and Shuter 1987; Walters 1987), it has long been clear that concepts such as maximum sustainable yield or optimal sustainable yield, which are necessary for derivation of quotas for individual stocks, do not always work well in an ecosystem context. For top predators, fishery quota systems are feasible (Koonce and Shuter 1987), even though their effectiveness may be limited by community responses to the variation in predator abundance (Walters 1987). In contrast, theoretical foundations of single-species quota derivation for secondary predators are nonexistent (Beddington and May 1977).

A second lesson of the history of percid management in Lake Erie is that even adopting conservative quotas (i.e., well below total allowable catch of a maximum sustainable yield or optimum sustainable yield) does not necessarily lead to stable fish communities. The quota derivation for walleye avoided the potential risks of overfishing by adopting a conservative strategy (Hatch et al. 1987). However, this conservative strategy could have initiated an oscillation in walleye abundance that now seems to entrain recent trends in Lake Erie. Viewed from the perspective of walleye only, this conservative strategy has produced walleye recovery, which must be judged as successful despite recent declining abundance. In the larger context of the ecosystem, however, this conservative strategy may be contributing to the decline of secondary predators, such as yellow perch, white perch, and smelt, making management of these species less successful. The challenge to fish managers is thus, to move from a species-by-species approach to an ecosystem approach as advocated by SGLFMP. Such a reorientation requires goals for community composition, as well as for harvest of various species.

Despite its mixed results, the walleye story in Lake Erie affirms the necessity of developing explicit institutional arrangements to coordinate resource management where multijurisdictional authorities fragment responsibility for management. Christie (1995) argues that the absence of such institutional arrangements often hinders implementation of an ecosystem approach. By creating a system of lakewide quotas for walleye and yellow perch, Lake Erie fishery managers have demonstrated operational approaches to coordinated management. Efforts under the terms of SGLFMP to develop fish community objectives for Lake Erie point the way to implementing an ecosystem approach, but face new impediments as fish managers grapple with water quality and habitat issues for which they have no management mandate. Target setting of phosphorus loading for Lake Erie did not consider direct effects on fish production (Charlton et al. 1993). Future coordination of fishery management and management of water quality, water quantity, and land-use development will be required to address wider ecological and social tradeoffs. Only in a wider context can the temptation to employ technological fixes to consequences of uncoordinated or unconstrained exploitation of ecosystems be tempered. SGLFMP and the quota management schemes offer templates upon which to build such broader coordination of management.

Finally, the history of percid management in Lake Erie points to the need to prepare for instabilities in the response of ecosystems to use. Ludwig et al. (1993) maintain that confronting uncertainty is central to the development of solutions to resource management and conservation problems. In addition to the fluctuations of walleye, which were induced at least in part by quota management, non-indigenous species have added to the fundamental instability of the Lake Erie ecosystem. White perch, rainbow smelt, and zebra mussels have all experienced major population expansions, and with the exception of zebra mussels, subsequent decline. Invasions of other species should be expected. Koonce et al. (1996) reasoned that the proliferation of non-indigenous species in Lake Erie is, in part, a consequence of large-scale degradation of near shore and tributary habitat favorable to native species. Prospects for future invasions, including the potential expansion of ruffe (*Gymnocephalus cernuus*) from Lake Superior, seem certain. An important concern for fish managers, therefore, is to begin assessing alternative configurations of the fish community that would prove more stable in the wake of continuing introductions. Such an assessment, by necessity, requires more general consideration with other management agencies of the contributions of water quality management decisions and habitat alterations to the instability of Lake Erie. Other uncertainties, such as climate variability and future human uses are also important, but the errors of past management, successful restoration of walleye, and population explosions of non-indigenous species have created a highly unpredictable ecosystem in Lake Erie. Unless managers find a way to include this uncertainty in their planning, they may find that, over the long term, their efforts result in more unintended than intended consequences.

Literature Cited

Baldwin, N. S., R. W. Saalfeld, M. A. Ross, and H. J. Buettner. 1979. Commercial fish production in the Great Lakes 1867–1977. Great Lakes Fishery Commission Technical Report 3, Ann Arbor, Michigan.

Beddington J. R. and R. M. May. 1977. Harvesting natural populations in a randomly fluctuating environment. Science 197:463–465.

Bertram, P.E. 1993. Total phosphorus and dissolved oxygen trends in the central basin of Lake Erie, 1970–1991. Journal of Great Lakes Research 19:224–236.

Britt, N. W. 1955. Stratification in western Lake Erie in summer of 1953: effects on the Hexagenia (Ephemeroptera) population. Ecology 36:239–244.

Carr, J. F. and J. K. Hiltunen. 1965. Changes in the bottom fauna of western Lake Erie from 1930 to 1961. Limnology and Oceanography 10:551–569.

Charlton, M. N., J. E. Milne, W. G. Booth, and F. Chiocchio. 1993. Lake Erie offshore in 1990: Restoration and resilience in the central basin. Journal of Great Lakes Resrarch 19:291–309.

Christie, W. J. 1995. The ecosystem approach to managing the Great Lakes: The new ideas and problems associated with implementing them. University of Toledo Law Review. 26:1–26.

Dermott, R. M. 1993. Distribution and ecological impact of Quagga mussels in the lower Great Lakes. In Proceedings of Third International Zebra Mussel Conference. Electric Power Research Institute Report No. TR-102077.

———, M. Munawar. 1993. Invasion of Lake Erie offshore sediments by Dreissena, and its ecological implications. Canadian Journal of Fisheries and Aquatic Sciences 50:2298–2304.

Dochoda, M. R. and J. F. Koonce. 1994. A perspective on progress and challenges under a joint strategic plan for management of Great Lakes Fisheries. University of Toledo Law Review 24:425–442.

Dolan, D. M. 1993. Point-source loadings of phosphorus to Lake Erie: 1986–1990. Journal of Great Lakes Research 19:212–223.

Great Lakes Fishery Commission. 1980. A Joint Strategic Plan for Management of Great Lakes Fisheries. Great Lakes Fishery Commission. Ann Arbor, Michigan.

Griffiths, R. W. 1993. Effects of zebra mussels (Dreissena polymorpha) on benthic fauna of Lake St. Clair. In Nalepa, T. F. and D. W. Schloesser eds. Zebra Mussels: Biology, Impacts, and Control. Lewis Publishers. Ann Arbor, Michigan. 810 p.

Hartman, W. L. 1973. Effects of exploitation, environmental changes, and new species on the fish habitats and resources of Lake Erie. Great Lakes Fishery Commission Technical Report 22, 43p.

Hatch, R. W., S. J. Nepszy, K. M. Muth, and C. T. Baker. 1987. Dynamics of the recovery of the Western Lake Erie walleye (Stizostedion vitreum vitreum) stock. Can. Journal of Fisheries and Aquatic Sciences 44:15–22.

Hilborn, R. 1979. Comparison of fisheries control systems that utilize catch and effort data. Journal of the Fisheries Research Board of Canada 36:1477–1489.

Jones, M. L., J. F. Koonce, and R. O'Gorman. 1993. Sustainability of hatchery-dependent salmonine fisheries in Lake Ontario: the conflict between predator demand and prey supply. Transactions of the American Fisheries Society 122:1022–1028.

Kitchell, J. F., D. J. Stewart, and D. Weininger. 1977. Applications of a bioenergetics model to yellow perch (Perca flavescens) and walleye (Stizostedion vitreum vitreum). Journal of the Fisheries Research Board of Canada 34:1922–1935.

Knight, R. L., F. J. Margraf, and R. F. Carline. 1984. Piscivory by walleye and yellow perch in western Lake Erie. Transactions of the American Fisheries Society 113:677–693.

Koonce, J. F. And B. J. Shuter. 1987. Influence of various sources of error and community interactions on quota management of fish stocks. Canadian Journal of Fisheries and Aquatic Sciences 44:61–67.

———, D. B. Jester, B. A. Henderson, R. W. Hatch, and M. L. Jones. 1983. Quota management of Lake Erie fisheries. Great Lakes Fishery Commission Special Publication 83–1. Ann Arbor, Michigan.

———, D. Busch, T. Czapla. (1996). Restoration of Lake Erie: Contribution of water quality and resource management. Canadian Journal of Fisheries and Aquatic Sciences 53 (Suppl. 1):105–112.

Krieger, K. A. and L. S. Ross. 1993. Changes in benthic macroinvertebrate community of the Cleveland Harbor area of Lake Erie from 1978 to 1989. Journal of Great Lakes Research 19:237–249.

Leach, J. H. 1993. Impacts of the zebra mussel (Dreissena polymorpha) on water quality and fish spawning reefs in Western Lake Erie. In Nalepa, T. F. and D. W. Schloesser eds. Zebra Mussels: Biology, Impacts, and Control. Lewis Publishers. Ann Arbor. 810 p.

———, S. J. Nepszy. 1976. The fish community in Lake Erie. Journal of the Fisheries Research Board of Canada 33:622–638.

Locci, A. B. 1988. Comparative Study of the Response of Simulated Temperate and Tropical Fish Communities to Fishery Exploitation and Environmental Variability. Ph.D Thesis. Case Western Reserve University, Cleveland, Ohio. 210 p.

Ludwig, D., R. Hilborn, and C. Walters. 1993. Uncertainty, resource exploitation, and conservation: Lessons from history. Science 260:17.

Mills, E. L., J. H. Leach, J. T. Carlton, and C. L. Secor. 1993. Exotic species in the Great Lakes: a history of biotic crises and anthropogenic introductions. Journal of Great Lakes Research 19:1–54.

Murdoch, W. W. 1973. The functional response of predators. Journal of Applied Ecology 10:335–342.

Nicholls, K. H. and G. J. Hopkins. 1993. Recent changes in Lake Erie (North shore) phytoplankton: Cumulative impacts of phosphorus loading reductions and the zebra mussel introduction. Journal of Great Lakes Research 19:637–647.

Parrish, D. L. and J. F. Margraf. 1990. Interactions between white perch (Morone americana) and yellow perch (Perca flavescens) in Lake Erie as determined from feeding and growth. Canadian Journal of Fisheries and Aquatic Sciences 47:1779–1787.

Parson, J. W. 1971. Selective food preferences of walleyes of the 1959 year class in Lake Erie. Transactions of the American Fisheries Society 100:474–485.

Price, J. W. 1963. A study of the food habits of some Lake Erie fish. Bulletin of the Ohio Biological Survey 2(1). 89p.

Rathke, D. E. and G. McRae. 1989. 1987 Report on Great Lakes Water Quality, Appendix B, Great Lakes Surveillance. Vol. III. Great Lakes Water Quality Board Report to the International Joint Commission. Windsor, Ontario.

Regier, H. A. and W. L. Hartman. 1973. Lake Erie's fish community: 150 years of cultural stresses. Science 180:1248–1255.

Shuter, B. J., J. F. Koonce, and H. A. Regier. 1979. Modeling the Western Lake Erie Walleye Population: A Feasibility Study. Great Lakes Fishery Commission Technical Report, Ann Arbor, Michigan.

Smith B.R. 1971. Sea lampreys in the Great Lakes of North America. In Hardisty, M. W. and L. C. Potter ed. The Biology of Lampreys. Vol. 1. Academic Press. New York. 207–247.

Trautman, M. B. 1981. The Fishes of Ohio. Ohio State University Press. Columbus, Ohio.

Vallentyne, J. R. and N. A. Thomas. 1978. Fifth Year Review of Canada-United States Great Lakes Water Quality Agreement. Report of Task Group III, A Technical Group to Report Phosphorus Loadings, to the Parties of the Great Lakes Water Quality Agreement of 1972. International Joint Commission. Windsor, Ontario.

Van Meter, H. D. and M. B. Trautman. 1970. An annotated list of the fishes of Lake Erie and its tributary waters exclusive of the Detroit River. Ohio Journal of Science 70:65–78.

Walters, C. J. 1987. Nonstationarity of production relationships in exploited populations. Canadian Journal of Fisheries and Aquatic Sciences 44:156–165.

Ware, D. M. 1982. Power and evolutionary fitness of teleosts. Canadian Journal of Fisheries and Aquatic Sciences 39:3–13.

Wu, Lin and D. A. Culver. 1991. Zooplankton grazing and phytoplankton abundance: an assessment before and after invasion of Dreissena polymorpha. Journal of Great Lakes Research 17:425–436.

CHAPTER 14

Lake Trout in the Great Lakes: Basinwide Stock Collapse and Binational Restoration

Michael J. Hansen

Introduction

The Laurentian Great Lakes contain the largest collection of surface freshwater in the world, and once supported commercial fisheries that were commensurate with their great size. In pre-settlement times, the lakes contributed substantially to the sustenance of native peoples that lived in the region. Following settlement by Europeans, the lakes sustained fisheries that initially supported local markets, and later, as preservation and transportation systems were developed, regional and national markets. Because of their great size, the Great Lakes were thought to be relatively immune to human impact, so fisheries developed unencumbered by regulations, and settlement occurred with little regard for cultural impacts on the lakes or their surrounding drainage basins. As a consequence of unbridled fishery development, introductions of exotic species, and alterations in aquatic habitats, most Great Lake fisheries underwent dramatic, and sometimes catastrophic, changes.

The lake trout (*Salvelinus namaycush*) was important to the human settlement of each of the Great Lakes, and underwent catastrophic collapses in each lake in the nineteenth and twentieth centuries. The timing of lake trout stock collapses were different in each lake, as were the causes of the collapses, and have been the subject of much scientific inquiry and debate. The purpose of this chapter is to summarize and review pertinent information relating to historical changes in Great Lakes lake trout stocks, binational efforts to restore those stocks, and progress toward stock restoration. The presentation will attempt to generalize patterns across the Great Lakes, rather than to focus within each lake. Lake specific analyses have been used to understand lake specific causes and effects, but there is continuing debate about some of these causes and effects. A basinwide review may suggest mechanisms for observed changes that are not evident by lake specific analysis.

Phenotypic Diversity of Lake Trout

The fish fauna of the Laurentian Great Lakes arose within the past ten thousand years, following the Pleistocene glaciation (Lawrie and Rahrer 1973), and includes 28 families, 71 genera, and 174 species (Bailey and Smith 1981). The ciscoes and whitefish (genus *Coregonus*) differentiated into the greatest number of species (Todd and Smith 1980; Smith and Todd 1984), and the lake trout differentiated into many morphological forms of widely-debated taxonomic status (Khan and Qadri 1970; Goodier 1981; Krueger and Ihssen 1995). In any case, there has been general agreement that lake trout formed discrete spawning stocks that used tributary streams, offshore shoals, and rocky shorelines throughout the Great Lakes, particularly in the upper three lakes (Berst and Spangler 1973; Lawrie and Rahrer 1973; Wells and McLain 1973).

Accounts of the nineteenth and early twentieth century fisheries of the Great Lakes often described several forms of lake trout, including a lean form that lived mostly in shallow water, a fat form that lived mostly in deep water, and an intergrade between the shallow-water and deep-water forms (Brown et al. 1981; Goodier 1981). Agassiz (1850) had described the siscowet, or fat form, as a subspecies of lake trout in Lake Superior, based on its deep body and rounded snout, as compared to the slimmer body and pointed snout of the lean form. Fat forms of lake trout were also described in northern Lake Michigan and northern Lake Huron, but were not regarded as siscowets in later reports (Brown et al. 1981). Jordan and Evermann (1904) stated that siscowet lake trout occurred in Lake Huron and Lake Erie, but made no reference to its occurrence in Lake Michigan. Brown et al. (1981) felt that persistent reports of fat lake trout in Lake Michigan during the 1800s indicated the presence of siscowets there. Eshenroder et al. (1995) argued that siscowet lake trout likely occurred throughout the upper Great Lakes because of post-glacial connections between the lakes.

Diverse lake trout forms persisted in the Great Lakes until stocks collapsed after 1900 (Krueger and Ihssen 1995). Restoration efforts (sea lamprey (*Petromyzon marinus*), control; stocking; and fishery regulation) saved a few stocks of offshore lake trout in Lake Superior, but virtually all lake trout were eliminated in the other lakes. Three lake trout forms still survive in Lake Superior, including lean lake trout that also inhabit most oligotrophic lakes in North America, and siscowet and humper lake trout that are endemic to the Great Lakes. Lean lake trout are slender with a low body fat content, have straight, pointed snouts, and inhabit depths <73m (<40 fathoms). Siscowet lake trout are robust with a high fat content, have blunt snouts, and inhabit basins 50m to 150m (27 to 82 fathoms) deep. Humper lake trout have large eyes, thin abdominal walls, intermediate fat content, and inhabit isolated shoals that are surrounded by basins >100m (>55 fathoms) deep.

Collapse of Great Lakes Lake Trout Stocks

Native peoples were the primary human users of Great Lakes fish resources prior to settlement by Europeans, and though they fished with spears, hooks, traps, and gill nets that were primitive by modern standards, their exploitation of local fish stocks may have been substantial (Lawrie and Rahrer 1973). Lake trout were important in these early fisheries, along with lake whitefish (*Coregonus clupeaformis*), and lake herring (*Coregonus artedi*), especially those that spawned in rivers where they were relatively easy to intercept. Early explorers of the seventeenth and eighteenth centuries also caught fish for their own subsistence. The size of these presettlement fisheries is not quantified, but probably amounted to only a small fraction of the size of postsettlement fisheries.

Europeans settled the Great Lakes basin from east to west, so their fisheries developed earlier in the lower lakes, Ontario and Erie, than in the upper lakes, Michigan, Huron, and Superior. In the lower lakes, lake trout were substantially reduced by 1900, though they increased during the 1920s in both lakes (Cornelius et al. 1995; Elrod et al. 1995). Total yields of lake trout from the lower lakes were never as great as in the upper lakes, because of the smaller size of the two lower lakes, and in the case of Lake Erie, the limited amount of habitat suitable for lake trout. In Lake Ontario, lake trout were one of the first species sought by early fisheries, but were rare by 1860 and virtually absent by 1885 (Koelz 1926). Lake trout increased in Lake Ontario in the 1920s, only to collapse in the 1930s and 1940s (Christie 1973). In Lake Erie, lake trout habitat was restricted to the eastern basin, so early lake trout fisheries were small compared to other species (Koelz 1926). Catches of lake trout from Lake Erie were greatest in the late 1800s, and have been very low since 1900.

In the upper Great Lakes, European colonization progressed more slowly in the late 1700s and early 1800s than in the lower lakes. Outposts and forts facilitated the development of fisheries engaged in the trade of salted fish in barrels. Lawrie and Rahrer (1973) estimated that the annual yield of such outpost fisheries in Lake Superior was 45,000kg to 91,000kg (100,000lbs to 200,000lbs.), principally of lake trout and lake whitefish. The fishery in Lake Superior grew slowly during this period, and peaked at about 450,000kg (1,000,000lbs.) in 1839, just before the failure of the American Fur Company, which owned and operated the major outposts. Similar fisheries operated in Lakes Huron and Michigan during the same period, and as in Lake Superior, were limited primarily by technologically primitive fishing vessels, fishing gears, shipping methods, and preservation methods (Berst and Spangler 1973; Wells and McLain 1973).

Innovations in harvesting and transportation methods in the late 1800s facilitated rapid growth of lake trout fisheries in the upper Great Lakes. Steam-powered fishing tugs were introduced in 1871 and steam-powered gill net lifters in 1890 (Berst and Spangler 1973; Lawrie and Rahrer 1973; Wells and McLain 1973). Fishing operations were therefore able to travel further away from ports of call, and were able to fish greater amounts of net than ever before. Marketing of fish was improved during the period by packing fish on ice taken from the lakes in the winter, and then transporting them to markets by freighter. Fisheries grew rapidly in each lake in the 1870s and 1880s, as the fisheries expanded from local, inshore fishing grounds to more distant, offshore fishing grounds (fig. 1).

Lake trout yield from Lakes Huron and Michigan declined slowly from the late 1800s through the early 1900s, and then declined sharply in each lake thereafter, because of fishery over-exploitation and sea lamprey predation (fig. 1). Lake Michigan generally accounted for greater lake trout yield than either Lakes Huron or Superior, and yielded over 4.1 million kilograms (9.0 million pounds) of lake trout in 1896, the largest single-year yield of lake trout from any lake on record. Yield was higher in Lake Michigan from 1929 to 1943, due to greater fishing intensity, as catch per unit of effort (CPE) was higher in both gill net and set hook fisheries in both Lakes Huron and Superior (Hile et al. 1951b). Lake trout yield was sustained through 1935 in Lake Huron, 1943 in Lake Michigan, and 1950 in Lake Superior, after which yield collapsed in each lake (Baldwin et

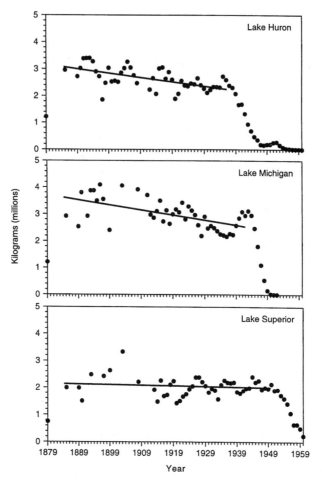

FIG. 1. *Yield of lake trout to commercial fisheries in each of the Laurentian Great Lakes (data from Baldwin et al. 1979).*

al. 1979). Lake trout yield fell from 2.7 million to 0.18 million kilograms (6.0 million to 0.39 million pounds) during the 1935 to 1947 period in Lake Huron, from 3.1 million to 0.16 million kilograms (6.9 million to 0.34 million pounds) from 1943 to 1949 in Lake Michigan, and from 2.1 million to 0.23 million kilograms (4.7 million to 0.50 million pounds) from 1950 to 1960 in Lake Superior. Lake trout stocks persisted in only two isolated locations in Lake Huron (Berst and Spangler 1973) and several, mostly offshore locations in Lake Superior (Lawrie 1978).

Causes of Stock Collapse

Settlement of the Great Lakes basin brought about massive changes in its fisheries through excessive fishery exploitation, introduction of deleterious non-native species, and alteration of supporting habitats (Berst and Spangler 1973; Christie 1973; Hartman 1973; Lawrie and Rahrer 1973; Wells and McLain 1973). The lake trout was one of the most highly valued Great Lakes fish species, so excessive fishery exploitation has been implicated in the demise of lake trout in all five of the

Great Lakes. The lake trout was also a prime target for the exotic sea lamprey, which invaded the lakes after 1900, so excessive sea lamprey-induced mortality has also been implicated in the demise of lake trout in each lake. Alteration of lake trout habitat has been less often blamed, though damming of rivers has been implicated in the demise of river spawning lake trout stocks.

The relative importance of fishery exploitation, sea lamprey predation, and habitat degradation on collapses of Great Lakes fish stocks remains ambiguous. For example, Langlois (1941) suggested that environmental stress resulting from poor land-use was the primary cause of declining fish yield in Lake Erie. Van Oosten (1948) challenged Langlois' claim, and asserted that fish yield in Lake Erie had declined largely because of unregulated fishing. Few disputed the role of fishing in the demise of lake trout stocks in the upper Great Lakes until Coble et al. (1990, 1992) reanalyzed existing fishery data and concluded that sea lampreys were the primary cause of stock collapses in Lakes Michigan and Huron. Eshenroder (1992) challenged Coble et al.'s assertion, because sea lampreys that first appeared in Lake Huron in the 1930s could not have been abundant enough to effect a decline in lake trout abundance by 1938, when lake trout stocks began to decline.

Fishery Exploitation

Early native people's fisheries may have affected local stocks of lake trout and other species that spawned in certain streams and rivers, but were unlikely to have been large enough to cause widespread depletions of any Great Lakes fish species, including lake trout. As commercial fisheries developed, however, they likely exerted increasingly greater impact on the species of interest. Koelz (1926) noted that lake trout were generally not sought when lake whitefish were readily caught, because lake whitefish were more valuable in commercial markets. For example, lake whitefish dominated fish yields in the late 1800s in each of the Great Lakes except Lake Erie, where lake herring predominated and lake whitefish were second. By 1922, however, catches of lake whitefish had declined to seventh among all Great Lakes species, and catches of lake trout had risen to third (behind ciscoes and blue pike (*Stizostedion vitreum glaucum*).

Lake trout were unimportant in Lake Ontario and Lake Erie fisheries in 1922, and were declining, though still relatively abundant, in the upper Great Lakes fisheries (Koelz 1926). For example, Koelz (1926 p. 564) noted that in Lake Michigan: "the species is apparently maintaining itself and is in no danger of extermination, though it is by no means as abundant as formerly." In Lake Superior, Koelz (1926 p. 574) described a more serious situation for lake trout:

> Trout are now less abundant than formerly, according to the testimony of the fishermen and as indicated by the census returns. On the American shore, census figures show that there has been a marked decrease since 1903. Though the registered amount of apparatus has been about the same during the period, the production given for each of the last three census years has been only a little over half of that recorded in 1903. In Canadian waters the fact that the amount of apparatus used is decreasing in the face of higher prices is a good indication of a decrease in abundance.

In Lake Huron, the species may have been better off, though Koelz (1926 p. 582) noted that "there are no longer any important fisheries for trout in sixty fathoms [110m] and deeper, and chub [deepwater ciscoes] nets set at these depths take relatively few small individuals."

In the absence of quantitative data on fishery catch and effort, qualitative descriptions of the impact of fishery exploitation on fish stocks, such as those provided by Koelz (1926), were often used to implicate fishing in the collapse of lake trout stocks in the Great Lakes. In Lakes Erie and Ontario, the role of fishing in the collapse of lake trout stocks will necessarily remain vague because stocks collapsed prior to the period when data describing fishery catch and effort were systematically collected (1931 in Ohio, 1946 in Ontario, 1950 in New York and Pennsylvania; Hile 1962). As in the lower lakes, catch reporting was instituted too late in Minnesota (1947), Illinois (1950), or Indiana (1951) (Hile 1962). In the upper Great Lakes, however, lake trout stocks collapsed after mandatory reporting of catch and effort was instituted for commercial fisheries in Michigan (1927) and Wisconsin (1936) (Hile 1962). The role of fishing in lake trout stock collapses in the upper Great Lakes can therefore be statistically evaluated by analyzing Michigan commercial fishery data.

Lake trout stocks collapsed first in Lake Huron, next in Lake Michigan, and last in Lake Superior, and in response, the lake trout fisheries shifted their effort from one lake to the next. In Lake Huron, gill net effort for lake trout was highest from 1929 to 1933 (Hile 1949; fig. 2), a period during which fishing effort for lake whitefish was correspondingly high (Van Oosten et al. 1946). The high level of fishing effort for lake whitefish was largely due to deep trap net fisheries, which caused the collapse of lake whitefish stocks in all areas of the lake after 1932 (Van Oosten et al. 1946). Deep trap nets were more effective than pound nets because they could be set in deep water where lake whitefish concentrated in summer, but were more destructive of undersized lake whitefish than gill nets in such situations because they were often constructed of smaller sized meshes (Van Oosten et al. 1946). To compensate, gill netters would have directed a greater portion of their effort toward lake trout, which were not as vulnerable to deep trap nets as lake whitefish (Van Oosten et al. 1946). The total portion of gill net effort targeted toward lake trout would have increased as lake whitefish stocks declined after 1932, though this shift in targeted effort was not discernable from statistics of the mixed-species gill net fishery (Hile 1949). Consequently, following the 1929 to 1932 period of increased fishing effort, lake trout abundance (i.e., catch-per-effort, CPE) fell gradually from 1933 through 1944, and then, more steeply from 1944 to 1949. Gill-net effort generally declined with abundance from 1933 to 1949, as effort shifted to Lake Michigan where densities were higher.

Lake trout stocks collapsed at different times in different areas of Lake Huron, but area-specific patterns strongly suggest a relationship between collapses of lake trout and lake whitefish. Lake trout stocks began to collapse first in the central part of the lake (MH–2, MH–3, MH–4), then in adjacent areas (MH–1, MH–5), and last, in the extreme southern end (MH–6) (figs. 3 and 4). Lake trout yield from MH–2 was highest (31% of the lake total), and abundance fell erratically after 1932, the year when lake whitefish stocks fell 50% (Van Oosten et al. 1946). In adjacent area MH–3, lake trout yield was next highest (23% of the lake total), and abundance fell after 1934, two years after lake whitefish declined 50% (Van Oosten et al. 1946). In MH–4 and MH–5, lake trout yields were lower (12% and 7% of the lake total), and abundance fell after 1934 and 1937, one and two years after lake whitefish declined 50% (Van Oosten et al. 1946). In MH–1 and MH–6, lake whitefish declined less precipitously, and lake trout stocks persisted longer than in other areas of Lake Huron. Lake trout yield was third highest (22% of the lake total) in MH–1, and abundance fell after

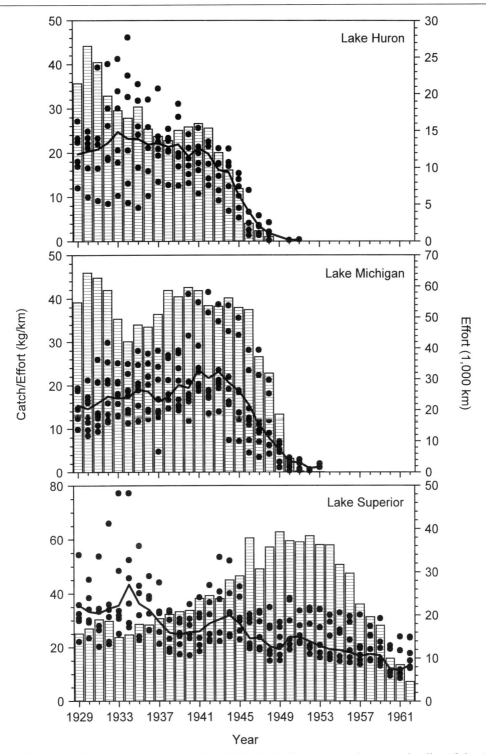

FIG. 2. *Effort (shaded bars) and catch-per-effort (CPE; line) of lake trout in large-mesh gill net fisheries of the Michigan waters of the upper Great Lakes, with CPE shown for each statistical district (dots) (data from Jensen and Buettner 1976).*

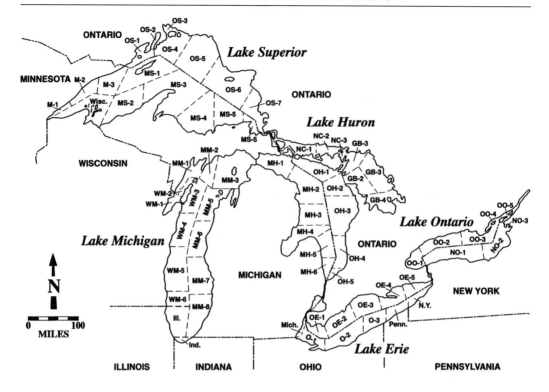

FIG. 3. *Fishery statistical districts of the Great Lakes (from Smith et al. 1961).*

1938, seven years after lake whitefish began to decline (Van Oosten et al. 1946). Lake trout yield was lowest (5% of the lake total) in MH–6, and abundance fell after 1944, nine years after lake whitefish began to decline (Van Oosten et al. 1946).

In Lake Michigan, gill net effort for lake trout was high from 1929 to 1933 (fig. 2), a period when deep trap net effort for lake whitefish was also high, as in Lake Huron (Van Oosten et al. 1946). Deep trap net operations caused lake whitefish stocks to decline in Lake Michigan after 1932, though not as severely as in Lake Huron (Van Oosten et al. 1946). Gill net fishing effort in Lake Michigan rose sharply through 1938, and remained high through 1946. Fishing effort increased in Lake Michigan from 1938 to 1946, as lake trout CPEs declined in Lake Huron and effort shifted into Lake Michigan. Lake trout stocks in Lake Michigan withstood this high fishing effort from 1929 through 1943, a period during which their abundance actually increased, but then collapsed between 1943 and 1951. Gill net effort in Lake Michigan declined quickly after 1946, as effort shifted into Lake Superior in pursuit of the last remaining lake trout stocks in the Great Lakes (Hile et al. 1951b).

Lake trout stocks also collapsed at different times in different areas of Lake Michigan, as they did in Lake Huron, and generally lagged behind preceding collapses of lake whitefish in the various areas by more than a decade. For example, lake whitefish stocks collapsed in every area of Lake Michigan after 1929 to 1931 (Van Oosten et al. 1946), but lake trout collapsed in every area after 1941 to 1943 (figs. 3 and 5). In Lake Michigan, the timing of both species' collapse was more uniform across areas and less related to the relative yield from each area than in Lake Huron. Hile

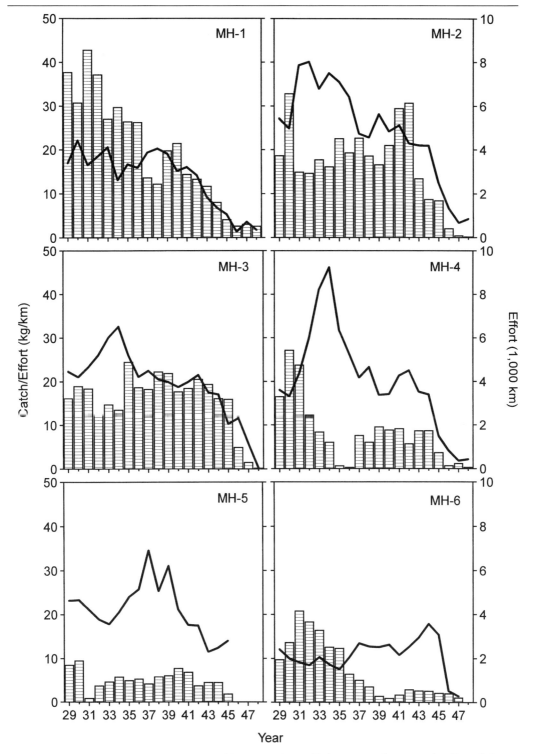

FIG. 4. *Effort (shaded bars) and catch-per-effort (CPE; line) of lake trout in large-mesh gill net fisheries of the six Michigan statistical districts of Lake Huron (data from Jensen and Buettner 1976; see figure 3 for location of statistical districts).*

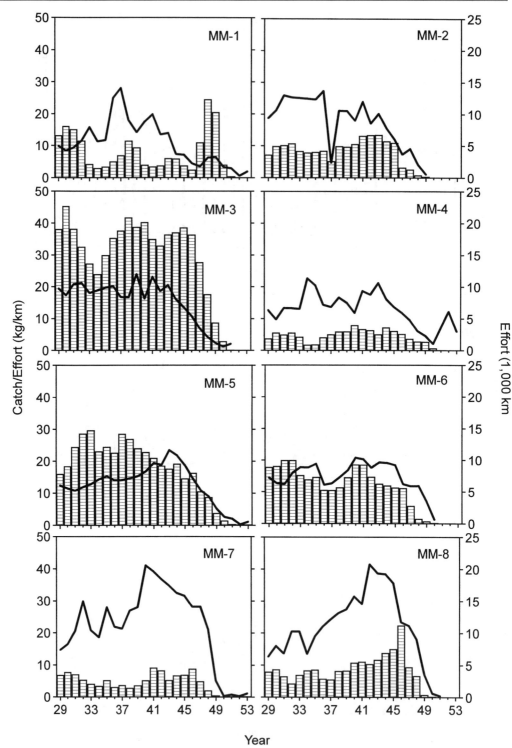

FIG. 5. *Effort (shaded bars) and catch-per-effort (CPE; line) of lake trout in large-mesh gill net fisheries of the eight Michigan statistical districts of Lake Michigan (data from Jensen and Buettner 1976; see figure 3 for location of statistical districts).*

et al. (1951a) noted that, though lake trout stocks began to collapse at approximately the same time throughout Lake Michigan, abundance reached low levels more rapidly in northern areas than in southern areas.

In Lake Superior, gill net effort for lake trout generally rose from 1929 through 1946, as lake trout stocks collapsed in Lake Huron and in Lake Michigan (fig. 2; Hile et al. 1951b). Gill net effort remained high from 1946 through 1954, but fell steeply thereafter. Lake trout abundance generally fell from 1929 through 1962, with cyclical peaks in 1933 and 1944. Gill net effort remained high in Lake Superior from 1946 to 1954, despite declining lake trout abundance, because Lake Superior was the last stronghold for lake trout in the Great Lakes. Lake trout fisheries were closed throughout the upper Great Lakes in 1962 to protect the last remaining, mostly remote stocks in Lake Superior.

Lake trout stocks exhibited erratic, persistent declines in abundance from 1929 to 1961 in all Michigan areas of Lake Superior, except at Isle Royale (MS–1) and adjacent to Wisconsin (MS–2) (figs. 3 and 6). Declining abundance was likely aided by high fishing effort, which increased in all areas but MS–1 in the 1940s. Fishing effort increased most sharply around the Keweenaw Peninsula (MS–3) and Marquette (MS–4), as the fisheries responded to the collapse of lake trout stocks in Lakes Huron and Michigan, and remained high through the early 1950s, before declining sharply. Unlike Lakes Huron and Michigan, lake trout stocks in Lake Superior did not decline precipitously, except in MS–1. In MS–1, lake trout stocks declined sharply after 1950, but not likely because of the fishery, because effort was exceptionally stable throughout the 1929 to 1956 period. In MS–2, lake trout stocks did not decline, perhaps because fishing effort was low during the 1929 to 1948 period, and declined thereafter.

Excessive fishery exploitation was blamed for the collapse of lake trout stocks in Lake Superior in the 1950s (Hile et al. 1951b), but not in Lakes Michigan (Hile et al. 1951a) or Huron (Hile 1949). In Lakes Huron and Michigan, yield corresponded well with abundance because fishing effort shifted across lakes in pursuit of remaining lake trout stocks. In contrast, yield in Lake Superior was unrelated to abundance because abundance was being depressed by increasing fishing intensity, which had nowhere else to move:

> The upward trend of fishing pressure and the high negative correlation between abundance and fishing intensity . . . invite the belief that the two may be related causally—that rising fishing pressure has brought about a decrease of abundance and that the fishermen in the face of this reduced availability have intensified their efforts in order to keep production at a good level. If so, a continuation of present trends until fishing becomes unprofitable is to be anticipated. On the other hand, it is possible that the rising fishing pressure has been merely the result of a natural decline in abundance but has not contributed materially to it. In this event, we can hope for a natural upturn of abundance which would relieve the precarious status of the fishery. Regardless of the causes that led to the present situation, the fact remains that the lake trout stocks . . . of Lake Superior are fast nearing a dangerously low level and are in poor condition to withstand the impending ravages of a growing population of sea lampreys (Hile et al. 1951b p. 311).

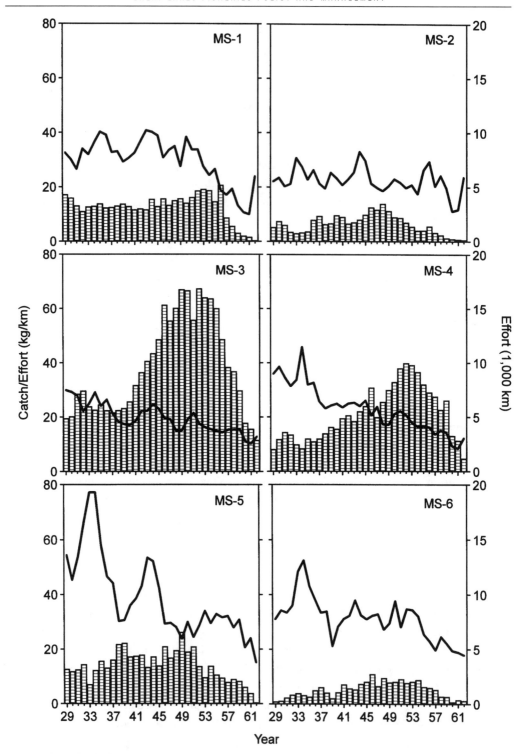

FIG. 6. *Effort (shaded bars) and catch-per-effort (CPE; line) of lake trout in large-mesh gill net fisheries of the six Michigan statistical districts of Lake Superior (data from Jensen and Buettner 1976; see figure 3 for location of statistical districts).*

Needless to say, a natural upturn of lake trout abundance did not occur in Lake Superior, as Hile et al. (1951b) hoped for, and the stocks continued to decline through the early 1960s. Pycha and King (1975) expanded the analysis of commercial fishery data through 1970, and into Wisconsin's waters of Lake Superior, and concluded that "intensive fishing, aided by the introduction of nylon gillnets, was the principal factor involved in the early years of the post-World War II decline of lake trout in Lake Superior." (Pycha and King 1975 p. 29)

Sea Lamprey Predation

Sea lampreys invaded Lake Ontario through the Erie Canal by 1835, but degraded stream conditions prevented them from becoming abundant until after the early 1900s (Smith 1995). Construction of the Welland Canal in 1829 permitted sea lampreys to reach Lake Erie from Lake Ontario, but the first sea lamprey was not found in Lake Erie until 1921, and the first spawning run was not observed until 1932 (Smith 1971). Colonization of the upper Great Lakes was more rapid, as sea lampreys were first found in Lake St. Clair in 1934, Lake Huron in 1937, Lake Michigan in 1936, and Lake Superior in 1946. The date of first encounter in Lake Michigan, however, suggests that sea lampreys made their way into Lake Huron before they were first found in Lake Michigan, perhaps as early as 1932. Spawning runs of sea lampreys were broadly monitored only in Lake Superior, but long-term monitoring of runs in the Ocqueoc River in Lake Huron and Hibbards Creek in Lake Michigan suggest that numbers of adult sea lampreys were too low to have caused initial declines of lake trout in any of the upper Great Lakes (fig. 7) (Smith 1968). In spite of this data, however, few researchers have agreed on the relative impacts of sea lamprey and fishing on lake trout stocks (see Coble et al. 1990 for a review).

The relative impact of fisheries and sea lampreys on lake trout stocks in the upper Great Lakes is unclear because of the relative timing of the arrival of sea lampreys in the lakes and the onset of stock declines. In Lake Huron, for example, Hile (1949) found that declining commercial fishing effort likely accounted for declining catches after 1935, rather than declining abundance of the stocks. Berst and Spangler (1973) also failed to establish a definitive link between fishing and lake trout stock collapse in Lake Huron, but nonetheless, attributed some of the blame for that collapse to overfishing. Simulation of Lake Huron lake trout stocks implicated fishing in the stock collapse, but implied that sea lampreys would also have collapsed the stocks in the absence of fishing (Lett et al. 1975). Coble et al. (1990, 1992) concluded that commercial fishery exploitation did not cause lake trout stocks to collapse, but Eshenroder (1992) concluded that they misinterpreted the time of arrival of sea lampreys relative to the time when overfishing began to depress lake trout stocks.

A critical point of any analysis of Lake Huron lake trout is the establishment of the year when lake trout stocks began to decline, relative to the year when sea lamprey arrived and became abundant enough to influence lake trout abundance. Clearly, lake trout were virtually gone from Lake Huron by the late 1950s, when fishery yield dropped to near zero (Baldwin et al. 1979), but the year when stocks began to decline is less clear. Coble et al. (1990) successively fit linear relationships to lake trout CPE statistics in the Michigan gill net and set hook fisheries in Lake Huron until the slope became negative and significant. They then reasoned that the stocks began to decline in the year when their test became negative and significant; 1938 in the set hook fishery and 1943 in

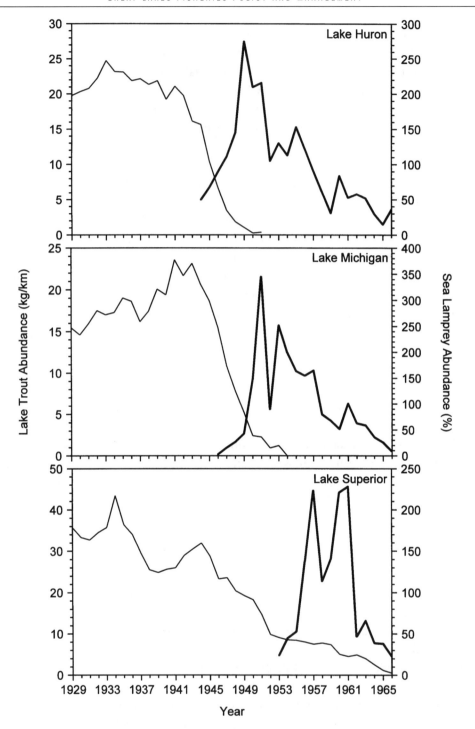

FIG. 7. *Lake trout catch-per-effort (CPE; line) in large-mesh gill net fisheries of the Michigan waters of the upper Great Lakes (data from Jensen and Buettner 1976), compared to indices of sea lamprey abundance in each lake (data from Smith 1968). Lake Superior lake trout CPE after 1950 was adjusted for the conversion from cotton to nylon twine as described by Pycha and King (1975).*

the gill net fishery. However, their use of a linear model to describe the relationship between CPE data for gill net and set hook fisheries was incorrect because the residuals from the linear model for these data clearly show that the errors are autocorrelated and curve linear. They also tied their linear regressions to the year 1929 (the first year of CPE data), in spite of the fact that CPE actually increased from 1929 to 1933, which therefore led to the erroneous result that stocks began to decline in 1943! By 1947, only four year later, Hile (1949) was able to find only eleven lake trout specimens in all of the fish houses on Lake Huron!

Sea lampreys were first found in Lake Huron in 1937, spawning in the Ocqueoc River, a tributary in the northern part of the lake. These spawning adults were likely the first progeny of the initial sea lampreys that colonized the lake in 1932 (Eshenroder 1992). Their abundance rose quickly thereafter; by 1944, when systematic trapping of adults began, nearly 4,500 were caught, and by 1949, more than 24,000 were caught (fig. 7). However, the decline of lake trout in Lake Huron began much earlier: after 1933 lakewide; after the 1933 to 1936 period in areas MH–1 through MH–5; and after 1945 in area MH–6 (figs. 3 and 4; Hile 1949). Consequently, lake trout stocks were declining in virtually all areas of Lake Huron before the first sea lamprey was reported in the lake. For some reason, Hile (1949) did not infer that fishing had caused lake trout stocks to decline in Lake Huron, even though a similar sequence of events in the lake whitefish fishery had led him and his colleagues to infer that fishing had caused lake whitefish stocks to decline a few years earlier (Van Oosten et al. 1946).

In Lake Michigan, sea lampreys were first found in 1936, but were not systematically monitored until 1946, and then only in Hibbards Creek, a tributary in the northern part of the lake on Wisconsin's Door Peninsula (fig. 7). Smith (1960) reasoned that the Hibbards Creek run was a reasonable surrogate for lakewide trends because of its consistency with data from several other streams that were monitored beginning in 1955. Sea lamprey abundance, as indicated by the Hibbards Creek spawning run, was very low at the time when lake trout stocks were in a state of collapse in Lake Michigan. Sea lampreys increased in Hibbards Creek from only 125 in 1946 to more than 12,000 in 1951, but by then, lake trout stocks were virtually extirpated throughout Lake Michigan (figs. 3 and 5; Eschmeyer 1957). By 1955, when three more streams were systematically trapped, the spawning run in Hibbards Creek had dropped to about 6,400, and more than 18,000 were trapped in the Bark, Cedar, and Sturgeon rivers.

In Lake Superior, lake trout abundance began to decline before sea lampreys were first observed or were numerous enough to have caused the collapse of lake trout stocks (fig. 7; Hile et al. 1951b; Pycha and King 1975). Jensen (1978) used a surplus production model to analyze commercial fishery data from central Michigan waters of Lake Superior, and confirmed that overfishing of lake trout stocks began in 1945, whereas sea lampreys were first reported in 1946. Coble et al. (1990) used regression analysis of trends in commercial fishery data, and determined that lake trout abundance began to decline lakewide as early as 1939 in the Michigan gill net fishery, 1949 in the Wisconsin gill net fishery, and 1950 in the Michigan hook fishery. Trends in gill net CPE suggest that lake trout stocks declined after 1929 in MS–3, and 1934 in MS–4, MS–5, and MS–6 (figs. 3 and 6). However, fishing cannot account for the collapse of stocks in MS–1, where lake trout abundance declined after 1950, and was unrelated to any change in fishing effort. Sea lampreys undoubtedly caused the decline of lake trout abundance in MS–1 after 1950 (Curtis et al. in press).

The most reasonable interpretation of the respective roles of fishing and sea lampreys in the demise of lake trout in the upper Great Lakes is that both were important. However, this interpretation is only clear when the lake trout fishery is viewed in context with the lake whitefish fishery and across the upper three lakes. The collapse of the lake whitefish fishery in Lake Huron in the mid 1930s put added pressure on lake trout stocks in Lake Huron as the fishery attempted to compensate for declining income. Lake trout stocks were unable to withstand this added pressure, however, and declined in the late 1930s and early 1940s. Sea lamprey colonized Lake Huron coincident with the collapse of lake trout stocks, thereby assuring the virtual extirpation of lake trout in the lake.

In Lake Michigan, lake whitefish had also declined, though not as severely as in Lake Huron, and markets for lake trout remained strong, so fishing effort for lake trout in Lake Michigan increased in the late 1930s and early 1940s (Hile et al. 1951a). As in Lake Huron, lake trout stocks in Lake Michigan were unable to withstand the added pressure, and declined in the late 1940s. Sea lampreys also colonized Lake Michigan as lake trout stocks collapsed, thereby assuring the extirpation of lake trout in that lake as well. Markets for lake trout remained good, so fishing effort for lake trout increased in Lake Superior in the late 1940s (Hile et al. 1951b). Again, lake trout stocks were unable to withstand the added fishing pressure, particularly because of the improved efficiency of nylon nets that were used after 1950 (Pycha 1962), and because the fishery could not shift to another lake as stocks collapsed. Once again, sea lampreys colonized Lake Superior as lake trout stocks collapsed, but chemical control of sea lamprey larvae in spawning streams took effect before lake trout stocks were completely eradicated.

Habitat Degradation

Habitat degradation has occurred to a certain extent in each Great Lake, including destruction of stream and inshore spawning habitat; eutrophication and consequent oxygen depletion of open waters; loading by organic contaminants and consequent disruption of spawning viability; and colonization by exotic species and consequent alteration of fish communities. Direct effects on lake trout from some of these habitat degradations have been difficult to impute, but the early history of Lake Ontario provides an eloquent portrayal of the impact of drainage degradation on stream and inshore spawning habitats for lake trout:

> The Lake Ontario watershed was settled early and developed thoroughly. It has been subjected to all of the stresses man applies to his water supplies. The alteration of the fish environment began with the clearing of the land. The streams became warmer and siltier and the near shore bottom sediments were probably changed in character. Dams were built in the streams in order to operate sawmills and gristmills. The sawmills ran until the land was cleared, and the water-powered gristmills ran until the drainage from the cleared land became too light in the fall to operate them. The dams not only warmed the water, but also blocked fish migrations—which in turn prevented reproduction and allowed increased exploitation (Christie 1973 p. 2).

Because of these degradations, stream-spawning fish stocks disappeared from Lake Ontario in the 1800s, including Atlantic salmon (*Salmo salar*) in the 1830s, ciscoes in the 1860s, and lake trout, lake whitefish, and burbot (*Lota lota*) in the 1890s (Christie 1973). Similar changes in stream and near shore spawning habitat also occurred in Lake Erie (Hartman 1973), and to a limited extent in the upper Great Lakes (Berst and Spangler 1973; Lawrie and Rahrer 1973; Wells and McLain 1973).

Eutrophication and consequent oxygen depletion of open waters was of greatest importance in Lake Erie (Hartman 1973). Eutrophication caused anoxia in the hypolimnion of Lake Erie's central and western basins in 1953, and subsequently, eliminated oxygen-sensitive burrowing mayfly nymphs (*Hexagenia* spp.), a major prey of many Lake Erie fishes, along with the oversummer sanctuary for cold stenotherms such as lake trout, lake herring, lake whitefish, and blue pike (Hartman 1973). Eutrophication has been less severe in the other Great Lakes, such that effects were only felt in near shore areas and major embayments where lake trout were not major components of fish communities (Berst and Spangler 1973; Christie 1973; Lawrie and Rahrer 1973; Wells and McLain 1973).

Organic contaminants have been suggested to cause reproductive problems in lake trout because of coherence of trends in ambient contaminant levels, egg hatchability, and fry survival; however, definitive proof of effects on lake trout reproduction in the wild has been lacking because several variables are frequently confounded in the field (Mac and Edsall 1991). Levels of PCBs and DDE in Lake Michigan in the early 1970s were high enough to increase mortality of lake trout fry (Berlin et al. 1981), and to increase preferred temperatures of fry (Mac and Bergstedt 1981), but did not affect egg hatching success (Mac et al. 1981), fry swimming performance (Rottiers and Bergstedt 1981), or vulnerability of fry to predation (Mac 1981). In 1980, lake trout fry hatched from eggs of Lake Michigan parents, the most highly contaminated among all Great Lakes adult lake trout, experienced elevated mortality at the swim up stage, compared to eggs from other Great Lakes origins (Mac et al. 1985; Mac 1988). Mortality rates of lake trout embryos and fry in the 1980s were higher from parents captured in years when, or locations where, PCBs were lowest (Mac et al. 1993). However, thiamin deficiency was found to be responsible for elevated mortality of lake trout fry in Lake Ontario, rather than contaminant burden (Fitzsimons et al. 1995).

Colonization of the Great Lakes basin by exotic species has altered fish communities in each lake, but has not apparently affected lake trout in a detectable way. The alewife (*Alosa pseudoharengus*) and sea lamprey colonized the Great Lakes basin in the same way, and alewives were first found in each lake within a decade of sea lampreys (Smith 1972). The alewife became dominant in Lakes Huron, Michigan, and Ontario, where temperatures were moderate and top predators were virtually nonexistent, but not in Lake Superior, where temperatures were too cold, or in Lake Erie, where abundant predators existed (Smith 1972). Abundance of native planktivores declined in Lakes Huron, Michigan, and Ontario as the alewife became dominant through competition for zooplankton or through predation on pelagic larvae (Smith 1972). Native planktivores were subsequently replaced by alewives as the dominant prey of reintroduced lake trout in Lakes Huron (Johnson et al. 1995), Michigan (Stewart and Ibarra 1991) and Ontario (Brandt 1986).

The rainbow smelt (*Osmerus mordax*), in contrast to the alewife, colonized the upper Great Lakes basin from Crystal Lake, which drains into Lake Michigan, and Lake Ontario from the St. Lawrence River (Smith 1972). The rainbow smelt was therefore found before either the alewife or the sea lamprey in Lakes Michigan, Huron, and Superior, but after both species in Lakes Erie and Ontario (Smith 1972). The rainbow smelt became dominant only in Lakes Erie and Superior, where the alewife did not become dominant (Smith 1972), but overfishing rather than competition with, or predation by rainbow smelt was the likely cause of lake herring population declines in both Lake Erie (Van Oosten 1930) and Lake Superior (Selgeby 1982). The rainbow smelt subsequently became the dominant prey for lake trout in both Lake Erie (Forage Task Group 1993) and Lake Superior (Conner et al. 1993).

Stock Restoration Efforts

The sea lamprey invasion of the upper Great Lakes led to interagency cooperation in fishery management and research. The first of such efforts were formalized as the Great Lakes Lake Trout and Sea Lamprey committees in 1946 (Smith 1971). The two committees were merged in 1952, renamed the Great Lakes Fishery Committee in 1953, and then replaced by the Great Lakes Fishery Commission in 1955, which was authorized by the Convention on Great Lakes Fisheries between the United States and Canada to undertake sea lamprey control and coordinate fishery research and management. The United States Fish and Wildlife Service and the Fisheries Research Board of Canada were contracted to perform sea lamprey control. In 1966, the sea lamprey control program in Canada was transferred to the Department of Fisheries (later, the Department of Fisheries and Oceans).

Under the aegis of the Great Lakes Fishery Commission, Lake Committees for each of the Great Lakes, made up of fishery managers from each management agency, were formed to coordinate fishery management on each lake. Each Lake Committee appointed a Lake Technical Committee made up of fishery researchers from each agency to serve as technical advisors on matters of fishery science. The Lake Technical Committees drafted lake trout restoration or rehabilitation plans for each lake that set targets for lake trout stock restoration or rehabilitation, prescribed management actions to enhance recruitment and survival, suggested pertinent topics for fishery research, and provided standards for coordinating stock assessment programs. The lake trout restoration plans for each of the Great Lakes were completed by the mid 1980s.

Stock Restoration Targets

Lake trout restoration targets were developed for each lake based on an understanding of historically sustainable yield, the need to enhance recruitment through stocking, and the need to limit mortality through sea lamprey control and fishery regulation. Restoration plans for each lake had targets for sustainable yield that anticipated a future state that would be at least reasonably similar in lake trout productivity to some historic condition. Targets for stocking were virtually all driven by early evaluations in Lake Superior (see below), but were adapted to each lake based on the distribution and amount of lake trout spawning habitat. Targets for mortality were based on a review by Healy (1978), which suggested that lake trout populations could only remain self-sustaining if total annual mortality was at or below 50%, but were adapted to each lake based on growth and maturity rates for lake trout.

In Lake Superior, the goal of lake trout restoration is to achieve a sustained annual yield of two million kilograms (4 million pounds) from naturally reproducing stocks (LSLTTC 1986). An intermediate objective was to restore recruitment in the best habitat by setting stocking priorities, and to reduce total mortality to levels consistent with rehabilitation. The plan therefore focused on strategies for ensuring that the goal was achieved: stocking yearling lake trout at an average size of 40 to 55 per kilogram (18 to 25 per pound.), derived from wild strains that spawned in the stocking areas, at a rate of 232 to 347 per square kilometer (600 to 900 per square mile) of lake trout habitat; and limiting total annual mortality to 42% (indexed as 50% on ages seven to eleven lake trout caught in 114mm, 4.5in, stretch-mesh gill nets). Stocking rates were reduced for Michigan's waters in 1988, Wisconsin's waters in 1989, and Ontario's waters in 1991 because of increased natural recruitment (Hansen 1988, 1989; Powell and Atkinson 1991).

In Lake Michigan, the goal of lake trout restoration is to achieve a self-sustaining lake trout population able to yield an annual harvest projected conservatively at 500,000 to 700,000 fish weighing 1.1 million kilograms (2.5 million pounds.; LMLTTC 1985). Interim objectives were: within ten years, achieve larger spawning populations that are subject to no more than 40% annual mortality; within fifteen years, demonstrate routinely in trawl and gill net surveys the presence of lake-produced young of several year classes in refuges and high priority rehabilitation areas; and within twenty years, show that spawning stocks of hatchery-origin are being augmented by significant numbers of wild spawners, and that the abundance of wild recruits is accelerating toward a level that will eliminate the need for stocking. The goal was to be achieved by emphasizing the planting of young lake trout in refuges and high priority rehabilitation areas, over the most historically productive spawning grounds, and with concurrent controls on sea lamprey populations and fisheries.

In Lake Huron, the goal of lake trout restoration is to rehabilitate the lake trout populations such that the adult spawning stocks are comprised of several year classes, and are capable of sustaining themselves at relatively stable levels by natural reproduction, and produce usable annual surpluses (LHLTTC 1985). Interim objectives included: by 1990, achieve increased broodstocks subject to no more than 45% total annual mortality in areas where sea lamprey wounding rates exceed 5%, or to no more than 40% total annual mortality in areas where sea lamprey wounding rates are less than 5%; and by 2010, develop self-sustaining lake trout populations that are capable of providing 450,000 kilograms (1 million pounds) of usable surplus in U.S. waters and 450,000 kilograms (1 million pounds) of usable surplus in Canadian waters. The goal was to be achieved by emphasizing the planting of yearling lake trout in highest priority rehabilitation zones over the most promising spawning grounds, and by implementing appropriate controls on sea lamprey populations and fisheries.

In Lake Erie, the goal for lake trout restoration is to restore a naturally reproducing lake trout population in the eastern basin that will eventually yield an annual harvestable surplus (LTTG 1985). Interim objectives included: by 1991, limit total annual mortality to less than 40%; by 2000, build an adult stock of about seventy-five thousand individuals, with adult females that average 7.5 years of age and produce ten-thousand yearling lake trout annually; and by 2020, attain an annual harvestable surplus of 50,000 kg (110,000 lb.) of lake trout and an annual production of 200,000, yearling lake trout through natural reproduction. The goal was to be achieved by: annually stocking two-hundred thousand yearlings of suitable strains; maximizing recruitment of stocked fish; restricting total annual survival to the maximum allowable level; and maximizing reproductive potential of lake trout.

In Lake Ontario, the goal for lake trout restoration is to establish a lake trout population that sustains itself and produces a useable annual surplus (LTS 1983). An interim objective for the year 2000 was to develop a population of 0.5 to 1.0 million adults with females that average 7.5 years old and produce 100,000 yearlings annually. The goal was to be achieved by: stocking 1.25 million yearlings annually in both U.S. and Canadian waters; maximizing recruitment of stocked fish by optimizing cultural techniques and stocking procedures; fostering an annual survival rate of 60% to 65% by suppressing sea lampreys and restricting fishing; and maximizing natural reproduction by identifying and mitigating factors that limit spawning and survival of early life stages. Several different genetic strains were to be stocked initially, and upon evaluation of their performance, stocking of those strains that did not perform well would be reduced or discontinued.

Management Actions

Fishery managers had a relatively short time to react to the collapse of lake trout stocks in the upper Great Lakes, and most managers were convinced that sea lamprey predation was the cause of those stock collapses, rather than overharvest (Hile 1949; Hile et al. 1951a), so lake trout stocks were mostly driven to extirpation before a selective sea lamprey toxicant was discovered in 1957 (Smith 1971; Smith et al. 1974). As described above, lake trout yields collapsed in about one decade in Lake Huron (1935 to 1947), a half-decade in Lake Michigan (1943 to 1949), and one decade in Lake Superior (1950 to 1960). State and federal fishery management agencies in Lake Superior initiated stocking in 1950, implemented sea lamprey control measures in 1953, and closed lake trout fisheries in 1962, before stocks had completely disappeared (Pycha and King 1975). Unfortunately, lake trout were extirpated in the remaining Great Lakes, except for two small isolated stocks in Lake Huron, before any of these fishery management actions were undertaken (Berst and Spangler 1973; Christie 1973; Hartman 1973; Wells and McLain 1973). Management of lake trout in the Great Lakes has therefore focused on the restoration of stocks, even in Lake Superior where some native stocks persisted through the ravages of overfishing and sea lamprey predation.

Stocking

Stocking of hatchery-reared juvenile lake trout was initiated in the early 1950s in Lake Superior to stave off the complete demise of failing stocks (Lawrie and Rahrer 1973; Lawrie 1978). Targets for lake trout stocking programs in the Great Lakes were largely based on analyses of the survival of these early plantings in Lake Superior. For example, natural reproduction of lake trout was reestablished in Keweenaw Bay, Lake Superior, during the late 1960s and the early 1970s when 347 yearlings were stocked per square km (1.5 per acre) of habitat (area <73m, <40 fathoms) for five years, followed by 232 yearlings per square km (1.0 per acre) thereafter (LSLTTC 1986). Another analysis of stocked lake trout survival indicated that yearlings survived best when stocked at an average weight of 23g to 27g (17 to 20 fish per lb.) (Pycha and King 1967).

Lake trout stocking in Lake Superior averaged 330,000 per year since 1951 in Wisconsin, 796,000 per year since 1952 in Michigan, 949,000 per year since 1957 in Ontario, and 333,000 per year since 1962 in Minnesota (Hansen et al. 1995a). These stocking rates were generally lower than recommended because of limited hatchery capacity: 105 yearlings per square km (0.42 per acre) in Wisconsin, 124 per square km (0.50 per acre) in Michigan, 153 per square km (0.62 per acre) in Ontario, and 543 per square km (2.20 per acre) in Minnesota. Yearling lake trout dominated the releases in each jurisdiction, 81% in Wisconsin, 91% in Michigan, 89% in Ontario, and 88% in Minnesota. Yearlings reared in Ontario and Michigan hatcheries were 8g to 10g (0.28oz to 0.35oz) lighter than those reared in U.S. federal hatcheries and 14g to 17g (0.49oz to 0.60oz) lighter than those reared in Minnesota and Wisconsin hatcheries. Yearlings increased 13g (0.46 ounces) per year in average weight from 1964 to 1983 in Lake Superior, but were lighter than the target size of 23g to 27g per fish (17 to 20 fish per lb.) in most years. Only the yearlings reared in Minnesota and Wisconsin hatcheries exceeded the target size of 23g to 27g per fish (17 to 20 fish per lb.).

In Lake Michigan, lake trout stocking began in 1965, when 1.1 million yearlings were stocked (Holey et al. 1995). Stockings increased steadily until the early 1970s, and leveled off at 2.4 million yearlings thereafter. As in Lake Superior, the number of lake trout stocked in Lake Michigan has generally been fewer than the number targeted (3.5 million) because of limited hatchery capacity. From 1964 through 1984, prior to implementation of the lake trout restoration plan, 53% of the lake trout stocked into Lake Michigan were put into what are now designated as secondary or deferred rehabilitation areas to maximize recreational fishing opportunities, whereas only 9% were put into what are now designated as refuges. Since the implementation of the lake trout restoration plan in 1985, however, 90% of lake trout were put into refuges (56%) and primary rehabilitation areas (34%), where lake trout are protected from harvest, whereas only 10% were put into secondary, or deferred, rehabilitation areas, where lake trout are less protected from harvest.

In Lake Huron, where sea lamprey were more numerous than in the other Great Lakes, lake trout reintroduction was to be based on the stocking of splake (hybrid of the brook trout and lake trout, *Salvelinus fontinalis* × *S. namaycush*) because they would grow faster and mature earlier than lake trout in the face of sea lamprey mortality, which tended to be more severe on larger, older fish than on smaller, younger fish (Eshenroder et al. 1995). Ontario stocked splake from 1969 to 1977, backcross of splake and lake trout since 1978, and lake trout since 1981. Michigan stocked splake only in 1970 to 1972, backcross in 1973, and lake trout thereafter. From 1969 through 1993, releases by both jurisdictions totaled 3.6 million splake, 11.6 million backcross, and 27.6 million lake trout, which averaged (1.7 million per year) well below the target rate (4.1 million per year) because of limited hatchery capacity.

In Lake Erie, lake trout restoration was not begun until political furor over relatively high contaminant levels in Lake Ontario allowed lake trout to be diverted from Lake Ontario into Lake Erie in 1978 and 1979 (Cornelius et al. 1995). Lake trout stocked in those two years survived well, and thereby, encouraged fishery managers to develop a lake trout restoration program for Lake Erie. Stockings since 1978 averaged 193,550 per year, which was very close to the number recommended in the lake trout restoration plan for the lake (LTTG 1985). Since 1978, all lake trout have been stocked in the eastern basin of the lake where all of the lake trout habitat is found (Cornelius et al. 1995).

In Lake Ontario, annual stocking of hatchery lake trout began with the release of 66,000 yearlings of the 1972 year class in the United States and 194,000 yearlings of the 1975 year class in Canada (Elrod et al. 1995). Most hatchery fish were stocked as yearlings because they survived better than fingerlings; in the United States fingerlings exceeded 0.4 million fish in only three years, and in Canada fingerlings totaled only 41,400 in 1987 and 195,100 in 1992. In the United States, yearling stockings averaged about one million each year for the 1979 to 1985 year classes, decreased to 0.75 million for the 1986 to 1990 year classes, and decreased again to 0.5 million for the 1991 to 1992 year classes due to a reduced water supply at the hatchery. In Canada, stockings of yearlings averaged about 0.4 million fish for the 1979 to 1982 year classes, increased to 0.85 million for the 1985 year class, and ranged between 0.95 and 1.12 million for the 1986 to 1991 year classes. Because of limited hatchery capacity, stockings in neither the United States nor Canada have ever achieved the target stocking rate of 1.25 million yearlings per year in each jurisdiction (LTS 1983).

Sea Lamprey Control

The earliest attempts to control sea lampreys were by mechanical and electrical weirs and traps (Smith and Tibbles 1980). In 1950 and 1951, mechanical weirs were installed in twelve tributaries to Lake Huron, seven tributaries to Lake Michigan, and two tributaries to Lake Superior. Electrical weirs were tested in 1952, and 162 were installed on tributaries of Lake Superior by 1960. Electrical weirs were used to monitor changes in sea lamprey populations in varying numbers of streams after 1958, but it is unclear whether they effected a significant reduction in sea lamprey populations in any of the Great Lakes. Permanent barriers have also been built on several streams in each lake. Other types of barriers are being tested for integration into the control program, including velocity barriers that are designed to block slow swimming sea lampreys without blocking faster swimming fish, inflatable barriers that are designed to inflate only during periods of sea lamprey spawning migrations, and portable traps that are designed to intercept significant portions of particular spawning runs.

In the search for a chemical means of controlling sea lamprey, about 6,000 possible chemicals were screened before researchers selected 3-trifluoromethyl-4-nitrophenol (Smith and Tibbles 1980). The use of TFM, as it is more commonly known, began in Lake Superior in 1958, Lakes Michigan and Huron in 1960, Lake Ontario in 1971, and Lake Erie in 1986 (Smith and Tibbles 1980; Pearce et al. 1980; Cornelius et al. 1995). Within each lake basin, streams that are most productive of sea lamprey larvae are treated more often than those that are less productive, so the control program requires routine monitoring of sea lamprey larvae. Over the years, most streams that produce sea lampreys have been treated at least once, and many have been treated numerous times. Releases of sterile male sea lampreys, which can theoretically reduce the number of successful spawnings by sea lamprey, are being tested in Lake Superior and the St. Mary's River for integration into the overall control program.

Fishery Regulation

Lake trout fisheries throughout the upper Great Lakes were closed in 1962, when it became apparent that the use of TFM had effectively reduced sea lamprey abundance in Lake Superior, and lake trout could, therefore, be saved from the sea lamprey (Pycha and King 1975). Since then, fishery regulations have focused on protecting lake trout from commercial exploitation, while allowing limited exploitation by recreational fisheries (Cornelius et al. 1995; Elrod et al. 1995; Eshenroder et al. 1995; Hansen et al. 1995a; Holey et al. 1995). Various strategies have been used to protect lake trout from commercial fisheries, including: prohibition of lethal gears such as large-mesh gill nets;

mandatory release, dead or alive, of all lake trout caught while fishing for other species; use of biologically based catch quotas, in conjunction with mandatory tagging of all fish caught; prohibition of fishing during the lake trout spawning season; establishment of primary restoration areas where commercial fishing is prevented but recreational fishing is permitted; and establishment of refuge areas where all fishing is prohibited.

Commercial fishery regulation in the upper Great Lakes was made more complicated in the 1970s when state and federal courts authorized treaty-based fishing by various native tribes in Michigan and Wisconsin (Eshenroder et al. 1995; Hansen et al. 1995a; Holey et al. 1995). Tribal fishing under the 1836 Treaty was affirmed in 1971 for all Michigan waters of the upper Great Lakes, including eastern Lake Superior (Hansen et al. 1995a), northern Lake Michigan (Holey et al. 1995), and northern Lake Huron (Eshenroder et al. 1995). Tribal fishing under the 1842 Treaty was affirmed in 1971 for western Michigan waters, and in 1972 for eastern Wisconsin waters of Lake Superior. Portions of tribal waters in each lake were reserved for maximizing lake whitefish yields, and were designated as deferred areas for lake trout rehabilitation because lake trout suffer excessive non-target mortality in large-mesh gill nets (Jensen 1991), the preferred gear in tribal fisheries (Hansen et al. 1995a). Tribal fishing effort, mostly with large-mesh gill nets, increased rapidly in the ensuing decade in eastern Wisconsin and in the 1980s in western Michigan.

Regulation of recreational fisheries has been through open seasons, daily bag limits, and size limits, but with little coordination among agencies. In Lake Superior, angling is prohibited in October and November (except in Michigan), the daily bag limit is three, and the minimum size limit is 152mm (6in) in Minnesota, 254mm (10in) in Michigan, 381mm (15in) in Wisconsin (only one can exceed 635mm, 25in), and no length limit in Ontario. In Lake Michigan, angling is prohibited from Labor Day to Memorial Day (Labor Day to May 1 in Michigan), the daily bag limit is two (three in Indiana), and the minimum length limit is 254mm (10in; 610mm, 24in, in Grand Traverse Bay and Michigan waters north of the 45th parallel). In Lake Huron, angling is prohibited from Labor Day to May 1 (November in Ontario), the daily bag limit is three, and the minimum length limit is 254mm (10in; no length limit in Ontario). In Lake Erie, angling is regulated by daily bag limits of one fish per day in New York, two fish per day in Pennsylvania, and three fish per day in Ontario. In Lake Ontario, angling is prohibited from October through December, the daily bag limit is three, and lake trout must be smaller than 635 mm (25 in) or larger than 762mm (30in) in length (no length limit in Ontario).

Stock Restoration Progress

Lake trout stocks increased dramatically in each of the Great Lakes within a decade of the institution of stocking, chemical suppression of sea lampreys, and regulation of fisheries, but natural reproduction was restored only in Lake Superior (Eshenroder et al. 1984). The widespread reproductive failure of hatchery-origin lake trout led the Great Lakes Fishery Commission to convene a Conference on Lake Trout Research (CLAR) in 1983 that was intended to recommend priorities for lake trout research, identify hypotheses to be tested, outline the associated experimental designs, and encourage a sharing of tasks among the various agencies and institutions (Eshenroder et al. 1984). In 1994, a decade after CLAR, another conference was convened to critically evaluate lake trout restoration in the Great Lakes (RESTORE), case history papers reviewed lake trout restoration in each lake, while other papers chronicled the results of topical research, some of which stemmed directly from the earlier CLAR proceedings. The ensuing discussion derives largely from the case histories presented at RESTORE (Cornelius et al. 1995; Elrod et al. 1995; Eshenroder et al. 1995; Hansen et al. 1995a; Holey et al. 1995).

Abundance

Since the 1960s, lake trout stocks in the Great Lakes have been monitored by spring or summer assessment fishing. The first assessment fishery put into place was in Michigan waters of Lake Superior, where deposed commercial fishermen were employed to fish for lake trout in a standardized manner immediately after the fishery was closed in 1962 (Pycha and King 1975). Assessment fisheries were implemented in 1975 in parts of Lakes Huron (Johnson et al. 1995) and Michigan (Holey et al. 1995), Lake Ontario in 1980 (Elrod et al. 1995), and Lake Erie in 1984 (Cornelius et al. 1995). Methods of conducting lake trout assessment were standardized within each lake, but varied among the lakes (Table 1). In general, lake trout density was indexed annually at fixed locations within each lake using standardized gill nets. Depictions of trends in lake trout density (catch per kilometer of gill net) below were adapted from Hansen et al. (1995a) for Lake Superior, Holey et al. (1995) for Lake Michigan, Johnson et al. (1995) for Lake Huron, Cornelius et al. (1995) for Lake Erie, and Elrod et al. (1995) for Lake Ontario.

In Lake Superior, lake trout were more abundant in Michigan than in Wisconsin or Minnesota (fig. 8). In Michigan, abundance of stocked lake trout increased sharply in the late 1960s and remained high through the 1970s, but declined sharply in the 1980s, and remained scarce after 1988. Wild lake trout increased slowly in Michigan in the 1970s and early 1980s, and declined slowly thereafter. In Wisconsin, stocked lake trout increased sharply in the 1960s, but declined sharply in the early 1970s, and more slowly thereafter. Wild lake trout increased slowly in Wisconsin after 1970. In Minnesota, stocked lake trout increased slowly in the 1970s and remained high through the 1980s, but declined thereafter. Wild lake trout increased slowly in Minnesota in the 1980s, but declined in the early 1990s. Densities of stocked lake trout declined in Michigan and Wisconsin because of increased large-mesh gill net fishing effort, but declined in Minnesota because of predation by increased density of wild lake trout (Hansen et. al.).

In Lake Michigan, lake trout stocking inshore produced higher densities in western waters than in eastern waters, and stocking offshore produced densities similar to those in western waters (fig. 9). Densities were lowest in northern Michigan waters where tribal, commercial, and sport fishing effort was highest (Holey et al. 1995). The changeover in stocking in the mid 1980s from low-priority areas to high-priority areas and refuges led to declining abundance in most low-priority areas and increasing abundance in high-priority and refuge areas (Holey et al. 1995). Lake trout densities in the 1990s were highest in Grand Traverse Bay, inshore waters off the Leelanau Peninsula in Michigan, inshore areas off southern Door County in Wisconsin, and offshore waters over the Sheboygan Reef complex near the Wisconsin-Michigan border (Holey et al. 1995). Within the Sheboygan Reef complex, lake trout densities were highest on the shallowest reef (Sheboygan Reef) and lowest on the deepest reef (Milwaukee Reef; Holey et al. 1995). Lake trout densities on

TABLE 1

Configurations of gill net assessment fisheries in the Great Lakes

LAKE	LENGTH OF GANG (M)	MESH SIZES IN GANG (MM)	INCREMENTS OF MESHES (MM)	LENGTH OF PANELS (M)	FISHING SEASON
Superior	varied	114	na	na	May
Huron	274	51–152	12.7	30.5	May–June
Michigan	244	64–152	12.7	30.5	May–August
Erie	152	38–152	12.7	15.2	August
Ontario	137	51–152	12.7	15.2	September

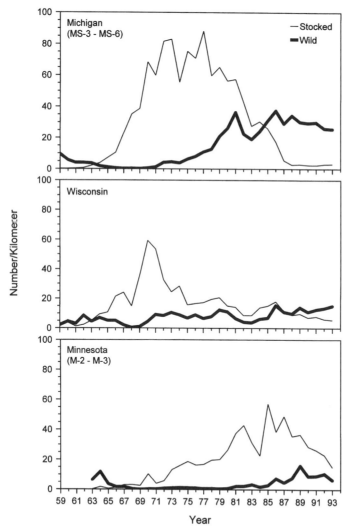

FIG. 8. *Density of stocked and wild lake trout (geometric mean catch per km of 114mm stretch measure gill net) in Michigan, Wisconsin, and Minnesota waters of Lake Superior during 1959–1993 (adapted from Hansen et al. 1995).*

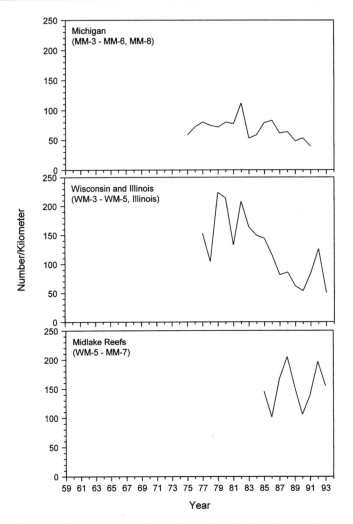

FIG. 9. *Density of stocked lake trout (catch per km of graded-mesh stretch-measure gill net) in in-shore Michigan, Wisconsin, and Illinois waters and offshore midlake reef waters of Lake Michigan during 1975–1993 (adapted from Holey et al. 1995).*

the Sheboygan Reef were among the highest recorded in the Great Lakes because the area was designated as a refuge for lake trout and stocking rates were high (Holey et al. 1995).

In Lake Huron, lake trout stocking caused densities to build rapidly in the late 1970s in both Michigan waters of the lake and southern Georgian Bay (fig. 10). Stocking produced higher densi-ties in northern Michigan waters (MH–1) in the late 1970s than in other areas (MH–2, MH–3, MH–4), but stocks declined or remained stable in all areas in the 1980s and early 1990s (Eshenroder et al. 1995). In MH–1, lake trout densities rose sharply in the late 1970s, but fell sharply in the early 1980s, and more gradually thereafter because of increased tribal fishing, which led to abandon-ment of lake trout restoration in much of the area (Eshenroder et al. 1995). In MH–2, lake trout densities increased slowly in the late 1970s, and stabilized thereafter in spite of increased stocking

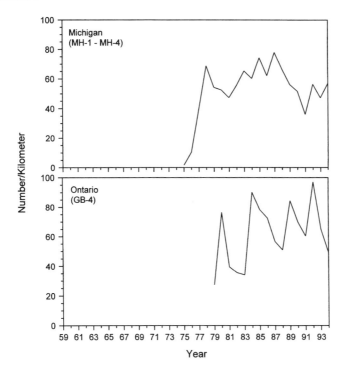

FIG. 10. *Density of stocked lake trout (catch per km of graded-mesh stretch-measure gill net) in in-shore Michigan and Ontario waters of Lake Huron during 1975–1994 (adapted from Johnson et al. 1995).*

because of increased production of sea lampreys in the St. Mary's River in the 1980s (Eshenroder et al. 1995). In MH–3 and MH–4, lake trout densities increased in the early 1980s and declined in the late 1980s, in concert with stocking rates (Eshenroder et al. 1995).

In eastern Lake Erie, lake trout densities increased slowly from 1985 through 1990, and then declined from 1991 through 1993 (fig. 11). Lake trout density increased from 1986 through 1990 because of the success of sea lamprey treatments after 1986, but declined after 1990 for uncertain reasons (Cornelius et al. 1995). Water clarity increased nearly three-fold after 1990, and may have caused lake trout catchability to decrease due to increased gear avoidance (Cornelius et al. 1995). Survival of stocked lake trout may also have declined because of predation by increased densities of mature lake trout (Cornelius et al. 1995).

In Lake Ontario, densities of lake trout in New York waters were relatively low from 1980 to 1982, but rose sharply in 1983 as fish recruited from the first stocking of one million yearlings (fig. 12). Lake trout density peaked in New York waters in 1986 and declined slowly thereafter, largely due to a decrease in survival of stocked fish during their first year in the lake (Elrod et al. 1993). The reduced stocking rate for the 1986 to 1990 year classes also contributed to lower catches of fish from 1989 to 1992 (Elrod et al. 1995). Density of mature females in New York waters was very low from 1980 to 1984, increased in 1985 and 1986, and stabilized thereafter (Elrod et al. 1995). In Ontario waters, lake trout density increased from 1985 to 1988, and then fluctuated without trend through 1992 (Elrod et al. 1995), most likely because survival of hatchery lake trout in their first

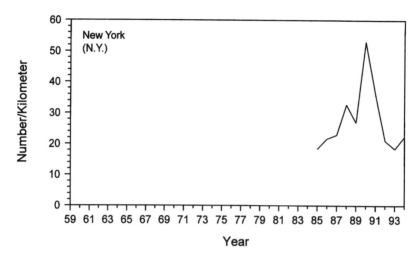

FIG. 11. *Density of stocked lake trout (catch per km of graded-mesh stretch-measure gill net) in New York waters of the eastern basin of Lake Erie during 1985–1994 (adapted from Cornelius et al. 1995).*

year or two in the lake declined in a manner similar to that in the United States (Elrod et al. 1993). Density of mature female lake trout in Ontario waters increased from 1985 to 1987, fell in 1988, increased sharply in 1989, fell again in 1990 and 1991, and rose again in 1992 (Elrod et al. 1995).

Mortality

Simulation of lake trout population dynamics have suggested that Great Lakes lake trout can withstand mortality from sea lampreys and fisheries, but only if sea lampreys are suppressed and fisheries are regulated (Lett et al. 1975; Walters et al. 1980). Lett et al. (1975) suggested that a linear increase in sea lamprey abundance caused a geometric decrease in lake trout abundance, but that lake trout could coexist with sea lampreys if large lake trout were not removed by commercial fisheries, and if sea lampreys were suppressed. Walters et al. (1980) suggested that either fishery exploitation or sea lamprey predation could have caused lake trout stocks to collapse in Lake Superior. Stocks declined in response to sea lamprey predation, but recovered in the absence of fishing when sea lampreys were suppressed. Similarly, stocks declined in response to fishing, but recovered rapidly in the absence of sea lampreys when stocking was instituted.

Sea Lamprey Predation

The use of TFM to control sea lampreys reduced spawning runs of sea lampreys 80% to 90% within four to six years of the first round of treatments in each of the upper Great Lakes (Smith and Tibbles 1980). In Lake Superior, sea lamprey control was effective enough to allow lake trout restoration to proceed (Hansen et al. 1995a), but sea lampreys still kill large numbers of lake trout, particularly in inshore areas of western Lake Superior (Hansen et al. 1994b). From 1968 to 1978, sea lamprey mortality on lake trout in inshore Michigan waters was 0.16 to 0.66, while fishing mortality (u) was only 0.13 to 0.16 (Pycha 1980). To allow for further advancement in lake trout restoration, the goal for sea lamprey control in Lake Superior is to reduce sea lampreys 50% from current levels by the

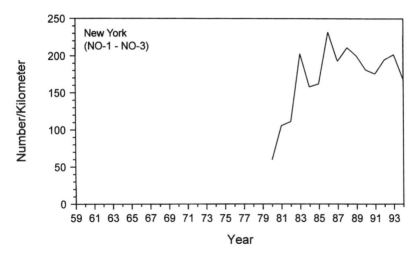

FIG. 12. *Density of stocked lake trout (catch per km of graded-mesh stretch-measure gill net) in New York waters of Lake Ontario during 1980–1994 (adapted from Elrod et al. 1995).*

year 2000, and 90% by the year 2010 (Busiahn 1990). Integrated control of sea lampreys with chemicals, barriers, and sterile male releases will likely be needed to achieve these goals (Sawyer 1980; Smith and Tibbles 1980).

In Lake Michigan, sea lamprey populations were suppressed lower than in either Lake Huron or Lake Superior (Holey et al. 1995). Nonetheless, sea lampreys caused high mortality on lake trout in Wisconsin waters of Green Bay following their colonization of the Peshtigo River in the early 1970s (Moore and Lychwick 1980). Prior to 1972, water quality in the Peshtigo River was so poor that sea lampreys could not reproduce, but pollution abatement from 1972 to 1974 caused water quality to improve to the point where sea lampreys colonized the river and caused excessive mortality on lake trout in Green Bay (Moore and Lychwick 1980). By the 1977 to 1978 period, sea lamprey-induced mortality was only 5% in Lake Michigan's main basin, but was 14% to 31% in Wisconsin waters of the lake (Wells 1980). During 1986 to 1992, wounding rates on lake trout (number per 100 fish) were lower in Lake Michigan (2.7 to 5.4) than in Lake Superior (4.1 to 9.0) or Lake Huron (8.0 to 21.5; Holey et al. 1995).

In Lake Huron, sea lampreys have caused higher mortality on lake trout than in any of the other Great Lakes, because of high sea lamprey production in the St. Mary's River (Eshenroder et al. 1995). Sea lampreys were suppressed to relatively low abundance in Lake Huron in the early 1970s, when lake trout restoration began, but increased in numbers in the 1980s, as recruitment of bloaters increased, which afforded newly metamorphosed sea lampreys a viable source of prey. Sea lamprey mortality on large lake trout ($\geq$735mm; 29in) is more than twice as high in northern Lake Huron, adjacent to the discharge of the St. Mary's River, than in the main basin to the south, and lower still in southern Georgian Bay. Prior to instituting a sea lamprey control program in Lake Huron, sea lamprey-induced mortality on lake trout in South Bay, on Manitoulin Island, was so severe that essentially no lake trout survived past ages six and seven (Budd et al. 1969). Sea lamprey-induced mortality on lake trout was equally high in northern Lake Huron from 1991 to

1993; few lake trout survived past age six because total mortality from sea lampreys, fishing, and natural causes was 70% from ages five to six (Eshenroder et al. 1995).

In Lake Erie, sea lamprey populations remained small until the 1980s when stocking of salmonids (*Salmonidae*) was instituted and stream water quality improved in response to pollution abatement programs (Cornelius et al. 1995). Sea lamprey populations increased and survival of stocked lake trout was quite low in the early 1980s (Cornelius et al. 1995). In response to the completion of the first chemical treatment cycle for sea lampreys from 1986 to 1987, wounding rates on lake trout (wounds per 100 fish) fell sharply, from twenty prior to 1988 to only one in 1989, and average survival of lake trout improved steadily, from 26% in the 1981 to 1987 period, to 32% in the 1988 to 1992 period (Cornelius et al. 1995).

In Lake Ontario, trends in numbers of lake trout killed by sea lampreys from 1982 to 1992 indicate that substantial numbers of sea lamprey transformers immigrated from untreated tributaries to Oneida Lake and from untreated tributaries to Lake Erie, in addition to those produced in Lake Ontario tributaries (Elrod et al. 1995). Estimated numbers of lake trout killed by sea lampreys fell from 84,700 in 1982–1984 to 35,700 in 1985–1987, following TFM treatments of Oneida Lake tributaries in May and June 1984, and declined again to 23,400 from 1988 to 1992, following TFM treatments of Lake Erie tributaries in fall 1986 and spring 1987 (Elrod et al. 1995).

Fishery Exploitation

The effectiveness of commercial and recreational fishery regulations varied greatly across the Great Lakes, but generally contributed to excessive total annual mortality in most areas. In Lake Superior, state and tribal fishery managers failed to control commercial large-mesh gill net fisheries in Michigan and Wisconsin, so lake trout restoration stalled in eastern Wisconsin, near the Keweenaw Peninsula in western Michigan, and around Grand Marais and Whitefish Bay in eastern Michigan (Hansen et al. 1995a). Large-mesh gill net fishing effort was recently constrained in eastern Wisconsin, but has not yet been effectively regulated in waters around the Keweenaw Peninsula, where abundance of wild lake trout continues to decline, or in eastern Michigan waters, where a consent order between Michigan and three tribes of Chippewa-Ottawa native peoples forgoes lake trout restoration in favor of large-mesh gill net fisheries for lake whitefish (Peck and Schorfhaar 1991).

In Lake Michigan, fishery exploitation caused excessive total annual mortality in most areas of the lake throughout the 1970s and 1980s (Holey et al. 1995). In Michigan waters, total annual mortality greatly exceeded acceptable rates each year from 1975 to 1989, and were regarded as the primary limitation on successful lake trout restoration (Rybicki 1991). A large portion of northern Michigan and Wisconsin waters is classified as deferred for lake trout rehabilitation, so that tribal and state commercial fisheries for lake whitefish can be maximized (Holey et al. 1995). Annual mortality of lake trout in the high-priority area in Wisconsin was also too high in the 1970s and early 1980s, but fell closer to acceptable levels in the late 1980s. The lowest rates of annual mortality on lake trout occur in Illinois, where fishery harvest of lake trout is relatively insignificant.

In Lake Huron, fishery exploitation works in concert with sea lamprey predation to cause excessive mortality on lake trout throughout the northern and central portions of the lake (Eshenroder et al. 1995). As in eastern Lake Superior and northern Lake Michigan, the northern portion of Lake Huron has been deferred for lake trout rehabilitation so that tribal commercial fisheries for lake whitefish can be maximized. As a consequence of this fishing mortality, in conjunction with excessive sea lamprey predation, annual mortality on lake trout is so high in these waters that few lake trout survive past age five. Fishing exploitation is lower in other areas, but still contributes to excessive total annual mortality in the area nearest to the deferred zone.

In Lake Erie, fishery exploitation of lake trout is generally low, though total mortality still exceeds acceptable levels (Cornelius et al. 1995). Much of the pattern of total mortality derives from year classes that were subjected to high sea lamprey predation. Total annual mortality rates will likely decline in the future as these older year classes phase out of the population in eastern Lake Erie.

In Lake Ontario, the relative impact of angler harvest on lake trout mortality in U.S. waters increased from 1982 to 1992, relative to the impact of sea lamprey predation (Elrod et al. 1995). Estimated numbers of lake trout killed by sea lampreys exceeded angler harvest in 1984, but angler harvest and hooking mortality accounted for 64% of estimated lake trout deaths from 1985 to 1992, while sea lamprey predation made up 28%. Unknown sources accounted for 8% of all lake trout deaths during the same period. Annual survival rates of lake trout varied among strains because of different depth distributions; the Superior and Clearwater strains were distributed shallower, in closer association with sea lampreys, and therefore, had lower survival than the Seneca strain, which was distributed deeper. However, survival rates of all three strains were low in both the United States and Canada.

Natural Reproduction

Stocked lake trout aggregate and spawn in many areas of the Great Lakes, but have not yet rebuilt self-sustaining populations anywhere but in Lake Superior. In Lake Superior, stocking caused the rapid build-up of inshore stocks that reproduced in all areas with widely distributed inshore spawning habitat (Hansen et al. 1995a). Wild lake trout replaced stocked lake trout in most areas of Lake Superior, through reproduction by stocked lake trout in inshore areas, remnant wild lean lake trout in offshore areas, and siscowet and humper lake trout in most deeper areas. Densities of wild lake trout in Michigan waters were lower and more variable from 1979 to 1993, than from 1929 to 1943 (Hansen et al. 1995b). Survival of stocked lake trout, at least those stocked at average sizes of 18g to 25g (18–25 per pound.), declined sharply in the late 1970s and 1980s (Hansen et al. 1994a). Survival of stocked lake trout declined in Wisconsin in the 1970s and Michigan in the 1980s, due to increased tribal large-mesh gill net fishing, and in Minnesota in the late 1980s due to predation by increased densities of wild lake trout (Hansen et. al.). Stocked lake trout are now much less abundant than wild lake trout throughout Michigan waters, and are declining elsewhere (Hansen et al. 1994b).

In Lake Michigan, stocked lake trout produced yearling and older progeny in only two areas, though viable eggs have been collected in many areas (Holey et al. 1995). Production of wild lake trout in Grand Traverse Bay and Platte Bay, Michigan, was evidenced by higher than expected percentages of unclipped fish in assessment catches (Rybicki 1991). In Grand Traverse Bay, the 1976 year class was composed of 13% wild fish, and the 1981 year class, of 7% wild fish, after accounting for fin clipping error. In Platte Bay, the 1983 year class was composed of 4% wild fish, after accounting for fin clipping error. This sporadic recruitment of wild fish, however, could not withstand the excessive fishing mortality in these areas, and disappeared from the lake trout populations in both areas.

In Lake Huron, only two isolated wild lake trout stocks survived the depredations of overfishing and sea lamprey predation (McGregor Bay and Parry Sound, Ontario; Berst and Spangler 1973), and stocked lake trout or backcross reproduced only in South Bay, Ontario, and Thunder Bay, Michigan (Eshenroder et al. 1995). Wild (unclipped) lake trout compose about 15% of the population in Thunder Bay, Michigan, but do not appear to be replacing stocked fish as the primary reproductive stock because mortality probably limits adult abundance (Johnson and VanAmberg 1995). Stocked lake trout produced fry on Six-Fathom Bank in both 1993 and 1994, but wild yearlings have not yet been recovered in the area (C. Bowen, U.S. Geological Survey–Biological Resources Division, personal communication).

In Lake Erie, lake trout restoration has only been underway since the mid 1980s, so natural reproduction of stocked populations is unlikely at present. Nonetheless, lake trout stocks in eastern Lake Erie appear to be protected from excessive mortality from either sea lampreys or fishing, and lake trout spawning habitat has not been observably degraded (Cornelius et al. 1995). In 1994, an unclipped juvenile lake trout of wild origin was captured (F. Cornelius, New York Department of Environmental Conservation, personal communication), which may indicate the first sign of natural reproduction.

In Lake Ontario, naturally produced lake trout fry were captured each spring from 1986 to 1990 (Marsden et al. 1988; Marsden and Krueger 1991), and mixed-stock analyses showed that these fry were produced by parental populations composed of 78% Seneca strain and 15% Superior strain fish (Marsden et al. 1993). In spite of this fry production, however, the incidence of unmarked juvenile lake trout in bottom trawl catches showed no detectable contribution of naturally produced fish to the population through 1993 (Elrod et al. 1995). In 1994, wild young-of-the-year and yearling lake trout were caught in bottom trawls at several locations in Ontario, which indicates survival past the fry stage for both the 1993 and 1994 year classes (J. Elrod, U.S. Geologcal Survey-Biological Resources Division, personal communication).

Prognosis for the Future

After more than thirty-five years of binational effort, lake trout restoration in the Great Lakes remains elusive. Stocking rates have generally been too low, particularly in the upper lakes, and mortality rates have generally been too high for lake trout populations to be restored in any lake. Progress has been greatest in Lake Superior, but even there, further progress appears to be inhibited by the same forces that caused stocks to collapse in the first place: excessive fishing and sea lamprey predation. Suppression of sea lampreys below their current level is needed, particularly in Lake Huron, but will depend on the development and integration of alternative control methods (sterile male releases and trapping) into the existing control program (chemicals and barrier dams). More restrictive regulation of fisheries is needed in every lake. Stocking rates can likely be increased, particularly in Lakes Huron and Michigan as stocking is abandoned in Lake Superior in favor of natural reproduction, but hatcheries cannot provide enough fish to balance current mortality rates.

Though more difficult than originally thought, lake trout restoration appears achievable in each of the Great Lakes. In Lake Superior, more stringent regulation of large-mesh gill net fisheries, in conjunction with current levels of sea lamprey control, appears to be all that is needed to restore stocks throughout Michigan waters. Lake trout management efforts in the rest of the lake appear sufficient to restore populations. In Lake Michigan, stocking in refuges is building densities of adult lake trout in areas with high quality spawning habitat that should provide adequate conditions for successful reproduction. In Lake Huron, stocking in the Six-Fathom Bank area has already produced fry under similar conditions. These large offshore reef complexes produced significant portions of the historic production in both lakes, and could therefore fuel lake trout recovery in each lake. In Lakes Erie and Ontario, wild juvenile lake trout were caught for the first time in 1994, which indicates good prospects for the future of lake trout restoration in the lower lakes.

Self-sustaining lake trout stocks that result from successful restoration programs will provide a bell-weather of ecosystem quality in each of the Great Lakes (Ryder and Edwards 1985; Marshall et al. 1987; Edwards et al. 1990). Prior to human settlement, the lake trout was the top predator in all areas of the Great Lakes, except inshore areas, large embayments, and western Lake Erie. The lake trout, therefore, integrated energy flow within most of the open waters of the Great Lakes, and served as an ideal indicator of the health of Great Lakes aquatic ecosystems. Perturbations to Great Lakes aquatic ecosystems can be measured in terms of changes in the biodiversity and abundance of lake trout populations. Increased genetic diversity and abundance of lake trout stocks in each Great Lakes should indicate improved health of the associated aquatic ecosystems.

Acknowledgments

I thank the many individuals who served on the Lake Superior Technical Committee since 1985, especially Richard G. Schorfhaar, Mark P. Ebener, James H. Selgeby, Donald R. Schreiner, and Stephen T. Schram, who tirelessly participated in many hours of conversation that educated me beyond my years. Richard L. Pycha provided an invaluable written legacy of lake trout research, and an indefatigable willingness to write about and discuss his personal observations of the lake trout. Wayne R. MacCallum and James W. Peck provided invaluable comments on the draft manuscript. This paper is contribution 930 of the U.S. Geological Survey, Biological Resources Division, Great Lakes Science Center.

Literature Cited

Agassiz, L. 1850. Lake Superior: its physical character, vegetation, and animals. Gould, Kendall, and Lincoln, Boston, Massachussetts.

Bailey, R. M., and G. R. Smith. 1981. Origin and geography of the fish fauna of the Laurentian Great Lakes Basin. Canadian Journal of Fisheries and Aquatic Sciences 38:1539–1561.

Baldwin, N. S., R. W. Saalfeld, M. A. Ross, and H. J. Buettner. 1979. Commercial fish production in the Great Lakes 1867–1977. Great Lakes Fishery Commission Technical Report 3, Ann Arbor, Michigan.

Berlin, W. H., R. J. Hesselberg, and M. J. Mac. 1981. Growth and mortality of fry of Lake Michigan lake trout during chronic exposure to PCB's and DDE. United States Fish and Wildlife Service Technical Paper 105:11–22.

Berst, A. H., and G. R. Spangler. 1973. Lake Huron: the ecology of the fish community and man's effects on it. Great Lakes Fishery Commission Technical Report 21, Ann Arbor, Michigan.

Brandt, S. B. 1986. Food of adult lake trout and salmon in Lake Ontario. Journal of Great Lakes Research 12:200–205.

Brown, E. H., Jr., G. W. Eck, N. R. Foster, R. M. Horrall, and C. E. Coberly. 1981. Historical evidence for discrete stocks of lake trout (*Salvelinus namaycush*) in Lake Michigan. Canadian Journal of Fisheries and Aquatic Sciences 38:1747–1758.

Budd, J. C., F. E. J. Fry, and P. S. M. Pearlstone. 1969. Final observations on the survival of planted lake trout in South Bay, Lake Huron. Journal of the Fisheries Research Board of Canada 26:2413–2424

Busiahn, T. R., editor. 1990. Fish community objectives for Lake Superior. Great Lakes Fishery Commission, Special Publication 90–1, Ann Arbor, Michigan.

Christie, W. J. 1973. A review of the changes in the fish species composition of Lake Ontario. Great Lakes Fishery Commission Technical Report 23, Ann Arbor, Michigan.

Coble, D. W., R. E. Bruesewitz, T. W. Fratt, and J. W. Scheirer. 1990. Lake trout, sea lampreys, and overfishing in the upper Great Lakes: a review and reanalysis. Transactions of the American Fisheries Society 119:985–995.

———. 1992. Decline of lake trout in Lake Huron. Transactions of the American Fisheries Society 119:550–554.

Conner, D. J., C. R. Bronte, J. H. Selgeby, and H. L. Collins. 1993. Food of salmonine predators in Lake Superior, 1981–87. Great Lakes Fishery Commission Technical Report 59, Ann Arbor, Michigan.

Cornelius, F. C., K. M. Muth, and R. Kenyon. 1995. Lake trout rehabilitation in Lake Erie: a case history. Journal of Great Lakes Research 21(Supplement 1):65–82.

Curtis, G. L., J. H. Selgeby, and R. G. Schorfhaar. in press. Decline and recovery of lake trout populations near Isle Royale, Lake Superior, 1929–1990. Transactions of the American Fisheries Society.

Edwards, C. J., R. A. Ryder, and T. R. Marshall. 1990. Using lake trout as a surrogate of ecosystem health for oligotrophic waters of the Great Lakes. Journal of Great Lakes Research 16:591–608.

Elrod, J. H., C. P. Schneider, and D. A. Ostergaard. 1993. Survival of lake trout stocked in U.S. waters of Lake Ontario. North American Journal of Fisheries Management 13:775–781.

———, R. O'Gorman, C. P. Schneider, T. H. Eckert, T. Schaner, J. N. Bowlby, and L. P. Schleen. 1995. Lake trout rehabilitation in Lake Ontario. Journal of Great Lakes Research 21(Supplement 1):83–107.

Eschmeyer, P. H. 1957. The near extinction of lake trout in Lake Michigan. Transactions of the American Fisheries Society 85:102–119.

Eshenroder, R. L. 1992. Decline of lake trout in Lake Huron. Transactions of the American Fisheries Society 121:548–550.

———, N. R. Payne, J. E. Johnson, C. H. Bowen, II, and M. P. Ebener. 1995. Lake trout rehabilitation in Lake Huron. Journal of Great Lakes Research 21(Supplement 1):108–127.

———, T. P. Poe, and C. H. Oliver. 1984. Strategies for rehabilitation of lake trout in the Great Lakes: proceedings of a conference on lake trout research, August 1983. Great Lakes Fishery Commission Technical Report 40, Ann Arbor, Michigan.

Fitzsimons, J. D., S. Huestis, and B. Williston. 1995. Occurrence of a swim-up syndrome in Lake Ontario lake trout in relation to contaminants and cultural practices. Journal of Great Lakes Research 21(Supplement 1):227–285.

Forage Task Group. 1993. Consumption of rainbow smelt by walleye and salmonine fishes in the central and eastern basins of Lake Erie. Pages 81–104 *in* Lake Erie Committee 1993 Annual Meeting, March 24–25, 1993, Ann Arbor, Michigan.

Goodier, J. L. 1981. Native lake trout (*Salvelinus namaycush*) stocks in the Canadian waters of Lake Superior prior to 1955. Canadian Journal of Fisheries and Aquatic Sciences 38:1724–1737.

Hansen, M. J. 1988. Report to the Lake Superior Committee by the Lake Superior Technical Committee. Pages 9–22 *in* Lake Superior Committee 1988 Annual Meeting, Great Lakes Fishery Commission, March 15–16, 1988, Ann Arbor, Michigan.

———, 1989. Report to the Lake Superior Committee by the Lake Superior Technical Committee. Pages 43–64 *in* Lake Superior Committee 1989 Annual Meeting, Great Lakes Fishery Commission, March 15, 1989, Ann Arbor, Michigan.

———, 1994. Dynamics of the recovery of lake trout (*Salvelinus namaycush*) in U.S. waters of Lake Superior. Ph.D. dissertation. Michigan State University, East Lansing. 114 pp.

———, M. P. Ebener, R. G. Schorfhaar, S. T. Schram, D. R. Schreiner, and J. H. Selgeby. 1994a. Declining survival of lake trout stocked in U. S. waters of Lake Superior during 1963–1986. North American Journal of Fisheries Management 14:395–402.

———, J. D. Shively, and B. L. Swanson. 1994b. Lake trout. Pages 13–34 *in* M. J. Hansen, editor. The State of Lake Superior in 1992. Great Lakes Fishery Commission, Special Publication 94–1, Ann Arbor, Michigan.

———, J. W. Peck, R. G. Schorfhaar, J. H. Selgeby, D. R. Schreiner, S. T. Schram, B. L. Swanson, W. R. MacCallum, M. K. Burnham-Curtis, G. L. Curtis, J. W. Heinrich, and R. J. Young. 1995a. Lake trout (*Salvelinus namaycush*) populations in Lake Superior and their restoration during 1959–1993. Journal of Great Lakes Research 21(Supplement 1):152–175.

———, R. G. Schorfhaar, J. W. Peck, J. H. Selgeby, and W. W. Taylor. 1995b. Abundance indices for determining the status of lake trout restoration in Michigan waters of Lake Superior. North American Journal of Fisheries Management 15:830–837.

———, M. P. Ebener, R. G. Schorfhaar, S. T. Schram, D. R. Schreiner, J. H. Selgeby, and W. W. Taylor. 1996. Causes of declining survival of lake trout stocked in U.S. waters of Lake Superior in 1963–1986. Transactions of the American Fisheries Society 125:831–843.

Hartman, W. L. 1973. Effects of exploitation, environmental changes, and new species on the fish habitats and resources of Lake Erie. Great Lakes Fishery Commission Technical Report 22, Ann Arbor, Michigan.

Healey, M. C. 1978. Dynamics of exploited lake trout populations and implications for management. Journal of Wildlife Management 42:307–328.

Hile, R. 1949. Trends in the lake trout fishery of Lake Huron through 1946. Transactions of the American Fisheries Society 76:121–147.

———, 1962. Collection and analysis of commercial fishery statistics in the Great Lakes. Great Lakes Fishery Commission Technical Report 5, Ann Arbor, Michigan.

———, P. H. Eschmeyer, and G. F. Lunger. 1951a. Decline of the lake trout fishery in Lake Michigan. United States Fish and Wildlife Service Fishery Bulletin 60:77–95.

———, 1951b. Status of the lake trout fishery in Lake Superior. Transactions of the American Fisheries Society 80:278–312.

Holey, M. E., R. W. Rybicki, G. W. Eck, E. H. Brown, Jr., J. E. Marsden, D. S. Lavis, M. L. Toneys, T. N. Trudeau, and R. M. Horrall. 1995. Progress toward lake trout restoration in Lake Michigan. Journal of Great Lakes Research 21(Supplement 1):128–151.

Jensen, A. L. 1978. Assessment of the lake trout fishery in Lake Superior, 1929–1950. Transactions of the American Fisheries Society 107:543–549.

———, 1991. Multiple species fisheries with no ecological interactions: two-species Schaefer model applied to lake trout and lake whitefish. ICES Journal of Marine Science 48:167–171.

———, H. J. Buettner. 1976. Lake trout, whitefish, chubs, and lake herring yield and effort data for state of Michigan waters of the upper Great Lakes: 1929–1973. Michigan Sea Grant Technical Report 52, Ann Arbor, Michigan.

Johnson, J. E., G. M. Wright, D. M. Reed, C. A. Bowen, II, N. R. Payne. 1995. Status of the cold-water fish community in 1992. Pages 21–72 *in* M. P. Ebener, editor. The state of Lake Huron in 1992. Great Lakes Fishery Commission, Special Publication 95–2, Ann Arbor, Michigan.

———, J. P. VanAmberg. 1995. Evidence of natural reproduction in western Lake Huron. Journal of Great Lakes Research 21(Supplement 1):253–259.

Jordan, D. S., and B. W. Evermann. 1904. American food and game fishes. The Nature Library, Volume 5. Doubleday, Page and Company, New York, New York.

Khan, N. Y., and S. U Qadri. 1970. Morphological differences in Lake Superior lake char. Journal of the Fisheries Research Board of Canada 27:161–167.

Koelz, W. 1926. Fishing industry of the Great Lakes. Pages 553–617 *in* Appendix XI to the Report of the U.S. Commissioner of Fisheries for 1925, Department of Commerce, Bureau of Fisheries Document No. 1001, Washington, D.C.

Krueger, C. C., and P.E. Ihssen. 1995. Review of genetics of lake trout in the Great Lakes: history, molecular genetics, physiology, strain comparisons, and restoration management. Journal of Great Lakes Research 21(Supplement 1):348–363.

Langlois, T. H. 1941. Two processes operating for the reduction in abundance or elimination of fish species from certain types of water areas. Transactions of the North American Wildlife Conference 6:189–201.

Lawrie, A. H. 1978. The fish community of Lake Superior. Journal of Great Lakes Research 4:513–549.

———, J. F. Rahrer. 1973. Lake Superior: a case history of the lake and its fisheries. Great Lakes Fishery Commission Technical Report 19, Ann Arbor, Michigan.

Lett, P. F., F. W. H. Beamish, and G. J. Farmer. 1975. System simulation of the predatory activities of sea lampreys (*Petromyzon marinus*) on lake trout (*Salvelinus namaycush*). Journal of the Fisheries Research Board of Canada 32:623–631.

LHLTTC (Lake Huron Lake Trout Technical Committee). 1985. A provisional lakewide management plan for lake trout rehabilitation in Lake Huron. Pages 103–113 *in* B. Menovske, editor. Lake Huron Committee 1985 Annual Meeting Minutes. Great Lakes Fishery Commission, Ann Arbor, Michigan.

LMLTTC (Lake Michigan Lake Trout Technical Committee). 1985. A draft lakewide management plan for lake trout rehabilitation in Lake Michigan. Pages 139–150 *in* B. Menovske, editor. Lake Michigan Committee 1985 Annual Meeting Minutes. Great Lakes Fishery Commission, Ann Arbor, Michigan.

LSLTTC (Lake Superior Lake Trout Technical Committee). 1986. A lake trout rehabilitation plan for Lake Superior. Pages 47–62 *in* S. D. Morse, editor. Lake Superior Committee 1986 Annual Meeting Minutes. Great Lakes Fishery Commission, Ann Arbor, Michigan.

LTTG (Lake Trout Task Group). 1985. A strategic plan for the rehabilitation of lake trout in eastern Lake Erie. Pages 89–109 *in* D. Einhouse and D. McLennan, editors. Lake Erie Committee 1985 Annual Meeting Minutes. Great Lakes Fishery Commission, Ann Arbor, Michigan.

LTS (Lake Trout Subcommittee). 1983. A joint plan for the rehabilitation of lake trout in Lake Ontario. Great Lakes Fishery Commission, Ann Arbor, Michigan.

Mac, M. J. 1981. Vulnerability of young lake trout to predation after chronic exposure to PCB's and DDE. United States Fish and Wildlife Service Technical Paper 105:29–32.

———, 1988. Toxic substances and survival of Lake Michigan salmonids: field and laboratory approaches. Pages 389–401 *in* M. S. Evans, editor. Toxic contaminants and ecosystem health: a Great Lakes focus. John Wiley & Sons, Inc., New York.

————, R. A. Bergstedt. 1981. Temperature selection by young lake trout after chronic exposure to PCB's and DDE. United States Fish and Wildlife Service Technical Paper 105:33–35.

————, W. H. Berlin, and D. V. Rottiers. 1981. Comparative hatchability of lake trout eggs differing in contaminant burden and incubation conditions. United States Fish and Wildlife Service Technical Paper 105:8–10.

————, C. C. Edsall. 1991. Environmental contaminants and the reproductive success of lake trout in the Great Lakes: an epidemiological approach. Journal of Toxicology and Environmental Health 33:375–394.

————, J. G. Seelye. 1985. Survival of lake trout eggs and fry reared in water from the upper Great Lakes. Journal of Great Lakes Research 11:520–529.

————, T. R. Schwartz, C. C. Edsall, and A. M. Frank. 1993. Polychlorinated biphenyls in Great Lakes lake trout and their eggs: relations to survival and congener composition 1979–1988. Journal of Great Lakes Research 19:752–765.

Marsden, J. E., and C. C. Krueger. 1991. Spawning by hatchery-origin lake trout (*Salvelinus namaycush*) in Lake Ontario: data from egg collections, substrate analysis, and diver observation. Canadian Journal of Fisheries and Aquatic Sciences 48:2377–2384.

Marsden, J. E., C. C. Krueger, P. M. Grewe, H. L. Kincaid, and B. May. 1993. Genetic comparison of naturally spawned and artificially propagated lake trout fry: evaluation of a stocking strategy for species rehabilitation. North American Journal of Fisheries Management 13:304–317.

Marsden, J. E., C. C. Krueger, and C. P. Schneider. 1988. Evidence of natural reproduction by stocked lake trout in Lake Ontario. Journal of Great Lakes Research 14:3–8.

Marshall, T. R., R. A. Ryder, C. J. Edwards, and G. R. Spangler. 1987. Using the lake trout as an indicator of ecosystem health—application of the dichotomous key. Great Lakes Fishery Commission Technical Report 49, Ann Arbor, Michigan.

Moore, J. D., and T. J. Lychwick. 1980. Changes in mortality of lake trout (*Salvelinus namaycush*) in relation to increased sea lamprey (*Petromyzon marinus*) abundance in Green Bay, 1974–78. Canadian Journal of Fisheries and Aquatic Sciences 37:2052–2056.

Pearce, W. A., R. A. Braem, S. M. Dustin, and J. J. Tibbles. 1980. Sea lamprey (*Petromyzon marinus*) in the lower Great Lakes. Canadian Journal of Fisheries and Aquatic Sciences 37:1802–1810.

Peck, J. W., and R. G. Schorfhaar. 1991. Assessment and management of lake trout stocks in Michigan waters of Lake Superior, 1970–1987. Michigan Department of Natural Resources, Fisheries Research Report 1956, Ann Arbor, Michigan.

Powell, M. J., and J. Atkinson. 1991. Lake Superior Lake Trout Stocking Review. Ontario Ministry of Natural Resources, Unpublished Report, Toronto.

Pycha, R. L. 1962. The relative efficiency of nylon and cotton gill nets for taking lake trout in Lake Superior. Journal of the Fisheries Research Board of Canada 19:1085–1094.

————, 1980. Changes in mortality of lake trout (*Salvelinus namaycush*) in Michigan waters of Lake Superior in relation to sea lamprey (*Petromyzon marinus*) predation, 1968–78. Canadian Journal of Fisheries and Aquatic Sciences 37:2063–2073.

————, G. R. King. 1967. Returns of hatchery-reared lake trout in southern Lake Superior, 1955–1962. Journal of the Fisheries Research Board of Canada 24:281–298.

————, 1975. Changes in the lake trout population of southern Lake Superior in relation to the fishery, the sea lamprey, and stocking, 1950–1970. Great Lakes Fishery Commission Technical Report 28, Ann Arbor, Michigan.

Rottiers, D. V., and R. A. Bergstedt. 1981. Swimming performance of young lake trout after chronic exposure to PCB's and DDE. United States Fish and Wildlife Service Technical Paper 105:23–26.

Rybicki, R. W. 1991. Growth, mortality, recruitment, and management of lake trout in eastern Lake Michigan. Michigan Department of Natural Resources, Fisheries Research Report 1979, Ann Arbor, Michigan.

Ryder, R. A., and C. J. Edwards, editors. 1985. A conceptual approach for the application of biological indicators of ecosystem quality in the Great Lakes basin. International Joint Commission and Great Lakes Fishery Commission, Windsor, Ontario.

Sawyer, A. J. 1980. Prospects for integrated pest management of the sea lamprey (*Petromyzon marinus*). Canadian Journal of Fisheries and Aquatic Sciences 37:2081–2092.

Selgeby, J. H. 1982. Decline of lake herring (*Coregonus artedii*) in Lake Superior: an analysis of the Wisconsin herring fishery, 1936–1978. Canadian Journal of Fisheries and Aquatic Sciences 39:554–563.

Smith, B. R. 1971. Sea lampreys in the Great Lakes of North America. Pages 207–247 in M. W. Hardisty and I. C. Potter, editors. The Biology of Lampreys, Volume 1. Academic Press, New York, New York.

————, J. J. Tibbles. 1980. Sea lamprey (*Petromyzon marinus*) in Lakes Huron, Michigan, and Superior: history of invasion and control, 1936–1978. Canadian Journal of Fisheries and Aquatic Sciences 37:1780–1801.

————, B. G. H. Johnson. 1974. Control of the sea lamprey (*Petromyzon marinus*) in Lake Superior. Great Lakes Fishery Commission Technical Report 26, Ann Arbor, Michigan.

Smith, G. R, and T. N. Todd. 1984. Evolution of species flocks of fishes in north temperate lakes. Pages 45–68 in A. A. Echelle and I. Kornfield, editors. Evolution of fish species flocks. University of Maine at Orono Press.

Smith, S. H. 1968. Species succession and fishery exploitation in the Great Lakes. Journal of the Fisheries Research Board of Canada 25:667–693.

———, 1972. Factors of ecologic succession in oligotrophic fish communities of the Laurentian Great Lakes. Journal of the Fisheries Research Board of Canada 29:717–730.

———, 1995. Early changes in the fish community of Lake Ontario. Great Lakes Fishery Commission Technical Report 60, Ann Arbor, Michigan.

Stewart, D. J., and M. Ibarra. 1991. Predation and production by salmonine fishes in Lake Michigan, 1978–88. Canadian Journal of Fisheries and Aquatic Sciences 48:909–922.

Todd, T. N., and G. R. Smith. 1980. Differentiation in *Coregonus zenithicus* in Lake Superior. Canadian Journal of Fisheries and Aquatic Sciences 37:2228–2235.

Van Oosten, J. 1930. The disappearance of the Lake Erie cisco—a preliminary report. Transactions of the American Fisheries Society 60:204–214.

———, 1948. Turbidity as a factor in the decline of Great Lakes fishes with special reference to Lake Erie. Transactions of the American Fisheries Society 75:281–322.

———, Hile, and F. W. Jobes. 1946. The whitefish fishery of Lakes Huron and Michigan with special reference to the deep-trap-net fishery. U.S. Department of the Interior, Fish and Wildlife Service, Fishery Bulletin 40:297–394.

Walters, C. J., G. Steer, and G. Spangler. 1980. Responses of lake trout (*Salvelinus namaycush*) to harvesting, stocking, and lamprey reduction. Canadian Journal of Fisheries and Aquatic Sciences 37:2133–2145.

Wells, L. 1980. Lake trout (*Salvelinus namaycush*) and sea lamprey (*Petromyzon marinus*) populations in Lake Michigan, 1971–78. Canadian Journal of Fisheries and Aquatic Sciences 37:2047–2051.

———, A. L. McLain. 1973. Lake Michigan: man's effects on native fish stocks and other biota. Great Lakes Fishery Commission Technical Report 20, Ann Arbor, Michigan.

CHAPTER 15

Pacific Salmonines in the Great Lakes Basin
John F. Kocik and Michael L. Jones

Introduction

Pacific salmonines, species of the genus *Oncorhynchus*, are indigenous to both coastal and land-locked drainages of the North Pacific rim, and are derived from a single ancestral stock (Stearley and Smith 1993). Overfishing, habitat destruction, and hatchery supplementation have led to over-all declines in abundance and the extirpation of many unique populations in their native range (Nehlson et al. 1991). However, because of their popularity as food and sportfish, fishery manag-ers have introduced Pacific salmonines extensively outside their native range (MacCrimmon 1971; Behnke 1992; Groot and Margolis 1991). The Great Lakes region has been the focus of numerous introduction attempts. Fishery managers have introduced seven Pacific salmonines into the Great Lakes: chinook salmon (*Oncorhynchus tshawytscha*), coho salmon (*O. kisutch*), cutthroat trout (*O. clarki*), kokanee salmon (*O. nerka*), masu salmon (*O. masu*), pink salmon (*O. gorbuscha*), and steel-head (*O. mykiss*) (Parsons 1973; Kruger et al. 1985). Five species, chinook salmon, coho salmon, pink salmon, kokanee salmon, and steelhead, are extant. Pacific salmonines have become a promi-nent element of Great Lakes ecosystems. They were pursued by 46% of U.S. Great Lakes anglers with a species preference, making them a significant component of fisheries with total expendi-tures of $1.3 billion (U.S.) in 1991 (Anonymous 1991). The ecological role of Pacific salmonines is also important, although variable between lakes (Eck and Brown 1985; Collins 1988; Jones et al. 1993; Hansen 1994). Thus, understanding fisheries management in the Great Lakes requires an understanding of the history, ecology, and management of Pacific salmonines.

Pacific salmonines occupy a unique place in ecological, social, economic, and political de-bates regarding fishery management in the Great Lakes (Gale 1987). Views on the role of Pacific salmonines in Great Lakes management run along a continuum; at the extremes are what might be called the enhancement view and the restoration view. The enhancement view considers the Great Lakes as dysfunctional ecosystems requiring intensive management actions such as Pacific salmonine stocking. The enhancement view regards Pacific salmonines as a primary species of management concern, and as a management success today and into the future (Gale 1987). The restoration view considers the Great Lakes as recovering ecosystems requiring management to restore ecosystem health—primarily through rehabilitation of native species. In this view, lake trout (*Salvelinus namaycush*) are a measure of ecosystem health (Edwards et al. 1990) and Pacific salmonines are temporary components whose prominence is inconsistent with long-term management strategies.

Despite polarized views, substantial common ground exists on issues such as water quality, contaminants, habitat protection, and control of sea lamprey (*Petromyzon marinus*). Furthermore, we believe that the polarized views are being driven back towards more reconcilable positions by an increased understanding of Great Lakes ecosystem structure and function, the dynamic nature of present ecosystems, and public expectation regarding management. In this context, Pacific salmonines have a role in both present and future Great Lakes ecosystems—a role that needs to be redefined in the context of sustainable fisheries.

Our goals in this chapter are to describe U.S. and Canadian Pacific salmonine management, and to assess, within the limits of current knowledge, the position of these fish in Great Lakes ecosystems. We have organized this chapter to first provide a historical context of stocking trends and a brief description of fisheries. We then summarize what is known about the ecology of Pacific salmonines in the Great Lakes. Given this background, we examine the relationship of Pacific salmonines to four important management issues. The chapter concludes with research needs and our perspective of the future of Pacific salmonines in the Great Lakes.

Great Lakes Introductions and Recent Stocking Trends

Overview

Following the example of Parsons (1973), the history of Great Lakes salmonine populations can be roughly divided into three periods: indigenous (before 1850), transitional (1850 to 1965), and modern (1966 to present). Lake troutand brook trout (*Salvelinus fontinalis*) were indigenous to all five Great Lakes. Atlantic salmon (*Salmo salar*) occurred only in Lake Ontario (Scott and Crossman 1973). Life history strategies utilized by these species are important indicators of the ecological role of indigenous salmonines and provide the most tangible junction between riverine and lacustrine ecosystems within the basin. Following McDowall's (1987) model of anadromy referring to crossing a saltwater barrier, we have elected to use the term potamodromous for a similar strategy in the Great Lakes and present a glossary of life-history types used in this chapter in table 1. Using these definitions, indigenous populations of brook trout were primarily riverine, lake trout were primarily lacustrine, and Atlantic salmon were potamodromous (table 1). In the lakes, these species were at or near the top of the food web, and Brandt (1986a) hypothesized that lake trout acted as keystone predators. The historical abundance of potamodromous populations relative to total piscivore abundance in the lakes is unclear. While historic records indicate that potamodromous populations of lake trout and brook trout occurred, distribution and abundance are unknown because these stocks were extirpated earliest, and quantitative records are scarce (Loftus 1958; Goodier 1981; Behnke 1994; Smith 1995; Coon 1999). However, Atlantic salmon were abundant in Lake Ontario (Webster 1982) and were likely an historically important potamodromous component of the lacustrine food web.

TABLE 1

Glossary of life history strategies related to salmonines in the Great Lakes

(Adapted from McDowall (1987) and Trotter (1989)).

TERM	DEFINITION
anadromous	migratory across a saline-freshwater boundary, use rivers for breeding and juvenile nursery habitat, use ocean for most of adult growth
lacustrine	entire life-history completed within a lake
potamodromous	migratory within freshwater
	lake-dwelling: use rivers for breeding and juvenile nursery areas and lakes for most of adult growth, analog to anadromous (primary definition in this chapter)
	stream-dwelling: complete life history in rivers but migrate between river systems
riverine	entire life history completed in a stream

During the transitional period (1850 to 1965), Great Lakes ecosystems were permanently altered by multiple factors including exploitation, watershed modification, pollution, and invasions of marine species (Smith 1995). The impact of these changes on native salmonines was the extirpation of Atlantic salmon, the loss of most populations of lake trout, and the loss of numerous populations of brook trout (Dymond 1966; Hansen 1999; Behnke 1994). Faced with declining native salmonine stocks, early fishery managers responded by attempting to rehabilitate native stocks—frequently with donor stocks from other regions. Secondary actions included introducing other salmonine species (Parsons 1973). The primary goal of transition period stocking was to restore salmonine fisheries with either native or introduced stocks. In the transitional period, managers first attempted to introduce chinook salmon (1873), steelhead (1876), brown trout (*Salmo trutta*) (1883), cutthroat trout (1895), masu salmon (1929), coho salmon (1933), and kokanee salmon (1950; Parsons 1973; Kruger et al. 1985). In addition, pink salmon were unintentionally introduced in Lake Superior in 1956 (Nunan 1967). As the transitional period ended, steelhead, pink salmon, and brown trout (primarily riverine) were becoming established in the Great Lakes watershed, and stocking programs for other introduced species had ceased.

During the modern period, management of the Great Lakes changed dramatically as U.S. states and Ontario began to take a more active role. To understand these decisions in their mid 1960s context, we consulted Dr. Howard Tanner, who was then Chief of Fisheries in Michigan. Dr. Tanner stated that two influential policy changes occurred at this time. First, Michigan decided to more actively manage its share of Great Lakes waters. Prior to this decision, federal agencies were responsible for most management activities occurring in Great Lakes waters. Secondly, state management emphasis was focused upon enhancing Great Lakes sportfishing opportunities. The ensuing management goal was to introduce a popular game fish well suited to Great Lakes waters. Concurrent increases in Pacific Northwest coho and chinook salmon abundance provided an opportunity for Michigan to import eggs of these popular game fish. These introductions created popular sportfishing opportunities, and other states and Ontario soon followed suit.

Compared to early attempts, the success of modern stocking programs has been attributed to the choice of stocking locations (cold-water streams), the use of smolts (less demanding on stream

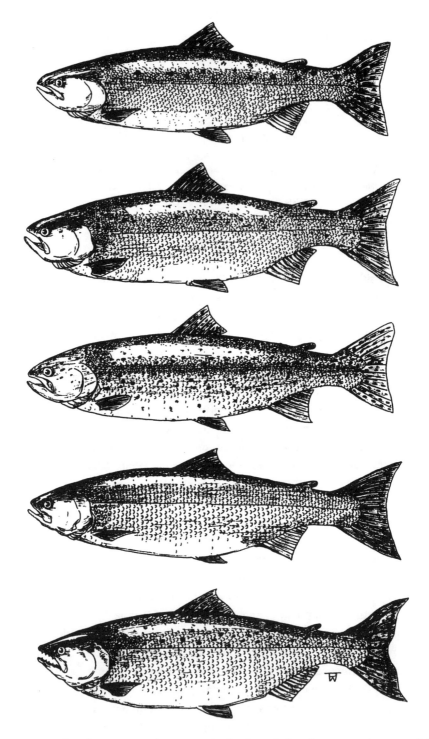

FIG. 1. *The five Pacific salmonine species present in the Great Lakes, from top to bottom, pink salmon* (O. gorbuscha), *coho salmon* (O. kisutch), *steelhead* (O. mykiss), *sockeye salmon* (O. nerka), *and chinook salmon* (O. tshawytscha), *original artwork by P. W. Thomas.*

resources), and changing ecosystems (Parsons 1973). Great Lakes ecosystems early in the modern period were characterized by decreasing sea lamprey abundance, increasing water quality, low native piscivore populations, and abundant forage fishes (Brown 1999; Hansen 1999; Beeton et al. 1998). These factors resulted in high growth and survival rates of coho and chinook salmon. During the modern period, stocking added coho, chinook, and kokanee salmon to the salmonine community. However, the abundance of these three species has declined in recent years for various reasons noted later in this chapter. Because the history of each species is different, and important to understanding its status, we highlight the stocking history of the five Pacific salmonines extant in the Great Lakes by species (fig. 1).

Steelhead

Steelhead were first introduced to the Great Lakes in 1876 with the release of McCloud River, California fish into Michigan's Au Sable River, a tributary of Lake Huron (MacCrimmon 1971; MacCrimmon and Gots 1972). Five years later, Canada introduced steelhead into Lake St. Clair using a brood stock obtained from Michigan. These plantings were the first of a succession of introductions to enhance salmonine fisheries that included many donor stocks (table 2; MacCrimmon and Gots 1972). The stocked steelhead reproduced in cold-water tributaries, and naturalized populations developed in all five Great Lakes by the early 1900s (table 2; MacCrimmon and Gots 1972; Biette et al. 1981). The genetic origin of naturalized populations is unclear because of multiple donor stocks and lack of donor-stock-specific assessments (Dueck 1994).

Steelhead are still stocked extensively throughout the Great Lakes. From 1975 to 1993, more than 157 million steelhead, primarily yearling smolts, have been stocked into the Great Lakes (fig. 2). Most broodfish are derived from naturalized stocks. However, summer-run strains from the Pacific Northwest have been introduced in recent years to extend fishing opportunities for stream anglers into summer months and to supplement open lake fisheries (Seelbach et al. 1994).

Kokanee Salmon

The Ontario Ministry of Natural Resources (OMNR) made the most concerted effort to establish kokanee salmon in the Great Lakes (Collins 1971; Parsons 1973). Their goal was to establish self-supporting kokanee salmon populations for commercial and sport fishing, and secondarily, as a forage species for lake trout (Collins 1971). Between 1964 and 1970, 14.4 million kokanee salmon were stocked in tributaries and lacustrine waters of Lakes Huron and Ontario. The majority of fish released were stocked as fry. Various donor stocks were used for these plants (table 2). Stocking yielded annual returns of several hundred adults in Lake Ontario and 1,100 to 9,000 adults in Lake Huron (Collins 1971). No kokanee salmon have been stocked in the Great Lakes since 1970. The few kokanee salmon remaining in the Great Lakes are descendents of fish introduced during this six-year stocking effort.

Pink Salmon

In 1956, nearly 22,000 odd-year stock pink salmon fry from British Columbia's Lakelse River were introduced into Lake Superior by the Ontario Ministry of Natural Resources (Schumacher and Eddy 1960; Nunan 1967). Except for about five hundred fry intentionally planted in Lake Superior

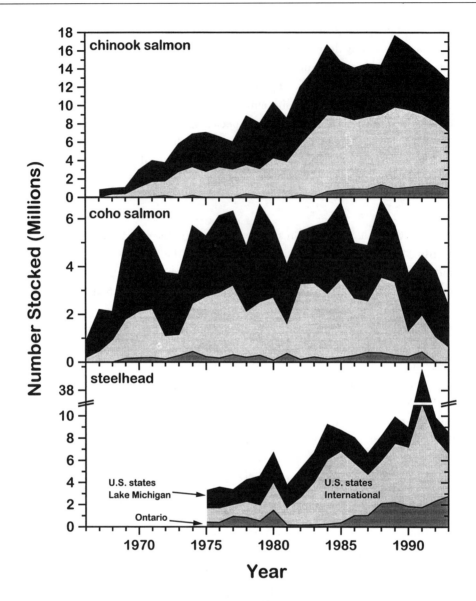

FIG. 2. *Recent stocking history of chinook salmon (1967–1993), coho salmon (1966–1993), and steelhead (1975–1993) in the five Great Lakes by Ontario (dark grey), U.S. states in Lake Michigan (black), and U.S. states in Lakes Erie, Huron, Ontario, and Superior (light gray). Data taken from GLFC stocking database through 1988 and individual state records through 1993. Steelhead stocking records prior to 1975 have not been assembled.*

by a hatchery worker, this was not a planned release but a consequence of the disposal of excess fry from unsuccessful Hudson and James Bay introductions. Approximately 21,000 fry were discarded down a sewer that led into the Current River, a Lake Superior tributary (Nunan 1967). From these small releases, pink salmon became established in the Great Lakes. All pink salmon presently in the lakes are of wild origin derived from the release of 22,000 fish.

TABLE 2

Partial list of Pacific salmonine stocks introduced in the Great Lakes since 1850.

SPECIES	STOCKS	CITATIONS

rainbow trout (steelhead)	British Columbia
	  Kamloops
	California
	  McCloud, Mount Shasta, Redwood
	Oregon
	  Rogue, Siletz, Willamette
	Washington
	  Columbia River, Puget Sound, Skamania,
	  Donaldson Strain
		MacCrimmon and Gots (1972) Biette et al. (1981)
		Peck et al. (1994)

kokanee salmon	British Columbia
	  Meadow Creek, Eagle River
	Colorado
	  Colorado River
	Montana
	  Flathead Lake, Swan River, McDonald Creek
	Washington
	  Brannian Creek
		Collins (1971)

pink salmon	British Columbia Lakelse River	Gharrett and Thomason (1987)

coho salmon	British Columbia
	  Big Qualcum River, Capilano River
	Oregon-Washington
	  Columbia River, Skagit River
		Parsons (1973)
		Stewart (1989)

chinook salmon	Alaska
	  Swan River
	Oregon-Washington
	  Columbia River, Donaldson Strain
		Parsons (1973)

Coho Salmon

The first attempt to stock coho salmon in the Great Lakes was made in 1933 with the release of 41,000 fingerlings in two Ohio tributaries of Lake Erie (Parsons 1973). No returns were reported from this attempt. Stocking did not resume until 1966, when the State of Michigan stocked coho salmon from the Columbia River into Lakes Michigan and Superior (table 2). By 1968, coho salmon had been stocked into all the Great Lakes, and by 1969 returns to weirs (approximately 15%) were sufficient to maintain hatchery production without external egg sources (Parsons 1973; Seelbach, Institute of Fisheries Research, MDNR, personal communication). Coho salmon in the Great Lakes are typically stocked as sixteen-month smolts. From 1966 to 1993, more than 133 million coho salmon were stocked in the five Great Lakes (fig. 2).

Chinook Salmon

Early attempts to introduce chinook salmon into the Great Lakes focused on Lake Ontario. The first introduction into Lake Ontario occurred in 1873, and by 1933, a total of 9.3 million fish were released. Nearly two million chinook salmon were stocked in the other four lakes (Parsons 1973). Managers selected fry for most of the stocking and released about half the fish directly into lacustrine environments. No naturalized populations developed, and, due to low returns, chinook salmon stocking was abandoned in 1933 (Parsons 1973).

Attempts to introduce chinook salmon resumed in 1967, when the State of Michigan introduced chinook salmon into tributaries of Lakes Michigan and Superior (Parsons 1973). Chinook salmon are typically planted in tributaries as five-month-old smolts. In Lake Michigan, stocking prior to 1970 yielded angler returns of about 13% of stocking rates (Parsons 1973). As a result, chinook stocking programs spread to Lakes Huron (1968), Ontario (1969), and Erie (1970). By 1977, all U.S. states were stocking chinook salmon in Great Lakes waters under their jurisdiction. Efforts in Canada were more restricted. In 1971, the OMNR began stocking chinook salmon in Lake Ontario and has continued this program. Canadian stocking of Lakes Huron and Superior began in 1985 and 1988 and are unique because they are run by angling and conservation groups through Community Fisheries Involvement Programs (CFIP). From 1967 to 1993, a total of 259 million chinook salmon have been stocked in the Great Lakes.

Allocation of Stocking by Country, Lake and Species

We compiled chinook salmon, coho salmon, and steelhead stocking records from the Great Lakes Fishery Commission (GLFC 1989) stocking database (1966 to 1988) and by contacting agencies that stocked Pacific salmonines (1989 to 1993). Steelhead stocking records prior to 1975 have not been assembled (GLFC 1989). Pacific salmonine stocking has included only these three species since the termination of kokanee salmon stocking in 1970 (Parsons 1973). Our analyses determined that from 1966 to 1993, more than half a billion Pacific salmonines were stocked in Great Lakes waters. Of this total, 93% were stocked by U.S. states and 7% by the Province of Ontario or CFIP. The choice of species and the numbers stocked in each basin are the result of agency management decisions and cold-water habitat available in the lake. Thus, our summary details the allocation of stocking over this period and highlights general use of Pacific salmonine stocking as a management tool in each lake and in all five lakes (basinwide).

Ontario has stocked coho salmon only into Lakes Superior and Ontario. The province briefly stocked coho salmon into Lake Superior from 1969 to 1971, accounting for less than 1% of coho salmon stocked into that lake from 1969 to 1991 (fig. 2). In Lake Ontario, efforts were more intense; Ontario accounted for nearly 44% of the total coho salmon stocked between 1969 and 1991. Since 1992, no coho salmon have been stocked by Ontario into any of the Great Lakes. U.S. states stock coho salmon into all five lakes with 75% going to Lakes Michigan (54%) and Erie (21%) combined (fig. 2).

Chinook salmon are stocked in all five Great Lakes. Ontario has been planting chinook salmon into Lake Ontario since 1971. CFIP programs in Lakes Huron and Superior began chinook salmon stocking programs in 1985 and 1988. In these three lakes, Ontario accounts for nearly 10% of the total stocked but less than 5% of basinwide totals. Of all chinook salmon stocked, U.S. states stock 41% into these three lakes, while 54% are stocked in Lakes Michigan and Erie (fig. 2).

Steelhead plants from 1975 to 1993 were made by both U.S. states (62%) and Ontario (13%) in Lakes Erie, Huron, Ontario, and Superior. The remaining 25% were stocked by U.S. states in Lake Michigan (fig. 2). In the four international lakes, Ontario accounted for nearly 18% of the steelhead stocked.

These stocking numbers highlight a significant difference in management emphasis between U.S. and Canadian fishery management agencies. Again, it is important to reinforce that Pacific salmonine stocking activities in the United States are conducted by the states. Federal stocking activities target lake trout. The management philosophy of U.S. states has emphasized put-grow-and-take fisheries for introduced Pacific salmonines. The objective of this management philosophy is to maximize angling opportunities by producing abundant fish that are popular with anglers (Keller et al. 1990). Since 1966, this approach has resulted in the growth of economically important, fishing-related industries and has increased public expectations for these fisheries (Gale 1987).

The Province of Ontario has used Pacific salmonines to supplement fisheries, but has done so to a lesser extent than its U.S. counterparts. Salmonine management in Canada has focused more on the development and maintenance of self-sustaining populations, especially lake trout, in keeping with the province's strategic plan for fisheries (OMNR 1992). This approach is illustrated by Canada's first attempt to introduce a Pacific salmonine, the kokanee salmon. Ontario chose kokanee salmon because they were "adapted naturally to freshwater environments" (Collins 1971 p. 1858). The stated objective of the program was to "establish kokanee salmon populations in Lake Huron without continuing hatchery support" (Collins 1971 p. 1858). When natural sustainability could not be met, the program was ended. Despite this underlying philosophy, Ontario has a long history of stocking steelhead in its Great Lakes waters and has been actively stocking chinook salmon in Lake Ontario. Furthermore, angler involvement in Ontario's CFIP has led to increased private stocking of Pacific salmonines into Lakes Huron and Superior.

Comparisons of annual stocking intensities (number per total surface area) reveal substantial differences among lakes. From 1983 to 1993, chinook salmon, coho salmon, and steelhead total stocking intensity was highest in Lake Ontario (251 fish-km^2), followed by Lakes Michigan (190 fish-km^2), Erie (144 fish-km^2), Huron (102 fish-km^2) and Superior (69 fish-km^2) (figs. 3 and 4). Among the three species, chinook salmon comprise 69%, 59%, and 58% of the total Pacific salmonines stocked into Lakes Ontario, Huron, and Michigan respectively (table 3). In Lake Erie, coho salmon are the predominant stocked Pacific salmonine (45%), followed closely by steelhead (40%). In Lake Superior, steelhead dominate recent plants, comprising 57% of fish stocked. Recent fry stocking efforts in Lake Superior tributaries, especially from 1983 to 1985 and in 1991, increased the number of steelhead stocked as compared with previous years, where yearlings were primarily stocked (Close and Anderson 1992).

Great Lakes Salmonine Fisheries

Historically, Great Lakes cold-water fisheries were primarily commercial operations targeting lake trout and coregonines (Brown et al. 1999). Recreational angling in the Great Lakes occurred but was limited in scope (Keller et al. 1990; Bence and Smith 1999). In Great Lakes tributaries, potamodromous and stream-resident species were targets of commercial and recreational fisheries (Bence and Smith; Brown et al. 1999). Many of these fisheries declined between 1850 and 1949.

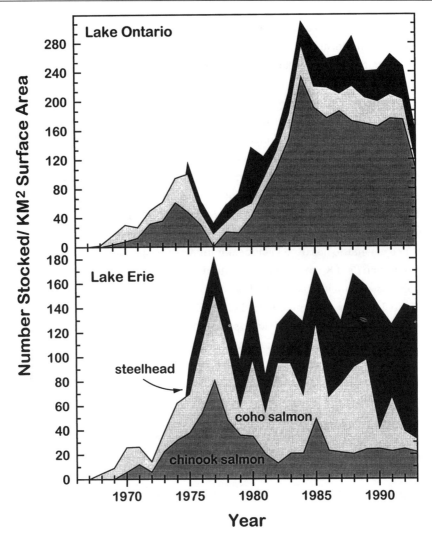

FIG. 3. *Stocking densities for chinook salmon (1967–1993, dark gray), coho salmon (1966–1993, light grey), and steelhead (1975–1993, black) in the lower Great Lakes.*

Pacific salmonines have played a pivotal role in the rehabilitation and transformation of these fisheries. By 1898, steelhead and riverine brown trout were becoming popular with anglers, fostering renewed interest in stream fisheries following the decline of native fish stocks (Kruger et al. 1985; Keller et al. 1990). With the introduction of coho and chinook salmon in the late 1960s, Great Lakes fisheries were altered dramatically. Early Pacific salmonine fisheries occurred mostly in near shore areas and rivers. Most coho and chinook salmon caught in rivers were taken by snagging (i.e., foul-hooking with large treble hooks) because anglers were inexperienced in fishing for them and such methods were legal (Parsons 1973). As Great Lakes anglers became accustomed to fishing for Pacific salmon, techniques were modified and methods changed. Offshore and river techniques

TABLE 3

Average number and density (per km^2 surface area) of Pacific salmonines
stocked each year by species and lake from 1983 to 1993.

| | AVERAGE NUMBER (1 X 10^6) | | | | AVERAGE DENSITY | | | |
LAKE	CHINOOK SALMON	COHO SALMON	STEELHEAD	TOTAL	CHINOOK SALMON	COHO SALMON	STEELHEAD	TOTAL
Erie	0.64	1.25	1.82	3.70	24.76	48.49	70.65	143.90
Huron	3.47	0.29	2.17	5.93	58.20	7.62	36.43	102.25
Michigan	6.33	2.54	2.13	11.00	109.16	43.78	36.68	189.63
Ontario	3.37	0.65	0.88	4.90	172.39	33.13	45.13	250.64
Superior	1.13	0.27	1.44	2.85	14.55	3.22	50.88	68.66

were acquired from Pacific Ocean anglers and modified for the region. Snagging and other types of foul-hooking have been greatly reduced through regulations and changing angler attitudes, and very few waters remain where this is a legal fishing method. As of 1991, the U.S. Great Lakes recreational fishery had a total value in excess of $1.3 billion (U.S.) and participants were estimated at 2.6 million (Anonymous 1991). Including multiple responses, a total of 12.3 million days of fishing were targeted at salmonines in the Great Lakes; 7.1 million of these were directed toward Pacific salmonines (Anonymous 1991). Of anglers studied in 1988, 75% were from the United States and 25% from Canada (Talhelm 1988). Sport angling for all species of Pacific salmonines is popular in both riverine and lacustrine systems. The valuation and role of Pacific salmonines in Great Lakes recreational fisheries are analyzed in greater detail by Bence and Smith (1999).

Pacific salmonines are generally not targeted by U.S. commercial fisheries because their possession is restricted. However, tribal fisheries for Pacific salmonines (primarily chinook salmon) began with the 1985 Consent Agreement and occur primarily in Lake Huron, with smaller fisheries in Lakes Superior and Michigan (Brown et al. 1999). In Lake Huron, tribal fisheries have grown from 1% of the total U.S. harvest of chinook salmon in 1986 to nearly 43% by 1993 as stocking efforts were refocused to support these fisheries (Wright 1994). In addition, approximately 30% of the chinook salmon harvested in Lake Huron are taken in weirs licensed by the State of Michigan. These fish are excess to Michigan's egg collection requirements and are sold by private companies that operate these weirs. Canadian commercial fisheries for Pacific salmonines are limited to incidental quotas for pink salmon in Lakes Huron and Superior. Catches typically fall far below quotas. For example, the Lake Huron pink salmon quota averaged 102,000kg from 1985 to 1987 while the average harvest was only 7,400kg (McNeil et al. 1988).

Pacific Salmonines in Great Lakes Ecosystems

Life History and Ecology

Pacific salmonines in their native range are primarily anadromous. In the Great Lakes, this basic strategy has been retained with the Great Lakes replacing marine habitat. Consequently, Pacific salmonines in the Great Lakes are more appropriately termed potamodromous (table 1). Pacific salmonines species and populations exhibit considerable variation in life history patterns throughout their native range (Healey 1991; Groot and Margolis 1991). A subset of this variation occurs in Great Lakes populations (Biette et al. 1981) with a few noteworthy additions.

Pink salmon in the Great Lakes use streams minimally, spawning in September and October and outmigrating as emergent fry (33 mm) in mid April to early May (Bagdovitz et al. 1986; Kocik et al. 1991). The juveniles disperse from river mouths quickly but remain common in shallow near shore waters, particularly at night (Collins 1988; Kocik 1988). Pink salmon typically spend sixteen or seventeen months in the lake, returning to spawn at age two (Bagdovitz et al. 1986). Spawning has been documented primarily in small second- or third-order streams, but pink salmon also spawn in large rivers (e.g., St. Mary's River). Observations of lacustrine shoal spawning have been made but reproductive success has not been determined (Kocik and Washbush, Michigan State University, unpublished data). Deviations from the typical two-year life cycle, which results in separate even-year and odd-year stocks, are rare in their native range (Anas 1959; Turner and Bilton 1968). However, the incidence of three-year-old pink salmon in Lake Superior was sufficient to establish modestly sized even-year populations (Kwain and Chappel 1978; Wagner and Stauffer 1980). Wagner and Stauffer (1982) determined that three-year-old pink salmon resulted from delayed maturation due to slow growth in Lake Superior's cold, oligotrophic waters. Even-year runs spread to Lakes Huron and Michigan as well but have remained small compared with odd-year runs in all lakes (Kocik et al. 1991).

Kokanee salmon also make limited use of streams in the Great Lakes. Adults return to tributaries and lacustrine shoals to spawn from late August to the first week of October (Collins 1971). Successful spawning in both riverine and lacustrine systems was documented by Collins (1971). Upon emergence from redds early in the spring, river-spawned fry moved directly into Great Lakes waters. Most documented returns of kokanee salmon in the Great Lakes were age-two adults or precocious males returning at age one (Collins 1971). Because of declining abundance, research on the ecology of these fish has stopped.

Chinook salmon in the Great Lakes typically enter rivers to spawn in late August through September with fry emerging in May. Chinook salmon spawning is greatest in larger streams (>5m wide) with production being positively related to high water velocities (Carl 1982). Juvenile chinook salmon reside in natal streams until late May, when smoltification occurs, and outmigration begins, continuing through mid July (Zaft 1992). This early migration pattern corresponds to the ocean-type life history typical of southern populations on the North American Pacific coast and is consistent with Columbia River donor stocks (Parsons 1973; Healey 1991). Elliott (1994) found that age-zero chinook salmon smolts disperse from river mouths quickly, occupying shallow near shore waters at night. Stream-type chinook salmon, those remaining in the tributary streams through their first winter, are also observed in the Great Lakes (Carl 1984; Kerr et al. 1988; Zaft 1992). However, Zaft (1992) found that 13% or less of juvenile production was stream-type in a Lake Michigan tributary. Two life-history variations have been observed: Kwain and Thomas (1984) observed spring-spawning chinook salmon in a Lake Superior tributary, and Powell and Miller (1990) documented shoal (i.e., lacustrine) spawning of chinook salmon in Lake Huron. Neither study quantified reproductive success. Chinook salmon typically return to spawn at ages three through five in the Great Lakes, although some males may return at age two.

Coho salmon and steelhead in the Great Lakes exhibit life history strategies that include more extended use of stream habitats. Coho salmon spawn from September through January, and fry emerge from the gravel in May or June (Carl 1983). Coho salmon typically migrate to the lakes the following May or June as age-one+ smolts. They spend eighteen months in the lake before returning to spawn, although some precocious fish return after only six months in the lakes.

Steelhead migration into tributaries starts in September and continues through May. Most steelhead spawn in March, April, or May, although winter spawning has been noted in some tributaries (Biette et al. 1981). Steelhead fry emerge from their redds in June or July, and remain in nursery streams until outmigration as age-1+, 2+, or 3+ smolts (Seelbach 1993). Steelhead typically spend two or three years in the Great Lakes, although a few will return after one or four years (Seelbach 1993).

Some steelhead and coho salmon parr move downstream out of some Great Lakes tributaries before their first winter (M. L. Jones, unpublished data). A similar phenomenon occurs in Pacific Ocean populations where pre-smolt fish are presumed to be unable to survive because they are not yet physiologically adapted to saltwater (Chapman 1962). In the Great Lakes, however, the stress associated with saltwater transition is absent. As such, the possibility exists that these migrants could survive to maturity. Numerous anecdotal accounts exist of coho salmon and steelhead parr using Great Lakes near shore habitats (L. Stanfield, OMNR, Picton; D. Barton, University of Waterloo, Ontario; J. George, OMNR, Thunder Bay, personal communications), although a systematic survey has yet to be completed. Scale analyses from adult steelhead suggests that one to three years of stream residency is typical (Biette et al. 1981; Seelbach 1993), but the slow growth rates likely to be exhibited by age-zero steelhead feeding on invertebrates in Great Lakes near shore environments could easily confound these interpretations. This strategy is also interesting because it suggests the potential for a local adaptation of these species in just over one hundred years for steelhead and in less then thirty years for coho salmon.

Information on the lacustrine feeding ecology of juvenile Pacific salmonines during their first growing season is extremely limited. A diet study of age-zero chinook salmon in near shore Lake Michigan habitats indicates that they fed primarily on terrestrial insects, larval fish, larval aquatic insects, and zooplankton during early summer (Elliot 1994). By late summer, the diet was dominated by small fish. Post-smolt coho salmon in northern Lake Michigan fed primarily on insects and crustaceans (Peck 1974). Detailed information on the diet of pink salmon and steelhead upon entering the Great Lakes, as well as broad scale geographic studies, are lacking.

During their second summer in the Great Lakes, Pacific salmonines become vulnerable to angling and assessment gears and more is known of their diet. While their diets vary somewhat between lakes, Pacific salmonine adults are primarily piscivores. Chinook, coho, and pink salmon feed heavily on alewife (*Alosa pseudoharengus*) and rainbow smelt (*Osmerus mordax*) (Brandt 1986b; Kocik and Taylor 1987; Collins 1988; Diana 1990; Elliott 1993). Steelhead also feed upon these fish, but utilize invertebrates, particularly insects, to a greater extent (Rand et al. 1993).

Natural Production of Pacific Salmonines

Natural reproduction of steelhead (1898), pink salmon (1959), kokanee salmon (1967), coho salmon (1967), and chinook salmon (1973) was documented very soon after their introduction. However, the magnitude of natural reproduction of Pacific salmonines in the Great Lakes is still poorly understood. Studies that have monitored Pacific salmonine natural production over time (Carl 1983; Seelbach 1987a, 1987b; Kocik et al. 1991) indicate that natural recruitment is highly variable but not more so than in their native range. Some estimates have been made in various streams and in specific regions of Lakes Michigan, Superior, and Ontario (Stauffer 1972; Kwain 1981; Carl 1982; Peck 1992; Zaft 1992; Hesse 1994; OMNR 1994). However, quantitative estimates of the contribution of natural reproduction on a lakewide and basinwide basis are lacking. Information on the reproductive success of each species does provide clues as to their productive potential and underscores connections between riverine and lacustrine ecosystems of the Great Lakes.

Small populations of kokanee salmon developed in Lake Huron following introduction attempts, but no successful spawning was documented in Lake Ontario. In Lake Huron, straying resulted in the development of ephemeral populations in unstocked tributaries. Low return rates and insufficient natural reproduction were documented in final assessments in 1975 (J. Collins, OMNR, Tehkummah, personal communication). However, small numbers (<10) of kokanee salmon have been reported in Blue Jay Creek, a Lake Huron tributary, as recently as 1993 (John Collins, OMNR, Tehkummah, personal communication). The current status of kokanee salmon in the Great Lakes is best described as rare in Lake Huron and absent in the other lakes.

Pink salmon populations in the Great Lakes basin are entirely self-sustaining. Spawners were first detected in 1959 in two Minnesota tributaries of Lake Superior. Pink salmon increased in abundance and spread to Lakes Huron (1969), Michigan (1973), Erie (1979), and Ontario (1979; Emery 1981). Pink salmon occur throughout Lake Superior and in the northern tributaries of the other Great Lakes. Abundance has fluctuated greatly in U.S. tributaries, building up to 10,000 to 30,000 fish in various rivers by the mid 1980s then declining to 200 to 3,000 fish (Kocik et al. 1991). In Canadian tributaries, abundance has been lower, but was believed to be more stable (Kelso and Noltie 1990). Recent data indicate that pink salmon abundance has continued a significant decline in the upper Great Lakes and Lake Erie, and they are only rarely reported in Lake Ontario (Kocik et al. 1991; J. Bowlby, OMNR, Picton, personal communication; J. Colonnello, OMNR, Simcoe, personal communication; Lake Huron Management Unit, Owen Sound, OMNR, unpublished data). Pink salmon populations in the Great Lakes are relatively homogenous but analyses suggest that some divergence may be occurring in these newly founded populations (Gharrett and Thomason 1987). Great Lakes pink salmon populations were found to be less variable than their donor stock, indicating genetic drift and possibly adaptively distinct lineages produced by selection in fifteen generations (Gharrett and Thomason 1987).

Successful reproduction of chinook salmon was not documented until 1973, but it likely occurred from the onset of spawning runs in the late 1960s (Carl 1982). In western Lake Superior, naturally produced chinook salmon were estimated to comprise between 50% and 90% of the population from 1989 to 1992 (Peck et al. 1994). In 1979, Carl (1982) estimated that 630,000 chinook salmon smolts were produced in Michigan's lower peninsula tributaries to Lake Michigan. This estimate equaled 27% of the total number of chinook salmon stocked by Michigan and 13% of the lakewide total. Zaft (1992) assessed production of chinook salmon in Michigan's Pere Marquette River from 1988 to 1990 and found that smolt production ranged from 52,000 to 100,000 fish. Furthermore, Elliot (1994) determined that natural production accounts for 45% to 66% of chinook salmon populations in Lake Michigan's northeast basin, and from 27% to 37% in the southeast basin. These numbers represent an increased contribution of wild chinook salmon to overall production in more recent years (Elliot 1994 v. Carl 1982), and may be due to declines in survival of stocked fish or increases in wild chinook salmon production.

Spawning of coho salmon in Lake Superior was first documented in October and November of 1967 by Peck (1970). Adult coho salmon were documented in thirty-five unstocked rivers, one of which was more than 300km from the stocking site (Peck 1970). This author also documented age-zero coho salmon in fourteen of these rivers with maximum densities of 950parr/ha. To encourage natural reproduction early in the program, 17,000 spawners were transferred to seven Lake Michigan and three Lake Huron tributaries (Keller et al. 1990). At present, natural reproduction contributes substantially to coho salmon populations in some areas of the Great Lakes. In Lake Superior waters near Marquette, Michigan, approximately 94% of the coho salmon sport catch consists of naturally spawned fish (Peck 1992). Production of 253 coho salmon smolts/ha has been reported in a Lake Michigan tributary, where attempts were made to exclude spawners (Seelbach 1985). In a relatively unproductive Lake Superior tributary, coho salmon production as high as 572 smolts/ha has been documented (Seelbach and Miller 1993). In Lake Ontario, wild coho juveniles are common in suitable tributary streams, with densities comparable to those cited above, and estimates of wild coho salmon contribution to the sport harvest have ranged as high as 50% (M.L. Jones, unpublished data).

Steelhead have been the most successful naturally producing Pacific salmonine in the Great Lakes. Wild steelhead populations occur throughout the Great Lakes, and many have persisted for nearly one hundred years (MacCrimmon and Gots 1972). New populations are still being established; a depressed stock in a Lake Superior tributary was rehabilitated with stocking from 1983 to 1985, resulting in the development of self-sustaining wild populations in only seven years (Peck 1994). Their persistence and adaptability to the Great Lakes basin is further emphasized by genetic differences that have developed between stocks found in tributaries separated by relatively small distances (Krueger and May 1987; Krueger et al. 1994). These differences may be the result of selection for local adaptations (MacCrimmon and Gots 1972; Seelbach and Miller 1993). They have developed in approximately twenty generations despite continued supplemental stocking of hatchery fish, often of non-local origin. In seven Michigan tributaries to Lake Michigan, wild fish comprised 100% of the population in unstocked rivers, 93% in stocked trout rivers, and 60% in two stocked marginal trout rivers (Seelbach and Whelan 1988). Rand et al. (1993) estimated that the contribution of wild smolts to populations in Lakes Michigan and Ontario ranged from 6% to 44% and from 18% to 33% respectively. Most sport-caught steelhead in Lake Superior are of natural origin (Peck et al. 1994).

Issues for Current and Future Management of Pacific Salmonines

Pacific salmonines currently occupy an ecologically and economically important position in Great Lakes ecosystems. The hatchery-based put-grow-take recreational fishery is important, and economic incentives for its perpetuation are considerable (Gale 1987). Because Pacific salmonine recruitment is artificially maintained through stocking, important questions have arisen concerning limits to stocking rates, particularly regarding the ability of the lakes' pelagic prey fish communities to sustain the predation pressure imposed by stocked salmonines. Concerns have also been raised about the impact on indigenous species of hatchery-supported, non-native Pacific salmonines. On the other hand, the establishment and expansion of naturalized stocks, particularly of steelhead, is an indication of improvements to the habitat quality of Great Lakes tributaries. These naturalized populations have been proposed as indicators of progress towards improving the health of the Great Lakes ecosystem. The presence of persistent toxic contaminants in the Great Lakes has been one of the most troubling issues in the basin during the past twenty-five years. Pacific salmonines, because of their position at the top of the food chain, their ability to achieve large maximum sizes, and their popularity as a sport-caught (and consumed) fish, have played a prominent role in this controversy. Finally, the development of Pacific salmon sport fisheries has been partly at the expense of allocation of Great Lakes fishes to commercial fishing operations

Sustainability of Artificially Maintained Fisheries

Total salmonine (Pacific salmonines, brown trout, and lake trout) abundance in the Great Lakes has increased dramatically during the past twenty years. By the mid 1980s, the total biomass of these top-level predators had returned to and perhaps even exceeded, historical levels (Stewart and Ibarra 1991; Jones et al. 1993). In all four of the deeper lakes and in eastern Lake Erie, hatchery-derived salmonines dominated piscivore biomass. Pacific salmonines represent a substantial proportion of total salmonine stocking in each lake. Of total salmonine stocking from 1968 to 1988, Pacific salmonines comprised 90% of the total for Lake Erie, 70% for Lakes Huron, Michigan, and Ontario, and 34% for Lake Superior (GLFC 1989).

Through the 1970s, the extreme abundance of introduced forage fish species such as alewife and rainbow smelt resulted in exceptional growth and survival rates for stocked salmonines. This led to calls for increasing levels of stocking as increasing numbers of sport anglers, charter-boat operators, and entrepreneurs (motels, restaurants, tackle shops, chambers of commerce, etc.) began to invest in Great Lakes salmon fisheries. However, within a natural ecosystem there will be limits to the level of stocking that can be sustained—limits determined by the balance between the productivity at lower trophic levels and the bioenergetic demands imposed by the total piscivore population (Jones et al. 1993). There is no controversy concerning whether this limit exists and therefore, that stocking rates cannot be allowed to increase unchecked. Nonetheless, accurately quantifying the sustainable stocking levels requires knowledge of trophodynamic interactions for which there are very little data (Stewart and Ibarra 1991; Jones et al. 1993; Negus 1995). As a result, the determination of Pacific salmonine stocking levels has been controversial for the two lakes with the highest stocking rates: Michigan and Ontario.

Stewart (1980) and Stewart et al. (1981) examined the issue of salmonine stocking limits in Lake Michigan. To derive estimates of population-level prey consumption rates, they used a bioenergetics modeling approach combining observed or estimated growth and survival rates, diet information, and assumptions about species metabolic rates (Kitchell et al. 1977). They concluded that total salmonine consumption could be as much as 20% to 33% of the annual alewife production in Lake Michigan. Their analysis is theoretically sound, but they were required to make numerous assumptions, which undoubtedly significantly affected their overall consumption

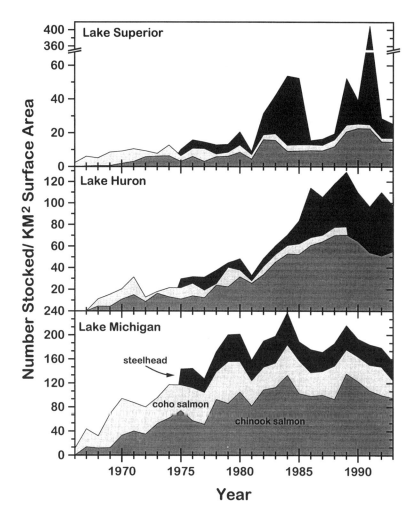

FIG. 4. *Stocking densities for chinook salmon (1967–1993, dark gray), coho salmon (1966–1993, light gray), and steelhead (1975–1993, black) in the upper Great Lakes.*

estimates. For example, they assumed that stocked salmonine survival rates were fixed and independent of size-at-stocking, an assumption that is known to be incorrect for steelhead (Seelbach 1987a; Ward et al. 1989). To compare their consumption estimates to prey supply, they assumed a conversion factor to translate trawl catches first to total biomass estimates and then to a production estimate, a conversion that they admitted was highly uncertain (Stewart et al. 1981, p. 757). In view of the substantial uncertainties associated with their calculations, their estimate of a 20% to 33% predation rate is certainly not so low as to eliminate the possibility of an excess of predator demand. At the same time, their evidence could not be construed as a convincing indication that the stocking rates of the late 1970s were excessive.

Soon after their analysis was completed, alewife abundance in Lake Michigan began to decline rapidly. Eck and Brown (1985) suggested that these declines were due to a series of colder than normal winters. However, the declining pattern, and more significantly, the failure of the alewife populations to rebound from winter mortalities, was consistent with the prediction that predator demand had exceeded levels that could be sustained by alewife. But a consensus was not reached, and state fisheries agencies around Lake Michigan did not take collective action to reduce predator demand. As shown in figure 4, Lake Michigan Pacific salmonine stocking rates increased to a peak in 1984, declined from 1985 to 1987, and then recovered almost to 1984 levels by 1989.

By the mid 1980s, alewife declines had leveled off and the indigenous bloater (*Coregonus hoyi*) had replaced alewife as the dominant planktivore in the lake (Eck and Wells 1987). Despite this change, alewife continued to be the primary prey of chinook salmon (Kogge 1985, cited by Rybicki et al. 1990; Jude et al. 1987; Stewart and Ibarra 1991), the predator responsible for the greatest prey demand of all the stocked salmonines (Stewart and Ibarra 1991; Rand et al. 1993). Stewart and Ibarra (1991), updating earlier bioenergetics models and applying them to more recent data, estimated that 76,000 tons of alewife were consumed annually by salmonine predators. This estimate was combined with a whole-lake acoustic estimate of alewife biomass to yield a predation estimate of 55% (Brandt et al. 1991). Again, this predation estimate is both uncertain (although perhaps less so than the earlier figure because of more accurate prey fish abundance estimates) and not far below a level that would be considered excessive for a prey fish population. The evidence of a predator demand–prey supply problem remained equivocal.

In 1988, chinook salmon survival began to decline, coincident with an outbreak of bacterial kidney disease. By 1993, the sport harvest of chinook salmon out of Lake Michigan had been reduced by 90% from the peak seen in 1987, although stocking rates had only been reduced by 5% from 1987, or 30% from the 1989 peak stocking year. There is no convincing evidence of a causal linkage between the changes that occurred to the prey fish community during the early 1980s and the precipitous declines in chinook salmon survival in the late 1980s. Chinook salmon growth decline during the period of low alewife abundance would be strong evidence of a prey supply problem; however, growth data are scarce prior to the mid 1980s, and that which has been reported (e.g., Stewart and Ibarra 1991) is confounded by the absence of age composition information or size-related biases inherent in sport fishery catch data. Recent evidence from known-aged fish (coded-wire tag returns) indicates that growth rates of ages three and four chinook salmon have increased subsequent to the declines in abundance from 1988 to 1990 (M. Toneys, Wisconsin Department of Natural Resources, personal communication; J. Wesley, University of Michigan, personal communication). This information suggests the possibility of a density dependent growth response. Nevertheless, there remains much uncertainty regarding the true causal connections among the fish community changes that have occurred in Lake Michigan since the late 1970s. This uncertainty, along with differences in strategic priorities among agencies, has prevented the five management agencies with jurisdiction over Lake Michigan's fisheries (Michigan, Indiana, Illinois, Wisconsin, and the Chippewa-Ottawa First Nation) from agreeing upon a lakewide course of action.

On Lake Ontario, chinook salmon stocking rates peaked in 1984 at levels (234 km^{-2}), exceeding those in Lake Michigan. O'Gorman et al. (1987) suggested that alewife populations in Lake Ontario were vulnerable to excessive salmonine predation, particularly in the face of climate-mediated, winter-mortality events. They also presented evidence that coho salmon and brown trout growth was related to alewife abundance. As alewife abundance continued an irregular pattern of decline during the period of 1986 to 1992, fishery managers in New York and Ontario called for an independent assessment of the status of the Lake Ontario pelagic-prey-fish community. An expert panel concluded that increases in salmonine stocking, coincident with declines in secondary production due to the success of basinwide phosphorus removal programs, had led to a situation in which predator demand for alewife exceeded sustainable levels (Anonymous 1992). Their conclusions were consistent with the findings of a series of Adaptive Environmental Assessment and Management (Holling 1978) modeling workshops (Jones et al. 1993; Koonce and Jones 1994). The workshop discussions concluded that there were considerable risks associated with the Lake Ontario stocking levels at that time but also stressed the uncertainty surrounding the entire issue. After extensive public consultation, both agencies agreed to stocking cuts sufficient to reduce predator demand by 50% over two years, beginning in 1993. Time will tell whether this action is judged a success. Management objectives for Lake Ontario are complex and evolving, and the same objectives are not held by all stakeholders. At this time, the decision to reduce stocking will likely be viewed as a success by most parties if a major alewife and Pacific salmonine collapse is avoided and if alewife do not return to the excessive levels of abundance witnessed prior to 1980. The bilateral nature of the stocking-reduction strategy on Lake Ontario and the deliberations that led to this decision provide a striking example of interagency and international cooperation between a Canadian province and an American state.

While Lakes Michigan and Ontario have been the focus of much of the debate regarding sustainability of hatchery-based Pacific salmonine fisheries, similar concerns have been raised about the other Great Lakes. In eastern Lake Erie, rainbow smelt populations have declined considerably in recent years; these fish are both the object of a valuable commercial fishery (in Canada) and the primary prey of salmonines in eastern Lake Erie (P. Ryan, Ontario Ministry of Natural Resources, personal communication). As such, fishery managers are faced with difficult tradeoffs to reconcile between competing uses of rainbow smelt as a commercial species and as a forage base supporting stocked Pacific salmonines and lake trout rehabilitation. In Lake Huron, total salmonine stocking rates have not reached the levels seen in Lakes Michigan and Ontario, but concerns have nevertheless been expressed that too little is known about the sustainable upper limit to predator demand (M. Ebener, Chippewa-Ottawa Treaty Fisheries Management Authority, personal communication). As in Lake Michigan, declines in alewife and rainbow smelt in Lake Huron have been offset by large increases in bloater abundance. In both lakes, declines in alewife abundance have been suggested as being causally linked to increases in bloater, the mechanism being a relaxation of predation by alewife on bloater larvae (Crowder 1980; Brown et al. 1987; Crowder et al. 1987; Eck and Wells 1987). Finally, in Lake Superior, rainbow smelt abundance has declined considerably since the 1970s, while lake herring (*Coregonus artedi*) abundance has recovered substantially (though not to historic levels) since the mid 1980s (Hansen 1994). Despite these changes, rainbow smelt continue to dominate the diet of Lake Superior salmonines (Conner et al. 1993). Bioenergetic calculations suggest that in local areas where Pacific salmonine abundance is relatively high (e.g., Minnesota waters), prey demand may be substantial (Negus 1995).

The prey fish communities of each of the Great Lakes either are presently in, or have recently passed through, a state of significant transition. The future trajectories for these communities are, therefore, difficult to predict, although there is some suggestion of a general trend towards communities with a stronger benthic component, as was the case historically (Regier and Kay 1996). Because of this uncertainty, it is difficult to predict what the future sustainable levels of Pacific salmonine abundance will be. There is little doubt that Pacific salmon rely more heavily on pelagic prey species such as alewife and rainbow smelt than do other predator species, most notably lake trout (Stewart and Ibarra 1991; Eck and Brown 1985; O'Gorman et al. 1987). Consequently, it is likely that the future sustainable levels of production of stocked Pacific salmon will be linked to the fate of these two planktivores. It is also possible, if not likely, that continued high stocking rates in the short term may hasten community shifts (i.e., replacement of the pelagic alewife and rainbow smelt by the more benthic deepwater ciscoes (*Coregonus sp.*) and sculpins (*Myoxocephalus sp.*)) that will ultimately limit the productive potential of the Great Lakes for Pacific salmonines.

Effects of Pacific Salmonines on Native Species

An examination of the potential effects of Pacific salmonines on Great Lakes native fishes requires an appropriate context. Pacific salmonines have been present in the Great Lakes for over one hundred years. During this time, the physical environment and the biological communities of all of the Great Lakes have undergone profound changes. Within the lakes themselves, invading species such as sea lamprey and alewife have contributed to a fundamental alteration of fish community structure. In the watershed, dam construction, land development (urbanization, agriculture, and forestry), and water uses (withdrawls and discharges) have greatly altered the physical habitats of innumerable tributaries, thereby affecting access to and suitability of these systems for fishes. Many of these changes are not reversible due to the current priorities of society, so returning the Great Lakes basin to an historic state is not a practical management objective. Nevertheless, management agencies continue to view the conservation and restoration of native species a priority (OMNR 1992). It is important, therefore, to consider whether Pacific salmonines—naturalized or hatchery-derived—are likely to affect native species populations and to judge whether these effects are significant relative to the myriad of other factors that might affect these populations.

Stream Interactions

The brook trout is the only species of native riverine salmonine that remains in the Great Lakes today, although historically there were Atlantic salmon in Lake Ontario (Webster 1982) and river-run lake trout in Lake Superior (Loftus 1958). The genetic composition of native brook trout populations may have been altered by stocking of hatchery strains, although mitochondrial DNA evidence suggests that native genotypes persist in wild populations (Danzmann and Ihssen 1996). The decline or extirpation of native salmonine stocks has little or nothing to do with the establishment of Pacific salmonines; habitat destruction, overfishing, and sea lamprey are considered the principal causes. Pacific salmonines have subsequently established naturally reproducing populations in many of the streams formerly occupied by these species, however. Agencies attempting to reestablish native salmonine populations, such as Atlantic salmon in Lake Ontario and coaster (migratory) brook trout in Lake Superior, must, therefore, consider the potential constraints imposed by the presence of Pacific salmonines in these waters. Stream interactions are possible at the spawning and early rearing life-history stages.

There is very little direct evidence for effects of spawning Pacific salmonines on native salmonines. Native populations of brook trout sometimes utilize similar habitats to pink salmon for spawning, and Kocik (unpublished data) observed mature pink salmon males attacking male brook trout and displacing them from females in spawning condition in a Lake Huron tributary. Atlantic salmon and Pacific salmonines require similar substrate for spawning, so fall spawning Pacific salmon may compete with Atlantic salmon for redd sites. In addition, late-fall spawners (e.g., coho salmon) and spring spawners (e.g., steelhead) may superimpose upon Atlantic salmon redds (Stanfield et al. 1995). Quality spawning habitat is fully utilized in streams where spawning Pacific salmonines are abundant (M. L. Jones, unpublished data; Kocik et al. 1991; Seelbach 1993). Where these areas overlap with target areas for native species rehabilitation (e.g., Lake Ontario's north shore) the possibility of an antagonistic interaction between native and exotic salmonines needs to be further examined.

More evidence exists for effects of Pacific salmonines on juvenile native salmonines, but even in this case the evidence is indirect. In artificial stream experiments, Great Lakes coho salmon dominated brook trout of an equal size (Fausch 1984; Fausch and White 1986). Because coho salmon typically have a substantial size advantage, the researchers felt that they would negatively influence brook trout when stream resources were limited. Rose (1986) found that the growth rate of age-zero brook trout decreased after the emergence of steelhead in a Lake Superior tributary. He hypothesized that the growth reduction could increase overwinter mortality of brook trout, negatively influencing their abundance. The diet of brook trout overlaps substantially with hatchery steelhead but less so than with wild steelhead (Johnson 1981). Johnson (1981) concluded that the impact of hatchery steelhead smolts on brook trout would be minimal, primarily because the smolts will spend a relatively short time in the stream after planting (a few weeks). As well, microhabitat use by steelhead tends to overlap most with brook trout soon after emergence of the former, when both species utilize shallow, slow-moving marginal areas of streams (Rose 1986). Larger steelhead tend to utilize riffle habitats, while larger brook trout prefer pools (Cunjak and Green 1983; Larson and Moore 1985). Thus, competitive interactions between steelhead and brook trout are most likely to occur at a very early life stage.

Further insights into interactions between potamodromous and riverine species can be inferred from examining Pacific salmonine interactions with naturalized resident brown trout populations. Coho salmon were found competitively superior to brown trout in laboratory studies and, under resource limited conditions, would likely inhibit their growth and survival (Fausch and White 1986). Laboratory and field studies of competition between juvenile steelhead and brown trout suggests that the impacts of steelhead on brown trout are minimal; however, the presence of a healthy brown trout population appeared to inhibit the survival of juvenile steelhead (Kocik and Taylor 1994; Kocik and Taylor 1995). These studies again underscore the bidirectional connections between riverine and lacustrine ecosystems in the Great Lakes.

Like brook trout, Atlantic salmon have no evolutionary history of sympatry with Pacific salmonines. Jones and Stanfield (1993) assessed the effect of age-one and older steelhead, age-zero coho salmon, and age-zero and older brown trout on stocked Atlantic salmon fry in a Lake Ontario tributary. When the densities of these potential predators and competitors were reduced, the summer growth and survival of Atlantic salmon increased significantly. Attempts to reduce the abundance of age-zero steelhead in their experiments were unsuccessful, so their effect was not measured. Jones and Stanfield (1993) speculated that, given the high degree of microhabitat overlap between age-zero Atlantic salmon and age-zero steelhead, once the latter reach 40mm total length (both are riffle specialists), steelhead could have substantial effects on salmon fry growth and survival. In contrast, Hearn and Kynard (1986), in the only other field investigation of this interaction, did not find evidence of competition between age-zero cohorts of these two species.

To date, no studies have considered the effects of introduced Pacific salmonines on the non-salmonid fauna of Great Lakes tributaries. In the past, most of these species have not been highly valued because their presence does not have a direct economic impact (e.g., as game or bait fish). The recent global trend towards increasing concern for the conservation of endemic fauna, regardless of their present economic value, argues that these species should not, and probably will not, receive so little attention in the future.

A recent reassessment of the fish fauna of Shelter Valley Creek, a tributary to Lake Ontario, showed pronounced changes in the species composition at five sites between the early 1970s and the present day (M.L. Jones and T. Bugada, unpublished data). During this period, steelhead relative abundance increased substantially at all five sites from 6.6% to 15.9% of the total weight of fish caught in the 1970s, to 29.5% to 61.1% in 1993. During this period, several non-salmonid species (longnose dace (*Rhinichthys cataractae*), blacknose dace (*Rhinichthys atratulus*), white sucker (*Catastomus commersoni*), johnny darter (*Etheostoma nigrum*), and fantail darter (*Etheostoma flabellare*)) disappeared or declined dramatically in abundance. It would be inappropriate to conclude from these data that the declines in the non-salmonids were caused by increases in steelhead abundance. Changes to hydrology or stream temperatures resulting from afforestation of the watershed, or the construction of a sea lamprey barrier that occurred between the two sampling periods may be responsible for the trends in both trout and non-salmonids. Little is known about the habitat requirements of the species that declined during this time so that the potential for interspecific interactions between them and Pacific salmonines, particularly steelhead, is difficult to determine. These data, and the lack of other studies, make it difficult to rule out the possibility that a negative interaction could occur.

Lake Interactions

Interaction of Pacific salmonines with other lake-dwelling Great Lakes fish species has been a topic of intense interest and analysis. In contrast to the situation in rivers, little concern has been expressed regarding the effects of Pacific salmonines on native lacustrine species. Instead the emphasis, as discussed earlier, has been on maintaining an appropriate balance between the salmonines and their (primarily exotic) prey. Despite this interest, and largely because of logistic difficulties, there have been very few field investigations of these interactions. Modeling studies, which rely on field-derived diet data (e.g., Brandt 1986; Diana 1990; Elliot 1993), and whatever population assessment data are available (mostly unpublished agency reports), clearly suggest that hatchery-derived Pacific salmonines have the potential to restructure the prey fish community through their predatory pressure on preferred species, most notably alewife (Stewart and Ibarra 1991; Anonymous 1992; Jones et al. 1993). In fact, the major Pacific salmonine stocking programs that began on Lake Michigan in the mid 1960s were partly justified by a desire to suppress the alewife, at that time considered an undesirable pest species. It is ironic that the species that was once considered a major pest is now the object of management concern for its protection as the forage base for the economically important Pacific salmonine fishery.

Through their predatory pressure, Pacific salmonines may have indirect community effects as well, leading to an even greater irony. Reductions in alewife abundance, brought about by salmonine predation, may be the primary mechanism by which native prey fish species, such as bloater, deepwater sculpin (*Myoxocephalus thompsoni*), and lake herring, may be able to rebuild populations to historical levels (Crowder 1980; Eck and Wells 1987) and may even be the key to reestablishment of naturally reproducing lake trout stocks in some areas of the Great Lakes (Krueger et al. 1995; Jones et al. 1995). Alewife are known to prey on the larvae of each of these species; reductions in the predatory pressure imposed by alewife may release them from a recruitment bottleneck. Increases in the dominance of these native species signals a trend towards a system in which benthic production is more important, a trend that is unlikely to favor Pacific salmonines in the long run.

Pacific Salmonines as Indicators of Ecosystem Health

For the most part, attempts to define suitable indicators of ecosystem health have focused on offshore, pelagic organisms (Edwards et al. 1990; Edwards and Ryder 1990). These authors acknowledge, however, that the overall health of the Great Lakes depends not only on the quality of Great Lakes waters themselves, but also on the condition of the tributaries that supply these waters. The primarily pelagic fish species (lake trout and walleye (*Stizostedion vitreum vitreum*)) that have been adopted as indicators for the lakes themselves are of less utility as indicators of the health of these tributaries; other species that provide an indication of the health of these fluvial habitats would be a useful complement.

The relatively stringent habitat requirements of Pacific salmonines make them potentially useful yardsticks of improvements to the quality of riverine habitats in Great Lakes tributaries. All Pacific salmonines require cold, well-oxygenated water and clean gravel or cobble substrates for successful reproduction (Raleigh et al. 1984, 1986; Groot and Margolis 1991). Stream-reared juveniles (e.g., steelhead, coho salmon, and stream-type chinook salmon) require thermally stable streams with adequate cover and food supplies to survive and grow to smolt size (Raleigh et al. 1984, 1986; Groot and Margolis 1991). Pacific salmonines have also been widely introduced throughout the Great Lakes basin and, in the case of steelhead at least, have had time to colonize streams that contain habitats that meet their requirements. Together, their broad naturalized distribution within the Great Lakes basin and their sensitivity to habitat degradation suggest that Pacific salmonines may be suitable as indicators of the health of Great Lakes cold-water tributaries.

Lake Ontario provides a good illustration of how one of these species, steelhead, has been an indicator of progress toward improved habitat quality in tributaries along the north shore of the lake. As noted earlier, steelhead have been present in Lake Ontario since the early part of this century. Until the 1970s, however, they were never particularly abundant. The large increases in wild steelhead production in Lake Ontario that have occurred within the past twenty-five years have been made possible by changes in the quality of tributary stream habitats and increased access to these habitats due to dam removal or the construction of fishways.

Much of the habitat destruction that resulted in the loss of Atlantic salmon from Lake Ontario was related to deforestation and dam construction. It was not until the 1940s that efforts to reforest the rural lands began along the north shore of Lake Ontario (Richardson 1944). As these newly planted forests began to grow, the delivery of sediment to streams diminished, greater amounts of precipitation infiltrated into regional aquifers, which increased groundwater supply to the streams of the area, and riparian vegetation was reestablished along the stream banks. By the early 1970s, the quality of these streams had improved to the point that steelhead abundance was beginning to increase. In 1974, the opening of a fishway on the Ganaraska River created access for steelhead to one of the largest and highest quality watersheds in the area, and, within fifteen years, the annual run of adult steelhead into the Ganaraska had increased to over 15,000 fish (OMNR 1994). Implementation of sea lamprey control, the gradual establishment of a Lake Ontario adapted genotype, and the stocking programs of the 1970s and 1980s also contributed to the expansion of wild populations. However, without improvements to habitat, substantial increases in wild steelhead production would not have been seen.

While the presence of wild populations of Pacific salmonines indicates good habitat quality in cold-water streams, it does not necessarily imply that conditions have returned to a state similar to that which existed prior to European settlement of the area. A recent evaluation of the possible constraints to the successful reestablishment of Atlantic salmon in Lake Ontario suggests that the preferred rearing habitat of juvenile Atlantic salmon, namely riffle areas containing abundant coarse inorganic material (boulders) to provide cover, may be quite scarce in the basin (M. L. Jones, unpublished data). Because Atlantic salmon were historically very abundant in Lake Ontario during the early period of European settlement, this suggests one of two possibilities: either the Lake Ontario Atlantic salmon had developed adaptations for effectively utilizing other habitats (i.e., those with large organic debris as cover), or the substrate conditions at that time were quite different from those seen presently. Investigations of watershed sediment dynamics in other areas with similar physiography and history of human settlement (Trimble 1983) point to the latter possibility. The massive movement of upland soils into river valleys that is likely to have occurred during the relatively brief (20 to 40 years) period of deforestation may take an immensely longer period to be reversed due to the very limited capability (through lateral erosion) of a stream to eliminate material from its floodplain. Thus, it may be centuries before these Lake Ontario streams return to the geomorphological state they were in when Atlantic salmon were present.

This somewhat speculative example points to an important, and ultimately philosophical, question concerning goals for ecosystem health. There is now widespread agreement that the Great Lakes basin ecosystem has, in many respects, been irreversibly altered. Thus, a return to a pre-European ecosystem is not a realistic goal. If the condition we believe the Great Lakes to have been in historically is more useful as a benchmark than a target, what criteria do we use for determining a suitable target for ecosystem health? Perhaps a healthy ecosystem could be defined as one in which wild, self-sustaining populations of species are found whose habitat tolerances are individually narrow but collectively span the range of conditions expected for that ecosystem, given its physiographic and climatic setting. In the context of this definition, naturalized Pacific salmonines seem well-suited as a contemporary indicator of Great Lakes ecosystem health. These species also provide a context for evaluating biotic connections between the riverine and lacustrine systems in the Great Lakes, an aspect of their indigenous structure and function that is often forgotten.

Pacific Salmonines and Commercial Fisheries

The development of the economically important recreational Pacific salmonine fisheries in U.S. waters of the Great Lakes has clearly been associated with a substantial decline in the magnitude of commercial fishing operations in these waters. Other than where tribal commercial fisheries occur, commercial gill net operations have been largely curtailed in U.S. waters due to efforts of recreational fishing interest groups (Brown et al. 1999). There remains a pervasive perception among U.S. recreational fishing interests that commercial gill netting and recreational salmonine fisheries are incompatible.

Tribal commercial fisheries for Pacific salmonines other than steelhead occur in northern Lakes Huron, Michigan, and Superior. For the most part, these fisheries have not generated much controversy, with one noteworthy exception. As part of an agreement between the State of Michigan and the member tribes of the Chippewa Ottawa Treaty Fisheries Management Authority, the state agreed to stock up to 500,000 Pacific salmon annually into northern Lake Huron waters for tribal use. Because of attractive prices for the sale of chinook salmon roe, an intensive terminal gill net fishery has developed in the St. Martin Bay area of northwestern Lake Huron. Harvest levels approached 400,000kg in 1993. The controversy has arisen less from the magnitude of the harvest than from the hazards created by this fishery to recreational anglers and boaters. Regulations require the tribal fishers to use bottom-set gill nets. The nets are so tall, however, that their float-lines rise to the surface and create a hazard to navigation. The State of Michigan, the tribes, and representatives of recreational user groups are currently attempting to negotiate a mutually acceptable resolution to the problem, perhaps by restricting the timing or location (or both) of the gill net operations.

In Canadian waters, the situation is quite different. Viable commercial fishing operations, including gill netting, continue on all four Canadian Great Lakes. Many conflicts have arisen between commercial and recreational fishery interests, but they have generally centered around species other than Pacific salmonines such as lake trout and walleye. Generally, the magnitude of conflict has been related to the extent to which the commercial operations are believed to result in incidental catches of game fish. Lake trout and walleye have historically been much more vulnerable to incidental catch from commercial lake whitefish (*Coregonus clupeaformis*) and yellow perch (*Perca flavescens*) fisheries, respectively, than have Pacific salmonines. As a result, Pacific salmonines have not been central to the debate surrounding the allocation of Great Lakes fisheries to commercial versus recreational interests. Similarly, commercial operators have not argued that Pacific salmonines might be having economically (for them) undesirable effects through community-level interactions, despite the fact that they have raised this as an issue for lake trout and rainbow smelt in eastern Lake Erie. There they have argued that lake trout stocking and the consequent predatory demand imposed by this species has contributed to the recent declines in rainbow smelt abundance (P. Ryan, Ontario Ministry of Natural Resources, personal communication). The latter species is an economically important component of the Lake Erie commercial fishery. Finally, aboriginal fishing interests in Ontario have begun to express philosophical reservations regarding Pacific salmon in the Great Lakes because they are not native species. Currently, however, this concern is not a high priority for native groups.

In summary, there are few conflicts presently between commercial, tribal, and recreational fishing interests in either U.S. or Canadian waters of the Great Lakes that revolve around Pacific salmonines. In the United States, conflicts have been largely eliminated because of reduced commercial fishing effort, except in waters where tribal commercial fisheries occur. In Canada, commercial fishing operations have been allowed to continue, and most parties now tend to accept that incidental catch of Pacific salmonines by commercial fisheries are sufficiently limited as to pose little or no threat to the recreational fishery (e.g., McNeil et al. 1988). This situation is likely to persist as long as commercial fishing operations continue to be limited to areas where incidental catch of Pacific salmonines is minimal.

Pacific Salmonine Research Needs

A shortage of good research and assessment information on Pacific salmonines, especially chinook and coho salmon, has significantly constrained management of these species. Management decision-making, particularly in the case of determining the limits to sustainable stocking levels, has been greatly influenced by the absence of critical, yet seemingly basic, information such as growth and survival rates of these species. This uncertainty has forced agencies into adopting risk-averse strategies, as was the case for Lake Ontario, or into situations of unilateral decision-making because of a lack of scientific consensus, as has been the case to date in Lake Michigan. We hope that fishery managers and policy makers will recognize the consequences of these historical omissions and, in the future, direct more resources toward collecting important baseline data and maintaining these data in electronic formats that facilitate data exchange and cooperative research. As long as Pacific salmonines continue to be an important ecological and economic component of Great Lakes ecosystems, the collecting and sharing of these data will remain critical to their effective management.

Perhaps the most important research need is to establish and maintain long-term data sets that track the population dynamics of Pacific salmonines and the species with which they interact most strongly. Some of the most valuable fisheries information available in the Great Lakes today comes from the long-term stock assessment data sets collected principally by the U.S. Geological Resources Division. Seelbach's multiyear study of the dynamics of the Little Manistee River steelhead population provides us with, perhaps, the most insightful data on this species from the Great Lakes (Seelbach 1993). He, too, argues for a greater commitment to even longer-term data than his study provides. The importance of long-term ecological data sets has been widely and convincingly advocated (Franklin 1989; 1990) for numerous ecosystem management issues. In the Great Lakes case, these data sets would be of critical value for: tracking demographic trends in Pacific salmonines and relating these trends to other changes, such as trends in prey fish species; monitoring changes in native riverine species in streams where Pacific salmonines have recently invaded or where the native species are the object of restoration efforts; and assessing the longer-term effects (i.e., demographic, genetic, and epidemiological) of Pacific salmonine stocking programs on naturalized salmonine populations in both tributaries and lakes.

Not all research questions related to Pacific salmonine management require long-term studies for their resolution. Quantitative estimates of lakewide natural production of Pacific salmonines are lacking for all of the Great Lakes and would provide valuable information on the population-level consequences of lakewide adjustments to salmonine stocking rates. A multiple stream natural production assessment program has recently been implemented for Ontario waters of Lake Ontario (J. Bowlby, Ontario Ministry of Natural Resources, personal communication). Similar programs should be established elsewhere. Equally important is research into the extent to which Pacific salmonines have, or are likely to, adapt their life-history strategies to the Great Lakes environment in ways that are impossible or unobserved within their native range. Examples include the utilization of lacustrine near shore habitats for early rearing by steelhead and coho, spawning on lacustrine shoals, and alterations of the timing of spawning, both seasonally and with respect to age. All of these possible adaptations, and particularly the first two, have major implications regarding the ultimate productive potential of the Great Lakes for these species.

It is unrealistic to presume that all of the information requirements for Pacific salmonine management in the Great Lakes will, or even could, be met through expanded basic research and assessment programs. Management also has an important role to play in increasing our understanding of these species. By embracing a philosophy of adaptive management (Walters 1986), fisheries management decisions can be effectively used to increase knowledge in ways that small-scale research experiments cannot. The Lake Ontario stocking decision, while not explicitly an adaptive decision, can be considered one, so long as the consequences of the decision are rigorously monitored. The notion of adaptive probing is also being seriously considered for Lake Michigan (K. Smith, Michigan Department of Natural Resources, personal communication), where much could be gained from carefully designed management experiments. Adaptive management is much easier said than done, but it is hard to imagine circumstances where the alternative, in which opportunities to learn through management continue to be ignored, could reasonably be considered preferable.

The Future of Pacific Salmonines in the Great Lakes

Four species of Pacific salmonines, aided by humans, have successfully invaded the Great Lakes over the past one hundred years. Reproduction of three of the four species is supplemented by intensive stocking programs. A fifth species, the kokanee, was introduced several times, most recently from 1965 to 1975, but these introductions failed. The establishment of sustained spawning runs of wild chinook, coho, pink salmon, and steelhead suggests that these species are likely lasting additions to the fish fauna of the Great Lakes.

The popularity and economic significance of Pacific salmonines as the object of recreational fisheries on each of the Great Lakes means that pressure to maintain the production of catchable-sized fish at levels approaching those enjoyed during the 1980s will persist for the future. It is human nature to set the standard of performance equal to or greater than the highest levels enjoyed in the past. This commonly creates problems in managing and exploiting natural resources where natural variation leads to occasional periods of exceptional performance that do not represent the long-term capability of the system. Although our knowledge of sustainable levels of production for Pacific salmonines in the Great Lakes is far from certain (Jones et al. 1993), there are good reasons to believe that the peak levels of stocking, at least on Lakes Ontario and Michigan, exceed those that are likely to be sustainable in the future. Perhaps more important, the risks of ignoring this possibility are sufficiently great to question the wisdom of setting historical peak production levels as targets. On the other hand, arguments favoring the wholesale reduction, or even elimination, of stocking programs, fail to recognize the social and economic benefits accrued from the hatchery stocking programs. The joint decision recently taken by the Ontario and New York management agencies to reduce stocking substantially on Lake Ontario should be commended as a choice that is wise for being both risk-sensitive and informative. The decision could be appropriately described as adaptive (Walters 1986), provided a commitment to monitor the consequences of the decision is maintained.

Perhaps more worrying to the future of hatchery-based salmonine programs has been the continued problem with disease outbreaks in hatchery fish. Although it is unclear where bacterial kidney disease (BKD) lies in the chain of cause and effect, there is no doubt that this disease has been strongly associated with a dramatic decline in Lake Michigan chinook salmon survival. Furthermore, the disease has shown signs of spreading to other species (Eshenroder et al. 1995). More recently, hatchery production has also been seriously affected by the occurrence of early mortality syndrome (EMS), or swim-up syndrome, in all three species of Pacific salmonines reared in Great Lakes hatcheries, and a whirling disease outbreak in New York hatcheries in 1994 forced the destruction of an entire year's production of steelhead destined for Lake Ontario plantings. At present, causes of these disease outbreaks are poorly understood at best, and the recent trends raise serious questions about both the ability of hatcheries to maintain a stable supply of fish and the ecological effects of planting fish that may be disease vectors into natural systems.

Natural production of Pacific salmonines is likely to increase in the short term, due to increases in both the quality of existing spawning and rearing habitats, and the creation of access to habitats upstream of artificial barriers. Interannual variation in natural production due to abiotic effects, such as harsh winters or dry summers, may mask these trends, but agency and public commitments to rehabilitating degraded stream habitats throughout the Great Lakes argue well for naturalized populations of these potamodromous species. The ultimate productive potential of the Great Lakes for naturally produced potamodromous salmonines is unknown, particularly given the uncertain potential of lacustrine habitats for production of juveniles.

However, there are significant concerns associated with trends toward increased natural salmonine production. First, if contaminants continue to be a public and environmental health concern in the Great Lakes, potamodromous species provide a mechanism for transport of these substances into areas that would otherwise be unaffected (Merna 1986; Giesy et al. 1994). Second, the spread of Pacific salmonine populations to new areas may have negative effects on existing native species, including brook trout, on efforts to reintroduce extirpated species such as Atlantic salmon, and on the stability of naturalized fish populations. Besides Great Lakes ecosystems, the potential exists for these fish to colonize the Atlantic Ocean and tributaries through the St. Lawrence River, although such dispersal has not been documented to date. Also, the establishment of self-sustaining naturalized populations of Pacific salmon with evidence of genetic differentiation (Gharrett and Thomason 1987; Krueger and May 1987; Krueger et al. 1994) raises the possibility of conflicts between conservation of naturalized stocks and hatchery supplementation. Concern for the genetic effects of hatchery supplementation has been widely raised as an issue for the management of Pacific salmonines in their native range (e.g., Carvalho 1993). Whether wild stocks of non-native species deserve protection from such genetic effects may ultimately become a philosophical issue, although there is no reason in principle why these stocks, so long as they are considered valuable, should not be subjected to the same stock management practices that are advocated for native populations. Management agencies need to determine their position on these naturalized species and develop the appropriate policies concerning supplemental stocking.

In the long term, the fate of Pacific salmonines in the Great Lakes is even less certain. Great Lakes fish communities remain in a period of significant transition, and the trend may be toward a greater predominance of benthic production, a state more like the historical community structure of these lakes. Declines in alewife and rainbow smelt and increases in burbot (*Lota lota*), bloater, lake whitefish, and deepwater sculpins (*Myoxocephalus thompsoni*) are all indications of this trend. It is too early to say with confidence that the trend is a long-term directional one, although the pattern is consistent with the hypothesis that the Great Lakes are moving towards a successional state more like that which existed historically (Eshenroder and Burnham-Curtis 1999). If the trend does continue, though, it will likely be to the detriment of species that depend on pelagic production, such as Pacific salmonines. A plausible forecast for the Great Lakes biota is one in which Pacific salmonines continue to be a quantitatively significant component of the ecosystem and its management, but not the dominant component that has existed in several lakes for the past two decades.

Literature Cited

Anas, R. E. 1959. Three-year-old pink salmon. Journal of the Fisheries Research Board of Canada 16:91–94.

Anonymous. 1991. U.S. Department of the Interior, Fish and Wildlife Service and U.S. Department of Commerce, Bureau of the Census. 1991. National Survey of Fishing, Hunting, and Wildlife-Associated Recreation. U.S. Government Printing Office, Washington, DC 1993. 174 pp.

———, 1992. Status of the Lake Ontario offshore pelagic fish community and related ecosystem in 1992. Report of an ad hoc task group to the Lake Ontario Committee, Great Lakes Fishery Commission, Ann Arbor, Michigan.

Bagdovitz, M. S., W. W. Taylor, W. C. Wagner, S. P. Nicolette and G. R. Spangler. 1986. Pink salmon populations in the U.S. waters of Lake Superior, 1981–1984. Journal of Great Lakes Research 12: 72–81.

Beeton, A. M., C. E. Sellinger, and D. F. Reid. 1999. An introduction to the Laurentian Great Lakes ecosystem. Pages 3–54 *in* W. W. Taylor and C P. Ferreri eds. Great Lakes Fisheries Policy and Management: A Binational Perspective. Michigan State University Press, East Lansing, Michigan.

Behnke, R. J. 1992. Native trout of western North America. American Fisheries Society Monograph 6, American Fisheries Society, Bethesda, MD.

Behnke, R. C. 1994. Coaster brook trout and evolutionary "significance". Trout autumn: 59–60.

Bence, J. R., and K. D. Smith. 1999. An overview of recreational fisheries of the Great Lakes. Pages 259–306 *in* W. W. Taylor and C P. Ferreri eds. Great Lakes Fisheries Policy and Management: A Binational Perspective. Michigan State University Press, East Lansing, Michigan.

Biette, R. M., D. P. Dodge, R. L. Hassinger and T. M. Stauffer. 1981. Life history and timing of migrations and spawning behavior of rainbow trout (*Salmo gairdneri*) populations of the Great Lakes. Canadian Journal of Fisheries and Aquatic Science 38: 1759–1771.

Brandt, S. B. 1986a. Disappearance of the deepwater sculpin (*Myoxocephalus thompsoni*) from Lake Ontario: the keystone predator hypothesis. Journal of Great Lakes Research 12: 18–24.

———, 1986b. Food of trout and salmon in Lake Ontario. Journal of Great Lakes Research 12: 200–205.

———, D. M. Mason, E. V. Patrick, R. L. Argyle, L. Wells, P. A. Unger, and D. J. Stewart. 1991. Acoustic measures of the pelagic planktivores in Lake Michigan. Canadian Journal of Fisheries and Aquatic Sciences 48:894–908.

Brown, E. H., T. R. Busiahn, M. L. Jones, and R. L. Argyle. 1999. Allocating Great Lakes forage bases in response to multiple demand. Pages 355–394 *in* W. W. Taylor and C P. Ferreri eds. Great Lakes Fisheries Policy and Management: A Binational Perspective. Michigan State University Press, East Lansing, Michigan.

———, R. L. Argyle, N. R. Payne, and M. E. Holey. 1987. Yield and dynamics of destabilized chub (*Coregonus spp.*) populations in Lakes Michigan and Huron, 1950–84. Canadian Journal of Fisheries and Aquatic Sciences 44(Suppl. 2): 371–383.

Brown, R. W., M. Ebener, and T. Gorenflo. 1999. Great Lakes commercial fisheries: Historical overview and prognosis for the future. Pages 307–354 *in* W. W. Taylor and C. P. Ferreri eds. Great Lakes Fisheries Policy and Management: A Binational Perspective. Michigan State University Press, East Lansing, Michigan.

Carl, L. M. 1982. Natural reproduction of coho salmon and chinook salmon in some Michigan streams. North American Journal of Fisheries Management 2: 375–380.

———, 1983. Density, growth, and change in density of coho salmon and rainbow trout in three Lake Michigan tributaries. Canadian Journal of Zoology 61: 1120–1127.

———, 1984. Chinook salmon (*Oncorhynchus tshawytscha*) density, growth, mortality, and movement in two Lake Michigan tributaries. Canadian Journal of Zoology 62: 65–71.

Carvalho, G. R. 1993. Evolutionary aspects of fish distribution-genetic variability and adaptation. Journal of Fish Biology 43 (SA):53–73.

Chapman, D. W. 1962. Aggressive behavior in juvenile coho salmon as a cause of emigration. Journal of the Fisheries Research Board of Canada 19(6):1047–1080.

Close, T. L. and C. S. Anderson. 1992. Dispersal, density-dependent growth, and survival of stocked steelhead fry in Lake Superior tributaries. North American Journal of Fisheries Management 12: 728–735.

Collins, J. J. 1971. Introduction of kokanee salmon (*Oncorhynchus nerka*) into Lake Huron. Journal of the Fisheries Research Board of Canada 28: 1857–1871.

———, 1988. Changes in the North Cannel fish community, with emphasis on pink salmon (*Oncorhynchus gorbuscha* Walbaum). Hydrobiologica 163: 195–213.

Conner, D. J., C. R. Bronte, J. H. Selgeby, and H. L. Collins. 1993. Food of salmonine predators in Lake Superior, 1981–87. Great Lakes Fisheries Commission Technical Report 59, Ann Arbor, Michigan.

Coon, T. G. 1999. Ichtyofauna of the Great Lakes basin. Pages 55–71 *in* W. W. Taylor and C. P. Ferreri eds. Great Lakes Fisheries Policy and Management: A Binational Perspective. Michigan State University Press, East Lansing, Michigan.

Crowder, L. B. 1980. Alewife, rainbow smelt and native fishes in Lake Michigan: competition or predation. Environmental Biology of Fishes 5:225–233.

———, M. E. McDonald, and J. A. Rice. 1987. Understanding recruitment of Lake Michigan fishes: The importance of size-based interactions between fish and zooplankton. Canadian Journal of Fisheries and Aquatic Sciences 44(Suppl. 2): 141–147.

Cunjak, R. A., and J. M. Green. 1983. Habitat utilization by brook char (*Salvelinus fontinalis*) and rainbow trout (*Salmo gairdneri*) in Newfoundland streams. Canadian Journal of Zoology 61:1214–1219.

Danzmann, R. G. and P.E. Ihssen. 1996. A phylogeographic survey of brook trout (*Salvelinus fontinalis*) in Algonquin Park, Ontario, based on mitochondrial DNA variation. Molecular Ecology, in press.

Diana, J. S. 1990. Food habits of angler-caught salmonines in western Lake Huron. Journal of Great Lakes Research 16: 271–278.

Dueck, L. A. 1994. Population divergence of introduced rainbow trout (*Oncorhynchus mykiss*) in the Lake Ontario watershed, based on the mitochondrial genome. Master's thesis, University of Guelph, Guelph, Ontario. 173 p.

Dymond, J. R. 1966. The Lake Ontario salmon (*Salmo salar*). Unpublished monograph, edited by H. H. MacKay, University of Toronto, Toronto, Ontario.

Eck, G. W. and E. H. Brown, Jr. 1985. Lake Michigan's capacity to support lake trout (*Salvelinus namaycush*) and other salmonines: an estimate based on the status of prey populations in the 1970s. Canadian Journal of Fisheries and Aquatic Sciences 42:449–454.

———, L. Wells. 1987. Recent changes in Lake Michigan's fish community and their probable causes, with emphasis on the role of the alewife (*Alosa pseudoharengus*). Canadian Journal of Fisheries and Aquatic Sciences 44(Suppl. 2):53–60.

Edwards, C. J., R. A. Ryder, and T. R. Marshall. 1990. Using lake trout as a surrogate of ecosystem health for oligotrophic waters of the Great Lakes. Journal of Great Lakes Research. 16:591–608.

———, (ed.) 1990. Biological surrogates of mesotrophic ecosystem health in the Laurentian Great Lakes. Report to the Great Lakes Science Advisory Board, International Joint Commission, Windsor, Ontario.

Elliott, R. F. 1994. Early life history of Chinook salmon in Lake Michigan. Michigan Federal Aid Project F–53–R–472. 23 pp

———, 1993. Feeding habits of Chinook salmon in eastern Lake Michigan. Master's Thesis. Michigan State University, East Lansing, Michigan. 79 pp.

Emery, L. 1981. Range extension of pink salmon (*Oncorhynchus gorbuscha*) into the lower Great Lakes. Fisheries 6: 7–10.

Eshenroder, R. L., and M. K. Burnham-Curtis. 1999. Species succession and sustainability of the Great Lakes fish community. Pages 145–184 *in* W. W. Taylor and C P. Ferreri eds. Great Lakes Fisheries Policy and Management: A Binational Perspective. Michigan State University Press, East Lansing, Michigan.

———, E. J. Crossman, G. K. Meffe, C. H. Olver, and E. P. Pister. 1996. Lake trout rehabilitation in the Great Lakes: an evolutionary, ecological, and ethical perspective. Journal of Great Lakes Research 21 (Suppl. 1):518–529.

Fausch, K. D. 1984. Profitable stream positions for salmonids: relating specific growth rate to net energy gain. Canadian Journal of Zoology 62: 441–451.

———, R. J. White. 1986. Competition among juveniles of coho salmon, brook trout, and brown trout in a laboratory stream, and implication for Great Lakes tributaries. Transactions of the American Fisheries Society 115: 363–381.

Franklin, J. F. 1989. Importance and justification of long-term studies in ecology. Pages 3–19 *in* G. E. Likens, ed., Long-term Studies In Ecology: Approaches And Alternatives. Springer-Verlag, New York.

———, 1990. Contributions of the long-term ecological research program. Bioscience 40: 509–523.

Gale, R. P. 1987. Resource miracles and rising expectations: A challenge to fishery managers. Fisheries 12(5):8–13.

Gharrett, A. J. and M. A. Thomason. 1987. Genetic changes in pink salmon (*Oncorhynchus gorbuscha*) following their introduction into the Great Lakes. Canadian Journal of Fisheries and Aquatic Science 44:787–792.

Giesy, J. P., and 15 coauthors. 1994. Contaminants in fishes from Great Lakes-influenced sections and above dams of three Michigan rivers. II: Implications for health of mink. Archives of Environmental Contamination and Toxicology. 27:213–223.

GLFC (Great Lakes Fishery Commission). 1989. Fish Stocking Database Management System. A R-Base Software Program

Goodier, J. L. 1981. Native lake trout (*Salvelinus namaycush*) stocks in the Canadian waters of Lake Superior. Canadian Journal of Fisheries and Aquatic Sciences 38: 1724–1737.

Groot, C. and Margolis, L. 1991. Pacific salmon Life Histories. University of British Columbia Press, Vancouver, British Columbia, Canada.

Hansen, M. J. 1999. Lake trout in the Great Lakes: Basinwide stock collapse and binational restoration. Pages 417–453 *in* W. W. Taylor and C. P. Ferreri eds. Great Lakes Fisheries Policy and Management: A Binational Perspective. Michigan State University Press, East Lansing, Michigan.

———, (ed.). 1994. The state of Lake Superior in 1992. Great Lakes Fishery Commission, Special Publication 94–1., Ann Arbor, MI.

Healey, M. C. 1991. Life history of chinook salmon (*Oncorhynchus tshawytscha*). Pages 313–396 *in* C. Groot, and L. Margolis, eds. Pacific salmon Life Histories. University of British Columbia Press, Vancouver, British Columbia, Canada.

Hearn, W. E., B. E. Kynard. 1986. Habitat utilization and behavioral interaction of juvenile Atlantic salmon (*Salmo salar*) and rainbow trout (*S. gairdneri*) in tributaries of the White River of Vermont. Canadian Journal of Fisheries and Aquatic Sciences 43: 1988–1998.

Hesse, J. A. 1994. Contribution of hatchery and natural chinook salmon to the Eastern Lake Michigan fishery, 1992–1993. Master's Thesis, Michigan State University, East Lansing, Michigan. 82 pp.

Holling, C. S. [ed.]. 1978. Adaptive environmental assessment and management. Wiley and Sons, New York.

Johnson, J. 1981. Food Interrelationships of Coexisting Brook Trout, Brown Trout and Yearling Rainbow Trout in Tributaries of the Salmon River, New York. New York Fish and Game Journal 28: 88–99.

Jones, M. L., J. F. Koonce, and R. O'Gorman. 1993. Sustainability of hatchery-dependent salmonine fisheries in Lake Ontario: the conflict between predator demand and prey supply. Transactions of the American Fisheries Society 122:1002–1018.

———, L. W. Stanfield. 1993. Effects of exotic juvenile salmonines on growth and survival of juvenile Atlantic salmon (*Salmo salar*) in a Lake Ontario tributary. Pages 71–79 *in* Gibson, R. J. and R. E. Cutting, editors. Production of Juvenile Atlantic Salmon, *Salmo Salar*, in Natural Waters . National Research Council of Canada, Ottawa, Ontario, No. 118.

———, nine coauthors. 1995. Limitations to lake trout (*Salvelinus namaycush*) rehabilitation in the Great Lakes imposed by biotic interactions occurring at early life stages. Journal of Great Lakes Research 21 (Suppl. 1):505–517.

Jude, D. J., F. J. Tesar, S. F. Deboe, and T. J. Miller. 1987. Diet and selection of major prey species by Lake Michigan salmonines. 1973–1982. Transactions of the American Fisheries Society 116:677–691.

Keller, M., Smith, K. D. and Rybicki, R. D. 1990. Review of salmon and trout management in Lake Michigan. Fisheries Special Report Number 14. Michigan Department of Natural Resources, Charlevoix, Michigan.

Kelso, J. R. M. and D. B. Noltie. 1990. Abundance of spawning Pacific salmon in two Lake Superior streams, 1981–1987. Journal of Great Lakes Research 16: 209–215.

Kerr, S. J., D. Gibson, and A. C. McKee. 1988. A study of chinook salmon, *Oncorhynchus tshawytscha*, smolting activity in the Pottawatomi River during the spring of 1987. Ontario Ministry of Natural Resources, Owen Sound, Ontario.

Kitchell, J. F., D. J. Stewart, and D. Weininger. 1977. Applications of a bioenergetics model to yellow perch (*Perca flavescens*) and walleye (*Stizostedion vitreum vitreum*). Journal of the Fisheries Research Board of Canada 34:1922–1935.

Kocik, J. F. 1988. Population parameters and abundance of pink salmon in the upper Great Lakes. Masters thesis, Michigan State University, East Lansing, Michigan. 64 pp.

————, W. W. Taylor. 1987. Diet and movements of age 1+ pink salmon in western Lake Huron. Transactions of the American Fisheries Society 116: 628–633.

————, 1994. Summer survival and growth of brown trout with and without steelhead under equal total salmonine densities in an artificial stream. Transactions of the American Fisheries Society 123: 931–938.

————, 1995. Effect of juvenile steelhead, *Oncorhynchus mykiss*, and age-zero and age-one brown trout, *Salmo trutta*, survival and growth in a sympatric nursery stream. Canadian Journal of Fisheries and Aquatic Sciences 52: 105–114.

————, W. C. Wagner. 1991. Abundance, size, and recruitment of pink salmon (*Oncorhynchus gorbuscha*) in selected Michigan tributaries of the upper Great Lakes, 1984–1988. Journal of Great Lakes Research 17: 203–213.

Koonce, J. F. and M. L. Jones. 1994. Sustainability of intensively managed fisheries of Lake Michigan and Lake Ontario: Final report of the SIMPLE task group. Great Lakes Fishery Commission. Ann Arbor, Michigan.

Krueger, C. C. and B. May. 1987. Genetic comparisons of naturalized rainbow trout populations among Lake Superior tributaries: differentiation based on allozyme data. Transactions of the American Fisheries Society 116: 795–806.

————, D. L. Perkins, R. J. Everett, D. R. Schreiner and B. May. 1994. Genetic variation in naturalized rainbow trout (*Oncorhynchus mykiss*) from Minnesota Tributaries to Lake Superior. Journal of Great Lakes Research 20: 299–316.

————, Perkins, D. L., Mills, E. L., and Marsden, J. E. 1995. Predation by Alewife on lake trout fry in Lake Ontario: role of an exotic species in preventing restoration of a native species. Journal of Great Lakes Research 21 (Suppl. 1):458–469.

Kruger, K. M., W. W. Taylor and J. R. Ryckman. 1985. Angler use and harvest in the Pere Marquette River near Baldwin, Michigan. Michigan Academician 317–330.

Kwain, W. 1981. Population dynamics and exploitation of rainbow trout in Stokely Creek, eastern Lake Superior. Transactions of the American Fisheries Society 110: 210–215;.

————, J. A. Chappel 1978. First evidence for even-year spawning pink salmon (*Oncorhynchus gorbuscha*) in Lake Superior. Journal of the Fisheries Research Board of Canada 35: 1373–1376.

————, E. Thomas. 1984. The first evidence of spring spawning by chinook salmon in Lake Superior. North American Journal of Fisheries Management 4:227–229.

Larson, G. L. and S. E. Moore. 1985. Encroachment of exotic rainbow trout into stream populations of native brook trout in the southern Appalachian mountains. Transactions of the American Fisheries Society 114:195–203.

Loftus, K. H. 1958. Studies on river-spawning populations of lake trout in eastern Lake Superior. Transactions of the American Fisheries Society. 87:259–277.

MacCrimmon, H. R. 1971. World distribution of rainbow trout (*Salmo gairdneri*). Journal of the Fisheries Research Board of Canada 28:663–704.

————, B. L. Gots. 1972. Rainbow trout in the Great Lakes. Ontario Ministry of Natural Resources 66 pp.

McDowall, R. M. 1987. The occurrence and distribution of diadromy among fishes. American Fisheries Society Symposium 1:1–13.

McNeil, F. I., N. R. Payne, and E. J. DeLaPlante. 1988. Estimated catches of rainbow trout, lake trout, and Pacific salmon in commercial gill nets, Lake Huron, 1979–1987. Ontario Ministry of Natural Resources, Lake Huron Fisheries Assessment Unit Report 1–88. Owen Sound, Ontario.

Merna, J. W. 1986. Contamination of stream fishes with chlorinated hydrocarbons from eggs of Great Lakes salmon. Transactions of the American Fisheries Society 115: 69–74.

Nehlsen, W., J. E. Williams and J. A. Lichatowich. 1991. Pacific salmon at the crossroads: stocks at risk from California, Oregon, Idaho, and Washington. Fisheries 16:4–21.

Negus, M. T. 1995. Bioenergetics modeling as a salmonine management tool applied to Minnesota waters of Lake Superior. North American Journal of Fisheries Management 15:60–78.

Nunan, P. J. 1967. Pink salmon in Lake Superior. Ontario Fish and Wildlife Review 6: 8–13.

O'Gorman, R., R. A. Bergstedt, and T. H. Eckert. 1987. Prey fish dynamics and salmonine predator growth in Lake Ontario, 1978–84. Canadian Journal of Fisheries and Aquatic Sciences 44(Suppl. 2): 390–403.

Ontario Ministry of Natural Resources (OMNR). 1992. Strategic plan for Ontario fisheries. Queens Printer for Ontario. Toronto, Ontario.

————, 1994. Lake Ontario Fisheries Unit: 1993 annual report. OMNR. Picton, Ontario

Parsons, J. W. 1973. History of salmon in the Great Lakes, 1850–1970. Technical Papers of the Bureau of Sport Fisheries and Wildlife 68: 80 pp.

Peck, J. W. 1970. Straying and natural reproduction of coho salmon, *Oncorhynchus kisutch*, planted in a Lake Superior tributary. Transactions of the American Fisheries Society 99:591–595.

————, 1974. Migration, food habits, and predation on yearling coho salmon in a Lake Michigan tributary and bay. Transactions of the American Fisheries Society 103:10–14.

————, 1992. The sport fishery and contribution of hatchery trout and salmon in Lake Superior and its tributaries at Marquette, Michigan, 1984–87. Michigan Department of Natural Resources Fisheries Research Report 1975: 62 pp.

————, 1994. Rehabilitation of a Lake Superior steelhead population by stocking yearling smolts. Michigan Department of Natural Resources Fisheries Research Report 2012: 16 pp.

————, W. R. MacCallum, S. T. Schram, D. R. Schreiner, and J. D. Shively. 1994. Other salmonines *In* M. J. Hansen, editor. *The State of Lake Superior in 1992*. Great Lakes Fishery Commission Special Publication 94–1. pp 35–52

Powell, M. J. and M. Miller. 1990. Shoal spawning by chinook salmon in Lake Huron. North American Journal of Fisheries Management 10:242–244.

Raleigh, R. F., T. Hackman, R. C. Solomon, and P. C. Nelson. 1984. Habitat suitability information: rainbow trout. United States Department of the Interior. Fish and Wildlife Service Report FWS/OBS–82/10.60.

————, W. J. Miller, and P. C. Nelson. 1986. Habitat suitability index models and instream flow suitability curves: chinook salmon. United States Department of the Interior. Fish and Wildlife Service Biological Report 82(10.122).

Rand, P. S., D. J. Stewart, P. W. Seelbach, M. L. Jones and L. R. Wedge. 1993. Modeling steelhead population energetics in Lakes Michigan and Ontario. Transactions of the American Fisheries Society 122: 977–1001.

Regier, H. A. and J. J. Kay. 1996. An heuristic model of transformations of the aquatic ecosystems of the Great Lakes–St. Lawrence River basin. Journal of Aquatic Ecosystem Health. 5:3–21.

Richardson, A. H. 1944. The Ganaraska watershed: A study in land use with recommendations for rehabilitation of the area in the post-war period. Ontario Department of Planning and Development, Toronto, Ontario.

Rose, G. A. 1986. Growth decline in subyearling brook trout (*Salvelinus fontinalis*) after emergence of rainbow trout (*Salmo gairdneri*). Canadian Journal of Fisheries and Aquatic Sciences 43: 187–193.

Rybicki, R. W., P. J. Seelbach, and W. M. Wagner. 1990. Biology of salmonids. Pages 138–194 *in* M. Keller, K. D. Smith, and R. W. Rybicki, editors. Review of salmon and trout management in Lake Michigan. Michigan Department of Natural Resources Fisheries Special Report No. 14. Charlevoix, Michigan.

Schumacher, R. E. and S. Eddy. 1960. The appearance of pink salmon, *Oncorhynchus gorbuscha* (Walbaum), in Lake Superior. Transactions of the American Fisheries Society 89: 371–373.

Scott, W. B. and E. J. Crossman. 1973. Freshwater fishes of Canada. Fisheries Research Board of Canada Bulletin 184: 966 pp.

Seelbach, P. W. 1985. Smolt migration of wild and hatchery-raised coho and Chinook salmon in a tributary of northern Lake Michigan. Fisheries Research Report No. 1935. Michigan Department of Natural Resources, Lansing, Michigan.

————, 1987a. Smolting success of hatchery-raised steelhead planted in a Michigan tributary of Lake Michigan. North American Journal of Fisheries Management 7:223–231.

————, 1987b. Effect of winter severity on steelhead smolt yield in Michigan: an example of the importance of environmental factors in determining smolt yield. American Fisheries Symposium 1: 441–450.

————, 1993. Population biology of steelhead in a stable-flow, low-gradient tributary of Lake Michigan. Transactions of the American Fisheries Society 122:179–198.

————, J. I. Dexter, and N. D. Ledet. 1994. Performance of steelhead smolts stocked in southern Michigan warm-water rivers. Fisheries Research Report No. 2003. Michigan Department of Natural Resources, Lansing, Michigan.

————, B. R. Miller. 1993. Dynamics in Lake Superior of hatchery and wild steelhead emigrating from the Huron River, Michigan. Fisheries Research Report No. 1993. Michigan Department of Natural Resources, Lansing, Michigan.

————, G. E. Whelan. 1988. Identification and contribution of wild and hatchery steelhead stocks in Lake Michigan tributaries. Transactions of the American Fisheries Society 117: 444–451.

Smith, S. H. 1995. Early ecologic changes in the fish community of Lake Ontario. Great Lakes Fishery Commission Technical Report 60, Ann Arbor, Michigan.

Stanfield, L. S., M. L. Jones, and J. N. Bowlby. 1995. A conceptual framework for Atlantic salmon restoration in Lake Ontario. Ontario Ministry of Natural Resources, Picton, Ontario.

Stauffer, T. M. 1972. Age, growth, and downstream migration of juvenile rainbow trout in a Lake Michigan tributary. Transactions of the American Fisheries Society 101:18–28.

Stearley, R. F. and G. R. Smith. 1993. Phylogeny of the Pacific trouts and salmon (Oncorhynchus) and genera of the family salmonidae. Transactions of the American Fisheries Society 122: 1–33.

Stewart, D.J. 1980. Salmonid predators and their forage base in Lake Michigan: a bioenergetics modeling synthesis. Doctoral dissertation. University of Wisconsin, Madison, Wisconsin, USA.

————, M. Ibarra. 1991. Predation and production by salmonine fishes in Lake Michigan, 1978–88. Canadian Journal of Fisheries and Aquatic Sciences 48:909–922.

————, J. F. Kitchell, and L. B. Crowder. 1981. Forage fish and their salmonid predators in Lake Michigan. Transactions of the American Fisheries Society 110:751–763.

Talhelm, D. R. 1988. The international Great Lakes sport fishery of 1980. Great Lakes Fishery Commission Special Publication 88–4, Ann Arbor, Michigan.

Trimble, S. W. 1983. A sediment budget for Coon Creek basin in the Driftless Area, Wisconsin, 1853–1977. American Journal of Science 283:454–474.

Trotter, P. C. 1989. Coastal cutthroat trout: a life history compendium. Transactions of the American Fisheries Society 118:463–473.

Turner, C. E. and H. T. Bilton. 1968. Another pink salmon in its third year. Journal of the Fisheries Research Board of Canada 25:1993–1996.

Wagner, W. C., and T. M. Stauffer. 1980. Three-year-old pink salmon in Lake Superior tributaries. Transactions of the American Fisheries Society 109(4):458–460.

———, 1982. Distribution and abundance of pink salmon in Michigan tributaries of the Great Lakes, 1967–1980. Transactions of the American Fisheries Society 111:523–526.

Walters, C. J. 1986. Adaptive management of renewable resources. MacMillan, New York.

Ward, B. R., P. A. Slaney, A. R. Facchin, and R. W. Land. 1989. Size-biased survival in steelhead trout (*Oncorhynchus mykiss*): back-calculated lengths from adult's scales compared to migrating smolts at the Keogh River, British Columbia. Canadian Journal of Fisheries and Aquatic Sciences 46:1853–1858.

Webster, D. A. 1982. Early history of the Atlantic salmon in New York. New York Fish and Game Journal 29:26–44.

Wright, G. M. 1994. State of the Lake Report Status of Chinook salmon. Great Lakes Fishery Commission Lake Huron Committee Meeting Report, Ann Arbor, Michigan.

Zaft, D. J. 1992. Migration of wild Chinook and coho salmon smolts from the Pere Marquette River, Michigan. Master's Thesis Michigan State University, E. Lansing, Michigan 87 pp.

CHAPTER 16

Ascent, Dominance, and Decline of the Alewife in the Great Lakes: Food Web Interactions and Management Strategies

Robert O'Gorman and Thomas J. Stewart

Introduction

The alewife (*Alosa pseudoharengus)* is an anadromous marine clupeid indigenous to the east coast of North America, from Newfoundland to North Carolina (Scott and Crossman 1973). It colonized the Great Lakes via navigation canals over an eighty-year period (Miller 1957; Ihssen et al. 1992), becoming first a well-publicized nuisance because of large die-offs in spring (Pritchard 1929; Greenwood 1970), then a low-value commercial resource (Hanson 1987), and finally, as prey for stocked salmonines, the key species supporting a multimillion dollar sport fishery (Talhelm 1988). Initial management efforts focused on reducing the abundance of alewives, first by commercial harvest, and later, by stocking salmonines to eat them. Then, as a valuable sport fishery developed, managers focused on maintaining sufficient numbers to sustain the piscivores, and finally, as alewife numbers waned, management efforts turned to conserving the diminished populations to preserve the sport fishery.

Although the alewife eventually fueled an economically important sport fishery, its colonization of the Great Lakes was not without injury. By the 1980s, decades of research had firmly implicated the alewife in the decline of many important Great Lakes fishes (Smith 1970, 1972; Crowder 1980; Stewart et al. 1981), and by the 1990s, contemporary research was suggesting that the alewife was a major impediment to the costly, binational effort to restore self-sustaining populations of lake trout (*Salvelinus namaycush*) (Jones et al. 1995; Fisher et al. 1996). How the alewife was able to reduce endemic fish populations is not completely known, but most evidence suggests that they did it by eating the endemic fish larvae (Brandt et al. 1987; Eck and Wells 1987; Brown et al. 1987). More recently, alewife were shown to eat the fry of hatchery-reared lake trout that spawned in Lake Ontario (Kreuger et al. 1995), and to produce a thiamine deficiency in mature lake trout that ate mostly alewife (Fisher et al. 1996). The offspring of the thiamine deficient lake trout suffer high mortality at an early life stage from a syndrome that appears linked to thiamine deficiency (Fitzsimons 1995). Thus, the alewife could be contributing to the failure of stocked lake trout to become self-reproducing, not only by eating young lake trout soon after they hatch, but also by causing a potentially lethal vitamin deficiency among the young lake trout.

We chronicle the ascent, dominance, and decline of the alewife in the Great Lakes and track the gradual accumulation of knowledge on the fish's effect on the aquatic community. Changes in management strategies for alewife are followed, and the current management dilemma is framed in light of the alewife's effect on indigenous fishes and the changing biota and trophic status of the Great Lakes.

Ascent

Lake Ontario

The first record of alewives in the Great Lakes was from Lake Ontario in 1873 (Bean 1884). But judging from the alewife's abundance later in that decade, it most certainly was present in the lake some years earlier (Smith 1970). Alewives had access to Lake Ontario from the ocean via the St. Lawrence River, and landlocked populations were within the Lake Ontario–St. Lawrence River drainage that might have been relics of the Champlain Sea invasion in the late Pleistocene (Radforth 1944). But prior to the establishment of alewife in Lake Ontario, there were no records of landlocked populations within the Lake Ontario–St. Lawrence River drainage, and there were few records of alewives in the 483 km stretch of river between Lake Ontario and the city of Quebec (Bean 1884; Smith 1995). Furthermore, a study of allozyme variation among alewives from the Great Lakes, Cayuga Lake (Seneca River drainage), and the Canadian Maritimes strongly suggests that the alewife is a recent invader, and that the most likely route of entry was not the St. Lawrence River, but rather the section of the Erie Canal that linked the Hudson-Mohawk River system, which drained into the Atlantic Ocean, with the Oneida-Seneca-Oswego River system, which drained into Lake Ontario (Ihssen et al. 1992). Although the section of the Erie Canal that connected the two drainages was completed in the early 1800s (Smith 1985), the enlargement and rerouting of the canal in the mid 1800s were probably the key changes needed for the alewife to successfully traverse it.

In the late 1870s, commercial fishermen reported that alewives were very common in northwestern Lake Ontario (Pritchard 1929), and by the late 1880s, they were abundant (Wright 1891). From 1876 to 1886 the city of Toronto, Ontario was asked to take measures to bury alewives that succumbed and washed ashore in spring, and in 1892, residents of Burlington, Ontario, at the extreme western end of the lake, complained of the smell and cost of removing dead alewives that washed ashore in summer (Pritchard 1929). Alewives were also abundant at the far eastern end of the lake where a fishery based at Pillar Point, New York caught sufficient numbers of alewives in 1885 to make five hundred gallons of oil and sixty-three tons of fertilizer (Smith and Snell 1891). Reduction of alewives for oil soon became unprofitable, however, because the oil content of alewives dropped when their abundance peaked (Smith 1892; Smith 1995).

We have no way of gauging the size of the population of alewives in Lake Ontario until about one hundred years after they colonized the lake. Alewives were generally not sought by commercial fisherman, so there are no annual records of commercial landings and the concept of routine assessment of fish abundance had yet to emerge, so there are only occasional, qualitative records from government agencies and universities. For example, Dymond et al. (1929) reported that alewives were common throughout the lake from 1927 to 1928. We do know, however, that noticeable mortalities of alewives occurred sporadically for more than a century (Wright 1891; Pritchard 1929; Graham 1956), and although they may not be an altogether reliable indicator of abundance, they would be unlikely to cause public comment if the alewife's population density was not relatively high (Christie 1973).

In the 1960s and 1970s, alewives apparently were the most abundant fish in the lake. Wells (1969) reported that alewives were, by far, the most common fish caught with an experimental bottom trawl at several locations in southern Lake Ontario in fall 1966. In 1968, bottom trawls fished in northeastern and northwestern Lake Ontario to evaluate the feasibility of a commercial trawl fishery caught mainly alewives, whereas from 1969 to 1970 they caught mainly rainbow smelt (*Osmerus mordax*) followed by alewives (Thurston 1975; Howell 1975). Finally, a synoptic survey of the lake's fishery resources, conducted in 1972, found that alewives were the dominant species in the main lake basin (Christie and Thomas 1981).

Lake Erie

Alewives were first reported from Lake Erie in 1931 (Dymond 1932). They most likely colonized Lake Erie from Lake Ontario by moving through the Welland Canal (Ihssen et al. 1992). Although the Niagara River drains Lake Erie into Lake Ontario, passage of fish upstream is blocked by Niagara Falls. The Welland Canal, connecting Lakes Erie and Ontario, was constructed to allow ships to bypass the falls, and it apparently allowed the alewife to do so also. Although the Welland Canal was completed in 1829 (Smith 1970), the decisive factor that resulted in fishes traversing the 44.4km canal system was probably the complete elimination of water flow to the canal from the Grand River in the early 1920s (Ashworth 1986). After this, any fish swimming upstream from Lake Ontario would eventually arrive in Lake Erie with no possibility of being detoured into the inflowing water of the Grand River.

Alewives were common in Lake Erie by 1942 (Smith 1968), and by 1950, they were abundant (Miller 1957). They remained noticeably abundant through 1954 (Miller 1957). But the 1950 to 1954 period was apparently the alewife's longest period of abundance in this lake. In subsequent years, despite an occasional strong year class (Leach and Nepszy 1976), alewives were not considered particularly numerous, nor were they thought to be responsible for changes in Lake Erie's fish community (Smith 1968; Hartman 1973, 1988). The failure of alewife to become abundant in Lake Erie for more than short periods of time has been attributed to a large population of predators and to the limited amount of deep water where the fish concentrate in winter to avoid cold water (Smith 1968). The temperatures at which alewives are found in Lakes Michigan and Ontario in winter suggest that the fish cannot tolerate temperatures at or lower than 1°C (Wells 1968; Bergstedt and O'Gorman 1989). In Lake Erie, water in the western and central basins is isothermal at about 1°C by late December and the deeper eastern basin is nearly isothermal at 1°C in winter (Hartman 1973).

Die-offs of alewives in Lake Erie were not reported with the regularity that accompanied the alewife's colonization of Lake Ontario. Conspicuous die-offs occurred in 1952, 1967, and in the mid 1970s (A. Larsen and R.G. Ferguson, personal communication cited in Smith 1970; Leach and Nepszy 1976). Another die-off occurred in northeastern Lake Erie in February 1992, as large numbers of dead alewives were seen floating down the Niagara River (O'Gorman, personal observation). The sporadic reports of mortalities generally coincided with periods of high alewife abundance, suggesting that only when abundance was high did sufficient numbers of fish die to arouse comment (i.e., annual mortalities of small numbers of fish when population levels were low went unnoticed). Alternatively, die-offs simply may not have occurred at low population levels.

Lake Huron

Once alewives reached Lake Erie, they had unobstructed access to Lakes Huron and Michigan via the Detroit-St. Clair River system, and by 1933, some of the fishes had used this route to reach northeastern Lake Huron (MacKay 1934; Miller 1957). However, it wasn't until the mid 1950s that alewives became abundant in Lake Huron (Miller 1957; Scott 1963; Smith 1970). Only one spawning alewife was taken by an experimental fishery on South Bay, Manitoulin Island in 1951, but by 1954 the catch had ballooned to about one metric ton (t) (Scott 1963). During the 1957 to 1963 period, the catch routinely exceeded 9t, suggesting that alewives were a dominant member of the fish community in the northeastern part of the lake during that period (Coble 1967). In the main lake basin, alewives were the most frequently caught fish in experimental gill nets from 1967 to 1968 (Spangler and Collins 1992).

Lake Michigan

The first report of alewives in Lake Michigan was from the northern part of the lake in 1949, and by 1953, alewives had spread throughout most of the lake (Miller 1957). By 1956, commercial netters in Green Bay, northwestern Lake Michigan, were catching large quantities of alewives in pound nets fished for other species (Wells and McLain 1973), and by the winter of 1956 to 1957, commercial operators in southern Lake Michigan were complaining about alewives fouling their gill nets set for deepwater ciscoes (*Coregonus* spp.) (Miller 1957). Wells and McLain (1973) characterized the population increase of alewives in the late 1950s and early 1960s as explosive. The catch of alewives per hour of experimental trawling in northern Lake Michigan was 0.4 in 1955, and 915 in 1961 (Brown 1972). Similarly, the catch per hour in southern Lake Michigan was 0 in 1954, and 582 in 1960. Commercial landings increased from 100t in 1957, to 2,151t in 1962, and reached 19,000t in 1967 (Baldwin et al. 1979). Landings of the low-value alewife might well have been even higher if market demand had been stronger. An index of relative abundance of adult alewives (age-two and older) in southern and central Lake Michigan derived from catches with experimental and commercial bottom trawls indicate a nearly five-fold increase in biomass from 1963 to 1966 (Brown 1968).

Alewives clearly became the most abundant fish in Lake Michigan during the 1960s. They made up 48% to 76% (by weight) of bottom trawl catches during exploratory surveys of northern Lake Michigan from 1963 to 1965 (Reigle 1969b, 1969d) and 51% (by weight) of bottom trawl catches during exploratory surveys of southern Lake Michigan from 1962 to 1965 (Reigle 1969c). During surveys in the southern portion of the lake, the proportion of alewives in the catch increased each year from 17% in 1962, to 74% in 1965. Pound nets fished in Milwaukee Harbor from May through August 1968 caught 100t of alewife, but only about 0.3t of suckers, Catostomidae, and common carp (*Cyprinus carpio)* and less than a thousand individuals of other species (Greenwood 1970).

Lake Superior

Alewives were first reported from Lake Superior in 1954 (Miller 1957), but they never became abundant there (Lawrie and Raher 1973). Although catches of a few thousand individuals have occurred sporadically through the years, alewives were not abundant enough to be routinely captured in survey netting (Reigle 1969a; Bronte et al. 1991) and were rarely present in stomachs of salmonines (Conner et al. 1993). The main reason alewives never established a substantial population in Lake Superior is that it is the northernmost, and thus, the coldest of the Great Lakes (Smith 1968, 1972; Bronte et al. 1991).

Reasons for the Alewife's Ascent to Dominance

Miller (1957) suspected that the rapid spread and increase in abundance of the alewife in the early 1950s was due to changes in the fish communities caused by the sea lamprey (*Petromyzon marinus*), another marine invader. The sea lamprey killed large numbers of lake trout, the main deepwater piscivore in all the Great Lakes except Lake Erie, and contributed to eliminating lake trout from all but the northernmost portion of their historic range in the 1940s and 1950s (Hile et al. 1951; Eschmeyer 1957). Although there is some debate over the relative importance of sea lamprey predation and overfishing in the destruction of the lake trout populations (Coble et al. 1990, 1992; Eshenroder 1992; Hansen 1999), there is little doubt that sea lampreys were highly influential. Smith (1970) examined the changes in the native fish communities of the Great Lakes that preceded colonization by the alewife and concluded that "...without exception the alewife has failed to appear or to become abundant in any of the Great Lakes during its dispersal when the lake was densely populated with large predators." (Smith 1970, p. 755). Alewives are a favorite prey of lake trout, they were eaten by the native lake trout in Lake Ontario (Dymond 1928), and are prominent in the diet of stocked lake trout in Lakes Ontario, Huron, and Michigan (Eck and Wells 1983; Brandt 1986; Diana 1990; Elrod and O'Gorman 1991). Thus, the relaxation of predation pressure that accompanied the extreme reduction of lake trout in Lakes Huron and Michigan seems quite clearly to have been the pivotal change in fish community structure that allowed rapid colonization and explosive growth in numbers of alewives in the two lakes. In Lake Superior, the severe reduction of lake trout in the 1950s no doubt facilitated colonization there (Smith et al. 1974). But in Lake Ontario, the native lake trout were not eliminated until about 1950, nearly ninety years after alewife colonized the lake (Christie 1973). However, the numbers of another abundant native piscivore, the Atlantic salmon (*Salmo salar*), declined sharply in the 1860s (Huntsman 1944), and this must have relaxed predation pressure enough to allow the alewife to successfully colonize Lake Ontario in the late 1860s, and to proliferate in the 1870s.

In Lake Erie, *Stizostedion spp.* were the dominant piscivores in the historic fish community, and Smith (1970) contended, on the basis of reduced commercial landings, that they declined in the late 1920s, just prior to the first sighting of the alewife in 1931. In our view, however, the decline of commercial landings was not of sufficient magnitude or duration to suggest a major relaxation of predation pressure. From 1920 to 1927 the *Stizostedion* catch was composed of about 75% blue pike (*S. vitreum glaucum*), 16% sauger (*S. canadense*), and 9% walleye (*S. vitreum vitreum*) (Applegate and Van Meter 1970; Baldwin et al. 1979). From 1928 to 1929, landings of blue pike and sauger declined about 50% from the 1920 to 1927 average, and landings of walleye declined about

20%. But from 1930 to 1931, landings of both blue pike and walleye were 26% higher than from 1920 to 1927, whereas landings of sauger were 35% lower. The synchronous fall and rise of landings may well have been due to a series of weak year classes in these closely related species, although diminished fishing effort, or fishing success from 1928 to 1929 would be an equally plausible explanation for the brief dip in landings. At any rate, had the *Stizostedion* populations actually suffered a short-term decline in the late 1920s, predation pressure would not have been relaxed due to the collapse of lake herring (*Coregonus artedii*), a pelagic planktivore like the alewife. Lake herring landings fell from about 9,525t in the early 1920s to about 454t in the late 1920s, and except for a brief resurgence in the mid 1940s, never recovered (Baldwin et al. 1979; Leach and Nepszy 1976). Excessive harvest by commercial netters is widely believed to be the reason for the collapse of lake herring in Lake Erie (Van Oosten 1930; Regier et al. 1969; Leach and Nepszy 1976).

We suggest that, rather than a relaxation of predation pressure, a more plausible explanation for the appearance of alewives in Lake Erie was that: collapse of the enormous populations of lake herring left a vacant niche for the alewife to exploit; elimination of water flowing into the Welland Canal from the Grand River in 1921 increased the number of alewives arriving from Lake Ontario (Ashworth 1986); and a lack of harsh winters from 1921 to 1930 improved survival of the colonizers (Bolsenga and Norten 1993; see below). Predation probably helped regulate the alewife population until the late 1950s, when blue pike, sauger, and walleye populations collapsed (Applegate and Van Meter 1970; Baldwin et al. 1979). Predation by walleye has been shown to reduce alewife numbers in the Great Lakes. During the late 1970s, resurgence of walleye populations in Lake Erie's western basin and in Lake Ontario's Bay of Quinte played a major role in local reductions of alewife abundance (Ridgway et al. 1990; Knight and Vondracek 1993). In spite of predation, however, the numbers of planktivorous alewives in Lake Erie surged in the early 1950s, and we attribute this to warmer than normal weather and to nutrient enrichment increasing planktonic production. In the open waters of central Lake Erie, total phosphorous apparently doubled between 1942 and 1958 (Hartman 1973).

Warm weather during the 1950s favored the spread of alewives throughout the upper Great Lakes, a region clearly on the northern edge of their distribution. Winters in the basin were all warmer than normal from 1949 to 1958, when alewives were increasing, and the summers of 1952 to 1955 were also warmer than normal (Bolsenga and Norton 1993). Massive mortalities of alewives often coincide with colder than average winters (Colby 1973; O'Gorman and Schneider 1986; O'Gorman et al. 1987), and weak year classes of alewives have been associated with colder than average growing seasons (Eck and Wells 1987). In most winters, alewives in the Great Lakes are exposed to temperatures between 1°C and 3°C (Mortimer 1971; Hartman 1973; Rodgers 1987), temperatures that they avoid in the ocean by migrating southward, and in severe winters, alewives in the Great Lakes may be exposed to temperatures at or lower than 1°C, temperatures that apparently can be lethal (Bergstedt and O'Gorman 1989). Temperature, however, is not the only factor that determines the severity of winter mortalities. In winters with similar air temperatures, mortality will be less severe in a population of alewives that are fat, or in good condition, than in a population of alewives that are skinny, or in poor condition (Brown 1972; Colby 1973; Bergstedt and O'Gorman 1989). In addition, factors other than condition apparently influence the severity of winter mortalities as the alewife population in Lake Ontario survived the 1993 to 1994 winter,

Chronological list of important events in the history of alewife in the Great Lakes

DECADE	LAKE	EVENT
1870	Ontario	First record of alewives, 1873.
1880	Ontario	Alewives are abundant.
1930	Erie	First record of alewives, 1931.
	Huron	First record of alewives, 1933.
1940	Michigan	First record of alewives, 1949.
1950	Erie	Alewives briefly abundant, 1950–1954.
	Huron	Alewives increase rapidly, mid 1950s.
	Superior	First record of alewives, 1954.
	Michigan	Alewives increase rapidly, late 1950s.
1960	Ontario	Alewives dominate fish community.
	Huron	Alewives dominate fish community.
	Michigan	Alewives dominate fish community; adult alewife biomass increases nearly 5-fold, 1963–1966; die-off reduces numbers by more then 50%, 1967.
1970	Ontario	Alewives dominate fish community; massive winter mortality, 1976–1977.
	Erie	Alewives briefly abundant, early 1970s.
	Huron	Alewives begin to decline.
	Michigan	Alewives increase, early 1970s, and then decrease, late 1970s.
1980	Ontario	Alewives abundant, early 1980s; alewives decline, late 1980s.
	Huron	Alewives continue to decline.
	Michigan	Alewives decline six-fold, 1981–1983; alewives <15% of planktivorous fish biomass, 1987–1989.
1990	Michigan	Commercial fishery for alewives closed, 1991.
	Ontario	Adult alewife biomass in 1990–1994, 42% lower than in 1980–1984.

one of the coldest of this century, even though the fish were in relatively poor condition (O'Gorman, personal observation). Perhaps mortality severity is influenced by the duration of exposure to low temperatures, or by how rapidly temperatures decline in the fall. Finally, exposure to stressfully low temperatures in winter may not kill the fish immediately, but it may result in them dying the subsequent spring (Colby 1971, 1973).

Colder than average growing seasons most likely result in poor recruitment because growth of young-of-year (YOY) alewives is reduced and size-dependent overwinter survival favors the larger young (Eck and Brown 1985; Eck and Wells 1987). Indeed, there is a threshold size (about 60mm, total length) that YOY alewives must achieve if they are to survive their first winter in the Great Lakes (Brown 1972; Argyle 1982; O'Gorman, personal observation). Why small YOY alewives fail to survive winters in the Great Lakes is not known. Condition, or body weight of adult alewives, as well as growth of YOY alewives varied inversely with population size in Lakes Michigan and Ontario (Brown 1972; O'Gorman et al. 1987). So, the initial wave of colonizers would not have been as dependent on favorable weather for continued expansion as would an established population. Nevertheless, normal, or warmer than normal, temperatures would have certainly facilitated the alewife's colonization of the upper lakes and would explain the rapid expansion of the initial populations after the collapse of lake trout.

Dominance and Decline

Alewives dominated fish communities in Lakes Huron and Michigan for a decade or more, and for an even longer time in Lake Ontario (table 1). Their dominance was ended by a reversal of the same factors that allowed the fish to colonize and proliferate: piscivore populations were reestablished by reducing sea lamprey numbers and releasing massive numbers of hatchery-reared fish, climate became cooler, and trophic status was lowered by reducing nutrient enrichment. The record during the period of dominance and decline is most complete for Lakes Michigan and Ontario, so we will focus our discussion on those two lakes. In Lake Huron, alewives began declining in the early 1970s and the decline continued into the 1980s (Henderson and Brown 1985; Hartman 1988). Harsh winters and intensified predation from increasingly numerous salmonines were cited as reasons for the decline.

Lake Michigan

Alewife numbers peaked in the mid 1960s in Lake Michigan and then declined sharply in 1967 when more than half of the fish perished during a die-off that started in winter and continued through midsummer (Brown 1968, 1972). Reductions in alewife condition, growth, and food (zooplankton) in the mid 1960s prompted Brown (1972) to speculate that abundance at the population peak was apparently beyond the carrying capacity of the lake. Indeed, alewife numbers never returned to levels close to those of the 1960s peak (Brown et al. 1999). After 1967, the population slowly rose to another, albeit lower, peak from 1973 to 1975 (Hatch et al. 1981), and then began an irregular decline that culminated in a six-fold decline from 1981 to 1983, from which the population never recovered (Eck and Wells 1987; Stewart and Ibarra 1991). By 1984, alewife abundance had slid to less than 20% of that after the 1967 die-off (Kitchell and Crowder 1986), and the bloater (*Coregonus hoyi)* had become the dominant planktivorous fish in Lake Michigan (Eck and Wells 1987). A whole-lake measure of planktivorous-fish biomass in 1987 and 1989 showed that the biomass of alewives made up only 9% to 14% of the combined biomass of alewives, bloater, and rainbow smelt (Brandt et al. 1991; Argyle 1992).

Failure of alewives to return to nuisance levels of abundance after the 1967 die-off was most certainly due to cropping by piscivorous salmonines with some assistance from the commercial fishery in the years immediately after the die-off. Commercial harvest of alewives exceeded salmonine consumption of alewives from 1967 to 1968, and was equal to salmonine consumption in 1969 (fig. 1). But in the ensuing years, salmonine consumption exceeded commercial harvest by increasingly large margins. Large-scale releases of hatchery-reared salmonines, which began in 1965, quickly established populations of piscivores, and these populations expanded rapidly due to increases in the number of fish planted. Total plantings increased from an annual average of 3.4 million fish from 1965 to 1968, to 8.6 million from 1969 to 1972, 12.1 million from 1973 to 1977, and 15.4 million from 1978 to 1984 (Eck and Wells 1987). As a result, the estimated annual average consumption of alewives by salmonines steadily rose from about 25,000t from 1970 to 1972, to about 64,000t from 1980 to 1988 (Stewart et al. 1981; Stewart and Ibarra 1991). In contrast, landings by the commercial fishery in the 1970s fluctuated without trend from about 13,000t to 22,000t, and sank to about 5,000t in the late 1980s (fig. 1). Commercial landings were strongly

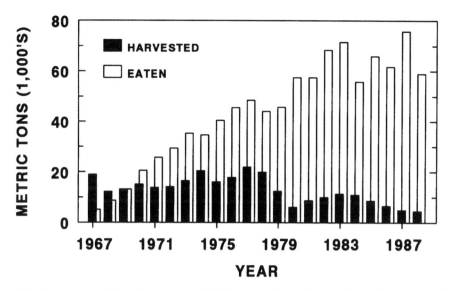

FIG. 1. *Alewives harvested by the commercial fishery and eaten by stocked salmonines, Lake Michigan, 1967–1988. Commercial harvest is from Baldwin et al. (1979), Hanson (1987), and Brown et al. (1999) and salmonine consumption is from Stewart et al. (1981) and Stewart and Ibarra (1991).*

influenced by market demand and only weakly linked to the size of the alewife population (Hanson 1987).

Reasons for the decline of alewife after the mid 1970s are a matter of controversy. Crowder and Binkowski (1983) suggested that competition from a native planktivore, the bloater (*Coregonus hoyi*), was important but only after alewives had begun to decline and bloaters had begun to increase. Eck and Wells (1987), however, argued against this hypothesis noting that the alewife successfully colonized Lake Michigan when bloaters were abundant. Eck and Brown (1985) hypothesized that colder than average growing seasons from 1976 to 1982 did not allow very many YOY alewife to exceed the threshold size needed to survive the winter and the resultant reduced recruitment was responsible for declining alewife numbers. Eck and Wells (1987) later presented evidence for this scenario. Kitchell and Crowder (1986), however, held that predation by salmon and trout was substantially, if not entirely, responsible. Stewart and Ibarra (1991) suggested that intraspecific processes and weather may have been the dominant forces controlling alewife numbers before 1982, but after that, heavy predation may have become dominant. Mean annual total phosphorus in Lake Michigan waters fell from about 7 mg L^{-1} in 1976 to about 4 mg L^{-1} in 1985 (Hartig et al. 1991), so the declining trophic status of the lake may also have played a role.

Lake Ontario

Alewife were probably the most abundant fish in the open waters of Lake Ontario for decades prior to the 1960s and 1970s, when surveys demonstrated their dominance (Wells 1969; Christie and Thomas 1981). Accurate records of changes in alewife abundance start in 1961 in Canadian waters of the northeastern basin (Christie et al. 1987) and in 1977 in U.S. waters (O'Gorman and

Schneider 1986; O'Gorman et al. 1987; Jones et al. 1993). Smoothed population trends in the northeastern basin show numbers building to a strong peak in the mid 1960s followed by a decline, a minor population peak in the mid 1970s followed by a decline, and low abundance in the late 1970s. These population trends are strikingly similar to those in Lake Michigan during the 1960s and much of the 1970s, suggesting that climatic conditions may have been the primary factor influencing alewife numbers in these two Great Lakes during most of those two decades. Low numbers of alewives in Lake Ontario in the late 1970s was clearly linked to a massive die-off in the unusually harsh winter of 1976 to 1977 (O'Gorman and Schneider 1986).

The trends of alewife abundance in Lakes Michigan and Ontario diverged in the 1980s. In Lake Michigan, alewives continued to decline, whereas in Lake Ontario, they rose to high levels by 1980 and remained abundant until 1986, when alewife biomass started an irregular decline (fig. 2). The discrepancy between lakes in the timing of alewife population declines was probably due, in large part, to differences between lakes in the build up of piscivore populations, particularly the chinook salmon (*Oncorhynchus tshawytscha)* population. Per fish stocked, chinook salmon consumed more alewives during their life cycle in the lake than did other salmonines (Stewart et al. 1981). Large-scale releases of salmonines in Lake Ontario started in 1968, three years after large-scale releases in Lake Michigan, and annual releases per unit surface area from 1973 to 1977 were about half those of Lake Michigan (Eck and Wells 1987; Stewart and Ibarra 1991; LeTendre and Savoie 1992). From 1975 to 1979, the number of chinook salmon stocked per unit surface area in Lake Ontario was about one-third of that stocked in Lake Michigan. Even though stocking rates were similar in both lakes for chinook salmon and for all salmonines combined in 1982, in Lake Ontario, poor control of the sea lamprey prior to 1985 undoubtedly retarded establishment of piscivore populations there (Elrod et al. 1995). By 1984, stocking per unit surface area of chinook salmon, and of all salmonines combined in Lake Ontario exceeded that in Lake Michigan by about 50%, and by 1988, improvements in control of sea lamprey had stabilized lamprey numbers at a lower level than in the early 1980s.

During the early 1990s, the alewife population in Lake Ontario was relatively stable at a level below the peak of the early 1980s. From 1990 to 1994, spring trawling assessments in U.S. waters indicated that the number of adult alewives (age two and older) was only 16% lower than from 1980 to 1984, but that the biomass of adult alewives was 42% lower than from 1980 to 1984 (fig. 2). Clearly, adult alewives were, on average, smaller in the early 1990s than they were in the early 1980s. This change was due to: sharp reductions in the numbers of the largest adult alewives in the population (Jones et al. 1993); and alewives in the 1990s being in worse condition (i.e., thinner) than alewives in the 1980s (Rand et al. 1994; U.S. Geological Service–Lake Ontario Biological Station, unpublished data). The number of large alewives was reduced by rising predation rates, accelerating removal of adults, while at the same time zooplankton was declining, slowing the growth of juveniles and resulting in smaller fish recruiting to the adult portion of the population. The thinner alewives were probably due to declines in zooplankton which, in turn, was probably due to reductions in phosphorous inputs to the lake. Aggressive nutrient abatement programs had reduced total phosphorous, midlake in spring, from a 1973 peak of 30.6 mg L^{-1} to 12.8 mg L^{-1} by 1982, and to 10.2 mg L^{-1} by 1989 (Stevens and Neilson 1987; Johannsson et al. 1991a). However, prior to about 1982 and 1983, summer phytoplankton was probably nitrogen limited, so only after

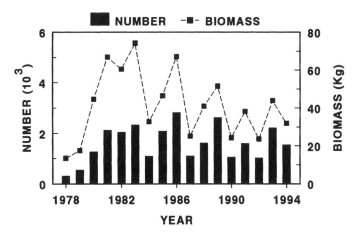

FIG. 2. *Stratified mean catch of adult alewives (age 2 and older) with 12m (headrope) bottom trawls in the U.S. waters of Lake Ontario, 1978–1994.*

those years would further declines in phosphorus be expected to lead to reductions in phytoplankton (Hartig et al. 1991) and, by corollary, zooplankton. Dobson (1994) reported that median values of total phosphorus in Lake Ontario's near-surface waters in the spring of 1991 and 1992 were about 25% lower than in the spring of 1982 and 1983.

Food Web Interactions

In the late 1800s, soon after the alewife colonized Lake Ontario, fishermen realized that alewives were being eaten by a variety of fishes. During this period, chroniclers of the commercial fisheries reported that growth and abundance of some predators increased with the rise in alewife numbers (numerous authors cited in Smith 1970, 1995). However, the earliest scientific investigations of the alewife's place in the food web of Lake Ontario, of which we are aware, were not made until the late 1920s by Dymond (1928) and Pritchard (1929). They found that: alewives ate mainly microcrustacean zooplankton with the larger individuals supplementing their diet with insects when in shallow water and with benthic macroinvertebrates when in deep water, and that alewives were eaten heavily by burbot (*Lota lota*) and lake trout in the open lake. Pritchard (1929) reasoned that predation on alewife reduced the numbers of commercially valuable ciscoes (*Coregonus spp.*) being eaten. Thus, he thought that alewives were beneficial, particularly because they competed "in no serious way with other species," a notion that would be discounted in later years as our understanding of food web dynamics increased. However, the early investigators' observations that alewives occupied an intermediate level in the food chain would be supported by later studies that found that alewives were an important link in the transfer of energy from pelagic zooplankton to top piscivores (Hewett and Stewart 1989; Stewart and Ibarra 1991).

The function of alewives in the Lake Ontario food web did not change through the years, although the fish's food habits and importance in the diet of piscivores shifted in response to introductions of exotic species and extinction of native species. Alewives ate the same foods in

1972 as they did in the 1920s, but they expanded their diets in the 1980s and 1990s to include the zooplankter (*Bythotrephes cederstromi*) and the planktonic veliger larvae of dreissenid mussels after these exotics colonized Lake Ontario (Mills et al. 1992, 1995). The piscivore populations established in recent years by massive stocking programs rely more heavily on alewives for food than did the native piscivores (Brandt 1986; Elrod and O'Gorman 1991; Jones et al. 1993; Rand et al. 1994) because by the early 1970s, some native prey fishes, the deepwater sculpin (*Myoxocephalus thompsoni*) and deepwater ciscoes, were extinct and another native, the lake herring, was severely reduced (Christie 1973; Christie and Thomas 1981). This heavy reliance on alewife was true even though rainbow smelt, another candidate prey fish, had become abundant since colonizing the lake in 1929 (Greeley 1940; Bergstedt 1983). Although the presence of rainbow smelt certainly spared some alewives from being eaten by the hatchery-reared salmonines, this gain was offset somewhat by the consumption of young alewives by the larger smelt each fall (O'Gorman 1974).

In Lake Michigan, alewives were the primary prey of the initial releases of coho salmon (*Oncorhynchus kisutch*) in the late 1960s (Harney and Norden 1972), and they remained the primary prey through the 1970s (Jude et al. 1987). Alewives were also important in the diets of chinook salmon and lake trout through the 1970s (Eck and Wells 1983; Jude et al. 1987). When alewives fell from dominance in the early 1980s, coho and chinook salmon increased their consumption of other fishes and invertebrates, although alewives still made up more than 50% of the salmons' diet (Stewart and Ibarra 1991). Lake trout diet was about 50% alewife before and after alewife numbers declined.

Why alewives remained important in the diets of coho and chinook salmon and lake trout in Lake Michigan after the alewife population was greatly reduced is not fully understood. Alternative prey fishes such as rainbow smelt and bloaters were present, and bloaters in particular were very abundant (Eck and Wells 1987). But alewives, chinook salmon, and we presume, coho salmon are pelagic (Olson 1988), whereas adult bloaters are demersal (Wells 1968). So, similar spatial distributions could account for the continued heavy reliance on alewives for food by coho and chinook salmon, whereas different spatial distributions could account for the relatively few bloaters in the diet of salmon. As for lake trout, they are found deeper in the water column than chinook salmon (Olson 1988). The fact that lake trout diet did not change after alewife declined suggests that the trout were spending more time feeding at mid-depths, or that alewife were much easier to capture than bloaters.

Evidence of how alewives disrupted aquatic communities in the Great Lakes through food web interactions comes primarily from Lake Michigan, where numerous studies were conducted during the alewife's colonization, dominance, and decline. Wells (1970) found that zooplankton populations in the offshore waters of southeastern Lake Michigan shifted towards smaller species between 1954 and 1966 coincident with the rise in alewife numbers. He concluded that the shift in zooplankton size structure was probably due to size-selective predation by alewife. In 1968, after alewife numbers were reduced by the massive 1967 die-off, the composition of zooplankton populations shifted back towards that of 1954. When the Lake Michigan alewife population decreased in the late 1970s, *Daphnia* community structure in offshore waters shifted towards greater dominance of the larger *D. galeata mendotae*, and when the alewife population collapsed in the early 1980s, the largest member of the Daphnia community, *D. pulicaria*, rose to dominance (Evans and Jude 1986). In Lake Ontario, size-selective planktivory by alewives determined the size structure and species composition of the zooplankton community from 1981 to 1988, and eliminated the large cladoceran (*Bythotrephes cederstromi)* from the lake in all but two years from 1986 to 1995 (Johannsson et al. 1991b; Johannsson and O'Gorman 1991; O'Gorman et al. 1991; Mills et al. 1995; O'Gorman, personal observation).

In those Great Lakes where the alewife became dominant, a similar sequence of change occurred in the native fish communities, first the shallow-water planktivores declined, next the shallow-water piscivores increased and then declined, and finally, the deepwater planktivores declined (Smith 1970). The demonstrated ability of alewife to structure zooplankton communities in the Great Lakes made the alewife a strong competitor with native fishes, particularly the larval stages which were obligate planktivores; this was soon cited as one of two potential mechanisms by which alewives could have contributed to the decline of native fishes (Smith 1972; Wells and McLain 1973). The other potential mechanism was predation on the larvae and eggs of native fishes (Crowder 1980; Stewart et al. 1981). Eck and Wells (1987) used their data, as well as the published data of other investigators, to argue convincingly that the mechanism was not competition for food, but rather predation on larval fish. However, the negative effect of competition on survival may have been mediated through predation. Rice et al. (1987) showed that although bloater larvae were resistant to starvation, food deprivation reduced their growth and swimming ability, and thus, could increase or prolong the young bloaters susceptibility to predation. But confirmation of the hypothesis that alewives eat the larvae of native fishes in the Great Lakes was lacking until Brandt et al. (1987) demonstrated that alewives eat the larvae of yellow perch (*Perca flavescens)* in Lake Ontario. In the laboratory, yearling alewives were shown to ignore zooplankton prey to search for and eat larval bloaters (Luecke et al. 1990). Yellow perch and bloaters declined from recruitment failure during the period of alewife dominance in Lake Michigan but their reproductive success, like that of other native fishes, increased as alewife numbers waned (Smith 1970; Brown et al. 1987; Eck and Wells 1987).

Current Management Strategies

Periodic die-offs made the alewife's colonization of the Great Lakes apparent to the public, and ultimately led to large-scale attempts by fisheries managers to manipulate their numbers. Initially, managers attempted to reduce nuisance levels of alewife by direct harvest and development of commercial markets. Later, stocking of salmonines was seen as an effective means to suppress alewives while providing angling opportunities and associated economic benefits. The magnitude of the sport fishery that eventually developed, however, dwarfed even the most optimistic expectations. In 1985, the gross all-or-none value of the Great Lakes sport fishery was estimated at $3.4 billion (U.S.) and about $2 billion (U.S.) worth of goods and services were purchased for angling (Talhelm 1988). Much of this economic activity was due to the fisheries maintained by stocking salmonines; fisheries which, in large measure, relied on the alewife for their existence. As alewife numbers began to decline, coincident with the buildup of salmonine populations, managers became increasingly concerned about maintaining alewife abundance at levels sufficient to support the salmonine fishery.

The amount of effort expended to manage alewife varied among the lakes, generally in direct relation to the abundance of alewives. In Lake Superior, alewives are rare and management efforts are nil (Busiahn 1990). In Lake Erie, alewives are viewed as a minor member of a diverse fish community and there is no evidence that they have had a negative influence on the fish community (Hartman 1973; Leach and Nepszy 1976). Consequently, little effort has been made to regulate their numbers, although such measures are discussed whenever an increase in the population occurs. For example, reports of a high abundance of alewives in the early 1970s resulted in suggestions to use commercial fishing gear to reduce alewife numbers and derive economic benefits (J. Paine, Ontario Ministry of Natural Resources, personal communication). In Lake Huron, alewives maintain moderate populations, but they have not received much management attention, although stocking of salmonines in the lake has contributed to reducing alewife numbers there (Hartman 1988).

The most management effort deliberately directed at affecting alewife abundance has been in Lakes Michigan and Ontario (Tody 1979; Kerr and LeTendre 1991; Koonce and Jones 1994). In these lakes, the management of alewives eventually became linked to the management of chinook salmon (Stewart and Ibarra 1991; T. Stewart, personal observation). Compared to the other species of salmonines stocked, the chinook salmon is inexpensive to culture and grows quickly to a large size. Chinook salmon are usually released from hatcheries in the spring of their first year of life weighing 3g to 5g, whereas all other salmonines are usually released in the spring of their second year of life weighing 25g to 50g (LeTendre and Savoie 1992). Thus, with relatively little investment in hatchery space or fish food, natural resource agencies could easily raise large numbers of chinook salmon. Less than four years after stocking the salmon, anglers would be catching fish weighing 15kg or more (Hansen 1986), which were the largest fish that most of them had ever seen. Indeed, although it was the successful introduction of coho salmon that fueled the initial resurgence of the Great Lakes sport fishery in the 1960s, it was predominantly the accessibility, mystique, and excitement of angling for large chinook salmon that maintained the large-scale sport fisheries in Lakes Michigan and Ontario in the 1980s (Hansen et al. 1990; Kerr and LeTendre 1991). But per fish stocked, the chinook salmon consumed more prey fish during their life cycle in the lake than did the other species of hatchery-reared salmonines (Stewart et al. 1981) and stocking more chinooks than other species meant that the chinook salmon population had the highest demand for prey (Stewart and Ibarra 1991). In Lake Ontario, stocking practices in the 1980s led to a salmonine population in which two-thirds of the demand for adult alewife came from chinook salmon (Jones et. al 1993). The large predatory demand of each chinook salmon meant that any reduction in the number of chinooks stocked would result in a disproportionate decline in predation. So, managers saw reducing releases of chinook salmon as an efficacious way to reduce predation pressure in Lake Ontario when recruitment to the alewife population declined in the early 1990s. However, reducing the number of chinook salmon released quickly became a contentious issue because the salmon were highly valued by anglers.

Lake Michigan

Alewife were so spectacularly abundant in the mid 1960s, that they repeatedly clogged municipal and industrial water intakes (Greenwood 1970; Wells 1973). During the 1967 die-off, which was thought to involve more than 130,000t of fish, tons of dead and dying alewives clogged harbors and washed ashore, presenting a difficult and costly cleanup problem. The loss to industries, municipalities, and recreational interests was reportedly in excess of $100 million (U.S.) (Greenwood 1970). In response to the 1967 die-off in Lake Michigan, a joint state-federal investigation was undertaken in 1968, with the aim of evaluating methods of reducing the numbers of dead fish washing ashore and removing large numbers of live fish during spawning runs. Skimming nets up to 4,000ft long were towed on the surface in the open lake to collect dead, floating alewives, while pound nets were set in harbors to remove live alewives (Greenwood 1970).

Commercial harvest of alewives for fish meal and fertilizer was encouraged during this period of dominance. Annual landings from 1967 to 1977 averaged 16,400t (range: 12,333 to 21,952) (Baldwin et al. 1979). The industrial fishery, however, was beset by numerous problems, among them: low value of the catch, contamination by polychlorinated biphenyls and other chemicals, and demand that was dependent on the worldwide supply of fish meal and other products (Hanson 1987). Catches plummeted about 70% from 1977 to 1980 and never recovered (fig. 1). The last directed fishery for alewives was closed by management action in 1991 amid public concern over declining alewife numbers (Brown et al. 1999).

By the late 1960s, management agencies began to realize that reestablishment of predator populations could be effective in controlling alewives (Tody 1979). Indeed, Smith (1970) even argued that the suppression of alewife through the reestablishment of piscivore populations would lead to a restoration of Great Lakes fisheries. By this time though, piscivore populations had started to build in response to annual releases of hatchery-reared salmonines and effective sea lamprey control. Extraordinary returns to recreational fisheries were being reported for the recently initiated plantings of salmon, and alewives had not shown any sign of returning to nuisance levels. With the available science and the public clearly supporting them, Lake Michigan fishery managers envisioned being able to both control nuisance levels of alewives and create an exceptional recreational fishery (Tody 1979).

The 1970s were tumultuous years marked by the phenomenal growth of the Lake Michigan sport fishery. By the end of the decade, however, some concerns were expressed that the buildup of salmonine populations could create a demand for prey fish in excess of that available, and thus, destabilize the predator-prey balance (Stewart et al. 1981). Numbers of salmonines released annually had risen steadily through the 1970s, and had reached about 16 million by 1980. Fisheries managers did not retreat from this level of stocking in the early 1980s (Keller et al. 1990). Indeed, the average number of chinook salmon released from 1982 to 1984 was the highest on record. By the mid 1980s, changes in the structure of the fish community were clearly evident. Piscivorous salmonines continued to increase, alewives were declining, and bloaters and yellow perch were increasing (Eck and Wells 1987). Fisheries managers, apparently concerned about potential effects on the alewife population, began to slowly lower the stocking rate in 1985.

Managers had little incentive to change stocking policy. In Wisconsin waters of Lake Michigan from 1969 to 1985, angling effort increased ten fold, angling catch rates doubled, and angling total catch increased twenty fold; a multimillion dollar sport fishery was booming (Hansen et al. 1990). Prior to the mid 1980s, the perception was that more stocking meant more benefits from recreational fishing. Scientific uncertainty contributed to the delay in changing management direction. Although declines in alewife numbers were evident, the cause was debated (Crowder and Binkowski 1983; Kitchell and Crowder 1986; Eck and Brown 1985; Eck and Wells 1987). Also, concern about the potential for a shortage of alewives for salmonine prey was mitigated by the speculation that expanding populations of bloater and yellow perch might provide sufficient alternative prey to maintain system balance (Stewart et al. 1981; Eck and Wells 1987). Finally, declines in salmonine growth that Stewart et al. (1981) postulated would coincide with a food shortage were not evident (Hansen 1986).

By 1987, the States of Michigan and Wisconsin had reduced stocking levels in an attempt to conserve the alewife population (Keller et al. 1990), and alewives were considered a valuable resource (Kitchell and Hewett 1987; Kitchell et al. 1988). Projected stocking plans in Wisconsin and Michigan targeted a salmonine population that would consume 8.5% less prey than the salmonine population did from 1980 to 1984 (Keller et al. 1990). However, management focus shifted abruptly in the late 1980s when large chinook salmon began dying from bacterial kidney disease (BKD) in southern Lake Michigan (Stewart and Ibarra 1991). Some thought this was a fish disease problem that could be solved by improving hatchery practices, whereas others thought this was a prey deficit problem, with BKD expressing itself only because the chinook salmon were severely stressed by a lack of alewife food. The apparent dilemma for managers was: *do we stock more chinook salmon to offset the higher BKD-induced mortality, or do we stock less to alleviate the potential for nutritionally-induced stress?* The management rationale is not documented, but some agencies reduced the number of chinook salmon released, whereas other agencies increased the number released, with the result being that the total number of chinook salmon planted rose to its highest level in 1989 (Kocik and Jones 1999). Despite enhanced stocking, by 1993, the number of chinook salmon harvested by anglers had declined to a small fraction of former levels in all the states bordering Lake Michigan (Bence and Smith 1999). Some of the decline in harvest may have been due to reduced fishing effort, which followed a sharp reduction in angling success for chinook salmon in the late 1980s. Although biologists still debate the cause of chinook salmon's decline, the sequence of events in Lake Michigan suggest that the size (and thus value) of the salmonine sport fishery in the Great Lakes is linked to the fate of the alewife.

Lake Ontario

The chronology of change in alewife management strategies for Lake Ontario was similar to that of Lake Michigan. In Lake Ontario, alewife also reached nuisance levels soon after colonization. Although management programs to reduce alewife numbers by large scale harvest were never implemented, studies were undertaken from 1968 to 1970 to determine the economic feasibility of a commercial trawl fishery for rainbow smelt and alewife (Thurston 1975; Howell 1975). These studies were in response to a drastic decline in fish species traditionally sought by commercial fishers and an abundance of unutilized alewives and rainbow smelt.

Vigorous salmonine populations were established in Lake Ontario just as in Lake Michigan. Development of salmonine populations in Lake Ontario, however, appears to have lagged behind that in Lake Michigan by about 3 to 5 years. In Lake Ontario, large scale releases of salmonines, particularly chinook salmon, started later and rose more slowly (Eck and Wells 1987; LeTendre and Savoie 1992), and control of sea lamprey was poor before 1985 (Elrod et al. 1995). This small time lag may have contributed to Lake Ontario managers making a different policy decision than was made for Lake Michigan, because events in the upper lake heightened their awareness of the potential consequences of a large decline in alewife abundance.

In the mid 1980s, the State of New York and the province of Ontario agreed to cap stocking at eight million salmonines annually (Kerr and Letendre 1991). We believe that this management response was precipitated by three factors: (1) the number of salmonines stocked per unit surface area in Lake Ontario was 50% higher than that in Lake Michigan, and documented declines in the Lake Michigan alewife population were being attributed to excess predation by stocked salmonines (Kitchell and Crowder 1986); (2) the total harvest of fish from Lake Ontario by commercial and sport fisheries not only exceeded historical commercial landings (Montgomery 1986; Eckert 1986; Loftus et al. 1987), but also exceeded potential yields predicted by some empirical models (Leach et al. 1987), and management agencies were aware of the need to better understand the factors limiting fish yields from the Great Lakes, particularly from the artificially maintained food webs dominated by non-indigenous fishes (Lewis et al. 1987); and (3) at that level of stocking, an economically important recreational fishery had developed (Kerr and LeTendre 1991; Jones et al. 1993), so the risks associated with further increases in stocking outweighed the potential benefits.

Throughout the late 1980s, fisheries management efforts in Lake Ontario recognized the importance of maintaining alewife abundance. The instability of the prey base supporting the artificially maintained piscivores was acknowledged, and a management objective was established to maintain the aggregate biomass of principal prey species (alewife, rainbow smelt, and slimy sculpin (*Cottus cognatus*)) at 110kg/ha to support the hatchery-reared salmonines (Kerr and LeTendre 1991).

By the late 1980s, management agencies on Lake Ontario began to enhance prey assessment programs (Kerr and LeTendre 1991). At the same time, fisheries scientists and managers across the Great Lakes basin were questioning the long-term sustainability of fish communities maintained primarily by stocking. In response, the Great Lakes Fishery Commission sponsored the SIMPLE project (Sustainability of Intensively Managed Pelagic Lake Ecosystems, Koonce and Jones 1994). SIMPLE provided a synthesis of the scientific information available for Lakes Michigan and Ontario by way of a computer model of prey and predator populations (Jones et al. 1993). The model assisted in the development of management policies and contributed heavily to the decision to reduce the numbers of fish stocked in Lake Ontario. Because the model could produce growth trajectories of predator and prey populations, it became an invaluable tool for illustrating to the public the likely results of various management actions.

In 1992, the Lake Ontario Committee of the Great Lakes Fishery Commission convened a multidisciplinary task group of scientists to provide a technical evaluation of the Lake Ontario fish community and related ecosystem properties, in response to unusually poor survival of juvenile alewife and the failure of the condition of adult alewife to increase following population declines. The task group found that Lake Ontario had undergone significant declines in productivity during the 1980s, paralleling declines in phosphorus concentrations, and that prey populations were in the process of adjusting to the new trophic state. Predator demand implied by the current, high stocking rate probably exceeded that which could be supported if the expected declines in prey fish production occurred. They concluded that if predator demand was not eased, then either prey fish abundance would be driven lower, or piscivore growth and survival would decline, or (more likely) that all three would drop.

The report of the task group led to an extensive public consultation process by New York and Ontario about future management strategies for Lake Ontario. The principal focus of the public discussions was the fate of the alewife population, as it was likely to be affected by reduced productivity, current levels of predation, and a catastrophic overwinter mortality. The public was informed that the current level of stocking presented a high risk of a severe decline in alewife, and that if there was a catastrophic winter mortality in the future, the alewife population would, in all likelihood, fail to recover. The probable effects of various management options on alewife and salmonine populations, forecast by use of the computer model developed for SIMPLE (Jones et al. 1993; Koonce and Jones 1994), were outlined at the public consultations.

The stakes were high, for in Lake Ontario, unlike Lake Michigan, there was no alternative prey species waiting to proliferate should alewife numbers plummet. Rainbow smelt, the second most abundant planktivore in the main lake (O'Gorman et al. 1987), had declined along with the alewife. The bloater, which rose to dominance in Lake Michigan as the alewife declined, was extinct in Lake Ontario, as were the other species of deepwater coregonids (Todd and Smith 1992). The scientific consensus was that a catastrophic decline in alewives would most affect the chinook salmon which, as in the case of Lake Michigan, was the most valuable salmonine in the sport fishery, and the greatest consumer of alewife. In Lake Ontario, chinook salmon accounted for 66% of the predator demand for large alewife (Jones et al. 1993). Clear communication to the public of the scientific uncertainty associated with the SIMPLE model predictions and the explicit statements of risks and tradeoffs were the key to an effective dialogue with stakeholders. This was a good example of evolving approaches to fisheries management decision making (Hilborn et al. 1993). In the end, the consultation process resulted in the binational decision by New York and Ontario to reduce annual stocking of hatchery-reared salmonines so that predation demand was lowered by 50%. One of the changes in management strategies made to accomplish this was a 60% cut in the total number of chinook salmon stocked into Lake Ontario. Although there was broad public support for these difficult decisions, there was not unanimous endorsement of the new management scheme; there continue to be those who do not believe the scientific explanations, and who oppose the management action taken. But it is important to note that the conservation of alewives had become the focus of public debate that led to a major change in management of an economically significant Great Lakes fishery.

Future Management Directions

Biotic and abiotic changes are occurring at a rapid rate in the Great Lakes, introducing considerable uncertainty about how effective any of the past management schemes will be at manipulating alewife abundance. In the 1980s, a new wave of exotic species colonized the Great Lakes and in the 1990s, populations of these exotics expanded. The invaders included a predatory cladoceran, the spiny water flea (*Bythotrephes cederstromi*) (Bur et al. 1986); two bivalve mollusks, the zebra mussel (*Dreissena polymorpha*) (Hebert et al. 1989), and the quagga mussel (*D. bugensis*) (May and Marsden 1992; Spidle et al. 1994; Rosenberg and Ludyanskiy 1994); a percid, the ruffe

(*Gymnocephalus cernus*) (Pratt et al. 1992); and two gobioids, the tubenose goby (*Proterorhinus marmoratus*), and the round goby (*Neogobius melanostomus*) (Jude et al. 1992). Nutrient enrichment of the lakes has been curtailed by decades of aggressive abatement programs, and consequently, phosphorous concentrations have fallen (Scavia et al. 1986; Lean et al. 1990; Bertram 1993). Reductions in phosphorus improve water quality, but lower the potential for fish production. The effect on alewife numbers of the insertion of novel organisms at various trophic levels in a food web with reduced production potential is not known, and until it is, formulation of effective management strategies with reasonably predictable outcomes will be difficult.

An emerging issue which will undoubtedly pose a dilemma for Great Lakes fishery managers is that maintaining an alewife population to support recreational fishing objectives may be in conflict with objectives to restore indigenous fish species. For example, in all Great Lakes, fisheries managers are attempting to either reestablish or maintain self-reproducing populations of lake trout (Fetterolf 1984). Recently, alewives were shown to eat the naturally spawned fry of hatchery-reared lake trout in Lake Ontario (Krueger et al. 1995). Also, alewives are rich in thiaminase, a group of enzymes that can destroy thiamine, and the progeny of lake trout that eat mainly alewives were recently shown to suffer high mortality from a maternally transmitted syndrome that appears due to thiamine deficiency (Fitzsimons 1995; Fisher et al. 1996). Thus, the limited survival of the progeny of stocked lake trout past the fry stage in Lakes Michigan, Huron, and Ontario could be due, in part, to high mortality of the fry from alewife predation (Jones et al. 1995), from alewife induced thiamine deficiency (Fitzsimons 1995), or from both.

Attempts to manage alewife populations in the Great Lakes are likely to remain contentious in the near future, particularly on Lakes Michigan and Ontario where important recreational fisheries are still heavily dependent on the fish that eat alewives. In the long term, however, changes in preference of sport anglers, in response to species shifts in the fish community, may cause management focus to shift away from the alewife. For example, while the chinook salmon fishery in Lake Michigan declined, the sport fishery for rainbow trout (*Oncorhynchus mykiss*) was growing (Bence and Smith 1999). Central to the resolution of future conflicts over alewife management in the Great Lakes are mechanisms for integrating scientific knowledge, public consultation, and decision making (Lewis et al. 1987; Hilborn et al. 1993). The changes in the ecological role played by the alewife since it invaded the lakes illustrate the challenge of ecosystem management in the Great Lakes and demonstrate the need for an improved institutional framework for planning and implementing the concept of ecosystem management (Bocking 1994).

Acknowledgments

We thank J. H. Elrod, M. J. Hansen, J. Savino, and P. Smith for constructive comments on an earlier draft of this manuscript. This is contribution 929 of the USGS, Great Lakes Science Center, Ann Arbor, Michigan.

Literature Cited

Applegate, V. C. and H. D. Van Meter. 1970. A brief history of commercial fishing in Lake Erie. U.S. Department of the Interior, Bureau of Commercial Fisheries, Fishery Leaflet 630. 28 pp.

Argyle, R. L. 1982. Alewives and rainbow smelt in Lake Huron: midwater and bottom aggregations and estimates of standing stocks. Transactions of the American Fisheries Society 111:267–285.

Argyle, R. L. 1992. Acoustics as a tool for the assessment of Great Lakes forage fishes. Fisheries Research 14:179–196.

Ashworth, W. 1986. The late, Great Lakes. Alfred A. Knopf, Inc. New York.

Baldwin, N. S., R. W. Saalfeld, M. A. Ross, and H. J. Buettner. 1979. Commercial fish production in the Great Lakes 1867–1977. Great Lakes Fishery Commission Technical Report No. 3.

Bean, T. H. 1884. On the occurrence of branch alewife in certain lakes of New York. Pages 588–593, section 1 *in* G. B. Goode, editor. The fisheries and fishing industries of the United States. U.S. Commission of Fish and Fisheries, Washington, D.C.

Bence, J. R., and K. D. Smith. 1999. An overview of recreational fisheries of the Great Lakes. Pages 259–306 *in* W. W. Taylor and C. P. Ferreri, eds. Great Lakes Fisheries Policy and Management: A Binational Perspective. Michigan State University Press, East Lansing, Michigan.

Bergstedt, R. A. 1983. Origins of rainbow smelt in Lake Ontario. Journal of Great Lakes Research 9:582–583.

———, R. O'Gorman. 1989. Distribution of alewives in southeastern Lake Ontario in autumn and winter: a clue to winter mortalities. Transactions of the American Fisheries Society 118:687–692.

Bertram, P. E. 1993. Total phosphorus and dissolved oxygen trends in the central basin of Lake Erie, 1970–1991. Journal of Great Lakes Research 19:224–236.

Bolsenga, S. J., and D. C. Norton. 1993. Great Lakes air temperature trends for land stations, 1901–1987. Journal of Great Lakes Research 19:379–388.

Bocking, S. 1994. Visions of nature and society: A history of the ecosystem concept. Alternatives 20:12–18.

Brandt, S. B. 1986. Food of trout and salmon in Lake Ontario. Journal of Great Lakes Research 12:200–205.

———, D. M. Mason, D. B. MacNeill, T. Coates, and J. E. Gannon. 1987. Predation by alewives on larvae of yellow perch in Lake Ontario. Transactions of the American Fisheries Society 116:641–645.

———, D. M. Mason, E. V. Patrick, R. L. Argyle, L. Wells, P. A. Unger, and D. J. Stewart. 1991. Acoustic measures of the abundance and size of pelagic planktivores in Lake Michigan. Canadian Journal of Fisheries and Aquatic Sciences 48:894–908.

Bronte, C. R., J. H. Selgeby, and G. L. Curtis. 1991. Distribution, abundance, and biology of the alewife in U.S. waters of Lake Superior. Journal of Great Lakes Research 17:304–313.

Brown, E. H., Jr. 1968. Population characteristics and physical condition of alewives, *Alosa pseudoharengus*, in a massive die-off in Lake Michigan, 1967. Great Lakes Fishery Commission Technical Report No. 13.

———, 1972. Population biology of alewives, *Alosa pseudoharengus*, in Lake Michigan, 1949–70. Journal of the Fisheries Research Board of Canada 29:477–500.

———, R. L. Argyle, N. R. Payne, and M. E. Holey. 1987. Yield and dynamics of destabilized chub (*Coregonus* spp.) populations in Lakes Michigan and Huron, 1950–84. Canadian Journal of Fisheries and Aquatic Sciences 44(Supplement 2):371–383.

———, T. R. Busiahn, M. Jones, and D. Stewart. 1999. Allocating Great Lakes forage bases in response to multiple demand. Pages 355–394 *in* W. W. Taylor and C. P. Ferreri, eds. Great Lakes Fisheries Policy and Management: A Binational Perspective. Michigan State University Press, East Lansing, Michigan.

Bur, M. T., D. M. Klarer, and K. A. Krieger. 1986. First records of a European cladoceran, *Bythotrephes cederstromi*, in Lakes Erie and Huron. Journal of Great Lakes Research 12:144–146.

Busiahn, T. R. (ed.). 1990. Fish community objectives for Lake Superior. Great Lakes Fishery Commission Special Publication 90–1.

Christie, W. J. 1973. A review of the changes in the fish species composition of Lake Ontario. Great Lakes Fishery Commission Technical Report No. 23.

———, N. A. Thomas. 1981. Biology. Pages 327–340 *in* E. J. Aubert and T. L. Richards, editors. IFYGL-the international field year for the Great Lakes. National Oceanic and Atmospheric Administration, Great Lakes Environmental Research Laboratory, Ann Arbor.

———, K. A. Scott, P. G. Sly, and R. H. Strus. 1987. Recent changes in the aquatic food web of eastern Lake Ontario. Canadian Journal of Fisheries and Aquatic Sciences 44(Supplement 2):37–52.

Coble, D. W. 1967. The white sucker population of South Bay, Lake Huron, and effects of the sea lamprey on it. Journal of the Fisheries Research Board of Canada 24:2117–2136.

———, R. E. Bruesewitz, T. W. Fratt, and J. W. Scheirer. 1990. Lake trout, sea lampreys, and overfishing in the upper Great Lakes: a review and reanalysis. Transactions of the American Fisheries Society 119:985–995.

———, 1992. Reply to comments: Decline of lake trout in Lake Huron. Transactions of the American Fisheries Society 121:550–554.

Colby, P. J. 1971. Alewife dieoffs: why do they occur? LIMNOS 4:18–27.

———, 1973. Response of the alewives, *Alosa pseudoharengus*, to environmental change. Pages 163–198 *in* W. Chavin, editor. Responses of fish to environmental changes. Thomas, Springfield, Illinois.

Conner, D. J., C. R. Bronte, J. H. Selgeby, and H. L. Collins. 1993. Food of salmonine predators in Lake Superior, 1981–1987. Great Lakes Fishery Commission Technical Report 59.

Crowder, L. B. 1980. Alewife, smelt, and native fishes in Lake Michigan: competition or predation? Environmental Biology of Fishes 5:225–233.

———, F. P. Binkowski. 1983. Foraging behaviors and the interaction of alewife, *Alosa pseudoharengus*, and bloater, *Coregonus hoyi*. Environmental Biology of Fishes 8:105–113.

Diana, J. S. 1990. Food habits of angler–caught salmonines in western Lake Huron. Journal of Great Lakes Research 16:271–278.

Dobson, H. F. H. 1994. Lake Ontario water quality trends, 1969 to 1992: some observational nutrient-science for protecting a major and vulnerable source of drinking water. Contribution Number 94–58, National Water Research Institute, 867 Lakeshore Road, Burlington, Ontario.

Dymond, J. R. 1928. Some factors affecting the production of lake trout (*Cristivomer namaycush*) in Lake Ontario. University of Toronto Studies Biological Series 31. Publications of the Ontario Fisheries Research Laboratory 33:27–41.

———, 1932. Records of the alewife and steelhead (rainbow) trout from Lake Erie. Copeia 1:32–33.

———, J. L. Hart, and A. L. Pritchard. 1929. The fishes of the Canadian waters of Lake Ontario. University of Toronto Studies Biological Series 37. Publications of the Ontario Fisheries Research Laboratory 37:3–35.

Eck, G. W., and L. Wells. 1983. Biology, population structure, and estimated forage requirements of lake trout in Lake Michigan. U.S. Fish and Wildlife Service Technical Paper 111. 18 pp. Washington.

———, E. H. Brown, Jr. 1985. Lake Michigan's capacity to support lake trout (*Salvelinus namaycush*) and other salmonines: an estimate based on the status of prey populations in the 1970s. Canadian Journal of Fisheries and Aquatic Sciences 42:449–454.

———, L. Wells. 1987. Recent changes in Lake Michigan's fish community and their probable causes, with emphasis on the role of the alewife (*Alosa pseudoharengus*). Canadian Journal of Fisheries and Aquatic Sciences 44 (Supplement 2):53–60.

Eckert, T. H. 1986. Preliminary report 1985 Lake Ontario fishing boat census. Minutes of the Lake Ontario Committee Annual Meeting, Great Lakes Fishery Commission, Ann Arbor, Michigan.

Elrod, J. H. and R. O'Gorman. 1991. Diet of juvenile lake trout in southern Lake Ontario in relation to abundance and size of prey fishes, 1979–1987. Transactions of the American Fisheries Society 120:290–302.

———, R. O'Gorman, C. P. Schneider, T. H. Eckert, T. Schaner, J. N. Bowlby, and L. P. Schleen. 1995. Lake trout rehabilitation in Lake Ontario. Journal of Great Lakes Research 21(Supplement 1):83–107.

Eshenroder, R. L. 1992. Comments: Decline of lake trout in Lake Huron. Transactions of the American Fisheries Society 121:548–550.

Eschmeyer, P. H. 1957. The near extinction of the lake trout in Lake Michigan. Transactions of the American Fisheries Society 85:102–119.

Evans, M. S., and D. J. Jude. 1986. Recent shifts in Daphnia community structure in southeastern Lake Michigan: a comparison of the inshore and offshore regions. Limnology and Oceanography 31:56–67.

Fetterolf, C. M., Jr. 1984. Lake trout futures in the Great Lakes. Pages 163–170 *in* Wild trout III, Proceedings of the Symposium. F. Richardson and R. H. Hamre, eds. Trout Unlimited, Vienna, VA.

Fisher, J. P., J. D. Fitzsimons, G. F. Combs, Jr., and J. M. Spitsbergen. 1996. Naturally occurring thiamine deficiency causing reproductive failure in Finger Lakes Atlantic salmon and Great Lakes lake trout . Transactions of the American Fisheries Society 125:167–178.

Fitzsimons, J. D. 1995. The effect of B-vitamins on a swim-up syndrome in Lake Ontario lake trout. Journal of Great Lakes Research 21(Suppl. 1):286–289.

Graham, J. J. 1956. Observations on the alewife, *Pomolobus pseudoharengus* (Wilson), in fresh water. University of Toronto Studies Biological Series Publication 62, Publication of the Ontario Fisheries Research Laboratory 74, Toronto.

Greeley, J. R. 1940. Fishes of the watershed with annotated list. Pages 42–81 i *n* A biological survey of the Lake Ontario watershed. New York Conservation Department, Supplement to the 29th Annual Report (1939). Albany, New York.

Greenwood, M. R. 1970. 1968 state-federal Lake Michigan alewife die-off control investigation. Bureau of Commercial Fisheries, Fish and Wildlife Service, Ann Arbor, MI. 156 pp.

Hansen, M. J. 1986. Size and condition of trout and salmon from the Wisconsin waters of Lake Michigan, 1969–1984. Wisconsin Department of Natural Resources, Bureau of Fisheries Management, Fish Management Report 126, Madison.

———, P. T. Schultz, and B. A. Lasee. 1990. Changes in Wisconsin's Lake Michigan sport fishery, 1969–1985. North American Journal of Fisheries Management 10:442–457.

———, 1999. Lake trout in the Great Lakes: Basinwide stock collapse and binational restoration. Pages 417–453 *in* W. W. Taylor and C. P. Ferreri, eds. Great Lakes Fisheries Policy and Management: A Binational Perspective. Michigan State University Press, East Lansing, Michigan.

Hanson, F. B. 1987. Bioeconomic model of the Lake Michigan alewife (*Alosa pseudoharengus*) fishery. Canadian Journal of Fisheries and Aquatic Sciences 44(Supplement 2):298–305.

Harney, M. A., and C. R. Norden. 1972. Food habits of the coho salmon, *Oncorhynchus kisutch*, in Lake Michigan. Wisconsin Academy of Sciences, Arts and Letters 60:79–85.

Hartig, J. H., J. F. Kitchell, D. Scavia, and S. B. Brandt. 1991. Rehabilitation of Lake Ontario: the role of nutrient reduction and food web dynamics. Canadian Journal of Fisheries and Aquatic Sciences 48:1574–1580.

Hartman, W. L. 1973. Effects of exploitation, environmental changes, and new species on the fish habitats and resources of Lake Erie. Great Lakes Fishery Commission Technical Report 22, Ann Arbor, Michigan.

———, 1988. Historical changes in the major fish resources of the Great Lakes. Pages 103–131 *in* M. S. Evans, editor. Toxic contaminants and ecosystem health; a Great Lakes focus. John Wiley & Sons, Inc., New York.

Hatch, R. W, P. M. Haack, and E. H. Brown, Jr. 1981. Estimation of alewife biomass in Lake Michigan, 1967–1978. Transactions of the American Fisheries Society 110:575–584.

Hebert, P. D., B. W. Muncaster, and G. L. Mackie. 1989. Ecological and genetic studies on *Dreissena polymorpha* (Pallas): A new mollusc in the Great Lakes. Canadian Journal of Fisheries and Aquatic Sciences 46:1587–1591.

Henderson, B. A. and E. H. Brown, Jr. 1985. Effects of abundance and water temperature on recruitment and growth of alewife (*Alosa pseudoharengus*) near South Bay, Lake Huron, 1954–1982. Canadian Journal of Fisheries and Aquatic Sciences 42:1608–1613.

Hewett, S. W. and D. J. Stewart. 1989. Zooplanktivory by alewives in Lake Michigan: ontogenetic, seasonal, and historical patterns. Transactions of the American Fisheries Society 118:581–596.

Hile, R. P., P. H. Eschmeyer, and G. F. Lunger. 1951. Status of the lake trout fishery in Lake Superior. Transactions of the American Fisheries Society 80: 278–312.

Hilborn, R., E. K. Pikitch, and R. C. Francis. 1993. Current trends in including risk and uncertainty in stock assessment and harvest decisions. Canadian Journal of Fisheries and Aquatic Sciences 50:874–880.

Howell, H. D. 1975. Lake Ontario exploratory trawling program, 1969. Pages 1–23 *in* D. P. Kolenosky, editor. Exploratory trawling in Lake Ontario. Ontario Ministry of Natural Resources.

Huntsman, A. G. 1944. Why did Ontario salmon disappear? Transactions of the Royal Society of Canada, 3rd Series 38:83–102.

Ihssen, P. E., G. W. Martin, and D. W. Rodgers. 1992. Allozyme variation of Great Lakes alewife, *Alosa pseudoharengus*: genetic differentiation and affinities of a recent invader. Canadian Journal of Fisheries and Aquatic Sciences 49:1770–1777.

Johannsson, O. E. and R. O'Gorman. 1991. Roles of predation, food, and temperature in structuring the epilimnetic zooplankton populations in Lake Ontario, 1981–1986. Transactions of the American Fisheries Society 120:193–208.

———, K. M. Ralph, R. Dermott, M. A. Neilson, and E. S. Millard. 1991a. Trends in nutrient and biological variables in the offshore: a preliminary report of the Lake Ontario Biomonitoring Program. Lake Ontario Fisheries Unit 1990 Annual Report. Chapter 12.

———, E. L. Mills, and R. O'Gorman. 1991b. Changes in the near shore and offshore zooplankton communities in Lake Ontario: 1981–88. Canadian Journal of Fisheries and Aquatic Sciences 48:1546–1557.

Jones, M. L., J. F. Koonce, and R. O'Gorman. 1993. Sustainability of hatchery-dependent salmonine fisheries in Lake Ontario: the conflict between predator demand and prey supply. Transactions of the American Fisheries Society 122:1002–1018.

———, G. Eck, D. O. Evans, M. C. Fabrizio, M. H. Hoff, P. L. Hudson, J. Janssen, D. Jude, R. O'Gorman, and J. F. Savino. 1995. Limitations to lake trout rehabilitation in the Great Lakes imposed by biotic interactions occurring at early life stages. Journal of Great Lakes Research 21(Supplement 1):505–517.

Jude, D. J., F. J. Tesar, S. F. Deboe, and T. J. Miller. 1987. Diet and selection of major prey species by Lake Michigan salmonines, 1973–1982. Transactions of the American Fisheries Society 116:677–691.

———, R. H. Reider, and G. R. Smith. 1992. Establishment of Gobiidae in the Great Lakes basin. Canadian Journal of Fisheries and Aquatic Sciences 49:416–421.

Keller, M., K. D. Smith, and R. W. Rybicki (editors). 1990. Review of salmon and trout management in Lake Michigan. Michigan Department of Natural Resources. Fish Technical Report No. 14. Lansing, Michigan.

Kerr, S. J. and G. C. LeTendre. 1991. The state of the Lake Ontario fish community in 1989. Great Lakes Fishery Commission Special Publication 91-3, Ann Arbor, Michigan.

Kitchell, J. F. and L. B. Crowder. 1986. Predator-prey interactions in Lake Michigan: model predictions and recent dynamics. Environmental Biology of Fishes 16:205–211.

———, S. W. Hewett. 1987. Forecasting forage demand and yield of sterile chinook salmon (*Oncorhynchus tshawytscha*) in Lake Michigan. Canadian Journal of Fisheries and Aquatic Sciences 44(Supplement 2):384–389.

———, M. S. Evans, D. Scavia, and L. B. Crowder. 1988. Regulation of water quality in Lake Michigan: report of the food web workshop. Journal of Great Lakes Research 14:109–114.

Knight, R. L. and B. Vondracek. 1993. Changes in prey fish populations in Western Lake Erie, 1969–88, as related to walleye, *Stizostedion vitreum*, predation. Canadian Journal of Fisheries and Aquatic Sciences 50:1289–1298.

Kocik, J. F. and M. L. Jones. 1999. Pacific salmonines in the Great Lakes Basin. Pages 455–488 *in* W. W. Taylor and C. P. Ferreri, eds. Great Lakes Fisheries Policy and Management: A Binational Perspective. Michigan State University Press, East Lansing, Michigan.

Koonce, J. F. and M. L. Jones. 1994. Sustainability of the intensively managed fisheries of Lake Michigan and Ontario. Final report of the SIMPLE task group. Board of Technical Experts. Great Lakes Fishery Commission, Ann Arbor, Michigan.

Krueger, C. C., D. L. Perkins, E. L. Mills, and J. E. Marsden. 1995. Predation by alewife on lake trout fry in Lake Ontario: role of an exotic species in preventing restoration of a native species. Journal of Great Lakes Research 21(Supplement 1):458–469.

Lawrie, A. H. and J. F. Rahrer. 1973. Lake Superior-A case history of the lake and its fisheries. Great Lakes Fishery Commission Technical Report 60, Ann Arbor, Michigan.

Leach, J. H. and S. J. Nepszy. 1976. The fish community in Lake Erie. Journal of the Fisheries Research Board of Canada 33:622–638.

——, L. M. Dickie, B. J. Shuter, U. Borgmann, J. Hyman, and W. Lysack. 1987. A review of methods for prediction of potential fish production with application to the Great Lakes and Lake Winnipeg. Canadian Journal of Fisheries and Aquatic Sciences 44(Supplement 2):471–485.

Lean, D. R. S., M. A. Neilson, R. J. J. Stevens, and A. Mazumder. 1990. Response of Lake Ontario to reduced phosphorus loading. Verh. Internat. Verein. Limnol. 24:420–425.

LeTendre, G. C. and P. J. Savoie. 1992. Lake Ontario stocking and marking program for 1991. Lake Ontario Committee 1992 Annual Meeting, Great Lakes Fishery Commission, Ann Arbor, Michigan.

Lewis, C. A., D. H. Schupp, W. W. Taylor, J. J. Collins, and R. W. Hatch. 1987. Predicting Great Lakes fish yields: tools and constraints. Canadian Journal of Fisheries and Aquatic Sciences 44(Supplement 2):411–416.

Loftus, D. H., C. H. Olver, E. H. Brown, P. J. Colby, W. L. Hartman, and D. H. Shupp. 1987. Partitioning potential fish yield from the Great Lakes. Canadian Journal of Fisheries and Aquatic Sciences 44(Supplement 2):417–424.

Luecke, C., J. A. Rice, L. B. Crowder, S. E. Yeo, and F. P. Binkowski. 1990. Recruitment mechanisms of bloater in Lake Michigan: an analysis of the predatory gauntlet. Canadian Journal of Fisheries and Aquatic Sciences 47:524–532.

MacKay, H. H. 1934. Record of the alewife from Lake Huron. Copeia 2:97.

May, B. and J. E. Marsden. 1992. Genetic identification and implications of another invasive species of dreissenid mussel in the Great Lakes. Canadian Journal of Fisheries and Aquatic Sciences 49:1501–1506.

Miller, R. R. 1957. Origin and dispersal of the alewife, *Alosa pseudoharengus*, and the gizzard shad, *Dorosoma pseudoharengus*, in the Great Lakes. Transactions of the American Fisheries Society 86;97–111.

Mills, E. L., R. O'Gorman, J. DeGisi, R. F. Heberger, and R. A. House. 1992. Food of the alewife (*Alosa pseudoharengus*) in Lake Ontario before and after the establishment of *Bythotrephes cederstroemi*. Canadian Journal of Fisheries and Aquatic Sciences 49:2009–2019.

——, R. O'Gorman, E. F. Roseman, C. Adams, and R. W. Owens. 1995. Planktivory by alewife (*Alosa pseudoharengus*) and rainbow smelt (*Osmerus mordax*) on microcrustacean zooplankton and dreissenid (Bivalvia: Dreissenidae) veligers in southern Lake Ontario. Canadian Journal of Fisheries and Aquatic Sciences 52:925–935.

Montgomery, D. 1986. Commercial harvests in Canadian and American waters of Lake Ontario. Great Lakes Fishery Commission, Minutes of the Lake Ontario Committee Annual Meeting, 1986:164–172.

Mortimer, C. H. 1971. Large scale oscillatory motions and seasonal temperature changes in Lake Michigan and Lake Ontario. University of Wisconsin-Milwaukee, Center for Great Lakes Studies, Special Report 12, parts I and II, Milwaukee.

O'Gorman, R. 1974. Predation by rainbow smelt (*Osmerus mordax*) on young-of-the-year alewives (*Alosa pseudoharengus*) in the Great Lakes. The Progressive Fish-Culturist 36:223–224.

——, C. P. Schneider. 1986. Dynamics of alewives in Lake Ontario following a mass mortality. Transactions of the American Fisheries Society 115:1–14.

——, R. A. Bergstedt, and T. H. Eckert. 1987. Prey fish dynamics and salmonine predator growth in Lake Ontario, 1978–84. Canadian Journal of Fisheries and Aquatic Sciences 44(Supplement 2):390–403.

——, E. L. Mills, and J. DeGisi. 1991. Use of zooplankton to assess the movement and distribution of alewives (*Alosa pseudoharengus*) in south-central Lake Ontario in spring. Canadian Journal of Fisheries and Aquatic Sciences 48:2250–2257.

Olson, R. A., J. D. Winter, D. C. Nettles, and J. M. Haynes. 1988. Resource partitioning in summer by salmonids in south-central Lake Ontario. Transactions of the American Fisheries Society 117:552–559.

Pratt, D. M., W. H. Blust, and J. H. Selgeby. 1992. Ruffe, *Gymnocephalus cernus*: Newly introduced in North America. Canadian Journal of Fisheries and Aquatic Sciences 49:1616–1618.

Pritchard, A. L. 1929. The alewife (*Pomolobus pseudoharengus*) in Lake Ontario. University of Toronto Studies Biological Series Publication 33, Publication of the Ontario Fisheries Research Laboratory 38.

Radforth, I. 1944. Some considerations on the distribution of fishes in Ontario. Contribution of the Royal Ontario Museum of Zoology 25:1–116.

Rand, P. J., B. F. Lantry, R. O'Gorman, R. W. Owens, and D. J. Stewart. 1994. Energy density and size of pelagic prey fishes in Lake Ontario, 1978–1990: implications for salmonine energetics. Transactions of the American Fisheries Society 123:519–534.

Regier, H. A., V. C. Applegate, and R. A. Ryder. 1969. The ecology and management of the walleye in western Lake Erie. Great Lakes Fishery Commission Technical Report 15, Ann Arbor, Michigan.

Reigle, N. J., Jr. 1969a. Bottom trawl explorations in Lake Superior, 1963–65. U.S. Fish and Wildlife Circular 294, 25 pp.

———, 1969b. Bottom trawl explorations in Green Bay, 1963–65. U.S. Fish and Wildlife Circular 297, 14 pp.

———, 1969c. Bottom trawl explorations in southern Lake Michigan, 1962–65. U.S. Fish and Wildlife Circular 301, 35 pp.

———, 1969d. Bottom trawl explorations in northern Lake Michigan, 1963–65. U.S. Fish and Wildlife Circular 318, 21 pp.

Rice, J. A., L. B. Crowder, and F. P. Binkowski. 1987. Evaluating potential sources of mortality for larval bloater (*Coregonus hoyi*): starvation and vulnerability to predation. Canadian Journal of Fisheries and Aquatic Sciences 44:467–472.

Ridgway, M. A., D. A. Hurley, and K. A. Scott. 1990. Effects of winter temperature and predation on the abundance of alewife (*Alosa pseudoharengus*) in the Bay of Quinte, Lake Ontario. Journal of Great Lakes Research 16:11–20.

Rodgers, G. K. 1987. Time of onset of full thermal stratification in Lake Ontario in relation to lake temperatures in winter. Canadian Journal of Fisheries and Aquatic Sciences 44:2225–2229.

Rosenberg, G. and M. L. Ludyanskiy. 1994. A nomenclature review of *Dreissena* (Bivalvia: Dreissenidae), with identification of the quagga mussel as *Dreissena bugensis*. Canadian Journal of Fisheries and Aquatic Sciences 51:1474–1484.

Scavia, D., G. L. Fahnenstiel, M. S. Evans, D. J. Jude, and J. T. Lehman. 1986. Influence of salmonine predation and weather on long-term water quality trends in Lake Michigan. Canadian Journal of Fisheries and Aquatic Sciences 43:435–443.

Scott, W. B. 1963. A review of the changes in the fish fauna of Ontario. Transactions of the Royal Canadian Institute 34:111–125.

———, E. J. Crossman. 1973. Freshwater fishes of Canada. Fisheries Research Board of Canada Bulletin 184. 966 pp.

Smith, B. R., J. J. Tibbles, and B. G. H. Johnson. 1974. Control of sea lamprey (*Petromyzon marinus*) in Lake Superior, 1953–70. Great Lakes Fishery Commission Technical Report 26, Ann Arbor, Michigan.

Smith, C. L. 1985. The inland fishes of New York state. New York State Department of Environmental Conservation, Wolf Rd, Albany, NY. 522 pp.

Smith, H. M. 1892. Report on an investigation of the fisheries of Lake Ontario. U.S Fish Commission Bulletin 10(1890):177–215.

———, M. M. Snell. 1891. Fisheries of the Great Lakes in 1885. Appendix 1, Part VIII, The fisheries of Lake Ontario. Pages 296–328 *in* Report of the commissioner for 1887. U.S. Commission of Fish and Fisheries. Government printing office, Washington, D.C.

Smith, S. H. 1968. Species succession and fishery exploitation in the Great Lakes. Journal of the Fisheries Research Board of Canada 25:667–693.

———, 1970. Species interactions of the alewife in the Great Lakes. Transactions of the American Fisheries Society 99:754–765.

———, 1972. Factors of ecologic succession in oligotrophic fish communities of the Laurentian Great Lakes. Journal of the Fisheries Research Board of Canada 29:717–730.

———, 1995. Early changes in the fish community of Lake Ontario. Great Lakes Fishery Commission Technical Report 60.

Spangler, G. R. and J. J. Collins. 1992. Lake Huron fish community structure based on gill-net catches corrected for selectivity and encounter probability. North American Journal of Fisheries Management 12:585–597.

Spidle, A. P., J. E. Marsden, and B. May. 1994. Identification of the Great Lakes quagga as *Dreissena bugensis* from the Dneiper River, Ukraine, on the basis of allozyme variation. Canadian Journal of Fisheries and Aquatic Sciences 51:1474–1484.

Stevens, R. J. J. and M. A. Neilson. 1987. Response of Lake Ontario to reductions in phosphorus load, 1967–82. Canadian Journal of Fisheries and Aquatic Sciences 44:2059–2068.

Stewart, D. J. and M. Ibarra. 1991. Predation and production by salmonine fishes in Lake Michigan, 1978–88. Canadian Journal of Fisheries and Aquatic Sciences 48:909–922.

———, J. F. Kitchell, and L. B. Crowder. 1981. Forage fishes and their salmonid predators in Lake Michigan. Transactions of the American Fisheries Society 110:751–763.

Talhelm, D. R. 1988. Economics of Great Lakes fisheries: a 1985 assessment. Great Lakes Fishery Commission Technical Report 54, Ann Arbor, Michigan

Thurston, L. 1975. Lake Ontario exploratory trawling, 1968. Pages 1–23 *in* D. P. Kolenosky, ed. Exploratory trawling in Lake Ontario. Ontario Ministry of Natural Resources.

Tody, W. H. 1979. Utilization of predator-prey relationships in fisheries management. Pages 361–364 *in* H. Clepper, editor. Predator-prey systems in fisheries management. Sport Fishing Institute, Washington, D.C.

Todd, T. N., and G. R. Smith. 1992. A review of differentiation in Great Lakes ciscoes. Pages 261–267 *in* T. N. Todd and M. Luczynski, editors. Biology and Management of Coregonid Fishes. Polskie Archiwum Hydrobiologii 39(3–4).

Van Oosten, J. 1930. The disappearance of the Lake Erie cisco-a preliminary report. Transactions of the American Fisheries Society 60:204–214.

Wells, L. 1968. Seasonal depth distribution of fish in southeastern Lake Michigan. U.S. Fish and Wildlife Service Fishery Bulletin 67:1–15.

———, 1969. Fishery survey of U.S. waters of Lake Ontario. Pages 51–57 *in* Limnological survey of Lake Ontario, 1964. Great Lakes Fishery Commission Technical Report 14, Ann Arbor, Michigan

———, 1970. Effects of alewife predation on zooplankton populations in Lake Michigan. Limnology and Oceanography 15:556–565.

———, A. L. McLain. 1973. Lake Michigan: Man's effects on native fish stocks and other biota. Great Lakes Fishery Commission Technical Report No. 20, Ann Arbor, Michigan

Wright, R. R. 1891. Preliminary report on the fish and fisheries of Ontario. Pages 419–476, *in* Ontario Game and Fish Commission, Commissioner's Report, 1892. 483 pp. Ontario.

CHAPTER 17

Lake Sturgeon: A Unique and Imperiled Species in the Great Lakes

Nancy A. Auer

Introduction

There are nineteen species in the Genus *Acipenser,* yet the lake sturgeon (*Acipenser fulvescens*) is the only one which lives its entire life within freshwater. Lake sturgeon once ranged throughout the Mississippi River, the Laurentian Great Lakes, and the Hudson Bay drainages (Harkness and Dymond 1961). Lake sturgeon spawn in rapids and high flow sections of Great Lakes tributaries. Spawning along rocky lake shorelines or reefs is suspected, but remains undocumented (Harkness and Dymond 1961). Once abundant members of the Great Lakes fish community, lake sturgeon were, and continue to be, commercially valuable, and can provide tremendous sport fishing opportunities. These benthic feeding fishes hold a low, but intrinsic, position in the trophic food web of aquatic systems. Of the twenty-seven species of sturgeon known worldwide, nine are native to North America, however, only the lake sturgeon is endemic to the Great Lakes basin. Considered relicts, fossil evidence suggests sturgeons existed one hundred to two hundred million years ago. Sturgeons are classified as Chondrosteans, the original ray-finned fishes and retain many of the characteristics of primitive fishes.

Unique Attributes of Lake Sturgeon

Lake sturgeon forage on lake and river bottoms, feeding on chironomid larvae, molluscs, mayfly nymphs, caddisfly larvae (Harkness 1923), crustaceans, and fish (Harkness and Dymond 1961). They possess the primitive characters of a shark-like heterocercal tail, bony scutes along their head, back and sides, a cartilaginous skeleton, and a toothless, protrusible mouth. Lake sturgeon are the largest freshwater fish in the Great Lakes basin. Adults captured in the Sturgeon River, Michigan of the Lake Superior watershed, between 1987 and 1994 ranged from 110cm to 180cm TL and 7kg to 41kg (Auer 1999). An almost identical range in size was observed by Priegel and Wirth (1975) for Lake Winnebago lake sturgeon captured from 1955 to 1967. The largest lake sturgeon on record include an individual caught in Batchewana Bay, Lake Superior, in 1922 which was 228.6cm TL and 140.6kg, and another found in Lake Michigan in 1943 at 241.3cm TL and 140.6kg

(Scott and Crossman 1973). Great size is not unusual for this fish; between 1900 and 1959 there were twenty-four reports of lake sturgeon greater than 188cm TL and 62.6kg from Great Lakes waters (Harkness and Dymond 1961). Lake sturgeon are slow-growing and long-lived, and can reach ages of 100 to 150 years (Scott and Crossman 1973). The specimen from Batchewana Bay, Lake Superior, was believed to be 100 years old.

Lake sturgeon not only have unique physical features, but they also have some unusual life history strategies. Unlike many fishes, lake sturgeon require fifteen to twenty-five years to reach maturity and are intermittent spawners (Priegel and Wirth 1977). Depending on location, males may spawn every one to two, two to three, or three to ten years, while females may spawn every three to four, four to six, or three to fourteen years in Lake Winnebago, Wisconsin, southern Lake Superior and Lake Nipissing, Ontario, respectively (Love 1972; Lyons and Kempinger 1992; Auer 1999). These fish prefer to spawn in upper reaches of rivers among large rocks in rapids or areas of high flow (Harkness and Dymond 1961). The adhesive eggs cling to clean surfaces, the undersides of large cobble and boulders, and among clean gravel (Auer unpublished data).

Lake sturgeon possess many of the life history strategies of a k-selected species (Stearns 1976); long life, delayed maturation, and large size. However, the typical characteristics of production of a few large ova, parental care, and small reproductive effort do not seem to apply. A 36kg female lake sturgeon may produce 85,000 eggs at spawning, yet spawn only once in six years. The overall reproductive effort, therefore, may more closely resemble that of a k-selected species, as energetics applied toward reproduction on a yearly basis are small for lake sturgeon, and instead of a few large ova, these fish periodically spawn one large batch of many small ova.

Lake Sturgeon Fisheries in the Great Lakes

Native Americans in the Great Lakes region valued the lake sturgeon for food, oil, and leather before such products from lake sturgeon were appreciated by European settlers (Holzkamm and Wilson 1988). Additionally, Hudson Bay Company records indicate that the Ojibwa supplied the company with isinglass (a clarifying agent made from the swim bladder of lake sturgeon) from 1823 to 1885 (Holzkamm and McCarthy 1988). Early explorers along the Ontonagon River, Michigan observed that Native Americans relied upon lake sturgeon almost entirely for their subsistence (Schoolcraft 1970). Kohl (1956) described lake sturgeon as the daily bread of Native Americans living near Lake Superior in 1860, because it was abundant, could be caught throughout the year, was wholesome, and had a very agreeable taste. Early Native Americans fished for lake sturgeon using spears (Kohl 1956) or they built weirs and use gaff hooks to take trapped fish (Schoolcraft 1970; Holzkamm and Wilson 1988).

Lake sturgeon were not considered a commercial species by settlers in the area before 1860 (Harkness and Dymond 1961) to 1875 (Koelz 1926), but were rather thought to be a nuisance, their sharp scutes often damaging cotton nets set for the more valued lake trout and whitefish (Smith 1968; Wells and McLain 1973). By 1880 smoked lake sturgeon flesh was found to be a good substitute for smoked halibut and a growing market began for sturgeon meat, as well as an increased European demand for caviar. Gill nets, pound nets, and set-lines were the primary gears used in capture of lake sturgeon (Smith and Snell 1889; Koelz 1926). The first records of commercial catch of lake sturgeon occurred in 1880 (Tower 1908), when 3,428,029kg were taken from Lakes Michi-

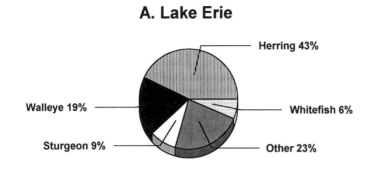

A. Lake Erie

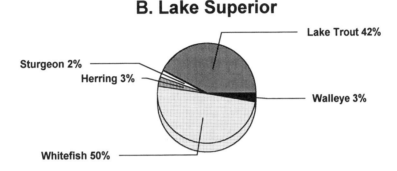

B. Lake Superior

FIG. 1. *Composition of Commercial Fish Catch from 1885. A. Lake Erie and B. Lake Superior.*

gan and Erie, 63% of the total sturgeon catch taken in the United States. In 1885, the total catch for these two lakes was 2,991,038kg (Baldwin et al. 1979). Prince (1905) specified eleven products obtained from lake sturgeon which included: caviar; isinglass; fresh, smoked, or salted flesh; oil; fertilizer; leather; and delicacies from other portions of the fish (brain, notochord and belly). Many products were used locally, however, much of the caviar produced was shipped to Hamburg, Germany, Russia or England (Smith and Snell 1889; Prince 1905).

Commercial harvest data for the Great Lakes is available from the early 1880s to the present (Baldwin et al. 1979; OMNR 1972a, 1972b, 1976, 1981). In 1885, lake sturgeon composed 7% of the total combined U.S. and Canadian commercial fish catch. Lake sturgeon, at that time, ranked among the five most abundant fishes in the commercial catch, with lake herring, whitefish, walleye, and lake trout (Baldwin et al. 1979). In Lake Erie, in 1885, the commercial catch included 11,478,802kg herring; 5,120,690kg walleye, 2,352,823kg lake sturgeon, and 1,686,485kg whitefish, while in Lake Superior that year the catch of fish included 2,348,741kg whitefish, 1,995,840kg lake trout, 147,420kg herring, 128,822kg walleye, and 101,606kg lake sturgeon (Baldwin et al. 1979; fig. 1).

The commercial harvest of lake sturgeon for the Great Lakes and tributaries from 1880 through 1994 is presented in table 1. These data reflect an increased fishing effort in the 1930s due to the Depression and a need to supplement diets with cheap protein sources, and a decrease in effort in

TABLE 1

Combined United States and Canadian Commercial Harvest of Lake Sturgeon (kg) for the five Great Lakes, Lake St. Clair, the St. Lawrence River through Lac St. Pierre and their tributaries grouped in five-year intervals.

FIVE-YEAR INTERVAL	LAKE SUPERIOR	LAKE MICHIGAN	LAKE HURON	LAKE ST. CLAIR	LAKE ERIE	LAKE ONTARIO	ST. LAWRENCE	TOTAL
1880–1884			327,950	97,070	501,230	126,550		1,052,800
1885–1889	271,250	915,820	1,608,010	274,430	3,753,080	347,010		7,169,600
1890–1894	218,180	795,160	1,507,310	378,300	1,916,000	383,740		5,198,690
1895–1899	92,080	213,650	1,094,990	171,460	1,110,420	200,950		2,883,550
1900–1904	43,090	133,360	406,880	130,640	519,820	75,300		1,309,090
1905–1909	17,690	100,700	141,520	84,370	198,680	33,110		576,070
1910–1914	28,120	26,310	115,210	107,960	275,340	9,070		562,010
1915–1919	17,240	27,670	99,340	42,180	175,540	19,500		381,470
1920–1924	28,580	18,600	70,770	34,020	73,480	15,880	126,049	367,379
1925–1929	2,720	9,520	56,240	29,940	102,060	29,940	262,340	492,760
1930–1934	7,250		45,360	28,580	90,720	33,560	408,820	614,290
1935–1939	4,080		31,300	19,500	70,760	32,660	436,311	594,611
1940–1944	4,990		21,320	15,870	49,440	26,760	84,324	202,704
1945–1949	2,720		30,390	9,980	57,160	18,140	396,621	515,011
1950–1954	6,350	4,540	48,540	19,960	32,210	24,490	396,659	532,749
1955–1959	11,790	2,720	74,390	29,480	33,562	21,320	407,188	580,450
1960–1964	11,340	4,540	42,640	28,120	8,160	21,770	424,697	541,267
1965–1969	5,444	4,080	42,634	28,580	3,174	5,444	394,751	484,107
1970–1974	1,450		27,220		539	61,690	184,291	275,190
1975–1979	1,900		5,927		3,788	73,790	255,134	340,539
1980–1984	2,399		23,127	48	2,103	24,820	316,000	368,497
1985–1989	635		27,663	1,437	67		558,462	588,264
1990–1994			23,095	1,819			599,931	624,845

For the St. Lawrence River methods of reporting harvest changed in 1985 and data from 1941, 1970, 1971 and 1985 were not included in Table 1. Data were taken from Baldwin et al. 1979, OMNR 1972a and b, 1976, 1981, and personal communication with individuals in the Ontario Ministry of Natural Resources: R.T. Thomson, Lake Superior Management Unit, 435 James Street South, Suite 221, Thunder Bay, Ontario P7E 6E3; L.Mohr and S. Munroe, Lake Huron Management Unit, 611 Ninth Avenue East, Owen Sound, Ontario N4K 3E4; D. MacLennan, Lake St. Clair Fisheries Assessment Unit, R. R., Tilbury, Ontario, N0P 2L0; P. Dietz, Lake Erie Fisheries Assessment Unit, R.R. 2, Wheatley, Ontario, N0P 2P0; P. Smith, Lake Ontario Manager; 1 Richmond Blvd., Napanee, Ontario K7R 3M8 and J. Robitaille, Bureau D'Ecologie Appliquee, 4950, Blvd de la Rive-Sud, Bureau 104, Levis Quebec G6V 4Z6.

the early 1940s due to World War II, and a loss of fishermen and materials to the war effort. These data also illustrate the tremendous reduction in catch of lake sturgeon throughout the Great Lakes region by the early 1900s, and the failure of stocks to rebuild to historic levels (fig. 2).

Reasons for the Decline

Three factors have been implicated as major reasons for the decline of sturgeons worldwide (Rochard et al. 1990); physical impacts on spawning and nursery habitats, barriers to migration, and effects of fishing. These factors have all been important in the decline of the lake sturgeon in the Great Lakes.

Physical Impacts on Nursery and Spawning Areas

Because of their small size and the absence of fully formed, sharp, protective bony scutes, the greatest mortality for lake sturgeon is thought to occur during the first years of life. Habitat needs and preferences, and movements of newly hatched and juvenile lake sturgeon have not been well studied. Newly hatched lake sturgeon larvae begin drifting downstream seventeen to twenty-four days after peak periods of spawning (Kempinger 1988; LaHaye et al. 1992). Larvae have been observed to drift for distances of 150m to 12.8km, 19km, and 43km by Kempinger (1988), LaHaye et al. (1992) and Auer (unpublished data), respectively. Industrial water use, water level manipulation, and bridge construction and repair during such periods of drift may affect survival of young. There is one study on movement and habitat selection of juvenile lake sturgeon (Thuemler 1988). This author concluded that habitat selection may depend on stock adaptation, and that young from differing stocks may have different habitat preferences.

Human-induced impacts on the physical environment of spawning grounds have been more intensively studied. Modifications in flow regimes below hydroelectric facilities have affected sedimentation, water temperatures (Graham 1985), and oxygen concentrations (Brousseau and Goodchild 1989) in spawning areas below these barriers. Dewatered sections of rivers allow eggs, newly hatched larvae, and even adult fish to become exposed to air, direct sunlight, and predators (Brousseau and Goodchild 1989). In many cases, barrier dams have been placed at historic spawning grounds simply because natural rapids or falls are areas of greatest hydrologic head (table 2). Construction and placement of some barriers has destroyed the natural spawning grounds preferred by lake sturgeon in many large rivers.

The rapid decline in lake sturgeon abundance throughout the Great Lakes basin is illustrated with commercial catch records for the United States and Canada (table 1, fig. 2). Sawmills, discharging sawdust waste into many rivers in Michigan, Wisconsin and other Great Lake states in the early 1800s, prior to records of catch, had perhaps the earliest impacts on fish in these systems (Gates et al. 1983). A sawmill was first built on the Menominee River, Wisconsin in 1832, and by 1867, there were several mills located on both sides of this river (Emich 1987). These mills deposited great quantities of sawdust into the river and large rafts of sawdust could be seen floating in the waters of Green Bay. Emich (1987) states that the sawdust in the water ruined the fishery in the area, and by 1870, no whitefish runs remained in the rivers. Such waste discharges probably degraded spawning sites used by lake sturgeon, however, no early documentation concerning lake sturgeon has been found.

Thuemler (1988) mentions water quality degradation as a probable reason for the disappearance of lake sturgeon from the uppermost section of the Menominee River (Sturgeon Falls to Chalk Hills Dam). Harkness and Dymond (1961) cite examples of pulp and paper mill waste burying lake sturgeon spawning beds on the Rainy and Spanish Rivers, Ontario. The harvesting of lumber probably had a small impact on lake sturgeon habitat, but the log driving and discharge of mill effluent severely degraded aquatic habitat (Brousseau and Goodchild 1989).

Nutrient-rich river mouths are believed to be important nursery areas for young lake sturgeon, yet this hypothesis remains untested by research. The river mouths of many Great Lake tributaries have been dredged to create shipping channels, filled to support development, or

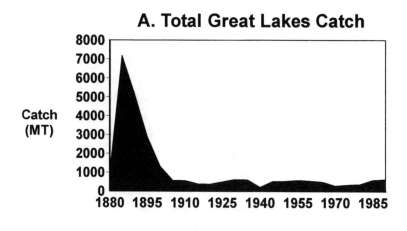

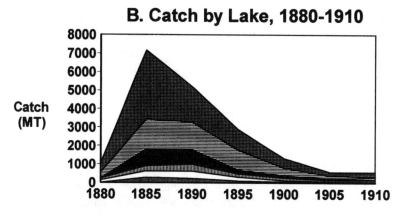

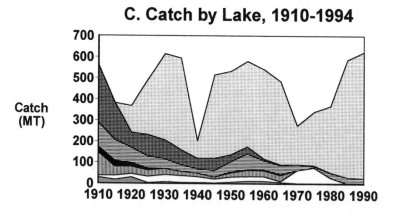

FIG. 2. *Commercial Catch of Lake Sturgeon from 7 regions in the Great Lakes, United States and Canada combined. A. Total Catch 1880–1994. B. 1880–1910. C. 1910–1994.*

TABLE 2

Great Lake tributaries known to be used by lake sturgeon and current or natural barriers (NB) to movement

Isolated stocks have no access to the Great Lakes. Dam locations are for first structure encountered above the river mouth, in several cases there are more dams in system.

STATE/PROV.	LAKE	RIVER	BLOCKAGE AND DISTANCE	SOURCE
MN	Superior	St. Louis	Dam at 40 rkm NB 41 rkm	Schram et al. 1992.
WI	Superior	Bad River	NB 32 rkm	Shively and Kmiecik 1989
WI	Michigan	Wolf	Dam at 201 km, isolated stock	Kempinger 1988
WI	Michigan	Fox	14 locks/17 dams	D. Folz, pers. comm.
WI	Michigan	Menominee	Dam at 6 rkm, NB at 125 rkm, isolated stock	Thuemler 1985
WI	Michigan	Peshtigo	Dam at 14 rkm	FERC 1981
WI	Michigan	Oconto	Dam at 21 rkm	T. Thuemler, pers. comm.
MI	Superior	Ontonagon	Dam at 55 rkm	Blumer et al. 1991
MI	Superior	Sturgeon	Dam at 69 rkm	Auer 1996
MI	Michigan	St. Joseph	Dam at 37 rkm	Baker 1980
MI	Michigan	Manistique	Dam at 2 rkm	Bassett 1981
MI	Michigan	Manistee	Dam at 45 km	FERC 1981
MI	Michigan	Escanaba	Dam at 35 rkm	Blumer et al. 1991
MI	Michigan	Grand	NB at 64 rkm	Ylkanen, pers. comm.
MI	Huron	Black	Dam at river mouth, isolated stock	Hay-Chmielewski 1987
MI	Huron	Ausable	Dam at 14 rkm	Ylkanen, pers. comm.
MI/ON	St. Clair	St. Clair	No blockage	D. MacLennon, pers. comm.
OH	Erie	Cuyahoga	Historic, no current known population	Trautman 1981
OH	Erie	Maumee	Dam at 77 rkm	Trautman 1981
OH	Erie	Sandusky	Historic, no current known population	Smith and Cnoll 1889
NY	Ontario	Black	Dam at 2 rkm, NB 14 rkm	Carlson 1995
NY	St. Lawrence	Oswegatchie	Dam at 0.5 rkm, NB at 94 rkm	Carlson 1995
NY	St. Lawrence	Grasse	Dam at 12 rkm	Jolliff and Eckert 1971
NY	St. Lawrence	Raquette	Dam at 31 rkm	Jolliff and Eckert 1971
NY	St. Lawrence	St. Regis	Dam at 3 rkm, NB at 32 rkm	Carlson 1995
PQ	St. Lawrence	St. Lawrence	Robert Moses Dam isolated stock	Jolliff and Eckert 1971
PQ	St. Lawrence	Ottawa Rr.	Dam 37 rkm, isolated stock	Guenette et al. 1993

Personal Communication: Dan Folz, Wisconsin Dept. Nat. Res., P.O. Box 2565, Oshkosh, WI 54903; Don MacLennon, Lake St. Clair Fishery Assessment Unit, R.R. 2, Tilbury, ON, N0P 2L0; Bernie Ylkanen, Michigan Dept. Nat. Res., 1990 US 41 S., Marquette, MI 49855; Thomas Thuemler, Wisconsin Dept. Nat. Res., 1636 Industrial Parkway, Box 16, Marinette, WI 54143.

stabilized for harbor or recreational purposes. Both the Cuyahoga and Maumee Rivers in Ohio supported populations of lake sturgeon in the late 1800s (Trautman 1981), yet today these fish may be extinct in these systems.

Barriers to Migration

During the late 1800s and early 1900s many barrier dams were built on Great Lake tributaries. Most of these structures were built to generate power and were placed in sections of rivers with greatest change in gradient to capture the greatest energy from falling water. The placement of dams in these areas has either eliminated the spawning habitat or prevented the sturgeon from moving into areas of rapids and high-flow preferred for spawning. Lake sturgeon are believed to

TABLE 3

Lake sturgeon commercial fishing closure dates for the five Great Lakes,
Lake St. Clair and the St. Lawrence River to Lac St. Pierre

LAKE	STATE/PROVINCE	DATE OF CLOSURE
Superior	Minnesota	1928 to present
	Wisconsin	1928 to present
	Michigan	1928–1950, 1970 to present
	Ontario	1990
Michigan	Illinois	1929 to present
	Indiana	1929 to present
	Wisconsin	1929 to present
	Michigan	1929–1950, 1970 to present
Huron	Michigan	1928–1950, 1970 to present
	Ontario	no closure
Erie	Michigan	1929–1950, 1970 to present
	Ohio	1968
	New York	1972 to present
	Pennsylvania	1977 to present
	Ontario	1920–1921, 1984 to present
Ontario	New York	1972 to present
	Ontario	1984 to present
St. Clair	Michigan	1909 to present
	Ontario	1970–1979
St. Lawrence Rr.	New York	1972 to present
	Quebec	Lac St. Francis only, 1987

return to natal areas to spawn, and fidelity to particular river systems is high (Lyons and Kempinger 1992). These authors found that only 3 of 203 fish (1.5%) chose a different river for spawning in a subsequent year.

In many areas throughout their range, lake sturgeon are now isolated in sections of rivers or lakes due to placement of more than one barrier structure (Basset 1981; Thuemler 1985). McKinley et al. (1993) have shown that nonesterified fatty acids (an energy source for metabolism) of lake sturgeon reflect nutritional status of these fish, and that lake sturgeon downstream of four hydroelectric facilities had lower lipid reserves than those in upstream sections. These authors suggested that sturgeon above the dams had more abundant and diverse food sources under the stable water levels which occurred above the facilities. No assessment of these food sources was made to verify these conclusions.

Effects of Fishing

The decline in commercial harvest presented in figure 2 may reflect the effects of exploitation; certainly the early fishery had some impact on lake sturgeon abundance. Unlicensed catch also affected lake sturgeon abundance. In 1894, a Canadian officer seized and destroyed 48km of set lines with thousands of large, unbaited grapnel hooks set to snag sturgeon in Lake St. Clair (Prince

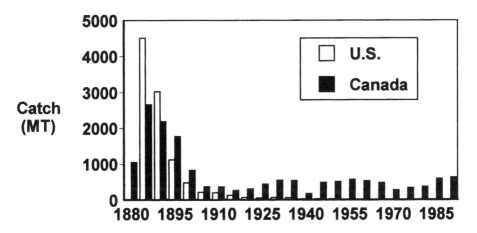

FIG. 3. *Commercial Catch of Lake Sturgeon from the United States and Canadian Waters of the Great Lakes.*

1905). A 56% to 90% decline in catch occurred over a seven to ten year span in most Great Lakes, and these declines were viewed as permanent because continued fishing from 1952 to 1956 (fig. 2) produced only a fraction (0.3% to 6%) of historic highs (Harkness and Dymond 1961). The decline may have been a direct affect of exploitation, or the catch may have reflected a decrease in abundance due to reduced spawning and egg survival resulting from dam construction, possible water quality changes, or a combination of these factors. Lake sturgeon are curious fish, often attracted to structures. This habit makes them vulnerable to modern fishing techniques, and they are easily captured using pound or gill nets.

Barriers prevent adult fish from reaching spawning grounds, and these fish are harvested without making a contribution to maintaining the stock. By 1929, most Great Lake states had prohibited commercial harvest of lake sturgeon because harvests had decreased to 20% or less of the catch reported in the early 1900s (table 1), and closure might protect remaining stocks. The fishery on the St. Lawrence River has provided the highest yield of any Great Lake area since 1920 (fig. 2). The Provinces of Quebec and Ontario have not prohibited commercial harvest except for short intervals or in very recent years (table 3). The harvest of lake sturgeon declined in Ontario inland waters, Lake of the Woods, Lake Nipigon, Lake Nipissing, and Great Lake waters at about the same time, 1895. In the "Draft Management Strategy for Lake Sturgeon in Ontario" all southern Ontario districts viewed lake sturgeon as rare or endangered warranting special protection, yet reasons for allowing the continued harvest in some regions and only closing fisheries in other areas in very recent years (table 3) were not stated (E. Iwachewski, Ontario Ministry of Natural Resources-Thunder Bay, personal communication).

Lack of Rebound

The commercial harvest of lake sturgeon declined so severely by the 1920s that most fishery professionals assumed the species was at or near extinction. Tower (1908) wrote an article for *Popular Science Monthly,* and chose this ominous title: "The Passing of the Sturgeon: a Case of

Unparalleled Extermination of a Species." In another article "Species Succession and Fishery Exploitation in the Great Lakes," Smith (1968, p.669) stated: "The lake sturgeon undoubtedly will continue to live in the Great Lakes at low abundance until influenced adversely by the changing environment, or by a fishery for another species that raises the incidental catch above the limits of its biological yield." These types of statements reflect a belief that the lake sturgeon was headed for extinction. Although commercial fisheries for lake sturgeon were closed, this attitude of resignation and lack of public interest kept the lake sturgeon from being considered as a species for rehabilitation. In the United States, the lake sturgeon has essentially been ignored since its decline in abundance and closure of commercial fisheries in the late 1920s. The lake sturgeon in the Great Lakes waters of the Provinces of Ontario and Quebec has been subject to a small, but continuous commercial harvest (fig. 3).

Unfortunately, even in the early 1990s, the lake sturgeon has attracted little attention. For example, although mentioned as part of the historic fish community of Lake Ontario, Kerr and LeTendre (1991) in their "State of the Lake" report fail to include lake sturgeon in their list of important species for the lake when discussing their first objective of maintaining viable populations of indigenous fish species. Lake sturgeon are indigenous to Lake Ontario, and historically provided a portion of the Great Lakes commercial harvest (fig. 2). Despite the lack of attention to rehabilitation by fishery managers, this species has persisted, and a few populations may be rebuilding in some areas of the Great Lakes. Gaston and Lawton (1990) suggest that thermal springs or deep cold lakes may buffer environmental variation and allow isolated populations of some freshwater fishes to persist. The size and physical characteristics of the Great Lakes, combined with the unusual life history strategies of this species, may have worked together to allow lake sturgeon to survive in some locations despite the great adversity of intense fishing and environmental perturbation. Currently, self-sustaining stocks of lake sturgeon are known from Lake St. Clair and the St. Clair River, Lake Superior, and Lake St. Francis in the St. Lawrence River. Lake sturgeon have access to deep, cool water in each of these systems. Some lake sturgeon may have sought refuge in deep water from environmental disturbances caused by industrial development and eutrophication occurring in shallow water. Populations that had access to deep, cool waters may have persisted, while those confined to shallow water systems perished.

Current Distribution and Abundance

Throughout the Great Lakes region, tributaries known to be used by lake sturgeon have been identified in the literature or through personal communication with fishery managers. Table 2 summarizes tributaries in the Great Lakes, primarily U.S., for which there is documentation that lake sturgeon have, or are utilizing the system, and includes the present type of barrier if applicable. A comprehensive summary of the major tributaries and historic spawning grounds of lake sturgeon in the Canadian region of the Great Lakes has not been undertaken. Harkness and Dymond (1961) provide some limited information on rivers used by lake sturgeon throughout their entire North American range, but recent population assessment and evaluation of status of these historic stocks is lacking.

Issues Facing Sturgeon Management

The lake sturgeon is under legal protection in seven of eight states which border the Great Lakes. Wisconsin considers the species one of special concern (Johnson 1987). Houston (1987) summarized the status of lake sturgeon in Canada as being neither endangered nor even rare, but decline and extirpation may occur if stock assessment and habitat protection are not soon considered. Lake sturgeon are widespread in Canada, not only in the Great Lakes region but in tributaries of the Hudson Bay drainage. In many areas of Canada, populations appear sufficient to allow a modest managed fishery (Houston 1987). The commercial harvest for lake sturgeon in the Canadian waters of Lakes Erie and Ontario has been closed since 1984, was recently closed for Lake Superior in 1990, but remains open in Lakes Huron, St. Clair, and the lower St. Lawrence River (table 1; fig. 3). In 1994, a single commercial set-line operation was licensed in Lake St. Clair, and as many as one hundred commercial operators were active in Lake Huron. Abundance of lake sturgeon is believed, by Canadian biologists, to be sufficient to support these fisheries as the 1994 harvest was below quotas set for Lake Huron. In the lower St. Lawrence River, sixty-three licensed commercial fishermen operated between Lac St. Louis and Lac St. Pierre in 1984 (Dumont et al. 1987). Lake sturgeon harvested by Canadian fishermen may use spawning grounds located in other management jurisdictions, but information on such habitat use has not been collected or verified.

Stock Identification and Genetic Diversity

Low genetic variability is common among most North American sturgeons, yet stocks of one species develop genetic differences as they adapt to specific environments (Leggett 1977). Many water level control structures and power facilities were built in the late 1800s and early 1900s, and some of these barriers have caused stocks of sturgeon to be isolated for almost one hundred years. White sturgeon, *(A. transmontanus)*, isolated in the upper sections of the Columbia and Fraser Rivers were found to have lower mitochondrial DNA diversity than white sturgeon located further downstream (Brown et al. 1992). Whether a sufficient time interval has allowed for accumulation of genetic differences or drift for isolated populations of lake sturgeon has only recently been investigated (Ferguson et al. 1993; Guenette et al. 1993).

Stock genetics of lake sturgeon are only now being studied. Guenette et al. (1992) chose to examine several stocks of lake sturgeon from the St. Lawrence River area for possible morphological differences. Their investigation concluded that stock separation based solely on morphology was not possible using their forty-one characteristics. These same authors later examined mitochondrial DNA variation among lake sturgeon collected from five sites in the St. Lawrence River area of Quebec and found that sturgeon from these locations were homogeneous with regard to mtDNA analysis (Guenette et al. 1993). Recent work by Ferguson et al. (1993) also indicated that there was little genetic differentiation among lake sturgeon taken from various sites within the Moose River watershed (James Bay drainage), yet these authors felt there may be differentiation between fish from the three basins, and hypothesize that lake sturgeon of differing glacial races have used separate routes in colonizing their current range (Ferguson et al. 1993).

In the Great Lakes basin, some lake sturgeon populations have been extirpated, and reintroduction through stocking is being considered by several management agencies. Using radio

TABLE 4

Upstream migration distances for lake sturgeon in eight
Great Lake tributaries and location of spawning site

RIVER SYSTEM	STATE	DISTANCE UPSTREAM (RKM)	SPAWNING SITE	SOURCE
Wolf	Wisconsin	210*	Shawano dam	Kempinger 1988
Bad	Wisconsin	32	Natural rapids	Shively and Kmiecik 1989
Menominee	Wisconsin	125**	Natural rapids and 4 dams	Thuemler 1988
Rainy	Minnesota	50–68	Natural rapids	Holzkamm et al. 1988
Ontonagon	Michigan	36**	Natural rapids	Schoolcraft 1970
Sturgeon	Michigan	69*	Natural rapids and 1 dam	Auer 1996b
Des Prairies	Quebec	19*	Power station	LaHaye et al. 1992
L'Assomption and Ouareau	Quebec	37	Natural rapids	LaHaye et al. 1992

* Barrier Dam terminates migration.   ** Historic information only, no sturgeon currently found spawning at the natural barriers.

telemetry, Thuemler (1988) investigated movements of juvenile (avg. 48cm TL) lake sturgeon captured below the Grand Rapids dam on the Menominee River, and than released into a 39km free-
flowing section above the point of a capture. These juveniles remained in riverine portions of the
river for the duration of the study (October 2 to 21) and were not found in impounded water above
dams. In contrast, juveniles (avg. 30.5cm TL) reared in a hatchery from eggs of a lake dwelling
stock, also released in the same area, quickly moved downstream and most were located in impounded waters. Thuemler's (1998) findings suggest that stock origin should be considered when
selecting eggs for rearing and reintroduction purposes. Yet Ferguson et al. (1993) and Guenette et
al. (1993) conclude that there is little genetic difference among stocks within a drainage. These
limited studies do not provide a definitive solution to the problem. Genetic research on lake sturgeon tissues collected from areas throughout their entire range is now in progress at the University of Guelph (M. Ferguson, University of Guelph, personal communication) and Ohio State
University (T. Cavendar, Ohio State University, personal communication). It is hoped that these
investigations will clarify current concerns regarding stock genetics and subsequent implications
for stocking and management of lake sturgeon.

Range Restrictions and Habitat Degradation

Lake sturgeon are one of the largest freshwater fishes in North America, and large-bodied organisms often require large areas in which to range for optimal growth and survival (Soule 1987). The
relationship between body size and range has been shown to be positive for fishes (Peters 1983;
McAllister et al. 1986), and holds for several different species of sturgeon (Auer 1996a). Data on
unrestricted spawning migration distances for lake sturgeon are sketchy (table 4), and absolute
range information does not exist. Some lake sturgeon in the Great Lakes have moved tremendous
distances. Each spring, spawning lake sturgeon are tagged in the Wolf River, Wisconsin. These fish
return to Lake Winnebago, but are blocked from further out-migration to Green Bay and Lake Michigan by seventeen locks and fourteen dams on the Fox River. However, two fish, still retaining tags,
were recovered in Lakes Erie and Huron after five and eight years at large, respectively (Dan Folz,

Wisconsin Department of National Resources, personal communication). Auer (1999) has found post-spawning lake sturgeon to disperse quickly from the spawning site with individuals located 70km to 265km from the site within a few months of spawning.

Due to barrier structures, some lake sturgeon populations have persisted in restricted environments with limited area and habitat (i.e., 34km of the Menominee River between Grand Rapids and Upper Scott dams, Thuemler 1988). However, isolated populations of any organism are more prone to catastrophe, genetic drift and environmental and demographic uncertainty (Shaffer 1981). Lake sturgeon in fragmented systems may spawn, but larvae and juveniles can drift or swim downstream, and out of the populated segment. Recruitment to the isolated stock is lost, and individuals returning to natal rivers to spawn are blocked below the barriers.

Many barrier dams were placed at the mouths of river systems, preventing lake sturgeon from utilizing spawning grounds. Should these migration routes be reopened, lake sturgeon could once again use areas of rapids or other historic spawning grounds, provided these areas have not been altered or destroyed by development or industry. This phenomenon has been observed on the Peshtigo River in Wisconsin. With improved water quality, large lake sturgeon returned to the river from Green Bay only to be trapped and injured in turbines of a power facility located there.

These fish could have been randomly searching for a place to spawn, or they could have originated from an earlier stock isolated above the dam when it was built and now seek their natal spawning grounds. Power facility personnel constructed a large mesh grid to prevent the fish from moving into the turbines, but this is only a partial solution to the problem. The dam still acts as a barrier to lake sturgeon reaching adequate spawning habitat.

The building of dams for, and operation of hydroelectric facilities has had a tremendous impact on lake sturgeon. Not only do the structures block migration routes and cover spawning grounds, but in many cases, facility operation produces water level fluctuations on seasonal and periodic scales (Brousseau and Goodchild 1989). Many facilities have operated in a peaking mode, generating power and discharging water at high flows during hours of peak demand (0800hr to 1700hr; Auer 1996b). Once demand diminished, water discharge was dramatically cut, and water flows rapidly decreased. On the Sturgeon River, Michigan, a small hydroelectric facility discharged 360 cfs to 630 cfs ($10.2m^3$/s to $17.8m^3$/s) during peaking operation, yet dropped within thirty minutes to 15 cfs ($0.42m^3$/s) when operation was terminated (Auer 1996b). These interrupted and fluctuating flows create difficulties for adult fish moving onto the spawning grounds and reduce survival of eggs and larval lake sturgeon. Eggs become exposed to drying, increased water temperature, and reduced oxygen concentrations, while newly-hatched lake sturgeon larvae may be trapped in dewatered stream bed material, and exposed to drying and temperature stresses as well (Brousseau and Goodchild 1989).

Hydroelectric facilities release water when high-water events, storms, and spring snow melt, begin to fill reservoirs beyond designed capacity. These spill events affect lake sturgeon spawning downstream of dam sites. Erosion of the streambed and banks can occur when large volumes of water are spilled from reservoirs to maintain water levels, and sediment loads are carried onto downstream sites (Brousseau and Goodchild 1989). Often these spill events occur in the spring, and the adhesive eggs of lake sturgeon can be scoured from the substrate in areas of high flow, or buried by sediment in areas of low flow. Spawning adults and newly-hatched young may also be washed over spillways in high flow events (Brousseau and Goodchild 1989). Reservoir drawdowns can occur during winter as natural flows decrease, but power demand remains unchanged. Winter drawdowns may reduce available winter habitat for lake sturgeon isolated above barrier dams, but drawdown effects on lake sturgeon stocks remain unstudied (Brousseau and Goodchild 1989).

When prevented from utilizing historic spawning grounds located upstream of barrier structures, lake sturgeon have been found to spawn directly below the barriers. Lake sturgeon spawning below the Shawano Dam in the Wolf River, Wisconsin have left spawned eggs in mats as much as 15cm thick (Kempinger 1988). Survival of eggs and larvae from such clumping remains unstudied. Lake sturgeon, spawning in areas of rapids, have not been observed to leave such large masses of eggs (Auer unpublished data).

Fishery Conflicts

All commercial fishing operations for lake sturgeon in the U.S. waters of the Great Lakes have been discontinued, yet commercial fisheries still operate in the Canadian waters of Lakes Huron, St. Clair, and the St. Lawrence River. There is no evidence that present commercial fisheries are depleting stocks. However, lake sturgeon harvested in one location may spawn in an area of another management jurisdiction. Lake sturgeon spawn in the New York waters below the Robert Moses Dam on the St. Lawrence River. Larvae and adults of this population have unrestricted movement into the Canadian waters of Lac St. Francis, 16km downstream of the dam (Jolliff and Eckert 1971). Spawning success and harvest rates will require careful monitoring by management agencies in both countries if this stock of lake sturgeon is to remain self-sustaining.

Once remnant Great Lakes populations receive protection and are allowed to rebuild in number, or populations are reestablished in areas in which they were extirpated, some stocks may be able to support a managed fishery. Determining exploitation rates for lake sturgeon stocks will depend on the type of fishery, and the abundance and age composition of the stock. Several isolated stocks in Wisconsin and Michigan have been managed for sport fisheries since the mid 1900s. Every legal lake sturgeon harvested from Wisconsin waters has been registered with the Department of Natural Resources since 1983. This enables managers to assess mortality, observe changes in population structure, and adjust size, bag limits, and season if necessary. This data on sport fishing impacts for the Winnebago and Menominee populations (Priegel and Wirth 1975; Folz and Meyers 1985; Kornely 1988) could be applied to larger systems once stocks have rebuilt to harvestable levels. Data on commercial and recreational harvest is also available for lake sturgeon populations in northern Ontario rivers (Olver 1987). Before reinstating any type of fishery for lake sturgeon in the Great Lakes, reproductive success must be assured, stocks must be identified, abundance monitored, and stocks allowed to rebuild if needed.

Throughout the Great Lakes, several Native American tribes actively pursue commercial fisheries. Eleven tribes are located in Minnesota, Wisconsin, and Michigan. Fisheries of these tribes are monitored by the Great Lake's Indian Fish and Wildlife Commission, and none of the fisheries target lake sturgeon, yet lake sturgeon are taken in incidental catch. The St. Regis/Mohawk tribe, located in New York, Ontario, and Quebec, has commercial fisheries in the St. Lawrence River which target lake sturgeon. Native American fishing in Canadian waters of the Great Lakes is not monitored, so harvest information is unavailable; however, the Ontario Ministry of Natural Resources considers the harvest to be low.

Although commercial fishing has been reduced or terminated throughout much of the Great Lakes, recreational fisheries have continued. Stocks of lake sturgeon in Lakes Winnebago, Poygan, Winneconne, Butte Des Morts, the Menominee River (Wisconsin), and in Black, Burt, Mullet, and Indian Lakes (Michigan) have provided sport fishing opportunities (either spear, hook and line, or both) (Priegel and Wirth 1975; Baker 1980; Kornely 1988). Speared lake sturgeon will fight, and quite commonly the spear pulls out, allowing the fish, often mortally wounded, to escape (Rickey 1987). Spearing is coming under scrutiny by fishery professionals in Michigan drafting management plans. Spear fishing may be phased out, perhaps to be replaced by catch and release barbless hook and line fisheries (B. Ylkanen, Michigan Department of Resources, personal communication)

By-Catch and Poaching

The impact or magnitude of the commercial, Native American, and sport fishery incidental catch on Great Lakes lake sturgeon stocks has not been investigated. In Lake Superior, commercial fishermen have reported over twenty incidental catches of lake sturgeon over an eight year period, all fish were returned alive (Auer 1999). Native American fishermen are also encouraged to return incidental catches, however some fish are taken.

Poaching is a threat to lake sturgeon throughout their entire range. Russian, Chinese, and Iranian sturgeon previously provided much of the world's caviar. With the collapse of the U.S.S.R., and economic development and overfishing in China, many sturgeon species in Europe and Asia are in jeopardy (Birstein 1993). As the supply of caviar in Europe and Asia diminishes, other sources will be sought. Paddlefish (K. Graham, Missouri Department of Conservation, personal communication); white sturgeon (A. Cohen, National Marine Fisheries Service, personal communication) and lake sturgeon in the Winnebago and Menominee systems (T. Thuemler, Wisconsin Department of National Resources, personal communication) have been subject to organized poaching. Egg laden lake sturgeon are the most vulnerable to poaching while concentrated on, or migrating to the spawning grounds. The implementation of high fines and the patrol of spawning grounds during the spawning season has helped discourage the illegal taking of lake sturgeon in Wisconsin. Such actions allow managers to reduce unnatural mortality and maintain stock size. Lake sturgeon stocks may come under an increased threat of poaching as markets for caviar remain lucrative and supplies of other caviar sources dwindle.

Sturgeon flesh is sold fresh or smoked and commands prices of $4.50 to $10.00 per kg (U.S.) (Houston 1987). Lake sturgeon caviar has had recent wholesale values of $13.00 to $16.50 per kg (Houston 1987). In 1994, a caviar company in Chicago, Illinois was paying fishermen $18.00 per kg for eggs. Wholesale prices for resulting caviar were $54.00 per kg and retail prices as high as $136.00 per kg. Gonads in mature lake sturgeon females account for 21.5% of the total body weight (Cuerrier 1966). A 36 kg female would hold about 8 kg in gonadal products. At $18.00 per kg the eggs, sold for caviar, would be worth about $144.00. Perhaps half of the remaining weight, 14kg, of the fish would be usable flesh. If sold for $10.00 per kg, the flesh could bring about $140.00, therefore, a single female would be worth about $300.00 to the fisherman with much higher prices realized in whole-sale and retail markets.

Sea Lamprey Management

Sea lamprey control efforts focus on sterile male programs, chemical treatments, and barriers to migration. The latter two directly affect lake sturgeon. Structures built to prevent spawning migrations of sea lamprey also prevent spawning migrations of lake sturgeon. This method of control should not be employed on river systems known to support lake sturgeon. Chemical control using TFM (3-trifluoromethyl-4-nitrophenol) can be toxic to lake sturgeon (D. Johnson, U.S. Fish and Wildlife Service, personal communication). Sea lamprey have a greater sensitivity to TFM, and careful application can adequately kill larval lamprey without harming juvenile lake sturgeon (D. Johnson, U.S. Fish and Wildlife Service, personal communication). Lamprey treatment sched-ules, changed from spring and early summer to fall can prevent chemical stress to young and drifting lake sturgeon larvae and should now be common practice in known lake sturgeon waters. Sea Lamprey treatment of the Sturgeon River, Michigan is now conducted in the fall. This change from spring treatment occurred after a bioassay in 1989 revealed lake sturgeon tolerance to TFM was dependent on pH, and juvenile lake sturgeon mortality occurred after 12hour exposure to TFM concentrations greater than 2.4mg per L (D. Johnson, U.S. Fish and Wildlife Service, personal communication).

Contaminants

Lake sturgeon, which feed on organisms living near the bottom of lakes and streams, are more likely to come in contact with contaminated sediments and organisms than are pelagic fishes. Since lake sturgeon can live to be sixty to one hundred years old, they can accumulate more con-taminants than short-lived fishes. Oil was one of the products harvested from lake sturgeon in early fisheries (Prince 1905; Holzkamm and Wilson 1988). Contaminants accumulate in the lipids of organisms (WDNR 1990), and since lake sturgeon are oily fish, contaminant accumulation can be a problem.

Twenty lake sturgeon, captured from 1979 to 1987 in Lake Wisconsin (outside the Great Lakes watershed), were analyzed for PCBs (polychlorinated biphenyls). The U.S. Food and Drug Admin-istration has set 2.0ppm as the tolerance level for human consumption of fish flesh. Nine fish, 135cm to 170cm TL, had levels of this contaminant at or above safe levels for human consumption (2.0ppm to 5.1ppm) (Larson 1988). The Wisconsin Department of Natural Resources has had lake sturgeon, taken from Lake Superior and Green Bay and its tributaries between 1985 and 1990, evaluated for contaminants. Three lake sturgeon (65 to 132 cm TL) from Lake Superior had PCB levels below 0.3ppm, while five fish (129cm to 221cm TL) from Green Bay and the Peshtigo River had levels between 3.9ppm and 34.0ppm. Of eleven fish (89–147cm TL) taken from the Menominee River, all but one had PCB levels below 1.0ppm (T. Thuemler, Wisconsin Department of Natural Resources, personal communication).A 1994 no-consumption advisory based on mercury levels which exceed 1.0ppm has also been established for lake sturgeon captured from the river mouth to the first dam on the Menominee, Oconto and Peshtigo Rivers of Wisconsin (WDNR 1994).

TABLE 5

Downstream movement of lake sturgeon after tagging at spawning locations in four North American river systems

RIVER SYSTEM	STATE/PROVINCE	DOWNSTREAM DISTANCE(RKM)	TAGGING SITE	SOURCE
Groundhog	Ontario	26	LaDuke rapids	Nowak and Jessop 1987
Kenogami	Ontario	130	Ogahalla rapids	Sandilands 1987
Black	Michigan	20	Kleber dam	Hay–Chmielewski 1987
Sturgeon	Michigan	70–265	Rapids 1 km below dam	Auer 1999
Des Prairies and L'Assomption	Quebec	138	250 m below Des Prairies Dam, rapids at Joliette	Dumont et al. 1987

Future Outlook

Fish Passage

To date, no fish passage facility in the United States or Canada has been designed or built specifically to pass sturgeon. The large body size of migrating adults can prevent this fish from using conventional fish ladders, which have narrow slots, high baffles, and short distances between steps. A joint and innovative study involving the Wisconsin and Michigan Departments of Natural Resources; The U.S. Fish and Wildlife Service, National Biological Survey Conte Anadromous Fish Research Center, Turners Falls Massachusetts; Guelph University; and the Department of Fisheries and Oceans Freshwater Institute in Winnipeg, Manitoba began in the fall of 1993, and is focusing on designing and building the first fish passage specifically for lake sturgeon on the Menominee River, Wisconsin. It is hoped that this technology will eventually be feasible for similar impounded systems known to be used by lake sturgeon throughout the Great Lakes.

Rearing, Stocking, and Aquaculture

One of the first North American sturgeons for which culture and rearing techniques have been developed and refined is the white sturgeon, (*A. transmontanus*) (Conte et al. 1988). Several firms which rear and culture white sturgeon to supply flesh, caviar, and even juveniles for the tropical fish trade have now prospered in California (B. Moore, Sea Farm Washington, Inc., personal communication).

Successful rearing and culture techniques have also been developed for lake sturgeon (Anderson 1984; Ceskleba et al 1985). Lake sturgeon are raised at Wild Rose State Fish Hatchery, Wisconsin, and Wolf Lake State Fish Hatchery, Michigan, for stocking and experimental purposes. In Michigan, rearing lake sturgeon to a stocking size of 13cm was estimated to cost $2.40 per fish (R. Poynter, Wolf Lake State Fish Hatchery, personal communication). At this time, stocking lake sturgeon is an uncommon practice. Not only is rearing expensive, but success of stocking efforts has not been evaluated. Mark Twain Lake, Missouri was stocked in 1984 and 1986, and Pool 24 of the Mississippi River in Missouri was stocked in 1988 and 1990 with fish reared from eggs obtained from the Lake Winnebago population (Graham 1986). The St. Louis River, Duluth, Minnesota has also been stocked with lake sturgeon, reared from eggs from the Lake Winnebago source, from 1983 to 1992 by the Minnesota and Wisconsin Departments of Natural Resources (J. Lindgren, Minnesota Department of Natural Resources, personal communication). Without knowledge of population genetics and an adequate understanding of native stock locations and abundance throughout the Great Lakes region, there is concern over the widespread stocking of this species. Stocked populations may compete for food and habitat with native populations. There is concern that stocking, which originates from a single broodstock source, may compromise natural genetic variation.

Research Needs for Future Management

Those individuals working on management plans for lake sturgeon have begun to identify areas needing further research to ensure effective management. One of the most important needs is for a clear idea of present stock abundance, location, and genetic integrity within the Great Lakes basin. Protection and enhancement of populations will not be effective if little is known about population size, sex and age structure, mortality rates, and location or habitat requirements. Opportunities to reintroduce lake sturgeon through stocking and aquaculture may be lost until the importance of genetic integrity among Great Lake stocks is established.

Movement and range requirements of adult lake sturgeon remain unclear. Six studies identified upstream migration movements of lake sturgeon and barriers encountered during spawning migrations (table 4). Four studies reveal downstream movement patterns after fish were tagged on spawning sites (table 5). This data on spawning migration distances illustrate how little is known about movements, and how few lake sturgeon populations in the Great Lakes are free-ranging. A comparative study of egg survival between natural spawning beds and eggs spawned below barrier dams is essential to the understanding of affects barriers have on lake sturgeon recruitment.

Development of fish passage facilities in systems now obstructed by barriers has only recently been considered. More research is needed on facilitating upstream passage of adults as well as safe downstream passage of adults and young.

Establishing habitat and water quality preferences and movement patterns of early-life and juvenile stages of lake sturgeon is also necessary to insure recruitment to existing stocks, and help determine if sufficient habitat exists in areas proposed for reintroductions. Developing methods and sampling equipment which collect various life stages of lake sturgeon yet do not destroy habitat or other fish species is a significant research priority.

Besides basic questions concerning life history of the species, many new research needs arise once stocking and reestablishing stocks is considered. The survival of stocked individuals and extent of reproduction provided by these individuals has not been well documented.

Stocking of non-native fishes in rivers utilized by lake sturgeon may cause unnatural mortality of newly-hatched young. Some lake sturgeon waters are stocked in the spring with salmonids which may feed on drifting larvae. No research has been done on the effects of exotic species on adult, juvenile or larval lake sturgeon. There is concern that lake sturgeon imprint to natal streams when very young. Stocked fry may grow to maturity, wander and not spawn, or spawn in suboptimal environments because these fish did not imprint to the stocked location.

Some recovery plans call for habitat rehabilitation or enhancement, yet habitat preferences for the species remain largely undefined. Habitat enhancement can not occur until habitat preferences are defined, and then careful experiments must be conducted to assess success of enhancing habitat.

There is a need to educate the public to recognize lake sturgeon as a valuable member of the Great Lakes fish community. The public should be involved in defining future sport fishing, aquaculture opportunities, and commercial fishing development.

Coordinating Agency Management Programs

Most remaining populations of lake sturgeon are endangered or threatened, and since these potamodromous fish live within interjurisdictional management areas, cooperation of federal, state, and tribal agencies, as well as private organizations and institutions will be needed if the lake sturgeon is to be given an opportunity to make a comeback in the Laurentian Great Lakes. Coordinating assessment, communication, management, enforcement of regulations, and research will not be an easy task, considering the objectives of eight U.S. states, two Canadian provinces, over twelve U.S. Native American tribes, and many Canadian First Nations.

The U.S. Fish and Wildlife Service has developed a management framework for conservation of sturgeons and paddlefish of the United States (Booker et al. 1993). In this framework, the Service's role in conservation management is defined, and the agency takes a leadership role in coordination and administration of management, research and networking responsibilities. The service also proposes coordinating culture and stocking programs throughout the United States and standardizing fishing regulations for interjurstictional stocks.

Several states and one province have drafted management plans. These plans target either restoration of entire systems with the feasibility of reestablishing a native species such as lake sturgeon (Lake Champlain, Vermont, Moreau and Parrish 1994), or general protection and restoration of the species throughout a state (Lake Sturgeon Management Plan, Michigan Department of Natural Resources; Lake Sturgeon Management, Wisconsin Department of Natural Resources) or province (Management Strategy for Lake Sturgeon in Ontario, Ontario Ministry of Natural Resources).

These plans have several similar objectives. The most common objective is to conserve and protect existing populations, allow them to build to self-sustaining levels, and to insure adequate spawning, juvenile and adult habitat for these populations to survive. A second objective is to reestablish lake sturgeon to areas from which they have been extirpated, and in which they would have a reasonable chance of becoming self-sustaining. Stocking and habitat restoration are common goals of this objective, and often habitat access and appropriate water level regimes must be addressed because many areas have been impacted by barrier or water level control structures. Some plans suggest establishing stocks in suitable areas within the native range of the lake sturgeon, but which did not originally support this species, and some suggest examining areas outside the native range, which might support populations solely for sport or trophy fisheries, thus reducing fishing pressure on native stocks. Most plans mention an eventual desire to reach a harvestable situation with some stocks, either as regulated sport catches or quota based commercial fisheries. Educating the public about this species and future management goals was also commonly addressed in management plans.

Perhaps the greatest opportunity the coordination of protection and management of lake sturgeon in the Great Lakes basin occurs when efforts are considered on a lakewide basis. In the "Strategic Vision of the Great Lakes Fishery Commission for the Decade of the 1990s" the commission states "The conservation of biological diversity through rehabilitation of native fish populations, species, communities, and their habitats has a high priority" (GLFC p. 15, 1992). In 1994, the Lake Superior Technical Committee formed the Lake Sturgeon Subcommittee, and charged this committee with developing a plan for the protection, reestablishment or enhancement of depleted stocks of lake sturgeon in Lake Superior. Individuals from the U.S. Fish and Wildlife Service; Michigan, Minnesota, and Wisconsin Departments of Natural Resources; Ontario Ministry of Natural Resources; local academic institutions, and several Native American communities are members and resource persons to this committee.

Many changes have occurred in the Great Lakes basin over the last one hundred years, and many lessons have been learned in fishery management. It is time now to apply that knowledge to protection and restoration of this valuable native species. With time and effort, the lake sturgeon can be restored to greater prominence in the Great Lakes ecosystem, and possibly sustain some aboriginal, sport, and commercial fisheries.

Literature Cited

Auer, N. A. 1996a. Importance of habitat and migration to sturgeons with emphasis on lake sturgeon. Canadian Journal Fisheries and Aquatic Sciences 53(Suppl. 1):152–160.

Auer, N. A. 1996b. Response of spawning lake sturgeon to change in hydroelectric facility operation. Transactions of the American Fisheries Society 125(1):66–77.

Auer, N. A. 1999. Population characteristics and movements of lake sturgeon in the Sturgeon River and Lake Superior. Journal of Great Lakes Research 15(2):282–293.

Anderson E. R. 1984. Artificial propagation of lake sturgeon Acipenser fulvescens (Rafinesque), under hatchery conditions in Michigan. Michigan Department of Natural Resources, Research Report No. 1898.

Baker, J. P. 1980. The distribution, ecology and management of the lake sturgeon (Acipenser fulvescens Rafinesque) in Michigan. Michigan Department of Natural Resources, Research Report No. 1883.

Baldwin, N. S., R. W. Saalfeld, M. A. Ross, and H. J. Buettner 1979. Commercial fish production in the Great Lakes 1867–1977. Great Lakes Fishery Commission Technical Report No. 3, Ann Arbor, Michigan.

Bassett, C. 1981. Management plan for lake sturgeon (Acipenser fulvescens) in the Indian River and Indian Lake, Alger and Schoolcraft counties, Michigan. U.S. Forest Service, Manistique District.

Birstein, V. J. 1993. Sturgeons and paddlefishes: threatened fishes in need of conservation. Conservation Biology 7(4):773–787.

Blumer, S. P., W. W. Larson, R. J. Minnerick, C. R. Whited, and R. L. LeuVoy. 1991. Water Resources Data, Michigan-Water Year 1990. U.S. Geological Survey Water-Data Report MI–90–1.

Booker, W. V., F. A. Chapman, S. P. Filipek, L. K. Graham, A. A. Rakes, J. L. Rasmussen, K. J. Semmens, R. A. St. Pierre and T. J. Smith. 1993. Framework for the management and conservation of paddlefish and sturgeon species in the United States. U.S. Fish and Wildlife Service, Washington, D.C.

Brousseau, C. S. and G. A. Goodchild. 1989. Fisheries and yields in the Moose River Basin, Ontario. Pages 145–158 in D. P. Dodge, ed. Proceedings of the International Large River Symposium. Canadian Special Publication of Fisheries and Aquatic Sciences 106.

Brown, J. R., A. T. Beckenbach and M. J. Smith 1992. Influence of Pleistocene glaciations and human intervention upon mitochondrial DNA diversity in white sturgeon (Acipenser transmontanus) populations. Canadian Journal of Fisheries and Aquatic Sciences 49:358–367.

Carlson, D. M. 1995. Lake sturgeon waters and fisheries in New York State. Journal of Great Lakes Research 21(1):35–41.

Ceskleba, D. G., S. AveLallemant and T. F. Thuemler. 1985. Artificial spawning and rearing of lake sturgeon, Acipenser fulvescens, in Wild Rose State Fish Hatchery, Wisconsin, 1982–1983. Environmental Biology of Fishes 14(1):79–85.

Conte, F. S., S. I. Doroshov, P. B. Lutes and E. M. Strange. 1988. Hatchery manual for the white sturgeon Acipenser transmontanus Richardson with application to other North American Acipenseridae. Cooperative Extension Service, University of California-Davis, Division of Agriculture and Natural Resources Publication 3322.

Cuerrier, J. P. 1966. L'Esturgeon de lac *Acipenser fulvescens* RAF. de la region du Lac St. Pierre au cours de la periode du frai. Le Naturaliste Canadiense 93(4):279–334.

Dumont, P., R. Fortin, G. Desjardins and M. Bernard. 1987. Biology and exploitation of lake sturgeon in the Quebec waters of the Saint-Laurent River. Pages 57–76 *in* C. H. Olver, ed. Proceedings of a Workshop on the Lake Sturgeon (*Acipenser fulvescens*). Ontario Fisheries Technical Report Series No. 23.

Emich, H. L. 1987. City of Marinette Centennial Program and History, 1887–1987. Marinette, Wisconsin.

FERC (Federal Energy Regulatory Commission). 1981. Hydropower sites of the United States developed and undeveloped. Office of Electric Power Regulation. Washington, D.C.

Ferguson, M. M., L. Bernatchez, M. Gatt, B. R. Konkle, S. Lee, M. L. Malott and R. S. McKinley. 1993. Distribution of mitochondrial DNA variation in lake sturgeon (*Acipenser fulvescens*) from the Moose River basin, Ontario, Canada. Journal of Fish Biology 43(A):91–101.

Folz, D. J. and L. S. Meyers. 1985. Management of the lake sturgeon, *Acipenser fulvescens*, population in the Lake Winnebago system, Wisconsin. Pages 135–146 *in* F. P. Binkowski and S. I. Doroshov, eds. North American sturgeons: Biology and aquaculture potential. Kluwer Academic Publishers, Norwell, Netherlands.

Gates, D. M., C. H. D. Clarke, and J. T. Harris. 1983. Wildlife in a Changing Environment. Pages 52–80 *in* S. L. Flader, ed. The Great Lakes forest: An environmental and social history. University of Minnesota Press, Minneapolis, Minnesota.

Gaston, K. J. and J. H. Lawton 1990. The population ecology of rare species. Journal of Fish Biology 37(A):97–104.

Graham, K. 1986. Reintroduction of lake sturgeon into Missouri. Missouri Department of Conservation., Jefferson City, Missouri.

Graham, P. J. 1985. Effects of small hydropower development on fisheries. Pages 11–16 *in* F. W. Olson, R. G. White and R. H. Hamre, eds. Symposium on Hydropower and Fisheries. American Fisheries Society, Bethesda, Maryland.

GLFC (Great Lakes Fishery Commission). 1992. Strategic Vision of the Great Lakes Fishery Commission for the Decade of the 1990s. Ann Arbor, Michigan.

Guenette, S. E. Rassart and R. Fortin. 1992. Morphological differentiation of lake sturgeon (*Acipenser fulvescens*) from the St. Lawrence River and Lac des Deux Montagnes (Quebec, Canada). Canadian Journal of Fisheries and Aquatic Sciences 49:1959–1965.

———, R. Fortin and E. Rassart. 1993. Mitochondrial DNA variation in lake sturgeon (*Acipenser fulvescens*) from the St. Lawrence River and James Bay drainage basins in Quebec, Canada. Canadian Journal of Fisheries and Aquatic Sciences 50.059 GG1.

Harkness, W. J. K. 1923. The rate of growth and the food of the lake sturgeon (*Acipenser fulvescens* Lesueur). University of Toronto Biological Studies Series 24, Publication of the Ontario Fisheries Research Lab 18.

———, J. R.Dymond. 1961. The lake sturgeon. Ontario Department of Lands and Forests.

Hay-Chmielewski, E. M. 1987. Habitat preferences and movement patterns of the lake sturgeon (*Acipenser fulvescens*) in Black Lake Michigan. Michigan Department of Natural Resources, Fisheries Division Research Report No.1949.

Holzkamm, T. E. and M. McCarthy. 1988. Potential fishery for lake sturgeon (*Acipenser fulvescens*) as indicated by the returns of the Hudson's Bay Company Lac la Pluie District. Canadian Journal of Fisheries and Aquatic Sciences 45(5):921–923.

———, W. Wilson 1988. The sturgeon fishery of the Rainy River Ojibway bands. Smithsonian Columbus Quincentenary Program, "Seeds of the Past".

———, V. P. Lytwyn and L. G. Waisberg. 1988. Rainy river sturgeon: Ojibway resource in the fur trade economy. Canadian Geographer 32(3):194–205.

Houston, J. J. 1987. Status of the lake sturgeon, *Acipenser fulvescens*, in Canada. Canadian Field Naturalist 101(2):171–185.

Johnson, J. E. 1987. Protected fishes of the United States and Canada. American Fisheries Society, Bethesda, Maryland.

Jolliff, T. M. and T. H. Eckert. 1971. Evaluation of present potential sturgeon fisheries of the St. Lawrence River and adjacent waters. N.Y. Department of Conservation, Cape Vincent Fisheries Station.

Kempinger, J. J. 1988. Spawning and early life history of lake sturgeon in the Lake Winnebago system, Wisconsin. American Fisheries Society Symposium 5:110–122.

Kerr, S. J. and G. C. LeTendre 1991. The state of the Lake Ontario fish community in 1989. Great Lakes Fishery Commission Special Publication 91–3, Ann Arbor, Michigan.

Koelz, W. 1926. Fishing industry of the Great Lakes. U.S. Commercial Fisheries Report for 1925:553–617.

Kohl, J. G. 1956. Kitchi-Gami, Wanderings Round Lake Superior. Ross and Haines, Inc, Minneapolis, Minnesota.

Kornely, G. W. 1988. Lake sturgeon creel survey of Menominee River Wisconsin-Michigan Boundary water, 1981–1984. Wisconsin Department of Natural Resources Fisheries Management Report 134.

LaHaye, M., A. Branchaud, M. Gendron, R. Verdon, and R. Fortin. 1992. Reproduction, early life history, and characteristics of the spawning grounds of the lake sturgeon (*Acipenser fulvescens*) in Des Prairies and L'Assomption rivers, near Montreal, Quebec. Canadian Journal of Zoology 70:1681–1689.

Larson, T. 1988. The lake sturgeon fishery of Lake Wisconsin, 1978–1985. Wisconsin Department of Natural Resources Fisheries Management Report 136.

Leggett, W. C. 1977. The ecology of fish migrations. Annual Reviews in Ecology and Systematics 8:285–308.

Love G. F. 1972. The lake sturgeon (*Acipenser fulvescens*) of Lake Nipissing—1971 study. Nipissing-Temagami Fisheries Management Unit, North Bay District, Ontario.

Lyons, J. and J. J. Kempinger. 1992. Movements of adult lake sturgeon in the Lake Winnebago system. Wisconsin Department of Natural Resources, Research Publication-RS-156 92.

McAllister, D. E., S. P. Platania, F. W. Schueler, M. E. Baldwin and D. S. Lee. 1986. Ichthyofaunal patterns on a geographic grid. Pages 17–51 *in* C. H. Hocutt and E. O. Wiley, eds. The Zoogeography of North American Freshwater Fishes. John Wiley and Sons, New York

McKinley, R. S., T. D. Singer, J. S. Ballantyne and G. Power. 1993. Seasonal variation in plasma nonesterified fatty acids of lake sturgeon (*Acipenser fulvescens*) in the vicinity of hydroelectric facilities. Canadian Journal of Fisheries and Aquatic Sciences 50:2440–2447.

Moreau, D. A. and D. L. Parrish. 1994. A study of the feasibility of restoring lake sturgeon to Lake Champlain. University of Vermont, Vermont Cooperative Fish and Wildlife Research Unit, Technical Report 9.

Nowak, A. M. and C. S. Jessop. 1987. Biology and management of the lake sturgeon (*Acipenser fulvescens*) in the Groundhog and Mattagami Rivers, Ontario. Pages 20–32 *in* C. H. Olver, ed. Proceedings of a Workshop on the Lake Sturgeon (*Acipenser fulvescens*). Ontario Fisheries Technical Report Series No. 23.

OMNR (Ontario Ministry of Natural Resources). 1972a. Ontario Commercial Fish Industry Statistics on Landings, 1961–1965.

———, 1972b. Ontario Commercial Fish Industry Statistics on Landings, 1966–1970.

———, 1976. Ontario Commerical Fish Industry Statistics on Landings, 1971–1975.

———, 1981. Ontario Commerical Fish Industry Statistics on Landings, 1976–1980.

Olver, C. H., ed. 1987. Proceedings of a Workshop on the Lake Sturgeon (*Acipenser fulvescens*). Ontario Fisheries Technical Report Series No. 23, Ontario Ministry of Natural Resources.

Peters, R. H. 1983. The ecological implications of body size. Cambridge Univer. Press., Cambridge.

Priegel, G. R. and T. L. Wirth 1975. Lake sturgeon harvest, growth and recruitment in Lake Winnebago, Wisconsin. Wisconsin Department of Natural Resources, Technical Bulletin No. 83.

———, 1977. The lake sturgeon: Its life history, ecology and management. Wisconsin Department of Natural Resources Publication 4–3600(77).

Prince, E. E. 1905. The Canadian sturgeon and caviar industries. Sessional Paper No. 22.

Rickey, D. 1987. Lake Huron spotlight: sturgeon spearing. Great Lakes Fisherman 17(2):61.

Rochard, E., G. Castelnaud and M. Lepage 1990. Sturgeons (Pisces:Acipenseridae); threats and prospects. Journal of Fish Biology 37(A):123–132.

Sandilands, A. P. 1987. Biology of the lake sturgeon (*Acipenser fulvescens*) in the Kenogami River, Ontario. Pages 36–46 *in* C. H. Olver, ed. Proceedings of a Workshop on the Lake Sturgeon (*Acipenser fulvescens*). Ontario Fisheries Technical Reptort Series No. 23, Ontario Ministry of Natural Resources.

Scott, W. B. and E. J. Crossman. 1973. Freshwater Fishes of Canada. Fisheries Research Board of Canada, Bulletin 184.

Schoolcraft, H. R. 1970. Narrative journals of travels from Detroit northwest through the Great chain of American lakes to the sources of the Mississippi river in the year 1820. Arno Press and N.Y. Times.

Schram, S. T., T. L. Margenau, and W. H. Blust. 1992. Population biology and management of the walleye in western Lake Superior. Wisconsin Department of Natural Resources, Technical Bulletin 177.

Shaffer, M. L. 1981. Minimum population sizes for species conservation. Bioscience 31:131–134.

Shively, J. D. and N. Kmiecik. 1989. Inland fisheries enhancement activities within the ceded territory of Wisconsin during 1988. Administrative Report 89–1. Great Lakes Indian Fish and Wildlife Commission.

Smith, S. H. 1968. Species succession and fishery exploitation in the Great Lakes. Journal of the Fisheries Research Board of Canada 25(4):667–693.

Smith, H. M. and M. M. Snell. 1889. Fisheries of the Great Lakes in 1885. Report of the Commission for 1887. U.S. Commission of Fish and Fisheries.

Soule, M. E., ed. 1987. Viable populations for conservation. Cambridge University Press, Cambridge.

Stearns, S. C. 1976. Life history tactics: A review of the ideas. The Quarterly Review of Biology 51(1):3–47.

Thuemler, T. F. 1985. The lake sturgeon, *Acipenser fulvescens*, in the Menominee River, Wisconsin-Michigan. Environmental Biology of Fishes 14(1):73–78.

———, 1988. Movements of young lake sturgeons stocked in the Menominee River, Wisconsin. American Fisheries Society Symposium 5:104–109.

Tower, W. S. 1908. The passing of the sturgeon: A case of the unparalleled extermination of a species. The Popular Science Monthly. 73:361–371.

Trautman, M. B. 1981. The Fishes of Ohio. Ohio State University Press, Columbus, Ohio.

Wells, L., and A. L. McLain. 1973. Lake Michigan: Man's effects on native fish stocks and other biota. Great Lakes Fishery Commission Technical Report No. 20, Ann Arbor, Michigan.

WDNR (Wisconsin Department of Natural Resources). 1990. Wisconsin Department of Natural Resources, Health guide to eating Wisconsin sport fish. Wisconsin Division of Health, 1 page.

———, 1994. Wisconsin Department of Natural Resources, Health guide for people who eat sport fish from Wisconsin waters. Wisconsin Division of Health, Publication-IE-019 4/94 Rev.

PART 5

Outlook for the Future

CHAPTER 18

Great Lakes Fisheries Futures: Balancing the Demands of a Multijurisdictional Resource

C. Paola Ferreri, William W. Taylor, and John M. Robertson

Introduction

The Laurentian Great Lakes comprise an expansive and complex ecosystem that is an integral component of the region's environmental health, economic well being, and general quality of life. Ranking among the fifteen largest fresh water lakes in the world, the Great Lakes collectively hold 20% of the world's surface fresh water supply (Donahue 1991). The diversity of habitat types found in the lakes, ranging from deep oligotrophic basins to shallow eutrophic embayments, supports a wide variety of fish species (Hartman 1988). There are 179 species of native and introduced fishes representing 29 families in 18 orders in the Great Lakes (Coon 1999).

Before European settlement of the Great Lakes region, the relatively few native peoples who lived along the shores fished for subsistence (Hartman 1988). With the influx of European settlers, commercial fisheries for species such as lake trout (*Salvelinus namaycush*), lake whitefish (*Coregonus clupeaformis*), lake herring (*Coregonus artedii*), yellow perch (*Perca flavescens*), and walleye (*Stizostedion vitreum*) became important. The fishery resources of the Great Lakes were adversely affected by the major landscape changes that occurred as a result of continued population growth. The successive impacts of human population growth around the Great Lakes basin, including overexploitation, habitat degradation, and the invasion/introduction of exotic species caused commercial fisheries to collapse (Christie 1974). Production of high value species declined from fourteen million pounds in 1940, to twelve million in 1955, to seven million in 1965 (Tanner et al. 1980). The negative economic impact of the degraded state of the Great Lakes led to the initiation of a fishery rehabilitation program that aimed to restore valuable fish populations. At the time, fishery managers focused their attention on single species management, with little understanding of ecosystem dynamics and their relationship to fish production. Thus, during the early 1960s when alewives were dying in droves during the spring and littering beaches with their corpses during the summer, fisheries managers only thought of reducing this nuisance species through the introduction of a new suite of predators: coho (*Oncorhynchus kisutch*) and chinook (*O. tshawytscha*) salmon. Although several earlier attempts, dating back to the late 1800s, to introduce these predators failed, the 1965 attempt was a huge success (Tanner et al. 1980). With the introduction of coho and chinook salmon, the priority of the Great Lakes fishery management changed from the historical commercial fishing industry to the newly created sport fishing industry (Tanner et al. 1980). The current total economic impact of the Great Lakes sport and commercial fishery on the basin's economy has been estimated at $2.3 to $4.3 billion annually (Talhelm 1988; Fedler and Nickum 1993).

Unfortunately, we do not fully understand why the coho and chinook salmon did so well during the late 1960s through the early 1980s, as our primary focus was on the salmon and their catch rate, rather than on understanding the production dynamics of these fishes in relation to the ecosystem in which they reside. Our ignorance of the linkages between salmon production and ecosystem processes became painfully obvious when our Great Lakes salmon populations collapsed during the mid 1980s and no obvious single species solutions were apparent. This forced the fisheries management community to move their management strategy away from a single species focus to an ecosystem focus, where ecosystem responses lead to enhanced and sustainable fishery resources. Additionally, managers also better understand the importance of the social dimension in designing management strategies, as the collapse of the salmon fisheries had a significant effect on the local economies of shoreline communities and the charter boat industry. Thus, management agencies such as the Michigan Department of Natural Resources set up lake advisory councils to receive input from interested publics and other agencies regarding where management of the Great Lakes should go in the future. These actions culminated a shift in fisheries management philosophy away from single species management towards a more holistic view that included an understanding of the physical, chemical, biological, and social environments that affect the sustainability of our Great Lakes fishery resources.

Currently, fishery managers are struggling to balance a wide variety of demands on the Great Lakes ecosystem that impact their fishery resources. These demands revolve around three major issues: non-indigenous species, allocation of fishery resources, and habitat/water quality. The Aquatic Nuisance Species Task Force (1993) defined a non-indigenous species as "any species or other viable biological material that enters an ecosystem beyond its historic range as defined by the area occupied at the time of European colonization of North America." In general, the introduction of a species to an ecosystem often alters the structure and function of natural communities making the system less stable and less predictable (Herbold and Moyle 1986). Non-indigenous species have been intentionally and accidentally introduced into the Great Lakes ecosystem with varying degrees of impact (Mills et al. 1993). Many non-indigenous species have had severe effects on the Great Lakes ecosystem. For example, sea lamprey played a key role in the collapse of lake trout and other valuable fisheries in the Great Lakes (Smith and Tibbles 1980); zebra mussels (*Dreissena polymorpha*) have detrimentally affected many native mussels (Mills et al. 1993), as well as having negatively impacted local communities through the fouling of water intake pipes (Claudi and Mackie 1994). Purple loosetrife (*Lythrum salicaria*) has competitively displaced many native aquatic plants and threatens prime marsh habitat for waterfowl, which is also important fish spawning and nursery habitat (Rawinski and Malecki 1984). On the other hand, the introduction of Pacific salmon positively aided the start of the Great Lakes fishery rehabilitation process and created an extremely valuable recreational fishery.

The second issue faced by fishery managers deals with the allocation of fishery resources. As the fishery is allocated to more than one user group, managers must make decisions about how to achieve allocation goals on a sustainable basis. Currently, Great Lakes fishery managers allocate the fishery resource to a diverse set of user groups, including the recreational fishery, commercial fishery, Native American fishery, and other sources of mortality not related to fishing, such as impingement or entrainment at hydroelectric power plants.

The last set of issues deals with habitat and water quality that play an integral role in determining the productivity of a fishery. Loss of wetlands as spawning areas and nursery habitats, as well as sedimentation of spawning habitat in streams due to land use practices can negatively impact fish populations and lead to their decline. In addition, contaminants have been found to impede the survival of fish at early life stages in the laboratory and may affect fish recruitment in the Great Lakes (Mac et al. 1994).

The impact of non-indigenous species, allocation decisions, and habitat degradation are rarely localized to a single region or to a single fishery; instead these impacts often cause disturbance in the entire ecosystem. Consequently, these issues must be dealt with at an ecosystem level, with a management approach that recognizes the interdependence of physical, chemical, biological, and social processes in the Great Lakes basin. The resulting strategy of ecosystem management is a comprehensive and integrated approach to natural resource management (Bocking 1994). However, ecosystem management is often hindered by problems that arise due to multiple jurisdiction and self interest.

Governance of the Great Lakes Ecosystem

Management of the Great Lakes fisheries at an ecosystem level is complex, not only because of the expansiveness and diversity of the ecosystem itself, but also because of the intricate governance structure involved. Governance is defined as the exercise of authority and control (Francis 1990). The political boundaries of the Great Lakes make it obvious that governance is shared by two federal governments, the United States and Canada. In addition, eight states and one province (two provinces if the St. Lawrence River watershed is included) share in the responsibility for the management of Great Lakes resources (Caldwell 1994). These in turn, can assign rights and responsibilities to local (i. e., municipal) governments. In both countries, governing structures have been created above the municipal level, but below the state or province level. Additionally, international commissions that answer directly to both federal governments, such as the International Joint Commission and the Great Lakes Fishery Commission have been created to facilitate decision making regarding the Great Lakes (Francis 1990). Thus, the system of governance in the Great Lakes basin is at least four levels deep (fig. 1; Francis 1990) including over 650 local to international jurisdictional units (Caldwell 1994).

The existence of this intricate, multijurisdictional governance arrangement can impose a stress on the ecosystem that we refer to as jurisdictional stress. This type of stress arises when the number of specific demands and interests represented by each of the governance structures and public interest groups exceeds the natural ability of the ecosystem to accommodate these demands. Conflicts arise as to whose demand is the most important, often leading to the implementation of

Great Lakes Governance

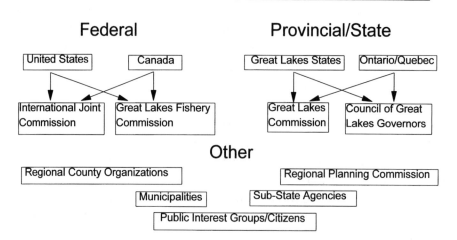

Federal

| United States | Canada |

| International Joint Commission | Great Lakes Fishery Commission |

Provincial/State

| Great Lakes States | Ontario/Quebec |

| Great Lakes Commission | Council of Great Lakes Governors |

Other

| Regional County Organizations | | Regional Planning Commission |

| Municipalities | Sub-State Agencies |

| Public Interest Groups/Citizens |

FIG. 1. *Example of the multiple layers of governance involved in making decisions about fisheries management in the Great Lakes.*

management actions that contradict or overlap each other. The solution for this type of stress is to focus on management of the ecosystem rather than on the management of individual jurisdictions. By focusing on the ecosystem, individual interests become subordinate to the collective interest in the welfare of the ecosystem and the sustainability of its natural resources. Thus, jurisdictional boundaries become secondary to the ecosystem boundaries, allowing management practices to be developed on a holistic rather than in a fragmented, jurisdictional basis. Without this holistic view, management actions narrowly focused on one component of the system can often counteract other management initiatives designed for some other component of the ecosystem.

Collaboration Using Consensus

To successfully implement fisheries management within an ecosystem framework, all parties with a vested interest in the Great Lakes ecosystem need to be brought together in a common forum, where the interests of each party are considered and opportunities for collaborative management of the resource are sought. Pinkerton (1989) coined the term "comanagement" to refer to a variety of formal and informal agreements between different interest groups and various levels of government aimed at producing collaborative management of local fisheries. In essence, comanagement agreements impose shared decision making among all interested parties, leading to a management scenario where benefits are greater for cooperation than for opposition or competition. Through these collaborative management agreements, the parties learn to optimize mutual benefits and to plan cooperatively (Pinkerton 1989). For example, the Strategic Great Lakes Fisheries Management Plan (GLFC 1997) is an attempt to establish collaborative fisheries management in the Great Lakes. It was adopted by fishery management agencies at all levels of governance in an effort to coordinate activities that optimize social and ecological benefits (GLFC 1997).

However, the agreement has not been successful in shifting the focus of the cooperating agencies from their jurisdictional boundaries to a more holistic view of the Great Lakes ecosystem.

Collaborative management agreements can be derived through several methods, ranging from informal decision making meetings to court-mandated negotiated settlements. Formal consensus building has become the preferred method for resolving problems and policy questions among private and public decision making groups. Decision making by consensus has been successfully implemented in fields ranging from health and medicine (Fink et al. 1984) to education (Garcia 1986) to aquatic resource management (Buzan 1981; McKinney 1992). The complexity of the governance system and the numerous demands to be balanced through the management of Great Lakes fisheries make the consensus approach to decision making appealing.

Typically, a consensus agreement is the result of the convergence of two or more points of view. The focus of the consensus strategy is to bring all parties to a mutually satisfactory solution to the problem at hand (Garcia 1986). The strategy encourages all parties to freely express their own point of view to enable the group to constructively explore all dimensions of the problem (Tjosvold and Field 1985; Garcia 1986). Members of the group need to have an overriding commitment to solving the problem, allowing for convergence upon a mutually acceptable solution. In this strategy, individuals argue their points, not necessarily to win the argument, but to enhance the development of a strong solution to the problem at hand (Garcia 1986). Consensus is measured by the degree to which the members of the group agree with, subscribe to, or endorse the final solution (Rossi and Bark 1985).

The use of consensus for decision making creates a sense of ownership, satisfaction, and commitment among the participants (Garcia 1986; Tjosvold and Field 1985). In managing Great Lakes fisheries, consensus would foster the creation of partnerships allowing for the resource to be managed for the benefit of all partners. Although unanimous agreement between the parties on all issues would be preferred (i. e., true consensus), this is often not possible due to the complexity of the issues. Several studies have found that when facing a judgmental problem, one that does not have a single, clear solution, groups tend to reach a modified consensus (Tjosvold and Field 1985; Rossi and Berk 1985; Kirchler and Davis 1986). Modified consensus occurs when members do not necessarily agree to every aspect of the decision, but do agree to support the group's decision without reservation. Consensus building among the multiple jurisdictions with interests in the Great Lakes will be critical to the success of ecosystem management. If all of the partners can agree on an overriding management focus, there is less risk that different agencies will institute management practices that work against each other. However, due to the complexity of the Great Lakes ecosystem and governance structure, there are many impediments to consensus building that must be overcome if ecosystem management is to be successful.

Impediments to Consensus Building

One impediment to consensus building arises from the mere size and diverse use of the Great Lakes ecosystem that leads to difficult and complex management questions. The number of interested parties involved in each decision gives rise to part of the complexity. Getting all the interested parties with a stake in the outcome of the decision process to the table can be a difficult task by itself. Managing such a large group of people and trying to address everyone's concerns can be equally as difficult, and often presents a formidable obstacle to consensus building. Even within agencies, the complexity of the issues at hand may require the cooperation of several branches of the agency. Each branch has its own set of priorities to which time and resources are allocated. It can be difficult to redirect these efforts to address a new issue, even if everyone agrees that it is important. Another aspect of the complexity of management issues in the Great Lakes deals with the types of management questions being asked. In general, these questions tend to be complex and difficult with no easy answers to be found. Much of the complication arises from the fact that many of these questions pose a tradeoff between short-term and long-term benefits. Short-term benefits are easy to define and readily accessible (e. g., number of pounds of fish to be harvested this year), while long-term benefits (e. g., sustainability of the fishery) are often difficult to define and quantify. Additionally, the political atmosphere is such that management agencies are often reactive rather than proactive, making it difficult for managers to give up short-term benefits for potentially greater long-term benefits that will not be realized for five or more years. The complexity of the management issues tends to make decision makers uncomfortable, as managers put themselves at the risk of losing their credibility and reputation by taking a chance on giving up today's known benefit if the action does not result in a greater benefit in the future.

The issues facing the Great Lakes ecosystem are complicated, and decisions made have the potential to affect many constituent groups. Although decision makers are willing to attend meetings that present a win-win situation, they often avoid those that present a difficult tradeoff. Unfortunately, it appears that the reward system in most agencies is based on creating happy publics rather than on providing leadership in the stewardship of the Great Lakes. Hence, decision makers will often send representatives to key meetings rather than attending themselves. This practice creates many instances where the people gathering at the decision table are not empowered either to make the decisions necessary or to freely share information bringing the consensus building process to a halt.

The decisions made by resource management agencies have the potential to affect many different sectors of society. However, identifying and reaching all affected constituent groups can be very difficult. In most cases, managers can identify and respond to the desires of relatively small groups who are well organized and publicly express their concerns. Other stakeholders, including the general public, are usually not as well organized, making it difficult for fishery managers to discern their needs and desires. This lack of knowledge about the general public poses a significant obstacle to consensus building, as the concerns and input of a large number of people cannot be easily considered.

In addition to the obstacles posed by the complexity of the issues, consensus building can be hindered by the multijurisdictional nature of the governance structure. Even an in-house agency perspective can pose obstacles to consensus building. Performing ecosystem management in the Great Lakes requires agency personnel to think beyond the boundaries often imposed by their agency mission because of geopolitical boundaries. However, it is often difficult to overcome local politics; issues that are important in a basin-wide sense may not be a high priority within one's own jurisdiction. Thus, assigning money and resources to these issues is often difficult because of the lack of public support for such a decision. The budget that ultimately defines the policy for

Great Lakes management agencies is allocated to those issues perceived as critical on the local level, leaving few resources to deal with long-term issues that affect other jurisdictions. The difficulty dealing with long-term, basinwide issues leaves agencies and their managers without a strategic vision or plan for the future of the resource. This lack of a common vision across the basin poses another major obstacle to consensus building, as managers cannot agree on management priorities for their interconnected ecosystem.

Other interjurisdictional obstacles to consensus building include the lack of a common information base and the lack of a common value system. In general, most agencies collect data to support efforts within their own jurisdiction. Often, the proprietary nature of data collection impedes the establishment of a common data base. Another problem with establishing a common data base is the fact that no common protocol exists for data collection and reporting throughout the basin. Therefore, comparing data collected by different agencies can be, at best, a very difficult and time consuming task. As a result, most decision makers are working without full knowledge of the problem as their agency only has information on an isolated portion of the ecosystem.

Suggestions for the Future: ACME

The problems faced by fishery managers in the Great Lakes are less fishery management problems than problems resulting from an entire suite of divergent and changing uses of the watershed and the five Great Lakes. However, all of these ecosystem problems affect the productivity of the fishery resources within the Great Lakes. In order to fully implement the needed ecosystem approach to managing Great Lakes fisheries, consensus decisions must become easier to attain. Here, we propose a management strategy that will facilitate consensus building and provide a goal-focused process for fisheries management in the Great Lakes ecosystem. This strategy can be referred to as ACME, the Adaptive Collaborative Management of Ecosystems. ACME is a consensus building process that can be used within the existing institutional framework in the Great Lakes basin. In particular, the Great Lakes Fishery Commission and the International Joint Commission provide the beginnings of a forum in which the ACME process can take place. Currently, the Great Lakes Fishery Commission focuses on fisheries of common concern to the two nations and on the control of the exotic sea lamprey (Dochoda 1991), while the International Joint Commission focuses on water quality issues (IJC 1988). However, the activities of the two agencies need to be more directly linked than they are currently in order to foster an holistic ecosystem view (fig. 1).

The purpose of using the ACME process is to shift the focus of those participating in the decision making process from any single jurisdiction within the ecosystem to the Great Lakes ecosystem as a whole. This process would use an adaptive management approach where aggressive experimentation is coupled with careful evaluation allowing managers to modify management strategies and activities as feedback from the ecosystem is received (Krueger and Decker 1993). Collaboration among the many institutions and groups responsible for Great Lakes management will be essential to the success of ecosystem management. Governance structures participating in the ACME process will be committed to sharing information and ideas so that all decision makers can base their decisions on the best knowledge available. Consensus decisions are facilitated by the pooling of information by members of the decision making group (Carpenter and Kennedy 1985; Stasser and Titus 1985) and by creating win-win situations (Garcia 1986).

In order to facilitate the ACME process, several fundamental shifts in the way that Great Lakes fisheries agencies operate must occur. First, managing the Great Lakes ecosystem must become the top priority for the group. Currently, the multijurisdictional nature of most Great Lakes fisheries issues lead to the perception that these issues are too complex to address. This perception causes managing with an ecosystem perspective to become less important than managing within political boundaries. To facilitate ecosystem management, this perception must be overcome at the highest levels of Great Lakes management agencies. Making Great Lakes ecosystem management a high priority will necessitate the formulation of a vision and strategic plan. This vision should be written so that the endpoints for the ecosystem reach beyond the career of management officials. Achieving agreement on where the ecosystem should be in fifty years is often easier than agreeing on what should be done right now. However, once the vision has been agreed upon by all of the interested parties in the basin, everyone will be working towards a common goal that will facilitate collaboration and consensus building allowing for constant progress towards the goal.

Along with the shift to making management of the Great Lakes ecosystem a priority, the ACME process compels agencies to realize that the resource is their ultimate customer, and that they need to work for the resource. Therefore, participants in the ACME process should espouse the belief that what is best for the resource will ultimately be best for those served by the resource. A healthy Great Lakes ecosystem will enhance the predictability and sustainability of its fishery resources that will benefit all user groups.

Finally, management agencies and the stakeholders need to recognize that consensus building takes an enormous amount of energy and commitment. The individuals involved in generating the consensus should be rewarded for their efforts. Agencies should facilitate rather than impede these consensus building exercises by making it part of the assignment of the individual working on the issue. Assigning time and resources to consensus building will allow individuals to spend the required amount of effort toward that end, rather than trying to accomplish a consensus on a volunteer basis.

To facilitate consensus building, the needs and desires of the all stakeholders, including the general public should be determined. Further research must be done to define these needs and desires and to determine how these groups are affected by decisions made for enhanced Great Lakes fishery resources. Social science techniques, such as the use of focus groups, can be utilized to gain the necessary insight into stakeholder needs and desires. By so doing, all stakeholders will be better engaged in the decision making process for the Great Lakes ecosystem. In this way, everyone who can possibly be affected by a decision will be encouraged to have input into the decision process. This protocol should help to make people feel a greater degree of ownership of the decision-making process and the resulting decisions. Active involvement in the decision-making process lends credibility to the solution and promotes cooperation in the implementation of agreements (Carpenter and Kennedy 1985).

Conclusion

The future sustainability of our Great Lakes fishery resources depends on our ability to manage these ecosystems effectively. Fisheries management is an integrative process that requires an understanding of the biology of the species of interest, the productive capacity of the ecosystem, and the socioeconomic expectations of the public. Balancing the diverse demands placed on our fishery resources and their ecosystems will be difficult without an effective consensus-building process. The ACME process will enhance collaboration between Great Lakes governance structures and stakeholders that should result in a common vision for healthy and productive ecosystems.

Acknowledgments

We would like to thank the following individuals for their insightful comments regarding agency perspectives and decision making in the Great Lakes: Gavin Christie (Great Lakes Fishery Commission), Ron Desjardine (Department Fisheries and Oceans, Canada), Mark Holey (U.S. Fish and Wildlife Service), and Mike Hansen (National Biological Survey).

Literature Cited

Aquatic Nuisance Species Task Force. 1993. Findings, Conclusions and Recommendations of the Intentional Introductions Policy Review-A Report to Congress. Washington, D. C.

Bocking, S. 1994. Visions of nature and society: a history of the ecosystem concept. Alternatives 20:12 19.

Buzan, B. 1981. Negotiating by consensus: developments in technique at the United Nations Conference on the law of the sea. The American Journal of International Law 75:324–348.

Caldwell, L. K. 1994. Disharmony in the Great Lakes Basin: Institutional Jurisdictions Frustrate the Ecosystem Approach. Alternatives 20:26–32.

Carpenter, S. and W. J. D. Kennedy. 1985. Managing environmental conflict by applying common sense. Negotiation Journal 1985:149–161.

Christie, W. J. 1974. Changes in the fish species composition of the Great Lakes. Journal of Fisheries Research Board Canada 31:827–854.

Claudi, R. and G. L. Mackie. 1994. Practical Manual for Zebra Mussel Monitoring and Control. CRC Press, Boca Raton, Florida.

Coon, T. G. 1999. Ichtyofauna of the Great Lakes. Pages 55–71 in W. W. Taylor and C. P. Ferreri eds. Great Lakes Fisheries Policy and Management: A Binational Perspective. Michigan State University Press, East Lansing, Michigan.

Dochoda, M. R. 1991. Meeting the challenge of exotics in the Great Lakes: The role of an international commission. Canadian Journal of Fisheries and Aquatic Sciences 48(Suppl. 1):171–176.

Donahue, M. J. 1991. Water resources and policy. Pages 57–71 in W. A. Testa, editor. The Great Lakes Economy: Looking North and South. Federal Reserve Bank of Chicago, Chicago, IL.

Fedler, J. A. and D. M. Nickum. 1993. The 1991 economic impact of sport fishing in Michigan. Sport Fishing Institute, Washington, D. C.

Fink, A., J. Kosecoff, M. Chassin, R. H. Brook. 1984. Consensus methods: Characteristics and guidelines for use. American Journal of Public Health 74(9):979–983.

Francis, G. 1990. Flexible governance. Pages 195–207 in C. J. Edwards and H. S. Regier, editors. An Ecosystem Approach to the Integrity of the Great Lakes in Turbulent Times. Great Lakes Fishery Commission Special Publication 90–4, Ann Arbor, Michigan.

Garcia, A. 1986. Consensus decision making promotes involvement, ownership, satisfaction. NASSP Bulletin 20:50–52.

GLFC (Great Lakes Fishery Commission). 1980. A joint strategic plan for management of Great Lakes fisheries. Great Lakes Fishery Commission, Ann Arbor, Michigan.

Hartman, W. L. 1988. Historical changes in the major fish resources of the Great Lakes. Pages 103–131 in M. S. Evans, editor. Toxic Contaminants and Ecosystem Health: A Great Lakes Perspective. John Wiley and Sons, New York.

Herbold, B. and P. B. Moyle. 1986. Introduced species and vacant niches. American Naturalist 128:751–760.

IJC (International Joint Commission). 1988. Revised Great Lakes Water Quality Agreement of 1978. International Joint Commission, United States and Canada. 130pp.

Kirchler, E. and J. H. Davis. 1986. The influence of member status differences and task type on group consensus and member position change. Journal of Personality and Social Psychology 51(1):83–91.

Krueger, C. C., and D. J. Decker. 1993. The process of fisheries management. Pages 33–55 in C. C. Kohler and W. A. Hubert, editors. Inland Fisheries Management in North America. American Fisheries Society, Bethesda, Maryland.

Mac, M. J., T. R. Schwartz, C. C. Edsall, and A. M. Frank. 1994. Polychlorinated biphenyls in Great Lakes lake trout and their eggs: Relations to survival and congener composition 1979–1988. Journal of Great Lakes Research 19:752–765.

McKinney, M. 1992. Designing a dispute resolution system for water policy and management. Negotiation Journal 1992:153–163.

Mills, E. L., J. H. Leach, J. T. Carlton, and C. L. Secor. 1993. Exotic species in the Great Lakes: A history of biotic crises and anthropogenic introductions. Journal of Great Lakes Research 19(1):1–54.

Pinkerton, E. 1989. Attaining better fisheries management through comanagement-prospects, problems, and propositions. Pages 3-33 in E. Pinkerton, ed. Cooperative Management of Local Fisheries. University of Columbia Press, Vancouver, British Columbia.

Rawinski, T. J. and R. A. Malecki. 1984. Ecological relationships among purple loosetrife, cattails, and wildlife at Montezuma National Wildlife Refuge. New York Fish and Game Journal 31:81–87.

Rossi, P. H. and R. A. Berk. 1985. Varieties of normative consensus. American Sociological Review 50:333–347.

Smith, B. R. and J. J. Tibbles. 1980. Sea lamprey (Petromyzon marinus) in Lakes Huron, Michigan, and Superior: History of invasion and control. Canadian Journal of Fisheries and Aquatic Sciences 37:1780–1801.

Stasser, G. and W. Titus. 1985. Pooling of unshared information in group decision making: Biased information sampling during discussion. Journal of Personality and Psychology 48(6):1467–1478.

Tanner, H. A., M. H. Patriarche, and W. J. Mullendore. 1980. Shaping the world's finest freshwater fishery. Michigan Department of Natural Resources, Lansing, Michigan. 86pp.

Talhelm, D. R. 1988. Economics of Great Lakes fisheries: A 1985 assessment. Great Lakes Fishery Commission Technical Report No. 54.

Tjosvold, D. and R. H. G. Field. 1985. Effect of concurrence, controversy, and consensus on group decision making. The Journal of Social Psychology 125(3):355–363.

Contributors

Ray L. Argyle *(Chapter 12)*
USGS–Biological Resources Division
Great Lakes Science Center
1451 Green Road
Ann Arbor, MI 48105

Nancy A. Auer *(Chapter 17)*
Michigan Technical University
Department of Biology
1400 Townsend Dr.
Houghton, MI 49931-1295

Alfred M. Beeton *(Chapter 1)*
Great Lakes Environmental Research Lab
2205 Commonwealth Blvd.
Ann Arbor, MI 48105

James R. Bence *(Chapter 10)*
Department of Fisheries and Wildlife
13 Natural Resources Building
Michigan State University
East Lansing, MI 48824

Edward H. Brown, Jr. *(Chapter 12)*
USGS–Biological Resources Division
Great Lakes Science Center
1451 Green Road
Ann Arbor, MI 40105

Russell W. Brown *(Chapter 11)*
NOAA–National Marine Fisheries Service
Northeast Fisheries Science Center
166 Water Street
Woods Hole, MA 02543

Mary K. Burnham-Curtis *(Chapter 6)*
USGS–Biological Resources Division
Great Lakes Science Center
1451 Green Road
Ann Arbor, MI 48105

Thomas R. Busiahn *(Chapter 12)*
U.S. Fish & Wildlife Service
Ashland Fishery Resources Office
Ashland, WI 54806

Thomas G. Coon *(Chapter 2)*
Department of Fisheries and Wildlife
13 Natural Resources Building
Michigan State University
East Lansing, MI 48824

Margaret A. Dochoda *(Chapters 4, 7)*
Great Lakes Fishery Commission
2100 Commonwealth Blvd., Suite 209
Ann Arbor, MI 48105

Mark Ebener *(Chapter 11)*
Chippewa/Ottawa Treaty Fishery
    Management Authority
186 East Three Mile Road
Sault Ste. Marie, MI 49783

Randy L. Eshenroder *(Chapter 6)*
Great Lakes Fishery Commission
2100 Commonwealth Blvd., Suite 209
Ann Arbor, MI 48105

C. Paola Ferreri *(editor, Chapter 18)*
School of Forest Resources
207 Ferguson Building
Pennsylvania State University
University Park, PA 16802

Tom Gorenflo *(Chapter 11)*
Chippewa/Ottawa Treaty Fishery
    Management Authority
186 East Three Mile Road
Sault Ste. Marie, MI 49783

Richard Groop *(Chapter 3)*
Department of Geography
311 Natural Science Building
Michigan State University
East Lansing, MI 48824

Michael J. Hansen *(Chapter 14)*
University of Wisconsin–Stevens Point
174 College of Natural Resources
1900 Franklin Street
Stevens Point, WI 54481

Daniel B. Hayes *(Chapter 8)*
Department of Fisheries and Wildlife
13 Natural Resources Building
Michigan State University
East Lansing, MI 48824-1222

Raymond M. Hoff *(Chapter 9)*
Senior Research Scientist
Atmospheric Environment Service
R.R. #1
Egbert, ON L0L 1N0
Canada

Michael L. Jones[1] *(Chapters 12, 15)*
Ontario Ministry of Natural Resources
Aquatic Ecosystems Research
R.R. #4
Picton, ON K0K 2T0
Canada

Roger L. Knight *(Chapter 13)*
Ohio Division of Wildlife
Lake Erie Fisheries Unit
305 E. Shoreland Dr.
Sandusky, OH 44870

John F. Kocik *(Chapter 15)*
NOAA–National Marine Fisheries Service
Northeast Fisheries Science Center
166 Water Street
Woods Hole, MA 02543

Joseph F. Koonce *(Chapter 13)*
Department of Biology
Case Western Reserve University
Cleveland, OH 44106

Joseph H. Leach *(Chapter 7)*
Ontario Ministry of Natural Resources
Lake Erie Fisheries Station
Wheatley, ON N0P 2P0
Canada

Ana B. Locci *(Chapter 13)*
Department of Biology
Case Western Reserve University
Cleveland, OH 44106

Edward L. Mills *(Chapter 7)*
Department of Natural Resources
Cornell University Biological Field Station
900 Shackelton Point Road
Bridgeport, NY 13030

Robert O'Gorman *(Chapter 16)*
USGS–Biological Resources Division
Lake Ontario Biological Station
17 Lake Street
Oswego, NY 13126

Judith A. Orendorff[2] *(Chapter 5)*
OMNR–Aquatic Ecosystem Research Section
P.O. Box 5000
Maple, ON L6A 1S9
Canada

David F. Reid *(Chapter 1)*
Great Lakes Environmental Research Lab
2205 Commonwealth Blvd.
Ann Arbor, MI 48105

John M. Robertson[3] *(Chapter 18)*
Michigan Department of Natural Resources
Fisheries Division
P.O. Box 30452
Lansing, MI 48909

Richard A. Ryder *(Chapter 5)*
Ontario Ministry of Natural Resources
Productivity Unit
1265 Lakeshore Drive
Thunder Bay, ON P7B 5E4
Canada

Cynthia E. Sellinger *(Chapter 1)*
Great Lakes Environmental Research Lab
2205 Commonwealth Blvd.
Ann Arbor, MI 48105

Kelley D. Smith *(Chapter 10)*
Michigan Department of Natural Resources
Fisheries Division
P.O. Box 30446
Lansing, MI 48909

Thomas J. Stewart *(Chapter 16)*
Ontario Ministry of Natural Resources
Lake Ontario Fisheries Unit
R.R. #4
Picton, ON K0K 2T0
Canada

William W. Taylor *(editor, Chapter 18)*
Department of Fisheries & Wildlife
13 Natural Resources Building
Michigan State University
East Lansing, MI 48824

---

[1] Present address: Department of Fisheries and Wildlife, 13 Natural Resources Bldg., Michigan State University, East Lansing, MI 48824.

[2] Present address: Ontario Ministry of Natural Resources, 50 Bloomington Road West, R.R. #2, Aurora, ON L4G 3G8 Canada

[3] Present address: Michigan Department of Natural Resources, Forest Management Division, PO Box 30028, Lansing, MI